IET CONTROL, ROBOTICS AND SENSORS SERIES 119

Swarm Intelligence

Other volumes in this series:

Swarm Intelligence

Volume 1: Principles, current algorithms and methods

Edited by
Ying Tan

The Institution of Engineering and Technology

Published by The Institution of Engineering and Technology, London, United Kingdom

The Institution of Engineering and Technology is registered as a Charity in England & Wales (no. 211014) and Scotland (no. SC038698).

First published 2018

The Institution of Engineering and Technology
Michael Faraday House
Six Hills Way, Stevenage
Herts, SG1 2AY, United Kingdom

www.theiet.org

British Library Cataloguing in Publication Data
A catalogue record for this product is available from the British Library

ISBN 978-1-78561-627-3 (Volume 1 hardback)
ISBN 978-1-78561-628-0 (Volume 1 PDF)
ISBN 978-1-78561-629-7 (Volume 2 hardback)
ISBN 978-1-78561-630-3 (Volume 2 PDF)
ISBN 978-1-78561-631-0 (Volume 3 hardback)
ISBN 978-1-78561-632-7 (Volume 3 PDF)

Typeset in India by MPS Limited

Contents

Preface

This book entitled "swarm intelligence: principles and current algorithms and methods" consists of 20 chapters, focusing on the principle and the current approaches of swarm intelligence algorithms for presenting the latest achievements in the principle and typical algorithms in recent years. In this book, we mainly introduce the basic principles of typical swarm intelligence algorithms and some of their efficient improvements. Specifically, it not only includes the introductions and improvements of typical PSO, ACO and typical swarm intelligence algorithm as well as swarm robotics methods; furthermore, it also includes parts of the general swarm intelligent method researches until now.

It is well known that swarm intelligence (SI) is one of the most important and challenging paradigms in the umbrella of computational intelligence, focusing on the research of collective behaviors of a swarm in nature and/or social phenomenon, to solve complicated and difficult problems which cannot be handled by the traditional approaches. The swarm intelligence has attracted much more research interests in recent years. Thousands of papers are published each year for presenting new algorithms, new improvements and numerous applications in different domains. The SI has been successfully applied to many real-world applications in nearly all walks of life. Thus, it will remain a main research topic for the following decades. In the last decade, swarm intelligence has attracted many researchers around the world. Current publications in swarm intelligence are very separate, and there are only several conferences and journals specified for this topic, which makes it hard for people to share their ideas with other colleagues or follow-up the works from other researchers with common interests. So there is an immediately urgent need to collect and up-to-date the researching work together as a complete collection so as to lay a vast panorama of the swarm intelligence research before us timely. This book has just filled this gap in time.

However, there are too many new methods invented in this field each year, which makes it hard to follow for researchers, learners and practitioners in the field. It is the right time to publish a solid all in one reference presenting the latest knowledge, research and applications in this field systematically and thoroughly.

The aim of this book is to provide readers with a full view of the field of swarm intelligence from the latest principles and approaches to practical applications. Students will learn the principles and theories of typical swarm intelligence algorithm; scholars in the field will get inspired for further promising research directions; practitioners will find suitable methods for their applications of interest along with useful instructions.

This book is primarily intended for researchers, engineers, graduates and senior undergraduates with interests in swarm intelligence and its applications. It also serves as an essential resource for the professionals and practitioners with interests in swarm intelligence working in the fields of computer science, information technology, artificial intelligence, computational intelligence, bioengineering, physics, mathematics, social sciences, etc. It is also useful and helpful for the interested people in a broad discipline in science and engineering domains.

The structure of the contents of this book is organized by a variety of topics, from general survey, algorithms and improvements to typical applications, for swarm intelligence. Besides, it is also a useful reference for some technology developers and technicians in many technological companies as well as other interested readers who will find the chapters in this book useful in furthering their work or research exposure to pertinent topics in swarm intelligence.

The following describes each chapter of this book one by one concisely:

1. *Survey of swarm intelligence by Ying Tan*
   In this chapter, a systematic survey of swarm intelligence is given for showing the overview of swarm intelligence, along with some recent developments in swarm intelligence and swarm robotics. First, the concept of swarm intelligence is given. Then, some general researches and several classical algorithms and their developments are presented in details. After that, some novel algorithms and some applications are reviewed individually. Finally, some developments in swarm robotics are given.
2. *Generalization ability of swarm intelligence algorithms by Shi Cheng, Quande Qin, Bin Liu, Jiqiang Feng, Xiujuan Lei and Yuhui Shi*
   In this chapter, generalization ability of swarm intelligence algorithms solving problems with different number of dimensions is analyzed and discussed. Three algorithms, brain storm optimization in objective space (BSO-OS), fireworks algorithm (FWA) and particle swarm optimization algorithm, are selected as illustrations to explain the definition of algorithm's generalization ability. The performance of BSO, FWA and PSO algorithm on solving problems with different number of dimensions is analyzed. Based on the experimental results, the algorithm's generalization ability was measured by the results ratio of algorithms with the same settings on problems with different number of dimensions. This generalization ability measurement could be extended to problems with different components. Without analysis on the landscape of problems, this measurement could give a practical illustration of the generalization ability of algorithms for solving problems with different number of dimensions or different components. Based on the analysis on the generalization of algorithms and the hardness of problems, we could have a better understanding of the relationship between problems and algorithms and therefore design more effective algorithms to solve different problems.
3. *A unifying framework for swarm intelligence-based hybrid algorithms by Bin Xin, Yipeng Wang and Bo Liu*
   This chapter is aimed at giving a classification and an analysis of various swarm intelligence-based hybrid optimizers by the systematic taxonomy we proposed in a recent work. To date, it is impossible to list all kinds of hybrids within a chapter

of a book, since a plethora of hybrids exists in the literature (about 1,610,000 related results according to the Google search result with the index terms "hybrid algorithm" and "optimization"). Based on the taxonomy, a unifying framework for SIOA-based optimizers is established. Designers can gain an insight into various possibilities for hybrid design of swarm intelligence-based optimizers.

4. *Ant colony systems for optimization problems in dynamic environments by Yirui Wang, Shangce Gao and Yuki Todo*
   Ant colony optimization (ACO) is an intelligent bionic algorithm which simulates the foraging behavior of ant colony. The conventional ACOs mainly deal with the static optimization problems. In other words, the environment of problem maintains invariant. Actually, the most problems in reality are dynamic, namely, the changing environments. The ACO can use its robustness and self-adaptability to resolve dynamic problems properly. In this chapter, the ACO with neighborhood search is introduced to address dynamic traveling salesman problem and the ACO with improved K-means clustering algorithm, which uses three immigrants schemes including random immigrants, elitism-based immigrants and memory-based immigrants, is used for dynamic location routing problem. Several conventional ACOs and other heuristic algorithms are utilized to compare with new ACOs in the corresponding dynamic problems. The comparative experiments demonstrate that two novel ACOs are effective and efficient for respective dynamic optimization problems.
5. *Ant colony optimization for dynamic combinatorial optimization problems by Michalis Mavrovouniotis and Shengxiang Yang*
   The ant colony optimization (ACO) metaheuristic was inspired from the foraging behavior of real ant colonies. In particular, real ants communicate indirectly via pheromone trails and find the shortest path. Although real ants proved that they can find the shortest path when the available paths are known a prior, they may face serious challenges when some paths are made available after the colony has converged to a path. This is because the colony may continue to follow the current path rather than exploring the new paths in case a shorter path is available. For the ACO metaheuristic, the challenges are similar when applied to dynamic optimization problems (DOPs). Once the algorithm converges, it loses its adaptation capabilities and may have poor performance in DOPs. Several strategies have been integrated with ACO to address difficult combinatorial DOPs. Their performance proved that ACO is a powerful computational technique for combinatorial DOPs once enhanced. This chapter investigates the applications of ACO for combinatorial DOPs.
6. *Comparison of multidimensional swarm embedding techniques by potential fields by Youngha Hwang and Okan K. Ersoy*
   Multidimensional embedding is a technique useful for characterizing spectral signature relations in hyperspectral images. However, such images consist of disjoint similar spectral classes that are spatially sensitive, thus presenting challenges to existing graph-embedding methods. In this chapter, the multidimensional-embedding techniques are interpreted by the potential fields method, where a sum of attraction and repulsion potential functions is minimized to find optimal

energy configuration of embedding. Repulsion dominates at short distances, thereby emphasizing local relations. On the other hand, attraction dominates at long distances, stressing global relations. The formulations capture long-range- and short-range-distance-related effects often associated with living organisms and help to establish algorithmic properties that mimic mutual behavior for the purpose of dimensionality reduction. Widely used t-SNE (stochastic neighbor embedding) and its variations are compared in terms of the potential field methods, and their sources of weakness are discussed. As part of evaluation, the embedding maps are visualized, their trajectories are plotted and their semisupervised classifications are conducted for image scenes acquired by multiple sensors at various spatial resolutions over different types of objects.

7. *Inertia weight control strategies for PSO algorithms by Ahmad Nickabadi, Reza Safabakhsh and Mohammad Mehdi Ebadzadeh*
Due to the strong effect of the inertia weight on the performance of PSO, many researchers have investigated different inertia weight control approaches during the past decades, and many different strategies have been proposed. Various inertia weighting strategies can be categorized into three main classes: (1) constant or random, (2) time-varying and (3) adaptive inertia weights. This chapter presents the major existing inertia weight control strategies in the above three categories and discusses examples of each category in detail.
8. *Robot path planning using swarms of active particles by E. Jorge Ivan Sofrony and C. Helbert Eduardo Espitia*
There exists a wide range of techniques that generate collision free (optimal) trajectories, in an autonomous fashion, for mobile robotics. Probably the most popular technique is that of artificial potential fields (APF), where the robot is treated as a particle subject to a potential field that is generated by the obstacles and the goal position. The path generation problem is then treated as an optimization problem where gradient descent methods have been traditionally used. Particle swarm optimization has also been widely used to solve optimization problems, and although it has proven to be more efficient in the search of minima, it also suffers from early convergence, i.e., the swarm may get trapped in a local minima. Many of the modifications that cater for this weakness involve added complexity in the construction of the potential field and thus translates into ad-hoc algorithms. A particle swarm approach with the property of escaping local minima by forcing vortex-like dynamics when the gradient of the potential field is close to zero is proposed, proving to be more compact and intuitive than previously proposed algorithms.
9. *MAHM: a PSO-based multiagent architecture for hybridisation of metaheuristics by Givanaldo R. de Souza, Elizabeth F. G. Goldbarg, Anne M. de P. Canuto, Marco C. Goldbarg and Iloneide C. de O. Ramos*
Hybridization of metaheuristics is an important subject that has been explored by several researchers. Multiagent systems have been important tools to accomplish the task of hybridizing metaheuristics. In those multiagent approaches, however, each metaheuristic is performed separately and the potential of the hybridization is not fully explored. In order to bridge this gap, this chapter

presents MAHM (multiagent architecture for hybridization of metaheuristics), a multiagent approach for metaheuristics hybridization inspired on particle swarm optimization (PSO). Introduced as a metaheuristic, PSO can be viewed as a multiagent system once particles can be thought as agents that interact and work together to achieve specific goals. In this context, particles are identified to autonomous virtual entities with social ability, reactivity, and proactiveness. Each agent in MAHM may use different sets of predefined metaheuristics in different moments of the search to look for high-quality solutions of an optimization problem. Thus, several elements of different metaheuristics can be found running in the swarm every MAHM iteration. In order to show the potentiality of the proposed architecture, computational experiments were carried out with the traveling salesman (TSP) and the quadratic assignment problem (QAP), two important test grounds for algorithmic ideas.

10. *The critical state in particle swarm optimisation by Adam Erskine, Thomas Joyce and J. Michael Herrmann*
We will first set out backgrounds to PSO, and to criticality. Next, we discuss some previous attempts to explore the theoretical dynamics of PSO. The earliest of these offer some insight into the behaviors that may be expected from the algorithm, but they have typically ignored the multiplicative role of the random element of the update rules. These have often been replaced by their expectation values or simply viewed as a drunken walk about the predicted dynamics of the system, enhancing the local search capabilities. We show that in fact they contribute to the dynamics of the swarm's capabilities in a manner that alters its expected behavior. More recent analyzes consider the random elements by considering expectation and variance values of the update matrices. These predict a locus of parameter values that yield convergent swarms but we believe are still too conservative. In Section 10.2, we consider the products of random matrices, in particular those that represent the PSO algorithm. We derive the location of the locus of a and w values that give rise to swarms that will neither diverge nor collapse over time. This curve matches the observed behavior of an unconstrained PSO swarm, i.e., one where neither positional or velocity constraints are imposed on the dynamic of the swarm.

11. *Bounded distributed flocking control of nonholonomic mobile robots by Thang Nguyen, Hung M. La, Vahid Azimi and Thanh-Trung Han*
There have been numerous studies on the problem of flocking control for multiagent systems whose simplified models are presented in terms of point-mass elements. Meanwhile, full dynamic models pose some challenging problems in addressing the flocking control problem of mobile robots due to their nonholonomic dynamic properties. Taking practical constraints into consideration, we propose a novel approach to distributed flocking control of nonholonomic mobile robots by bounded feedback. The flocking control objectives consist of velocity consensus, collision avoidance and cohesion maintenance among mobile robots. A flocking control protocol which is based on the information of neighbor mobile robots is constructed. The theoretical analysis is conducted with the help of a Lyapunov-like function and graph theory. Simulation results

are shown to demonstrate the efficacy of the proposed distributed flocking control scheme.

12. *Swarming in forestry environments: collective exploration and network deployment by Micael S. Couceiro and David Portugal*
Landscape maintenance is a pivotal factor to reduce wildfire hazard potential. Being a difficult, time-consuming and generally expensive task, humans often face great difficulties to address the problem effectively. Considering its spatial distribution and exploration capabilities, swarm robotics presents itself as a natural solution to assist humans or other robotic agents in the reconnaissance of forestry environments and identification of fuel accumulation, such as forest debris. In this work, we propose an optimization framework for the exploration problem of forestry environments with a team of swarming robots, which also entails the deployment of a network infrastructure formed by the robots to assist agents during fire-prevention tasks. Experiments show the benefits of employing the multirobot coordination algorithm described, which provides an effective solution for the exploration task and guarantees a final formation which maximizes network coverage for mission information exchange, considering obstacles density and other interfering factors. Thus, the system proposed might assist human operators with relevant forestry information acquired during exploration and maintain a fail-safe network for communication while agents operate in the field.
13. *Guiding swarm behavior by soft control by Jing Han and Caiyun Wang*
Guiding swarm behavior is one of important problems in the area of swarm intelligence. While in many multiagent systems, the local rules for each agent is difficult or impossible to be redesigned. Even for artificial systems, it is not easy to design simple local rules for each agent which can lead the group emerge desired collective behavior. In this chapter, we introduce a new method called 'soft control' which does not need to change the underlying local rules of agents, but to add some special agents, called 'shills' which can be controlled by us, into the system. It is a nondestructive intervention method, because shills are still treated as normal agents by normal ones. The soft control method is effectively applied to three different multiagent system models to guide the swarm behavior: (1) guide consensus of a Vicsek-like flocking model by adding one intelligent shill; (2) promote cooperation for a group of players that play the end-unknown repeated prisoner's dilemma game by adding a cooperating team of shill plays; (3) add one or some shills to guide the collective opinion of the DeGroot model system.
14. *Agreeing to disagree: synergies between particle swarm optimisation and complex networks by Mihai Udrescu and Andrei Lihu*
Due to its numerous applications in fields like systems biology, medicine, technology, engineering or social sciences, the new science of complex networks has become extremely popular over the last couple of decades. Particle swarm optimization (PSO) and complex network (CN) science have significant common ground, as they both deal with a set of agents (particles for PSOs, vertices for CNs) which interact according to an underlying topology. Moreover, one of

the most important branches of CN science is represented by social networks, while PSO was originally conceived for optimization through the simulation of social behavior, thus further emphasizing the strong tie between the two scientific fields. A prominent problem in CNs is the opinion dynamics in social networks. To this end, opinion interaction models are tested in order to verify if they can recreate a realistic social behavior; an ubiquitous feature of such real-life social behaviors is persistent opinion disagreement. One of the most important characteristics of a social network which fosters disagreement is that it contains stubborn agents, namely, vertices which never change their opinion. We discuss models of disagreement, such as that with stubborn agents and tolerance-based models, then offer a new perspective in exploring the frontiers between network science and PSO.

15. *Ant colony algorithms for the travelling salesman problem and the quadratic assignment problem by Nikola Ivkovic*
The ant colony optimization (ACO) is a metaheuristic which has been successfully used to solve computationally difficult optimization problems, especially combinatorial optimization problems which belong to the class of NP-hard problems. This chapter explains ACO algorithms, their most important variants and hybridization with local optimization methods. Practical considerations for successful ACO implementation and parameter setting are given. The working of the algorithm is explained in details by using simple examples. The chapter ends with overview of research trends in ACO.
16. *A review of particle swarm optimization for multimodal problems by Jian-Ping Li, Shanxin Yuan, Qing Sheng Li, Bo Li and Yim Fun Hu*
Particle swarm optimizations (PSO) are population-based methods inspired from the flight of a flock of birds seeking food. After the development of over 20 years, PSOs have been become a major branch of evolutionary algorithms and have been successfully applied to solve many science and engineering optimization problems. Most of PSOs are designed to search one solution of a problem. However, many science and engineering optimization problems are complex and multimodal in nature. More and more researches are aiming to identify multiple global and local solutions of complex multimodal problems, and several competitions in recent international conferences had been set up to encourage researchers to develop more effective and efficient algorithms for exploring multiple solutions. There are several techniques in the literature which can be used by combining with an existing evolutionary algorithm to solve multimodal optimization problems. Those techniques are called Niching. The most commonly used niching techniques are crowding, fitness share, clustering and species conserving. PSO-related methods for multimodal problems are reviewed in this chapter, including hybrid PSO with other evolutionary algorithms. Additionally, the multimodal functions, including some challenge composition multimodal functions, are listed as references for researchers to test their new algorithms. The species conserving PSO (SCPSO) is described in detail and used to solve

some multimodal engineering optimization problems to demonstrate the power of niching in exploring multiple solutions.

17. *Decentralized control in robotic swarms by Hamed Rezaee and Farzaneh Abdollahi*
    Robotic swarms are groups of autonomous and self-organized mobile robots interacting under a communication network in order to accomplish cooperative tasks. This chapter will provide an overview on main outcomes on decentralized control in robotic swarms based on applications and the degree of autonomy in the view of stability and control theory. Based on application, we will category various problems in the area of robotic swarms such as swarming, formation and consensus/rendezvous. First-path planning and collision-avoidance approaches for mobile robots will be introduced. Then, swarming and swarm formation of mobile robots as most important problems in control of multirobot systems will be investigated. Finally, the consensus and rendezvous as important problems in multirobot systems will be studied. In each category, existing structures will be classified and explained.
18. *PSO in ANN, SVM and data clustering by Terje Kristensen and Fabien Guillaume*
    In this chapter, one gives an introduction to different kinds of particle swarm optimization (PSO) algorithms. One also introduces artificial neural networks (ANN), support vector machines (SVM) and evolutionary computing (EC) to show how PSO may be used to determine optimal parameters using an ANN or SVM regime, for classification of DNA-strings. In addition, PSO is used in the design of a SVM-based clustering algorithm. Ant colony optimization (ACO) algorithms are also introduced in the chapter. Using ACO algorithms has been of great success to solve many discrete optimization and NP-hard problems, for instance the traveling salesman problem (TSP). The behavior of ants is also used, for instance, to design an algorithm for data clustering. We want to develop later a similar application based on this clustering algorithm and compare it with the SVM one using PSO.
19. *Modelling of interaction in swarm intelligence focused on particle swarm optimization and social networks optimization by A. Niccolai, F. Grimaccia, M. Mussetta and R.E. Zich*
    In this chapter, the modelling of the interaction in swarm intelligence will be analyzed using differential and discrete models. Among the several computational intelligence techniques recently developed, swarm intelligence (SI) is one of the most popular. Based mainly on the development of particle swarm optimization (PSO) and ant colony optimization (ACO), SI focuses on the collective behavior of swarms in nature and social phenomena, to solve highly nonlinear optimization problems. The aim of this chapter is to provide an overview of the general interaction models in SI algorithms. In particular, the presented interaction analysis will focus on the well-known PSO technique, and it will be applied considering a recently developed population-based algorithm, namely, the social network optimization (SNO).

20. *Coordinating swarms of microscopic agents to assemble complex structures by Bruce J. MacLennan*
    This chapter addresses the problem of coordinating very large swarms of microscopic agents to assemble complex, hierarchically structured physical systems. The agents might be microscopic robots or genetically engineered microorganisms. The approach we use is a form of artificial morphogenesis, which studies the self-organized morphogenetic processes in the developing embryo, by which billions of cells cooperate to create physical form, and abstracts these processes to control microscopic agents to assemble desired physical structures. We use an approach based on the description of morphogenetic processes by partial differential equations, which ensures that our morphogenetic algorithms will scale up to arbitrarily large swarms (millions or more). We present several simulation experiments demonstrating the coordination of massive swarms to construct complex objects.

This book entitled "swarm intelligence: principles and current algorithms and methods" along with its sister two books entitled "swarm intelligence: noval algorithms and methods" and "swarm intelligence: applications" consist of a 3-volume book set which finally contains 64 chapters in total, being ready for the book entitled "swarm intelligence: from concepts to applications" published by the IET, and present the latest research achievements of many active researchers and academicians in the field of swarm intelligence all over the world. It is a collection of the most important research achievements in swarm intelligence thus far, covering the principles, approaches, developments and applications of swarm intelligence. It will provide a platform to hold high-quality state-of-art researches on swarm intelligence from many aspects including theoretical analysis, algorithm research, simulation implementation and real-world applications.

Because of the abundant and knowledged contents in this book set, we divided them into three parts individually published by three volumes. Volume 1 contains 20 chapters which are to present the principles and current algorithms and methods, which mainly introduces the basic principles of well-known swarm intelligence algorithms and efficient improvements from typical particle swarm optimization (PSO), ant colony optimization (ACO) and fireworks algorithm (FWA) as well as other swarm intelligence algorithms to swarm robotics methods. In Volume II, I include 17 chapters that are to report some novel algorithms and methods which are intended to present the front-edge research with newly proposed swarm intelligence algorithms. Finally, in Volume III, there are 27 chapters for showing the main applications in the field of swarm intelligence in which we want to present the latest real-world applications of swarm intelligence algorithms and related evolutionary algorithms. With these three volumes, we want to show the effectiveness and validation of swarm intelligence and encourage more researchers and practitioners to use them in their ongoing projects, as well as attract newcomers to this area.

The scope of this book set includes the principles, approaches, developments, analyzes and applications of typical swarm intelligence algorithms, which covers almost all aspects of swarm intelligence. Meanwhile, this book will present the latest research work in the field of swarm intelligence. Important newly proposed swarm intelligences are introduced in detail along with their theories and applications. Each of the chapters will be written by the inventor or most knowledgeable scholar of this swarm intelligence approach to ensure high-level expertise and comprehensiveness.

This book set will provide an abundant resource and a good platform for researchers in the community of swarm intelligence all over the world to share their innovative ideas. With such a large number of research work included here together, this book no doubt can attract much attention from a number of researchers; therefore it must play a very important role in swarm intelligence and push it forward remarkably.

About the editor

Dr. Ying Tan is a full professor of Peking University and director of Computational Intelligence Laboratory at Peking University. He is also a professor at Faculty of Design, Kyushu University, Japan. He worked at Columbia University as a senior research fellow in 2017 and at Chinese University of Hong Kong in 1999 and 2004–2005 as a research fellow, and he was as an electee of 100 Talent Program of Chinese Academy of Science (CAS) in 2005. As a visiting professor, he visited many universities including University of California at Santa Cruz, Kyushu University, Auckland University of Technology, City University of Hong Kong, etc. He is the inventor of Fireworks Algorithm (FWA). He serves as the Editor-in-Chief of International Journal of Computational Intelligence and Pattern Recognition (IJCIPR), the Associate Editor of IEEE Transactions on Evolutionary Computation (TEC), IEEE Transactions on Cybernetics (CYB), IEEE Transactions on Neural Networks and Learning Systems (NNLS), International Journal of Swarm Intelligence Research (IJSIR), International Journal of Artificial Intelligence (IJAI), etc. He also served as an Editor of Springer's Lecture Notes on Computer Science (LNCS) for 30+ volumes, and Guest Editors of several referred journals, including IEEE/ACM Transactions on Computational Biology and Bioinformatics, Information Science, Neurocomputing, Natural Computing, etc. He is the founder general chair of the ICSI International Conference series since 2010. He won the 2nd-Class Natural Science Award of China in 2009 and many other awards from academic communities. His research interests include swarm intelligence, swarm robotics, data mining, pattern recognition, intelligent information processing for information security and financial applications, etc. He has published more than 300 papers in refereed journals and conferences in these areas and authored/coauthored 15 books and 20+ chapters in book and received four invention patents.

List of Technical Reviewers

Niusha Abady	ODAS
Om Prakash Acharya	KIIT University, Bhubaneswar
Jose Aguilar	Universidad de Los Andes
Bestoun Ahmed	Dr. Bestoun S. Ahmed IDSIA Istituto Dalle Molle di Studi sull'Intelligenza Artificiale
Ahmad Almomani	Clarkson University
Subhabrata Banerjee	FIEM
Zafer Bingul	Kocaeli University
Ilyes Boulkaibet	University of Johannesburg
Sandip Chanda	Budge Budge Institute of Technology
Abhaya Chandra	TU kl
Walter Chen	National Taipei University of Technology
Shi Cheng	Shaanxi Normal University
Radu-Codrut David	Politehnica University of Timisoara
Massimo Donelli	University of Trento
Haibin Duan	Beijing University of Aeronautics and Astronautics
Dilson Espenchitt	COPPE UFRJ
Vitor Hugo Ferreira	Fluminense Federal University
Shangce Gao	University of Toyama
Debasish Ghose	Indian Institute of Science
Sotirios Goudos	Aristotle University of Thessaloniki
Roberto Gueleri	University of São Paulo
Jing Han	Academy of Mathematics and Systems Science
Thomas Hanne	University of Applied Sciences Northwestern Switzerland
J. Michael Herrmann	The University of Edinburgh
Weiwei Hu	Peking University
Chia-Ling Huang	KAINAN University
Nikola Ivkovic	Faculty of Organization and Informatics
Adnan Khan	MCS NUST
Salman Khan	University of Bahrain
Terje Kristensen	Bergen University College
Waroth Kuhirun	Kasetsart University
Hung La	University of Nevada
Xiujuan Lei	Shanxi Normal University
Jian-Ping Li	University of Bradford
Junzhi Li	Peking University

Andrei Lihu	Politehnica University of Timisoara
Bo Liu	Chinese Academy of Science
Wenju Liu	Institute of Automation, China Academy of Sciences
Mohammad Javad Mahmoodabadi	Sirjan University of Technology
Michalis Mavrovouniotis	Nottingham Trent University
Chilukuri Mohan	Syracuse University
Hiroyuki Mori	Meiji University
V. Mukherjee	Indian School of Mines, Dhanbad, Jharkhand, India
Marco Mussetta	Politecnico di Milano
Sreelaja N.K	University of South Australia
Rathindra Nath Biswas	A.J.C. Bose Polytechnic
Yan Pei	The University of Aizu
Wellington Pinheiro Dos Santos	Universidade Federal de Pernambuco
David Portugal	Ingeniarius, Ltd.
Amarjeet Prajapati	NIT Kurukshetra
Rabie Ramadan	Cairo University
Sivakumar Ramakrishnan	A.V.V.M. Sri Pushpam College (Autonomous)
Srikanth Reddy K	MNIT Jaipur
Hamed Rezaee	Amirkabir University of Technology
Givanaldo Rocha de Souza	IFRN
Jesús Ángel Román Gallego	USAL
Reza Safabakhsh	Amirkabir University of Technology
Chuan Shi	Beijing University of Posts and Telecommunications
Gopal Shyam	REVA College
Ana Paula Silva	Escola Superior de Tecnologia de Castelo Branco
Arlindo Silva	Escola Superior de Tecnologia do IPCB
Ganesan Sivarajan	Annamalai University
Petr Skobelev	Smart Solutions, Ltd
Adam C. Sniderman	University of Toronto
Jorge Sofrony	Universidad Nacional
Kavitha Sooda	B.M.S. College of Engineering
José A. Sáez	University of Salamanca
Fatemeh Tabib Mahmoudi	Shahid Rajaee Teacher Training University
Ying Tan	Peking University
Vedat Topuz	Marmara University
Milan Tuba	John Naisbitt University
Bin Xin	Beijing Institute of Technology
Seyed Hamid Zahiri	University of Birjand
Haijun Zhang	Harbin Institute of Technology Shenzhen Graduate School
Hong Zhang	Kyushu Institute of Technology

List of Additional Reviewers

A Appasani, Bhargav
C Couceiro, Micael
E Ebadzadeh, Mohammad Mehdi
F Freire, Renê Cruz
H Han, Jing
J John, Jason
K Kavanagh, Katie
M Martins, Gonçalo
N Nguyen, Thang
Niccolai, Alessandro
Nickabadi, Ahmad
S Sanangavarapu Mohan, Ananda
Shyam, Gopal
W Weng, Yuhao

Acknowledgments

I appreciate the contributors of each chapter for their great work that consists of the main components of this book set in such cutting-edge topics on swarm intelligence. In addition, I owe my gratitude to all the authors who promptly responded my call for chapter proposal, for their valuable participation which makes our work more competitive. I also would like to thank the valuable experts and reviewers for their constructive suggestions, and earnest and strict reviews to each chapter assigned to them, whose efforts and hard work guarantees such a high-level quality of this book.

For such a tedious editing workload, beside my own hard work and efforts, I have also received a plenty of help and support from my colleagues and graduate students. Without their help, I could not complete such a great work.

I am graceful to Val Moliere, Senior Commissioning Book Editor of The Institution of Engineering and Technology (IET), and Olivia Wilkins, Assistant Editor of the IET for their kind coordination and suggestions during the whole editing process of this book set. In addition, I am also grateful to Mr. Srinivasan N, an energetic editor of the MPS Limited, who has done a lot of editing work during the production of this book.

I want to thank everyone who gave me any assistance or help in the research of such an innovative idea and an amazing work on the title of swarm intelligence.

This book is dedicated to my wife Chao Deng and my daughter Fei Tan for their unconditional love; without their unselfish and salient supports and encouragements, it is impossible for me to make such great work a reality.

While working on this book, I was supported by the Natural Science Foundation of China (NSFC) under grant no. 61673025 and 61375119, and I was also partially supported by National Key Basic Research Development Plan (973) Project of China with grant no. 2015CB352302.

December 12, 2017 — Ying Tan
Beijing, China

Chapter 1
Survey of swarm intelligence

Ying Tan[1]

Abstract

In this chapter, a systematic survey of swarm intelligence is given to show the overview of swarm intelligence, along with some recent developments in swarm intelligence and swarm robotics. First, the concept of swarm intelligence is given. Then, some general researches and several classical algorithms and their developments are presented in detail. After that, some novel algorithms and their applications are reviewed individually. Finally, some developments in swarm robotics are given.

1.1 Introduction

Swarm intelligence is a rising research field under the concept of computational intelligence. Some researchers believe it is a promising approach toward artificial intelligence. In this paper, the concept of swarm intelligence and the tendency of this research field are first introduced. Then some recent developments of swarm intelligence are reviewed, including swarm intelligence optimization algorithms, modeling and theories and swarm robotics.

The emphasis of this survey is placed on recent advances in this field. Still, with thousands of papers published about swarm intelligence every year, it is inevitable that some important and interesting works are not included. Instead of a comprehensive survey, we suggest readers consider this paper as a rather brief index of this field.

1.1.1 Concept

The concept of swarm intelligence was first introduced in 1989 in the context of cellular robotic systems [1]. Then it has been rapidly spread and developed since the invention of particle swarm optimization (PSO) [2]. It is also quite closely related to other concepts like collective intelligence [3], multiagent system [4], nature/bioinspired computation [5], metaheuristic [6] and evolutionary computation [7].

[1]Department of Machine Intelligence, School of Electronics Engineering and Computer Science, Peking University, China

So far, there have been a number of proposed definitions for swarm intelligence. For examples:

1. The property of a system whereby the collective behaviors of (unsophisticated) agents interacting locally with their environment cause coherent functional global patterns to emerge [8].
2. Two or more individuals independently, or at least partially independently, acquire information, and these different packages of information are combined and processed through social interaction, which provides a solution to a cognitive problem in a way that cannot be implemented by isolated individuals [9].
3. Collective behavior of decentralized, self-organized systems, natural or artificial [10].

They look generally similar but actually refer to different things: property, population, behavior, etc.

It is said that the term "swarm intelligence" is difficult to define because the term "intelligence" is very obscure itself [11]. However, some essential properties of swarm intelligence have been widely accepted [12]:

1. It is composed of multiple individuals.
2. The individuals are relatively homogeneous.
3. The interactions among the individuals are based on simple behavioral rules.
4. The overall behavior of the system emerges from the interactions of individuals.

1.1.2 Tendency

So far, more than 12,000 papers have been published in the field of swarm intelligence. Figure 1.1 shows the number of papers the *topics* of which are related to swarm

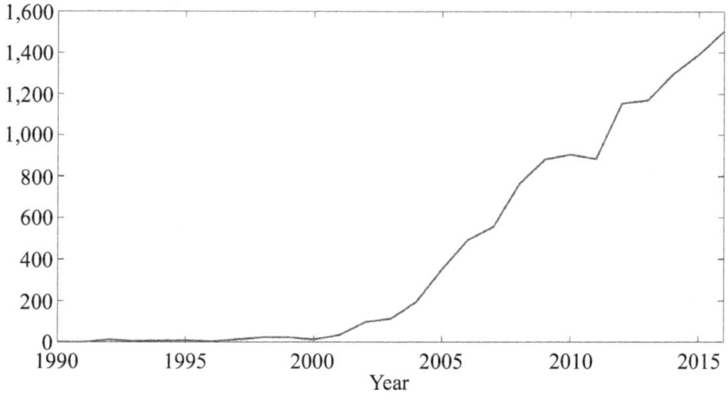

Figure 1.1 Number of papers about swarm intelligence published in each year

intelligence in each year.[1] It can be seen that the number of papers in this field has been growing dramatically in this century.

Table 1.1 shows the most highly cited papers related to swarm intelligence by the date of October 9, 2017.[2] In the history of swarm intelligence, PSO, ant colony optimization (ACO) and artificial bee colony (ABC) are the most popular algorithms. Besides, swarm robotics has also been a very hot spot.

Figure 1.2 shows the countries/territories from which scholars have published most papers. Chinese scholars have contributed about half papers in this field, followed by American and Indian scholars.

1.2 Swarm intelligence optimization algorithms

Swarm intelligence optimization algorithms are emerging technologies that search with plenty of intelligent agents which are usually analogous to natural swarm of populations, like ants, birds and bees. These algorithms obtain the optimal solution through cooperation and evolution, aiming to achieve complicated and efficient searching manners from simple individuals and limited communications. Swarm intelligence optimization algorithms have gradually received more and more attentions from academic researchers and practical optimization engineers for their significant advantages:

- **Low requirement for objective functions.**
- **Simplicity.**
- **Parallelism and distributed ability.**

In this section, we are going to discuss researches on classic Swarm Intelligence Optimization Algorithms (SIOAs), novel SIOAs and their applications.

1.2.1 Developments of existing SIOAs

1.2.1.1 Particle swarm optimization

PSO [16,17] was proposed by Kennedy and Elberhart in 1995 and became the most important swarm SIOA since then. PSO simulates the foraging behavior of birds: each individual updates its position iteratively, according to its former momentum, individual history and information from other individuals. Researches of PSO mainly focus on following aspects.

Algorithm improvements

Many researchers concentrate on improving the optimization ability of PSO. Since 2006, successive versions of standard PSO, SPSO2006, SPSO2007 and SPSO2011 [18], are proposed as baselines and available PSO algorithm improved by recent researches. There are some commonly used methods to improve PSO.

[1]Accessed from Web of Science [13], searching for TOPIC = "swarm intelligence."

[2]This list excludes papers that do not mention "swarm intelligence" specifically, such as the early papers of ant colony optimization [14,15] which have been cited for 5,300 and 3,777 times, respectively.

Table 1.1 Highly cited papers of swarm intelligence

Authors	Title	Times cited	Year published	Conference/journal
Kennedy, J; Eberhart, R	Particle swarm optimization	22,187	1995	IEEE International Conference on Neural Networks
Jadbabaie, A; Lin, J; Morse, AS	Coordination of groups of mobile autonomous agents using nearest neighbor rules	4,204	2003	IEEE Transactions on Automatic Control
Karaboga, Dervis; Basturk, Bahriye	A powerful and efficient algorithm for numerical function optimization: artificial bee colony (ABC) algorithm	1,812	2007	Journal of Global Optimization
Moreau, L	Stability of multiagent systems with time-dependent communication links	1,524	2005	IEEE Transactions on Automatic Control
Karaboga, D; Basturk, B	On the performance of ABC algorithm	1,422	2008	Applied Soft Computing
Liang, JJ; Qin, AK; Suganthan, Ponnuthurai Nagaratnam; *et al.*	Comprehensive learning particle swarm optimizer for global optimization of multimodal functions	1,403	2006	IEEE Transactions on Evolutionary Computation
Dorigo, M; Di Caro, G; Gambardella, LM	Ant algorithms for discrete optimization	1,395	1999	Artificial Life
Shi, YH; Eberhart, R	A modified particle swarm optimizer	1,209	1998	IEEE International Conference on Evolutionary Computation
Passino, KM	Biomimicry of bacterial foraging for distributed optimization and control	1,167	2002	IEEE Control Systems Magazine
Karaboga, Dervis; Akay, Bahriye	A comparative study of ABC algorithm	1,098	2009	Applied Mathematics and Computation

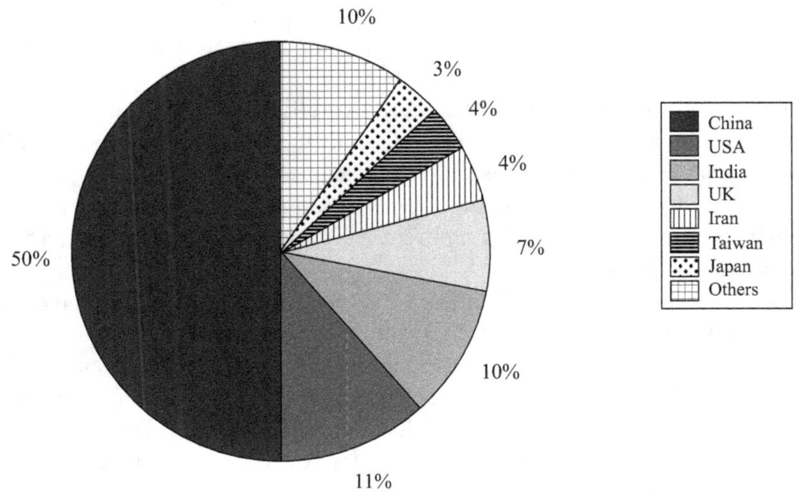

Figure 1.2 Number of papers about swarm intelligence published by authors from each country/territory

One method to build stronger version of PSO is to introduce mechanism from other fields like mathematics, physics and biology. A Quantum PSO [19,20] is proposed since 2004, and there are still researches proposing new PSO variations based on quantum behavior. Meng *et al.* proposed a new QPSO with special selection and mutation methods to solve economic load dispatch problem [21]. Dos Santos Coelho *et al.* proposed a novel quantum-inspired particle swarm optimization (QPSO) with Gaussian mutation operator for constrained engineering design problems [22]. Chaos is also introduced in PSO variations. Park *et al.* proposed an improved PSO with chaotic sequences for nonconvex economic dispatch problems [23]. Gandomi *et al.* combined an accelerated PSO and chaos and proposed chaotic APSO to enhance its global search ability [24]. Other introduced methods are fuzzy theory [25], opposition [26], biological clone [27], time-varying acceleration coefficients [28] and so on.

Another major approach to improve PSO is to create hybridization of PSO with other metaheuristic methods. PSO has been combined with genetic algorithm (GA) [29], ACO [30], ABC [31], differential evolution [32], simulated annealing [33] and other methods.

Some researchers focus on the topology structures of individuals in PSO and how it affects searching behaviors. Li proposed a PSO with ring topology to achieve niching without niching parameters [34]. Lim and Isa proposed PSO which gradually increases its topology connectivity to improve efficiency [35]. Tsujimoto *et al.* carried out an empirical research on the relation between network topology and search performance [36].

Theoretical analysis

Theoretical analysis of PSO mainly includes its parameter selection and convergence analysis.

Parameters of PSO impact on its optimization efficiency remarkably, researchers have employed many techniques to tune the parameters for different versions of PSO. Nickabadi *et al.* considered various strategies designed for the inertia weight parameter and proposed a new PSO with adaptive inertia weight [37]. Zheng *et al.* conducted convergence analysis for several simplified PSO and proposed efficacy parameter selection methods [38].

The convergence property of standard PSO was proved in [39–41]. And recently, there were convergence analysis of improved PSOs. For example, Lin implemented PSO for dynamic router node placement in wireless mesh network and analyzed convergence of PSO with constriction coefficient [42]. In practical applications, PSO might converge to global optimal or any local optimal positions. Different methods of analysis on how PSO behaves or how to guarantee convergence to global optimal is taken. For examples, Sun *et al.* conducted convergence analysis and parameters selection for PSO with inertia weight in [43].

Extensions to other optimization fields

PSO can be extended for some difficult optimization situations like multiobjective optimization [44], constrained optimization [45] or combinatorial optimization [46].

Applications and implementations

PSO can be directly implemented for many engineering problems, like photovoltaic (PV) system design [47,48], wind power forecasting [49] and so on.

Researchers also implemented PSO or modified PSO to tune parameters of other models. It was applied for clustering [50], support vector machine (SVM) [51] and fuzzy system [52].

Some other researchers studied how to exploit the parallel ability of PSO and implement it more efficiently on multicore central processing unit (CPU), graphics processing unit (GPU) or distributed systems. Cagnoni *et al.* implemented PSO with OpenCL and compared its performance between multicore CPU and GPU [53]. Mussi *et al.* evaluated the parallel ability of PSO within CUDA architecture [54]. Elamvazuthi *et al.* compared four parallel and distributed PSO in [55].

1.2.1.2 Ant colony optimization

ACO was proposed by Marco Dorigo in 1992 in his Ph.D. thesis. ACO simulates the foraging behavior of ants, which not only finds the targets scattered in the searching space but also finds the shortest path to them. Unlike PSO and most other SIAOs, individuals in ACO communicate through the environment by sensing and secreting pheromones (as shown in Figure 1.3).

We are going to exhibit researches of ACO on following aspects.

Algorithm analysis and improvements

Some fundamental researches have been done for ACO to improve its efficiency, understand its principles and extend it for more complicated problems. Dorigo and

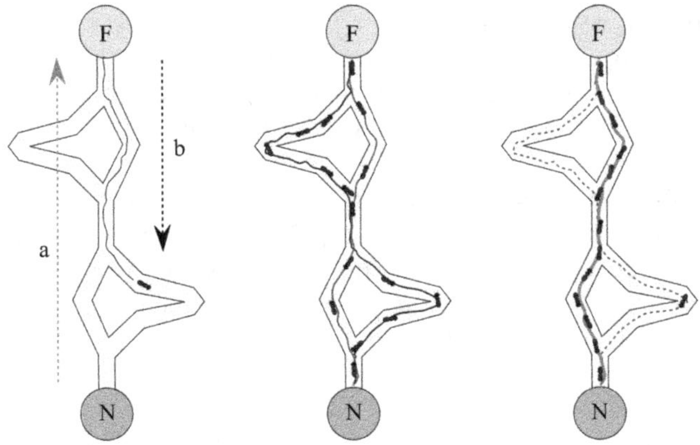

Figure 1.3 Convergence of ant colony

some researchers have created different variations of ACO since the first one, ant colony system, was proposed by Dorigo in 1996 [14]. They mainly focus on four components of ACO: transformation rules [15,56], updating rules of pheromones [57], local pheromones [58] and candidate list [59].

Recently, there are still improvements made upon the ACO framework. Stützle *et al.* proposed an online adaptation method of the parameters of ACO [60]. Zhao *et al.* introduced mutation operator from GA to get rid of local optimum [61]. Puris *et al.* proposed a two-stage methodology for ACO and proved its efficacy for traveling salesman problem (TSP) and quadratic assignment problem (QAP) problems [62].

Researches on fundamental theory of ACO started long ago. For example, Stützle and Dorigo have given a convergence proof of ACO for a class of ACO [63]. Theoretical analysis of ACO is also conducted in recent years. Huang *et al.* analyzed the convergence speed of ACO based on the absorbing Markov chain model and revealed the relationship between convergence time and pheromone rate [64]. Oliveira *et al.* conducted a detailed analysis of population-based ant colony optimization algorithm (P-ACO) for TSP and QAP problems and investigated the impact of a local search on P-ACO parameters and performance [65].

ACO is also combined with other optimization methods. Huang and Liao combined ACO and taboo search for the job shop scheduling problem [66]. Kiran *et al.* proposed a hybridization of ACO and PSO for an energy demand model to forecast energy demand [67].

Extensions and implementations

The behavior of ants is naturally defined on graphs, which makes ACO suitable for specific problems like TSP. ACO was also implemented for continuous or discrete optimization problems, or even more complex ones. Socha and Dorigo extended

ACO for continuous domains in [68]. In [69], Dorigo showed how to apply ACO for a number of discrete optimization problems. In recent years, ACO has been modified as different versions of multiobjective optimization methods for solving virtual machine placement in cloud computing [70], flow shop scheduling [71] and supply chain designing [72].

ACO was also implemented on different platforms. Cecilia *et al.* proposed a series of parallelism methods of ACO on GPU in [73]. Dawson *et al.* implemented ACO on GPU and applied a parallel Double-Spin Roulette selection and new pheromone update rules for TSP [74]. Tsutsui *et al.* designed and implemented an ACO with the MapReduce framework on Hadoop [75].

Applications

Thanks to the unique principle of ACO; it suits a group of problems which are hard for other optimization methods, like TSP. Researches that apply ACO on classic problems have begun long ago, and there are still works in this area. Hlaing and Khine proposed a new ACO system with well distribution strategy and information entropy which is conducted on the configuration strategy for updating the heuristic parameter in ACO for TSP [76]. Mavrovouniotis *et al.* proposed an ACO algorithm with immigrants schemes for the dynamic traveling salesman problem with traffic factors [77]. ACO is also frequently used for scheduling tasks. Li *et al.* proposed a load-balancing ACO algorithm for cloud task scheduling [78]. Xing *et al.* proposed a knowledge-based ACO algorithm for flexible job scheduling problems [79].

1.2.1.3 Fireworks algorithm

Fireworks Algorithm (FWA) is a relatively young swarm intelligence optimization algorithm compared with PSO and ACO. It was proposed by Tan *et al.* in 2010 [80]. Instead of natural creature community, FWA simulates the fireworks in the sky. Each individual in FWA is called a firework, which explodes and generates a group of sparks to search a local area. Fireworks converge by explosion and selection iteratively, and the core of FWA becomes the resource (sparks numbers and explosion amplitudes) assignment (see Figure 1.4).

As a new class of SIOA, FWA quickly becomes an active research area, and plenty of works have been done in different aspects of FWA like theoretical analysis, algorithm improvements and applications.

Theoretical analysis

As one of the most important theoretical topic of SIOAs, analysis of convergence and convergence rate for FWA was conducted in [81], in which Liu *et al.* considered FWA as a Markov process and analyzed its time complexity of convergence. Li *et al.* analyzed the information utilization in heuristic algorithms and proposed a novel guiding spark mutation methods for FWA [82].

Algorithm improvements

Researchers have improved the original FWA in many ways.

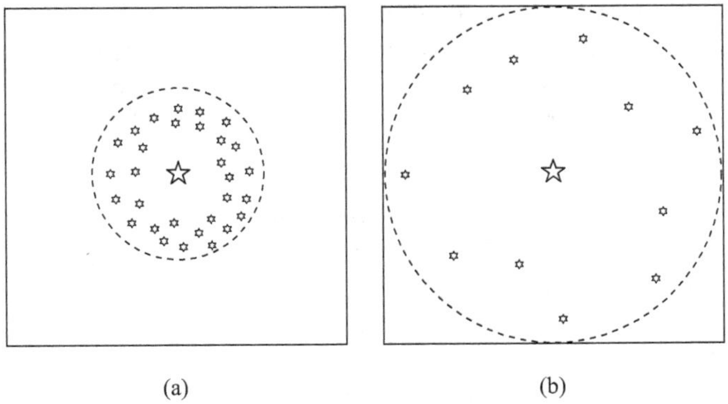

Figure 1.4 Explosion in FWA. (a) Explosion of good firework. (b) Explosion of bad firework

Most fundamental improvements of FWA were modifications on its core components, like explosion (including sparks assignment and explosion amplitudes) and mutation (generate additional sparks to accelerate optimization or increasing diversity). For explosion methods, Zheng *et al.* proposed a new version of FWA (dynFWA [83]) with dynamic explosion amplitude for core-firework (best individual) which changes according to whether the current best solution is improved. Li proposed a FWA with adaptive explosion amplitudes (AFWA [84]) that fits the local search conditions. Liu *et al.* designed a transfer function to replace the linear assignment of sparks number and explosion amplitudes along with the best individual selection (IFWABS [85]), and this method is improved further by Si with an adaptive transfer function (FWA-ATF [86]). For mutation operators, Zheng *et al.* first introduced Gaussian mutation in EFWA [87]. Yu *et al.* introduced a covariance mutation operator from covariance matrix adaptation-evolution strategy (CMA-ES) for FWA [88]. And Li *et al.* have proposed a novel guiding spark mutation operator [82]. In order to enhance the cooperation ability of fireworks in FWA, specific techniques have been invented. Zheng *et al.* proposed a cooperative framework for FWA [89] with crowdness-avoiding strategy, which means fireworks too close to the best firework are restarted.

FWA was also combined with other heuristic optimization methods. Yu *et al.* proposed a FWA with differential mutation for the CEC 2014 competition problems [90]. Zhang *et al.* combined FWA with a biogeography-based optimization methods to enhance information sharing among the population [91].

Implementations and extensions

As a swarm-based algorithm, FWA can be implemented in different parallel settings to improve computing efficiency. Ding *et al.* proposed a parallel version of FWA and implemented it on GPU since 2013 [92].

FWA is also extended for complicated optimization problems. Tan proposed a *s*-metric based multiobjective FWA for multiobjective problems [93]. Bacanin *et al.* applied FWA for portfolio optimization problems with constrains [94]. And Tan also proposed a discrete FWA for combinatorial optimization problems [95].

Applications
Except the applications in researches mentioned above, FWA has also been utilized for a large group of engineering and academic applications. Imran and Kowsalya implemented FWA to solve the distribution network reconfiguration problem for power system to minimize power loss an enhance voltage profile [96]. He *et al.* applied FWA to optimize the parameters of a local-concentration model for spam detection [97]. And Tuba *et al.* utilized FWA for multilevel image thresholding [98].

1.2.2 Novel SIOAs

1.2.2.1 Water wave optimization

Water wave optimization (WWO) is a young metaheuristic optimization method proposed by Zheng in 2015 [99]. WWO takes inspiration from shallow water wave models, that (for maximization problem) represents searching space as the seabed area and represents fitness as the water depth. Each individual in WWO is analogous to a wave which has a height and wavelength. And in each iteration, individuals in WWO updates by propagation, refraction and breaking operations. The convergence of WWO was analyzed in [100].

After the creation of WWO, Zheng and Zhang proposed a simplified wave optimization algorithm which leave out the refraction operator and introduced a population size reduction strategy [101]. Zhang proposed a WWO called VC-WWO [102] with variable population size and introduced a comprehensive learning mechanism for the refraction operator. Wu *et al.* proposed elite opposition-based strategy WWO [103] by a series of modifications including an elite opposition-based learning strategy, a local neighborhood search strategy and an improved propagation operator.

WWO has already been implemented in many practical optimization problems. Wu *et al.* utilized WWO for traveling salesman problem [104]. Yun proposed a WWO-based memetic algorithm to solve flow-shop scheduling problem [105]. And Zhao *et al.* used WWO for infinite impulse response (IIR) filters designing [106].

1.2.2.2 Brain storm optimization

Brain storm optimization (BSO) was inspired by one of the creative problem solving processes of human being, that is, the brainstorming process. In brainstorming, a group of people propose some ideas for a problem, and these ideas are evaluated and selected. New ideas are generated based on former ideas the person accepted iteratively and finally a good idea for the problem would be concluded. This process is summarized by BSO as grouping, replacing and creating operators.

The original BSO algorithm [107] was proposed in 2011 by Shi. Plenty of improvements on BSO were conducted in the following years. Zhan *et al.* proposed a modified BSO [108] which introduce a new simple grouping method and a novel idea difference strategy. Duan *et al.* proposed a BSO variation with predator–prey strategy [109] to enhance the population diversity. Zhou *et al.* improved BSO by a modified step-size strategy and a new individual generation and selection scheme [110].

Some parts of theoretical analysis of BSO were attempted. Cheng *et al.* conducted analysis of solution clustering and other properties of BSO in [111]. And Zhan *et al.* conducted investigation on the parameters of BSO and proposed a new variant [112].

BSO has already been used in optimal satellite formation reconfiguration [113], economic dispatch [114], wireless sensor networks deployment [115] and many other applications.

1.2.2.3 Whale optimization algorithm

Whale optimization algorithm (WOA [116]) is a newly proposed (in 2016) optimization algorithm inspired by the hunting method of humpback whales. Each iteration of the algorithm is combined a prey encircling operation and a bubble-net attacking operation.

Although WOA has only been proposed for a short time, there are already plenty of work done on this area. A Kaveh *et al.* proposed an enhanced WOA [117] which modified the original formulation of WOA. Trivedi *et al.* integrated WOA with a novel randomization technique termed adaptive technique [118]. Ling *et al.* proposed a new variation of WOA [119] based on a Levy flight trajectory for global optimization.

WOA has been applied in many areas. HJ Touma applied WOA for economic dispatch problem [120]. Aljarah *et al.* [121] and Bhesdadiya *et al.* [122] implemented WOAs to train neural networks, and Cherukuri *et al.* applied WOA for global maximum power point (MPP) tracking of PV system [123].

1.3 Swarm robotics

1.3.1 Introduction

1.3.1.1 Definition of swarm robotics

Swarm robotics is a new approach to the coordination of multirobot systems, which consist of large numbers of mostly simple physical robots [124]. It is supposed that a desired collective behavior emerges from the interactions between the robots and interactions of robots with the environment [125]. People have done lots of research in various areas of swarm robotics [126,127]. This approach emerged on the field of artificial swarm intelligence, as well as, but not limited to, the biological studies of insects, ants and other fields in nature, where swarm behavior occurs.

1.3.1.2 Characteristics of swarm robotics

Some researchers have summarized the properties of swarm robotics, such as robustness, flexibility and scalability [125]. In this section, advantages and characteristics of the swarm robotics system are presented through comparison with single robot.

To complete a sophisticated task, the single robot must be designed with complicated structure and control modules resulting in high cost of designing, construction and maintenance. Single robot is vulnerable especially when a small broken part of the robot may affect the whole system and difficult to predict what will happen. Swarm robotics can achieve the same ability through intergroup cooperation and takes the advantage of reusability of the simple agents and low cost of construction and maintenance. The advantages of swarm robotics are listed below.

- **Parallel:** The population size of swarm robotics is usually quite large and can deal with multiple targets in one task. This indicates the swarm can perform greatly in tasks involving multiple targets distributed in a vast range in the environment, and the search of the swarm would save time significantly. Some distributed algorithms can be designed accordingly [128].
- **Scalable:** All the interaction in the swarm is local, and this allows individuals to join or quit the task at any time without interrupting the whole swarm. The swarm can adapt to the changes of population through implicit task re-allocating schemes without the need of any external operations. This also indicates the system is adaptable for different size of population without any modification of the software or hardware which is very useful for real-life applications. To some extent, the scalability is related to the distributed control [129].
- **Stable:** Similar to scalability, swarm robotics systems will not be affected greatly even when part of the swarm quitted due to majeure factors. The swarm can still work toward the objective of the task although performances may degrade inevitably with fewer robots. This feature is especially useful for tasks under dangerous environments.
- **Economical:** As mentioned before, cost of swarm robotics is significantly low in designing, manufacturing and daily maintaining. The whole system is cheaper than a complex single robot even the robotics in the swarm can be hundreds or thousands. Since the individuals in the swarm can be massively produced while single robot requires precision machining.
- **Energy efficient:** Since individuals in the swarm are much smaller and simpler than a giant robot, the energy cost is far beyond the single robot comparing with the battery size. This means that the lifetime of the swarm is enlarged. In environments without fueling facilities or where wired electricity is forbidden, swarm robotics can be much useful than traditional single robot.

1.3.1.3 Fundamental tasks of swarm robotics

In past decades, swarm robotics is deployed in various scopes of applications [125] including odor localization, mobile sensor networking, medical operations, surveillance and search-and-rescue. The tasks of these applications are very sophisticated and hard to propose a direct solution. To solve these tasks, we propose several basic tasks of swarm robotics, such as flocking, navigating, obstacle avoidance, etc.

- **Flocking strategy and formation:** Flocking is widely observed in many nature swarms or even human beings. The Boids model, proposed by Reynolds [130], is a typical individual model for flocking behavior using distance metrics. The model is widely adopted in various applications including spacecraft, UAV, robot, etc. In these applications, group behaviors cannot be explicitly defined at group level, and the use of individual rules are adopted [131].

- **Directed flocking:** Besides the flocking strategy, direction control in flocking are the most concern in flocking researches and are widely adopted in navigation, migration and searching applications. Until now, there already exist a number of studies for directing the swarm with target positions and propagating information in the swarm.
- **Positioning and navigation:** In flocking and migration, positioning of the goal, nearby robots and various obstacles in the fields is also an important task. In the application taking place in the large outdoor environment, global positioning is expensive and requires more hardware, which is unaffordable for swarm robotics. Thus, local positioning in flocking should be specially focused.
- **Obstacle avoidance:** Obstacle avoidance is also considered an important basic task in the swarm robotics society. In most researches, some sort of potential functions are applied to robots. The swarm steers around the encountered obstacles according to the potential fields.

As to the tasks performed by swarm robotics, they also can be viewed from the angle of swarm behaviors. And the tasks or behaviors in swarm robotics include aggregation, flocking, foraging [132], object clustering and sorting, object assembly [133], navigation, path formation, deployment, collaborative manipulation, task allocation, order source localization, hole avoidance, decoy avoidance [134], rendezvous problem, MTT [135] and so on. Some other tasks such as human control [136], fault detection [137] and adversarial subversion [138] also attract the researchers. Some researchers [139] have also classified these collective behaviors into four categories: spatially organizing behaviors, navigation behaviors, collective decision-making [140,141], and other collective behaviors, which is comprehensive and reasonable. Some more specialized problems are also formalized and categorized [142].

1.3.2 Application scopes of swarm robotics

The studies on robotics application in target searching have grown substantially in recent years. It is more preferable when the working area is dangerous or inaccessible to human. Deduced from the advantages and characteristics introduced in previous section, several potential application scopes [125,143] of swarm robotics which are very suitable are listed below.

- **Tasks cover large areas:** Swarm robotics system is distributed and specialized for tasks requiring a large area of spaces, e.g. tasks cover large areas [144]. Robots in the swarm are distributed in the environment and can detect dynamic changes of the entire area, such as chemical leaks or pollution.
- **Tasks dangerous to robots:** Thanks to the scalability and stability, the swarm provides redundancy for dealing with dangerous tasks. Swarm can suffer loss of

robots to a great extent before the job has to be terminated. The robots are very cheap and are first choices in areas which will probably damage the workers.

- **Tasks require scaling population:** Workload in some tasks may change over time, and the swarm size should scale based upon the current workload for high efficiency in both time and economic. For example, in the task of clearing oil leakage after tank accidents, the swarm should maintain high population when the oil leaks fast at the beginning of the task and gradually reduce the robots when the leak source has been plugged and the leaking area is almost cleared.
- **Tasks require redundancy:** Robustness in swarm robotics systems mainly benefits from the redundancy of the swarm, i.e. removing some robots does not have a significant impact on the performance. Some tasks focus mainly on the result rather than the process, i.e. the system should make sure the task will complete successfully, mostly in the way of increasing redundancy.

1.3.3 Swarm robotics searching algorithms

1.3.3.1 Earlier progress of swarm robotics algorithms

In the earlier years of swarm intelligence research, scientists simulated the cooperative mechanisms in nature and explored the possibility of reproducing swam behaviors using artificial agents.

Self-organizing clustering observed in bacteria was one of the first swarm behaviors reproduced by scientists [145]. The individuals in the swarm are controlled by a simple rule: the possibility of joining or leaving a colony is conducted by the density nearby. In the experiment, 1,500 individuals in the swarm gradually clustered into three colonies without any prior information or external control.

A similar approach simulating ants behavior of clearing up graves are also proposed [146]. The task of the swarm is to collect all the items in the area together. There are no predefined storage spots available. Individual in swarm follows a simple and local rule to transport an item from spot of low-to-high densities only. Experiment shows the swarm completed the task for collecting 80 items without communication. They also explored that how these rules can impart on the result.

Currently, swarm robotic searching algorithms are the most concern by the researchers besides those basic tasks mentioned in previous section. In this section, we classify searching strategies in two types: inspired from swarm intelligence algorithms or inspired from other methods. These two types of algorithms vary from many aspects, such as searching schemes, ways to detect targets and information exchanges inside the swarm. As a distributed system, swarm robotics is very suited for tasks requiring lots of individuals, such as area coverage or multitarget search. Multitarget search in swarm robotics [147] is a process where a swarm of robots try to find and collect multiple targets distributed in a vast unknown environment.

1.3.3.2 Inspired from swarm intelligence algorithms

From the general point of view, swarm optimization algorithms share several similarities with swarm robotics searching, e.g. searching for the best points using a swarm of individuals. PSO is the swarm intelligence approach that adopted most in the swarm robotics due to the great similarity with flocking and searching schemes. Besides PSO, other methods have also inspired many successful approaches, such as ACO and glowworm swarm optimization. The scope of these approaches includes path finding [148], navigation, odor localization, etc. Couceiro *et al.* [149] proposed a Robotic Darwinian Particle Swarm Optimization (RDPSO) for solving the last issue. They divide the swarms into subswarms with dynamic topology updated in several iterations based on a reward and punishment mechanism. We divide the usage of swarm intelligence algorithms into three types:

- **Optimize Parameters:** The first type of searching algorithms inspires strategies from other methods with several parameters which are hard to optimize, such as neural network or heuristic schemes. Swarm intelligence algorithms are employed to optimize these parameters. Meng [150] proposed a collective construction task consisting of searching for randomly distributed building blocks and transporting these blocks to predefined locations. The method employed virtual pheromone trail for information exchange and task allocation for cooperative transportation. A modified PSO is proposed to balance the exploration and exploitation in their work.
- **Modeling individual behaviors:** Pugh and Martinoli [151] explored the possibility of adopting PSO strategies in swarm robotics searching directly. Each robot is regarded as a particle, and various neighborhood topologies and update strategies are verified. In their follow-up work [152], they designed an effective algorithm that allows a swarm of robots to work together to find targets. They proposed techniques inspired from PSO modified to mimic the swarm robotics search process. Analysis of parameters and setups in the model are also presented at an abstracted level. Group explosion strategy (GES) [153] is the first search strategy we proposed, which borrows the explosion schemes from the FWA [80], a swarm intelligence optimization algorithm inspired by the firework explosion. In GES, intragroup cooperation helps to converge quickly, while intergroup cooperation enhances the parallelism of the swarm to search multiple targets, which make GES a promising search strategy. However, several shortcomings are observed in our simulations especially when the population is very large. In GES, a random one is chosen if multiple robots in a group share the same fitness value, which can cause the robot to get stuck or fall back to worse regions [154].
- **Mixing two methods above:** Some algorithms try using the swarm intelligence model and optimizing parameters using swarm intelligence in the same time. Doctor, Venayagamoorthy and Gudise [155] proposed a method utilizing two layers of PSO for controlling unmanned mobile robots in target tracing applications. The robots are controlled by schemes in inner layer of the PSO, and parameters of inner layer are optimized by the outer layer. Signal intensity from targets is defined as the fitness to search for the swarm.

1.3.3.3 Inspired from other methods

Olfaction is a common ability that animals use in their everyday activities, such as hunting, mating, interacting and evading predators. Such schemes inspired from animals are widely used in swarm robotics applications localizing odor sources, which have attracted a growing interest in areas such as antiterrorists, locating toxic or harmful gas leakage, checking for contraband, exploring mineral resources in dangerous areas and search-and-rescue in collapsed building [156].

Cui *et al.* [157] proposed a biasing expansion swarm approach to collaboratively search and locate various number of emission sources in an unknown area using a swarm of simple robots. Jatmiko *et al.* [158] provided a model of odor-gated rheotaxis combined from chemotaxic and anemotaxic (upstream) methods to solve odor source localization problems. The combined model can achieve high accuracy in real life scenarios containing dynamic sources, random winds and obstacles.

Russell *et al.* [159] summarized and compared the implementation and evaluation of four chemotaxis algorithms which provide fast, simple and cost-effective solutions for olfaction-based searching applications in obstructive environments. They listed the details of the algorithms together with typical results of these algorithms obtained in both simulated and practical experiments. Besides olfaction, other searching applications and strategies are also proposed. Varela *et al.* [160] developed an algorithm aimed for coordination of a group of UAVs to monitor environment. The UAV swarm can locate the undesired phenomenon. The UAVs compare the average fitness of last five iterations with their neighbors and select the direction of the best neighbor to search in the next iteration. They validated the algorithm in real UAVs monitoring and industrial area.

Wu and Zhang [161] develop a switching strategy for locating a local minimum in an unknown noisy scalar field. Robots will switch to cooperative exploration only when they are not able to converge to a local minimum at a satisfying rate according to a cooperative filter. The switching strategy can result in faster convergence and is robust to unknown noises and communication delays.

A robust-satisficing approach based on info-gap theory is suggested as a solution for a spatial search-planning problem by Sisso *et al.* [162]. The swarm is given uncertain prior information data with severe errors. The proposed method shows great superior in robustness to the expected-utility maximizing strategy.

Lee and Ahn [163] proposed a foraging algorithm specially focused on energy efficiency. Through adding several temporal storage stations in the environment, the swarm can improve searching efficiency since robot will move shorter distance to the storage before next forage. The swarm is divided into two part, one part searches for the food and sends the food to the nearest station, while others transfer from the station to the nest. In this way, both time and energy efficiency are improved.

1.4 Conclusion

In this chapter, the concept of swarm intelligence is first discussed. Then some recent developments of swarm intelligence optimization algorithms and swarm robotics are surveyed. Some hot spots and research directions of swarm intelligence are pointed out.

References

[1] Gerardo Beni and Jing Wang. Swarm intelligence in cellular robotic systems. In *Robots and Biological Systems: Towards a New Bionics?*, pages 703–712. Springer, Berlin, Heidelberg, 1993.

[2] James F Kennedy, James Kennedy, Russell C Eberhart, and Yuhui Shi. *Swarm Intelligence*. Morgan Kaufmann, San Francisco, CA, 2001.

[3] Pierre Lévy. *Collective Intelligence*. Plenum/Harper Collins, New York, 1997.

[4] Michael Wooldridge. *An Introduction to Multiagent Systems*. John Wiley & Sons, 2009.

[5] Xin-She Yang. *Nature-Inspired Metaheuristic Algorithms*. Luniver Press, 2010.

[6] Ibrahim H Osman and James P Kelly. Meta-heuristics: an overview. In *Meta-Heuristics*, pages 1–21. Springer, Boston, MA, 1996.

[7] Thomas Bäck, David B Fogel, and Zbigniew Michalewicz. Handbook of evolutionary computation. *Release*, 97(1):B1, 1997.

[8] Eric Bonabeau, Marco Dorigo, and Guy Theraulaz. Swarm intelligence: from natural to artificial systems. *Quarterly Review of Biology*, 2003.

[9] Jens Krause, Graeme D Ruxton, and Stefan Krause. Swarm intelligence in animals and humans. *Trends in Ecology & Evolution*, 25(1):28–34, 2010.

[10] Wikipedia. Swarm intelligence—Wikipedia, the free encyclopedia, 2017. [Online; accessed 3-Oct-2017].

[11] Gerardo Beni. *From Swarm Intelligence to Swarm Robotics*, pages 1–9. Springer Berlin Heidelberg, Berlin, Heidelberg, 2005.

[12] Marco Dorigo and Mauro Birattari. Swarm intelligence. *Scholarpedia*, 2(9):1462, 2007. revision #138640.

[13] ISI. Web of science. www.isiknowledge.com, 2017.

[14] Marco Dorigo, Vittorio Maniezzo, and Alberto Colorni. Ant system: optimization by a colony of cooperating agents. *IEEE Transactions on Systems, Man, and Cybernetics, Part B (Cybernetics)*, 26(1):29–41, 1996.

[15] Marco Dorigo and Luca Maria Gambardella. Ant colony system: a cooperative learning approach to the traveling salesman problem. *IEEE Transactions on Evolutionary Computation*, 1(1):53–66, 1997.

[16] Russell Eberhart and James Kennedy. A new optimizer using particle swarm theory. In *Micro Machine and Human Science, 1995. MHS'95, Proceedings of the Sixth International Symposium on*, pages 39–43. IEEE, 1995.

[17] James Kennedy and Russell C Eberhart. Particle swarm optimization. In *IEEE International Conference on Neural Networks, 1995. Proceedings*, volume 4, pages 1942–1948, 2002.

[18] Particle swarm central. http://www.particleswarm.info.

[19] Jun Sun, Wenbo Xu, and Bin Feng. A global search strategy of quantum-behaved particle swarm optimization. In *Cybernetics and Intelligent Systems, 2004 IEEE Conference on*, volume 1, pages 111–116. IEEE, 2004.

[20] Jun Sun, Bin Feng, and Wenbo Xu. Particle swarm optimization with particles having quantum behavior. In *Evolutionary Computation, 2004. CEC2004. Congress on*, volume 1, pages 325–331. IEEE, 2004.

[21] Ke Meng, Hong Gang Wang, ZhaoYang Dong, and Kit Po Wong. Quantum-inspired particle swarm optimization for valve-point economic load dispatch. *IEEE Transactions on Power Systems*, 25(1):215–222, 2010.

[22] Leandro dos Santos Coelho. Gaussian quantum-behaved particle swarm optimization approaches for constrained engineering design problems. *Expert Systems with Applications*, 37(2):1676–1683, 2010.

[23] Jong-Bae Park, Yun-Won Jeong, Joong-Rin Shin, and Kwang Y Lee. An improved particle swarm optimization for nonconvex economic dispatch problems. *IEEE Transactions on Power Systems*, 25(1):156–166, 2010.

[24] Amir Hossein Gandomi, Gun Jin Yun, Xin-She Yang, and Siamak Talatahari. Chaos-enhanced accelerated particle swarm optimization. *Communications in Nonlinear Science and Numerical Simulation*, 18(2):327–340, 2013.

[25] Fevrier Valdez, Patricia Melin, and Oscar Castillo. An improved evolutionary method with fuzzy logic for combining particle swarm optimization and genetic algorithms. *Applied Soft Computing*, 11(2):2625–2632, 2011.

[26] Hui Wang, Zhijian Wu, Shahryar Rahnamayan, Yong Liu, and Mario Ventresca. Enhancing particle swarm optimization using generalized opposition-based learning. *Information Sciences*, 181(20):4699–4714, 2011.

[27] Junqi Zhang, Zhongmin Xiao, Ying Tan, and Xingui He. Hybrid particle swarm optimizer with advance and retreat strategy and clonal mechanism for global numerical optimization. In *Evolutionary Computation, 2008. CEC 2008. (IEEE World Congress on Computational Intelligence). IEEE Congress on*, pages 2059–2066. IEEE, 2008.

[28] Asanga Ratnaweera, Saman K Halgamuge, and Harry C Watson. Self-organizing hierarchical particle swarm optimizer with time-varying acceleration coefficients. *IEEE Transactions on Evolutionary Computation*, 8(3):240–255, 2004.

[29] Mohammad Hasan Moradi and M Abedini. A combination of genetic algorithm and particle swarm optimization for optimal DG location and sizing in distribution systems. *International Journal of Electrical Power & Energy Systems*, 34(1):66–74, 2012.

[30] Yannis Marinakis and Magdalene Marinaki. A hybrid multi-swarm particle swarm optimization algorithm for the probabilistic traveling salesman problem. *Computers & Operations Research*, 37(3):432–442, 2010.

[31] Xiaohu Shi, Yanwen Li, Haijun Li, Renchu Guan, Liupu Wang, and Yanchun Liang. An integrated algorithm based on artificial bee colony and particle swarm optimization. In *Natural Computation (ICNC), 2010 Sixth International Conference on*, volume 5, pages 2586–2590. IEEE, 2010.

[32] Hui Liu, Zixing Cai, and Yong Wang. Hybridizing particle swarm optimization with differential evolution for constrained numerical and engineering optimization. *Applied Soft Computing*, 10(2):629–640, 2010.

[33] Amin Jamili, Mohammad Ali Shafia, and Reza Tavakkoli-Moghaddam. A hybrid algorithm based on particle swarm optimization and simulated annealing for a periodic job shop scheduling problem. *The International Journal of Advanced Manufacturing Technology*, 54(1):309–322, 2011.

[34] Xiaodong Li. Niching without niching parameters: particle swarm optimization using a ring topology. *IEEE Transactions on Evolutionary Computation*, 14(1):150–169, 2010.

[35] Wei Hong Lim and Nor Ashidi Mat Isa. Particle swarm optimization with increasing topology connectivity. *Engineering Applications of Artificial Intelligence*, 27:80–102, 2014.

[36] Takahiro Tsujimoto, Takuya Shindo, Takayuki Kimura, and Kenya Jin'no. A relationship between network topology and search performance of PSO. In *Evolutionary Computation (CEC), 2012 IEEE Congress on*, pages 1–6. IEEE, 2012.

[37] Ahmad Nickabadi, Mohammad Mehdi Ebadzadeh, and Reza Safabakhsh. A novel particle swarm optimization algorithm with adaptive inertia weight. *Applied Soft Computing*, 11(4):3658–3670, 2011.

[38] Yong-Ling Zheng, Long-Hua Ma, Li-Yan Zhang, and Ji-Xin Qian. On the convergence analysis and parameter selection in particle swarm optimization. In *Machine Learning and Cybernetics, 2003 International Conference on*, volume 3, pages 1802–1807. IEEE, 2003.

[39] Frans Van Den Bergh. *An analysis of particle swarm optimizers*. PhD thesis, University of Pretoria, 2007.

[40] Maurice Clerc and James Kennedy. The particle swarm-explosion, stability, and convergence in a multidimensional complex space. *IEEE Transactions on Evolutionary Computation*, 6(1):58–73, 2002.

[41] Ioan Cristian Trelea. The particle swarm optimization algorithm: convergence analysis and parameter selection. *Information Processing Letters*, 85(6): 317–325, 2003.

[42] Chun-Cheng Lin. Dynamic router node placement in wireless mesh networks: a PSO approach with constriction coefficient and its convergence analysis. *Information Sciences*, 232:294–308, 2013.

[43] Xiang Sun, Da-Wei Zhou, and Xi-Wang Zhang. Convergence analysis and parameter selection of PSO model with inertia weight. *Computer Engineering and Design*, 31(18):4068–4071, 2010.

[44] DY Sha and Hsing-Hung Lin. A multi-objective PSO for job-shop scheduling problems. *Expert Systems with Applications*, 37(2):1065–1070, 2010.

[45] Konstantinos E Parsopoulos and Michael N Vrahatis. Particle swarm optimization method for constrained optimization problems. *Intelligent Technologies—Theory and Application: New Trends in Intelligent Technologies*, 76(1):214–220, 2002.

[46] Bassem Jarboui, Najeh Damak, Patrick Siarry, and Ahmed Rebai. A combinatorial particle swarm optimization for solving multi-mode resource-constrained project scheduling problems. *Applied Mathematics and Computation*, 195(1):299–308, 2008.

[47] Kashif Ishaque, Zainal Salam, Muhammad Amjad, and Saad Mekhilef. An improved particle swarm optimization (PSO)–based MPPT for PV with reduced steady-state oscillation. *IEEE Transactions on Power Electronics*, 27(8):3627–3638, 2012.

[48] Masafumi Miyatake, Mummadi Veerachary, Fuhito Toriumi, Nobuhiko Fujii, and Hideyoshi Ko. Maximum power point tracking of multiple photovoltaic arrays: a PSO approach. *IEEE Transactions on Aerospace and Electronic Systems*, 47(1):367–380, 2011.

[49] JPS Catalao, HMI Pousinho, and VMF Mendes. Hybrid wavelet-PSO-ANFIS approach for short-term wind power forecasting in Portugal. *IEEE Transactions on Sustainable Energy*, 2(1):50–59, 2011.

[50] Taher Niknam and Babak Amiri. An efficient hybrid approach based on PSO, aco and k-means for cluster analysis. *Applied Soft Computing*, 10(1):183–197, 2010.

[51] Abdulhamit Subasi. Classification of EMG signals using PSO optimized SVM for diagnosis of neuromuscular disorders. *Computers in Biology and Medicine*, 43(5):576–586, 2013.

[52] Zafer Bingül and Oğuzhan Karahan. A fuzzy logic controller tuned with PSO for 2 DOF robot trajectory control. *Expert Systems with Applications*, 38(1):1017–1031, 2011.

[53] Stefano Cagnoni, Alessandro Bacchini, and Luca Mussi. OpenCL implementation of particle swarm optimization: a comparison between multi-core CPU and GPU performances. In *EvoApplications*, pages 406–415. Springer, 2012.

[54] Luca Mussi, Fabio Daolio, and Stefano Cagnoni. Evaluation of parallel particle swarm optimization algorithms within the CUDA architecture. *Information Sciences*, 181(20):4642–4657, 2011.

[55] Leonardo Vanneschi, Daniele Codecasa, and Giancarlo Mauri. A comparative study of four parallel and distributed PSO methods. *New Generation Computing*, 29(2):129–161, 2011.

[56] Luca Maria Gambardella and Marco Dorigo. Solving symmetric and asymmetric TSPs by ant colonies. In *Evolutionary Computation, 1996, Proceedings of IEEE International Conference on*, pages 622–627. IEEE, 1996.

[57] Thomas Stützle and Holger H Hoos. Max–min ant system. *Future Generation Computer Systems*, 16(8):889–914, 2000.

[58] Marco Dorigo and Luca Maria Gambardella. A study of some properties of ant-q. In *International Conference on Parallel Problem Solving from Nature*, pages 656–665. Springer, 1996.

[59] Bernd Bullnheimer, Gabriele Kotsis, and Christine Strauß. Parallelization strategies for the ant system. In *High Performance Algorithms and Software in Nonlinear Optimization*, pages 87–100. Springer, Boston, MA, 1998.

[60] Thomas Stützle, Manuel López-Ibánez, Paola Pellegrini, *et al.* Parameter adaptation in ant colony optimization. In *Autonomous Search*, pages 191–215. Springer, 2011.

[61] Nan Zhao, Zhilu Wu, Yaqin Zhao, and Taifan Quan. Ant colony optimization algorithm with mutation mechanism and its applications. *Expert Systems with Applications*, 37(7):4805–4810, 2010.

[62] Amilkar Puris, Rafael Bello, and Francisco Herrera. Analysis of the efficacy of a two-stage methodology for ant colony optimization: case of study with TSP and QAP. *Expert Systems with Applications*, 37(7):5443–5453, 2010.

[63] Thomas Stützle and Marco Dorigo. A short convergence proof for a class of ant colony optimization algorithms. *IEEE Transactions on Evolutionary Computation*, 6(4):358–365, 2002.

[64] Han Huang, Chun-Guo Wu, and Zhi-Feng Hao. A pheromone-rate-based analysis on the convergence time of ACO algorithm. *IEEE Transactions on Systems, Man, and Cybernetics, Part B (Cybernetics)*, 39(4):910–923, 2009.

[65] Sabrina M Oliveira, Mohamed Saifullah Hussin, Thomas Stützle, Andrea Roli, and Marco Dorigo. A detailed analysis of the population-based ant colony optimization algorithm for the TSP and the QAP. In *Proceedings of the 13th Annual Conference Companion on Genetic and Evolutionary Computation*, pages 13–14. ACM, 2011.

[66] Kuo-Ling Huang and Ching-Jong Liao. Ant colony optimization combined with taboo search for the job shop scheduling problem. *Computers & Operations Research*, 35(4):1030–1046, 2008.

[67] Mustafa Servet Kıran, Eren Özceylan, Mesut Gündüz, and Turan Paksoy. A novel hybrid approach based on particle swarm optimization and ant colony algorithm to forecast energy demand of turkey. *Energy Conversion and Management*, 53(1):75–83, 2012.

[68] Krzysztof Socha and Marco Dorigo. Ant colony optimization for continuous domains. *European Journal of Operational Research*, 185(3):1155–1173, 2008.

[69] Marco Dorigo, Gianni Di Caro, and Luca M Gambardella. Ant algorithms for discrete optimization. *Artificial Life*, 5(2):137–172, 1999.

[70] Yongqiang Gao, Haibing Guan, Zhengwei Qi, Yang Hou, and Liang Liu. A multi-objective ant colony system algorithm for virtual machine placement in cloud computing. *Journal of Computer and System Sciences*, 79(8): 1230–1242, 2013.

[71] Betul Yagmahan and Mehmet Mutlu Yenisey. A multi-objective ant colony system algorithm for flow shop scheduling problem. *Expert Systems with Applications*, 37(2):1361–1368, 2010.

[72] Luis A Moncayo-Martínez and David Z Zhang. Multi-objective ant colony optimisation: a meta-heuristic approach to supply chain design. *International Journal of Production Economics*, 131(1):407–420, 2011.

[73] José M Cecilia, José M García, Andy Nisbet, Martyn Amos, and Manuel Ujaldón. Enhancing data parallelism for ant colony optimization on GPUs. *Journal of Parallel and Distributed Computing*, 73(1):42–51, 2013.

[74] Laurence Dawson and Iain Stewart. Improving ant colony optimization performance on the GPU using CUDA. In *Evolutionary Computation (CEC), 2013 IEEE Congress on*, pages 1901–1908. IEEE, 2013.

[75] Shigeyoshi Tsutsui and Noriyuki Fujimoto. Parallel ant colony optimization algorithm on a multi-core processor. In *International Conference on Swarm Intelligence*, pages 488–495. Springer, 2010.

[76] Zar Chi Su Su Hlaing and May Aye Khine. An ant colony optimization algorithm for solving traveling salesman problem. In *International Conference on Information Communication and Management*, volume 16, pages 54–59. 2011.

[77] Michalis Mavrovouniotis and Shengxiang Yang. Ant colony optimization with immigrants schemes for the dynamic travelling salesman problem with traffic factors. *Applied Soft Computing*, 13(10):4023–4037, 2013.

[78] Kun Li, Gaochao Xu, Guangyu Zhao, Yushuang Dong, and Dan Wang. Cloud task scheduling based on load balancing ant colony optimization. In *Chinagrid Conference (ChinaGrid), 2011 Sixth Annual*, pages 3–9. IEEE, 2011.

[79] Li-Ning Xing, Ying-Wu Chen, Peng Wang, Qing-Song Zhao, and Jian Xiong. A knowledge-based ant colony optimization for flexible job shop scheduling problems. *Applied Soft Computing*, 10(3):888–896, 2010.

[80] Ying Tan and Yuanchun Zhu. Fireworks algorithm for optimization. *Advances in Swarm Intelligence*, pages 355–364, 2010.

[81] Jianhua Liu, Shaoqiu Zheng, and Ying Tan. Analysis on global convergence and time complexity of fireworks algorithm. In *2014 IEEE Congress on Evolutionary Computation (CEC)*, pages 3207–3213. July 2014.

[82] Junzhi Li, Shaoqiu Zheng, and Ying Tan. The effect of information utilization: Introducing a novel guiding spark in the fireworks algorithm. *IEEE Transactions on Evolutionary Computation*, 21(1):153–166, 2017.

[83] Shaoqiu Zheng, Andreas Janecek, Junzhi Li, and Ying Tan. Dynamic search in fireworks algorithm. In *Evolutionary Computation (CEC), 2014 IEEE Congress on*, pages 3222–3229. IEEE, 2014.

[84] Junzhi Li, Shaoqiu Zheng, and Ying Tan. Adaptive fireworks algorithm. In *Evolutionary Computation (CEC), 2014 IEEE Congress on*, pages 3214–3221. IEEE, 2014.

[85] Jianhua Liu, Shaoqiu Zheng, and Ying Tan. The improvement on controlling exploration and exploitation of firework algorithm. In *International Conference in Swarm Intelligence*, pages 11–23. Springer, 2013.

[86] Tapas Si and Rituparna Ghosh. Explosion sparks generation using adaptive transfer function in firework algorithm. In *Signal Processing, Communication and Networking (ICSCN), 2015 3rd International Conference on*, pages 1–9. IEEE, 2015.

[87] Shaoqiu Zheng, Andreas Janecek, and Ying Tan. Enhanced fireworks algorithm. In *Evolutionary Computation (CEC), 2013 IEEE Congress on*, pages 2069–2077. IEEE, 2013.

[88] Chao Yu and Ying Tan. Fireworks algorithm with covariance mutation. In *Evolutionary Computation (CEC), 2015 IEEE Congress on*, pages 1250–1256. IEEE, 2015.

[89] Shaoqiu Zheng, Junzhi Li, Andreas Janecek, and Ying Tan. A cooperative framework for fireworks algorithm. *IEEE/ACM Transactions on Computational Biology and Bioinformatics*, 14(1):27–41, 2017.

[90] Chao Yu, Lingchen Kelley, Shaoqiu Zheng, and Ying Tan. Fireworks algorithm with differential mutation for solving the CEC 2014 competition

problems. In *Evolutionary Computation (CEC), 2014 IEEE Congress on*, pages 3238–3245. IEEE, 2014.

[91] Bei Zhang, Min-Xia Zhang, and Yu-Jun Zheng. A hybrid biogeography-based optimization and fireworks algorithm. In *Evolutionary Computation (CEC), 2014 IEEE Congress on*, pages 3200–3206. IEEE, 2014.

[92] Ke Ding, Shaoqiu Zheng, and Ying Tan. A GPU-based parallel fireworks algorithm for optimization. In *Proceedings of the 15th Annual Conference on Genetic and Evolutionary Computation*, pages 9–16. ACM, 2013.

[93] Ying Tan. S-metric-based multi-objective fireworks algorithm. In *Fireworks Algorithm*, pages 189–208. Springer, Berlin, Heidelberg, 2015.

[94] Nebojsa Bacanin and Milan Tuba. Fireworks algorithm applied to constrained portfolio optimization problem. In *Evolutionary Computation (CEC), 2015 IEEE Congress on*, pages 1242–1249. IEEE, 2015.

[95] Ying Tan. Discrete firework algorithm for combinatorial optimization problem. In *Fireworks Algorithm*, pages 209–226. Springer, Berlin, Heidelberg, 2015.

[96] A Mohamed Imran and M Kowsalya. A new power system reconfiguration scheme for power loss minimization and voltage profile enhancement using fireworks algorithm. *International Journal of Electrical Power & Energy Systems*, 62:312–322, 2014.

[97] Wenrui He, Guyue Mi, and Ying Tan. Parameter optimization of local-concentration model for spam detection by using fireworks algorithm. In *International Conference in Swarm Intelligence*, pages 439–450. Springer, 2013.

[98] Milan Tuba, Nebojsa Bacanin, and Adis Alihodzic. Multilevel image thresholding by fireworks algorithm. In *Radioelektronika (RADIOELEKTRONIKA), 2015 25th International Conference*, pages 326–330. IEEE, 2015.

[99] Yu-Jun Zheng. Water wave optimization: a new nature-inspired metaheuristic. *Computers & Operations Research*, 55:1–11, 2015.

[100] Bei Zhang and Yu-jun Zheng. Convergence analysis of water wave optimization algorithm. *Computer Science*, 4:009, 2016.

[101] Yu-Jun Zheng and Bei Zhang. A simplified water wave optimization algorithm. In *Evolutionary Computation (CEC), 2015 IEEE Congress on*, pages 807–813. IEEE, 2015.

[102] Bei Zhang, Min-Xia Zhang, Jie-Feng Zhang, and Yu-Jun Zheng. A water wave optimization algorithm with variable population size and comprehensive learning. In *International Conference on Intelligent Computing*, pages 124–136. Springer, 2015.

[103] Xiuli Wu, Yongquan Zhou, and Yuting Lu. Elite opposition-based water wave optimization algorithm for global optimization. *Mathematical Problems in Engineering*, 2017, 2017.

[104] Xiao-Bei Wu, Jie Liao, and Zhi-Cheng Wang. Water wave optimization for the traveling salesman problem. In *International Conference on Intelligent Computing*, pages 137–146. Springer, 2015.

[105] Xin Yun, Xiaoyi Feng, Xin Lyu, Shouyang Wang, and Bo Liu. A novel water wave optimization based memetic algorithm for flow-shop scheduling. In *Evolutionary Computation (CEC), 2016 IEEE Congress on*, pages 1971–1976. IEEE, 2016.

[106] Shuchen Zhao, Zhengyang Li, Xin Yun, Keyao Wang, Xin Lyu, and Bo Liu. IIR filters designing by water wave optimization. In *Control & Automation (ICCA), 2017 13th IEEE International Conference on*, pages 347–352. IEEE, 2017.

[107] Yuhui Shi. Brain storm optimization algorithm. In *International Conference in Swarm Intelligence*, pages 303–309. Springer, 2011.

[108] Zhi-hui Zhan, Jun Zhang, Yu-hui Shi, and Hai-lin Liu. A modified brain storm optimization. In *Evolutionary Computation (CEC), 2012 IEEE Congress on*, pages 1–8. IEEE, 2012.

[109] Haibin Duan, Shuangtian Li, and Yuhui Shi. Predator–prey brain storm optimization for dc brushless motor. *IEEE Transactions on Magnetics*, 49(10):5336–5340, 2013.

[110] Dadian Zhou, Yuhui Shi, and Shi Cheng. Brain storm optimization algorithm with modified step-size and individual generation. *Advances in Swarm Intelligence*, pages 243–252, 2012.

[111] Shi Cheng, Yuhui Shi, Quande Qin, and Shujing Gao. Solution clustering analysis in brain storm optimization algorithm. In *Swarm Intelligence (SIS), 2013 IEEE Symposium on*, pages 111–118. IEEE, 2013.

[112] Zhi-hui Zhan, Wei-neng Chen, Ying Lin, Yue-jiao Gong, Yuan-long Li, and Jun Zhang. Parameter investigation in brain storm optimization. In *Swarm Intelligence (SIS), 2013 IEEE Symposium on*, pages 103–110. IEEE, 2013.

[113] Changhao Sun, Haibin Duan, and Yuhui Shi. Optimal satellite formation reconfiguration based on closed-loop brain storm optimization. *IEEE Computational Intelligence Magazine*, 8(4):39–51, 2013.

[114] HT Jadhav, Utkarsh Sharma, Jay Patel, and Ranjit Roy. Brain storm optimization algorithm based economic dispatch considering wind power. In *Power and Energy (PECon), 2012 IEEE International Conference on*, pages 588–593. IEEE, 2012.

[115] Junfeng Chen, Shi Cheng, Yang Chen, Yingjuan Xie, and Yuhui Shi. Enhanced brain storm optimization algorithm for wireless sensor networks deployment. In *International Conference in Swarm Intelligence*, pages 373–381. Springer, 2015.

[116] Seyedali Mirjalili and Andrew Lewis. The whale optimization algorithm. *Advances in Engineering Software*, 95:51–67, 2016.

[117] A Kaveh and M Ilchi Ghazaan. Enhanced whale optimization algorithm for sizing optimization of skeletal structures. *Mechanics Based Design of Structures and Machines*, 45(3):345–362, 2017.

[118] Indrajit N Trivedi, Jangir Pradeep, Jangir Narottam, Kumar Arvind, and Ladumor Dilip. Novel adaptive whale optimization algorithm for global optimization. *Indian Journal of Science and Technology*, 9(38), 2016.

[119] Ying Ling, Yongquan Zhou, and Qifang Luo. Lévy flight trajectory-based whale optimization algorithm for global optimization. *IEEE Access*, 5:6168–6186, 2017.

[120] Haider J Touma. Study of the economic dispatch problem on IEEE 30-bus system using whale optimization algorithm. *International Journal of Engineering Technology and Sciences (IJETS)*, 5(1):11–18, 2016.

[121] Ibrahim Aljarah, Hossam Faris, and Seyedali Mirjalili. Optimizing connection weights in neural networks using the whale optimization algorithm. *Soft Computing*, 22(1):1–15, 2018.

[122] R Bhesdadiya, Pradeep Jangir, Narottam Jangir, Indrajit N Trivedi, and Dilip Ladumor. Training multi-layer perceptron in neural network using whale optimization algorithm. *Indian Journal of Science and Technology*, 9(19):28–36, 2016.

[123] Santhan Kumar Cherukuri and Srinivasa Rao Rayapudi. A novel global MPP tracking of photovoltaic system based on whale optimization algorithm. *International Journal of Renewable Energy Development*, 5(3): 225, 2016.

[124] Ying Tan. Swarm robotics: collective behavior inspired by nature. *Journal of Computer Science and Systems Biology*, 6:e106, 2013.

[125] Erol Şahin. Swarm robotics: from sources of inspiration to domains of application. In *International Workshop on Swarm Robotics*, pages 10–20. Springer, Berlin, Heidelberg, 2004.

[126] Ying Tan and Zhong-yang Zheng. Research advance in swarm robotics. *Defence Technology*, 9(1):18–39, 2013.

[127] Ying Tan. *Handbook of Research on Design, Control, and Modeling of Swarm Robotics*. IGI Global, 2015.

[128] Yu Zhou, Ron Goldman, and James McLurkin. An asymmetric distributed method for sorting a robot swarm. *IEEE Robotics and Automation Letters*, 2(1):261–268, 2017.

[129] Qi Lu, Melanie E Moses, and Joshua P Hecker. A scalable and adaptable multiple-place foraging algorithm for ant-inspired robot swarms. *arXiv preprint arXiv:1612.00480*, 2016.

[130] Craig W Reynolds. Flocks, herds and schools: a distributed behavioral model. *ACM SIGGRAPH Computer Graphics*, 21(4):25–34, 1987.

[131] Laura E Barnes, Mary Anne Fields, and Kimon P Valavanis. Swarm formation control utilizing elliptical surfaces and limiting functions. *IEEE Transactions on Systems, Man, and Cybernetics, Part B (Cybernetics)*, 39(6):1434–1445, 2009.

[132] Jie Li and Ying Tan. Triangle formation based multiple targets search using a swarm of robots. In *International Conference in Swarm Intelligence*, pages 544–552. Springer, Cham, 2016.

[133] Gregor HW Gebhardt, Kevin Daun, Marius Schnaubelt, Alexander Hendrich, Daniel Kauth, and Gerhard Neumann. Learning to assemble objects with a robot swarm. In *Proceedings of the 16th Conference on Autonomous Agents*

and MultiAgent Systems, pages 1547–1549. International Foundation for Autonomous Agents and Multiagent Systems, 2017.

[134] Zhongyang Zheng, Junzhi Li, Jie Li, and Ying Tan. Avoiding decoys in multiple targets searching problems using swarm robotics. In *Evolutionary Computation (CEC), 2014 IEEE Congress on*, pages 784–791. IEEE, 2014.

[135] Zhongyang Zheng and Ying Tan. Mobile target tracking of swarm robotics in unknown obstructive environment. *Handbook of Research on Design, Control, and Modeling of Swarm Robotics*, pages 397–420, IGI Global, 2016.

[136] Rohan Tiwari, Puneet Jain, Sachit Butail, Sujit P Baliyarasimhuni, and Michael A Goodrich. Effect of leader placement on robotic swarm control. In *Proceedings of the 16th Conference on Autonomous Agents and MultiAgent Systems*, pages 1387–1394. International Foundation for Autonomous Agents and Multiagent Systems, 2017.

[137] Hao Wei, Jonathan Ian Timmis, and Robert David Alexander. Evolving test environments to identify faults in swarm robotics algorithms. In *IEEE Congress on Evolutionary Computation 2017*. York, 2017.

[138] Navyata Sanghvi, Sasanka Nagavalli, and Katia Sycara. Exploiting robotic swarm characteristics for adversarial subversion in coverage tasks. In *Proceedings of the 16th Conference on Autonomous Agents and MultiAgent Systems*, pages 511–519. International Foundation for Autonomous Agents and Multiagent Systems, 2017.

[139] Manuele Brambilla, Eliseo Ferrante, Mauro Birattari, and Marco Dorigo. Swarm robotics: a review from the swarm engineering perspective. *Swarm Intelligence*, 7(1):1–41, 2013.

[140] Gabriele Valentini, Heiko Hamann, and Marco Dorigo. Self-organized collective decision-making in a 100-robot swarm. In *AAAI*, pages 4216–4217, 2015.

[141] Alexander Scheidler, Arne Brutschy, Eliseo Ferrante, and Marco Dorigo. The k-unanimity rule for self-organized decision-making in swarms of robots. *IEEE Transactions on Cybernetics*, 46(5):1175–1188, 2016.

[142] Gabriele Valentini, Eliseo Ferrante, and Marco Dorigo. The best-of-n problem in robot swarms: formalization, state of the art, and novel perspectives. *Frontiers in Robotics and AI*, 4(9), 2017.

[143] SD Xue and JC Zeng. Swarm robotics: a survey. *Pattern Recognition and Artificial Intelligence*, 21(2):177–185, 2008.

[144] Xiangyu Liu and Ying Tan. Adaptive potential fields model for solving distributed area coverage problem in swarm robotics. In *Advances in Swarm Intelligence – 8th International Conference, ICSI 2017, Fukuoka, Japan, July 27–August 1, 2017, Proceedings, Part II*, pages 149–157, 2017.

[145] Dario Floreano and Claudio Mattiussi. *Bio-Inspired Artificial Intelligence: Theories, Methods, and Technologies*. MIT Press, 2008.

[146] Abdelhak Chatty, Ilhem Kallel, Philippe Gaussier, and Adel M Alimi. Emergent complex behaviors for swarm robotic systems by local rules. In *Robotic Intelligence In Informationally Structured Space (RiiSS), 2011 IEEE Workshop on*, pages 69–76. IEEE, 2011.

[147] Jie Li and Ying Tan. The multi-target search problem with environmental restrictions in swarm robotics. In *Robotics and Biomimetics (ROBIO), 2014 IEEE International Conference on*, pages 2685–2690. IEEE, 2014.
[148] Eshwaran Vijaya Kumar, Mansimar Aneja, and Dipti Deodhare. Solving a path planning problem in a partially known environment using a swarm algorithm. *arXiv preprint arXiv:1705.03176*, 2017.
[149] Micael S Couceiro, Rui P Rocha, and Nuno MF Ferreira. A novel multi-robot exploration approach based on particle swarm optimization algorithms. In *Safety, Security, and Rescue Robotics (SSRR), 2011 IEEE International Symposium on*, pages 327–332. IEEE, 2011.
[150] Yan Meng and Jing Gan. A distributed swarm intelligence based algorithm for a cooperative multi-robot construction task. In *Swarm Intelligence Symposium, 2008. SIS 2008. IEEE*, pages 1–6. IEEE, 2008.
[151] Jim Pugh and Alcherio Martinoli. Multi-robot learning with particle swarm optimization. In *Proceedings of the Fifth International Joint Conference on Autonomous Agents and Multiagent Systems*, pages 441–448. ACM, 2006.
[152] Jim Pugh and Alcherio Martinoli. Inspiring and modeling multi-robot search with particle swarm optimization. In *Swarm Intelligence Symposium, 2007. SIS 2007. IEEE*, pages 332–339. IEEE, 2007.
[153] Zhongyang Zheng and Ying Tan. Group explosion strategy for searching multiple targets using swarm robotic. In *Evolutionary Computation (CEC), 2013 IEEE Congress on*, pages 821–828. IEEE, 2013.
[154] Zhongyang Zheng, Jie Li, Junzhi Li, and Ying Tan. Improved group explosion strategy for searching multiple targets using swarm robotics. In *Systems, Man and Cybernetics (SMC), 2014 IEEE International Conference on*, pages 246–251. IEEE, 2014.
[155] Sheetal Doctor, Ganesh K Venayagamoorthy, and Venu G Gudise. Optimal PSO for collective robotic search applications. In *Evolutionary Computation, 2004. CEC2004. Congress on*, volume 2, pages 1390–1395. IEEE, 2004.
[156] Ji-Gong Li, Jing Yang, Shi-Gang Cui, and Li-Hui Geng. Speed limitation of a mobile robot and methodology of tracing odor plume in airflow environments. *Procedia Engineering*, 15:1041–1045, 2011.
[157] Xiaohui Cui, C Tim Hardin, Rammohan K Ragade, and Adel Said Elmaghraby. A swarm approach for emission sources localization. In *Tools with Artificial Intelligence, 2004. ICTAI 2004. 16th IEEE International Conference on*, pages 424–430. IEEE, 2004.
[158] Wisnu Jatmiko, Kosuke Sekiyama, and Toshio Fukuda. A PSO-based mobile robot for odor source localization in dynamic advection-diffusion with obstacles environment: theory, simulation and measurement. *IEEE Computational Intelligence Magazine*, 2(2):37–51, 2007.
[159] R Andrew Russell, Alireza Bab-Hadiashar, Rod L Shepherd, and Gordon G Wallace. A comparison of reactive robot chemotaxis algorithms. *Robotics and Autonomous Systems*, 45(2):83–97, 2003.

[160] Gervasio Varela, Pilar Caamaño, Felix Orjales, Álvaro Deibe, Fernando López-Peña, and Richard J Duro. Swarm intelligence based approach for real time UAV team coordination in search operations. In *Nature and Biologically Inspired Computing (NaBIC), 2011 Third World Congress on*, pages 365–370. IEEE, 2011.

[161] Wencen Wu and Fumin Zhang. Robust cooperative exploration with a switching strategy. *IEEE Transactions on Robotics*, 28(4):828–839, 2012.

[162] Itay Sisso, Tal Shima, and Yakov Ben-Haim. Info-gap approach to multiagent search under severe uncertainty. *IEEE Transactions on Robotics*, 26(6):1032–1041, 2010.

[163] Jong-Hyun Lee and Chang Wook Ahn. Improving energy efficiency in cooperative foraging swarm robots using behavioral model. In *Bio-Inspired Computing: Theories and Applications (BIC-TA), 2011 Sixth International Conference on*, pages 39–44. IEEE, 2011.

Chapter 2

Generalization ability of swarm intelligence algorithms

Shi Cheng[1], Quande Qin[2], Bin Liu[3], Jiqiang Feng[4], Xiujuan Lei[1], and Yuhui Shi[5]

Abstract

In this chapter, generalization ability of swarm intelligence algorithms solving problems with different number of dimensions is analyzed and discussed. Three algorithms, brain storm optimization in objective space (BSO-OS), fireworks algorithm (FWA), and particle swarm optimization (PSO) algorithm, are selected as illustrations to explain the definition of algorithm's generalization ability. The performance of BSO-OS, FWA, and PSO algorithm on solving problems with different number of dimensions is analyzed. Based on the experimental results, the algorithm's generalization ability was measured by the results ratio of algorithms with the same settings on problems with different number of dimensions. This generalization ability measurement could be extended to problems with different components. Without analysis on the landscape of problems, this measurement could give a practical illustration of the generalization ability of algorithms for solving problems with different number of dimensions or different components. Based on the analysis on the generalization of algorithms and the hardness of problems, we could have a better understanding of the relationship between problems and algorithms, and therefore design more effective algorithms to solve different problems.

2.1 Introduction

Swarm intelligence is a set of search and optimization techniques [1,2]. To search a problem domain, a swarm intelligence algorithm processes a population of individuals. Different from traditional single-point based algorithms such as

[1] School of Computer Science, Shaanxi Normal University, China
[2] Department of Management Science, Shenzhen University, China
[3] School of Computer Science and Technology, Nanjing University of Posts and Telecommunications, China
[4] College of Mathematics and Statistics, Shenzhen University, China
[5] Department of Computer Science and Engineering, Southern University of Science and Technology, China

hill-climbing algorithms, each swarm intelligence algorithm is a population-based algorithm, which consists of a set of points (population of individuals). Each individual represents a potential solution to the problem being optimized. The population of individuals is expected to have high tendency to move towards better and better solution areas over iterations through cooperation and competition among themselves.

From small scale and/or low-dimensional problems to large scale and/or high-dimensional problems, an algorithm finds results with different quality. Generalization ability of an algorithm indicates the changing search ability of the algorithm on problems with different scale/dimensional numbers. With the increasing of computing power, nearly all algorithms could find good solutions on problems with lower dimensions and/or simple components. However, the performance of algorithms is getting worse on problems with higher dimensions or complex components. In addition, different algorithms could find solutions with different level, which indicates that some algorithms may more fit for some special problems. Thus, finding the relations between algorithms and problems is essential for swarm intelligence on solving various problems. Two definitions of generalization performance are proposed in coevolutionary learning [3].

Problem's difficulty or problem's hardness measure is a basic and hard problem in optimization field [4,5]. The fitness landscapes and difficulty in genetic programming were introduced in [6]. A number of predictive measures of problem's difficulty, which includes epistasis variance and fitness distance correlation, were reviewed in [4]. In [5], difficulty measures were rigorously defined for black-box optimization. The linkage information and neighborhood structures were utilized to measure problem's difficulty for genetic algorithm and estimation of distribution algorithms [7]. The network motifs were deployed to measure problem's difficulty for evolutionary algorithms in [8]. The difficulty measure for evolutionary program-induction algorithms was given in [9]. The hardness of problems and the complexity of black box problems are discussed in [10]. The discussions are based on the discrete problems, such as traveling salesperson problem, bin packing problem, etc. [10].

The aim of this paper is to give an analysis on the performance of different swarm intelligence algorithms on solving problems with different number of dimensions and to determine which kind of problems could be easily solved by a given algorithm and which are not. Most of the difficulty analyses are based on the theoretical analysis of problems, such as the fitness landscape of problems. Three algorithms, brain storm optimization in objective space (BSO-OS), fireworks algorithm (FWA), and particle swarm optimization (PSO) algorithm, are selected as illustrations to explain the definition of generalization ability of algorithms. Based on the experimental results, generalization ability of algorithms on various problems with different number of dimensions and problem's difficulty measure with different kinds of algorithms are proposed.

The brain storm optimization (BSO) algorithm is a young and promising algorithm in swarm intelligence. It is based on the collective behavior of human being, that is, the brainstorming process [11,12]. The solutions in BSO are diverging into several clusters. The new solutions are generated based on the mutation of one individual or a combination of two individuals. A comprehensive review on the state-of-the-art BSO research is given in [13,14]. In the original BSO algorithm, the computational resources are consumed a lot on the clustering strategy at each iteration. To reduce the computational burden, the BSO-OS algorithm was proposed, and the clustering strategy was replaced by a simple classification strategy based on the fitness values [15].

The FWA [16,17] and BSO [11,12] algorithms are two good examples of developmental swarm intelligence (DSI) algorithms [18]. The "good enough" optimum could be obtained through solutions diverging and converging in the search space. In FWA algorithm, mimicked by the fireworks exploration, the new solutions are generated by the exploration of existed solutions. While in BSO algorithm, the solutions are clustered into several categories, and new solutions are generated by the mutation of clusters or existed solutions. The capacity developing, i.e., the adaptation in search, is another common feature in these two algorithms.

The remaining of the paper is organized as follows: Section 2.2 reviews the basic BSO-OS algorithms, FWAs, and PSO algorithms. Section 2.3 gives a definition on the generalization ability of swarm intelligence algorithms on solving different kinds of benchmark functions. Experiments are conducted in Section 2.4, followed by analysis and discussion on the generalization ability of BSO-OS, FWA, and PSO algorithms in Section 2.5. Finally, Section 2.6 concludes with some remarks and future research directions.

2.2 Swarm intelligence algorithms

Three DSI algorithms are selected as illustrations to explain the definition of algorithm's generalization ability. These algorithms are BSO-OS, FWA, and PSO algorithm with global star and local ring structure.

2.2.1 Brain storm optimization algorithms

2.2.1.1 Original brain storm optimization

The BSO algorithm is based on the collective behavior of human being, that is, the brainstorming process [11,12]. The solutions in BSO are converging into several clusters. The best solution of the population will be kept if the newly generated solution at the same index is not better. New individual can be generated based on the mutation of one or two individuals in clusters. The exploitation ability is enhanced when the new individual is close to the best solution so far, while the exploration ability is enhanced when the new individual is randomly generated, or generated by individuals in two clusters.

The original BSO algorithm is simple in concept and easy in implementation [11,12]. The procedure of BSO algorithm is given in Algorithm 1. There are three strategies in this algorithm: the solution clustering, new individual generation, and selection [19].

Algorithm 1: The procedure of the original brain storm optimization algorithm

Initialization: Randomly generate n potential solutions (individuals), and evaluate the n individuals;
while *not find "good enough" solution or not reach the pre-determined maximum number of iterations* **do**
  Clustering: Cluster n individuals into m clusters by clustering algorithm;
  New individual generation: randomly select one or two cluster(s) to generate new individual;
  Selection: The newly generated individual is compared with the existing individual with the same individual index; the better one is kept and recorded as the new individual;
  Evaluate the n individuals;

Algorithm 2: The procedure of the BSO in objective space algorithm

Initialization: Randomly generate n potential solutions (individuals), and evaluate them;
while *have not found "good enough" solution or not reached the pre-determined maximum number of iterations* **do**
  Classification: Classify all solutions into two categories: the solutions with better fitness values as elitists and the others as normals;
  New individual generation: randomly select one or two individuals from elitists or normal to generate new individual;
  Solution disruption: re-initialize one dimension of a randomly selected individual and update its fitness value accordingly;
  Selection: The newly generated individual is compared with the existing individual with the same individual index; the better one is kept and recorded as the new individual;
  Evaluate all individuals;

2.2.1.2 Brain storm optimization in objective space

In the original BSO algorithm, a lot of computational resources are spent on the clustering strategy at each iteration. To reduce the computational burden, the BSO-OS algorithm was proposed, and the clustering strategy was replaced by a simple classification strategy based on the fitness values [15]. The procedure of the BSO in objective space algorithm is given in Algorithm 2.

2.2.1.3 New individual generation

The main difference between the original BSO and the BSO-OS is the new individual generation strategy. In the original BSO algorithm, individuals are clustered into several groups. For the BSO-OS algorithm, individuals are classified into two categories according to their fitness values. The procedure of new individual generation strategy

---

Algorithm 3: The new individual update operation

New individual generation: randomly select one or two individual(s) to generate new individual;
if *random value rand is less than a probability* $p_{elitist}$ **then** `/* generate a new individual based on elitists */`
  if *random value rand is smaller than a pre-determined probability* p_{one} **then**
    generate a new individual based on one randomly selected elitist;
  else
    two individuals from elitists are randomly selected to generate new individual;
else `/* generate a new individual based on normal */`
  if *random value rand is less than a pre-determined probability* p_{one} **then**
    generate a new individual based on one randomly selected normal;
  else
    two individuals from normal are randomly selected to generate new individual;
The newly generated individual is compared with the existing individual with the same individual index, the better one is kept and recorded as the new individual;

---

is given in Algorithm 3. Two parameters, probability $p_{elitist}$ and probability p_{one}, are used in this strategy. The new individuals are generated according to the functions (2.1) and (2.2).

$$x^i_{\text{new}} = x^i_{\text{old}} + \xi(t) \times \text{rand}() \tag{2.1}$$

$$\xi(t) = \text{logsig}\left(\frac{0.5 \times T - t}{k}\right) \times \text{rand}() \tag{2.2}$$

where x^i_{new} and x^i_{old} are the ith dimension of $\mathbf{x}_{\text{new}}$ and $\mathbf{x}_{\text{old}}$; rand() is a random function to generate uniformly distributed random numbers in the range [0, 1); and the value $\mathbf{x}_{\text{old}}$ is a copy of one individual or the combination of two individuals. The parameter T is the maximum number of iterations, t is the current iteration number, k is a coefficient to change logsig() function's slope of the step size function $\xi(t)$, which can be utilized to balance the convergence speed of the algorithm.

2.2.1.4 Transfer function

A transfer function logsig(a), which is given in (2.3), has been deployed in step size (2.2). The parameter k is used to adjust the function's slope. In the original BSO-OS algorithm, the maximum number of iterations T is set as 2,000, and the slope k is 25 [15]. Thus, the value $((0.5 \times T - t)/k)$ in (2.2) is linearly decreased from $((1{,}000 - 0)/25)$ to $((1{,}000 - 2{,}000)/25)$, i.e., it is mapped into a range $[-40, 40]$.

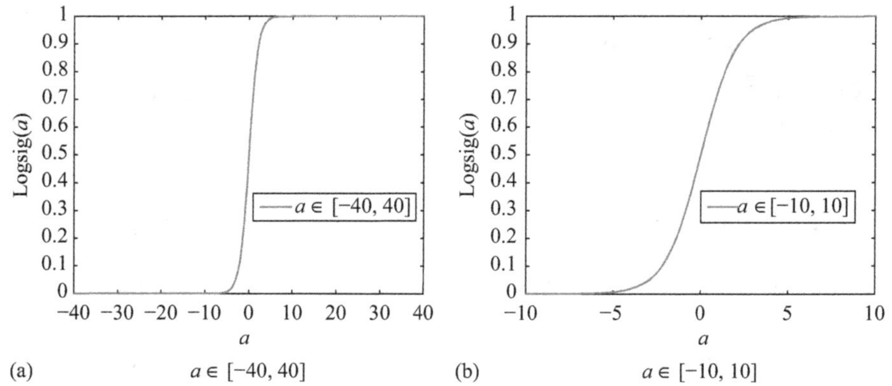

Figure 2.1 The function logsig(a) with different variable ranges: (a) $a \in [-40, 40]$ and (b) $a \in [-10, 10]$

Figure 2.1 gives the function logsig() with different variable ranges. It shows that, for function logsig() with variables in range $[-40, 40]$, nearly the half values are close to one, and the other half values are close to zero. In this paper, to have a smooth curve of step size, the value $((0.5 \times T - t)/k)$ is mapped into the range $[-10, 10]$.

$$\text{logsig}(a) = \frac{1}{1 + \exp(-a)} \tag{2.3}$$

2.2.1.5 Boundary constraint

The original BSO algorithm and BSO-OS algorithm lack of strategy to handle the boundary constraints. A new solution may be generated out of search space. The conventional boundary handling methods try to keep the solutions inside the feasible search space. The fitness value is only calculated when solutions are created in the search space. However, if a solution exceeds the boundary limit in one dimension at one iteration, that search information will be abandoned. Instead, a new solution will be reset to the particle in that dimension. The classic strategy is to set the solution at boundary when it exceeds the boundary. The equation of this strategy is as follows:

$$x_j = \begin{cases} x_{\max,j} & \text{if } x_j > x_{\max,j} \\ x_{\min,j} & \text{if } x_j < x_{\min,j} \\ x_j & \text{otherwise} \end{cases} \tag{2.4}$$

This strategy resets solutions in a particular point, i.e., the search boundary, which constrains solutions to explore in the search space limited by boundary.

2.2.2 Fireworks algorithm

FWA is a swarm intelligence optimization method that mimics the explosion process of fireworks [16,17,20]. The basic framework of FWA is given in Figure 2.2 [17].

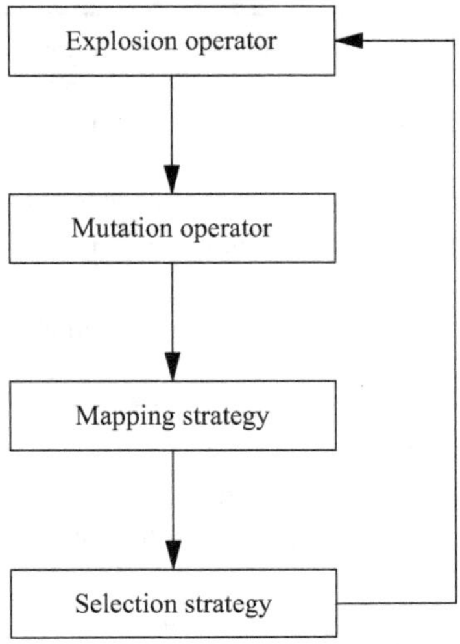

Figure 2.2 The framework of fireworks algorithm

Algorithm 4: Procedure of fireworks algorithm

Initialize n locations;
while *have not found "good enough" solution or not reached the pre-determined maximum number of iterations* **do**
  Set off fireworks at n locations;
  Generate sparks through **explosion** operator;
  Generate sparks through **mutation** operator;
  Obtain the locations of sparks;
  Mapping the locations into feasible search space;
  Evaluate the quality of the locations;
  Select n locations as new fireworks;

The procedure of FWA is given in Algorithm 4. There are four operators/strategies in FWA, which are explosion operator, mutation operator, mapping strategy, and selection strategy, respectively.

2.2.3 Particle swarm optimization algorithms

PSO emulates the swarm behavior, and the individuals represent points in the n-dimensional search space. A particle represents a potential solution. Each particle is associated with two vectors, i.e., the velocity vector and the position vector. For the purpose of generality and clarity, m represents the number of particles and n the number of dimensions. The position of a particle is represented as x_{ij}, and the

Algorithm 5: The basic procedure of particle swarm optimization

Initialization: Initialize velocity and position randomly for each particle in every dimension;
while *have not found "good enough" solution or not reached the pre-determined maximum number of iterations* **do**
  Calculate each particle's fitness value;
  Compare fitness value between current position and the best position in history (personal best, termed as *pbest*). For each particle, if the fitness value of the current position is better than *pbest*, then update *pbest* to be the current position;
  Select the particle which has the best fitness value among current particle's neighborhood, this particle is called the neighborhood best (termed as *nbest*);
  Update each particle's velocity and position, respectively;

velocity of a particle is represented as v_{ij}, i represents the ith particle, $i = 1, \ldots, m$, and j is the jth dimension, $j = 1, \ldots, n$.

The velocity and position update equations in canonical PSO algorithm are as follows [1,2,21,22]:

$$\mathbf{v}_i^{t+1} \leftarrow w\mathbf{v}_i^t + c_1\text{rand}()(\mathbf{p}_i^t - \mathbf{x}_i^t) + c_2\text{rand}()(\mathbf{p}_n^t - \mathbf{x}_i^t) \quad (2.5)$$

$$\mathbf{x}_i^{t+1} \leftarrow \mathbf{x}_i^t + \mathbf{v}_i^{t+1} \quad (2.6)$$

where w denotes the inertia weight and usually is less than 1 [23], c_1 and c_2 are two positive acceleration constants, rand() is a random function to generate uniformly distributed random numbers in the range [0, 1) and are different for each dimension and each particle, $\mathbf{x}_i$ represents the ith particle's position, $\mathbf{v}_i$ represents the ith particle's velocity, $\mathbf{p}_i$ is termed as personal best, which refers to the best position found by the ith particle, and $\mathbf{p}_n$ is termed as local best, which refers to the position found by the members in the ith particle's neighborhood that has the best fitness evaluation value so far.

The basic procedure of PSO is shown as Algorithm 5. A particle updates its velocity according to (2.5) and updates its position according to (2.6). The $c_1\text{rand}()(\mathbf{p}_i - \mathbf{x}_i)$ part can be seen as a cognitive behavior, while $c_2\text{rand}()(\mathbf{p}_g - \mathbf{x}_i)$ part can be seen as a social behavior.

2.3 Generalization ability of algorithms

The representation of problems is critical for the search performance of evolutionary computation algorithms. In [24], problems considered do not have a fitness landscape; the search space and fitness landscape are constructed based on the representation. Wolpert and Maceradv have proved that under certain assumptions, no algorithm is better than other one on average for all problems [25]. It means that different algorithms fit for different kinds of problems. Therefore, a good representation and an appropriate algorithm are completely necessary to solve a real-world problem.

From a practical approach, the computational resource consumption on a problem is not only related to the structure of the solved problem, but also with the chosen algorithm. It's difficult, if not impossible, to find an ideal algorithm for a real-world problem. The optimization result of the solved problem is related to the representation and the selected algorithm. In other words, the hardness of a problem is related to the chosen algorithm. The problem's difficulty is changed with the problem size or the components.

2.3.1 Problem with different settings

The x^* and $f(x^*)$ are used to represent global optimum and the corresponding fitness value, respectively. The fitness value of the best solution found by an algorithm in a run is denoted by $f(x_{\text{best}})$. The search error of this run is defined as $error = f(x_{\text{best}}) - f(x^*)$. For problems with small size or with small number of dimensions, a swarm intelligence algorithm could easily obtain good enough solutions. The results of search error on a problem with small size could be utilized as a baseline to measure generalization ability of the algorithm on the solving problem. For example, an algorithm obtained a search error result $error_1$ on problem with dimension n_1, and a result $error_2$ on the problem with dimension n_2 ($n_2 > n_1$). The generalization ability of algorithms could be defined in the following equation:

$$generalization = \frac{error_2 \times n_1}{error_1 \times n_2} \tag{2.7}$$

For the specific algorithm, the *generalization* measures the solved problem's difficulty changes from a lower dimension n_1 to a higher dimension n_2. In general, result on a problem will get worse with increasing of the problem's dimension. Thus, the value of *generalization* is in the range $[1, \infty)$. The closer to 1 the *generalization* is, the algorithm has a good generalization ability on the solved problem. On contrast, the closer to ∞ the *generalization* is, the algorithm has a bad generalization ability on the solved problem. The dimension changes is considered in (2.7). This definition could be used to measure difficulties that a problem with different increased dimension numbers.

2.3.2 Problem with different components

The generalization ability measure could also be extended to problems with different components. The results on a problem with one or several components could be utilized as a baseline to measure the difficulty of the whole problem. For example, an algorithm obtained a result $error_1$ on problem with a component c_1, and a result $error_{all}$ on the problem with n components $\mathscr{C}_{all}$ ($\mathscr{C}_{all} = c_1 + c_2 + \cdots + c_n$). The generalization ability of an algorithm on solving problems with several components and the problem's difficulty measure based on the component c_1 could be defined in the following equation:

$$generalization = \frac{error_{all}}{error_1 \times n} \tag{2.8}$$

For the specific algorithm, the value of *generalization* measures the problem's difficulty changes from one component c_1 to the whole problem $\mathscr{C}_{all}$. In general, the result on the whole problem will get worse than problem with part of components. Thus, the value of *generalization* is in the range $[1, \infty)$. The closer to 1 the *generalization* is, the components c_1 is harder to be solved among all components. On contrast, the closer to ∞ the *generalization* is, the components c_1 is easier to be solved.

The (2.9) could be used to compare the whole problem's difficulty and the problem with several components.

$$generalization = \frac{error_{all} \times m}{error_{1+\cdots+m} \times n} \tag{2.9}$$

The generalization ability definition of algorithm could be deployed to measure the hardness of different algorithms on the same problem. For real-world problems, it is difficult to determine the search error between the unknown real optimum and the optimal result found. The defined difficulty ratio will be affected by distance between real optimum and the optimum found. A hypothetical real optimum or boundary may be needed for the real-world problems.

2.4 Experimental study

The aim of experiment is not to compare the ability or the efficacy of BSO-OS algorithm, FWA, or PSO algorithms on different benchmark functions but to measure the difficulty of different algorithms on problems with different number of dimensions.

2.4.1 Benchmark test functions

The experiments have been conducted to test the benchmark functions listed in Table 2.1. Considering the generalization, five unimodal test functions and seven multimodal test functions are utilized. All functions are performed for 40 runs to ensure a suitable degree of statistical power necessary to compare the different approaches.

2.4.2 Parameter setting

In all experiments, the population in BSO-OS, FWA, and PSO algorithms has 50 individuals. Each algorithm runs 40 times, 10,000 iterations in every run. All other parameters of FWA are taken from [16]. The parameters of BSO-OS algorithm, except the slope k, are taken from [15]. All other parameters of PSO are taken from [26]. The parameters' setting for three algorithms is as follows:

1. BSO-OS algorithm:
   - $p_{elitist} = 0.1$; $p_{one} = 0.8$; $p_{disrupt} = 1.0$
   - slope $k = 500$.

Table 2.1 The benchmark functions used in our experimental study, where n is the dimension of each problem, S is the search range of each dimension, and $S \subseteq \mathbb{R}^n$. $f_{\min}$ is the minimum value of the function, which is the same for a problem with different number of dimensions

Name	Function	S	$f_{\min}$
Parabolic	$f_0(\mathbf{x}) = \sum_{i=1}^{n} x_i^2 + \text{bias}_0$	$[-100, 100]^n$	−450.0
Schwefel's P2.22	$f_1(\mathbf{x}) = \sum_{i=1}^{n} \lvert x_i\rvert + \prod_{i=1}^{n} \lvert x_i\rvert + \text{bias}_1$	$[-10, 10]^n$	−330.0
Schwefel's P1.2	$f_2(\mathbf{x}) = \sum_{i=1}^{n} \left(\sum_{k=1}^{i} x_k \right)^2 + \text{bias}_2$	$[-100, 100]^n$	450.0
Step	$f_3(\mathbf{x}) = \sum_{i=1}^{n} (\lfloor x_i + 0.5 \rfloor)^2 + \text{bias}_3$	$[-100, 100]^n$	330.0
Quartic Noise	$f_4(\mathbf{x}) = \sum_{i=1}^{n} i x_i^4 + \text{random}[0, 1) + \text{bias}_4$	$[-1.28, 1.28]^n$	−450.0
Rosenbrock	$f_5(\mathbf{x}) = \sum_{i=1}^{n-1} [100(x_{i+1} - x_i^2)^2 + (x_i - 1)^2] + \text{bias}_5$	$[-10, 10]^n$	180.0
Rastrigin	$f_6(\mathbf{x}) = \sum_{i=1}^{n} [x_i^2 - 10\cos(2\pi x_i) + 10] + \text{bias}_6$	$[-5.12, 5.12]^n$	−330.0
Noncontinuous Rastrigin	$f_7(\mathbf{x}) = \sum_{i=1}^{n} [y_i^2 - 10\cos(2\pi y_i) + 10] + \text{bias}_7$, $y_i = \begin{cases} x_i & \lvert x_i\rvert < \frac{1}{2} \\ \frac{\text{round}(2x_i)}{2} & \lvert x_i\rvert \geq \frac{1}{2} \end{cases}$	$[-5.12, 5.12]^n$	450.0
Ackley	$f_8(\mathbf{x}) = -20\exp\left(-0.2\sqrt{\frac{1}{n}\sum_{i=1}^{n} x_i^2}\right) - \exp\left(\frac{1}{n}\sum_{i=1}^{n}\cos(2\pi x_i)\right) + 20 + e + \text{bias}_8$	$[-32, 32]^n$	180.0
Griewank	$f_9(\mathbf{x}) = \frac{1}{4{,}000}\sum_{i=1}^{n} x_i^2 - \prod_{i=1}^{n}\cos\left(\frac{x_i}{\sqrt{i}}\right) + 1 + \text{bias}_9$	$[-600, 600]^n$	120.0
Generalized Penalized	$f_{10}(\mathbf{x}) = \frac{\pi}{n}\left\{10\sin^2(\pi y_1) + \sum_{i=1}^{n-1}(y_i - 1)^2 \times [1 + 10\sin^2(\pi y_{i+1})] + (y_n - 1)^2\right\} + \sum_{i=1}^{n} u(x_i, 10, 100, 4) + \text{bias}_{10}$, $y_i = 1 + \frac{1}{4}(x_i + 1)$, $u(x_i, a, k, m) = \begin{cases} k(x_i - a)^m & x_i > a, \\ 0 & -a < x_i < a \\ k(-x_i - a)^m & x_i < -a \end{cases}$	$[-50, 50]^n$	330.0
Schwefel	$f_{11}(\mathbf{x}) = \sum_{i=1}^{n} -x_i \sin(\sqrt{\lvert x_i\rvert}) + 418.9829n + \text{bias}_{11}$	$[-500, 500]^n$	−120.0

2. Fireworks algorithm:
   - $N = 5$; $M = 50$; AM = 2; BM = 40; MM = 5.
3. Particle swarm optimization algorithm:
   - $w = 0.72984$; $c_1 = c_2 = 1.496172$.

2.4.3 Experimental results

Tables 2.2–2.5 give the search error results of BSO-OS, FWA algorithms, PSO with star structure, and PSO with ring structure on 12 functions, respectively. Figures 2.3–2.6 display the performance of BSO-OS, FWA, and two PSO variants on solving different benchmark functions, respectively. The results are conducted on problems with four kinds of dimension settings, which are 30, 50, 100, and 200. The aim is to show the result changes of different algorithms on problems with different number of dimensions. It is not to compare the performance of BSO-OS algorithm, FWA, and PSO algorithms.

2.5 Analysis and discussion

The generalization ability of an algorithm is not to evaluate the performance of an algorithm on problems with certain dimension but to measure the difficulty changes of an algorithm on problems with different dimensions. Tables 2.6–2.9 give results of normalized difficulty measure of BSO-OS, FWA, and PSO algorithms on solving benchmark functions, respectively. The normalized values are varied in a large range when the dimensions increased from 30 to 200. Different algorithms obtain the largest value on solving different problems. For BSO-OS algorithm, the largest value is 184.2400 when the BSO-OS algorithm solving function f_{11}. For FWA algorithm, the largest value is 76.1653 when the FWA algorithm solving function f_4. For the PSO with star structure, the largest value is 2.456E+16 when the algorithm solving function f_2. For PSO with ring structure, the largest value is 4.844E+12 when the algorithm solving function f_{10}.

Figure 2.7 displays the difficulty changes for BSO-OS, FWA, and two PSO variants on solving different problems, respectively. It illustrates that problems are getting hard and hard with the increasing of dimensions. For algorithms with the same settings, the large value of hardness indicates that problems are going to be much harder when problems with the large number of dimensions.

For BSO-OS algorithm, the search results on f_0, f_3, f_{10}, and f_{11} are significantly getting worse than other functions. For FWA, the results on f_0, f_4, and f_9 are worse than others. This indicates that the hardness is not only a characteristic of problems but also related to the chosen algorithm.

The values of generalization ability measure are varied for different PSO algorithms. In general, the PSO with star structure obtains large values on solving unimodal functions; on the contrary, the PSO with ring structure obtains large values on solving multimodal functions. This may be due to that the PSO with star structure performs well on unimodal functions, and the PSO with ring structure performs

Table 2.2 Results of BSO in objective space algorithm solving benchmark functions with different dimension numbers. All algorithms are run for 40 times, where "best", "median", "worst," and "mean" indicate the best, median, worst, and mean of the best fitness values for all runs, respectively

Result		BSO in objective space				
Function	**Dimension**	**Best**	**Median**	**Worst**	**Mean**	**Std. dev.**
f_0	30	0.209837	0.580301	1.577664	0.643289	0.291868
	50	1.011012	3.790374	5.818549	3.708325	1.110167
	100	21.69920	35.36444	73.85534	35.64992	10.2613
	200	205.9324	369.7682	579.9516	370.2696	79.9965
f_1	30	0.169046	0.330898	0.569255	0.337617	0.093051
	50	0.686892	0.955796	1.643273	0.960657	0.178064
	100	2.865468	4.344910	5.710372	4.425545	0.612729
	200	16.12892	20.38410	25.43409	20.58539	2.119205
f_2	30	807.1557	2,274.281	5,647.022	2,636.902	1,089.290
	50	5,547.058	10,066.142	20,210.925	10,143.86	3,101.635
	100	21,973.14	40,917.12	68,163.49	40,802.72	10,638.19
	200	93,109.25	149,555.74	329,812.10	157,501.45	45,081.811
f_3	30	0	3	6	2.85	1.475635
	50	7	13	22	13.2	3.702701
	100	64	90	208	99.925	29.89095
	200	610	838	1097	821.35	125.4293
f_4	30	0.003672	0.022698	0.042056	0.023138	0.009698
	50	0.044996	0.090575	0.199930	0.093727	0.026787
	100	0.341780	0.583224	0.909355	0.566697	0.123648
	200	2.599941	3.558206	4.788286	3.518651	0.511945
f_5	30	16.93422	33.847736	172.0398	60.22175	42.0200
	50	63.0335	122.7471	265.0828	139.6795	50.6294
	100	330.9214	489.4711	823.16569	502.1855	96.7489
	200	1646.972	2252.212	3285.725	2357.723	369.9427
f_6	30	0.416640	1.388393	3.260824	1.414642	0.670178
	50	1.555191	6.350471	10.456115	6.283973	1.987473
	100	24.95245	38.23325	49.16030	37.51206	5.28087
	200	131.8728	155.7558	180.5220	155.4634	12.3300
f_7	30	0.426550	0.781479	2.393791	0.863062	0.400240
	50	1.683943	4.971352	8.276081	**5.081469**	1.482941
	100	31.19094	38.40965	49.93321	38.97856	4.63691
	200	115.0072	139.2571	168.4121	139.7471	11.3179
f_8	30	0.146062	0.306472	0.677666	0.328485	0.120134
	50	0.427103	0.816000	1.317300	0.834369	0.248823
	100	1.636474	2.141944	2.537777	2.118205	0.213663
	200	3.512959	3.715722	4.177061	3.736266	0.170590
f_9	30	0.151893	0.452037	0.761395	0.461278	0.147311
	50	0.546821	1.000853	1.056695	0.933778	0.125592
	100	1.120373	1.236722	1.575073	1.239222	0.079813
	200	2.762060	3.600078	5.257334	3.565025	0.608741
f_{10}	30	0.000307	0.001289	0.018580	0.001990	0.002848
	50	0.001718	0.004780	0.070557	0.010248	0.016764
	100	0.019253	0.089004	0.267716	0.101604	0.055587
	200	0.705357	1.080577	1.873520	1.119544	0.256962
f_{11}	30	0.901412	1.769673	3.913279	**2.013143**	0.907418
	50	3.647038	8.706800	29.83316	**10.53803**	5.384747
	100	92.05637	180.9894	458.7189	**189.4323**	74.86218
	200	1,745.1584	2,378.121	3,789.837	**2,472.676**	479.4473

Table 2.3 Results of fireworks algorithm solving benchmark functions with different dimension numbers. All algorithms are run for 40 times, where "best", "median", "worst", and "mean" indicate the best, median, worst, and mean of the best fitness values for all runs, respectively

Result		FWA solving benchmark functions				
Function	**Dimension**	**Best**	**Median**	**Worst**	**Mean**	**Std. dev.**
f_0	30	0.002357	0.005643	0.014456	0.006157	0.002497
	50	0.020368	0.038977	0.060787	0.038859	0.010155
	100	0.192331	0.359608	0.993684	0.379768	0.130699
	200	1.097181	2.06634	3.27736	2.08669	0.37848
f_1	30	0.131048	0.215762	0.391558	0.211181	0.053245
	50	0.595632	0.789250	1.107467	0.795623	0.125373
	100	2.004681	3.788736	4.819202	3.667816	0.672650
	200	4.422439	12.93191	19.38788	12.53630	3.222473
f_2	30	0.653721	3.010509	6.4977726	3.058899	1.443453
	50	0.256243	12.023407	58.44603	14.98579	11.08770
	100	9.705156	68.763572	165.3180	**72.59903**	37.09566
	200	68.01035	352.15093	1,134.6337	**392.34045**	234.8247
f_3	30	0	0	2	0.225	0.523808
	50	0	1	4	1.05	1.094303
	100	0	3	9	2.6	2.321637
	200	0	10	28	**10.475**	7.280066
f_4	30	0.007325	0.019505	0.038734	0.021384	0.008305
	50	0.051143	0.106483	0.228573	0.109319	0.040619
	100	0.400934	0.778728	1.700485	0.869485	0.301294
	200	2.261547	8.603377	38.61749	10.85812	8.159832
f_5	30	0.327751	27.01694	28.32994	16.96017	12.9613
	50	1.323782	8.508665	51.30070	**18.59891**	18.93635
	100	50.43790	110.6781	120.0602	**95.53180**	22.99670
	200	68.39149	289.2363	313.7366	**285.9893**	36.8288
f_6	30	0.074986	0.406972	2.691437	**0.692551**	0.695753
	50	0.073863	3.254909	8.480454	**3.329751**	2.291009
	100	0.008719	18.70481	34.80209	**16.46164**	10.22026
	200	9.392860	39.4997	92.5979	**44.58098**	23.6337
f_7	30	0.053777	0.379578	2.329029	**0.710945**	0.606785
	50	0.462282	5.452850	11.86414	5.486454	3.055328
	100	5.541884	21.31908	45.04252	**20.83864**	10.98438
	200	6.729064	43.8465	101.2184	**49.2168**	24.16501
f_8	30	0.017284	0.054225	0.397499	0.063616	0.056197
	50	0.033017	0.121202	0.491180	0.152402	0.101034
	100	0.130622	0.4953581	1.474574	**0.591435**	0.341860
	200	0.389816	0.890729	1.470779	**0.888777**	0.293218
f_9	30	0.000166	0.000372	0.002944	0.000437	0.000430
	50	0.000308	0.001959	0.113784	0.005876	0.017979
	100	0.006279	0.016196	0.058500	0.016745	0.009193
	200	0.024272	0.039489	0.063855	0.039949	0.009196
f_{10}	30	3.558E−11	9.028E−09	2.127E−07	3.460E−08	5.317E−08
	50	7.389E−13	3.153E−09	1.456E−07	**1.677E−08**	3.069E−08
	100	1.256E−11	1.002E−09	1.723E−08	**2.553E−09**	3.922E−09
	200	6.821E−13	3.694E−10	1.234E−08	**1.522E−09**	2.600E−09
f_{11}	30	3.025015	595.2057	1,213.786	574.6454	287.3841
	50	156.9476	1,469.0898	2,384.4171	1,361.7754	549.8476
	100	1,460.5972	6,333.5212	9,874.7195	6,221.4047	1,934.6737
	200	7,555.1047	24,914.803	30,099.201	23,259.730	5,279.074

Table 2.4 Results of PSO algorithm with global star structure solving benchmark functions with different dimension numbers. All algorithms are run for 40 times, where "best", "median", "worst", and "mean" indicate the best, median, worst, and mean of the best fitness values for all runs, respectively

Result		**PSO algorithm with global star structure**				
Function	**Dimension**	**Best**	**Median**	**Worst**	**Mean**	**Std. dev.**
f_0	30	0	5.684E−14	1.136E−13	**5.968E−14**	1.774E−14
	50	5.684E−14	1.705E−13	8.753E−12	3.723E−13	1.342E−12
	100	5.115E−13	5.400E−12	7.225E−07	1.817E−08	1.127E−07
	200	0.000300	0.051463	117.0674	8.879628	27.329776
f_1	30	5.684E−14	1.136E−13	1.210E−11	6.721E−13	2.256E−12
	50	2.273E−13	1.534E−12	9.515E−10	4.023E−11	1.577E−10
	100	6.275E−11	4.807E−09	0.000782	1.968E−05	0.000122
	200	0.0535081	1.6067625	22.519262	4.8258535	6.2781619
f_2	30	2.842E−13	4.547E−13	1.477E−12	**5.329E−13**	2.600E−13
	50	2.914E−06	4.348E−05	0.001463	**9.782E−05**	0.000226
	100	500.7058	957.4248	4,358.6327	1,238.6266	748.7982
	200	50,117.897	89,375.843	12,5132.001	87,262.934	19,247.680
f_3	30	0	0	82	3.525	14.78848
	50	0	10	1765	74.325	277.8051
	100	150	687	7339	1,412.125	2,063.8236
	200	1,966	11,924	33,994	13,048.125	8,040.03318
f_4	30	0.000663	0.001545	0.003126	**0.001610**	0.000711
	50	0.002141	0.005239	0.015852	**0.005929**	0.002793
	100	0.017701	0.041171	0.449424	0.066423	0.083962
	200	0.223564	1.684754	27.00333	3.369108	4.890259
f_5	30	1.534E−06	2.120397	72.92778	**5.604286**	15.522474
	50	0.006618	23.772137	78.438271	24.789296	23.462080
	100	19.026049	139.69599	238.01714	138.58872	49.921990
	200	213.84271	532.37553	8,563.21662	733.61886	1,259.4969
f_6	30	39.798311	78.601543	161.182370	86.038717	28.86472
	50	126.35946	197.9957	295.5009	199.264426	41.6134273
	100	362.1635	501.4566	676.5675	509.4410	72.954169
	200	909.39941	1,183.9942	1,420.7919	1,185.7987	97.32094
f_7	30	7.000000	54	132	53.9	31.93571
	50	23.00000	82.00000	239	93.40000	46.74280
	100	153.0000	310.0006	510.0000	319.86258	85.5076
	200	717.50016	983.51168	1,263.7549	989.369183	114.05331
f_8	30	5.684E−14	1.340421	2.579534	1.079306	0.840700
	50	1.027081	2.768558	5.417600	2.738344	0.934523
	100	4.219999	9.582494	17.166035	10.31868	3.903569
	200	15.936288	18.322041	18.84516	18.15455	0.561715
f_9	30	1.421E−14	0.009864	0.072997	0.016561	0.017872
	50	2.842E−14	0.007396	0.167881	0.027477	0.042940
	100	1.989E−13	0.174450	6.421892	0.457950	1.135783
	200	0.0032036	0.300309	27.159807	1.102331	4.19604973
f_{10}	30	5.684E−14	0.103669	1.970941	0.303680	0.446607
	50	1.136E−13	0.062201	1.993613	0.272546	0.445872
	100	3.979E−13	0.031110	0.920799	0.212371	0.283869
	200	0.8256429	2.945502	12.091627	3.455138	1.914228
f_{11}	30	1,658.7070	2,613.5152	4,300.0757	2,581.2254	515.0935
	50	3,168.4002	5,811.3073	9,746.8371	5,867.08078	1,280.43849
	100	10,118.7472	15,611.731	19,624.184	15,593.429	2,194.0101
	200	33,534.8059	39,681.9236	47,087.1921	39,898.1320	3,111.493040

Table 2.5 Results of PSO algorithm with local ring structure solving benchmark functions with different dimension numbers. All algorithms are run for 40 times, where "best", "median", "worst", and "mean" indicate the best, median, worst, and mean of the best fitness values for all runs, respectively

Result		**PSO algorithm with local ring structure**				
Function	**Dimension**	**Best**	**Median**	**Worst**	**Mean**	**Std. dev.**
f_0	30	5.684E−14	1.136E−13	1.705E−13	8.952E−14	3.084E−14
	50	1.136E−13	1.705E−13	2.273E−13	**1.747E−13**	3.454E−14
	100	2.842E−13	3.979E−13	6.252E−13	**4.305E−13**	7.614E−14
	200	1.250E−12	1.875E−12	6.423E−12	**2.076E−12**	8.775E−13
f_1	30	1.136E−13	1.705E−13	2.842E−13	**1.847E−13**	3.538E−14
	50	2.273E−13	3.410E−13	4.547E−13	**3.325E−13**	5.761E−14
	100	4.547E−13	7.389E−13	3.240E−12	**8.057E−13**	4.033E−13
	200	2.842E−12	6.309E−12	8.515E−11	**1.618E−11**	2.372E−11
f_2	30	2.729E−09	1.213E−07	2.708E−06	3.156E−07	5.599E−07
	50	0.824867	3.906257	20.937487	5.673151	4.855619
	100	18,263.8048	42,056.735	73,617.545	42,876.351	14,182.875
	200	175,965.8139	245,794.861	341,100.085	253,452.746	42,760.626
f_3	30	0	0	0	**0**	0
	50	0	0	0	**0**	0
	100	0	0	2	**0.35**	0.526782
	200	4	14	62	17.15	10.2116
f_4	30	0.001509	0.002870	0.004991	0.002922	0.000839
	50	0.004447	0.008213	0.012832	0.008319	0.002284
	100	0.014215	0.032322	0.050631	**0.033854**	0.007115
	200	0.124484	0.193760	0.291623	**0.191594**	0.038465
f_5	30	4.145E−07	9.235352	20.750470	10.028368	6.370256
	50	1.137488	39.443631	93.041226	44.043531	27.62117
	100	69.143924	171.10076	298.31675	165.6484	44.97973
	200	351.8936	485.8045	1,045.035	520.9299	126.5079
f_6	30	40.79328	65.66717	109.5160	67.463671	15.87436
	50	96.51091	171.27308	252.86066	174.31033	32.98838
	100	338.3561	442.1896	559.4544	444.3186	47.6045
	200	823.7839	1,033.800	1,233.757	1,026.297	110.959
f_7	30	32.00000	64.0000	98.00000	64.75031	16.95075
	50	125.07285	177.00187	266.00000	183.31232	33.51210
	100	348.0000	491.2500	640.2500	491.4526	53.9021
	200	912.5000	1,172.027	1,470.840	1,189.096	122.056
f_8	30	8.526E−14	1.421E−13	1.989E−13	**1.357E−13**	2.657E−14
	50	1.421E−13	2.557E−13	0.879051	**0.021976**	0.137241
	100	5.684E−13	2.109321	2.836222	1.842688	0.766436
	200	2.109321	3.126557	4.085244	3.108210	0.417764
f_9	30	1.421E−14	2.842E−14	0.007396	**0.000184**	0.001154
	50	2.842E−14	5.684E−14	6.963E−13	**7.638E−14**	1.038E−13
	100	5.684E−14	1.136E−13	1.918E−12	**1.602E−13**	2.842E−13
	200	3.126E−13	4.689E−13	0.017236	**0.000430**	0.002690
f_{10}	30	5.684E−14	1.136E−13	1.136E−13	**8.668E−14**	2.838E−14
	50	1.136E−13	1.705E−13	0.062201	0.004665	0.016383
	100	3.410E−13	7.958E−13	0.155736	0.010898	0.027450
	200	0.093752	2.471069	6.063923	2.799358	1.328801
f_{11}	30	2,658.6772	3,794.8463	4,669.7768	3,769.3159	429.0876
	50	5,686.735	7,509.0763	8,794.7300	7,448.2504	800.4772
	100	14,936.875	17,617.207	20,033.983	17,582.414	1,254.389
	200	33,929.591	40,064.457	45,760.337	39,958.598	2,629.163

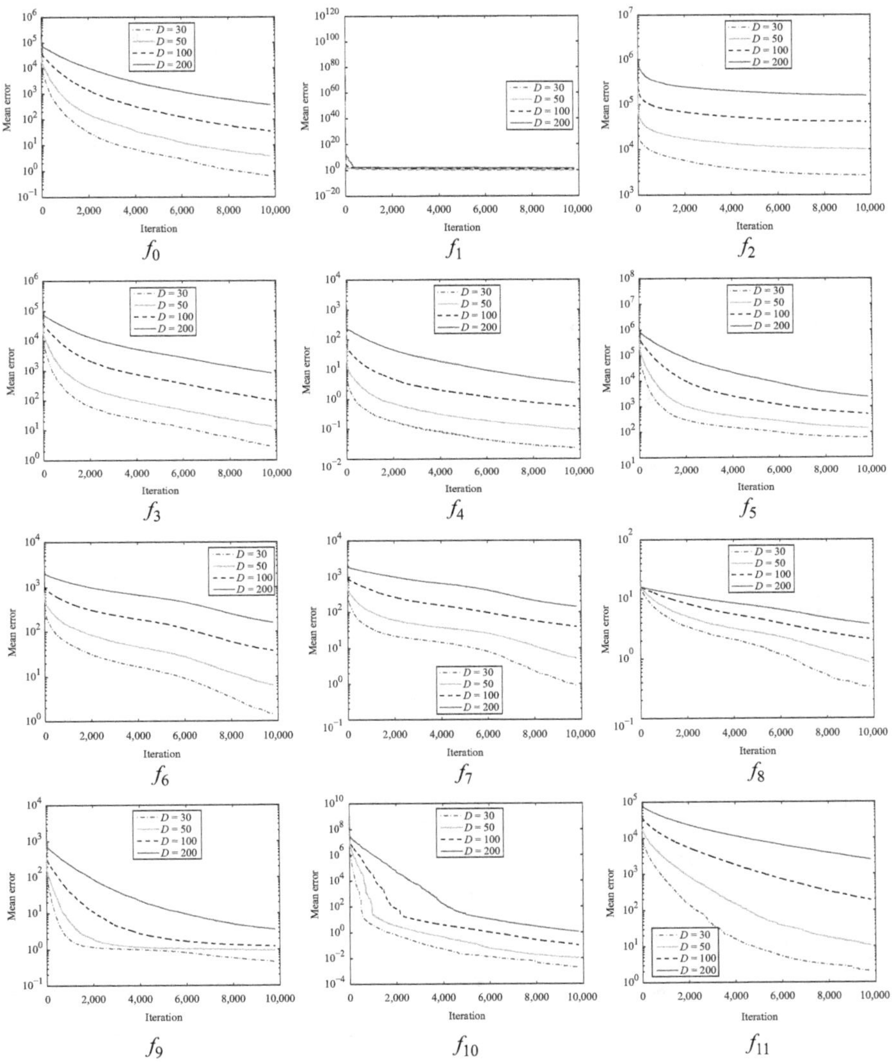

Figure 2.3 Performance of BSO in objective space algorithm solving different benchmark functions

well on multimodal functions. The values become large when these two PSO variants couldn't find good value on solving problems with large number of dimensions. The large values indicate that the generalization ability should be enhanced when the PSO variants solve problems with large number of dimensions. Otherwise, the good solution will be very difficult to be found.

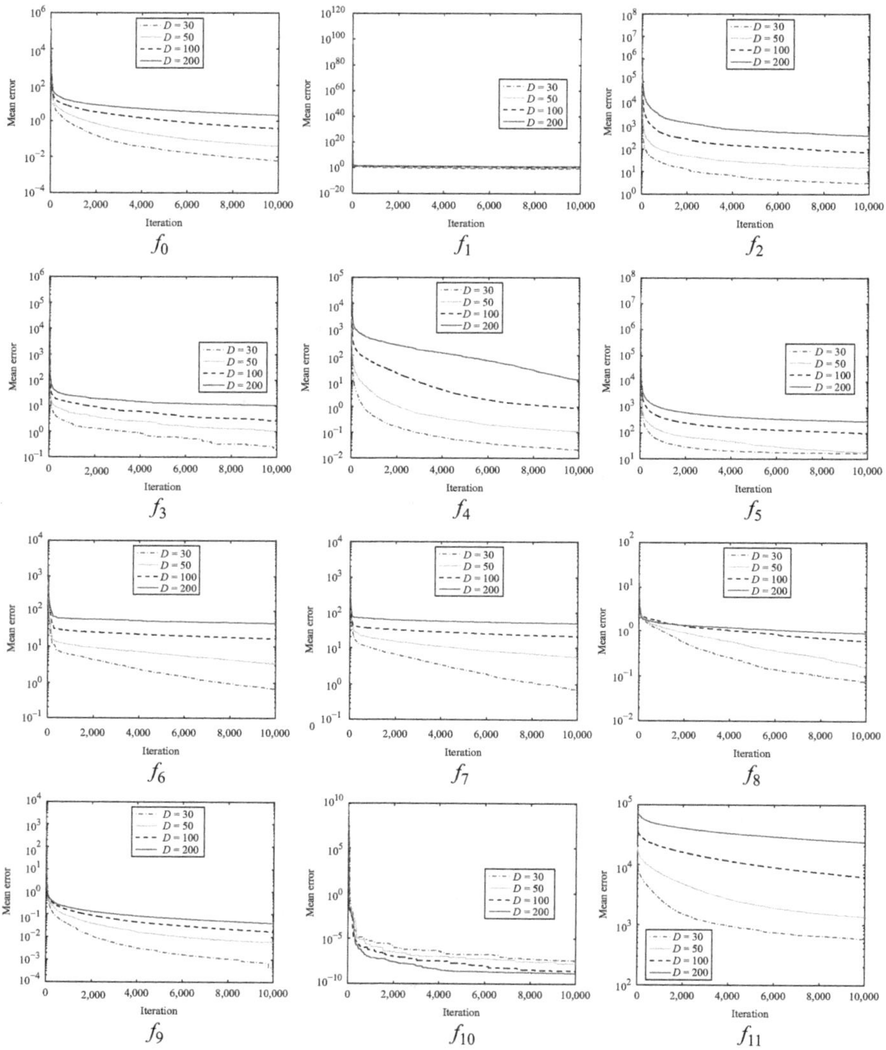

Figure 2.4 Performance of fireworks algorithm solving different benchmark functions

2.6 Conclusions

In this paper, we analyzed the performance of BSO-OS, FWA, and PSOs with star and ring structures on solving different benchmark functions. The search performance was significantly decreased with the increasing number of problem's dimensions. It indicates that problems are getting hard with the large dimension size.

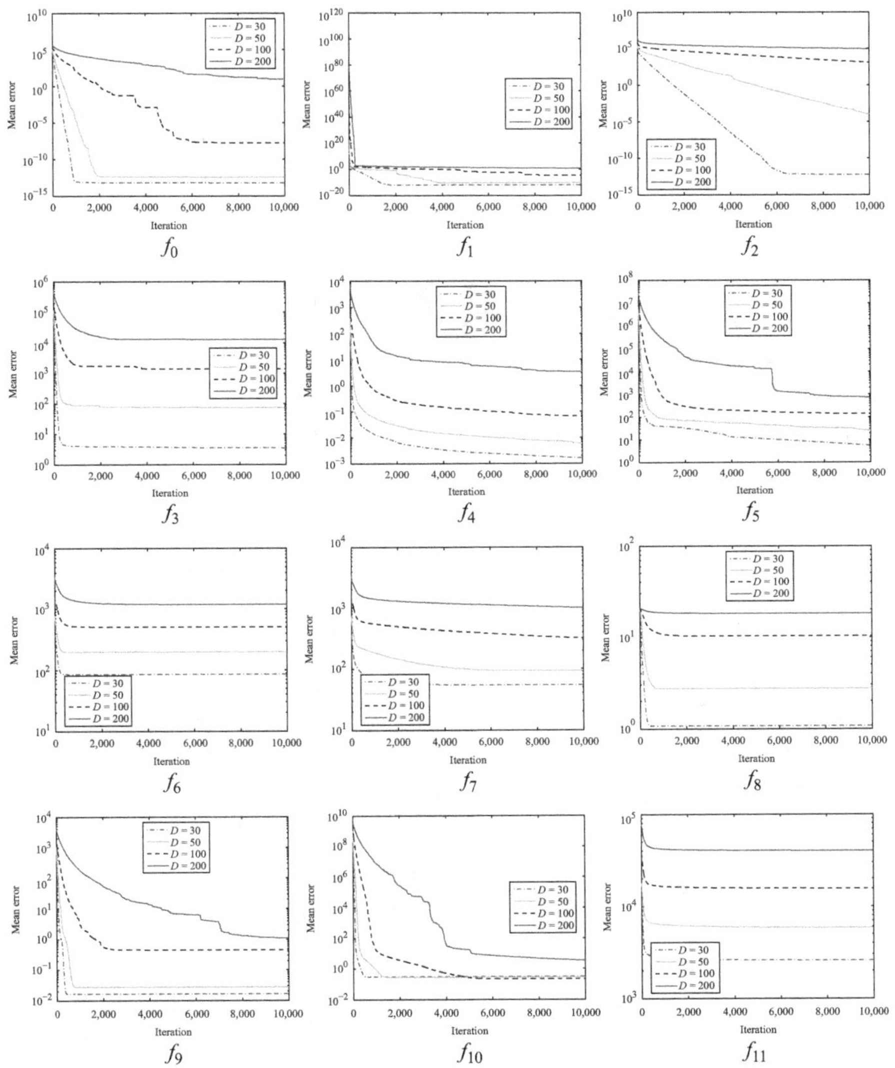

Figure 2.5 Performance of PSO algorithm with global star structure solving different benchmark functions

In the literature, the problem's difficulty measure was analyzed on the landscape of problems. In this paper, it is considered that the problem's difficulty or the generalization ability of algorithms is related to the representation of problems and the selected algorithms. The generalization ability of algorithms is defined in a practical way. It is measured by the result ratio of algorithms with the same settings on

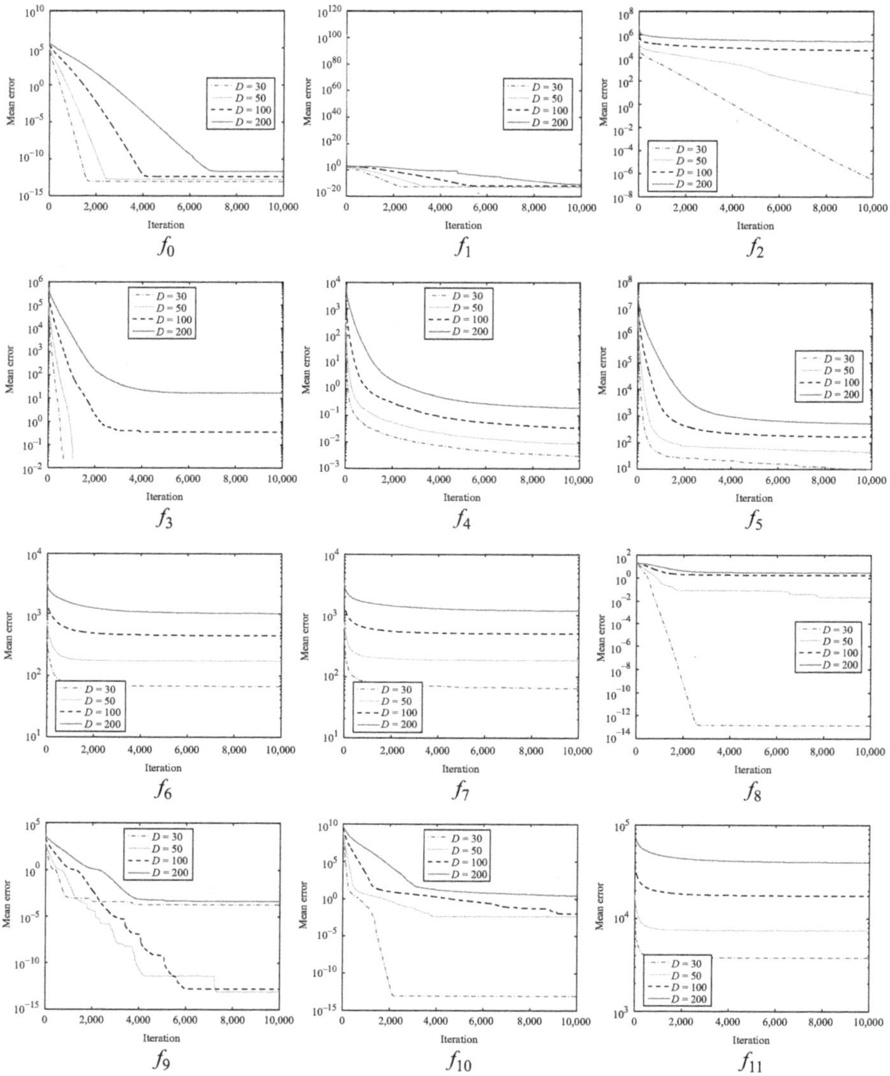

Figure 2.6 Performance of PSO algorithm with local ring structure solving different benchmark functions

problems with different number of dimensions. This definition could be extended to problems with different components. With the analysis on the hardness of problems, we can have more understanding on the relationship between problems and algorithms and can design more effective algorithms to solve different problems.

Table 2.6 Normalized difficulty measure of BSO in objective space algorithm solving benchmark functions

	BSO in objective space			
Function dim.	**30**	**50**	**100**	**200**
f_0	1.0000	3.4588	16.6255	86.3382
f_1	1.0000	1.7072	3.9325	9.1459
f_2	1.0000	2.3081	4.6421	8.9595
f_3	1.0000	2.7789	10.5184	43.2289
f_4	1.0000	2.4305	7.3476	22.8109
f_5	1.0000	1.3917	2.5017	5.8726
f_6	1.0000	2.6653	7.9551	16.4844
f_7	1.0000	3.5326	13.5489	24.2880
f_8	1.0000	1.5240	1.9345	1.7061
f_9	1.0000	1.2146	0.8059	1.1593
f_{10}	1.0000	3.0898	15.3172	84.3877
f_{11}	1.0000	3.1408	28.2293	**184.2400**

Table 2.7 Normalized difficulty measure of fireworks algorithm solving benchmark functions

	FWA solving benchmark functions			
Function dim.	**30**	**50**	**100**	**200**
f_0	1.0000	3.7868	18.5042	50.8370
f_1	1.0000	2.2605	5.2104	8.9044
f_2	1.0000	2.9394	7.1201	19.2393
f_3	1.0000	2.8000	3.4667	6.9833
f_4	1.0000	3.0673	12.1982	**76.1653**
f_5	1.0000	0.6580	1.6898	2.5294
f_6	1.0000	2.8848	7.1309	9.6558
f_7	1.0000	4.6303	8.7934	10.3841
f_8	1.0000	1.4374	2.7891	2.0956
f_9	1.0000	8.0677	11.4954	13.7125
f_{10}	1.0000	0.2908	0.0221	0.0066
f_{11}	1.0000	1.4219	3.2480	6.0715

Acknowledgments

This work is partially supported by National Natural Science Foundation of China under Grant Numbers 71402103, 61672334, 61273367, 61571238, and 61302158; the MOE Youth Foundation Project of Humanities and Social Sciences at Universities in China under grant 13YJC630123; The Youth Foundation Project of Humanities and Social Sciences in Shenzhen University under grant 14QNFC28, and the Fundamental Research Funds for the Central Universities under Grant GK201703062.

Table 2.8 Normalized difficulty measure of PSO algorithm with global star structure solving benchmark functions

	PSO algorithm with star structure			
Function dim.	**30**	**50**	**100**	**200**
f_0	1.0000	3.7429	9.133E+04	2.231E+13
f_1	1.0000	35.9142	8.784E+06	1.077E+12
f_2	1.0000	1.101E+08	6.972E+14	**2.456E+16**
f_3	1.0000	12.6510	120.1808	555.239
f_4	1.0000	2.2095	12.3769	313.892
f_5	1.0000	2.6539	7.4187	19.6354
f_6	1.0000	1.3895	1.7763	2.0673
f_7	1.0000	1.0397	1.7803	2.7533
f_8	1.0000	1.5222	2.8681	2.5230
f_9	1.0000	0.9954	8.2956	9.9842
f_{10}	1.0000	0.5384	0.2097	1.7066
f_{11}	1.0000	1.3637	1.8123	2.3185

Table 2.9 Normalized difficulty measure of PSO algorithm with local ring structure solving benchmark functions

	PSO algorithm with ring structure			
Function dim.	**30**	**50**	**100**	**200**
f_0	1.0000	1.1709	1.4426	3.4785
f_1	1.0000	1.0801	1.3086	13.1402
f_2	1.0000	1.078E+07	4.075E+10	1.204E+11
f_3	–	–	1.0000	24.5000
f_4	1.0000	1.7082	3.4757	9.8354
f_5	1.0000	2.6351	4.9553	7.7918
f_6	1.0000	1.5502	1.9758	2.2818
f_7	1.0000	1.6986	2.2769	2.7546
f_8	1.0000	9.716E+10	4.073E+12	3.435E+12
f_9	1.0000	2.490E−10	2.611E−10	0.3505
f_{10}	1.0000	3.229E+10	3.771E+10	**4.844E+12**
f_{11}	1.0000	1.1856	1.3993	1.5901

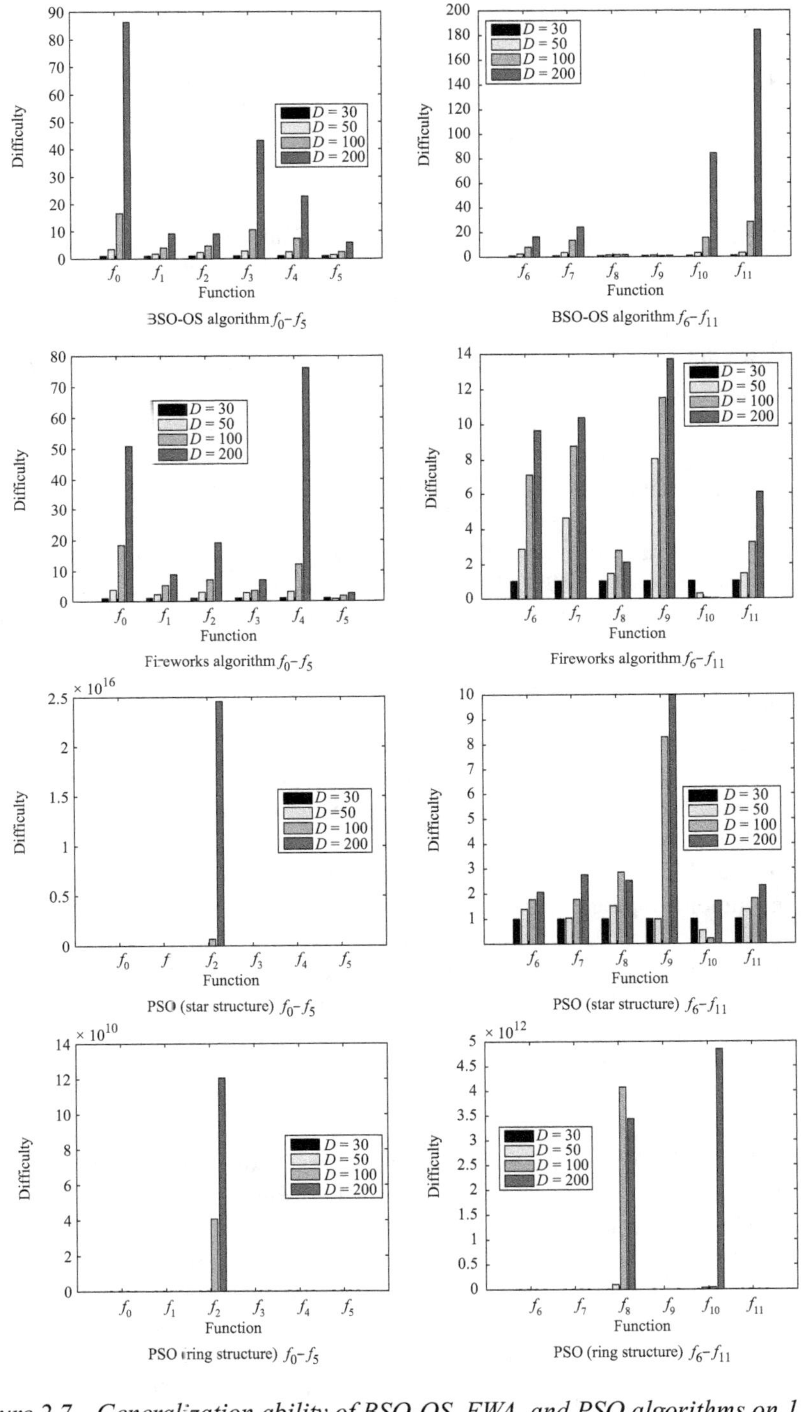

Figure 2.7 Generalization ability of BSO-OS, FWA, and PSO algorithms on 12 benchmark functions

References

[1] James Kennedy, Russell Eberhart, and Yuhui Shi. *Swarm Intelligence*. Morgan Kaufmann Publisher, San Francisco, CA, USA, 2001.

[2] Russell Eberhart and Yuhui Shi. *Computational Intelligence: Concepts to Implementations*. Morgan Kaufmann Publisher, San Francisco, CA, USA, 2007.

[3] Siang Yew Chong, Peter Tiño, and Xin Yao. Measuring generalization performance in coevolutionary learning. *IEEE Transactions on Evolutionary Computation*, 12(4):479–505, August 2008.

[4] Bart Naudts and Leila Kallel. A comparison of predictive measures of problem difficulty in evolutionary algorithms. *IEEE Transactions on Evolutionary Computation*, 4(1):1–15, 2000.

[5] Jun He, Colin Reeves, Carsten Witt, and Xin Yao. A note on problem difficulty measures in black-box optimization: Classification, realizations and predictability. *Evolutionary Computation*, 15(4):435–443, 2007.

[6] Kenneth E. Kinnear. Fitness landscapes and difficulty in genetic programming. In *Proceedings of 1994 IEEE Congress on Evolutionary Computation (CEC 1994)*, pages 142–147, Orlando, FL, 1994. IEEE.

[7] Mark Hauschild and Martin Pelikan. Advanced neighborhoods and problem difficulty measures. In *Proceedings of Annual Conference on Genetic and Evolutionary Computation (GECCO 2011)*, pages 625–632, Dublin, Ireland, 2011. ACM.

[8] Jing Liu, Hussein A. Abbass, David G. Green, and Weicai Zhong. Motif difficulty (MD): A predictive measure of problem difficulty for evolutionary algorithms using network motifs. *Evolutionary Computation*, 20(3):321–347, 2012.

[9] Mario Graff, Riccardo Poli, and Juan J. Flores. Models of performance of evolutionary program induction algorithms based on indicators of problem difficulty. *Evolutionary Computation*, 21(4):533–560, 2013.

[10] Ingo Wegener. *Complexity Theory: Exploring the Limits of Efficient Algorithms*. Springer, Berlin Heidelberg, 2005.

[11] Yuhui Shi. Brain storm optimization algorithm. In Ying Tan, Yuhui Shi, Yi Chai, and Guoyin Wang, editors, *Advances in Swarm Intelligence*, volume 6728 of *Lecture Notes in Computer Science*, pages 303–309. Springer, Berlin Heidelberg, 2011.

[12] Yuhui Shi. An optimization algorithm based on brainstorming process. *International Journal of Swarm Intelligence Research (IJSIR)*, 2(4):35–62, October–December 2011.

[13] Shi Cheng, Quande Qin, Junfeng Chen, and Yuhui Shi. Brain storm optimization algorithm: A review. *Artificial Intelligence Review*, 46(4):445–458, December 2016.

[14] Shi Cheng, Yifei Sun, Junfeng Chen, *et al.* A comprehensive survey of brain storm optimization algorithms. In *Proceedings of 2017 IEEE Congress on Evolutionary Computation (CEC 2017)*, pages 1637–1644, Donostia/San Sebastián, Spain, 2017. IEEE.

[15] Yuhui Shi. Brain storm optimization algorithm in objective space. In *Proceedings of 2015 IEEE Congress on Evolutionary Computation (CEC 2015)*, pages 1227–1234, Sendai, Japan, 2015. IEEE.

[16] Ying Tan and Yuanchun Zhu. Fireworks algorithm for optimization. In Ying Tan, Yuhui Shi, and Kay Chen Tan, editors, *Advances in Swarm Intelligence*, volume 6145 of *Lecture Notes in Computer Science*, pages 355–364. Springer, Berlin Heidelberg, 2010.

[17] Ying Tan. *Fireworks Algorithm: A Novel Swarm Intelligence Optimization Method*. Springer, Berlin Heidelberg, 2015.

[18] Yuhui Shi. Developmental swarm intelligence: Developmental learning perspective of swarm intelligence algorithms. *International Journal of Swarm Intelligence Research (IJSIR)*, 5(1):36–54, January–March 2014.

[19] Shi Cheng, Yuhui Shi, Quande Qin, and Shujing Gao. Solution clustering analysis in brain storm optimization algorithm. In *Proceedings of The 2013 IEEE Symposium on Swarm Intelligence (SIS 2013)*, pages 111–118, Singapore, 2013. IEEE.

[20] Shi Cheng, Quande Qin, Junfeng Chen, Yuhui Shi, and Qingyu Zhang. Analytics on fireworks algorithm solving problems with shifts in the decision space and objective space. *International Journal of Swarm Intelligence Research (IJSIR)*, 6(2):52–86, 2015.

[21] Yuhui Shi and Russell Eberhart. A modified particle swarm optimizer. In *Proceedings of the 1998 Congress on Evolutionary Computation (CEC1998)*, pages 69–73, 1998.

[22] Quande Qin, Shi Cheng, Qingyu Zhang, Li Li, and Yuhui Shi. Particle swarm optimization with interswarm interactive learning strategy. *IEEE Transactions on Cybernetics*, 46(10):2238–2251, 2016.

[23] Yuhui Shi and Russell Eberhart. Parameter selection in particle swarm optimization. In *Evolutionary Programming VII*, volume 1447 of *Lecture Notes in Computer Science*, pages 591–600. Springer, Berlin Heidelberg, 1998.

[24] Daniel Ashlock, Cameron McGuinness, and Wendy Ashlock. Representation in evolutionary computation. In Jing Liu, Cesare Alippi, Bernadette Bouchon-Meunier, Garrison W. Greenwood, and Hussein A. Abbass, editors, *Advances in Computational Intelligence*, volume 7311 of *Lecture Notes in Computer Science*, pages 77–97. Springer, Berlin Heidelberg, 2012.

[25] David H. Wolpert and William G. Macready. No free lunch theorems for optimization. *IEEE Transactions on Evolutionary Computation*, 1(1):67–82, April 1997.

[26] Daniel Bratton and James Kennedy. Defining a standard for particle swarm optimization. In *Proceedings of the 2007 IEEE Swarm Intelligence Symposium (SIS 2007)*, pages 120–127, April 2007.

Chapter 3
A unifying framework for swarm intelligence-based hybrid algorithms

Bin Xin[1,2,3], *Yipeng Wang*[1], *and Bo Liu*[4]

Abstract

This chapter is aimed at giving a classification and an analysis of various hybrid optimisers based on swarm intelligence optimisation algorithms (SIOAs) by the systematic taxonomy we proposed in a recent work. The taxonomy comprises five factors including the relationship between parent optimisers, hybridisation level, operation order, type of information transfer and type of transferred information. Based on the taxonomy, a unifying framework for SIOA-based optimisers is established. Some typical SIOA-based hybrids which are divided into two parts according to the combination patterns about global search and local search are analysed in accordance with the taxonomy. By the classification-based analysis, designers can gain an insight into various possibilities for hybrid design of SIOA-based optimisers.

3.1 Introduction

The concept of swarm intelligence (SI) derives from the observation of collective behaviours of a swarm in nature and/or social phenomenon. The swarm refers to a group of individuals that can communicate with each other directly or indirectly, and such swarm can solve complicated problems using decentralised control and self-organisation of relatively simple individuals [1,2]. Optimisation techniques inspired by SI (i.e. SI optimisation algorithms, SIOAs), which inherit the features of SI, are population-based stochastic methods mainly for solving combinatorial optimisation problems. Besides, SIOAs have become increasingly popular during the last decade, and plenty of optimisers have been proposed. Particle swarm optimisation (PSO), ant

[1]School of Automation, Beijing Institute of Technology, China
[2]Key Laboratory of Intelligent Control and Decision of Complex Systems, Beijing Institute of Technology, China
[3]Beijing Advanced Innovation Center for Intelligent Robots and Systems, Beijing Institute of Technology, China
[4]Academy of Mathematics and Systems Science, Chinese Academy of Sciences, China

colony optimisation (ACO), artificial bee colony (ABC), the artificial fish swarm algorithm, and the bacterial foraging optimisation algorithm (BFOA) are among the most representative SIOAs.

Inspired by various mechanisms in nature, numerous optimisation algorithms (OAs) have been proposed, including SIOAs. It seems that the pursuit of a 'better' or 'best' (sometimes only conceptually novel) optimiser continues unabated, though the interest in the best optimiser has subsided by the publication of the no free lunch theorems (NFLT) [3]. The NFLT mathematically proved that the average performance of any pair of algorithms A and B across all possible problems is the same. This fact implies that if an algorithm performs well on a certain class of problems, then it necessarily pays for that with degraded performance on the set of all remaining problems (as this would be the only way that all algorithms can have the same performance averaged over all the optimisation problems) [4]. Specifically, following different paradigms, various SIOAs have their own merit and demerit. Since a plethora of optimisers have been proposed, we are more concerned with the question:

Which SIOA has the best (or better) performance in solving a specific problem?

Researchers and engineers who aim to solve their domain-specific problems may be more interested in the following generalised question:

How to choose the most suitable optimiser in different cases to solve the problem in hand or specific problem instances?

No doubt that the choice of the 'best' SIOA is problem dependent, and even instance dependent. With limited SIOAs, we can only answer these questions by testing the performance of every SIOA to choose the best one. Moreover, some efforts to seek for the most suitable optimiser for specific problems have been made, including the research on adaptive algorithms and hyper-heuristics [5,6]. Nevertheless, we can overcome the limitation of their original frameworks by hybridisation. As a result, new optimisers with better performance are put forward, which contributes to the diversity of optimisers. Besides, emergence of various hybrids offers more choices for the optimisation problem solving and a better optimiser might be found. Hence, how to design a novel and excellent hybrid is a topic worthy of devotions.

In the past decade, hybrid optimisers have attracted persistent attention from scholars that are interested in design of optimisers and their applications [7–13]. As Raidl claimed in his unified view of hybrid meta-heuristics [13], it seems that choosing an adequate hybrid approach is determinant to achieve top performance in solving most difficult problems. A common template for hybridisation is provided by memetic algorithms (MAs), which combine the respective advantages of global search (GS) and local search (LS) [14]. Due to excellent performance, MAs have been favoured by many scholars in their research on different optimisation problems [14]. However, MAs only represent a special class in the family of hybrid optimisers. There are manifold possibilities of hybridising different optimisers, which follow diverse philosophies and paradigms. Admittedly, all population-based OAs suffer from large time cost because of their stochastic nature. In order to obtain the most of the advantage of SIOAs and to eliminate their disadvantages like premature convergence and large computational cost, many studies on the SIOAs-based hybridisation have been conducted in recent decades and this topic becomes increasingly attractive [15–22].

This chapter is aimed at giving a classification, and an analysis of various SIOA-based hybrid optimisers by the systematic taxonomy we proposed in a recent work [15]. To date, it is impossible to list all kinds of hybrids within a chapter of a book, since a plethora of hybrids exist in the literature (about 1,610,000 related results according to the Google search result with the index terms 'hybrid algorithm' and 'optimisation'). Based on the taxonomy, a unifying framework for SIOA-based optimisers is established. Designers can gain an insight into various possibilities for hybrid design of SIOA-based optimisers.

3.2 Taxonomy on hybridisation strategies

A favourable taxonomy of hybridisation strategies should be capable of differentiating different strategies as well as providing designers a convenient and efficient means of determining a hybridisation scheme. Several taxonomies can be found in the literature, which provide certain classification for the hybridisation of meta-heuristics [10,13,23] or MAs [14]. The taxonomy proposed by Talbi for the hybridisation of meta-heuristics considered solutions to design and implementation issues [10]. Both hierarchical and flat classifications were proposed by Talbi with respect to each issue. Besides, Jourdan *et al.* extended this taxonomy to the case of cooperation between meta-heuristics and exact methods [23]. Moreover, Raidl presented a similar taxonomy with the Talbi's without some flat classification mechanisms, which brings in a new classification mechanism, i.e. control strategy, to distinguish integrative hybridisation and collaborative hybridisation [13]. Nevertheless, a scheduler-based taxonomy was proposed by Krasnogor and Smith to classify different MAs, and it provides an effective approach to capture the interaction between LS and standard evolutionary operators (mutation, crossover and selection) [14].

In this section, we will introduce the taxonomy of hybridisation strategies we proposed in recent work [15], and the elements of a hybridisation strategy (i.e. hybridisation factors) will be presented in detail.

3.2.1 Hybridisation factors

Different hybrids can be differentiated by the relationship between parent optimisers, hybridisation level (HL), operation order (OO), type of information transfer (TIT) and type of transferred information (TTI).

3.2.1.1 Relationship between parent optimisers

The first classification criterion considers the relationship between parent optimisers for hybridisation. We mainly consider three kinds of parent relationships (PRs): collaboration, embedding and assistance.

Collaboration means that parent optimisers sample iteratively in the search space following their own paradigms, respectively, and they cooperate with each other during the search for optimum by sharing or exchanging accumulated information. Conversely, if the operating manners of parent optimisers are integrated into a novel unified manner for the iterative sampling and cannot be separated explicitly, then such relationship will be called as 'embedding'. In this case, it is hard to distinguish their contribution to fitness improvement since the parents are integrated into a holistic hybrid optimiser. In contrast, assistance refers to a very special hybridisation strategy in which one parent will not sample directly in search space but serve as an assistance of another parent. For example, one parent optimiser may be applied in the parameter space of another one under the assistance relationship. It should be noted that only if a point in search space undergoes fitness evaluation can it be treated as a sampling point. Usually, the assistant will follow its own search paradigm, which is a very important feature for distinguishing 'assistance' from 'embedding'. The following is a mathematical formulation of the three kinds of PR:

$$\textit{Collaboration}\begin{cases}\text{Parent A}, & f_{\mathrm{A}}: S_s^{n_{\mathrm{A}}(t)} \mapsto S_s^{n_{\mathrm{A}}(t+1)}[S_{ob}];\\ \text{Parent B}, & f_{\mathrm{B}}: S_s^{n_{\mathrm{B}}(t)} \mapsto S_s^{n_{\mathrm{B}}(t+1)}[S_{ob}].\end{cases} \tag{3.1}$$

$$\text{Assistance-Parent B} f_{\mathrm{B}}: S_s^{n_{\mathrm{A}}(t)} \xrightarrow{\text{Parent}_{\mathrm{A}} f_{\mathrm{A}}:\, S \mapsto S} S_s^{n_{\mathrm{B}}(t+1)}[S_{ob}] \tag{3.2}$$

$$\text{Embedding-Parent A and B}: f_{\mathrm{A}\oplus\mathrm{B}}: S_s^{n_{\mathrm{A}}(t)} \mapsto S_s^{n_{\mathrm{B}}(t+1)}[S_{ob}] \tag{3.3}$$

where f_{A} and f_{B} represent the mapping of parents A and B, respectively; S_s and S_{ob} are the solution space and the objective space, respectively; $S_s^{(n)} = \underset{1}{S_s} \times \underset{2}{S_s} \times \cdots \times \underset{n}{S_s}$; $n(t)$ is the population size (the optimiser adopts a static population sizing strategy if $n(t) =$ constant, and a dynamic strategy otherwise); S_{ob} indicates that solutions that are generated will go through objective evaluation. In the case of assistance, parent A as the assistant of parent B can work in different spaces, i.e. $S \in \{S_p, S_{op}, S_s, S_{it}, \ldots\}$ (where S_p is the parameter space, S_{op} is the operator (optimiser) space, and S_{it} is the information topology space), and the mapping f_{A} reflects the complete operation of parent A. The whole mapping in the case of assistance can be regarded as a composite mapping $f_{\mathrm{A}} \circ f_{\mathrm{B}}$, and the operations of parents can be differentiated. In the case of embedding, $f_{\mathrm{A}\oplus\mathrm{B}}$ represents a nonseparable hybrid mapping in which parents A and B are integrated into a new holistic operator.

3.2.1.2 Hybridisation level

The HL of the parent optimisers, which highly depends on their respective operating levels (OLs), refers to the OL at which they are hybridised, indicating the spatial assignment of parent optimisers. As illustrated in Figure 3.1, there are four kinds of HLs: component level, individual level, subpopulation level, and population level (in order from the lowest to the highest level). This classification is also applicable to the OL. The OL of a parent optimiser is determined by the object to which it applies according to hybridisation design, which can apply the optimiser without being required to use the same optimiser with any other objects. If the object is a whole

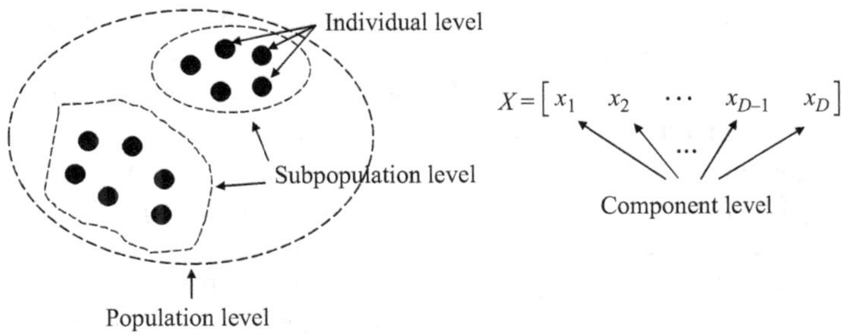

Figure 3.1 Hybridisation levels

population, we say the parent operates at population level. Likewise, a parent operates at subpopulation level, individual level or component level, if the corresponding object is a subpopulation, an individual or a solution component (or multiple components), respectively.

Furthermore, if parents work at the same OL, we term the resultant hybridisation as homogeneous level hybridisation (HOLH), and the OL of parents is regarded as their HL. In contrast, heterogeneous level hybridisation (HELH) means that the OLs of parents are different. For example, regarding the hybridisation between a population-based optimisation (PO) and a trajectory search method (TSM), i.e. single-point-based optimiser [12], in which the PO usually operates at population level, TSM will be regarded to be hybridised with the PO at individual level if a TSM manipulates only one solution at each generation. As a rule, the HL in HELH between two optimisers is determined by the lower OL of parents. Furthermore, some special HOLH can be termed as homogeneous-object hybridisation (HOOH), if parents share the same object to determine their OLs, which means that they are employed to evolve the same object at their HL. In contrast, heterogeneous-object hybridisation (HEOH) implies that the objects which determine the OLs of parents are different even though the objects may be of the same type (e.g. subpopulation type). For example, parents that operate on different subpopulations give birth to HEOH. Besides, undoubtedly we can categorise any HELH as HEOH.

3.2.1.3 Operating order

The OO of parent optimisers indicates the temporal assignment of parent optimisers, and three OO types are concerned here: no order, serial and parallel. It is worth noting that what the operating order means here is not the actual order of execution. Instead, it refers to the operating order decided by the information dependency between the parent optimisers. Generally, the parent optimisers can work in a sequential (alternate) or parallel order. Talbi terms the sequential OO type as relay, indicating the case where a group of meta-heuristics is applied one after another, each using the output of the previous as its input, acting in a pipeline fashion [10]. Das categorises hybrids based on the two OO types as serial hybridisation and parallel hybridisation, respectively [24]. Particularly, parents of embedding-based hybrids work together as an ensemble, regardless of the OO. Das terms embedding-based hybrids as implicit hybridisation [24].

For clarity of classification, the case, where the operating order of parents is tough to tell apart, is termed as 'no order' in our taxonomy.

Besides, the OO with respect to a hybrid should be canonically analysed at its HL, which is a crucial rule to determine the OO of parents since we can draw different conclusions on the OO from different angles. To avoid confusion, we will treat the analysis of OO regarding HOOH and HEOH in different ways. For HOOH, parents hold a parallel OO if and only if they can simultaneously operate on the same object that determines their OLs. If the parents evolve the object in a *relay* manner as Talbi claimed [10], we say they hold a sequential OO. For HEOH, parents usually hold a parallel OO as they operate on different objects, facilitating the implementation of parallel OO. However, parents may hold a sequential OO owing to some special interdependence between them. For example, in assistance-based hybrids, the master parent and its assistant usually operate on different objects, and the parents definitely hold a sequential OO since the implementation of the master relies on the information provided by its assistant.

In addition, we should focus on the interaction of parents regarding the parallel OO, including the information shared between parents and the occasions on which the information is shared. The parallel order can be carried out with the support of hardware architecture being capable of achieving parallel computing (e.g. single-instruction stream with multiple-data stream and multiple-instruction stream with multiple-data stream). A further discussion on the parallel architecture for the implementation of hybrid meta-heuristics can be found in [10].

3.2.1.4 Type of information transfer

The TIT indicates the interconnection structure of parent optimisers, concerning the information flow between them. Taking each parent optimiser as an 'individual (particle)', we can equate the TIT with the population topology of parent optimisers. Four typical interconnection structures are presented in Figure 3.2. Pertaining to the hybridisation between two parent optimisers, what concerns the classification of the TIT is the direction of information flow, either unidirectional (simplex TIT) or bidirectional (duplex TIT). Generally speaking, the duplex TIT leads to a stronger coupling between parents than the simplex TIT. In addition, if the parents in a duplex TIT transfer information of the same type, the TIT will be termed as the homogeneous TIT, and heterogeneous TIT otherwise.

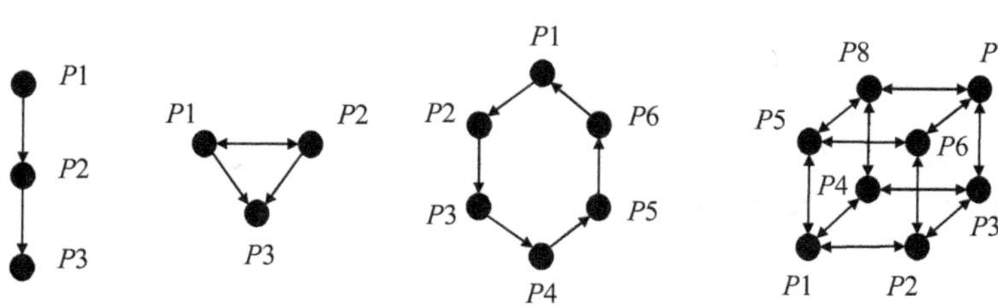

Figure 3.2 Typical interconnection structures

It should be stressed that the goal of information transfer in hybridisation is the information utilisation that is linked to interactions between parents. If the transferred information is merely shared but not utilised to affect the search behaviour of other parents, the parents will work independently. In this case, the hybridisation without interactions cannot bring the expected performance emergence since it can be easily derived that

$$\begin{aligned} Perf_{\mathrm{AB}}(m+n) &= Perf_{\mathrm{A}}(m) + Perf_{\mathrm{B}}(n) \\ &\leq \max\{Perf_{\mathrm{A}}(m+n), Perf_{\mathrm{B}}(m+n)\} \end{aligned} \tag{3.4}$$

where AB is the hybrid of the parents A and B, $Perf_{\mathrm{AB}}(m+n)$ is the performance index of AB with $m+n$ function evaluations (FEs), and $Perf_{\mathrm{A}}(m)$ and $Perf_{\mathrm{B}}(n)$ are the performance indices of A with m FEs and B with n FEs, respectively.

The aforementioned inequality holds for the optimisation of any problem, indicating that the hybrid AB cannot outperform both parents at the same time. Therefore, the hybridisation without interactions is meaningless, and the utilisation of transferred information is essential. To conclude, no interaction, no hybridisation!

3.2.1.5 Type of transferred information

As a crucial design element of hybrid optimisers, the TTI is the most complicated factor among the aforementioned hybridisation factors. It is not only related to specific optimisers, but also to the type of the problems to be solved. Common TTIs include current solutions, group of elite solutions, so-far-best solution (**gbest**), solution components and control parameters. Besides, the transferred information may be a combination of the TTIs mentioned earlier.

From the earlier discussion, it can be seen that the TTI is a complicated hybridisation factor with plentiful options, making it very hard to enumerate all TTIs. Instead, a coarse-grained classification of the TTI is provided, covering the following TTI types: (1) solutions, (2) fitness information, (3) solution components, (4) algorithm-induced in-betweens, (5) auxiliaries (e.g. Lagrange multipliers) and (6) control parameters.

The difference between solutions and algorithm-induced in-betweens lies in that solutions are expected to have gone through fitness evaluation but algorithm-induced in-betweens are merely intermediate variables that are introduced by specific optimisers and usually taken as ingredients for generation of new solutions.

3.2.2 Taxonomy

First of all, we will introduce some concise notations to express the candidate classes with respect to each factor as follows.

1. Parent relationship (PR)
   Collaboration: $\langle C\rangle$, Embedding: $\langle E\rangle$, Assistance: $\langle A\rangle$.
2. Hybridisation level (HL)
   Population level: $\langle P\rangle$, Subpopulation level: $\langle S\rangle$, Individual level: $\langle I\rangle$, Component level: $\langle C\rangle$.
3. Operating order (OO)
   Sequential order: $\langle S\rangle$, Parallel order: $\langle P\rangle$, No order: $\langle N\rangle$.

4. Type of information transfer (TIT)
   Simplex TIT: $\langle S \rangle$, Duplex TIT: $\langle D \rangle$.
   The simple classification of the TIT, here, is only appropriate for hybridisation between two parents. For hybridisation that involves more than two parents, designers will have more choices with the TIT in respect that the information topology of parents can be flexibly configured in a variety of ways (see Figure 3.2). In the generalised case, we can resort to a graph representation of the TIT.
5. Type of transferred information (TTI)
   Solutions: $\langle S \rangle$, Fitness information: $\langle F \rangle$, Solution components: $\langle S_c \rangle$, Auxiliaries: $\langle A \rangle$, Control parameters: $\langle C_p \rangle$, Algorithm-induced in-betweens: $\langle A_i \rangle$.

The six TTI types that are mentioned earlier are not mutually exclusive, that is to say, a hybridisation strategy may involve multiple TTI types simultaneously. For example, solutions $\langle S \rangle$ are usually transferred along with their fitness information $\langle F \rangle$. In some special cases, the fitness information may be transferred alone. For brevity, the transferred fitness information will not be labelled in the following TTI classification of different hybrids unless its corresponding solution is not transferred.

The five hybridisation factors that are mentioned earlier are either fine-grained or coarse-grained, covering primary elements to be considered in hybridisation design and exhibiting a hierarchical feature (see Figure 3.3). PR is a rough but fundamental classification criterion, grouping hybrids into three general categories—collaboration, embedding and assistance. The HL that is based on the OLs of parent optimisers helps one to differentiate the spatial assignment of parents, while the OO indicates the temporal assignment of parents. Both of them are crucial factors for the design of hybrid optimisers. In contrast, the TIT and TTI are mainly associated with details in hybridisation design. With this taxonomy, one can differentiate various hybridisation strategies in more detail and easily compare different hybridisation schemes. Due to the hierarchical feature of the taxonomy, one can also take it as a reference of a step-by-step procedure to design any hybrid OAs, by determining in turn PR, HL, OO, TIT and TTI. The template $\langle PR, HL, OO, TIT \rangle$ that incorporates

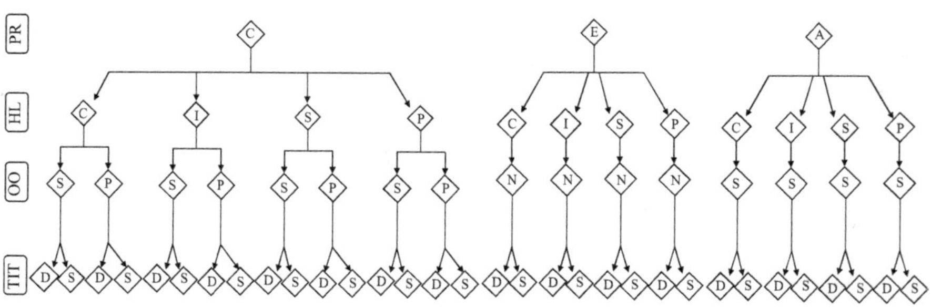

Figure 3.3 Taxonomy of hybridisation strategies. Each string connecting four hybridisation factors corresponds to a class of hybridisation strategies

the first four factors will be utilised to represent various sorts of hybridisation strategies. The reason that the TTI is not included in this template is that different TTI types in the classification of the TTI are not mutually exclusive for the construction of a hybridisation strategy. In the proposed taxonomy, the TTI will be treated as a complementary criterion for the detailed analysis of or to design a hybridisation strategy.

According to the classification of each factor in the template, there seem to be total $3 \times 4 \times 3 \times 2 = 72$ classes of hybridisation strategies. However, there exist some special connections (constraints) between the hybridisation factors, which makes it impossible to arbitrarily combine these factors. For example, parents in embedding- or assistance-based hybridisation cannot hold a parallel OO.

3.3 Previous SIOA-based hybrid optimisers

In this section, we will introduce and analyse some typical SIOA-based hybrids in the literature in accordance with the taxonomy aforementioned. Since the **tr**adeoff between **e**xplo**r**ation and **e**xplo**i**tation (Tr:Er&Ei) is a core of all kinds of optimisers, we primarily divide the hybrids into two parts according to the combination patterns that involve GS and LS. Besides, it is imperative to implement certain type of GS for solving complex problems, and SIOAs can be used for both GS and LS in solution space. Therefore, the hybrids considered here include two main patterns: ‘GS⊕GS’ and ‘GS⊕LS’.

In order to build a unified nomenclature and differentiate various SIOA-based hybrids with the same parents conveniently, we will name a given hybrid by combining the initial of the last name of each inventor of the hybrid into a suffix that follows the parent algorithms.

3.3.1 Hybrids based on SIOAs and GS methods

Although Tr:Er&Ei has been considered in SIOAs, the stagnation phenomenon caused by the weakness of a sole algorithm still could not be prevented. Therefore, the methods with GS ability can be combined with any SIOA to design a new optimiser. A rational motivation behind this hybridisation is that the two or more optimisers correspond to different landscapes, which may give birth to some shortcuts to escape from the local optima [25], even though both of them are implemented as GS. This concept can be demonstrated briefly as shown in Figure 3.4.

As we can see in Figure 3.4, algorithm A is likely to trap into local optimal point P, while P and Q in Figure 3.4(b) are closer. Consequently, switching A to B is conducive to escaping from local optima.

3.3.1.1 SIOA (GS) ⊕ SIOA (GS)

Although every SIOA is a population-based stochastic method with GS ability inherently, stagnation may appear during searching in use of any SIOA. Since different

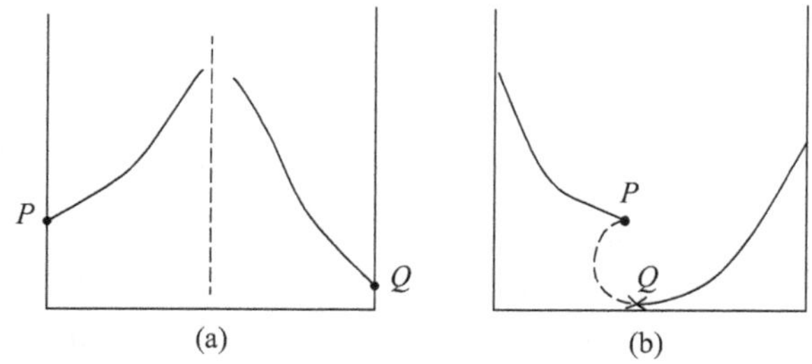

Figure 3.4 The landscapes of different algorithms for the same problem. Part (a) represents the landscape created by algorithm A for certain problem. Part (b) shows the landscape for algorithm B

SIOAs follow various paradigms, a hybrid that combines two or more different SIOAs is expected to have better performance than a sole SIOA. We will introduce some representative hybrids based on two different SIOAs.

El-Abd proposed a hybrid optimiser combining PSO and ABC in order to integrate the advantages of each other [26]. This optimiser is denoted by PSOABC-E, and it incorporates an ABC component into a standard PSO which updates the personal bests (**pbest**) of the particles in every iteration. More specifically, for every particle i in the swarm, its personal best $\mathbf{pbest}_i = \{pbest_{i,1}, pbest_{i,2}, \ldots, pbest_{i,D}\}$ will be updated by ABC after the main loop of PSO. The update equation of ABC can be described as follows:

$$pbest_{ij} = pbest_{ij} + \phi_{ij} \times (pbest_{ij} - pbest_{kj}) \tag{3.5}$$

where j is a randomly selected number in $[1, D]$ and D is the number of dimensions, and ϕ_{ij} is a random number uniformly distributed in the range $[-1, 1]$, and k is the index of a randomly chosen solution.

Besides, the $\mathbf{pbest}_i$ will be renewed if it has a better fitness. In this approach, the standard PSO is assigned as many times of FEs as ABC. Apparently, the ABC algorithm in PSOABC-E operates at component level. Hence, the hybridisation occurred at the component level. According to our taxonomy, the type of the hybridisation strategy in PSOABC-E is $\langle C, C, S, D\rangle$. Note that various other versions of PSOABC can also be found in the literature. And given the limited space available, these optimisers will not be covered here.

Abd-Elazim and Ali proposed a hybrid involving PSO and the BFOA [27], thus named PSOBFOA-EA. This optimiser is applied to design optimal power system stabilisers in a multi-machine power system. In PSOBFOA-EA, the personal best and global best information derived from PSO is employed to orient the search directions of the tumble behaviour for each bacterium. This hybrid is aimed at gaining benefits from both the PSO's ability to exchange social information and the BFOA's ability in finding a new solution by elimination and dispersal. The velocity and position update equations for the dth dimension of the ith particle in the swarm can be given as shown in (3.6):

$$\begin{aligned} v_{i,d}^{k1} &= \omega \times v_{i,d}^{k} + c_1 \times \text{rand1} \times (pbest_{i,d}^{k} - x_{i,d}^{k}) \\ &\quad + c_2 \times \text{rand2} \times (nbest_{i,d}^{k} - x_{i,d}^{k}) \\ x_{i,d}^{k+1} &= x_{i,d}^{k} + v_{i,d}^{k+1} \end{aligned} \tag{3.6}$$

where $X_i^k = [x_{i,1}^k, x_{i,2}^k, \ldots, x_{i,D}^k]$ and $V_i^k = [v_{i,1}^k, v_{i,2}^k, \ldots, v_{i,D}^k]$ represent the position and velocity of the ith particle at the kth generation ($i = 1, 2, \ldots, PS$; PS is the population size), respectively; $pbest_i^k = [pbest_{i,1}^k, pbest_{i,2}^k, \ldots, pbest_{i,D}^k]^T$ is the best position that has been found so far by the ith particle; $nbest_i^k = [nbest_{i,1}^k, nbest_{i,2}^k, \ldots, nbest_{i,D}^k]^T$ is the best position that has been found by particles in the neighbourhood of the ith particle; and rand1 and rand2 are the random numbers that are uniformly distributed in [0, 1]; ω is the so-called inertia weight; c_1 and c_2 are acceleration coefficients, which are also termed as the cognitive factor and the social factor, respectively. In PSOBFOA-EA, the position and velocity update equations of PSO will be adopted to update each bacterium (solution) in the population after the main loop of BFOA. According to our taxonomy, the type of the hybridisation strategy in PSOBFOA-EA is $\langle C, P, S, D \rangle$.

Mandloi and Bhatia designed a PSOACO (namely PSOACO-MB) to solve the symbol vector detection problem in large MIMO systems, which combines the distance based search of ants in ACO and the velocity-based search of particles in PSO [28]. In this hybrid, a new probability metric is generated by a weighted combination of the distance metric of ACO and the direction (velocity) metric of PSO. In the generation of the probability metric, they employed a sigmoid function to convert the distance and velocity into heuristic values. Besides, the velocity update of PSO is applied to replace the concept of pheromone update of ACO in PSOACO-MB. In fact, the velocity update of PSO is only used in the computation of the probability metric for the construction of the solution of each ant, and it does not generate new solutions directly. Thus, PSOACO-MB is an embedding-based hybrid. According to our taxonomy, the type of the hybridisation strategy in PSOACO-MB is $\langle E, P, N, D \rangle$.

In addition, the hybridisation strategies of the hybrids based on SIOA (GS) and EA (GS) are listed in Table 3.1.

Table 3.1 The hybridisation strategies of the hybrids based on 'SIOA (GS) ⊕ SIOA (GS)'

Name	Parent$_A$	Parent$_B$	PR	OL$_A$	OL$_B$	HL	OO	TIT	TTI
PSOABC-E	PSO	ABC	C	P	C	C(HELH)	S	D	S
PSOACO-MB	PSO	ACO	E	P	P	P(HOLH)	N	D	S
PSOBFOA-EA	PSO	BFOA	C	P	P	P(HOLH)	S	D	S

3.3.1.2 SIOA (GS) ⊕ evolutionary algorithms (GS)

Evolutionary algorithms (EAs) like genetic algorithm (GA), cultural algorithm (CA) and differential evolution (DE) algorithm with GS ability perform well in solving global optimisation problems. Thus, an EA can be hybridised with a certain SIOA to offer a novel and efficient search pattern and such a hybrid optimiser may be more capable for some specific problems. Several representative hybrids based on an SIOA and an EA have been proposed in recent decades.

Lin *et al.* presented a hybrid of cooperative PSO (CPSO) and CA, which is denoted as PSOCA-LCL [29]. In this hybrid, a collection of multiple swarms are used to optimise every component of the solution vector. Besides, the concept of belief space of CA is introduced into PSOCA-LCL to store the information of the individuals such that they can exchange information indirectly. Thus, all the parameters of CPSO are simultaneously tuned by the belief space of CA. Additionally, each position of the CPSO can be adjusted not only using the belief space that stores the paragons of each swarm, but also by searching around the local best solutions and the global best solution. According to our taxonomy, the type of the hybridisation strategy in PSOCA-LCL is $\langle A, P, S, D \rangle$.

Wang *et al.* also proposed a hybrid of DE and PSO (namely DEPSO-WYZ), which maintains three subpopulations that are evolved, respectively, by classical PSO and two DE variants [30]. The three subpopulations share the global best solution during their evolution. The DEPSO-WYZ combines the local optimisation ability of PSO and the global optimisation ability of DE to gain more optimisation precision and search efficiency. According to our taxonomy, the type of the hybridisation strategy in DEPSO-WYZ is $\langle C, S, P, D \rangle$.

Combining GA and PSO, Kao and Zahara put forward a hybrid method (denoted by PSOGA-KZ) for the global optimisation of multimodal functions [17]. PSOGA-KZ integrates the concept of evolving individuals originally modelled by GA with the concept of self-improvement of PSO, where individuals enhance themselves based on social interactions and their private cognition. When solving an N-dimensional problem, this hybrid initialises the population by generating $4N$ individuals randomly, and these individuals are ranked based on the fitness values, and the top $2N$ individuals are selected to used GA to create $2N$ new individuals by crossover and mutation operations. And PSO is applied on the remaining $2N$ individuals. It should be noted that the new $2N$ individuals created by GA are involved in the update of the global best particle and the personal best particles in PSO. Thus, PSO and GA operate at different subpopulations in parallel in PSOGA-KZ, and it is a heterogeneous level hybridisation and such hybridisation occurs at the subpopulation level. According to our taxonomy, the type of the hybridisation strategy in PSOGA-KZ is $\langle C, S, P, D \rangle$.

Juang proposed an evolutionary learning algorithm based on a hybrid of GA and PSO and is thus called PSOGA-J [18]. In PSOGA-J, the top half of the population regarding the fitness values are deemed as elites, and the other will be discarded. Besides, the elite individuals will be enhanced by PSO to generate one half of the generalised population. Then, a series of operations of GA, including tournament selection, crossover and mutation, will work with these enhanced elites to reproduce the other half. The elite selection will work with these individuals (enhanced by PSO or evolved by GA), and only the better half of the generalised population will enter the next generation and it will be regarded as the whole population both of PSO and GA. Hence, PSO and GA work with the whole population in order, and the hybridisation occurs at the population level. According to our taxonomy, the type of the hybridisation strategy in PSOGA-J is $\langle C, P, S, D\rangle$.

Ciornei and Kyriakides proposed a hybrid optimiser based on a special variant of ACO for continuous domains named API and a GA [31]. The hybrid is denoted by ACOGA-KC. In each iteration of ACOGA-KC, the sites forgotten in the API search and the best sites of every ant are added into the current population of GA to generate new solutions. The nest position can be updated by the best solution from GA and API. Thus, ACOGA-KC is a collaboration-based hybrid. In addition, ACO and GA in ACOGA-KC operate at the population level and the subpopulation level, respectively. As a consequence, the hybridisation of ACOGA-KC occurs at the subpopulation. According to our taxonomy, the type of the hybridisation strategy in ACOGA-KC is $\langle C, S, S, D\rangle$.

To enhance the efficiency of the ABC optimiser, Jatoth and Rajasekhar put forward a novel hybrid approach based on GA and ABC, which is named ABCGA-JR [32]. And this method is used for tuning a proportional integral speed controller in a vector-controlled permanent magnet synchronous motor drive. In this hybrid, GA is used to work with the whole population first to reproduce a new generation. Then, the solutions evolved by GA will be taken as the initial solutions (i.e. food sources) of ABC based on the fitness values. After each loop of ABCGA-JR, the best fitness value and optimum control parameters will be stored. Supposed the number of bees (i.e. the population) in ABC is $2n$, the population of ABC is initialised by taking n best solutions from GA. The best fitness value and control parameters are memorised after the ABC process. Thus, GA and ABC will be carried out successively in the main loop of ABCGA-JR, and the process continues till the best value obtained or predefined criterion is satisfied. Since GA work with the whole population and ABC only applies on the best half solutions (subpopulation), the hybridisation occurs at the subpopulation level. Moreover, the type of the hybridisation strategy in ABCGA-JR is $\langle C, S, S, D\rangle$ according to our taxonomy.

Liu *et al.* incorporated DE and PSO into the framework of the CA [33], and the resultant hybrid is named DEPSO-LWY here. A primary operation reflects the hybridisation of DE and PSO based on both embedding (DE's mutation is perturbed by PSO's velocity) and collaboration (PSO and perturbed DE are further hybridised at component level), which is shown in the following equation:

$$x_{i,d}^{k+1} = \begin{cases} x_{r1,d}^{k} + F \times (x_{r2,d}^{k} - x_{r3,d}^{k}) + v_{i,d}^{k+1}, & \text{if rand} < CR \text{ or } d = rn_i; \\ x_{i,d}^{k} + v_{i,d}^{k+1}, & \text{otherwise.} \end{cases} \tag{3.7}$$

where CR is the predefined crossover probability; rn_i is a number that is randomly selected from the index set $\{1, 2, \dots, D\}$ and D is the number of dimensions.

Besides, unlike most PSO variants, the particle positions and velocities are modulated according to four kinds of knowledge sources that are contained in the belief space of the CA. Particle positions will be drawn towards the global best position (i.e. **gbest**), and particle velocities will be confined by lower and upper bounds that are dependent on personal best particle positions. According to our taxonomy, the type of the hybridisation strategy in DEPSO-LWY is $\langle E, C, N, D\rangle$.

Hao *et al.* constructed another hybrid based on DE and PSO (namely DEPSO-HGH) [34]. In DEPSO-HGH, DE and PSO are regarded as two operators to generate candidate solutions, and they act on the level of dimensional components of individuals. Stated another way, each dimensional component of a candidate solution can be generated in either DE or PSO manner. The choice of generation manners depends on a predefined probability (CR) shown as follows:

$$\mathbf{DE}: y_{i,d}^{k+1} = \frac{x_{i,d}^k + x_{r1,d}^k}{2} + F \times (x_{r_1,d}^k - x_{i,d}^k) + x_{r_2,d}^k - x_{r_3,d}^k \tag{3.8}$$

$$u_{i,d}^{k+1} = \begin{cases} y_{i,d}^{k+1}, & \text{if rand} < CR; \\ z_{i,d}^{k+1}, & \text{otherwise.} \end{cases} \tag{3.9}$$

where F is the so-called scaling factor, which is a positive constant; $\mathbf{X}_i$ is the ith individual in the DE population $\{\mathbf{X}_i | i = 1, 2, \ldots, PS\}$; $r_1 \neq r_2 \neq r_3$ will be randomly (rand) chosen from $[1, PS]$; PS is the population size; $x_{i,d}$, $z_{i,d}$ and $u_{i,d}$ are the dth-dimensional components of the vectors $\mathbf{x}_i$, $\mathbf{z}_i$, $\mathbf{u}_i$, respectively; CR is the predefined crossover probability, which is usually set to a fixed value in (0, 1) or changes dynamically within (0, 1); rn_i is a number that is randomly selected from the index set $\{1, 2, \ldots, D\}$. According to our taxonomy, the type of the hybridisation strategy in DEPSO-HGH is $\langle C, C, S, D\rangle$.

Duan *et al.* proposed a hybrid algorithm integrating ABC and the quantum EA (QEA) for solving continuous optimisation problems [35]. In ABCQEA-DXX, the population is initialised randomly using the quantum chromosomes to represent the solutions. In each iteration, QEA, including roulette selection, quantum crossover and quantum mutation, is applied first on the quantum population to provide the initial positions of food sources (as well as employed bees) of ABC. Note that only the better solutions will replace the old ones in the crossover and mutation operations in QEA. Then, the onlooker phase and the scout bee phase of ABC are used to search for better solutions. In a word, QEA and ABC will alternately work till the stopping criterion is satisfied. Thus, ABCQEA-DXX is a collaboration-based hybrid and the hybridisation occurs at the population level. According to our taxonomy, the type of the hybridisation strategy in ABCQEA-DXX is $\langle C, P, S, D\rangle$. Moreover, the hybridisation strategies of the hybrids based on SIOA (GS) and EA (GS) are listed in Table 3.2.

3.3.1.3 SIOA (GS) ⊕ other meta-heuristics (GS)

Some other meta-heuristics also are inspired by the mechanisms in nature but they do not have the essential characteristics of neither SIOAs or EAs. However, they follow various paradigms and have good GS ability. Therefore, they are regarded as candidate parents for the SIOA-based hybrids.

Dedicated to pattern synthesis of conformal arrays, Bai *et al.* introduced a hybrid based on invasive weed optimisation (IWO) and PSO [36], named by PSOIWO-BXL. In this hybrid, IWO is applied first to generate the initial population (solutions), and seeds are produced randomly. Each seed is an initial solution which will grow into

Table 3.2 The hybridisation strategies of the hybrids based on 'SIOA (GS) ⊕ EA (GS)'

Name	Parent$_A$	Parent$_B$	PR	OL$_A$	OL$_B$	HL	OO	TIT	TTI
PSOCA-LCL	PSO	CA	A	P	P	P(HOLH)	S	D	C_p, S
PSOGA-KZ	PSO	GA	C	S	S	S(HEOH-&HOLH)	P	D	S
PSOGA-J	PSO	GA	C	P	P	P(HOLH)	S	D	S
ACOGA-KC	ACO	GA	C	P	S	S(HELH)	S	D	S
ACOGA-NBAA	ACO	GA	C	P	P	P(HOLH)	P	D	S
ABCGA-JR	ABC	GA	C	P	S	S(HELH)	S	D	S, C_p
DEPSO-LWY	PSO	DE	E	P	C	C(HELH)	N	D	S, A_i
DEPSO-WYZ	PSO	DE	C	S	S	S(HEOH-&HOLH)	P	D	S
DEPSO-HGH	PSO	DE	C	C	C	C(HOLH)	S	D	S_c
ABCQEA-DXX	ABC	QEA	C	P	P	P(HOLH)	S	D	S

a flowering plant (i.e. a weed) ranked in the colony according to the fitness value. Then PSO will be applied on these plants (solutions) to update these solutions (plants) in each iteration. And then, the flowering plants will produce new seeds. The number of seeds produced by each plant depends on the rank of its fitness value and decreases from the maximum possible seed production number $S_{\max}$ to the minimum $S_{\min}$. Thus, the dispersion method of the IWO algorithm will be applied on the plants updated by PSO to generate new seeds (solutions). The update of solutions in IWO is described as follows:

$$\sigma_{iter} = \left(\frac{iter_{\max} - iter}{iter_{\max}}\right)^n \times (\sigma_{\text{initial}} - \sigma_{\text{final}}) + \sigma_{\text{final}} \tag{3.10}$$

$$x_{i,d}^{k+1} = x_{i,d}^{k+1} + \text{rand} * \sigma_{iter} \tag{3.11}$$

where $iter_{\max}$ is the maximum number of iterations; σ_{initial} and σ_{final} are the initial and final standard deviations, respectively; and n is the nonlinear modulation index.

Moreover, new seeds will be compared with their parents based on the fitness value and their positions in the colony will be determined. The seeds with a higher rank will grow into flowering plants. On the contrary, those with a lower rank in the colony are eliminated to reach the maximum number of plants in the colony. In PSOIWO-BXL, PSO and IWO are carried out in series to update the solutions in the whole population. As a consequence, this is a collaboration-based hybrid and the hybridisation occurs at the population level. Moreover, the type of the hybridisation strategy in PSOIWO-BXL is $\langle C, P, S, D\rangle$ according to our taxonomy.

Wu *et al.* proposed a hybrid by incorporating the harmony search (HS) with ABC, thus called ABCHS-WQN, for solving global numerical optimisation problems [37]. In each iteration of ABCHS-WQN, the solutions in harmony memory are regarded as the positions of the food sources (i.e. possible solutions to the optimisation problem) of ABC, and the main loop of ABC will be carried out to update the solutions. Note that the harmony memory is a pool of solutions and corresponding fitness value that

Table 3.3 The hybridisation strategies of the hybrids based on 'SIOA (GS) ⊕ other meta-heuristics (GS)'

Name	Parent_A	Parent_B	PR	OL_A	OL_B	HL	OO	TIT	TTI
PSOIWO-BXL	PSO	IWO	C	P	P	P(HOLH)	S	D	S
ABCHS-WQN	ABC	HS	C	P	P	P(HOLH)	S	D	S
PSOSFLA-NAZ	PSO	SFLA	C	S	S	S(HEOH-&HOLH)	P	D	S

will be updated during the iteration. Then, a new harmony vector (i.e. a solution) is generated by an HS operator. If it is better than the worst solution in the current population, then it will replace the worst one of the population. According to our taxonomy, the type of the hybridisation strategy in ABCHS-WQN is $\langle C, P, S, D \rangle$.

Narimani *et al.* presented a new hybrid algorithm based on PSO and the shuffle frog leaping algorithms (SFLA) for solving the optimal power flow in power systems [38], thus called PSOSFLA-NAZ. In this hybrid, the whole population includes $2N$ solutions, N solutions for the modified PSO and the other N solutions for SFLA. In each iteration, PSO and SFLA work with different parts of the population in parallel, and the best solution for the next generation will be selected from the solutions generated by both algorithms. It is obvious that PSOSFLA-NAZ is a collaboration-based hybrid. Since PSO and SFLA in this hybrid operate at different subpopulations, the hybridisation occurs at the subpopulation level HEOH. According to our taxonomy, the type of the hybridisation strategy in PSOSFLA-NAZ is $\langle C, S, P, D \rangle$.

In addition, the hybridisation strategies of the hybrids based on SIOA (GS) and other meta-heuristics (GS) are listed in Table 3.3.

3.3.1.4 SIOA (GS) ⊕ simple heuristics (GS)

Since simple heuristics with special search patterns also have GS ability, some hybrids based on SIOAs and simple heuristics with GS ability are proposed to obtain better optimisers for specific problems.

Liu *et al.* proposed a hybrid incorporating chaotic search (CS) into a PSO with adaptive inertia to enrich the searching behaviour and avoid being trapped into local optimum [39]. The hybrid is denoted by PSOCS-LWJ. This hybrid performs PSO to update the position and velocity of each particle in the population at the beginning of each iteration. Then, the global best solution from PSO will be further improved and updated by CS. Note that CS is an explorative TSM actually since it does not exploit local information. Besides, in each iteration, only the top one-fifth best-performing solutions, in which the best solution is updated by CS, will enter the next iteration together with the other solutions generated randomly. According to our taxonomy, the type of the hybridisation strategy in PSOCS-LWJ is $\langle C, I, S, D \rangle$.

Besides, Wan *et al.* also developed a PSOCS, namely PSOCS-WWS, to solve the nonlinear bilevel programming problem [40]. In each iteration of PSOCS-WWS, CS is used to determine whether the M ($M = m + n$) particles in the population are

Table 3.4 The hybridisation strategies of the hybrids based on 'SIOA (GS) ⊕ simple heuristics (GS)'

Name	**Parent$_A$**	**Parent$_B$**	**PR**	**OL$_A$**	**OL$_B$**	**HL**	**OO**	**TIT**	**TTI**
PSOCS-LWJ	PSO	CS	C	P	I	I(HELH)	S	D	S
PSOCS-WWS	PSO	CS	C	S	S	S(HEOH-&HOLH)	P	D	S

Table 3.5 The hybridisation strategies of the hybrids based on 'SIOA (GS) ⊕ MP (GS)'

Name	**Parent$_A$**	**Parent$_B$**	**PR**	**OL$_A$**	**OL$_B$**	**HL**	**OO**	**TIT**	**TTI**
PSOBB-NGSC	PSO	BB	C	P	I	I(HELH)	S	D	S

feasible and those feasible particles (solutions) will be added into the feasible list and ranked based on the fitness values. Then, the positions and velocities of the *m* particles with better performance will be updated by PSO, and the other *n* particles with worse fitness will be improved by CS. Evidently, PSO and CS in PSOCS-WWS operate at different subpopulation, and the hybridisation occurs at the subpopulation level HEOH. Besides, the type of the hybridisation strategy in PSOCS-WWS is $\langle C, S, P, D\rangle$ according to our taxonomy. Furthermore, the hybridisation strategies of the hybrids based on SIOA (GS) and simple heuristics (GS) are listed in Table 3.4.

3.3.1.5 SIOA (GS) ⊕ mathematical programming methods (GS)

The branch-and-bound (BB) is a typical MP method with good GS ability. Thus, it can be hybridised with a certain SIOA to avoid the stagnation and premature convergence to a local solution.

Nema *et al.* proposed a hybrid algorithm for solving mixed discrete nonlinear programming problems, designed to efficiently combine PSO and BB [41]. This hybrid optimiser is denoted as PSOBB-NGSC. In this hybrid, BB is applied first to determine a feasible solution, which will be taken as an initial solution for PSO. To be specific, this solution plays the role of the global best solution for all the swarms in PSO at the beginning. Besides, once an improvement in the global best has been achieved by PSO, the BB process will be executed to determine a new feasible solution and update the global best, taking the improved global best from PSO as a starting point. That is to say, BB does not need to work in every single iteration of the hybrid algorithm but only when PSO has obtained a better solution. PSO resumes after the solution generation of the BB. It is evident that PSO operates at the whole population while the BB just works on a single solution (the global best). As a result, the hybridisation occurs at the individual level. According to our taxonomy, the type of the hybridisation strategy in PSOBB-NGSC is $\langle C, I, S, D\rangle$. The hybridisation strategies of the hybrids based on SIOA (GS) and MP (GS) are listed in Table 3.5.

3.3.2 Hybrids based on SIOAs and LS methods

Tr:Er&Ei is a core of all kinds of optimisers, for which MAs provide us with a common template for hybridisation to combine the respective advantages of GS and LS. For the same purpose, a flood of SIOA-based hybrids following different paradigms emerged in recent decades.

3.3.2.1 SIOA (GS) ⊕ SIOA (LS)

An SIOA can also be implemented as LS, thus hybridising different kinds of SIOAs is another choice in the hybridisation strategies considered in this section. Different from those hybrids described in the GS ⊕ GS pattern, the main roles of two (or more) different SIOAs in such hybrids are not the same (i.e. one for GS and another for LS).

Li *et al.* proposed a hybrid algorithm denoted by PSOABC-LWYZ, which combined PSO with ABC [42]. In this hybrid, three states are defined to describe the ageing degree of the individuals, including active, aged and dying states. During the iteration process of PSOABC-LWYZ, the ageing degree of the personal best solution (**pbest**) for each individual will be examined to determine which type of search phase from both PSO and ABC to implement. More specifically, PSO is performed on the active individuals to exploit a new solution using the position and velocity update equations of PSO. Besides, an aged individual will adopt the onlooker bee phase to explore a new possible solution in the search space different from PSO. Moreover, a dying individual that cannot be improved by PSO or the onlooker bee phase of ABC implements the modified scout bee phase to explore the whole search space. Furthermore, some parameters are defined to record the state of the individual. Since both PSO and ABC in PSOABC-LWYZ operate at the individual level, the hybridisation occurs at the individual level. According to our taxonomy, the type of the hybridisation strategy in PSOABC-LWYZ is $\langle C, I, S, D\rangle$.

Shelokar *et al.* proposed a hybrid algorithm based on PSO and ACO (namely PSOACO-SSJK) by applying PSO for GS and ACO to update the positions of particles to rapidly attain the feasible solution space [19]. In this hybrid, PSO and ACO will be applied in series to update the positions of the particles. After the initialisation of the population, PSO will work with the whole population to update the positions of the particles. And then, ACO works with the global best position from PSO using the simple pheromone-guided mechanism, which works as an LS for better solutions in the neighbourhood of the global best solution. Thus, ACO operates at the individual level. According to our taxonomy, the type of the hybridisation strategy in PSOACO-SSJK is $\langle C, I, S, D\rangle$. Moreover, the hybridisation strategies of the hybrids based on SIOA (GS) and SIOAs (LS) are listed in Table 3.6.

3.3.2.2 SIOA (GS) ⊕ evolutionary algorithms (LS)

EAs with LS also are taken into account to enhance the exploitation ability of the SIOAs, thus plenty of hybrid optimisers based on an EA and an SIOA arose recently.

Gong *et al.* hybridised DE/best/2/bin with a PSO variant, which removes traditional velocity terms and adopts timevariant acceleration coefficients [43]. Their

Table 3.6 The hybridisation strategies of hybrids based on 'SIOA (GS) ⊕ SIOA (LS)'

Name	Parent$_A$	Parent$_B$	PR	OL$_A$	OL$_B$	HL	OO	TIT	TTI
PSOABC-LWY	ABC	PSO	C	I	I	I(HEOH-&HOLH)	S	D	S
PSOACO-SSJK	PSO	ACO	C	P	I	I(HELH)	S	D	S

hybrid (namely DEPSO-GZQ) was proposed to solve multiobjective environmental/economic power dispatch problems. The motivation behind DEPSO-GZQ is an effective synergy of PSO's exploration ability and the exploitation ability of DE/best/2/bin. The DE and PSO parents of DEPSO-GZQ share and update an external archive, which records nondominated solutions that have been found so far. Particularly, the crowding distance, as an index of the solution diversity in multiobjective space, is utilised to differentiate the quality of nondominated solutions. According to our taxonomy, the type of the hybridisation strategy in DEPSO-GZQ is $\langle C, P, S, D \rangle$.

A hybrid based on DE and PSO (namely DEPSO-RJAMB) was proposed by Ramesh *et al.* to solve economic dispatch problems [44]. Its DE and PSO parents share and update the current positions of particles in their respective manners. Ramesh *et al.* claimed that DEPSO-RJAMB combines the vibrancy and explorative nature of PSO with the superior exploitative nature of hybrid DE (HDE). Note that the HDE is de facto a memetic DE that is proposed by Lin *et al.*, which hybridises DE/rand/1/bin with an LS method (steepest descent search) [45]. According to our taxonomy, the type of the hybridisation strategy in DEPSO-RJAMB is $\langle C, P, S, D \rangle$.

Besides, Xu and Gu proposed another DEPSO, namely DEPSO-XG, in which PSO and DE evolve the whole population alternately [46]. In particular, Xu and Gu incorporated the average **pbest** position and velocity of particles into the velocity regulation of the classical PSO. The DE operations in DEPSO-XG involve two populations: the original DE population and an extra population to record those trial vectors inferior to their target vectors. The 'inferior' vectors will be reused in a so-called prior crossover operation to generate intermediate vectors and update particle positions:

$$\begin{aligned} v_{i,d}^{k+1} &= \omega \times v_{i,d}^{k} + c_1 \times \text{rand}_1 \times (pbest_{i,d}^{k} - x_{i,d}^{k}) \\ &\quad + c_2 \times \text{rand}_2 \times (nbest_{i,d}^{k} - x_{i,d}^{k}) + c_3 \times \text{rand}_3 \times (\overline{pbest_{i,d}^{k}} - x_{i,d}^{k}) \\ &\quad + c_4 \times \text{rand}_4 \times (\overline{v_{i,d}^{k}} - x_{i,d}^{k}) \end{aligned} \tag{3.12}$$

where $\overline{pbest_{i,d}^{k}} = ((\sum_{i=1}^{PS} pbest_{i,d}^{k})/PS)$ and $\overline{v_{i,d}^{k}} = ((\sum_{i=1}^{PS} v_{i,d}^{k})/PS)$. DE is designed to enhance the exploitation in the neighbourhood of the current particle positions from PSO. Since PSO and DE evolve the whole population alternately, the DEPSO-XG is a collaboration-based hybrid and the hybridisation occurs at the population level. According to our taxonomy, the type of the hybridisation strategy in DEPSO-XG

Table 3.7 The hybridisation strategies of the hybrids based on 'SIOA (GS) ⊕ EA (LS)'

Name	Parent$_A$	Parent$_B$	PR	OL$_A$	OL$_B$	HL	OO	TIT	TTI
DEPSO-GZQ	PSO	DE	C	P	P	P(HOLH)	S	D	S
DEPSO-RJAMB	PSO	DE	C	P	P	P(HOLH)	S	D	S
DEPSO-XG	PSO	DE	C	P	P	P(HOLH)	S	D	S

is $\langle C, P, S, D \rangle$. Moreover, the hybridisation strategies of the hybrids based on SIOA (GS) and EA (LS) are listed in Table 3.7.

3.3.2.3 SIOA (GS) ⊕ other meta-heuristics (LS)

Except for SIOAs and EAs, there are other meta-heuristics with strong LS ability. Thus, a new hybridisation strategy is available to improve the performance of a certain SIOA.

Li *et al.* proposed a hybrid algorithm that introduces tabu search (TS) algorithm to perform LS in the ABC algorithm for solving the flexible job shop scheduling problems [47]. This hybrid is named as ABCTS-LQX. TS works as an LS and it is implemented to update the positions of food sources of employed bees, onlooker bees and scout bees, respectively. Since ABC works with the whole population while TS only operates at the individual level, the hybridisation occurs at individual level in ABCTS-LQX. Meanwhile, an external Pareto archive set is employed to record enough nondominated solutions for the problem considered. According to our taxonomy, the type of the hybridisation strategy in ABCTS-LQX is $\langle C, I, S, D \rangle$.

Zhang *et al.* combined PSO and TS to solve the multiobjective flexible job-shop scheduling problem (FJSP) with several conflicting and incommensurable objectives [48]. Thus, this hybrid algorithm is denoted as PSOTS-ZSLG here. In each iteration, PSO works on the whole population first, and the particles represent the assignment of operations to machines and the schedule of operations on each machine. In addition, each particle updates themselves using the position and velocity equations of PSO. Then, TS works as an LS to look for better solutions in the neighbourhood of the FJSP scheduling solutions originated from each position of the particle. In turn, PSO uses the best solutions improved by TS in the next iteration until it meets the termination condition. Since PSO works with the whole population while TS only operates at the individual level, the hybridisation occurs at individual level in PSOTS-ZSLG. According to our taxonomy, the type of the hybridisation strategy in PSOTS-ZSLG is $\langle C, I, S, D \rangle$.

Besides, based on PSO and the simulated annealing (SA) algorithm, Xia and Wu proposed an easily implemented hybrid approach (denoted by PSOSA-XW) for the multiobjective FJSP [49]. Similar to PSOTS-ZSLG, PSOSA-XW makes use of PSO to assign operations on machines and SA to schedule operations on each machine. During the process of this hybrid algorithm, PSO updates the solutions using the position and velocity equations and provides initial solutions for SA. Then, the new

Table 3.8 The hybridisation strategies of the hybrids based on 'SIOA (GS) ⊕ other meta-heuristics (LS)'

Name	Parent$_A$	Parent$_B$	PR	OL$_A$	OL$_B$	HL	OO	TIT	TTI
ABCTS-LQX	ABC	TS	C	P	I	I(HELH)	S	D	S
PSOTS-ZSLG	PSO	TS	C	P	I	I(HELH)	S	D	S
PSOGSA-MH	PSO	GSA	E	P	P	P(HOLH)	N	D	S
PSOSA-ZW	PSO	SA	C	P	I	I(HELH)	S	D	S
PSOSA-XW	PSO	SA	C	P	I	I(HELH)	S	D	S

solutions obtained by SA through an LS will be passed over to PSO and the main loop of the hybrid continues until the termination criterion is met. Since SA operates at individual level while PSO operates at population level, the hybridisation occurs at individual level in PSOSA-ZSLG. Moreover, the type of the hybridisation strategy in PSOSA-XW is $\langle C, I, S, D\rangle$ according to our taxonomy.

Mirjalili and Hashim proposed a hybrid population-based algorithm which combines PSO and the gravitational search algorithm (GSA) [50]. This hybrid algorithm is named as PSOGSA-MH. The basic idea of PSOGSA-MH is to combine the ability of social thinking (the global best, **Gbest**) in PSO with the LS ability of GSA. In this hybrid, the idea of GSA is embedded in the velocity update of PSO. In each iteration, the gravitational force, gravitational constant and resultant forces concept of GSA are used to calculate the accelerations of particles. And the velocity and position of the particle can be updated as follows:

$$\begin{aligned} V_i(t+1) &= \omega \times V_i(t) + c_1 \times \text{rand}_1 \times ac_i(t) + c_2 \times \text{rand}_2 \times (\mathbf{Gbest} - X_i(t)) \\ X_i(t+1) &= X_i(t) + V_i(t+1) \end{aligned} \tag{3.13}$$

where $ac_i(t)$ is the acceleration of particle i at iteration t, $V_i(t)$ the velocity of particle i at iteration t, $X_i(t)$ the velocity of particle i at iteration t, and **Gbest** is the best known solution. Hence, PSOGSA-MH is an embedding-based hybrid. According to our taxonomy, the type of the hybridisation strategy in PSOGSA-MH is $\langle E, P, N, D\rangle$.

Zhu and Wu proposed a hybrid algorithm which incorporates SA and PSO to improve the capability of seeking for the best allocation results [51]. This hybrid is named by PSOSA-ZW. In each iteration of PSOSA-ZW, PSO is executed to evolve the whole population first. If the fitness value of a particle (solution) is better than those in the last iteration, then it will be chosen as the initial solution for SA. Then SA serves as an LS to find better solutions and substitute this particle with a better solution. According to our taxonomy, the type of the hybridisation strategy in PSOSA-ZW is $\langle C, I, S, D\rangle$. Moreover, the hybridisation strategies of the hybrids based on SIOA (GS) and other meta-heuristics (LS) are listed in Table 3.8.

3.3.2.4 SIOA (GS) ⊕ simple heuristics (LS)

Some simple heuristics with strong LS ability have been considered to be hybridised with SIOAs for a desirable Tr:Er&Ei.

Table 3.9 The hybridisation strategies of the hybrids based on 'SIOA (GS) ⊕ simple heuristics (LS)'

Name	$Parent_A$	$Parent_B$	PR	OL_A	OL_B	HL	OO	TIT	TTI
PSODCS-ZGGS	PSO	DCS	C	P	I	I(HELH)	S	D	S

Zhang *et al.* proposed an efficient hybrid algorithm (namely PSODCS-ZGGS) to solve dynamic economic dispatch problems with valve-point effects, by integrating an improved bare-bones PSO with a local searcher called directionally CS (DCS) [52]. During the iterative process of PSODCS-ZGGS, PSO is implemented on the initial population at the beginning to serve as a basic level search. Then the hybrid performs DCS on the best solution obtained by PSO, which serves as an LS. Note that the improved bare-bones PSO updates the particles' positions using the Gaussian sampling instead of the particle velocity, which is shown as follows:

$$x_{i,d}^{k+1} = \mathcal{N}\left(\frac{pbest_{i,d}^{k} + gbest_{d}^{k}}{2}, |pbest_{i,d}^{k} - gbest_{d}^{k}|\right) \tag{3.14}$$

where $\mathcal{N}()$ is a Gaussian sampling function. Besides, an adaptive disturbance factor and a new genetic operator are also incorporated into the improved bare-bones PSO to enhance its search capability. Moreover, a heuristic constraint handling mechanism is introduced to modify infeasible particles (solutions). Since DCS is implemented on the global best position obtained from a series of improvement in the results of PSO, thus the hybridisation in PSODCS-ZGGS occurs at the individual level. According to our taxonomy, the type of the hybridisation strategy in PSODCS-ZGGS is $\langle C, C, S, D\rangle$. Besides, the hybridisation strategies of the hybrids based on SIOA (GS) and simple heuristics (LS) are listed in Table 3.9.

3.3.2.5 SIOA (GS) ⊕ mathematical programming methods (LS)

Since some MP methods can be used for LS, plenty of SIOA-based hybrids have incorporated the SIOA and the MP with LS ability to achieve a better Tr:Er&Ei.

Begambre and Laier proposed a hybrid optimiser based on the Nelder–Mead simplex method (NMSM) and PSO [53], and it is called PSONMSM-BL. In this hybrid, NMSM is employed to determine three control parameters (i.e. ω, c_1, c_2) of PSO. Moreover, a PSO is used to evaluate the solutions in NMSM. For the PSO loop, when the standard deviation of the last ten solutions is lower than 1E−06, the control parameters will be sent to the NMSM process starting a new NMSM iteration. According to our taxonomy, the type of the hybridisation strategy in PSONMSM-BL is $\langle A, P, S, D\rangle$.

Marrouchi *et al.* proposed a hybrid optimiser which combines PSO and the gradient method (GM) to solve the unit commitment problem [54], and this optimiser is named PSOGM-MHC. This hybrid mainly consists of two stages. In the first stage, these particles in the population are evolved by PSO to improve the solution quality with the social and personal information. And these particles are expected to present new search space for GM which serves as a local searcher in the second stage. That is to say, GM searches for better solutions in the neighbourhood of the final best solution from PSO. Since PSO and GM operate at population level and individual level, respectively, the hybridisation of this hybrid occurs at individual level. According to our taxonomy, the type of the hybridisation strategy in PSOGM-MHC is $\langle C, I, S, D\rangle$.

Ouyang *et al.* designed a hybrid optimiser in which the conjugate GM (CGM) is incorporated into the basic PSO [55], and this optimiser is named PSOCGM-OTZ. In PSOCGM-OTZ, each particle in the population is regarded as a point of the CGM. In each iteration, an LS by CGM is executed for the best solution in the population after the main loop of the original PSO. Thus, the elite individual generated by PSO will be replaced with the individual improved by CGM. According to our taxonomy, the type of the hybridisation strategy in PSOCGM-OTZ is $\langle C, I, S, D\rangle$.

A similar hybridisation strategy was taken in the hybrid optimiser proposed by Noel, which combines PSO and gradient-based LS (GBLS) algorithms to achieve faster convergence and better accuracy of final solutions without getting trapped in local minima [56]. This hybrid optimiser is called PSOGBLS-N, in which PSO is designed for global exploration while a gradient-based algorithm is applied for accurate local exploitation. To be specific, PSO updates the positions and velocities of the particles and locates a good local minimum (i.e. the global best) approximately at first, after which a gradient-based LS is carried out with the best solution has been found by PSO so far as its starting point. Next, the improved solution originated from GBLS will be used to update the global best information of PSO, and then PSO resumes. The cycle is repeated till any of the termination criterions is met. According to our taxonomy, the type of the hybridisation strategy in PSOGBLS-N is $\langle C, I, S, D\rangle$.

Basu proposed a hybrid optimiser integrating ABC with sequential quadratic programming (SQP) for solving dynamic economic dispatch problem of generation units considering valve-point effects [57]. Thus, this hybrid is named by ABCSQP-B. This ABCSQP-B incorporates ABC as a GS which can give a good direction to the optimal region and sequential quadratic programming as an LS procedure which is used to fine tune that region to reach the final solution. The ABCSQP-B method is divided into two stages. In the first stage, ABC is employed to update the solutions in the population. In the second stage, SQP is employed to improve the best solution found by ABC. According to our taxonomy, the type of the hybridisation strategy in ABCSQP-B is $\langle C, I, S, D\rangle$. Furthermore, the hybridisation strategies of the hybrids based on SIOA (GS) and MP (LS) are listed in Table 3.10.

After giving a classification and an analysis to these presentative SIOA-based hybrids, a unifying framework for SIOA-based optimisers is established and it is listed in Table 3.11. According to this table, designers can gain an insight into various possibilities for hybrid design of SIOA-based optimisers. A simple example is given to demonstrate the application of this unifying framework. First, we need to choose an SIOA (e.g. PSO) and another optimiser (e.g. GA) for hybridisation. Second, determine the relationship between the parent optimisers. As an example, 'collaboration' means that both PSO and GA are operated in the solution space. Third, we should select the HL of the hybrid optimiser, 'population level' for example. Thus, both PSO

Table 3.10 The hybridisation strategies of the hybrids based on 'SIOA (GS) ⊕ simple MP (LS)'

Name	$Parent_A$	$Parent_B$	PR	OL_A	OL_B	HL	OO	TIT	TTI
PSONMSM-BL	PSO	NMSM	A	P	P	P(HOLH)	S	D	C_p, S
PSOGM-MHC	PSO	GM	C	P	I	I(HELH)	S	D	S
PSOCGM-OTZ	PSO	CGM	C	P	I	I(HELH)	S	D	S
PSOGBLS-N	PSO	GBLS	C	P	I	I(HELH)	S	D	S
ABCSQP-B	ABC	SQP	C	P	I	I(HELH)	S	D	S

Table 3.11 A unifying framework for SIOA-based hybrids

<table>
<tr><th colspan="7">SIOA ⊕ GS or LS</th></tr>
<tr><td>PR</td><td colspan="2">Collaboration</td><td colspan="2">Embedding</td><td colspan="2">Assistance</td></tr>
<tr><td>HL</td><td>Population</td><td colspan="2">Subpopulation</td><td colspan="2">Individual</td><td>Component</td></tr>
<tr><td>OO</td><td colspan="2">Sequential</td><td colspan="2">Parallel</td><td colspan="2">No order</td></tr>
<tr><td>TIT</td><td colspan="3">Simplex</td><td colspan="3">Duplex</td></tr>
<tr><td>TTI</td><td>Solution</td><td>Fitness</td><td>S_c</td><td>Auxiliary</td><td>C_p</td><td>A_i</td></tr>
</table>

and GA operate at all the population. Then, decide the OO of the hybrid, such as 'sequential' which indicates that PSO and GA are implemented in sequence. Next, we ought to consider TIT, and as an example, 'simplex: $PSO \mapsto GA$' shows that the information only is transferred from PSO to GA. At last, the TTI of the hybrid also need to be determined, and we simply take 'solution' as an example. That is to say, the information of the solution in PSO will be sent to GA. Moreover, the type of the hybridisation strategy in this instance is $\langle C, P, S, S \rangle$. Therefore, we can obtain the hybridisation strategy that we can follow to design more details of a hybrid optimiser.

3.4 Discussion and future research on hybrid optimisation

As can be seen from the proposed SIOA-based hybrids, the researchers have considered multifarious hybridisation strategies in the design of a hybrid optimiser. Here, we make attempt to keep an open mind on different hybridisation strategies without bias or partiality. Besides, we do not intend to distinguish the performance differences of various hybridisation strategies through some benchmark tests, since an incomplete comparison may misguide other researchers into a prejudice against some hybridisation strategies. Undoubtedly, a successful application of certain hybridisation strategy may depend on many design elements, such as the selection of parent optimisers and other design details aforementioned. Exactly as is claimed at the outset, the chapter is aimed at unfolding a broader and systematic view of hybridisation strategies as well as at promoting the development of more efficient SIOA-based hybrid optimisers.

Exploration allows extensive search to determine which region is more likely to achieve the global optimum while exploitation refines the search and focuses on a special region in the search space [22]. Generally, a hybrid follows the hybridisation pattern of the MAs can generate a high-quality solution to the considered problem. However, how to incorporate LSs to ensure the expected problem-solving efficiency is still an open issue. See the MA tutorial of Krasnogor and Smith for more design issues on how to balance GS and LS [14]. Moreover, on account of this approved hybridisation philosophy, it is expected that the improvement of such a hybrid optimiser is reliable, whereas the performance and the efficiency of a hybrid that follow a 'GS⊕GS' pattern could not be guaranteed.

However, the 'GS⊕GS' hybridisation pattern also contributes to seeking for a more suitable optimiser for a specific problem. No matter which optimiser in SIOAs, EAs, MP methods, simple heuristics or other meta-heuristics, the essence of a stochastic search method is iterative sampling. They are distinguished from each other by the search patterns and the range. Even though the validity of a hybrid optimiser that follows the 'GS⊕GS' hybridisation pattern cannot be ensured, a rational motivation behind such hybridisation is that the two or more optimisers correspond to different landscapes, which can give birth to a new search paradigm, may suppress the stagnation during the iteration process that arises from the limitation of a sole search method [25].

Although we focus on the hybridisation of two parent optimisers in this chapter, which is the simplest case, the proposed taxonomy can also be utilised to analyse some intricate hybridisation that involves three or more parents. One can also follow the hierarchical framework that is shown in Figure 3.3 to determine primary hybridisation factors one by one to design a hybrid that involves three or more parent optimisers. When designing a hybrid optimiser, one should also consider the computation complexity that is possibly caused by the incorporation of too many optimisers. However, a sophisticated hybridisation that is based on multiple parents may be prospective for time-consuming high-dimensional numerical optimisation, since the computational cost that results from complicated hybridisation is usually insignificant in contrast to the inherent computation complexity and optimisation hardness of high-dimensional problems.

Besides, according to the analysis of existing SIOA-based hybrids, the hybridisation patterns $\langle C,P,?,?\rangle$ (especially, $\langle C,P,S,D\rangle$), $\langle C,S,?,?\rangle$ (especially, $\langle C,S,P,D\rangle$), $\langle C,I,?,?\rangle$ (especially, $\langle C,I,S,D\rangle$) have attracted more attention of researchers. It is expected that more efficient hybrids utilising the pattern $\langle C,C,?,?\rangle$ will emerge in the future. For example, the component-level adaptive or self-adaptive alternation of parents is a promising research line as it enriches the diversity of evolutionary operations to a large extent. In addition, the hybridisation of GS (especially, POs) and LS methods at component level is also a very interesting topic for future research.

Acknowledgements

This work was supported in part by the National Natural Science Foundation of China under grants nos. 61673058, U1609214, 71101139 and 61304215, in part by the Foundation for Innovative Research Groups of the National Natural Science Foundation of China under grant no. 61321002, and in part by the Research Fund for the Doctoral Program of Higher Education of China under grant no. 20131101120033.

References

[1] Dorigo M, Birattari M. Swarm intelligence. Scholarpedia. 2007;2(9):1462.
[2] Hinchey MG, Sterritt R, Rouff C. Swarms and swarm intelligence. Computer. 2007;40(4):111–113.
[3] Wolpert DH, Macready WG. No free lunch theorems for optimisation. IEEE Transactions on Evolutionary Computation. 1997;1(1):67–82.
[4] Caraffini F, Neri F, Picinali L. An analysis on separability for memetic computing automatic design. Information Sciences. 2014;265:1–22.
[5] Benveniste A, Wilson SS, Metivier M, *et al.* Adaptive Algorithms and Stochastic Approximations. Stochastic Modelling and Applied Probability. Springer Berlin Heidelberg; 2012.
[6] Burke EK, Gendreau M, Hyde M, *et al.* Hyper-heuristics: a survey of the state of the art. Journal of the Operational Research Society. 2013;64(12): 1695–1724.
[7] Blum C, Blesa Aguilera MJ, Roli A, *et al.*, editors. Hybrid Metaheuristics: An Emerging Approach to Optimization (Studies in Computational Intelligence Series 114). Springer Berlin Heidelberg; 2008.
[8] Blum C, Puchinger J, Raidl GR, *et al.* Hybrid metaheuristics in combinatorial optimization: A survey. Applied Soft Computing. 2011;11(6):4135–4151.
[9] Blesa MJ, Blum C, Cangelosi A, *et al.* Hybrid Metaheuristics: 10th International Workshop, HM 2016, Plymouth, UK, June 8–10, 2016, Proceedings. vol. 9668. Springer; 2016.
[10] Talbi EG. A taxonomy of hybrid metaheuristics. Journal of Heuristics. 2002;8(5):541–564.
[11] Gendreau M, Potvin JY. Metaheuristics in combinatorial optimization. Annals of Operations Research. 2005;140(1):189–213.
[12] Blum C, Roli A. Metaheuristics in combinatorial optimization: Overview and conceptual comparison. ACM Computing Surveys. 2003;35(3):268–308.
[13] Raidl GR. A unified view on hybrid metaheuristics. In: Proceedings of the 3rd International Workshop on Hybrid Metaheuristics. Berlin, Germany: Springer-Verlag; 2006. p. 1–12.
[14] Krasnogor N, Smith J. A tutorial for competent memetic algorithms: model, taxonomy, and design issues. IEEE Transactions on Evolutionary Computation. 2005;9(5):474–488.
[15] Xin B, Chen J, Zhang J, *et al.* Hybridizing differential evolution and particle swarm optimization to design powerful optimisers: A review and taxonomy.

IEEE Transactions on Systems, Man, and Cybernetics, Part C (Applications and Reviews). 2012;42(5):744–767.

[16] Deng W, Chen R, He B, *et al.* A novel two-stage hybrid swarm intelligence optimization algorithm and application. Soft Computing. 2012;16(10): 1707–1722.

[17] Kao YT, Zahara E. A hybrid genetic algorithm and particle swarm optimization for multimodal functions. Applied Soft Computing. 2008;8(2): 849–857.

[18] Juang CF. A hybrid of genetic algorithm and particle swarm optimization for recurrent network design. IEEE Transactions on Systems, Man, and Cybernetics, Part B (Cybernetics). 2004;34(2):997–1006.

[19] Shelokar PS, Siarry P, Jayaraman VK, *et al.* Particle swarm and ant colony algorithms hybridised for improved continuous optimization. Applied Mathematics and Computation. 2007;188(1):129–142.

[20] Das S, Abraham A, Konar A. Particle swarm optimization and differential evolution algorithms: technical analysis, applications and hybridization perspectives. In: Advances of Computational Intelligence in Industrial Systems. Springer, Berlin, Heidelberg; 2008. p. 1–38.

[21] Patel V, Tiwari A, Patel A. A comprehensive survey on hybridization of artificial bee colony with particle swarm optimization algorithm and ABC applications to data clustering. In: Proceedings of the International Conference on Informatics and Analytics. ACM; 2016. p. 65.

[22] Moslehi G, Mahnam M. A Pareto approach to multi-objective flexible job-shop scheduling problem using particle swarm optimization and local search. International Journal of Production Economics. 2011;129(1):14–22.

[23] Jourdan L, Basseur M, Talbi EG. Hybridizing exact methods and metaheuristics: A taxonomy. European Journal of Operational Research. 2009; 199(3):620–629.

[24] Das S. Nelder-Mead Evolutionary Hybrid Algorithms. Encyclopedia of Artificial Intelligence. 2009;3:1191–1196.

[25] Jones T. One Operator, One Landscape. Santa Fe, NM: Santa Fe Institute; 1995.

[26] El-Abd M. A hybrid ABC-SPSO algorithm for continuous function optimization. In: Proceedings of the 2011 IEEE Symposium on Swarm Intelligence; 2011. p. 1–6.

[27] Abd-Elazim SM, Ali ES. A hybrid particle swarm optimization and bacterial foraging for optimal power system stabilizers design. International Journal of Electrical Power & Energy Systems. 2013;46:334–341.

[28] Mandloi M, Bhatia V. A low-complexity hybrid algorithm based on particle swarm and ant colony optimization for large-MIMO detection. Expert Systems with Applications. 2016;50:66–74.

[29] Lin CJ, Chen CH, Lin CT. A hybrid of cooperative particle swarm optimization and cultural algorithm for neural fuzzy networks and its prediction applications. IEEE Transactions on Systems, Man, and Cybernetics, Part C (Applications and Reviews). 2009;39(1):55–68.

[30] Wang XC, Yang Q, Zhao YT. Research on hybrid PSODE with triple populations based on multiple differential evolutionary models. In: Proceedings of the 2010 International Conference on Electrical and Control Engineering. IEEE; 2010. p. 1692–1696.

[31] Ciornei I, Kyriakides E. Hybrid ant colony-genetic algorithm (GAAPI) for global continuous optimization. IEEE Transactions on Systems, Man, and Cybernetics, Part B (Cybernetics). 2012;42(1):234–245.

[32] Jatoth RK, Rajasekhar A. Speed control of PMSM by hybrid genetic artificial bee colony algorithm. In: Proceedings of the 2010 IEEE International Conference on Communication Control and Computing Technologies; 2010. p. 241–246.

[33] Liu S, Wang XY, You XM. Cultured differential particle swarm optimization for numerical optimization problems. In: Proceedings of the 3rd International Conference on Natural Computation. vol. 4. IEEE; 2007. p. 642–648.

[34] Hao ZF, Guo GH, Huang H. A particle swarm optimization algorithm with differential evolution. In: Proceedings of the 2007 International Conference on Machine Learning and Cybernetics. vol. 2. IEEE; 2007. p. 1031–1035.

[35] Duan HB, Xu CF, Xing ZH. A hybrid artificial bee colony optimization and quantum evolutionary algorithm for continuous optimization problems. International Journal of Neural Systems. 2010;20(1):39–50.

[36] Bai YY, Xiao SQ, Liu CR, *et al.* A hybrid IWO/PSO algorithm for pattern synthesis of conformal phased arrays. IEEE Transactions on Antennas and Propagation. 2013;61(4):2328–2332.

[37] Wu B, Qian CH, Ni WH, *et al.* Hybrid harmony search and artificial bee colony algorithm for global optimization problems. Computers & Mathematics with Applications. 2012;64(8):2621–2634.

[38] Narimani MR, Azizipanah-Abarghooee R, Zoghdar-Moghadam-Shahrekohne B, *et al.* A novel approach to multi-objective optimal power flow by a new hybrid optimization algorithm considering generator constraints and multi-fuel type. Energy. 2013;49:119–136.

[39] Liu B, Wang L, Jin YH, *et al.* Improved particle swarm optimization combined with chaos. Chaos, Solitons & Fractals. 2005;25(5):1261–1271.

[40] Wan ZP, Wang GM, Sun B. A hybrid intelligent algorithm by combining particle swarm optimization with chaos searching technique for solving nonlinear bilevel programming problems. Swarm and Evolutionary Computation. 2013;8:26–32.

[41] Nema S, Goulermas J, Sparrow G, *et al.* A hybrid particle swarm branch-and-bound (HPB) optimiser for mixed discrete nonlinear programming. IEEE Transactions on Systems, Man, and Cybernetics-Part A: Systems and Humans. 2008;38(6):1411–1424.

[42] Li ZY, Wang WY, Yan YY, *et al.* PS-ABC: A hybrid algorithm based on particle swarm and artificial bee colony for high-dimensional optimization problems. Expert Systems With Applications. 2015;42(22):8881–8895.

[43] Gong DW, Zhang Y, Qi CL. Environmental/economic power dispatch using a hybrid multi-objective optimization algorithm. International Journal of Electrical Power & Energy Systems. 2010;32(6):607–614.

[44] Ramesh V, Jayabarathi T, Asthana S, *et al.* Combined hybrid differential particle swarm optimization approach for economic dispatch problems. Electric Power Components and Systems. 2010;38(5):545–557.

[45] Lin YC, Hwang KS, Wang FS. Co-evolutionary hybrid differential evolution for mixed-integer optimization problems. Engineering Optimization. 2001;33(6):663–682.

[46] Xu W, Gu XS. A hybrid particle swarm optimization approach with prior crossover differential evolution. In: Proceedings of the First ACM/SIGEVO Summit on Genetic and Evolutionary Computation. ACM; 2009. p. 671–678.

[47] Li JQ, Pan QK, Xie SX. Flexible job shop scheduling problems by a hybrid artificial bee colony algorithm. In: Proceedings of the 2011 IEEE Congress of Evolutionary Computation. IEEE; 2011. p. 78–83.

[48] Zhang GH, Shao XY, Li PG, *et al.* An effective hybrid particle swarm optimization algorithm for multi-objective flexible job-shop scheduling problem. Computers & Industrial Engineering. 2009;56(4):1309–1318.

[49] Xia WJ, Wu ZM. An effective hybrid optimization approach for multi-objective flexible job-shop scheduling problems. Computers & Industrial Engineering. 2005;48(2):409–425.

[50] Mirjalili S, Hashim SZM. A new hybrid PSOGSA algorithm for function optimization. In: Proceedings of the 2010 International Conference on Computer and Information Application. IEEE; 2010. p. 374–377.

[51] Zhu LY, Wu JH. Hybrid particle swarm optimization algorithm for flexible task scheduling. In: Proceedings of the Third International Conference on Genetic and Evolutionary Computing. IEEE; 2009. p. 603–606.

[52] Zhang Y, Gong DW, Geng N, *et al.* Hybrid bare-bones PSO for dynamic economic dispatch with valve-point effects. Applied Soft Computing. 2014;18:248–260.

[53] Begambre O, Laier JE. A hybrid Particle Swarm Optimization–Simplex algorithm (PSOS) for structural damage identification. Advances in Engineering Software. 2009;40(9):883–891.

[54] Marrouchi S, Hessine MB, Chebbi S. New strategy based on combined use of Particle Swarm Optimization and Gradient methods to solve the unit commitment problem. In: Proceedings of the 15th IEEE International Conference on Environment and Electrical Engineering; 2015. p. 1362–1367.

[55] Ouyang AJ, Tang Z, Zhou X, *et al.* Parallel hybrid PSO with CUDA for ID heat conduction equation. Computers & Fluids. 2015;110:198–210.

[56] Noel MM. A new gradient based particle swarm optimization algorithm for accurate computation of global minimum. Applied Soft Computing. 2012;12(1):353–359.

[57] Basu M. Hybridization of bee colony optimization and sequential quadratic programming for dynamic economic dispatch. International Journal of Electrical Power & Energy Systems. 2013;44(1):591–596.

Chapter 4
Ant colony systems for optimization problems in dynamic environments

Yirui Wang[1], Shangce Gao[1], and Yuki Todo[2]

Abstract

Ant colony optimization (ACO) is an intelligent bionic algorithm which simulates the foraging behavior of ant colony. The conventional ACOs mainly deal with the static optimization problems. In other words, the environment of problem maintains invariant. Actually, the most problems in reality are dynamic, namely, the changing environments. The ACO can use its robustness and self-adaptability to resolve dynamic problems properly. In this chapter, the ACO with neighborhood search is introduced to address dynamic traveling salesman problem and the ACO with improved K-means clustering algorithm, which uses three immigrants schemes including random immigrants, elitism-based immigrants and memory-based immigrants, is used for dynamic location routing problem. Several conventional ACOs and other heuristic algorithms are utilized to compare with new ACOs in the corresponding dynamic problems. The comparative experiments demonstrate two novel ACOs are effective and efficient for respective dynamic optimization problems.

4.1 Introduction

The initial ant colony optimization (ACO) [1] was invented by Dorigo *et al.* to emulate foraging action of ants between nest and food. The core factor of algorithm is the pheromone ants release aiming to achieve the communication among their individuals. The pheromone on the path guides ants to select the best tour where the pheromone concentration is the highest. This mechanism has motivated people to make full use of ACO for processing non-deterministic polynomial (NP)-hard combinatorial optimization problems. Nowadays, the ACO as an effective algorithm has been successfully applied for a number of combinatorial optimization problems due

[1]Faculty of Engineering, University of Toyama, Japan
[2]Faculty of Electrical and Computer Engineering, Kanazawa University, Japan

to its intelligent search, global optimization, strong robustness, parallel computation, positive feedback and combination of other algorithms [2].

The traditional ACO is usually proposed to tackle the static optimization problems, such as traveling salesman problem (TSP), job-shop scheduling problem and graph coloring problem. These problems can be resolved well in the stationary environment. To be specific, ACO can obtain the better and better solution over iterations and gradually converge into a best-so-far solution in the fixed environment. Nevertheless, a majority of real-world problems belong to dynamic optimization with changing environments. Compared with static problems, it is more complex and difficult to address dynamic problems where the objective function, constraints and control parameters can vary during the execution of algorithm [3]. Meanwhile, the optimal solution of the problem can also alter with environmental changes. Consequently, for dynamic problems, the purpose of the algorithm is not to fast converge into an optimal solution anymore but to efficiently trace a new optimal solution in terms of existing information between previous and current environments. Even though the conventional ACO possesses strong robustness and self-adaption in the static environment, it is much inferior to handle dynamic environments only by itself [4]. The information from previous environments should be effectively utilized by ACO to accelerate the efficiency of the algorithm for tracing the optimal solution when environments change.

The first application of ACO was for solving static TSP [1]. In fact, dynamic TSP (DTSP) is even worth investigating and challenging. The dynamic changes not only reflect the actual environment but also influence the performance of ACO. Guntsch *et al.* designed a kind of DTSP regarding exchanging cities [5]. Eyckelhof *et al.* proposed a DTSP with respect to the changing cost of edges among cities [6]. Mavrovouniotis *et al.* introduced a type of DTSP composed of traffic factors [4]. Thus, several ACOs have been proposed to deal with different kinds of DTSPs. The straightforward approach is to reinitialize the pheromone of algorithm to respond to environmental changes within the adequate time, while it is usually not proper for the dynamic environments with noise [7]. Guntsch *et al.* presented two local restart schemes to handle the DTSP where the distribution of cities varied [8]. Two local restart schemes called η-strategy and τ-strategy just considered to reinitialize the pheromone of actual changing cities instead of global reinitialization. Population-based ACO (P-ACO) was proposed to resolve the DTSP with exchanging cities and show a distinctive structure of the algorithm for enhancing the diversity of population on account of its population list [5,9]. Furthermore, several variants of P-ACO such as fitness sharing P-ACO and simple crowding P-ACO have been proposed for reinforcing the performance of the algorithm in DTSP [10]. Besides, the ACO with immigrants schemes as a prospective approach was verified to obtain the superior performance for the DTSP with traffic factors [4].

Different from TSP, the location routing problem (LRP) is a more complicated logistical distribution system which contains location allocation and vehicle routing. Therefore, tackling LRP is to actually implement the combination of these two subproblems. At present, a number of LRPs are researched by numerous approaches. Baldacci *et al.* used dynamic programing and dual ascent approaches to resolve the

LRP which had multiple depots and vehicles [11]. Glicksman *et al.* investigated a classified LRP where the capacities of depots and vehicles were not considered and customers were categorized into several classes [12]. The purpose of the problem was the minimum total cost of routes and depots. Barreto *et al.* introduced some hierarchical and nonhierarchical clustering methods to address the capacitated LRP [13]. Several effective metaheuristics such as tabu search [14], iterative local search [15] and simulated annealing [16] have been applied for the LRP as well. Nowadays, a promising algorithm for the capacitated LRP is a multiple ACO which performs best on Barreto's instances [17]. The dynamic LRP (DLRP) indicates customers' demand, service time, traffic or capacities of depots and vehicles can alter in changing environments. Hence, it is necessary to resolve DLRP via an effective and sophisticated algorithm.

In this chapter, the ACO is used to address two dynamic optimization problems (DOPs) including DTSP and DLRP. The changing environments comprise dynamic traffic factors which represent the traffic jam. Two original variants of ACO are introduced to tackle DTSP and DLRP, respectively. First, the ACO with neighborhood search (NS–ACO) is utilized to deal with the DTSP constructed by random traffic factor. Neighborhood search can effectively optimize the local structure of solutions. Three search mechanisms which consist of swap, insertion and 2-opt operations are randomly implemented to update solutions. By this means, the diversity of solutions can be increased so that the algorithm has a high chance to fast obtain a better solution for adapting to dynamic environments. The experimental results demonstrate the superiority of NS–ACO compared with several conventional ACOs and the ACO with random immigrants. Second, the ACO with improved K-means clustering algorithm (KACO) which combines with three immigrants schemes involving random immigrants, elitism-based immigrants and memory-based immigrants is applied for DLRP with random and cyclic environments. The clustering approach can adaptively ascertain the positions of depots and service customers in each depot. Thereafter, the KACO with three immigrants achieves vehicle routes in DLRP. The experiments indicate the clustering algorithm significantly changes the construction of solutions to intensify the performance of immigrants. Consequently, KACO with three immigrants outperforms the ACO without K-means algorithm, several conventional ACOs and other heuristics such as simulated annealing and genetic algorithm, implying it is a prospective method for DLRP.

The rest of this chapter is organized as follows. Section 4.2 introduces two kinds of DOPs involving DTSP and DLRP. Section 4.3 presents five conventional ACOs and two novel ACOs, elaborating their basic principles and characteristics. Section 4.4 shows the experimental results and analyses for DTSP and DLRP. Section 4.5 gives some conclusions and points out some research directions.

4.2 Dynamic optimization problems

A DOP usually means its objective function, constraints and other parameters change over time. The environment of problem is diverse. This dynamic characteristic enhances the complexity of problem and indicates the realistic significance. In this chapter, two kinds of DOPs are described and resolved.

4.2.1 Dynamic traveling salesman problem

4.2.1.1 Stationary TSP

The traditional TSP is one of stationary NP-hard combinatorial optimization problems. It is depicted as follows: a salesman departs from home and must visit every city only once and finally returns to his home. The length of his route needs to be minimum. Mathematically, it is described by a completely connected graph $G = (V, E)$, where V is a group of vertices indicating cities and E is a group of edges denoting the connection between cities. The formula is shown as follow:

$$f(x) = \min \sum_{i=0}^{n} \sum_{j=0}^{n} d_{ij}\omega_{ij} \tag{4.1}$$

where n indicates the number of cities, and d_{ij} denotes the distance between cities i and j. $\omega_{ij} \in \{0, 1\}$ represents whether the $E(i,j)$ is in the tour.

4.2.1.2 Dynamic TSP

Compared with stationary TSP, DTSP possesses changing environments [4–6]. That is to say, it is more complicated to handle the tour when the environment is various during the execution of algorithm. Thus, the purpose for addressing DTSP is not to find the global optimal solution anymore but to trace it effectively with environmental changes.

In this chapter, the DTSP consists of random traffic factor. It considers the cost of edge between cities i and j, which is described as follows:

$$D_{ij}(t) = D_{ij} \times Z_{ij}(t) \tag{4.2}$$

where $D_{ij}(t)$ and $Z_{ij}(t)$ indicate the distance and random traffic factor between cities i and j in t moment, respectively. $Z_{ij} \in [Z_l, Z_h]$, where Z_l and Z_h are the lower and upper bounds of traffic factor. When each f iterations are satisfied, Z_{ij} will randomly change in terms of the magnitude m of environmental changes. The value of Z_{ij} manifests the degree of traffic jam between cities i and j. For instance, when the environment changes slightly, the value of m is low and Z_{ij} is closer to Z_l. On the contrary, the value of m is high and Z_{ij} is closer to Z_h over significant environmental changes.

In this way, the DTSP is collectively constructed by random traffic factor Z, the magnitude m and frequency f of environmental changes. Every changing environment is random and distinctive, which is not associated with previous environments. In other words, the environment may not necessarily occur once again.

4.2.2 Dynamic location routing problem

4.2.2.1 Conventional LRP

Generally, the LRP [18] is described as follows: a set of candidate depots is given, and several vehicles depart from determined depots to serve all the customers within

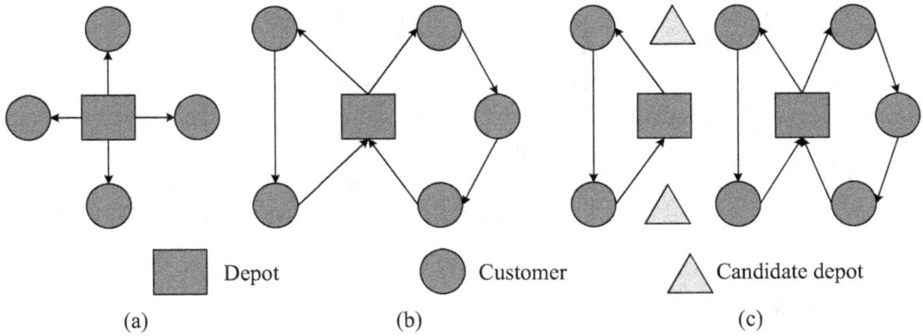

Figure 4.1 The routes displayed by LAP, VRP and LRP

certain time. The number and position of depots, optimal vehicle routes and allocation of customers' demand enable the total logistical cost to be minimum. The logistical cost contains the cost of depot construction, transportation, vehicles, time and so on. Therefore, LRP can be regarded as a logistical allocation model, integrating location allocation problem (LAP) with vehicle routing problem (VRP). For LRP, there is a serious matter where the solution space enhances exponentially along with the scale of problem increasing.

LAP is usually defined as follows: in the specific service range, the number and position of depots depend on the position of customers and allocation of commodities. The aim of LAP is to implement the minimum logistical cost in terms of those selected depots. For instance, Figure 4.1(a) shows the divergent routes between a depot and customers, which only considers the transportation from point to point, meaning each vehicle directly returns to the depot after serving one customer. This kind of transportation is just suitable for some special conditions such as the capacity of one vehicle is less than one customer's demand. Thus, in LAP, the total logistical cost can significantly enhance due to the limited transportation.

VRP has been an important logistical problem, depicting the whole vehicles depart from one depot to visit customers in order to achieve the shortest vehicle route or the minimum cost. The mathematical expression is $G = (C, M, d)$, where $C = \{c_0, c_1, \ldots, c_n\}$ represents customers and $M = \{(c_i, c_j), i \neq j\}$ indicates the linkage between customers i and j. Each vehicle whose capacity is Q sets off from one depot located in c_0 to satisfy n customers who have respective demands q_i. d_{ij} denotes the Euclidean distance between customers i and j. The constraint conditions of VRP are limited as (1) every customer is only visited by one vehicle; (2) every vehicle must depart from the depot and return at last and (3) every vehicle whose capacity is less than or equal to Q serves several customers, and $q_i \leq Q$. In VRP, the number of solutions will expand exponentially according to the number of customers. In Figure 4.1(b), a circle route is applied for improving the vehicle utilization rate and optimizing traveled routes. However, the depot is fixed and only one. In fact, the depot is a crucial factor to decide the entire logistical system. If neglecting it, the total cost cannot reach the optimum.

LRP combines LAP with VRP to establish a new combinatorial optimization problem for simulating a logistical distribution system, considering a set of depots, several vehicle routes and customers' demand to accomplish the minimum cost. As shown in Figure 4.1(c), the number and position of depots, optimal routes and service customers are simultaneously taken into account. Consequently, LRP integrating LAP with VRP has more complicated constraint conditions and shows reality, which can be generally categorized into three parts: locating depots, distributing demands and optimizing vehicle routes.

4.2.2.2 Dynamic LRP

The conventional LRP belongs to a static problem, whereas addressing a DLRP is more challenging [19]. This dynamic characteristic not only enhances the complexity of LRP but also brings new considerable circumstance for programing the vehicle routes of LRP. The dynamic environments can influence the space of solutions over time because a new optimal solution in new environment may be not the same as previous one. Accordingly, the goal of DLRP is to quickly track the optimal solution in terms of changing environments.

The changing environments are composed of random traffic factor and cyclic traffic factor. The random traffic factor Z_{ij} has been mentioned in DTSP, which indicates the traffic jam between customers i and j. The greater the value of Z_{ij}, the more serious the traffic jam between customers i and j. $Z_{ij} \in [Z_l, Z_h]$ and it randomly alters according to the frequency f and magnitude m of environmental changes. The eventual result is described by the expression $D_{ij} = D_{ij} \times Z_{ij}$.

The cyclic traffic factor possesses periodic characteristic which can enable former environments to reappear. Compared with random traffic factor, the cyclic traffic factor similar to the traffic of respective periods in one day is more actual. For example, the traffic jam is severe during commuting hours time, namely, Z_{ij} is closer to Z_h. While in other time, the traffic jam is expedite, i.e. Z_{ij} is closer to Z_l. The cyclic environments contain several cyclic traffic factors acted as the basic situations, and each environment reappears cyclically. The constructing way of cyclic traffic factor is identical to random traffic factor.

In this chapter, the DLRP has two dynamic environments including random and cyclic environments. Meanwhile, the constraint condition of DLRP is that each depot has sufficient capacity and each customer is just served by one vehicle. All the customers have the same demand, and the capacity of each vehicle is identical. In this way, determining the number and position of depots relies on the customers' distribution. Then vehicles depart from determined depots to construct traveled routes in dynamic environments so as to implement the overall minimum cost.

4.3 Ant colony optimization

Ants can process a great deal of complex activities via their mutual collaboration or division of labor since they are social beings. The collective strength shows ants have extraordinary self-adaptation for different environments. This kind of feature is decided by exchanging information called pheromone among ants. The pheromone guides ants to select the optimal path, and then ants deposit new pheromone on their traveling paths. Besides, the pheromone is evaporated over time. Therefore, a majority of ants follow one path, suggesting the pheromone concentration on this path is high and maybe attract more subsequent ants to enhance its concentration. This swarm behavior reveals a positive feedback phenomenon to let ants finally find an optimal route during successive iterations. In this chapter, five conventional ACOs and two improved ACOs are presented as follows.

4.3.1 Ant system

Ant system (AS) [2] was initially proposed by Dorigo M to simulate the foraging manner of ants and acquire a good result for tackling TSP. It is mainly constructed by search process and pheromone update.

In the search process, m ants are placed randomly in one city, and then each ant selects next city to visit according to the random possibility rule formulated as the following:

$$p_{ij}^{k}(t) = \frac{\tau_{ij}^{\alpha}(t)\eta_{ij}^{\beta}}{\sum_{l \in N_i^k} \tau_{il}^{\alpha}(t)\eta_{il}^{\beta}}, \quad j \in N_i^k \tag{4.3}$$

where p_{ij}^{k} indicates the possibility of ant k traveling from city i to city j. τ_{ij} denotes the pheromone between cities i and j. N_i^k represents the set of cities which are not visited by ant k in the city i, and the next city j is included in N_i^k. $\eta_{ij} = 1/d_{ij}$ manifests the heuristic information between cities i and j, where d_{ij} is the distance between two cites. α and β reflect the influence of pheromone and heuristic information on the performance of algorithm, respectively. Hence, balancing the relation between two parameters is essential.

In the pheromone update, ants release individual pheromone and the previous pheromone on the path is evaporated after the search process. Mathematically, it is described as the following:

$$\tau_{ij}(t+1) = (1-\rho)\tau_{ij}(t) + \sum_{k=1}^{m} \Delta\tau_{ij}^{k}(t) \tag{4.4}$$

where $\rho \in (0, 1]$ indicates the evaporation coefficient, which not only avoids the infinite increase of pheromone but also eliminates useless pheromone to accelerate the convergence of algorithm. $\Delta\tau_{ij}^{k} = Q/L_k$ is the pheromone released by ant k between cities i and j, where Q is a constant and L_k is the length of route constructed by ant k.

4.3.2 Elitist ant system

AS initially shows a good property in small-scale problems. Nevertheless, the performance of AS declines seriously with gradual expansion of scale of problems. Thus, the first improved version called elitist AS (EAS) [1] was proposed to reinforce the feature of algorithm. In terms of the entire framework of AS, EAS adds extra pheromone on the optimal path to attract more ants. To be specific, the search process of EAS is the same as that of AS, but in the pheromone update, EAS deposits more pheromone on the best-so-far path. The pheromone update is expressed as follows:

$$\tau_{ij}(t+1) = (1-\rho)\tau_{ij}(t) + \sum_{k=1}^{m} \Delta\tau_{ij}^{k}(t) + \Delta\tau_{ij}^{bs} \tag{4.5}$$

where $\Delta\tau_{ij}^{bs} = \sigma Q / L^{bs}$ indicates the extra pheromone released by the best-so-far ant. σ denotes the number of best-so-far ant and L^{bs} is the length of best-so-far path. In this way, EAS boosts the quality and convergency of solution, which overall outperforms AS.

4.3.3 MAX–MIN ant system

Stützle and Hoos presented MAX–MIN AS (MMAS) [20] to improve the characteristic of algorithm. MMAS principally revises four parts: (1) the iteration-best or best-so-far ant only deposits pheromone on its traveling routes; (2) the pheromone interval $[\tau_{min}, \tau_{max}]$ is set to effectively avoid the fast increase of pheromone and stagnation of algorithm; (3) the initialization of pheromone which is the upper bound of interval τ_{max} combines with a lower evaporation coefficient so that ants can expand the search space to obtain a better solution during the beginning of algorithm and (4) the pheromone will be reinitialized when the algorithm stays stagnation or cannot find a better solution in a long time.

The concrete change is shown in the pheromone update formulated as the following:

$$\tau_{ij}(t+1) = (1-\rho)\tau_{ij}(t) + \Delta\tau_{ij}^{bs}, \quad \tau_{ij} \in [\tau_{min}, \tau_{max}] \tag{4.6}$$

where $\Delta\tau_{ij}^{bs} = 1/L^{bs}$ indicates the releasing pheromone via the iteration-best or best-so-far ant. It should be noticed that the iteration-best ant can obtain better solutions with continuous iterations, while the best-so-far ant can enhance the convergency speed. The alternative depends on the scale of problem, and the total performance of MMAS is enhanced significantly.

4.3.4 Ant colony system

Ant colony system (ACS) [2] was proposed by Dorigo and Gambardella to handle the high-scale TSP. Its search mechanism and pheromone update strategy immensely boost the property of algorithm. New search method further improves the ability of ants finding solutions. Global and local pheromone update extend the search domain of ants.

In ACS, the new search mechanism called pseudorandom proportion rule is utilized to implement the approach of ants selecting next city j from city i, which is depicted as follows:

$$j = \begin{cases} \arg\max\limits_{l \in N_i^k} \{\tau_{il}(t)\eta_{il}^{\beta}\}, & q \leqslant q_0 \\ \arg\max\{p_{ij}^{k}(t)\}, & q > q_0 \end{cases} \tag{4.7}$$

where q is a random variable distributed uniformly in the interval $[0, 1]$. $q_0 \in [0, 1]$ is a parameter influencing the efficiency and result of algorithm. When $q \leqslant q_0$, ACS can accelerate convergency but has finite search space. When $q > q_0$, ACS expands search range to find better solutions according to the search process of AS. Hence, ACS not only enhances the efficiency of algorithm but also prompts the quality of solution by integrating two methods.

In the pheromone update, ACS uses global and local update methods. As for global pheromone update, the best-so-far ant is only permitted to release pheromone, and the evaporation is just executed on the best-so-far routes, which is shown as follow:

$$\tau_{ij}(t+1) = (1-\rho)\tau_{ij}(t) + \rho\Delta\tau_{ij}^{bs} \tag{4.8}$$

ACS introduces distinctive local pheromone update, elucidating each ant updates the path immediately after moving from city i to city j. The formulation is given as follow:

$$\tau_{ij}(t+1) = (1-\varphi)\tau_{ij}(t) + \varphi\tau_0 \tag{4.9}$$

where $\varphi \in (0, 1]$ is a parameter and τ_0 is the initial pheromone. The purpose of local update is to enable the subsequent ants to explore other paths different from the current traveling path so that all the ants avoid to progress toward the same path to trap into local optimum and have a high chance to acquire better solutions.

4.3.5 Population based ACO

Guntsch and Middendorf proposed a variant of ACO called population-based ACO (P-ACO) [9] which owned a population list whose size was k to store k iteration-best ants to update the pheromone on the path. In particular, there is no evaporation operation in P-ACO. P-ACO increases the diversity of population due to its population list. When ants in the population list are deleted, the corresponding deposited pheromone is also deleted. This kind of operation effectively decreases the computational complexity of algorithm.

The process of P-ACO is primarily described as follow: The method of routes constructed by ants is the same as ACS. Before the initial k iterations, the iteration-best ant is put in the population list in each iteration until the population list is full. From the beginning of $k+1$ iteration, the population list removes the earliest ant in terms of first-in–first-out scheme and new iteration-best ant enters the population list in each iteration. Meanwhile, the previous pheromone deposited by the removed ant is deleted and new population list continues to update pheromone. P-ACO sets a upper bound τ_{max} of pheromone to prevent the pheromone from increasing excessively.

4.3.6 ACO with neighborhood search

Ants usually construct several solutions but only containing one optimal solution, and the surplus of them are inferior. However, an inferior solution is not completely

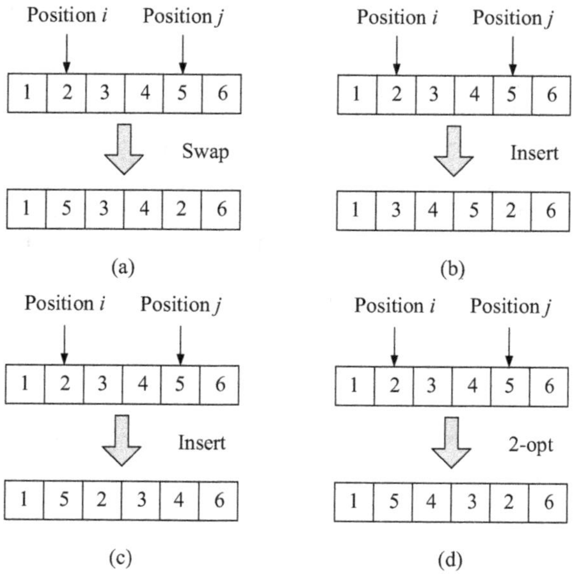

Figure 4.2 The illustrative graph of neighborhood search

useless because the most segments of solution maybe have been optimal and other segmental nonoptimal components just influence the overall quality of solution. Thus, modifying nonoptimal segments can remarkably enhance the result of solution. Based on this consideration, the ACO with neighborhood search (NS–ACO) [21] is proposed to locally optimize the component of solution to intensify performance for addressing DTSP.

NS–ACO adopts a short-term memory to save K iteration-best ants in current iteration to update pheromone in order to increase the diversity of population. The search process of NS–ACO is the same as AS, which utilizes (4.3) to structure tour. After the whole ants achieve respective tours, the neighborhood search begins to be executed for optimizing solutions. The neighborhood search contains swap, insertion and 2-opt three operations. Each operation is selected to realize the adjustment of solution with the 1/3 probability, and then new solution compares with the previous solution to retain the better one. Swap operation indicates two positions i and j of one solution are randomly selected and swap each other, as shown in Figure 4.2(a). Insertion operation first randomly chooses position i and position j of a solution. If $i < j$, then each position moves toward left one step from positions $i + 1$ to j and position i substitutes position j as displayed in Figure 4.2(b). Otherwise, from positions j to $i - 1$, each position moves toward right one step and position i substitutes position j as exhibited in Figure 4.2(c). 2-opt operation is to entirely reverse the segment between positions i and j randomly selected as Figure 4.2(d). By this means, neighborhood search can effectively rectify the component of solution to enhance the quality of solution.

After the neighborhood search is implemented, the short-term memory is used to conduct the pheromone update. In iteration t, the short-term memory releases the pheromone given as follow:

$$\tau_{ij}(t) = \tau_{ij}(t) + \Delta\tau_{ij}^k(t) \tag{4.10}$$

where $\Delta\tau_{ij}^k = (\tau_{max} - \tau_0)/K$, τ_{max} is the upper bound of pheromone and τ_0 is the initial value of pheromone. K indicates the size of short-term memory. $\tau_{max} = \tau_0 + \sum_{k=1}^{K} \Delta\tau_{ij}^k$. In iteration $t+1$, the short-term memory deletes all the ants in previous iteration to offer space for new K iteration-best ants. The previous releasing pheromone is deleted simultaneously, which is depicted as follows:

$$\tau_{ij}(t+1) = \tau_{ij}(t+1) - \Delta\tau_{ij}^k(t) \tag{4.11}$$

New generated short-term memory carries on adding the pheromone according to (4.10) after executing the deleting operation. In this way, there is no ant surviving in excess of one iteration and no evaporation operation. This pheromone update remains the pheromone concentration as well as improves the diversity of population.

4.3.7 Clustering ACO with immigrants schemes

The clustering ACO (KACO) with three immigrants schemes including random immigrants, elitism-based immigrants and memory-based immigrants is proposed to tackle the DLRP [22]. An improved K-means clustering algorithm is utilized to resolve the LAP to select the positions of depots and customers served by each depot, and then the KACO with three immigrants deals with the VRP in random and cyclic environments.

4.3.7.1 Clustering algorithm

Clustering is that a set of data are classified into several subsets where the data among subsets are distinctive, while the data in each subset have high similarity. K-means algorithm is a classical clustering approach whose objective function is the sum of distance between data and clustering center. Its principle is to minimize the objective function and implement the classification of data at the utmost. In K-means algorithm, the judgment of similarity depends on the sum of error squares $E_s = \sum_{i=1}^{k} \sum_{p \in X_i} ||p - m_i||^2$, where X_i indicates the class i and m_i represents the average value among data in the class i. p is the data. The E_s is less and less over iterations until invariant, and the clustering process ends.

Although the conventional K-means algorithm is useful for achieving clustering, the k value is necessary to be given initially. The selection of k value can directly influence the clustering effect. Therefore, an improved K-means algorithm called distance cost function [23] is adopted to adaptively determine the k value, which is described as follows:

$$F = \sum_{i=1}^{k} |m_i - m| + \sum_{i=1}^{k} \sum_{p \in X_i} |p - m_i| \tag{4.12}$$

where m indicates the average value among the whole data and other parameters are identical to E_s. The minimum of F reveals the best clustering performance and finds the optimal k value. Besides, the upper bound of k is set by the experimental rule $k \leqslant \sqrt{n}$ which can effectively decrease the search space of the best k value. The improved K-means algorithm can efficiently acquire the optimal k value owing to its simple structure and low-computational complexity. The result of clustering ultimately determines the position and number of depots as well as their surrounding customers. Then, the KACO with three immigrants schemes is implemented to address the DLRP.

4.3.7.2 Three immigrants schemes

Random immigrant scheme is to generate random ants to substitute the worst ants in the short-term memory k_{short} in terms of $r \times K_s$, where r indicates the replacement rate and K_s denotes the size of short-term memory. This method can resolve the DTSP in the quickly and significantly changing environments since the remarkable environmental changes lead to insufficient time to transmit information for ACO, while random immigrants enhance the diversity of population. It should be noted that too many random immigrants may bring in too much diversity so that ACO maybe converges into the local optimal solution.

Elitism-based immigrant scheme utilizes the best ant from the previous environments to substitute the worst ants in the short-term memory k_{short}. A mutation is executed to randomly swap two positions of a solution for improving the quality of solution according to a mutation rate p_m. In the slowly and slightly changing environments, elitism-based immigrants have enough time to transmit similar information from previous environment to current environment. It also should be noticed that a number of information which need to be transmitted maybe cause ACO to trap into local optimum.

Memory-based immigrant scheme uses not only the short-term memory k_{short} but also the long-term memory k_{long}. The k_{long} stores several optimal solutions from previous environments, which is initially full of random ants and then lets the optimal ant substitutes them or the closest ant. The closest metric is measured as follows:

$$M_{ij} = 1 - \frac{CE_{ij}}{N_l} \tag{4.13}$$

where CE_{ij} indicates the number of common paths between ants i and j. N_l means the number of total paths. The memory-based immigrant generated from the k_{long} replaces the worst ants in the k_{short} and also employs the mutation operation. The k_{long} is updated when the environment changes. Nevertheless, noise can interfere the detection of dynamic environment. To avoid the noisy influence, a time interval is set as $t_l = f + R$, where t_l is the next update time for k_{long} and f is the current iteration. $R \in [5, 10]$ indicates a random variable. Accordingly, the update of k_{long} has two patterns, including the detection of dynamic environment and the random implementation of t_l with no environmental change for a long time. Memory-based immigrant scheme which provides diverse optimal solutions shows the superiority in different environments, especially in cyclic environments which can be reappeared.

4.3.7.3 KACO with immigrants schemes

For DLRP, several constraint conditions such as the customers' demand and the capacity of each vehicle in each class should be taken into account. Consequently, within the service range of one depot, each ant is regarded as one vehicle. When the capacity of vehicle is empty, the vehicle automatically returns to the depot and departs again to serve customers who are not satisfied. After meeting all the customers, the vehicle routes are reasonably constructed in one class. In this way, several vehicle routes can be obtained in other depots.

In this chapter, the KACO with random immigrants, elitism-based immigrants and memory-based immigrants is called KACO–RI, KACO–EI and KACO–MI, respectively. In the search process, the whole ants are first arranged in the same depot. Then each ant departs from the depot to select the next visited customer according to (4.3), customer's demand and the capacity of vehicle. After the routes are established by all the ants, three immigrants schemes are applied for updating the k_{short}. Thereafter, the pheromone update is carried out by the k_{short}. The update method is the same as the above NS–ACO, and there is no repeated mention. The overall distance of vehicle routes in all the depots is deemed to be the objective value in DLRP. Eventually, the minimum distance illustrates the optimal vehicle routes in dynamic environments.

4.4 Experiments

4.4.1 The experiment for DTSP

4.4.1.1 Experimental settings

To illustrate the property of NS–ACO for resolving the DTSP, the comparative experiments among AS, EAS, MMAS, ACS, P-ACO, RIACO which uses the random immigrants [4] and NS–ACO are implemented. Six benchmark TSP instances adopted from TSPLIB are utilized to constitute the DTSP with random traffic factor. eil51 and st70 indicate the small-scale DTSP. kroA100 and kroA150 denote the medium-scale DTSP. kroA200 and pr299 signify the large-scale DTSP.

In DTSP, the upper and lower bounds of random traffic factor Z are set to 8 and 1, i.e. $Z \in [1, 8]$. The frequency f of environmental changes are set to 10 and 100, indicating the environment happens to quick or slow changes in each 10 or 100 iterations, respectively. The magnitude m of environmental changes are set to 0.1, 0.5 and 0.9, corresponding to the low, medium and high changes, respectively. Thus, for each DTSP, there are six dynamic environments forming six actual instances. In this experiment, the total 36 instances are constructed and handled by 7 ACOs. Each ACO is independently run 30 times and 1,000 iterations in each run to guarantee the accuracy of results. The calculation of experimental results are achieved by the formulation given as follows:

$$P^{bs} = \frac{1}{U}\sum_{t=1}^{U}\left(\frac{1}{V}\sum_{i=1}^{V} f_{ti}^{bs}\right) \tag{4.14}$$

Table 4.1 Parameter settings of seven ACOs

Algorithm	α	β	ρ	φ	q_0	K_s	r	M
AS	1	5	0.5	–	0	–	–	30
EAS	1	5	0.5	–	0	–	–	30
MMAS	1	5	0.6	–	0	–	–	30
ACS	1	5	0.5	0.5	0.9	–	–	30
P-ACO	1	5	–	–	0.9	4	–	30
RIACO	1	5	–	–	0	6	0.4	30
NS–ACO	1	5	–	–	0	6	–	30

where U and V mean the number of iterations and runs, respectively. f_{ti}^{bs} indicates the optimal solution in the tth iteration of ith run.

The parameters of each ACO are listed in Table 4.1. Several values of parameters are set in terms of the previous literatures [1,2,4,9,20]. M denotes the number of ants, and the initial pheromone is given to $\tau_0 = 1/L^{nn}$, where L^{nn} is the length of tour constructed by the nearest neighborhood heuristic.

4.4.1.2 Experimental analysis

Table 4.2 exhibits the experimental results among seven ACOs according to (4.14). The Wilcoxon signed ranks test at a level of significance $\alpha = 0.05$ is used to further analyze the performance of NS–ACO. Table 4.3 illustrates the significant difference between NS–ACO and other ACOs via the symbol, where the symbol + reveals the former outperforms the latter between two algorithms and the symbol – is contrary. symbol $\sim$ means no significance.

From Tables 4.2 and 4.3, it can be observed that NS–ACO always maintains the symbol + for whatever environments in eil51, st70, kroA100 and kroA150, implying NS–ACO outperforms other ACOs in small-scale and medium-scale DTSP and demonstrating the neighborhood search is an effective approach to locally optimize solutions for adapting to dynamic environments. Nevertheless, the results in $f = 10$, $m = 0.1$ of kroA200 and $f = 10$, $m = 0.1$ or 0.5 of pr299 display ACS outperforms NS–ACO in large-scale DTSP with quick and slight environmental changes on the basis of the symbol – of ACS. Since NS–ACO needs several iterations to successively optimize solutions in an extensive domain, it cannot be completely accomplished in large-scale DTSP with quickly changing environments. Furthermore, compared with NS–ACO, ACS effectively utilizes local and global pheromone update to track the optimal solution in slightly changing environments where the previous pheromone may be useful for guiding ants to quickly find the optimal solution. When the environment changes significantly, i.e. $m = 0.9$, the search mechanism of ACS cannot be suitable for adapting to environments. Thus, the performance of ACS declines. Besides, NS–ACO possesses sufficient iterations to update solution to enhance the quality of solution and outperforms other ACOs in slowly changing environments such as $f = 100$.

Table 4.2 The comparison of experimental results among seven ACOs for DTSP

Algorithm $f = 10, m$	eil51			st70			kroA100		
	0.1	0.5	0.9	0.1	0.5	0.9	0.1	0.5	0.9
AS	491.9	685.2	1,372.3	799.8	1,148.2	2,269.8	26,842.1	39,725.1	79,821.8
EAS	491.2	685.5	1,373.2	799.1	1,148.5	2,267.5	26,821.3	39,704.5	79,746.0
MMAS	490.7	719.1	1,412.1	824.4	1,209.4	2,308.5	27,625.0	41,597.2	80,927.0
ACS	492.6	687.5	1,446.5	790.5	1,148.5	2,363.9	26,376.3	38,624.8	80,633.2
P-ACO	495.4	731.1	1,515.0	799.6	1,225.1	2,451.6	28,319.1	46,993.5	95,517.2
RIACO	482.5	674.6	1,352.5	792.5	1,127.6	2,209.5	26,805.1	39,064.3	79,117.0
NS–ACO	478.1	669.0	1,338.1	779.6	1,111.8	2,180.3	25,971.5	37,880.8	76,284.7
$f = 100, m$	0.1	0.5	0.9	0.1	0.5	0.9	0.1	0.5	0.9
AS	469.6	639.2	1,273.6	772.3	1,051.1	2,017.5	25,373.1	35,871.9	73,447.1
EAS	468.7	637.4	1,270.8	771.3	1,050.8	2,010.6	25,324.7	35,838.3	73,233.2
MMAS	461.8	646.1	1,286.3	772.8	1,067.7	2,018.4	25,134.2	36,130.9	72,437.8
ACS	480.7	660.6	1,346.1	775.3	1,085.6	2,177.0	25,827.6	37,067.7	78,995.3
P-ACO	458.9	639.9	1,297.7	740.2	1,046.7	2,028.2	24,387.9	36,662.7	75,106.6
RIACO	454.4	627.2	1,247.8	751.9	1,033.7	1,959.3	24,758.0	34,968.6	71,477.7
NS–ACO	452.3	622.9	1,240.1	746.8	1,025.1	1,953.5	24,420.5	34,356.3	70,349.0

Algorithm $f = 10, m$	kroA150			kroA200			pr299		
	0.1	0.5	0.9	0.1	0.5	0.9	0.1	0.5	0.9
AS	34,596.9	51,683.5	102,192.3	40,086.5	60,403.6	118,856.8	70,807.0	111,697.1	219,845.0
EAS	34,559.8	51,638.9	102,193.9	40,101.4	60,361.1	118,829.0	70,816.1	111,787.6	219,868.0
MMAS	35,116.8	52,803.1	101,731.1	40,194.0	60,664.8	116,488.6	69,487.0	108,288.8	208,184.5
ACS	33,118.7	48,175.9	100,994.9	37,682.9	54,873.9	112,722.7	62,717.6	94,851.5	199,270.0
P-ACO	36,429.7	60,483.0	122,380.6	41,609.3	69,393.8	138,931.9	72,468.5	125,846.4	250,457.7
RIACO	34,623.8	49,977.3	99,867.5	40,063.6	57,503.1	114,114.7	69,267.9	102,961.5	204,917.9
NS–ACO	33,017.2	47,896.4	95,353.9	37,759.8	54,602.7	108,006.3	63,957.7	95,836.2	191,617.0
$f = 100, m$	0.1	0.5	0.9	0.1	0.5	0.9	0.1	0.5	0.9
AS	32,635.8	45,940.3	92,900.1	37,270.7	52,861.6	106,137.4	65,227.6	97,268.9	191,452.7
EAS	32,562.4	45,804.6	92,671.3	37,211.9	52,802.4	105,869.9	65,188.8	97,114.2	191,671.6
MMAS	32,561.8	45,578.8	89,411.9	36,360.5	51,880.2	102,142.5	62,584.2	92,862.7	178,715.9
ACS	32,674.3	46,203.2	97,857.3	36,870.1	52,690.8	108,795.9	61,933.1	93,555.3	190,426.9
P-ACO	31,914.2	46,336.8	93,095.9	35,475.9	52,746.4	105,596.6	62,334.9	95,370.2	187,557.8
RIACO	31,990.4	44,584.4	90,080.4	36,036.9	51,317.7	102,436.7	63,012.6	93,553.4	182,788.8
NS–ACO	31,300.7	43,493.0	87,680.1	34,999.2	49,906.0	99,384.8	60,223.7	89,988.3	175,112.3

To intuitively show the performance of each ACO, Figure 4.3 indicates the box-and-whisker diagrams of optimal solutions among seven ACOs in the $m = 0.5$, $f = 10$ or $f = 100$ of small-scale st70, $m = 0.1, f = 10$ or $f = 100$ of medium-scale kroA150 and large-scale pr299. The horizontal and vertical axes correspond to 7 ACOs and 30 independent optimal solutions, respectively. From Figure 4.3(a)–(d), the distribution of optimal solutions of NS–ACO obviously outperforms other ACOs, suggesting NS–ACO is superior in small-scale and medium-scale DTSP. Figure 4.3(e) exhibits ACS has the least distribution of solutions in large-scale DTSP with quick and slight changes, illustrating ACS is more proper than NS–ACO for this environment.

Table 4.3 Analytical results of Wilcoxon signed ranks test at a level of significance $\alpha = 0.05$ among seven ACOs

Algorithm	**eil51**			**st70**			**kroA100**		
$f = 10, m$	**0.1**	**0.5**	**0.9**	**0.1**	**0.5**	**0.9**	**0.1**	**0.5**	**0.9**
NS–ACO:AS	+	+	+	+	+	+	+	+	+
NS–ACO:EAS	+	+	+	+	+	+	+	+	+
NS–ACO:MMAS	+	+	+	+	+	+	+	+	+
NS–ACO:ACS	+	+	+	+	+	+	+	+	+
NS–ACO:P-ACO	+	+	+	+	+	+	+	+	+
NS–ACO:RIACO	+	+	+	+	+	+	+	+	+
$f = 100, m$	**0.1**	**0.5**	**0.9**	**0.1**	**0.5**	**0.9**	**0.1**	**0.5**	**0.9**
NS–ACO:AS	+	+	+	+	+	+	+	+	+
NS–ACO:EAS	+	+	+	+	+	+	+	+	+
NS–ACO:MMAS	+	+	+	+	+	+	+	+	+
NS–ACO:ACS	+	+	+	+	+	+	+	+	+
NS–ACO:P-ACO	+	+	+	+	+	+	+	+	+
NS–ACO:RIACO	+	+	+	+	+	+	+	+	+

Algorithm	**kroA150**			**kroA200**			**pr299**		
$f = 10, m$	**0.1**	**0.5**	**0.9**	**0.1**	**0.5**	**0.9**	**0.1**	**0.5**	**0.9**
NS–ACO:AS	+	+	+	+	+	+	+	+	+
NS–ACO:EAS	+	+	+	+	+	+	+	+	+
NS–ACO:MMAS	+	+	+	+	+	+	+	+	+
NS–ACO:ACS	+	+	+	−	+	+	−	−	+
NS–ACO:P-ACO	+	+	+	+	+	+	+	+	+
NS–ACO:RIACO	+	+	+	+	+	+	+	+	+
$f = 100, m$	**0.1**	**0.5**	**0.9**	**0.1**	**0.5**	**0.9**	**0.1**	**0.5**	**0.9**
NS–ACO:AS	+	+	+	+	+	+	+	+	+
NS–ACO:EAS	+	+	+	+	+	+	+	+	+
NS–ACO:MMAS	+	+	+	+	+	+	+	+	+
NS–ACO:ACS	+	+	+	+	+	+	+	+	+
NS–ACO:P-ACO	+	+	+	+	+	+	+	+	+
NS–ACO:RIACO	+	+	+	+	+	+	+	+	+

Figure 4.3(f) demonstrates NS–ACO recovers the best performance due to slow environmental changes.

Figure 4.4 depicts the convergence characteristics of solutions during seven ACOs in $m = 0.9$, $f = 10$ or $f = 100$ of kroA100, $m = 0.1$, $f = 10$ or $f = 100$ of eil51 and kroA200. The horizontal and vertical axes represent iterations and the best solutions, respectively. In order to distinctly observe the convergence results, five dynamic environments are only shown. For example, for $f = 10$, Figure 4.4(a), (c)

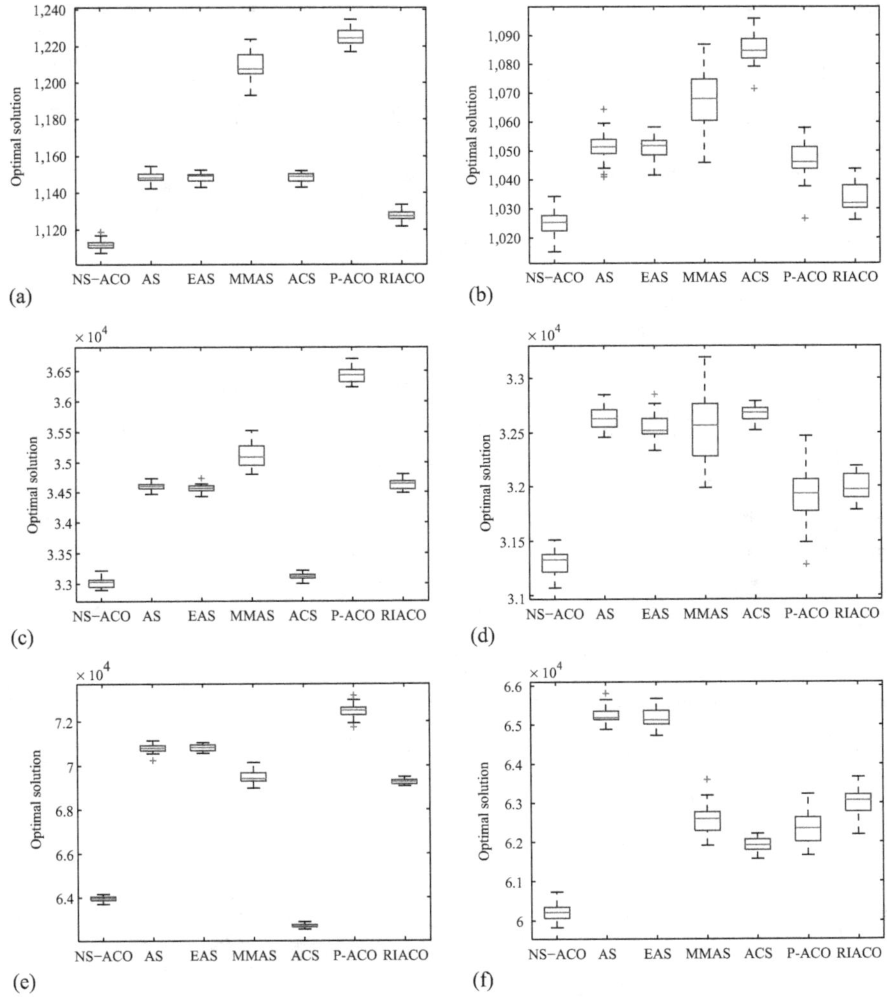

Figure 4.3 The box-and-whisker diagrams of optimal solutions found by seven ACOs in several DTSPs: (a) st70 $f = 10$ $m = 0.5$, (b) st70 $f = 100$ $m = 0.5$, (c) kroA150 $f = 10$ $m = 0.1$, (d) kroA150 $f = 100$ $m = 0.1$, (e) pr299 $f = 10$ $m = 0.1$ and (f) pr299 $f = 100$ $m = 0.1$

and (e) reveals the convergence of optimal solutions under the iterations from 350 to 400. For $f = 100$, Figure 4.4(b), (d) and (f) denote the convergence of optimal solutions under the iterations from 500 to 1,000. According to Figure 4.4(a)–(d), it can be seen that the convergence of NS–ACO is less than that of other ACOs, verifying the best property of NS–ACO in small-scale and medium-scale DTSP once again. Figure 4.4(e) and (f) elucidates the best convergence of ACS and NS–ACO under the large-scale DTSP with slight and quick or slow environmental changes, respectively.

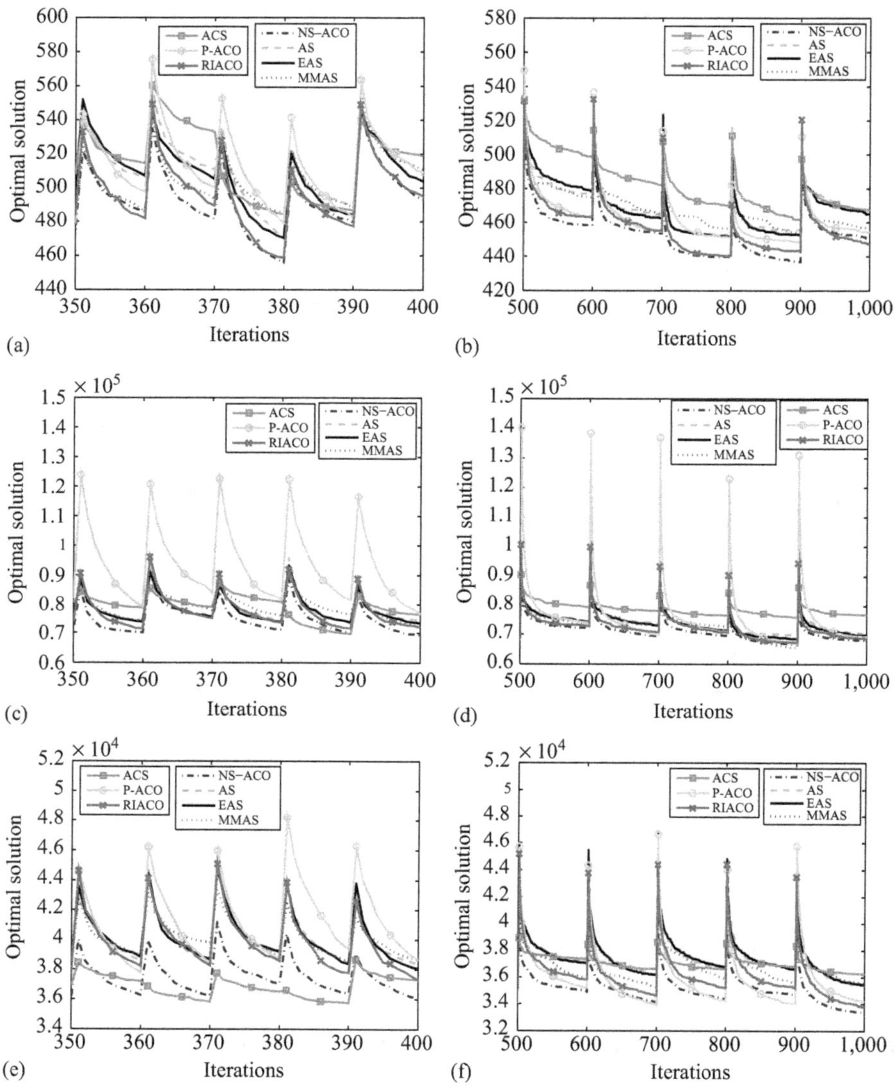

Figure 4.4 The convergence graphs of optimal solutions among seven ACOs for several DTSPs: (a) st70 $f = 10$ $m = 0.5$, (b) st70 $f = 100$ $m = 0.5$, (c) kroA150 $f = 10$ $m = 0.1$, (d) kroA150 $f = 100$ $m = 0.1$, (e) pr299 $f = 10$ $m = 0.1$ and (f) pr299 $f = 100$ $m = 0.1$

Based on the above experiment, NS–ACO has several characteristics shown as follows: (1) NS–ACO can effectively use neighborhood search to enhance the quality of solution in small-scale and medium-scale DTSP. (2) In large-scale DTSP, NS–ACO requires the sufficient time to optimize the structure of solutions within the extensive space. Thus, the performance of NS–ACO in slowly changing environments is better than that in quickly changing environments. (3) ACS outperforms NS–ACO via the global and local pheromone update in large-scale DTSP with quick and slight environmental changes. Nevertheless, for significant environmental changes, ACS is inefficient for tracing the optimal solution in terms of the previous pheromone and inferior to NS–ACO. (4) NS–ACO not only maintains the diversity of solutions but also manifests the remarkable robustness and self-adaptation by using the neighborhood search to address DTSP.

4.4.2 The experiment for DLRP

4.4.2.1 Experimental settings

The DLRPs are established by three benchmark TSP instances with random and cyclic traffic factors, where eil51, lin105 and kroB200 are employed to manifest the small-scale, medium-scale and large-scale city, respectively. The traffic factor Z is set in the interval $[1, 5]$, i.e. $Z_l = 1$ and $Z_h = 5$. The cyclic environments are composed of four cyclic traffic factors, indicating four distinctive circumstances formed by random traffic factor. The magnitude m of environmental changes are set to 0.1, 0.5 and 0.9 to denote low, medium and high changing degree, respectively. The frequency f of environmental changes are set to 10 and 100 to indicate the quick and slow changes, respectively. Consequently, in this experiment, there are 36 DLRPs constituted by two diverse frequencies, three different magnitudes and two kinds of traffic factors in three distinctive scales of cities.

The parameters of KACO with three immigrants are adopted as follows: $\alpha = 1$, $\beta = 5$, $K_s = 6$, $K_l = 4$, $r = 0.4$, $p_m = 0.02$, which have demonstrated the best performance for DTSP [4]. The initial pheromone is set to $\tau_0 = 1/L^{nn}$, where L^{nn} is the length of tour constructed by the nearest neighborhood heuristic. The number M of ants is 30. The capacity C of each vehicle is 1,000, and each customer's demand R is 100. The experiment is totally executed 30 times and 1,000 iterations in each time. In order to evaluate the performance of algorithms in dynamic environments, the experimental results are shown by the overall average optimal solution in all runs, described as follow:

$$P^{bs} = \sum_{k=1}^{K} \left[\frac{1}{U} \sum_{t=1}^{U} \left(\frac{1}{V} \sum_{i=1}^{V} f_{kti}^{bs} \right) \right] \tag{4.15}$$

where K is the number of classes. U and V indicate the iterations and executing times, respectively. f_{kti}^{bs} represents the optimal solution in the tth iteration of ith executing time of the class k.

4.4.2.2 Experimental analysis

To confirm the feature of clustering algorithm, the comparison between KACO and the ACO [4] without using the improved K-means algorithm (WKACO) with three immigrants is first conducted. The experimental results obtained by (4.15) in random and cyclic environments are listed in Tables 4.4 and 4.5, respectively. The corresponding statistical results of Wilcoxon signed ranks test at a level of significance $\alpha = 0.05$ in random and cyclic environments are shown in Tables 4.6 and 4.7.

Table 4.4 Experimental results of KACO and WKACO with three immigrants in random environments

Algorithm $f = 10, m$	eil51			lin105			kroB200		
	0.1	0.5	0.9	0.1	0.5	0.9	0.1	0.5	0.9
	7 depots and 7 vehicles			6 depots and 13 vehicles			11 depots and 25 vehicles		
KACO–RI	517.2	755.1	1,380.9	20,713.3	29,102.4	51,594.2	43,527.9	61,975.4	111,063.3
KACO–EI	521.6	770.1	1,414.5	21,535.5	30,127.4	53,101.2	45,701.1	63,522.2	113,602.3
KACO–MI	519.0	765.4	1,400.1	20,918.1	29,696.5	52,382.3	44,213.0	62,869.7	112,678.8
	1 depot and 6 vehicles			1 depots and 11 vehicles			1 depots and 20 vehicles		
WKACO–RI	658.0	924.3	1,663.3	34,177.7	49,952.9	88,443.5	85,574.1	125,959.6	217,836.5
WKACO–EI	658.3	922.0	1,655.1	34,229.2	49,193.3	87,102.2	84,291.5	123,394.3	214,484.3
WKACO–MI	654.9	923.4	1,658.5	33,615.5	49,160.1	87,095.5	83,569.8	123,443.2	214,323.3
$f = 100, m$	0.1	0.5	0.9	0.1	0.5	0.9	0.1	0.5	0.9
	7 depots and 7 vehicles			6 depots and 13 vehicles			11 depots and 25 vehicles		
KACO–RI	516.5	734.2	1,361.2	19,803.8	27,055.4	49,242.8	41,412.0	55,758.7	104,510.2
KACO–EI	518.3	743.6	1,382.6	20,418.0	28,186.4	51,217.2	42,808.7	57,635.6	108,124.0
KACO–MI	517.4	742.3	1,378.5	20,149.9	27,957.0	50,946.3	42,319.0	57,371.8	107,662.6
	1 depot and 6 vehicles			1 depots and 11 vehicles			1 depots and 20 vehicles		
WKACO–RI	616.3	842.9	1,566.2	32,171.4	44,855.4	80,428.1	80,823.5	112,251.2	198,169.6
WKACO–EI	618.0	853.2	1,587.6	32,621.6	45,776.7	80,864.7	81,009.3	112,241.8	196,741.3
WKACO–MI	617.5	856.8	1,585.0	32,486.4	45,531.7	80,725.2	80,516.8	112,835.8	197,165.1

From Tables 4.4 and 4.5, the utilization of depots and vehicles can be obviously observed. The number of vehicles depend on both the capacity of each vehicle and customers' demand, while the selection of depots are determined by the improved K-means algorithm. For KACO, the number of depots in eil51, lin105 and kroB200 are 7, 6 and 11 respectively. However, WKACO have only one depot. Furthermore, the number of vehicles determined by KACO in three different scales of cities are 7, 13 and 25, respectively, while WKACO correspondingly requires 6, 11 and 20 vehicles, respectively. It is evident that the number of depots obtained by KACO are clearly more than WKACO via the clustering algorithm, but the difference of number of vehicles are nearly slight. The cost of using depots and vehicles are almost one-time expenses; hence, the average length of vehicle routes can significantly influence the property of algorithms. In addition, Tables 4.4 and 4.5 show the total average solutions found by KACO are notably less than those found by WKACO in the whole DLRPs, meaning the clustering method is beneficial for enhancing the performance of algorithm.

Figure 4.5 shows the box-and-whisker diagrams of optimal solutions found by KACO and WKACO with three immigrants in $f = 10$, $m = 0.1$ of random and cyclic environments, where RI, EI and MI of the horizontal axis indicate random immigrants,

Table 4.5 Experimental results of KACO and WKACO with three immigrants in cyclic environments

Algorithm $f = 10, m$	eil51			lin105			kroB200		
	0.1	**0.5**	**0.9**	**0.1**	**0.5**	**0.9**	**0.1**	**0.5**	**0.9**
	7 depots and 7 vehicles			**6 depots and 13 vehicles**			**11 depots and 25 vehicles**		
KACO–RI	530.1	730.1	1,360.4	20,406.1	29,197.1	51,357.6	43,246.8	59,311.1	113,632.4
KACO–EI	535.7	742.8	1,395.7	20,905.4	30,306.3	52,918.9	45,072.2	61,111.8	116,241.1
KACO–MI	529.4	725.7	1,365.4	20,019.2	28,455.3	49,876.6	42,763.5	57,298.3	110,371.3
	1 depot and 6 vehicles			**1 depots and 11 vehicles**			**1 depots and 20 vehicles**		
WKACO–RI	665.4	894.6	1,669.6	34,505.9	49,603.5	85,318.7	85,586.0	126,594.6	219,573.9
WKACO–EI	665.0	893.6	1,662.0	34,449.4	48,652.2	83,642.5	85,066.5	123,822.0	215,919.8
WKACO–MI	648.5	848.2	1,611.7	33,162.3	46,660.8	79,927.3	82,000.2	117,606.5	207,429.4
$f = 100, m$	**0.1**	**0.5**	**0.9**	**0.1**	**0.5**	**0.9**	**0.1**	**0.5**	**0.9**
	7 depots and 7 vehicles			**6 depots and 13 vehicles**			**11 depots and 25 vehicles**		
KACO–RI	524.5	698.6	1,332.3	19,746.7	27,179.6	47,505.4	41,416.5	54,647.3	104,991.6
KACO–EI	525.9	705.4	1,355.1	20,112.7	28,089.4	49,549.5	42,628.8	56,835.5	108,866.5
KACO–MI	524.9	702.2	1,347.9	19,918.0	27,697.8	48,781.5	42,005.2	55,848.3	107,571.1
	1 depot and 6 vehicles			**1 depots and 11 vehicles**			**1 depots and 20 vehicles**		
WKACO–RI	627.2	824.7	1,570.2	32,261.4	44,587.5	78,289.7	81,042.8	111,642.7	200,979.5
WKACO–EI	630.1	835.5	1,582.7	32,790.7	45,001.2	78,562.7	81,102.7	111,363.8	201,458.9
WKACO–MI	627.5	827.6	1,575.7	32,264.2	44,670.0	77,626.3	80,042.2	109,612.5	199,717.1

elitism-based immigrants and memory-based immigrants, respectively, and the vertical axis signifies the average optimal solution P^{bs}. It is clear that the optimal solutions acquired by KACO are obviously less than those acquired by WKACO, and the difference between them gradually intensifies along with the increase of scale of city in terms of Figure 4.5, suggesting KACO outperforms WKACO, and especially in large-scale DLRP.

In Tables 4.6 and 4.7, the symbols + and − denote the former is superior and inferior to the latter between two comparative algorithms, respectively. The symbol ∼ means no significance. From Table 4.6, KACO–RI performs best in all random environments, while the performance of WKACO–RI observably descends in random environments especially with the quick environmental changes $f = 10$. Moreover, WKACO–RI reacts poorly for resolving the large-scale DLRP. The reason may be two parts. On the one hand, the improved K-means algorithm optimizes the construction of solutions to reinforce not only the performance of random immigrants but also the diversity of population so that KACO–RI is appropriate for searching in random environments. Nevertheless, the construction of solutions obtained by WKACO–RI is much more complicated and has no benefit for fast achieving the optimal solution. To be specific, KACO–RI utilizes the diversity of population and optimized solution

Table 4.6 The Wilcoxon signed ranks test of KACO and WKACO with three immigrants in random environments

Algorithm	**eil51**			**lin105**			**kroB200**		
$f = 10, m$	**0.1**	**0.5**	**0.9**	**0.1**	**0.5**	**0.9**	**0.1**	**0.5**	**0.9**
KACO–RI:KACO–EI	+	+	+	+	+	+	+	+	+
KACO–EI:KACO–MI	−	−	−	−	−	−	−	−	−
KACO–MI:KACO–RI	−	−	−	−	−	−	−	−	−
WKACO–RI:WKACO–EI	∼	−	−	∼	−	−	−	−	−
WKACO–EI:WKACO–MI	−	∼	+	−	∼	∼	−	∼	∼
WKACO–MI:WKACO–RI	+	∼	+	+	+	+	+	+	+
$f = 100, m$	**0.1**	**0.5**	**0.9**	**0.1**	**0.5**	**0.9**	**0.1**	**0.5**	**0.9**
KACO–RI:KACO–EI	+	+	+	+	+	+	+	+	+
KACO–EI:KACO–MI	−	∼	−	−	−	−	−	−	−
KACO–MI:KACO–RI	−	−	−	−	−	−	−	−	−
WKACO–RI:WKACO–EI	∼	+	+	+	+	∼	∼	∼	−
WKACO–EI:WKACO–MI	∼	∼	∼	∼	∼	∼	−	∼	∼
WKACO–MI:WKACO–RI	∼	−	−	−	−	∼	∼	∼	+

Table 4.7 The Wilcoxon signed ranks test of KACO and WKACO with three immigrants in cyclic environments

Algorithm	**eil51**			**lin105**			**kroB200**		
$f = 10, m$	**0.1**	**0.5**	**0.9**	**0.1**	**0.5**	**0.9**	**0.1**	**0.5**	**0.9**
KACO–RI:KACO–EI	+	+	+	+	+	+	+	+	+
KACO–EI:KACO–MI	−	−	−	−	−	−	−	−	−
KACO–MI:KACO–RI	+	+	−	+	+	+	+	+	+
WKACO–RI:WKACO–EI	∼	∼	−	∼	−	−	−	−	−
WKACO–EI:WKACO–MI	−	−	−	−	−	−	−	−	−
WKACO–MI:WKACO–RI	+	+	+	+	+	+	+	+	+
$f = 100, m$	**0.1**	**0.5**	**0.9**	**0.1**	**0.5**	**0.9**	**0.1**	**0.5**	**0.9**
KACO–RI:KACO–EI	+	+	+	+	+	+	+	+	+
KACO–EI:KACO–MI	−	−	−	−	−	−	−	−	−
KACO–MI:KACO–RI	−	−	−	−	−	−	−	−	−
WKACO–RI:WKACO–EI	+	+	+	+	+	∼	∼	∼	∼
WKACO–EI:WKACO–MI	∼	+	∼	−	−	−	−	−	−
WKACO–MI:WKACO–RI	∼	∼	∼	∼	∼	+	+	+	+

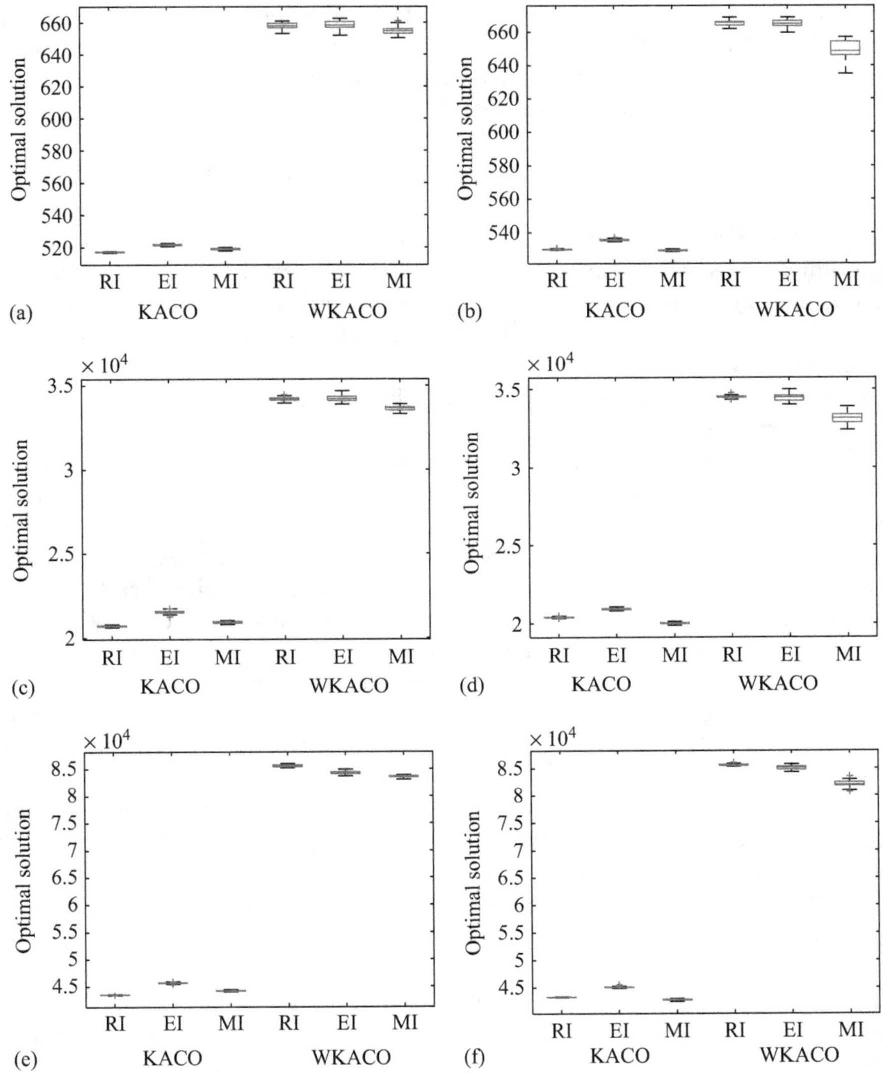

Figure 4.5 The box-and-whisker diagrams of optimal solutions found by KACO and WKACO with three immigrants in several DLRPs: (a) eil51 random environment (f = 10 m = 0.1), (b) eil51 cyclic environment (f = 10 m = 0.1), (c) lin105 random environment (f = 10 m = 0.1), (d) lin105 cyclic environment (f = 10 m = 0.1), (e) kroB200 random environment (f = 10 m = 0.1) and (f) kroB200 cyclic environment (f = 10 m = 0.1)

structure to search more areas for discovering a better solution, whereas the solution structure of WKACO–RI is unsuccessful for enhancing the diversity and contrarily damages the current pheromone on the optimal path so that WKACO–RI has insufficient time to trace the optimal solution. On the other hand, the diversity formed by random immigrants is more suitable than transmitting previous information in random environments. Accordingly, KACO–RI is superior to KACO–EI and KACO–MI in random environments.

Table 4.7 depicts the statistical results of Wilcoxon signed ranks test between two algorithms in cyclic environments. Because the similar environmental information can be used by the memory-based immigrants in quickly cyclic environments, the memory-based immigrants outperform random immigrants and elitism-based immigrants owing to the special function of long-term memory k_{long}. When the previous environment occurs once again, the memory-based immigrants can adopt the optimal solution stored in k_{long} to fast match the current environment. This mechanism can effectively intensify the efficiency of algorithm for quickly cyclic environments. However, there is an exception in $f = 10$, $m = 0.9$ of eil51, where KACO–RI performs best. This is because the diversity of random immigrants is more proper for small-scale city with quick and significant environmental changes. Besides, KACO–RI is superior to KACO–EI and KACO–MI in slowly cyclic environments $f = 100$ because the diversity generated by KACO–RI expands the search area and the algorithm acquires adequate time to ultimately find a better solution. That is to say, the new solutions obtained by KACO–RI may be better than those in k_{long} under slowly changing environments. Meanwhile, WKACO–RI exhibits the similar property between slowly random and cyclic environments in terms of its complex solution structure.

In Figures 4.6 and 4.7, the convergence characteristics of optimal solutions obtained by KACO with three immigrants are displayed in $m = 0.9$, $f = 10$ or $f = 100$ of random environments and $m = 0.5, f = 10$ or $f = 100$ of cyclic environments. The horizontal and vertical axes manifest the iterations and optimal solutions, respectively. Figure 4.6 demonstrates the performance of KACO–RI is the best in random environments. It can be observed that the KACO–MI outperforms KACO–RI and KACO–EI in quickly cyclic environments according to Figure 4.7(a), (c) and (e), while in Figure 4.7(b), (d) and (f), the performance of KACO–RI is superior to others in slowly cyclic environments. Eventually, from Figures 4.6 and 4.7, it is also obvious that the performance of KACO–EI is the worst in the whole environments.

KACO primarily utilizes clustering method to resolve the determination of depots in consideration of each customer's position and demand. The improved K-means algorithm optimizes the construction of solution to enhance the property of immigrants. Thus, to intuitively show the performance of clustering method, Figures 4.8 and 4.9 are plotted to illustrate the total lengths of vehicle routes in lin105 with $m = 0.9$, $f = 10$ of random environments and kroB200 with $m = 0.5, f = 10$ of cyclic environments according to KACO and WKACO with three immigrants, respectively, where d denotes the depot. From two figures, it can be declared that the total lengths of vehicle routes obtained by KACO are much less than those by WKACO, suggesting

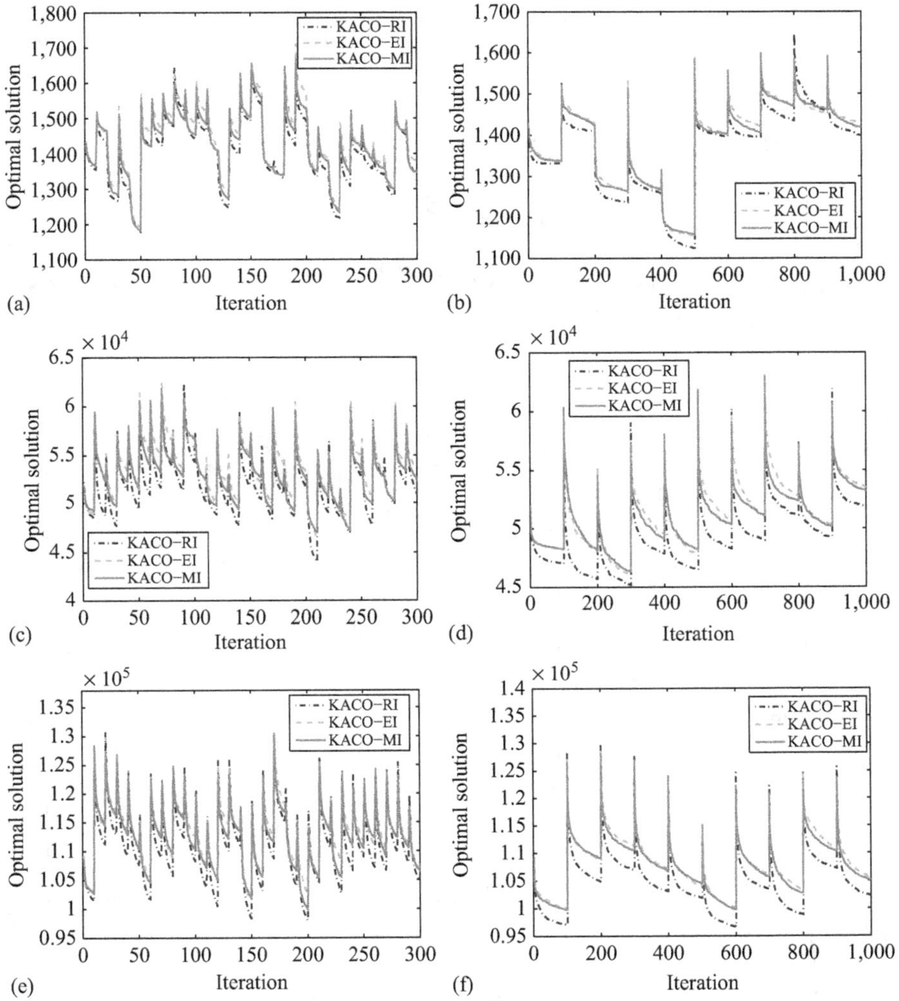

Figure 4.6 The convergence curves of optimal solutions obtained by KACO with three immigrants in random environments: (a) eil51 (f = 10 m = 0.9), (b) eil51 (f = 100 m = 0.9), (c) lin105 (f = 10 m = 0.9), (d) lin105 (f = 100 m = 0.9), (e) kroB200 (f = 10 m = 0.9) and (f) kroB200 (f = 100 m = 0.9)

the clustering algorithm not only remarkably diminishes the overall length of vehicle routes but also changes the structure of solution to reinforce the performance of immigrants for gaining a better solution.

To further elaborate the characteristics of KACO with three immigrants, the comparative experiment between KACO and other heuristics including AS, EAS, MMAS, ACS, P-ACO, SA [16] and GA [24] is carried out. SA forms the initial

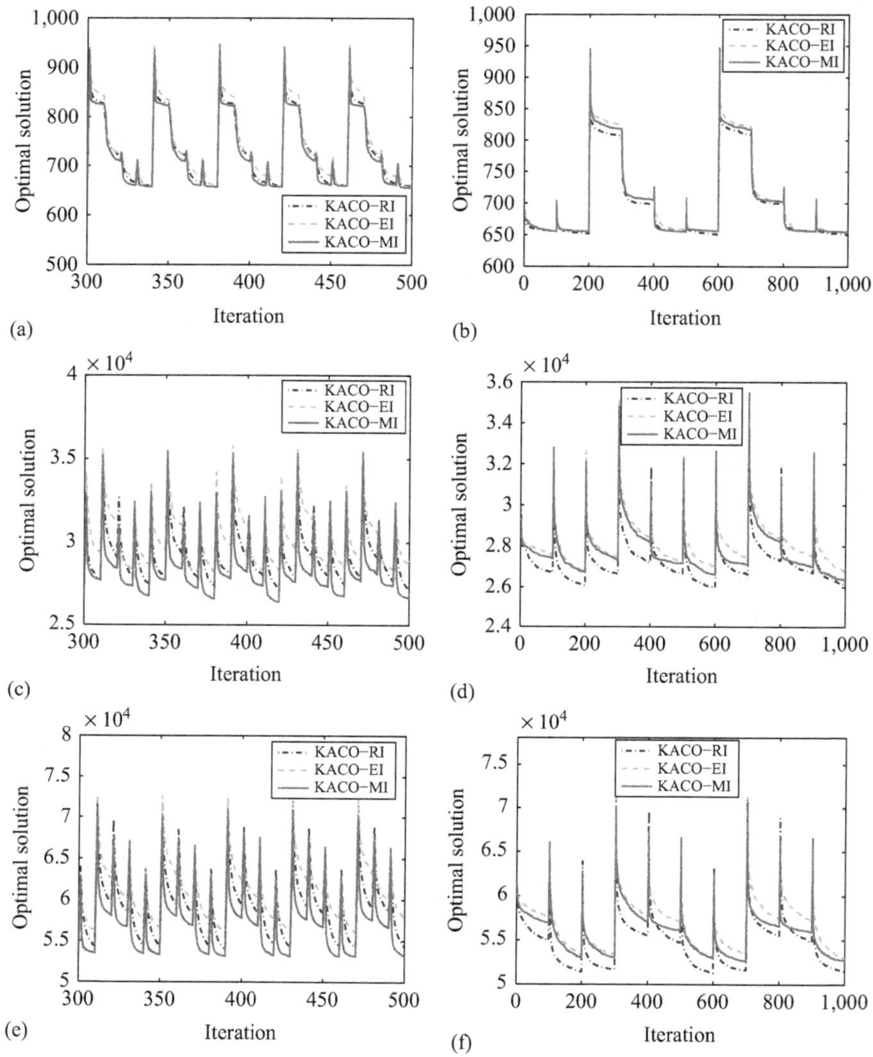

Figure 4.7 The convergence curves of optimal solutions obtained by KACO with three immigrants in cyclic environments: (a) eil51 ($f = 10$ $m = 0.5$), (b) eil51 ($f = 100$ $m = 0.5$), (c) lin105 ($f = 10$ $m = 0.5$), (d) lin105 ($f = 100$ $m = 0.5$), (e) kroB200 ($f = 10$ $m = 0.5$) and (f) kroB200 ($f = 100$ $m = 0.5$)

population via the greedy algorithm and uses local search to improve solutions for resolving the DLRP. For GA, the initial population is randomly generated and the new offspring is produced by the crossover between two parents. Thereafter, the mutation is implemented to randomly swap two positions of genes in the new offspring three

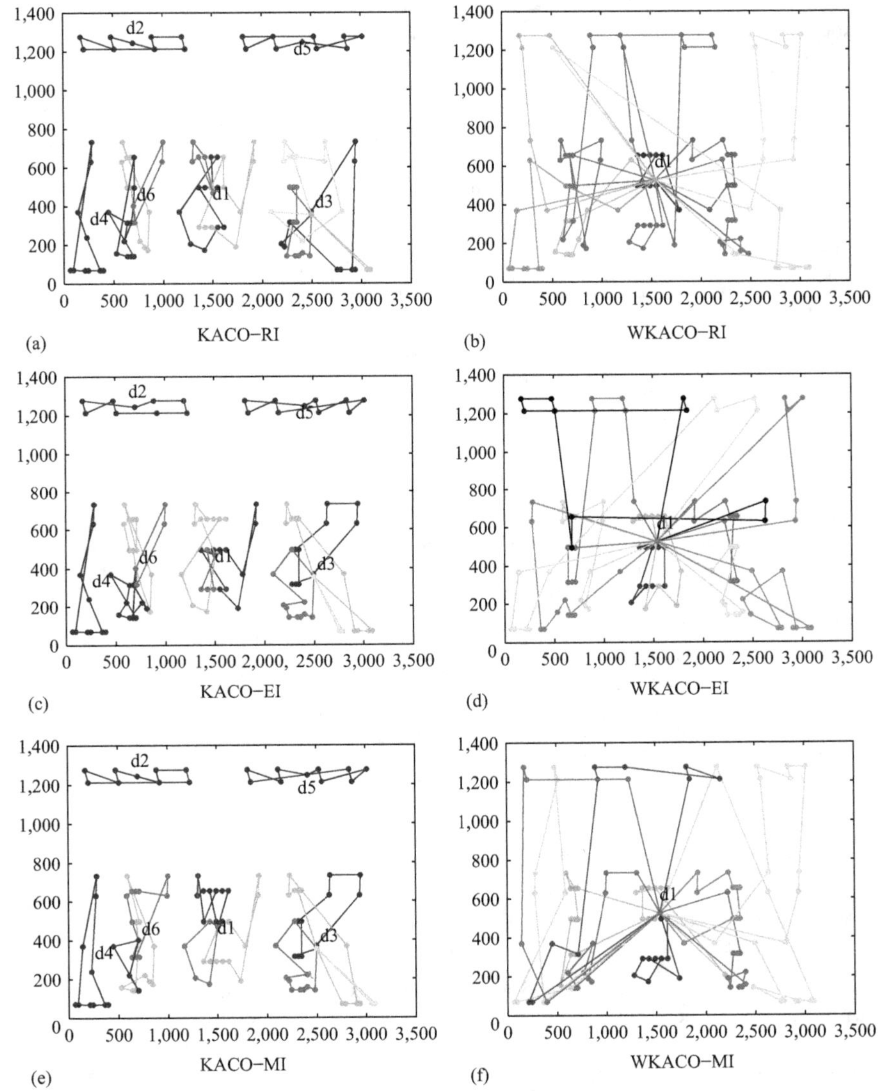

Figure 4.8 *The vehicle routes constructed by KACO and WKACO with three immigrants in lin105 with random environments: (a) total length = 47,096 vehicles = 13, (b) total length = 94,254 vehicles = 11, (c) total length = 49,653 vehicles = 13, (d) total length = 94,156 vehicles = 11, (e) total length = 49,590 vehicles = 13 and (f) total length = 91,472 vehicles = 11*

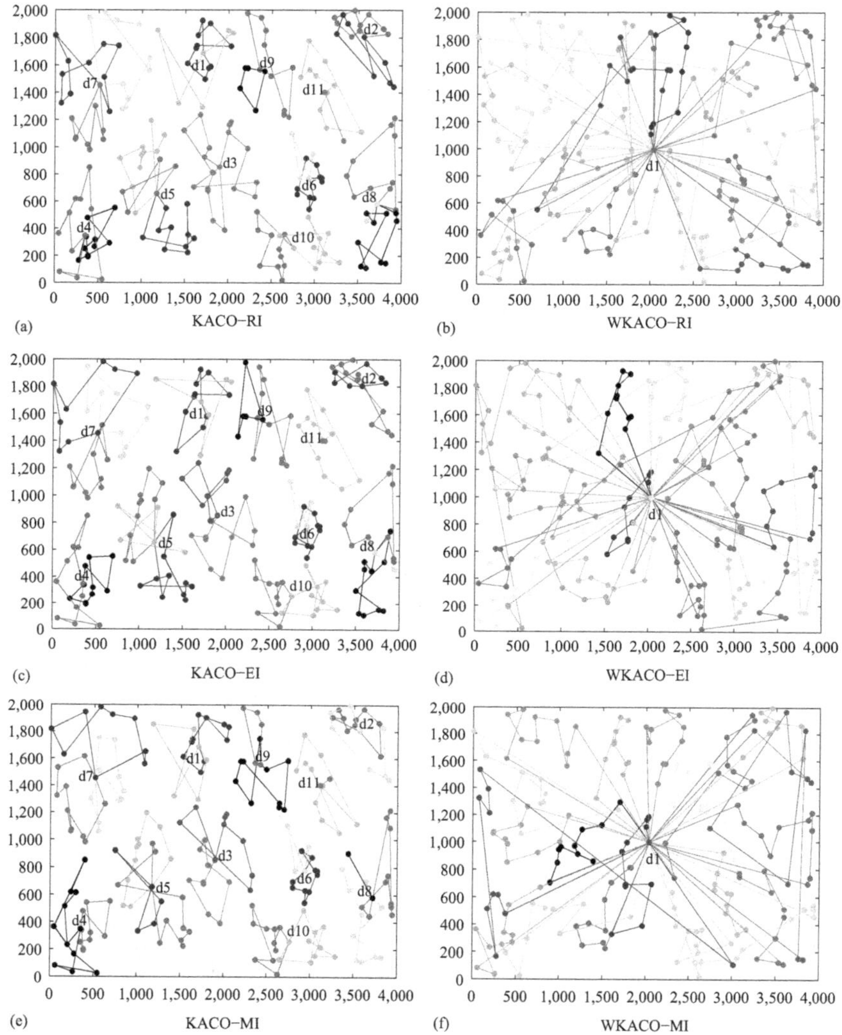

Figure 4.9 The vehicle routes constructed by KACO and WKACO with three immigrants in kroB200 with cyclic environments: (a) total length = 60,120 vehicles = 25, (b) total length = 121,172 vehicles = 20, (c) total length = 62,559 vehicles = 25, (d) total length = 117,157 vehicles = 20, (e) total length = 57,660 vehicles = 25 and (f) total length = 114,727 vehicles = 20

times in terms of a mutation rate. The parameter settings are listed in Table 4.8. In SA, T_0 and T_f indicate the beginning and ending temperature, respectively. γ denotes the cooling coefficient. K is the Boltzmann constant. N manifests the tolerable time where there is no better solution. I signifies iterations in every temperature. In GA,

Table 4.8 Parameter settings for several ACOs, SA and GA

Algorithm	α	β	ρ	φ	q_0	K_s	K_l	r	M
KACO–RI	1	5	–	–	–	6	–	0.4	30
KACO–EI	1	5	–	–	–	6	–	0.4	30
KACO–MI	1	5	–	–	–	6	4	0.4	30
AS	1	5	0.5	–	–	–	–	–	30
EAS	1	5	0.5	–	–	–	–	–	30
MMAS	1	5	0.6	–	–	–	–	–	30
ACS	1	5	0.1	0.1	0.9	–	–	–	30
P-ACO	1	5	–	–	0.9	–	4	–	30
	T_0	T_f	K	γ	N	I			
SA	3,000	$1e^{-8}$	1/9	0.95	100	5,000			
	P_c	P_m	P_l	T					
GA	0.8	0.05	30	1,000					

P_c and P_m represent the crossover and mutation rate, respectively. P_l is the scale of population. T indicates the iterations. The whole algorithms are also implemented 30 runs and 1,000 iterations in each run. For DLRP, SA and GA adopt the global restart mechanism to adapt to the environmental changes.

Table 4.9 displays the experimental results between KACO with three immigrants and other heuristics in lin105 with random and cyclic environments. Tables 4.10 and 4.11 indicate the corresponding Wilcoxon signed ranks test among all the algorithms. In Table 4.10, it is obvious that the performances of KACO with three immigrants are superior to other heuristics. From Table 4.11, it can be observed that ACS outperforms other heuristics due to its global and local search mechanism in random and cyclic environments. P-ACO which is inferior to ACS can perform best with slow environmental changes or quick and slight environmental changes because the population list can increase the diversity of solutions for adapting to dynamic environments. Meanwhile, except for ACS, the performance of EAS is superior with quick and significant environmental changes because the quick changes interfere the pheromone of optimal routes so that EAS avoids trapping into the local convergence. In slowly changing environments, MMAS is only superior to AS, illustrating that the pheromone interval is not suitable for quick environmental changes. This is because the pheromone on the path requires adequate time to volatilize. Even then, the previous pheromone may provide useless guideline for tracing new optimal solution in current environment.

To observe the convergence property of all the algorithms, Figures 4.10 and 4.11 are plotted to indicate the convergence curve of each algorithm in $m = 0.5$, $f = 10$ or $f = 100$ of random environments and $m = 0.9$, $f = 10$ or $f = 100$ of cyclic environments, respectively. From Figure 4.10, it is clear that the KACO with three immigrants can quickly trace the optimal solution and obviously outperform

Table 4.9 Experimental results among several algorithms in lin105 with random and cyclic environments

Algorithm $f = 10, m$	Random environments			Cyclic environments		
	0.1	**0.5**	**0.9**	**0.1**	**0.5**	**0.9**
KACO–RI	20,713.3	29,102.4	51,594.2	20,406.1	29,197.1	51,357.6
KACO–EI	21,535.5	30,127.4	53,101.2	20,905.4	30,306.3	52,918.9
KACO–MI	20,918.1	29,696.5	52,382.3	20,019.2	28,455.3	49,876.6
AS	23,446.5	33,068.7	57,637.8	22,539.6	33,267.9	57,390.8
EAS	23,170.3	32,968.8	57,076.5	22,181.9	33,226.7	57,280.2
MMAS	23,581.6	35,493.0	59,118.9	23,579.7	35,057.0	59,884.7
ACS	21,059.3	30,571.5	54,799.0	21,105.0	30,773.7	54,169.8
P-ACO	23,016.1	33,743.9	58,795.3	21,842.7	33,151.7	59,189.2
SA	23,839.4	38,868.2	60,855.7	23,545.5	37,748.2	61,649.1
GA	38,942.6	68,985.0	113,645.2	38,746.5	68,567.5	114,889.3
$f = 100, m$	**0.1**	**0.5**	**0.9**	**0.1**	**0.5**	**0.9**
KACO–RI	19,803.8	27,055.4	49,242.8	19,746.7	27,179.6	47,505.4
KACO–EI	20,418.0	28,186.4	51,217.2	20,112.7	28,089.4	49,549.5
KACO–MI	20,149.9	27,957.0	50,946.3	19,918.0	27,697.8	48,781.5
AS	23,116.3	32,508.6	57,723.5	22,908.4	31,967.2	57,197.3
EAS	22,792.2	31,724.5	56,880.9	21,766.7	31,491.5	55,864.9
MMAS	23,047.7	32,012.5	57,406.2	21,851.7	31,586.4	55,967.0
ACS	20,564.8	29,124.6	53,159.7	20,617.8	28,942.7	51,729.0
P-ACO	22,245.9	31,468.8	56,587.9	21,503.2	30,901.8	55,267.3
SA	23,198.1	38,377.1	59,350.4	23,323.9	38,778.2	59,644.4
GA	36,853.1	64,540.1	105,967.7	36,779.7	63,258.6	106,895.1

other ACOs in dynamic environments. Figure 4.11 shows the performance among KACO with three immigrants, SA and GA. The observation that the performance of SA is just better than that of GA demonstrates the initial population generated by the greedy method is quicker to acquire a near-optimal solution than that by the random method. Unfortunately, the performances of SA and GA are much worse than those of ACOs owing to their various search mechanisms and maladjustment for dynamic environments. Therefore, the experiments manifest KACO with three immigrants has advantage to track the optimal solution for DLRP, meaning it is an efficient approach.

Based on the above discussion and analysis, KACO with three immigrants schemes has several properties described as follows: (1) KACO using the clustering algorithm obviously outperforms WKACO without utilizing the clustering algorithm since the improved K-means algorithm optimizes the construction of solution to intensify the influence of three immigrants schemes for DLRP; (2) KACO with random immigrants can effectively process the random environments and slowly cyclic environments due to the diversity generated by random immigrants; (3) KACO with

Table 4.10 The comparison of Wilcoxon signed ranks test between KACO with three immigrants and other heuristics in lin105 with random and cyclic environments

Algorithm m	Random environments						Cyclic environments					
	$f = 10$			$f = 100$			$f = 10$			$f = 100$		
	0.1	0.5	0.9	0.1	0.5	0.9	0.1	0.5	0.9	0.1	0.5	0.9
KACO–RI:AS	+	+	+	+	+	+	+	+	+	+	+	+
KACO–RI:EAS	+	+	+	+	+	+	+	+	+	+	+	+
KACO–RI:MMAS	+	+	+	+	+	+	+	+	+	+	+	+
KACO–RI:ACS	+	+	+	+	+	+	+	+	+	+	+	+
KACO–RI:P-ACO	+	+	+	+	+	+	+	+	+	+	+	+
KACO–RI:SA	+	+	+	+	+	+	+	+	+	+	+	+
KACO–RI:GA	+	+	+	+	+	+	+	+	+	+	+	+
KACO–EI:AS	+	+	+	+	+	+	+	+	+	+	+	+
KACO–EI:EAS	+	+	+	+	+	+	+	+	+	+	+	+
KACO–EI:MMAS	+	+	+	+	+	+	+	+	+	+	+	+
KACO–EI:ACS	−	+	+	+	+	+	+	+	+	+	+	+
KACO–EI:P-ACO	+	+	+	+	+	+	+	+	+	+	+	+
KACO–EI:SA	+	+	+	+	+	+	+	+	+	+	+	+
KACO–EI:GA	+	+	+	+	+	+	+	+	+	+	+	+
KACO–MI:AS	+	+	+	+	+	+	+	+	+	+	+	+
KACO–MI:EAS	+	+	+	+	+	+	+	+	+	+	+	+
KACO–MI:MMAS	+	+	+	+	+	+	+	+	+	+	+	+
KACO–MI:ACS	+	+	+	+	+	+	+	+	+	+	+	+
KACO–MI:P-ACO	+	+	+	+	+	+	+	+	+	+	+	+
KACO–MI:SA	+	+	+	+	+	+	+	+	+	+	+	+
KACO–MI:GA	+	+	+	+	+	+	+	+	+	+	+	+

memory-based immigrants can primely handle the quickly cyclic environments in terms of the long-term memory; (4) KACO with elitism-based immigrants performs worst during three immigrants even if it can transmit the previous environmental information; (5) KACO with three immigrants compared with other heuristics is verified to be an efficient and superior method for DLRP.

4.5 Conclusions

In this chapter, two types of dynamic combinatorial optimization problems including DTSP and DLRP composed of traffic factors are investigated. The experiments manifest the superiority of NS–ACO for DTSP and KACO with three immigrants schemes for DLRP, respectively. These can be summarized as follows: (1) For DTSP, NS–ACO applies the neighborhood search involving swap, insertion and 2-opt operations for

Table 4.11 The comparison of Wilcoxon signed ranks test during other heuristics in lin105 with random and cyclic environments

Algorithm	**Random environments**						**Cyclic environments**					
m	$f = 10$			$f = 100$			$f = 10$			$f = 100$		
	0.1	**0.5**	**0.9**	**0.1**	**0.5**	**0.9**	**0.1**	**0.5**	**0.9**	**0.1**	**0.5**	**0.9**
AS:EAS	−	∼	−	−	−	−	−	∼	∼	−	−	−
AS:MMAS	+	+	+	∼	−	∼	+	+	+	−	−	−
AS:ACS	−	−	−	−	−	−	−	−	−	−	−	−
AS:P-ACO	−	+	+	−	−	−	−	∼	+	−	−	−
AS:SA	∼	+	+	∼	+	+	+	+	+	+	+	+
AS:GA	+	+	+	+	+	+	+	+	+	+	+	+
EAS:MMAS	+	+	+	+	+	+	+	+	+	∼	∼	∼
EAS:ACS	−	−	−	−	−	−	−	−	−	−	−	−
EAS:P-ACO	−	+	+	−	∼	∼	−	∼	+	−	−	−
EAS:SA	+	+	+	∼	+	+	+	+	+	+	+	+
EAS:GA	+	+	+	+	+	+	+	+	+	+	+	+
MMAS:ACS	−	−	−	−	−	−	−	−	−	−	−	−
MMAS:P-ACO	−	−	−	−	−	−	−	−	−	−	−	−
MMAS:SA	∼	+	+	∼	+	+	∼	+	+	+	+	+
MMAS:GA	+	+	+	+	+	+	+	+	+	+	+	+
ACS:P-ACO	+	+	+	+	+	+	+	+	+	+	+	+
ACS:SA	+	+	+	+	+	+	+	+	+	+	+	+
ACS:GA	+	+	+	+	+	+	+	+	+	+	+	+
P-ACO:SA	+	+	+	+	+	+	+	+	+	+	+	+
P-ACO:GA	+	+	+	+	+	+	+	+	+	+	+	+
SA:GA	+	+	+	+	+	+	+	+	+	+	+	+

enhancing the quality of solution in dynamic environments. The experimental results indicate NS–ACO outperforms other ACOs in small-scale and medium-scale DTSPs, whereas the performance of NS–ACO in slow changing environments is better than that in quick changing environments. The overall framework of NS–ACO reinforces both the diversity of population and the exploring ability of solutions. (2) For DLRP, KACO with three immigrants called KACO–RI, KACO–EI and KACO–MI is used, respectively. The algorithm first employs an improved K-means clustering method to adaptively establish several depots and corresponding service customers and then utilizes random immigrants, elitism-based immigrants and memory-based immigrants to tackle the vehicle routes in dynamic environments. The experiments demonstrate that the utilization of clustering algorithm can notably enhance the performance of KACO and adjust the construction of solution to further perfect the property of immigrants. KACO–RI shows the best performance in random environments and slowly cyclic environments. KACO–MI is the fittest for quickly cyclic environments. Even

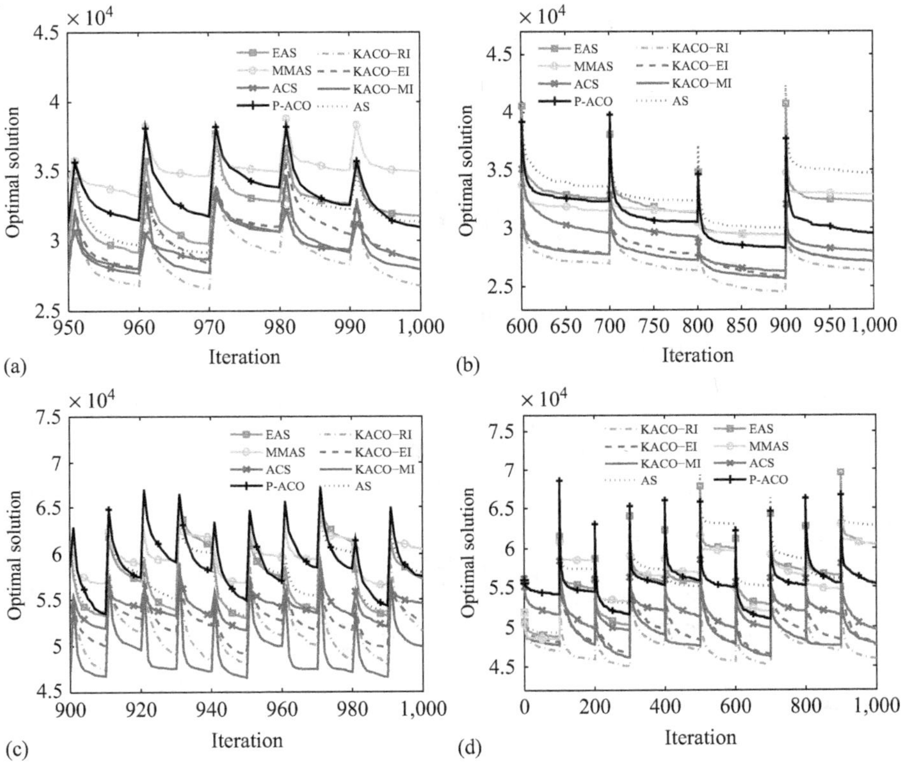

Figure 4.10 The convergence curves of optimal solutions obtained by KACO with three immigrants and other ACO algorithms in lin105 with random and cyclic environments: (a) random environment ($f = 10$ $m = 0.5$), (b) random environment ($f = 100$ $m = 0.5$), (c) cyclic environment ($f = 10$ $m = 0.9$) and (d) cyclic environment ($f = 100$ $m = 0.9$)

though the performance of KACO–EI is the worst among three immigrants, it still outperforms other conventional ACOs. The comparison between KACO and other heuristics obviously denotes the validity and high efficiency of KACO with three immigrants schemes for handling DLRP.

This chapter brings new inspiration to resolve dynamic combinatorial optimization problems. For the future work, several researches worth concerning as follows: (1) Some variants of TSP and LRP in dynamic environments should be considered. For instance, the TSP with time windows and the LRP with different customers' demand, various capacities of vehicles and depots, service time and so on in changing environments can attempt to be addressed by ACO; (2) new methods or more advanced immigrants schemes should be proposed to further enhance the searching ability of ant to track the optimal solution with dynamic changes and (3) DOPs may be handled by other heuristics. Besides ACO, evolutionary algorithms with some schemes might be another effective approach for adapting to dynamic environments.

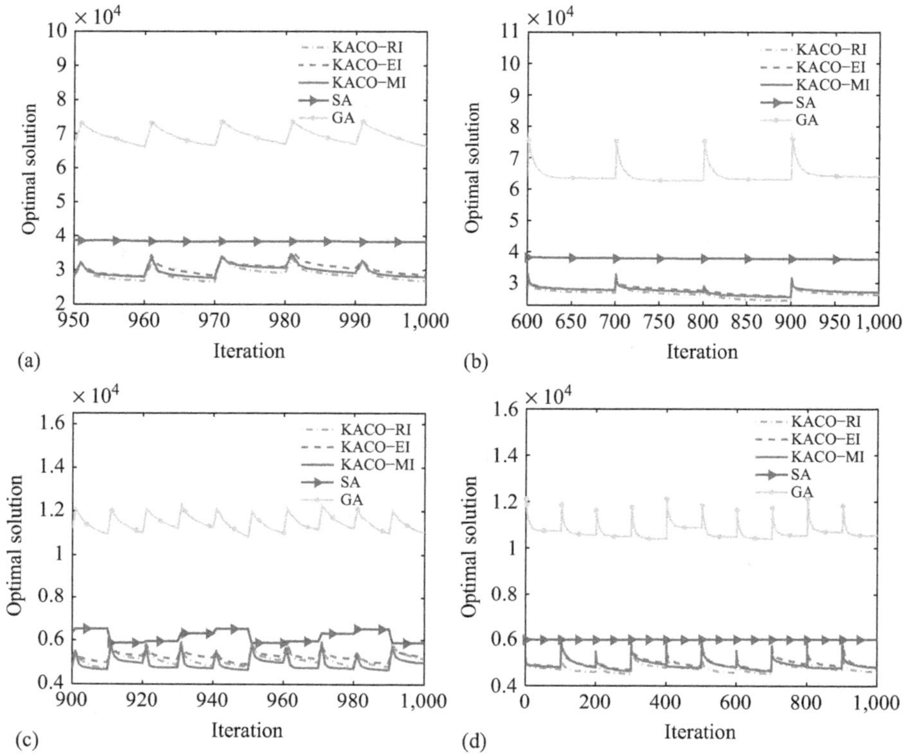

Figure 4.11 The convergence curves of optimal solutions obtained by KACO with three immigrants, SA and GA in lin105 with random and cyclic environments: (a) random environment ($f = 10$ $m = 0.5$), (b) random environment ($f = 100$ $m = 0.5$), (c) cyclic environment ($f = 10$ $m = 0.9$) and (d) cyclic environment ($f = 100$ $m = 0.9$)

References

[1] Dorigo M, Maniezzo V, Colorni A. Ant system: optimization by a colony of cooperating agents. IEEE Transactions on Systems, Man, and Cybernetics, Part B: Cybernetics. 1996;26(1):29–41.

[2] Dorigo M, Birattari M, Stutzle T. Ant colony optimization. IEEE Computational Intelligence Magazine. 2006;1(4):28–39.

[3] Cruz C, González JR, Pelta DA. Optimization in dynamic environments: a survey on problems, methods and measures. Soft Computing. 2011;15(7): 1427–1448.

[4] Mavrovouniotis M, Yang S. Ant colony optimization with immigrants schemes for the dynamic travelling salesman problem with traffic factors. Applied Soft Computing. 2013;13(10):4023–4037.

[5] Guntsch M, Middendorf M. Applying population based ACO to dynamic optimization problems. In: International Workshop on Ant Algorithms. Springer; 2002. p. 111–122.

[6] Eyckelhof CJ, Snoek M. Ant systems for a dynamic TSP. In: International Workshop on Ant Algorithms. Springer; 2002. p. 88–99.

[7] Stützle T, Hoos HH. MAX–MIN ant system. Future Generation Computer Systems. 2000;16(8):889–914.

[8] Guntsch M, Middendorf M. Pheromone modification strategies for ant algorithms applied to dynamic TSP. In: Workshops on Applications of Evolutionary Computation. Springer; 2001. p. 213–222.

[9] Guntsch M, Middendorf M. A population based approach for ACO. In: Applications of Evolutionary Computing. Springer; 2002. p. 72–81.

[10] Angus D. Niching for population-based ant colony optimization. In: e-Science and Grid Computing, 2006. e-Science'06. Second IEEE International Conference on. IEEE; 2006. p. 115.

[11] Baldacci R, Mingozzi A, Wolfler Calvo R. An exact method for the capacitated location-routing problem. Operations Research. 2011;59(5):1284–1296.

[12] Glicksman H, Penn M. Approximation algorithms for group prize-collecting and location-routing problems. Discrete Applied Mathematics. 2008;156(17):3238–3247.

[13] Barreto S, Ferreira C, Paixao J, *et al.* Using clustering analysis in a capacitated location-routing problem. European Journal of Operational Research. 2007;179(3):968–977.

[14] Albareda-Sambola M, Díaz JA, Fernández E. A compact model and tight bounds for a combined location-routing problem. Computers & Operations Research. 2005;32(3):407–428.

[15] Derbel H, Jarboui B, Hanafi S, *et al.* An iterated local search for solving a location-routing problem. Electronic Notes in Discrete Mathematics. 2010;36:875–882.

[16] Yu VF, Lin SY. A simulated annealing heuristic for the open location-routing problem. Computers & Operations Research. 2014;62(2):288–299.

[17] Ting CJ, Chen CH. A multiple ant colony optimization algorithm for the capacitated location routing problem. International Journal of Production Economics. 2013;141(1):34–44.

[18] Laporte G, Nobert Y, Arpin D. An exact algorithm for solving a capacitated location-routing problem. Annals of Operations Research. 1986;6(9):291–310.

[19] Chan Y, Carter WB, Burnes MD. A multiple-depot, multiple-vehicle, location-routing problem with stochastically processed demands. Computers & Operations Research. 2001;28(8):803–826.

[20] Stützle T, Hoos H. MAX–MIN ant system and local search for the traveling salesman problem. In: Evolutionary Computation, 1997, IEEE International Conference on. IEEE; 1997. p. 309–314.

[21] Wang Y, Xu Z, Sun J, *et al.* Ant colony optimization with neighborhood search for dynamic TSP. In: International Conference in Swarm Intelligence. Springer; 2016. p. 434–442.

[22] Gao S, Wang Y, Cheng J, *et al.* Ant colony optimization with clustering for solving the dynamic location routing problem. Applied Mathematics and Computation. 2016;285:149–173.

[23] Yang S, Li Y, Hu X, *et al.* Optimization study on k value of K-means algorithm. Systems Engineering – Theory & Practice. 2006;2:012.

[24] Mohammed MA, Ahmad MS, Mostafa S, *et al.* Using genetic algorithm in implementing capacitated vehicle routing problem. In: Computer & Information Science (ICCIS), 2012 International Conference on. vol. 1. IEEE; 2012. p. 257–262.

Chapter 5

Ant colony optimization for dynamic combinatorial optimization problems

Michalis Mavrovouniotis[1] *and Shengxiang Yang*[2]

Abstract

The ant colony optimization (ACO) meta-heuristic was inspired from the foraging behaviour of real ant colonies. In particular, real ants communicate indirectly via pheromone trails and find the shortest path. Although real ants proved that they can find the shortest path when the available paths are known a prior, they may face serious challenges when some paths are made available after the colony has converged to a path. This is because the colony may continue to follow the current path rather than exploring the new paths in case a shorter path is available. For the ACO meta-heuristic, the challenges are similar when applied to dynamic optimization problems (DOPs). Once the algorithm converges, it loses its adaptation capabilities and may have poor performance in DOPs. Several strategies have been integrated with ACO to address difficult combinatorial DOPs. Their performance proved that ACO is a powerful computational technique for combinatorial DOPs once enhanced. This chapter investigates the applications of ACO for combinatorial DOPs.

5.1 Introduction

Ant colony optimization (ACO) meta-heuristic is one of the mainstream swarm intelligence algorithms [1]. ACO algorithms were first applied to difficult combinatorial optimization problems such as the travelling salesman problem (TSP) [2], vehicle routing problem (VRP) [3] and job-shop scheduling problem (JSP) [4] with static environment. However, many real-world problems have dynamic environment in which their objective function, problem constraints, problem instance, optimization variables may vary over time. In this chapter, we investigate ACO's applications for combinatorial dynamic optimization problems (DOPs).

[1]School of Science and Technology, Nottingham Trent University, United Kingdom
[2]School of Computer Science and Informatics, De Montfort University, United Kingdom

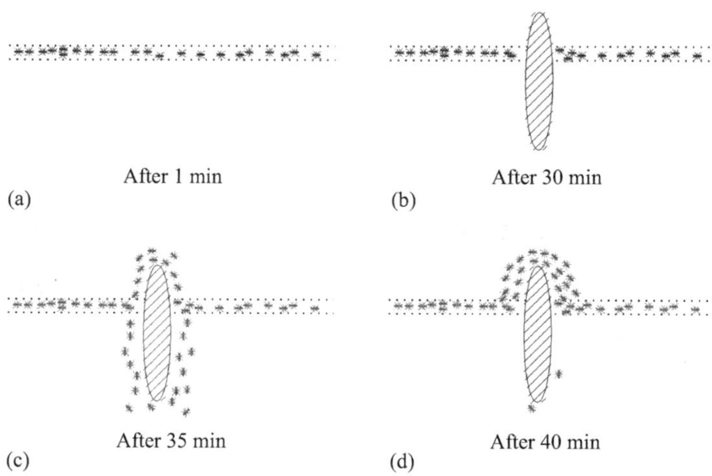

Figure 5.1 In this experiment, the colony has already converged to the shortest path from their nest to food sources. After 30 min, an obstacle interrupts their path. The ants will successfully find the best alternative shortest path after a few minutes

The motivation of applying ACO to DOPs comes from the observation of the foraging behaviour of real ants [5]. Ants lay pheromone as they are moving to mark their way back to the nest but also other ants in the colony can use them to find locations of food sources. Ants are more likely to be attracted to areas (or paths) that have higher intensity of pheromone. Therefore, ants can discover the shortest path between food sources and their nest. The shortest path will contain a trail of high concentration of pheromones.

More interestingly, ants are capable of adapting dynamic changes when their shortest path becomes infeasible [5,6]. For example, in Figure 5.1(a), the colony is moving on the shortest path that connects its nest with the food source. After 30 min, an obstacle appeared and blocks the current path of the colony in Figure 5.1(b). Therefore, the ants will start exploring the alternative paths generated on the top and at the bottom of the obstacle as shown in Figure 5.1(c). After a few minutes, the ants will converge to the path at the top as shown in Figure 5.1(d). This is because the path at the top which is the shortest will receive more pheromone more quickly than the path at the bottom within the given time.

ACO meta-heuristic has been widely used to address DOPs due to its intrinsic characteristics. While ACO is optimizing a particular problem, it maintains a pheromone matrix that contains information of the current environment. When a change occurs, the key idea is to use the pheromone trails from previously optimized environments to speed up the re-optimization process rather than starting optimizing from scratch. Although the adaptation strategy is working, sometimes the previous pheromone trails may misguide the colony.

Since ACO has been originally designed for stationary optimization problems (e.g. to converge fast into an optimum or near-optimum solution), it faces a serious challenge when tackling DOPs via transferring pheromone trails. This is because after a change, the pheromone trails of the previous environment may bias the population to search into an old optimum, making it difficult to track the changing optimum. As a result, ACO may not adapt well to dynamic changes once the colony converges into an optimum. Several strategies have been integrated with ACO applications to enhance its performance that include (a) increase diversity after change, (b) maintain diversity during execution, (c) memory schemes, (d) multi-colony schemes and (e) hybridizations.

In this chapter, we describe and discuss these applications of ACO on different DOPs. The rest of the paper is organized as follows. Section 5.2 describes how a DOP can be generated for different combinatorial problems. Section 5.3 gives a generic description of the ACO meta-heuristic and then describes three main ACO algorithms that are used in DOPs. Section 5.4 describes the different applications of ACO in DOPs, which are classified into different categories. Section 5.5 describes real-world applications with a dynamic environment in which ACO was found useful. Finally, Section 5.6 concludes this chapter with different direction for future work.

5.2 Dynamic optimization problems

5.2.1 DOP definition

Many static combinatorial optimization problems are *NP*-hard, that is it is strongly believed that they cannot be solved to optimality within polynomial computation time [7]. A combinatorial DOP can be intuitively defined as a sequence of static problem instances that need to be optimized [8]. The main characteristics of a DOP are usually defined by the frequency and magnitude of the environmental changes. In other words, the former and latter characteristic corresponds to the speed and degree at which the environment of the problem changes, respectively.

An environmental change may involve the objective function, input variables, problem instance and constraints. Formally, a DOP can be defined as follows:

$$\text{DOP} = \text{optimize } \phi(\mathbf{x}, t) \text{ subject to } X(t) \subseteq S, \quad t \in P,$$

where S is the search space, t is the time, $\phi : S \times P \rightarrow R$ is the objective function that assigns a value in R to each possible solution $\mathbf{x} \in S$ and $X(t)$ is the set of feasible solutions $\mathbf{x} \in X(t) \subseteq S$ at time t [9,10]. Each feasible solution $\mathbf{x}$ consists of optimization variables $\mathbf{x} = \{x_1, \ldots, x_n\}$, where variables x_i are discrete that take values $x_i \in \mathbf{D}_i = \{v_i^1, \ldots, v_i^{|\mathbf{D}_i|}\}, i = 1, \ldots, n$. Every solution $\mathbf{x} \in X(t)$ has a a set of neighbours $N(\mathbf{x}) \subseteq X(t)$ where $N : X \rightarrow 2^X$. A local optimum solution is a feasible solution $\mathbf{x}'$ such that $\phi(\mathbf{x}', t) \leq \phi(\mathbf{x}, t), \forall \mathbf{x} \in N(\mathbf{x})$ for minimization problems or $\phi(\mathbf{x}', t) \geq \phi(\mathbf{x}, t), \forall \mathbf{x} \in N(\mathbf{x})$ for maximization problems, respectively.

The global optimum is a feasible solution $\mathbf{x}^*$ that minimizes (or maximizes) the objective function $\phi(\mathbf{x}^*, t) \leq \phi(\mathbf{x}, t),\ \forall \mathbf{x} \in X(t)$ for minimization problems or $\phi(\mathbf{x}^*, t) \geq \phi(\mathbf{x}, t),\ \forall \mathbf{x} \in X(t)$ for maximization problems, respectively.

5.2.2 Dynamic characteristics

Apart from the magnitude and frequency of change, DOPs are also characterized by the following aspects:

- Predictability: whether the dynamic changes are predictable or not
- Visibility: whether the dynamic changes are known to the optimization algorithm
- Detectability: whether the dynamic changes are detectable or not. If they are known, there is no need to detect them
- Cyclicity: whether previous dynamic changes will reappear again or not
- Time linkage: whether the current solution found by an optimization algorithm will affect the future behaviour of the problem or not [11,12]
- Constrained: whether the dynamic changes affect the constraints or not [13]

5.2.3 Generate dynamic environments

A dynamic environment is generated by changing the stationary base of combinatorial optimization problem over time. A dynamic environment consists of two important parameters (e.g. f and m) that define the frequency and magnitude of change. The frequency of change is typically expressed as algorithmic iterations. It can be also expressed as function evaluations or real time. A lower value of f means faster dynamic changes. The magnitude of change is highly problem-specific and depends on the components of the optimization problem. Usually, the magnitude of change is defined as $m \in [0, 1]$ where a higher value of m means significant dynamic changes. The environmental period index for each dynamic change is defined as $T = \lceil t/f \rceil$ where t is the algorithmic iteration count and f is the frequency of change.

In the following sections, the dynamic versions of different combinatorial optimization problems are expressed.

5.2.3.1 Travelling salesman problem

The TSP is a fundamental combinatorial problem. The classical TSP is modelled by a fully connected weighted graph $G = (N, A)$, where N is a set of n nodes (e.g. cities) and A is a set of arcs links, where n represents the size of a problem instance. Each link $(i,j) \in A$ is associated with a non-negative value $d_{ij} \in \mathbb{R}^+$, which represents the distance between cities i and j. The objective of the problem is to find the shortest Hamiltonian cycle that starts from one city and visits each of the other cities once before returning to the starting city.

There are several dynamic TSP (DTSP) variations in the literature such as changing the topology by replacing cities [14] and changing the weights of the links between cities [15]. The magnitude of change can be expressed as the number cities that will be replaced (e.g. $\lceil mn \rceil$) or the number of weights that will be modified (e.g. $\lceil mn(n-1) \rceil$) where m is the magnitude of change defined above, and n is the number of cities. DTSPs have a wide range of real-world applications in telecommunication networks [16], robotics [17] and transportation [18].

5.2.3.2 Vehicle routing problem

The VRP is direct variation of the TSP above. Similarly, it is modelled with the same weighted graph G, but it contains a special node that represents the central depot. In addition, each node (e.g. customer) $i \in N$ is associated with a quantity q_i of some goods that need to be delivered by K vehicles. The objective of the problem is to find the shortest routes for a fleet of K vehicles with capacity Q all starting from and ending at the depot satisfying the demands of all customers. For the central depot, the quantity is always set to zero.

There are also several dynamic VRP (DVRP) variations in the literature such as the VRP in which new customer orders arrive [19] and the VRP where the weights of the links between customers change [20]. Similarly with the DTSP above, the magnitude of change can be expressed as the number of customer demands change or the number of weights that will be modified. The real-world applications of DVRP appear in transportation and logistics [21].

5.2.3.3 Job-shop scheduling problem

The JSP is a classical combinatorial problem which can be described as follows. Given a set M of machines and a set J of jobs that consists of an ordered sequence of operations that is associated with non-negative processing time. The operations need to be assigned to the machines for processing. The objective of the problem is to minimize the completion times of all operations.

A dynamic JSP (DJSP) variation exist in which jobs with new operations may arrive [22]. The magnitude of change may be expressed as the number of new jobs that need to be rescheduled. Hence, the current schedule on the machines needs to be re-scheduled to include the new operations. Real-world applications of DJSP may appear in re-scheduling trains [23].

5.2.3.4 Binary-encoded functions

The XOR DOP generator [24–26] can construct dynamic environments from any binary-encoded stationary function $\phi(\mathbf{x})$ ($\mathbf{x} \in \{0, 1\}^n$) by a bitwise XOR operator, where n is the binary encoding length. The dynamics for any binary-encoded function (e.g. *OneMax*, *Royal Road* and *Deceptive*) can be formulated as follows:

$$\phi(\mathbf{x}, t) = \phi(\mathbf{x} \oplus \mathbf{M}(T)),$$

where '$\oplus$' is the XOR operator (i.e. $1 \oplus 1 = 0$, $1 \oplus 0 = 1$, $0 \oplus 0 = 0$), $\mathbf{M}(T)$ is the XORing mask that occurs incrementally and is defined as follows:

$$\mathbf{M}(T) = \mathbf{M}(T - 1) \oplus \mathbf{B}(T),$$

where $\mathbf{B}(T)$ is an intermediate binary template randomly created with exactly $\lceil mn \rceil$ ones.

5.2.3.5 Knapsack problem

The Knapsack problem (KP) is another classical combinatorial problem which can be described as follows. Given a set of items $i \in I$, where each item i is associated with a weight a_i and a profit b_i. The objective of the problem is to select a subset of items to fill up a knapsack of limited size to maximize the total profit of the chosen items.

A dynamic KP (DKP) variation can be constructed by changing the weights of items or the profits or the knapsack capacity [27]. The magnitude of change can be defined as how many problem components will change. The XOR DOP described above can also generate DKPs since the KP can be represented in a binary space [28].

5.3 Ant colony optimization meta-heuristic

The ACO meta-heuristic has been developed to address combinatorial optimization problems [29–31]. To apply ACO on the combinatorial problems described above, a construction graph needs to be defined, and a pheromone model needs to be associated with the construction graph. Let $C_G = (C, L)$ define the construction graph that artificial ants can search for solutions. C represents the set of components and L represents the set of connections that connect the components (i.e. $C = N$ and $L = A$, where N and A are the set of nodes and arcs, respectively) of graph G defined in Section 5.2.3. Basically, the construction graph C_G needs to be mapped to the considered combinatorial optimization problem.

In particular, for DTSPs, the construction graph is identical with the weighted graph $G = (N, A)$ that represents the problem. One component in C corresponds to one city in N. Similarly, for DVRPs, we can use one component for each customer and for the depot. For DJSPs, the construction graph contains one component for every job operation plus two dummy components that define where the scheduling task starts and where it ends. For DKPs, the components in the construction graph correspond to the set of items (i.e. $C = I$), and L defines the connection of the items. For binary-encoded problems, the construction graph components correspond to the only two possible values the problem can take (i.e. 1 or 0).

The pheromone model is mapped with the components in the construction graph. For DTSPs and DVRPs, the pheromone trails τ_{ij} are associated with the components i and j and represent the desirability of visiting component j after component i. Similarly, the pheromone trails in DJSPs refer to the desirability of choosing operation j directly after operation i. For binary-encoded problems generated by XOR DOP, pheromone trails are associated only to the choices of 1 and 0 value. For DKPs, the pheromone trails τ_i are associated with the components individually and refer to the desirability of adding item i. A list of important ACO applications of combinatorial DOPs is given in Table 5.1.

The main procedures for ACO are illustrated in Algorithm 1 and are described in the following sections.

Table 5.1 ACO's applications for combinatorial DOPs

Authors	Reference
Travelling salesman problem	
Angus and Hendtlass	[34]
Guntsch and Middendorf	[14,35]
Eyckelhof and Snoek	[15]
Mavrovouniotis and Yang	[8,36–40]
Melo *et al.*	[41,42]
Mavrovouniotis *et al.*	[33,43]
Guntsch *et al.*	[44]
Mavrovouniotis *et al.*	[45]
Liu	[46]
Silva and Runkler	[47]
Vehicle routing problem	
Euchi *et al.*	[48]
van Veen *et al.*	[49]
Mavrovouniotis and Yang	[20,50,51]
Toklu *et al.*	[19,52]
Gao *et al.*	[53]
Yang *et al.*	[54]
Montemanni *et al.*	[21]
Ankerl and Hämmerle	[55]
Job-shop scheduling problem	
Zhou *et al.*	[22,56]
Lorpunmanee *et al.*	[57]
Knapsack problem	
Randall	[27]
Binary-encoded functions	
Fernandes *et al.*	[58]
Mavrovouniotis and Yang	[59]

Algorithm 1: ACO meta-heuristic

$t \leftarrow 0$
SetParameters
InitializePheromoneTrails
while (termination condition *not* satisfied) **do**
  ConstructSolutions
  $\mathbf{x}^{ib} \leftarrow$ FindIterationBest
  if $(\phi(\mathbf{x}^{ib}, t) < \phi(\mathbf{x}^{bs}, t))$ **then**
    $\mathbf{x}^{bs} \leftarrow \mathbf{x}^{ib}$
  end if
  ApplyLocalSearch %optional
  PheromoneUpdate
  $t \leftarrow t + 1$
end while

5.3.1 Construct solutions

A colony of artificial ants construct solutions according to some problem-depended criteria. Each ant builds an independent solution by continuously adding a problem component. An ant's (partial) solution is initially empty and it is filled at each construction step until a feasible solution is completed. The feasibility of the solution depends on the constraints of the combinatorial optimization problem. Each ant must select a component from a particular finite set that ensures no violation of any problem-specific constraint.

In addition, the selection of each problem component at each construction step is guided by pheromone trails which are associated to each component and some problem-specific heuristic information (if available). A stochastic policy is applied in which the component that has the highest pheromone trails and best heuristic information has a higher probability to be selected.

5.3.2 Local search

Although a local search scheme is optional in ACO, it is recommended to improve the solution quality. In fact, it has been shown that an ACO with local search significantly outperforms an ACO without local search in both static [32] and dynamic environments [33]. However, an extensive usage of local search may not be very efficient for DOPs especially when the time between changes is short. This is because the computation time will be increased significantly since local search operators are usually computationally expensive. In addition, local search is a highly problem-specific procedure.

5.3.3 Pheromone update

Once the ants complete their construction task, they will then need to update the pheromone values that are associated with the problem-specific components. In this way, ants mark their constructed solution. Initially, all the pheromone values are set with an equal amount. The aim is to increase the pheromone values of those components that belong to promising quality solutions (according to the objective of the optimization problem) and decrease the rest of the pheromones. In other words, the colony starts with a high explorative phase which is gradually decreased with the update of the pheromones.

Pheromone evaporation is responsible for decreasing all the existing pheromones with a small factor. Therefore, the components that are not frequently selected by ants will contain less amount of pheromone.

5.3.4 Main ACO algorithms

5.3.4.1 *MAX–MIN* ant system

One of the state-of-the-art ACO variations is the *MM*AS [32]. A colony of ω ants read pheromones in order to construct their solutions and write pheromones to store their solutions. Each ant k uses a probabilistic rule to choose the next node to visit. The decision rule of the kth ant to move from node i to node j is defined as follows:

$$p_{ij}^{k} = \frac{\left[\tau_{ij}\right]^{\alpha}\left[\eta_{ij}\right]^{\beta}}{\sum_{l \in N_i^k}\left[\tau_{il}\right]^{\alpha}\left[\eta_{il}\right]^{\beta}}, \quad \text{if } j \in N_i^k, \tag{5.1}$$

where τ_{ij} and η_{ij} are the existing pheromone trail and heuristic information available a priori between nodes i and j, respectively, N_i^k is the set of feasible components for ant k to select, α and β are the two parameters, which determine the relative influence of τ_{ij} and η_{ij}, respectively.

The pheromone trails in *MM*AS are updated by applying evaporation as follows:

$$\tau_{ij} \leftarrow (1-\rho)\,\tau_{ij}, \quad \forall (i,j), \tag{5.2}$$

where ρ is the evaporation rate which satisfies $0 < \rho \leq 1$, and τ_{ij} is the existing pheromone value. After evaporation, the best ant deposits pheromone as follows:

$$\tau_{ij} \leftarrow \tau_{ij} + \Delta\tau_{ij}^{best}, \quad \forall (i,j) \in \mathbf{x}^{best}, \tag{5.3}$$

where $\Delta\tau_{ij}^{best} = 1/C^{best}$ is the amount of pheromone that the best ant deposits and C^{best} defines the solution quality of tour $\mathbf{x}^{best}$. The best ant that is allowed to deposit pheromone may be either the best-so-far, in which case $C^{best} = C^{bs}$, or the iteration-best, in which case $C^{best} = C^{ib}$, where C^{bs} and C^{ib} are the solution quality of the best-so-far and the iteration best ant, respectively. The best-so-far ant is a special ant that may not necessarily belong to the current colony of ants as the iteration best ant. Both update rules are used in an alternate way in the implementation [60].

The lower and upper limits τ_{min} and τ_{max} of the pheromone trail values are imposed. The τ_{max} value is bounded by $1/(\rho C^{bs})$, where C^{bs} is initially the solution quality of an estimated optimal tour and later on is updated whenever a new best-so-far ant solution quality is found. The τ_{min} value is set to $\tau_{min} = \tau_{max}/2n$.

Since only the best ant is allowed to deposit pheromone, the colony may quickly converge towards the best solution found in the first iteration. This behaviour is known as stagnation behaviour where all ants construct the same solution from the early stages of the optimization process. Therefore, the pheromone trails are occasionally reinitialized to the τ_{max} value to increase exploration. For example, whenever the stagnation behaviour occurs or when no improved solution is found for a given number of iterations, the pheromone trails are reinitialized.

5.3.4.2 Ant colony system

Another state-of-the-art ACO algorithm is the ant colony system (ACS) [29]. This variation uses a different decision rule from the *MM*AS as follows. With a probability q_0, the kth ant will choose the city with the highest probability as follows:

$$j = \arg\max_{l \in N_i^k}\{[\tau_{il}]^{\alpha}[\eta_{il}]^{\beta}\}.$$

With a probability $(1 - q_0)$, a probabilistic selection is made using the same decision rule with *MM*AS described above. The q_0 parameter satisfies $0 \leq q_0 \leq 1$ and basically regulates the degree of exploration.

In ACS, only the best-so-far ant is allowed to deposit pheromone. Also the pheromone evaporation occurs only on the pheromone trails of the best-so-far solution rather than globally as in *MM*AS. The pheromone update policy at each iteration for ACS is defined as follows:

$$\tau_{ij} \leftarrow (1-\rho)\tau_{ij} + \rho\Delta\tau_{ij}^{bs}, \quad \forall(i,j) \in \mathbf{x}^{bs},$$

where $\mathbf{x}^{bs}$ is the best-so-far solution. In addition, to this pheromone update which occurs at the end of each iteration, pheromone trails are updated locally while the ants are constructing solutions. The local pheromone update when an ant just traversed arc (i,j) is defined as follows:

$$\tau_{ij} \leftarrow (1-\xi)\tau_{ij} + \xi\tau_0,$$

where ξ is a parameter that satisfies $0 < \xi < 1$ and τ_0 is the initial pheromone value.

The local pheromone update policy does not allow stagnation behaviour because whenever an ant visits an arc, it becomes less desirable. Therefore, it allows an increase in exploration of other arcs that have not been visited yet.

5.3.4.3 Population-based ant colony optimization

The population-based ACO (P-ACO) uses a population (explicit memory), where the best ant from each iteration t is stored in the population [14]. The population is directly associated with the pheromone trails (e.g. whenever an ant *in* enters the population) a positive pheromone update is performed as follows:

$$\tau_{ij} \leftarrow \tau_{ij} + \Delta\tau_{ij}^{in}, \quad \forall(i,j) \in \mathbf{x}^{in}, \tag{5.4}$$

where $\Delta\tau_{ij}^{in}$ is the constant pheromone amount to be deposited, and $\mathbf{x}^{in}$ is the tour of the ant entering the population. Whenever an ant *out* leaves the population, due to the limited size of $M(t)$, a negative pheromone update is performed as follows:

$$\tau_{ij} \leftarrow \tau_{ij} - \Delta\tau_{ij}^{out}, \quad \forall(i,j) \in \mathbf{x}^{out}, \tag{5.5}$$

where $\Delta\tau_{ij}^{out}$ is the constant amount to be removed, and $\mathbf{x}^{out}$ is the tour of the ant leaving the population. In this way, outdated pheromone trails are eliminated directly, and pheromone evaporation is not needed as in *MM*AS and ACS. Note that the constant amounts of pheromones are always $\Delta\tau_{ij}^{out} = \Delta\tau_{ij}^{in}$. Therefore, the pheromone values can never drop below the initial pheromone value.

5.4 ACO for DOPs

A straightforward method to address DOPs is to consider each dynamic change as the arrival of a new problem instance that needs to be optimized from scratch. With an ACO algorithm, this method can be implemented by re-initializing all the pheromone trails with an equal amount (usually the initial value).

Considering the *NP*-hardness of combinatorial DOPs, such strategy would not be practical. A more efficient method is to transfer knowledge from previously optimized problem instances. Typically, the previously optimized problem instance with the

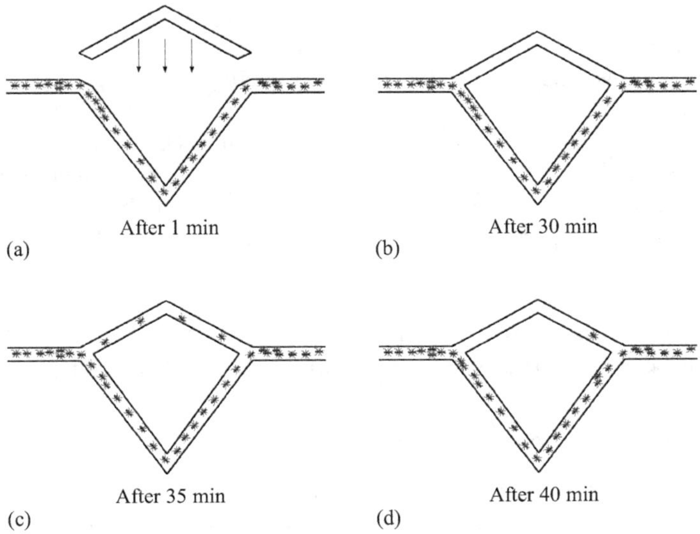

Figure 5.2 *In this experiment, the colony of ants initially has only one out of the two paths available. After 30 min where the colony has generated a trail of pheromone on the longer branch, the shortest branch on the top becomes available. Once the shortest branch on the top was made available, the ants continued to the use the longer branch at the bottom*

newly generated problem instance will have similarities since the dynamic changes (in most real-world applications) are small to medium [61]. Therefore, when a dynamic change occurs, the pheromone trails of the previously optimized problem instance can be directly transferred to the newly generated problem instance, rather than being re-initialized, to speed up the re-optimization process.

However, some of the transferred pheromone trails may misguide the optimization process to the new environment. In fact, this is the case with real ants [5]. For example, in the experiments shown in Figure 5.2(a), the colony is offered a single path between the nest and food sources for the first 30 min. After 30 min, another path becomes available (e.g. an environmental change occurs), which happens to be shorter than the previous one; Figure 5.2(b). Only a few ants explored the shorter path sporadically as presented in Figure 5.2(c), but the colony was trapped on the longer path (e.g. the path from the old environment) as presented in Figure 5.2(d). This is because the pheromone trails generated on the longer path from the previous environment are very high, whereas the shortest path has (initially no pheromone) lower concentration of pheromone from the few ants that explored it. However, the pheromone concentrated on the shorter branch is not enough to bias the colony to select the shorter branch.

The evaporation of pheromone may favour the selection of the shorter path because the strength of existing pheromone trails will gradually drop. But, since in reality pheromone evaporates too slowly, the colony is not capable to forget the previously converged path and discover the new shorter path.

In other words, once the colony converges into a particular path, it is difficult to escape from it when a change occurs due to the high concentration of pheromone. Analogously, a similar case occurs with the stagnation behaviour in ACO meta-heuristic when addressing dynamic changes. In contrast, the pheromone evaporation in an ACO meta-heuristic can be tuned accordingly: using a high evaporation rate for fast adaptation [37]. Apart from the pheromone evaporation, which is useful in the adaptation of ACO in dynamic changes, several strategies were proposed (e.g. the P-ACO algorithm) to address this issue.

The strategies integrated with ACO algorithms (any of the main algorithms described in Section 5.3.4) are categorized and described in the following subsections.

5.4.1 Increasing diversity after a change

Many strategies in DOPs depend on the detection of dynamic changes. The detection is typically performed by re-evaluating a solution and whenever a change to its fitness occurs then a dynamic change is detected. The strategies in this category perform actions whenever a change is detected to increase diversity. In case the dynamic changes are known, there is no need to explicitly detect them [14]. The actions performed when a change occurs are categorized as follows.

The pheromone trails of ACO algorithms are re-initialized with an equal amount whenever a change occurs (e.g. for DTSPs and DVRPs [15,36,37]). This corresponds to a complete restart of the algorithm that optimizes each dynamic change from scratch. Since the initial phase of an algorithm is highly explorative, the diversity is increased significantly. However, all previous knowledge gained is destroyed.

Other strategies aim to simply increase the diversity and maintain the knowledge gained, simultaneously, when a change occurs. Partial restart strategies using local information have been proposed, such as the η-strategy and τ-strategy [35,44], which take into account where the dynamic change actually occurred (e.g. which cities are added/removed when a change occurs in DTSPs). The aim of both strategies is to give a higher degree of re-initialization to the pheromone trails closer to the offended areas. This corresponds to a partial restart of the algorithm because knowledge is maintained and could speed up the optimization process. These strategies can be applied only if the locations of changes are known a priori. In case they are not available, it may require extensive computational efforts to explicitly detect them.

A similar partial restart strategy has been proposed for the DJSP [22,56]. The key idea for this strategy is to exclude jobs that will not have the opportunity to execute before a new job arrives (i.e. a dynamic change occurs). Therefore, the re-scheduling of the jobs will include only jobs that can be executed within a given period assuming that the period of change is known a prior or the problem is decomposed into time slices. The pheromone trails are transferred when a new dynamic change occurs.

Differently, the previous pheromone trails are modified proportionally to the maximum pheromone trail value for DTSPs [34]. Similarly, a ‘shaking technique’ has been proposed to regulate the previous pheromone trails [15]. All pheromone trails are modified using a logarithmic formula. The pheromone values closer to the initial pheromone value are deducted slightly, whereas higher pheromone values are deducted severely.

The above strategies performed modification via the pheromone trails directly. In contrast, the best solution of the previous environment is repaired for DKPs, since it became infeasible by the change, using a deconstruction technique [27]. The pheromone trails are affected indirectly according to the changes made to the solutions during the repair. This way, partial knowledge of the previous environment is preserved. A repair of the best solution is also performed when a change occurs in DTSPs [44].

5.4.2 Maintain diversity during the execution

There are strategies that do not necessarily require the detection of change to increase diversity. Diversity is maintained during the execution. In fact, unmodified ACO algorithms were applied to the DJSP [57] and binary-encoded problems [58,59]. Because pheromone evaporation helps to forget unused pheromone trails, it helps the adaptation process when a dynamic change occurs. Another way is to gather statistics from the pheromone trails to detect stagnation and re-initialize the pheromone trails [60].

Based on the pheromone evaporation effect, the rate ρ was adapted during the optimization process [37] for DTSPs and DVRPs to speed up the adaptation. The idea is to increase the evaporation rate when the algorithm is approaching stagnation behaviour to eliminate pheromone trails faster. Since the adaptive method requires a fixed step size, a self-adaptive evaporation rate was later proposed to address this issue [40]. In particular, the evaporation rate is discretized to a set of values with which pheromone trails are associated. The ants are responsible to select the appropriate rate at each iteration.

The evaporation rate is a global ACO parameter. Other ACO parameters (e.g. the local ACO parameters α and β) were self-adapted [55]. These two parameters control the weight of pheromone trails and heuristic information, respectively. Each ant uses a different set of these parameters, which initially are the same. This way enables ants to explore different locations of the search space and, thus, maintain diversity.

In contrast, instead of adapting the control parameters, either local or global described above, the decision rule of ACO was modified instead in [46], where a rank-based non-linear selection pressure function was integrated to the decision rule. Although the integration enhances diversity, it causes the introduction of new parameters that need to be optimized.

A very popular strategy integrated with ACO is the immigrants scheme [8,20,37–39,53]. The general idea is to introduce immigrant ants, either randomly or using the best from the previous environment, to deposit pheromone trails. This way, diversity is maintained. However, if too many random immigrant ants are generated, they may disturb the optimization process.

5.4.3 Memory schemes

Another way to transfer knowledge from previous environments is to use an external memory. Typically, promising solutions are stored in the memory and reused when a dynamic change occurs. Again, this type of strategies require the detection of change to update the memory or even repair it.

One of the most popular ACO for DTSPs is the P-ACO [14]. P-ACO maintains a memory of the best ants that is used for updating the pheromones. When a dynamic change occurs, the ants in the memory are repaired heuristically, and thus, the pheromone values are modified. Similarly, with the η-strategy and τ-strategy mentioned above, the location of changes is required for the heuristic repair.

The memory-based immigrants ACO [8,20,50] uses a limited size memory to store the best representative solution from several previously optimized environments. The idea is to generate immigrant ants based on the currently best memory to transfer knowledge and increase diversity simultaneously.

5.4.4 *Multi-colony schemes*

The use of separate colonies can naturally increase and maintain diversity. The idea is to allocate different colonies into different areas in the search space.

A multi-colony ACO algorithm, where each colony uses a separate pheromone table in an attempt to maximize the search area explored, was proposed for the DTSP [45] and DVRP [49,54]. An explicit method to keep the colonies into different areas was not applied, but the results showed that it performs better than a single colony ACO algorithm.

In contrast, a multi-colony ACO algorithm, in which different colonies, called castes, use the same pheromone table, was proposed in [41,42]. Each caste uses different parameter settings that correspond to different behaviour for each caste that can cover different location of the search space.

5.4.5 *Hybridizations*

The strategies in this category emphasizes more on the hybridization of the ACO algorithm with a local search method. Usually, algorithms enhanced by local search, known as memetic algorithms, provide significantly better solutions with the price of increasing the computational time. The idea is to let the main algorithm provide an initial solution for local search operators to optimize. But, since local search operators generate strong exploitation, the algorithms need to maintain diversity.

The P-ACO described above was improved in [36,51], where several local search steps based on simple and adaptive inversions are applied to the iteration best ant. In addition, diversity is enhanced with the *triggered* random immigrants entering the memory, whenever all ants stored in the memory are identical. Recently, another memetic-based ACO algorithm was proposed in [33], which uses a local search operation whenever a new best solution is found. Whenever the algorithm is entering stagnation behaviour (very often because of the local search), the pheromone trails are re-initialized to the maximum value to maintain diversity.

5.4.6 *Discussion*

Each aforementioned strategy proposed to enhance the adaptation capabilities of ACO in dynamic environments has advantages and disadvantages. The strategies

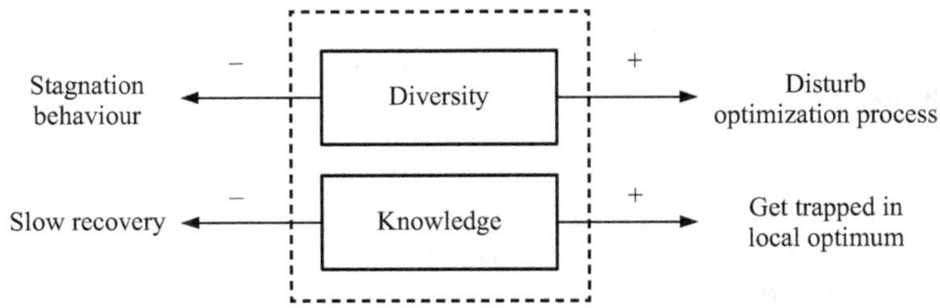

Figure 5.3 Effects of tuning diversity maintenance and knowledge transfer of ACO algorithms in DOPs

that fall into the categories of *increasing diversity after a change* and *maintaining diversity during the execution* can well address the diversity loss issue, but it is difficult to know the optimal level for diversity maintenance in order to design efficient diversity mechanisms. The strategies that fall into the category of *memory schemes* can help the colony to quickly adapt to dynamic changes, but they are effective only in environments with small dynamic changes. The strategies that fall into the category of *multi-colony schemes* can also well address the diversity loss issue, but the optimal number of colonies, the optimal number of ants for each colony and the search area of each colony are very hard to determine.

The balance of *diversity maintenance* and *knowledge transfer* should always be configured for handling environmental changes effectively as shown in Figure 5.3. Once the ACO algorithm converges (e.g. reaches the stagnation behaviour), it loses its adaptation capabilities because of the diversity lost issue. This issue can be addressed by increasing or maintaining the diversity. However, if too much diversity is generated, it may disturb the optimization process because of too much randomization. Within ACO, knowledge can be transferred via pheromone trails (e.g. the pheromone trails of the previous environment can be transferred and used in the new environment). However, if exactly the same pheromone trails are transferred, it may take some time for the ants to escape from old optimum and track the new one when a dynamic change occurs. On the contrary, if too much knowledge is transferred, the optimization process may start close to a poor local optimum and get stuck there.

Furthermore, these two aspects are also conflicting because if diversity is not maintained during the optimization process, the algorithm will not be very flexible to utilize the knowledge transferred when a change occurs [1].

5.5 ACO in real-world DOP applications

In this section, real-world applications of ACO for DOPs are discussed. There are four different interpretations of the definition 'real-world applications' classified as follows [62]:

1. Applications that are actually used in the industry.
2. Applications that use real data for experimentation.

3. Applications that are tested on models/simulators of a real-world problem.
4. Other applications based on artificial benchmarks (e.g. constraint-handling and multi-objective optimization).

Among the classification, only the case of category 1 is clearly a real-world application, whereas the remaining categories are analogies of real-world applications. In this section, we refer to real-world applications as those applications that are actually used in some industries or are tested in simulations with real-world data.

An ACO algorithm was used to solve the on-line fuel oil distribution problem [21]. The application enables new orders to be assigned to vehicles while they have already left the depot. Basically, the working day is divided into a series of several static problems that are optimized by ACO separately. Pheromone conservation of the previous static problem was used to prevent starting the optimization from scratch for the next problem. Time slices will contain the new orders that will be assigned to the vehicles.

An ACO algorithm was proposed to address dynamic changes in the production systems [63]. A pheromone matrix is maintained to store the operations. Whenever a new job arrives, the pheromone values are re-initialized. Their experiments on a real-world data collection showed that the proposed ACO method outperforms the manual method and a traditional priority-rule method.

Another ACO was used to reschedule trains at junctions and stations when train delays occur [23]. A new feasible schedule was generated based on the previous infeasible schedule using a path-preserving heuristic.

A conventional ACO algorithm was used to address the cash machine problem [18]. Basically, the problem is that a cash logistics company needs to fill the cash containers according to their money levels for a number of cash machines of a specific bank in Frankfurt. The results showed that ACO is able to automatically output solutions with the minimum travelling time, considering the distances between the cash machines and their money levels. However, there are no comparisons with any other automatic or manual methods.

Finally, ACO was used to solve the problem of path planning for mobile robots by avoiding obstacles during the operation [64]. The workspace (search space) is discretized in a matrix of 50×50 nodes, where the mobile robot is able to navigate and build paths.

5.6 Conclusions

ACO is successfully applied for many combinatorial DOPs due to its intrinsic characteristic. Specifically, the pheromone trails of the previously optimized environment can be re-used to the newly generated environment. Adapting to dynamic changes contributes to the speed up of the re-optimization process.

This chapter describes the most important ACO applications. The strategies used to enhance ACO algorithms to cope with dynamic changes are classified as follows: (a) increasing diversity after a change, (b) maintaining diversity during the execution,

(c) memory schemes, (d) multi-colony schemes and (e) hybridizations. Although every strategy in each category has advantages and disadvantages, their key idea is to maintain the balance between knowledge transfer and diversity maintenance.

Although the research area in ACO for DOPs has been growing in the last decade, it still has several gaps that can be addressed in the future, which are summarized as follows:

- Experimentation protocols: Still there is not any agreed computational experimentation protocol yet to compare ACO algorithms in DOPs. For example, how the frequency of change must be defined (time, algorithmic iterations, algorithmic evaluations) or which performance measurements should be used.
- Modelling real-world scenarios: Most research focuses on academic benchmarks. It will be interesting to consider modelling more real-world benchmarks and apply ACO algorithms to solve them. The number of existing real-world application at the moment are far below the expectations. Some examples have been discussed in Section 5.5.
- Consideration of other prospectives: Typically, ACO algorithms track the moving optimum. Recently, a new perspective on DOPs has been introduced, known as robust optimization over time [65], where the aim of the algorithm is to find the solution that its quality if acceptable to a predefined number of dynamic changes.
- Theoretical development: Usually, the performance of ACO algorithms in DOPs is analysed empirically, rather than theoretically, due to the mathematical challenges. Although theoretical works exist, the research area is still very limited. Theoretical development should be enhanced significantly in order to have deep understanding why and how ACO works in DOPs.

Acknowledgement

This work was supported by the Engineering and Physical Sciences Research Council (EPSRC) of UK under Grant EP/K001310/1.

References

[1] M. Mavrovouniotis, C. Li, and S. Yang. A survey of swarm intelligence for dynamic optimization: Algorithms and applications. *Swarm and Evolutionary Computation*, 33:1–17, 2017.

[2] M. Dorigo, V. Maniezzo, and A. Colorni. Ant system: Optimization by a colony of cooperating agents. *IEEE Transactions on Systems, Man, and Cybernetics, Part B: Cybernetics*, 26(1):29–41, 1996.

[3] L. M. Gambardella, E. D. Taillard, and C. Agazzi. MACS-VRPTW: A multi-colony ant colony system for vehicle routing problems with time windows. In *New Ideas in Optimization*, pages 63–76, 1999.

[4] A. Colorni, M. Dorigo, M. Maniezzo, and M. Trubian. Ant system for job-shop scheduling. *JORBEL – Belgian Journal of Operations Research, Statistics and Computer Science*, 34(1):39–53, 1994.

[5] S. Goss, S. Aron, J. L. Deneubourg, and J. M. Pasteels. Self-organized shortcuts in the argentine ant. *Naturwissenschaften*, 76(12):579–581, 1989.

[6] R. Beckers, J. L. Deneubourg, and S. Goss. Trails and u-turns in the selection of a path by the ant *Lasius niger*. *Journal of Theoretical Biology*, 159(4):397–415, 1992.

[7] M. Garey and D. Johnson. *Computer and intractability: A guide to the theory of NP-completeness*. Freeman, San Francisco, 1979.

[8] M. Mavrovouniotis and S. Yang. Ant colony optimization with immigrants schemes for the dynamic travelling salesman problem with traffic factors. *Applied Soft Computing*, 13(10):4023–4037, 2013.

[9] C. Cruz, J. R. Gonzalez, and D. A. Pelta. Optimization in dynamic environments: A survey on problems, methods and measures. *Soft Computing*, 15(7):1427–1448, 2011.

[10] S. Yang, Y. Jiang, and T. Nguyen. Metaheuristics for dynamic combinatorial optimization problems. *IMA Journal of Management Mathematics*, 24(4): 451–480, 2013.

[11] P. Bosman. Learning, anticipation and time-deception in evolutionary online dynamic optimization. In *Proceedings of the 2005 Genetic and Evolutionary Computation Conference*, pages 39–47, 2005.

[12] T. Nguyen, X. Yao. Dynamic time-linkage problems revisited. In M. Giacobini, A. Brabazon, S. Cagnoni, *et al.*, editors, *Applications of Evolutionary Computing*, volume 5484 of *Lecture Notes in Computer Science*, 735–744, Springer, Berlin, Heidelberg, 2009.

[13] T. Nguyen, X. Yao. Continuous dynamic constrained optimization-the challenges. *IEEE Transactions on Evolutionary Computation*, 16(6):769–786, 2012.

[14] M. Guntsch and M. Middendorf. Applying population based ACO to dynamic optimization problems. In M. Dorigo, G. Di Caro, and M. Sampels, editors, *Ant Algorithms*, volume 2463 of *Lecture Notes in Computer Science*, pages 111–122. Springer, Berlin, Heidelberg, 2002.

[15] C. Eyckelhof and M. Snoek. Ant systems for a dynamic TSP. In M. Dorigo, G. Di Caro, and M. Sampels, editors, *Ant Algorithms*, volume 2463 of *Lecture Notes in Computer Science*, pages 88–99. Springer, Berlin, Heidelberg, 2002.

[16] D. Apiletti, E. Baralis, and T. Cerquitelli. Energy-saving models for wireless sensor networks. *Knowledge and Information Systems*, 28(3):615–644, 2011.

[17] L. Gueta, R. Chiba, J. Ota, T. Arai, and T. Ueyama. A practical and integrated method to optimize a manipulator-based inspection system. In *Robotics and Biomimetics, 2007. ROBIO 2007. IEEE International Conference on*, pages 1911–1918, December 2007.

[18] C. Silva and T. Runkler. Ant colony optimization for dynamic traveling salesman problems. In *ARCS Workshops*, pages 259–266, 2004.

[19] N. Toklu, R. Montemanni, and L. Gambardella. A multiple ant colony system for a vehicle routing problem with time windows and uncertain travel times. *Journal of Traffic and Logistics Engineering*, 2(1):52–58, 2014.

[20] M. Mavrovouniotis and S. Yang. Ant algorithms with immigrants schemes for the dynamic vehicle routing problem. *Information Sciences*, 294:456–477, 2015.

[21] R. Montemanni, L. M. Gambardella, A. E. Rizzoli, and A. V. Donati. Ant colony system for a dynamic vehicle routing problem. *Combinatorial Optimization*, 10:327–343, 2005.

[22] R. Zhou, H. P. Lee, and A. Nee. Applying ant colony optimization algorithm to dynamic job shop scheduling problems. *International Journal of Manufacturing Research*, 3(3):301–320, 2008.

[23] J. Eaton, S. Yang, and M. Mavrovouniotis. Ant colony optimization with immigrants schemes for the dynamic railway junction rescheduling problem with multiple delays. *Soft Computing*, 20(8):2951–2966, 2016.

[24] S. Yang. Non-stationary problem optimization using the primal-dual genetic algorithm. In *The 2003 Congress on Evolutionary Computation CEC '03*, volume 3, pages 2246–2253, December 2003.

[25] S. Yang and X. Yao. Experimental study on population-based incremental learning algorithms for dynamic optimization problems. *Soft Computing*, 9(11):815–834, November 2005.

[26] S. Yang and X. Yao. Population-based incremental learning with associative memory for dynamic environments. *IEEE Transactions on Evolutionary Computation*, 12(5):542–561, 2008.

[27] M. Randall. A dynamic optimisation approach for ant colony optimisation using the multidimensional knapsack problem: Recent advances in artificial life. *Advances in Natural Computation*, 3:215–226, 2005.

[28] S. Yang. Genetic algorithms with memory- and elitism-based immigrants in dynamic environments. *Evolutionary Computation*, 16(3):385–416, 2008.

[29] M. Dorigo and L. M. Gambardella. Ant colony system: A cooperative learning approach to the traveling salesman problem. *IEEE Transactions on Evolutionary Computation*, 1(1):53–66, 1997.

[30] M. Dorigo and T. Stützle. *Ant colony optimization*. MIT Press, Cambridge, MA, 2004.

[31] M. Dorigo. Ant colony optimization. *Scholarpedia*, 2(3):1461, 2007.

[32] T. Stützle and H. Hoos. MAX–MIN ant system and local search for the traveling salesman problem. In *IEEE International Conference on Evolutionary Computation*, pages 309–314, 1997.

[33] M. Mavrovouniotis, F. M. Müller, and S. Yang. Ant colony optimization with local search for the dynamic travelling salesman problems. *IEEE Transactions on Cybernetics*, 47(7):1743–1756, 2017.

[34] D. Angus and T. Hendtlass. Ant colony optimisation applied to a dynamically changing problem. In T. Hendtlass and M. Ali, editors, *Developments in Applied Artificial Intelligence*, volume 2358 of *Lecture Notes in Computer Science*, pages 618–627. Springer, Berlin, Heidelberg, 2002.

[35] M. Guntsch and M. Middendorf. Pheromone modification strategies for ant algorithms applied to dynamic TSP. In E. J. W. Boers, editor, *Applications of*

Evolutionary Computing, volume 2037 of *Lecture Notes in Computer Science*, pages 213–222. Springer, Berlin, Heidelberg, 2001.

[36] M. Mavrovouniotis and S. Yang. A memetic ant colony optimization algorithm for the dynamic travelling salesman problem. *Soft Computing*, 15(7): 1405–1425, 2011.

[37] M. Mavrovouniotis and S. Yang. Adapting the pheromone evaporation rate in dynamic routing problems. In A. Esparcia-Alcázar, editor, *Applications of Evolutionary Computation*, volume 7835 of *Lecture Notes in Computer Science*, pages 606–615. Springer, Berlin, Heidelberg, 2013.

[38] M. Mavrovouniotis and S. Yang. Memory-based immigrants for ant colony optimization in changing environments. In C. Di Chio, S. Cagnoni, C. Cotta, *et al.*, editors, *Applications of Evolutionary Computation*, volume 6624 of *Lecture Notes in Computer Science*, pages 324–333. Springer, Berlin, Heidelberg, 2011.

[39] M. Mavrovouniotis and S. Yang. Interactive and non-interactive hybrid immigrants schemes for ant algorithms in dynamic environments. In *2014 IEEE Congress on Evolutionary Computation (CEC)*, pages 1542–1549, July 2014.

[40] M. Mavrovouniotis and S. Yang. Ant colony optimization with self-adaptive evaporation rate in dynamic environments. In *2014 IEEE Symposium on Computational Intelligence in Dynamic and Uncertain Environments (CIDUE)*, pages 47–54, December 2014.

[41] L. Melo, F. Pereira, and E. Costa. Multi-caste ant colony algorithm for the dynamic traveling salesperson problem. In M. Tomassini, A. Antonioni, F. Daolio, and P. Buesser, editors, *Adaptive and Natural Computing Algorithms*, volume 7824 of *Lecture Notes in Computer Science*, pages 179–188. Springer, Berlin, Heidelberg, 2013.

[42] L. Melo, F. Pereira, and E. Costa. Extended experiments with ant colony optimization with heterogeneous ants for large dynamic traveling salesperson problems. In *2014 14th International Conference on Computational Science and Its Applications (ICCSA)*, pages 171–175, June 2014.

[43] M. Mavrovouniotis, F. M. Müller, and S. Yang. An ant colony optimization based memetic algorithm for the dynamic travelling salesman problem. In *Proceedings of the 2015 Genetic and Evolutionary Computation Conference (GECCO15)*, pages 49–56, 2015.

[44] M. Guntsch, M. Middendorf, and H. Schmeck. An ant colony optimization approach to dynamic TSP. In *Proceedings of the 2001 Genetic and Evolutionary Computation Conference*, pages 860–867, 2001.

[45] M. Mavrovouniotis, S. Yang, and X. Yao. Multi-colony ant algorithms for the dynamic travelling salesman problem. In *2014 IEEE Symposium on Computational Intelligence in Dynamic and Uncertain Environments (CIDUE)*, pages 9–16, December 2014.

[46] J. Liu. Rank-based ant colony optimization applied to dynamic traveling salesman problems. *Engineering Optimization*, 37(8):831–847, 2005.

[47] K. Socha. ACO for continuous and mixed-variable optimization. In M. Dorigo, M. Birattari, C. Blum, L. Gambardella, F. Mondada, and T. Stützle, editors, *Ant*

Colony Optimization and Swarm Intelligence, volume 3172 of *LNCS*, pages 25–36. Springer, Berlin, Heidelberg, 2004.

[48] J. Euchi, A. Yassine, and H. Chabchoub. The dynamic vehicle routing problem: Solution with hybrid metaheuristic approach. *Swarm and Evolutionary Computation*, 21:41–53, 2015.

[49] B. van Veen, M. Emmerich, Z. Yang, T. Bäck, and J. Kok. Ant colony algorithms for the dynamic vehicle routing problem with time windows. In J. M. Ferrández Vicente, J. R. Álvarez Sánchez, F. de la Paz López, and F. J. Toledo Moreo, editors, *Natural and Artificial Computation in Engineering and Medical Applications: 5th International Work-Conference on the Interplay Between Natural and Artificial Computation, IWINAC 2013, Mallorca, Spain, June 10–14, 2013. Proceedings, Part II*, pages 1–10. Springer, Berlin, Heidelberg, 2013.

[50] M. Mavrovouniotis and S. Yang. Ant colony optimization with memory-based immigrants for the dynamic vehicle routing problem. In *2012 IEEE Congress on Evolutionary Computation (CEC)*, pages 2645–2652, June 2012.

[51] M. Mavrovouniotis and S. Yang. Dynamic vehicle routing: A memetic ant colony optimization approach. In A. Uyar, E. Ozcan, and N. Urquhart, editors, *Automated Scheduling and Planning*, volume 505 of *Studies in Computational Intelligence*, pages 283–301. Springer, Berlin, Heidelberg, 2013.

[52] N. Toklu, R. Montemanni, and L. Gambardella. An ant colony system for the capacitated vehicle routing problem with uncertain travel costs. In *2013 IEEE Symposium on Swarm Intelligence (SIS)*, pages 32–39, April 2013.

[53] S. Gao, Y. Wang, J. Cheng, Y. Inazumi, and Z. Tang. Ant colony optimization with clustering for solving the dynamic location routing problem. *Applied Mathematics and Computation*, 285:149–173, 2016.

[54] Z. Yang, M. Emmerich, and T. Bäck. Ant based solver for dynamic vehicle routing problem with time windows and multiple priorities. In *2015 IEEE Congress on Evolutionary Computation (CEC)*, pages 2813–2819, May 2015.

[55] M. Ankerl and A. Hämmerle. Applying ant colony optimisation to dynamic pickup and delivery. In R. Moreno-Díaz, F. Pichler, and A. Quesada-Arencibia, editors, *Computer Aided Systems Theory – EUROCAST 2009*, volume 5717 of *Lecture Notes in Computer Science*, pages 721–728. Springer, Berlin, Heidelberg, 2009.

[56] R. Zhou, A. Nee, and H. P. Lee. Performance of an ant colony optimisation algorithm in dynamic job shop scheduling problems. *International Journal of Production Research*, 47(11):2903–2920, 2009.

[57] S. Lorpunmanee, M. Sap, A. Abdullah, and C. Chompoo-Inwai. Ant colony optimization for dynamic job scheduling in grid environment. *International Journal of Computer and Information Engineering*, 1(8):469–476, 2007.

[58] C. Fernandes, A. Rosa, and V. Ramos. Binary ant algorithm. In *Proceedings of the 9th Annual Conference on Genetic and Evolutionary Computation, GECCO '07*, pages 41–48, New York, NY, USA, 2007. ACM.

[59] M. Mavrovouniotis and S. Yang. Applying ant colony optimization to dynamic binary-encoded problems. In A. Mora and G. Squillero, editors,

EvoApplications 2015: Applications of Evolutionary Computation, volume 9028 of *Lecture Notes in Computer Science*, pages 845–856. Springer, Berlin, Heidelberg, 2015.

[60] T. Stützle and H. H. Hoos. MAX–MIN ant system. *Future Generation Computer Systems*, 16(8):889–914, 2000.

[61] J. Branke. Memory enhanced evolutionary algorithms for changing optimization problems. In *Proceedings of the 1999 Congress on Evolutionary Computation*, volume 3, pages 1875–1882, 1999.

[62] Z. Michalewicz. Quo vadis, evolutionary computation? In J. Liu, C. Alippi, B. Bouchon-Meunier, G. W. Greenwood, and H. A. Abbass, editors, *Advances in Computational Intelligence: IEEE World Congress on Computational Intelligence, WCCI 2012, Brisbane, Australia, June 10–15, 2012. Plenary/Invited Lectures*, pages 98–121. Springer, Berlin, Heidelberg, 2012.

[63] A. Vogel, M. Fischer, H. Jaehn, and T. Teich. Real-world shop floor scheduling by ant colony optimization. In M. Dorigo, G. D. Caro, and M. Sampels, editors, *Ant Algorithms*, volume 2463 of *Lecture Notes in Computer Science*, pages 268–273. Springer, Berlin, Heidelberg, 2002.

[64] M. Garcia, O. Montiel, O. Castillo, R. Sepúlveda, and P. Melin. Path planning for autonomous mobile robot navigation with ant colony optimization and fuzzy cost function evaluation. *Applied Soft Computing*, 9(3):1102–1110, 2009.

[65] X. Yu, Y. Jin, K. Tang, X. Yao. Robust optimization over time – A new perspective on dynamic optimization problems. In *2010 IEEE Congress on Evolutionary Computation (CEC)*, pages 3998–4003, 2010.

Chapter 6

Comparison of multidimensional swarm embedding techniques by potential fields

Youngha Hwang[1] *and Okan K. Ersoy*[1]

Abstract

Multidimensional embedding is a technique useful for characterizing spectral signature relations in hyperspectral images. However, such images consist of disjoint similar spectral classes that are spatially sensitive, thus presenting challenges to existing graph embedding methods. In this chapter, the multidimensional embedding techniques are interpreted by the potential fields method, where a sum of attraction and repulsion potential functions is minimized to find optimal energy configuration of embedding. Repulsion dominates at short distances, thereby emphasizing local relations. On the other hand, attraction dominates at long distances, stressing global relations. The formulations capture long-range- and short-range-distance-related effects often associated with living organisms and help to establish algorithmic properties that mimic mutual behavior for the purpose of dimensionality reduction. Widely used tSNE (stochastic neighbor embedding) and its variations are compared in terms of the potential field methods and their sources of weakness are discussed. As part of the evaluation, the embedding maps are visualized, their trajectories are plotted, and their semisupervised classifications are conducted for image scenes acquired by multiple sensors at various spatial resolutions over different types of objects.

6.1 Introduction

For readers who are not familiar with the dimensionality reduction in hyperspectral imaging, the background, previous approaches to the dimensionality reduction, and merit of the potential field methods or force field framework are repeated from the published article [1].

6.1.1 Background

Hyperspectral remote sensing is concerned with the extraction of information from objects or scenes lying on the Earth surface, based on their radiance acquired by

[1]Electrical and Computer Engineering, Purdue University, USA

airborne or spaceborne sensors [2]. The images are characterized by high spectral resolution that generates several hundreds of bands capturing the electromagnetic reflectance properties of different materials. However, having many bands (or channels) poses several challenges to conventional land-cover studies, which include classification algorithms due to the curse of dimensionality. Other additional challenges include the inherent nonlinear characteristics that stem from bidirectional reflectance distribution function (BRDF) [3]. BRDF often leads to variations in spectral reflectance of different classes as a function of position in the landscape, depending on the local topology. Furthermore, for imagery that is acquired over coastal environments (e.g., coastal wetlands), the variable presence of water in pixels as a function of position in landscape presents more sources of nonlinearities. Thus, recent research efforts have identified dimensionality reduction and manifold learning techniques as key for preprocessing hyperspectral images.

6.1.2 Review of previous approaches

Dimensionality reduction methods bring better compression, better visualization, and extraction of useful classifier input features. Widely used techniques include linear formulations that are easy to implement such as principal component analysis [4], multidimensional scaling (MDS) [5], and local Fisher discriminant analysis [6].

Further developments in hyperspectral sensing enabled the acquisition of greater details about objects on the earth's surface. Thus, linear dimensionality reduction techniques are challenged. Nonlinear methods include the following methods. The locally linear embedding (LLE) [7] represents the neighborhood relations by a linear transformation that best reconstruct each data point from its neighbors. The Laplacian eigenmaps (LE) [8] exploit the correspondence between the Laplace Beltrami operator on a manifold and the connections to the heat equation. It aims to devise a geometrically motivated algorithm to find a representation for data sampled in a low-dimensional manifold embedded in a higher dimensional space. Both LLE and LE solution spaces consist of the trailing eigenvectors of the linear transformation and the Laplacian matrices, respectively. Isometric feature Mapping (Isomap) [9] assumes linearity on the local neighborhoods of data and proposes their piecewise combination to form global nonlinear structures. It first solves LLE for locality and applies MDS for the global map.

Very few conventional methods have been successfully used in the analysis of hyperspectral data. According to a comparative study that evaluated the effect of various nonlinear manifold learning algorithms when reducing the dimension of hyperspectral data [10], conventional dimensionality reduction methods fail to take advantage of the disjoint class structure in hyperspectral imagery. When embedding algorithms cannot discriminate dissimilar objects or handle disjoint classes, we say that it suffers from the crowding problem [11]. In hyperspectral imaging, the crowding problem appears as collapsing maps toward the center of the embedding space [1]. This phenomenon causes embedding algorithms to fail to establish discriminative boundaries that are necessary for classification accuracy.

6.1.3 Review of merits in force field framework

The crowding problem can be handled by a general embedding framework that can be used as a platform for developing new dimensionality reduction models. The new dimensionality reduction framework is reviewed as multidimensional artificial field embedding (MAFE) [1] that finds a minimum energy configuration state of the high-dimensional neighborhood graph. The force field intuition from mechanics inspired the framework. Pair-dependent attraction and repulsion potential functions are designed to embody long- and short-range force effects. The potentials are superimposed to generate an odd function that invokes a pairwise interaction force between pairs of maps in the embedding space.

Probabilistic methods can be interpreted as MAFE because they minimize a sum of Kullback–Leibler divergence between two distributions of high-dimensional space and lower dimensional embedding space. The attractive potential comes from the distribution of high-dimensional space and the repulsive potential from that of low-dimensional space. For example, the Stochastic Neighbor Embedding (SNE) [12] represents each object by a mixture of widely separated low-dimensional factors capturing some of the local structure and establishes global formations of clusters of similar maps. Student t-distribution based stochastic neighbor embedding (tSNE) [11] is a variation of SNE, which assumes the probability relations in the lower dimensional space to be inversely proportional to the distance between pairs of maps. Spherical stochastic neighbor embedding (sSNE) [13] is a variation of the SNE that the lower dimensional maps are modeled by the Exit distribution. sSNE maps the high-dimensional data to unit l_2-norm nonlinear manifold; thus, it captures well important regularities of the data.

6.1.4 Contributions

This chapter reviews nonlinear multidimensional embedding in the framework of potential field or force field based on the paper [1] and original papers that were cited in it. MAFE formulation is repeated from the paper. However, we added newly explicit connection of multidimensional embedding to the swarm intelligence that was missing from the paper. Specifically, we found that MAFE is indeed an application of swarm intelligence combined with the potential field methods. In addition, interpretation of neighbor embedding methods by MAFE is improved in its clarity and conciseness. Interpretations of sSNE and large margin nearest neighbor (LMNN) are newly added in this chapter. LMNN learns a Mahalanobis distance metric for k-nearest neighbor (kNN) classification by semidefinite programming [14]. The metric is trained with the goal that the kNN always belong to the same class while examples from different classes are separated by a large margin.

Experiment method followed the approach of [1] in visualization, trajectories of embedding map, and semi-supervised classification. However, tSNE combined with the bilateral similarity function was not discussed before. Its classification accuracy actually surpasses that of a MAFE model with the bilateral similarity function, whereas the MAFE model combined with the bilateral similarity function turned out to be better than tSNE with the photometric similarity in [1]. The bilateral similarity combines the photometric similarity with the geometric closeness. Spatially sensitive kernel function was used together with the photometric similarity for computing high-dimensional neighborhood graphs. The kernel function encodes spatial details that are essential for maintaining disjoint spectral classes in hyperspectral embedding. On the other hand, the photometric similarity weighs pixel values as a function of the photometric differences between the center pixel and its neighbor pixels. In addition, missed comparison to sSNE and LMNN is added. The performance of sSNE turns out better than the MAFE model.

The rest of this chapter is organized as follows. Section 6.2 reviews swarm intelligence and potential field methods widely used in robot planning studies. Swarm robotics based on potential field methods are discussed as an analogy to multi-dimensional embedding by force field framework. Section 6.3 provides a brief review of the force field intuition and presents a graph embedding approach based on the intuition. Section 6.4 discusses the interpretation of the existing neighbor embedding methods within the potential field framework. Then, Section 6.5 recapitulates potential forms used in formulating MAFE model along with the bilateral similarity kernel to add spatial sensitivity to the neighborhood graph. Section 6.6 presents the experimental results. Then the discussion follows to conclude the chapter.

6.2 Swarm intelligence and potential field methods

6.2.1 Swarm intelligence

The term of *swarm intelligence* was first reportedly used in [15] when it discussed "intelligent" behavior of cellular robotic systems (CRS). CRS consist of collections of autonomous, non-synchronized, non-intelligent robots cooperating to achieve global tasks. It operates on a finite n-dimensional cellular space under distributed control. No system-wide centralized mechanism, synchronous clock, or shared memory is assumed. Limited communication exists only between adjacent robots. Swarm intelligence is defined in systems of non-intelligent robots exhibiting collectively intelligent behavior evident in the ability to *unpredictably* produce specific (not in a statistical sense) ordered patterns of matter in the external environment. Here *unpredictable* is meant as globally "intractable" or "externally not-representable." On the other hand, swarm intelligence is defined more broadly in [16] by any attempt to design algorithms or distributed problem-solving devices inspired by the collective behavior of social insect colonies and other animal societies.

The application of swarm intelligence principles to collective robotics can be termed "swarm robotics." Swarm robotics is an approach to the coordination of multi-robot systems that consist of large numbers of relatively simple robots which take its inspiration from social insects [17]. The most remarkable characteristic of swarm robots is the ability to work cooperatively to achieve a common goal.

6.2.2 Review of potential field methods

This subsection reviews the force field formulation in robot planning studies, so the contents are repeated here to enhance readers' understanding from [18]. Potential field method, also called force field framework [1], was proposed as a robot planning method. It treats the robot represented as a point in configuration space as a particle under the influence of an *artificial potential field* U whose local variations are expected to reflect the "structure" of the free space denoted by $\mathbf{q}$. The potential function is typically defined over free space as the sum of an *attractive* potential pulling the robot toward the goal configuration and a *repulsive* potential pushing the robot away from the obstacles. Motion planning is performed in an iterative fashion. At each iteration, the artificial force [Equation (6.1)] induced by the potential function at the current configuration is regarded as the most promising direction of motion, and path generation proceeds along this direction by some increment.

$$\mathbf{F}(\mathbf{q}) = -\nabla\mathbf{U}(\mathbf{q}) \tag{6.1}$$

The potential field was originally developed as an on-line collision avoidance approach, applicable when the robot does not have a prior model of the obstacles but senses them during motion execution [19]. Emphasis was put on real-time efficiency, rather than on guaranteeing the attainment of the goal. Potential field methods are often referred to as "local methods." This comes from the fact that most potential functions are defined in such a way that their values at any configuration do not depend on the distribution and shapes of the obstacles beyond some limited neighborhood around the configuration. However, one could construct an "ideal" potential function with a single minimum (at the goal configuration) in the connected subset of the free space containing the goal configuration, this function could be regarded as some kind of "global" information about free space.

6.2.3 Swarm intelligence and potential field methods

There have been attempts to combine swarm robotics and potential field methods for unmanned aerial vehicles (UAVs). A strategy for organizing swarms of unmanned vehicles into a formation was presented by utilizing artificial potential fields that were generated from normal and sigmoid functions [20,21]. These functions construct the surface on which swarm members travel, controlling the overall swarm geometry and the individual member spacing. Several basic swarming laws for UAVs were developed for both two-dimensional (2D) plane and three-dimensional (3D) space [22]. Effects of these basic laws on the group behavior of swarms of UAVs were studied. Algorithms were developed and successfully tested for the two types of collision avoidance, group-wise and individual. Formation control analysis and design problems for UAV swarm systems to achieve time-varying formations were investigated [23]. To achieve predefined time-varying formations, formation protocols are presented for UAV swarm systems first, where the velocities of UAVs can be different when achieving formations. Then, consensus-based approaches are applied to deal with the time-varying formation control problems for UAV swarm systems.

6.3 Review of formulation of MAFE

Force field framework or potential field methods are widely used in robotics studies [18,24] in which related models are described to study the stability and motion planning of robots. Application of the framework to graph-based embedding was made by an analogy of objects moving in a coordinate space [1]. Specifically, an attractive potential pulls the objects in the lower dimensional space to the prior model, high-dimensional neighbor graph, while a repulsive potential push the objects away from the other objects except themselves. It is exactly the same as utilization of the artificial potential fields for formation control of swarm robots that was discussed in the previous section.

The following contents of this section are iterated to help readers' understanding from [1]. MAFE proposes each data point to be associated with a vertex of a high-dimensional neighborhood graph that has a corresponding map in the lower dimensional space. Each embedding map has both attractive and repulsive force that is dynamically determined by varying distances between the images of the data. So the dimensionality reduction problem is posed as finding minimum energy configuration or the graph topology that preserves the neighborhood.

6.3.1 Graph embedding framework

Graph embedding can be regarded as a planning of simultaneous motions of embedded maps in the lower dimensional space by the superposition of potential functions that are defined for all objects in the high-dimensional space. All the objects except itself are regarded as its obstacles.

Let $\mathscr{G} = (\mathscr{E}, \mathscr{V})$ be a finite undirected graph with vertices $\mathscr{V}$ and edges $\mathscr{E}$. Elements of the edges can be thought as ideal springs. Let $\mathscr{S} = \{(w_{ij}, k_{ij})\}$ be the spring properties between each vertex i and j for all edges $(v_i, v_j) \in \mathscr{E}$, where w_{ij} is the normalized length without any force between the vertices in $\mathscr{V}$ or equivalently in $\mathbf{Y} = \{\mathbf{y}_1, \mathbf{y}_2, \ldots, \mathbf{y}_N\}$ with $\mathbf{y}_i \in \mathbb{R}^d$ and $k_{ij} = 1$ is the spring force constant. $\mathscr{S}$ is called a neighborhood spring graph and denoted by $\mathscr{G}_{\mathscr{S}}$.

Embedding of $\mathscr{G}_{\mathscr{S}}$ is an assignment of vertices into an m-dimensional Euclidean space $\mathbb{R}^m$, where $\mathbb{R}^m$ is a lower dimensional space. Let $\mathbf{Z} = \{\mathbf{z}_1, \mathbf{z}_2, \ldots, \mathbf{z}_N\}$, be the images of $\mathscr{G}_{\mathscr{S}}$ by the embedding and $\mathbf{z}_i \in \mathbb{R}^m$ is the position of the vertex image. Then, the problem of dimensionality reduction becomes that of finding the minimum energy embedding that is controlled by the structure in $\{w_{ij}\}$. Such embedding is desired to preserve pairwise distances of the neighborhood graph $\mathscr{G}_{\mathscr{S}}$.

The graph embedding framework is interpreted as follows: each map $\mathbf{z}_i \in \mathbf{Z}$ is assumed to be moving simultaneously and to be aware of the forces that exist and the positions of all the other maps, $\{\mathbf{z}_j\}_{j \neq i}$. Assuming such motion changes in a continuous time, the velocity determined by the effect of group members on each map i at position $\mathbf{z}_i$ is described by

$$\dot{\mathbf{z}}_i = \sum_{j \neq i} F^{ij}(\mathbf{z}_i - \mathbf{z}_j), \quad i = 1, \ldots, N, \tag{6.2}$$

where $F^{ij}(\mathbf{z}_i - \mathbf{z}_j)$ denotes pairwise symmetric interactions between the ith and jth maps. The symmetry of the function follows from the law of motion in physics. Force on $\mathbf{z}_i$ is made of attraction term and repulsion as follows:

$$F^{ij}(\mathbf{z}_i - \mathbf{z}_j) = (\mathbf{z}_i - \mathbf{z}_j)\left(F_r^{ij}(\|\mathbf{z}_i - \mathbf{z}_j\|) - F_a^{ij}(\|\mathbf{z}_i - \mathbf{z}_j\|)\right), \tag{6.3}$$

where F_r^{ij} and F_a^{ij} denote the magnitude of the repulsion force and that of the attraction force, respectively.

6.3.2 Force field functions for embedding

We assume that at large distances the attraction dominates and at short distances the repulsion dominates, while in between there is a unique distance where the attraction and the repulsion balance. Embedding forces $F^{ij}(\mathbf{z}_i - \mathbf{z}_j)$ should obey the following force field properties.

1. There exists a pair equilibrium distance $\hat{w}_{ij}$ where $F_r^{ij}(\hat{w}_{ij}) = F_a^{ij}(\hat{w}_{ij})$, else For $\|\mathbf{z}_i - \mathbf{z}_j\| > \hat{w}_{ij}$, $F_a^{ij}(\|\mathbf{z}_i - \mathbf{z}_j\|) > F_r^{ij}(\|\mathbf{z}_i - \mathbf{z}_j\|)$ or For $\|\mathbf{z}_i - \mathbf{z}_j\| < \hat{w}_{ij}$, $F_a^{ij}(\|\mathbf{z}_i - \mathbf{z}_j\|) < F_r^{ij}(\|\mathbf{z}_i - \mathbf{z}_j\|)$.
2. F^{ij} is an odd function, that is, $F^{ij}(-(\mathbf{z}_i - \mathbf{z}_j)) = -F^{ij}(\mathbf{z}_i - \mathbf{z}_j)$.
3. There exist pair-dependent positive functions or potentials U_a^{ij} and U_r^{ij} such that

$$\begin{aligned}\nabla_{\mathbf{z}_i} U_a^{ij}(\|\mathbf{z}_i - \mathbf{z}_j\|) &= F_a^{ij}(\|\mathbf{z}_i - \mathbf{z}_j\|)(\mathbf{z}_i - \mathbf{z}_j)\\ \nabla_{\mathbf{z}_i} U_r^{ij}(\|\mathbf{z}_i - \mathbf{z}_j\|) &= F_r^{ij}(\|\mathbf{z}_i - \mathbf{z}_j\|)(\mathbf{z}_i - \mathbf{z}_j)\end{aligned}$$

U_a^{ij} and U_r^{ij} are artificial attraction and repulsion potentials. The term $(\mathbf{z}_i - \mathbf{z}_j)F_r^{ij}(\|\mathbf{z}_i - \mathbf{z}_j\|)$ represents the repulsion force and the term $-(\mathbf{z}_i - \mathbf{z}_j)F_a^{ij}(\|\mathbf{z}_i - \mathbf{z}_j\|)$ represents the attraction force. Note that the potential functions do not dictate the trajectories of the maps explicitly. Instead, the interactions of every map with its neighbors are described by a superposition of potentials so that the position of the map adapts to the changes of other maps. Thus, the resultant forces on each individual map can be written as

$$\dot{\mathbf{z}}_i = -\sum_{j\neq i}\left(\nabla_{\mathbf{z}_i} U_a^{ij}(\|\mathbf{z}_i - \mathbf{z}_j\|) - \nabla_{\mathbf{z}_i} U_r^{ij}(\|\mathbf{z}_i - \mathbf{z}_j\|)\right). \tag{6.4}$$

Let $U_i(\mathbf{Z})$ be the potential on the map i be the superposition of the attraction potential and the repulsion potential.

$$U_i(\mathbf{Z}) = \sum_{\mathbf{j}\neq \mathbf{i}}\left(\mathbf{U}_\mathbf{a}^{\mathbf{ij}}(\|\mathbf{z}_\mathbf{i} - \mathbf{z}_\mathbf{j}\|) - \mathbf{U}_\mathbf{r}^{\mathbf{ij}}(\|\mathbf{z}_\mathbf{i} - \mathbf{z}_\mathbf{j}\|)\right). \tag{6.5}$$

Then, the above framework can be written with the potentials on each individual map i as

$$\dot{\mathbf{z}}_i = -\nabla_{\mathbf{z}_i} U_i(\mathbf{Z}). \tag{6.6}$$

The total superposed potential function on $\mathcal{G}_\mathcal{S}$ is defined by

$$U(\mathbf{Z}) = \sum_{\mathbf{i}=1}^{\mathbf{N}} \mathbf{U}_\mathbf{i}(\mathbf{Z}). \tag{6.7}$$

So new embedding maps can be solved by the optimization problem of the form

$$\mathbf{Z}^* = \arg\min_{\mathbf{Z}} \mathbf{U}(\mathbf{Z}), \tag{6.8}$$

where $\mathbf{Z}^*$ denotes the optimal $\mathbf{Z}$.

6.4 Examples of swarm embedding by potential fields

This section reviews interpretation of SNE [12], Student-*t* stochastic neighbor embedding (t-SNE) [11], sSNE [13], and LMNN [14].

6.4.1 Stochastic neighbor embedding

SNE is an embedding that preserves neighbor identities on lower dimensional manifolds. Edge weights or pairwise dissimilarities of neighbor vertices in high-dimensional space are not necessarily symmetric in SNE. However, if the dissimilarities, d_{ij}^2, are computed by the scaled squared Euclidean distances between the high-dimensional points $\mathbf{y}_i$, $\mathbf{y}_j$, then they are symmetric as shown below

$$d_{ij}^2 = \frac{\|\mathbf{y}_i - \mathbf{y}_j\|^2}{\sigma_i^2}, \tag{6.9}$$

where σ_i is either set by hand or found by a binary search for the value of σ_i that makes the entropy of the distribution over neighbors equal to $\log k$. Here, k is the effective number of local neighbors or perplexity and is chosen by hand [12].

The edge weights or the probability that i would pick j as its neighbor is assumed by the Gaussian distribution,

$$w_{ij} = \frac{\exp(-d_{ij}^2)}{\sum_{k\neq i}\exp(-d_{ik}^2)}. \tag{6.10}$$

In the lower dimensional space, we also use Gaussian neighborhoods. Thus, the embedding graph weight or the induced probability that points i would pick j as its neighbor is given as a function of the lower dimensional images $\mathbf{z}_i$ of all the data,

$$\hat{w}_{ij} = \frac{\exp(-\|\mathbf{z}_i - \mathbf{z}_j\|^2)}{\sum_{k\neq i}\exp(-\|\mathbf{z}_i - \mathbf{z}_k\|^2)}. \tag{6.11}$$

The aim of the embedding is to match these two distributions w_{ij} and $\hat{w}_{ij}$ as well as possible. This is achieved by minimizing a cost function that is a sum of Kullback–Leibler divergences ($KL(w_i||\hat{w}_i||)$) between the original (w_i) and induced ($\hat{w}_i$) distributions over neighbors for each data.

$$U = \sum_i KL(w_i||\hat{w}_i) = \sum_i\sum_{j\neq i} w_{ij}\log\left(\frac{w_{ij}}{\hat{w}_{ij}}\right). \tag{6.12}$$

SNE is shown to be an improvement over LLE [7], since it does not allow widely separated data points collapsed as near neighbors in the low-dimensional space. Its cost function cleanly enforces both keeping the images of nearby objects nearby and keeping the images of widely separated objects relatively far apart [12]. The gradient of the cost function or the potential U is given as

$$\dot{\mathbf{z}}_i = -\frac{\partial U}{\partial \mathbf{z}_i} = -4\sum_j (\mathbf{z}_i - \mathbf{z}_j)(w_{ij} - \hat{w}_{ij}), \tag{6.13}$$

which indicates that a sum of forces pulling $\mathbf{z}_i$ toward $\mathbf{z}_j$ or pushing it away depending on whether j is observed to be a neighbor more or less often than desired. The attractive force comes from $-w_{ij}(\mathbf{z}_i - \mathbf{z}_j)$ and the repulsion force from $\hat{w}_{ij}(\mathbf{z}_i - \mathbf{z}_j)$. At longer distances, the attraction forces dominate and the repulsion forces are negligible, while the opposite happens with close ranges. Ignoring the terms independent of $\hat{w}_{ij}$ simplifies the potential as

$$\tilde{U} = -\sum_j w_{ij} \log \hat{w}_{ij} \tag{6.14}$$

$$= \sum_j w_{ij} \|\mathbf{z}_i - \mathbf{z}_j\|^2 + \log \sum_{k\neq i} \exp\left(-\|\mathbf{z}_i - \mathbf{z}_k\|^2\right), \tag{6.15}$$

where the first term $\sum_j w_{ij}\|\mathbf{z}_i - \mathbf{z}_j\|^2$ represents the attraction potential and the repulsive potential is shown by the second term $\sum_j \log \sum_{k\neq i} \exp\left(-\|\mathbf{z}_i - \mathbf{z}_k\|^2\right)$.

6.4.2 t-Stochastic neighbor embedding

tSNE [11] is a variation of the SNE that the lower dimensional maps are modeled by the Student t-distribution of the degree of freedom one or the Cauchy distribution. tSNE maps the high-dimensional data to a 2D or 3D map. It significantly reduced the tendency to crowd points together in the center of the map against SNE thanks to the heavy-tailed nature of the Cauchy distribution.

$$\hat{w}_{ij} = \frac{(1 + \|\mathbf{z}_i - \mathbf{z}_j\|^2)^{-1}}{\sum_{k\neq i}(1 + \|\mathbf{z}_i - \mathbf{z}_k\|^2)^{-1}}. \tag{6.16}$$

The gradient of the sum of Kullback–Leibler divergences is given as

$$\dot{\mathbf{z}}_i = -\frac{\partial U}{\partial \mathbf{z}_i} = -4\sum_j (\mathbf{z}_i - \mathbf{z}_j)(w_{ij} - \hat{w}_{ij})(1 + \|\mathbf{z}_i - \mathbf{z}_j\|^2)^{-1}. \tag{6.17}$$

The attractive force comes from $-w_{ij}(\mathbf{z}_i - \mathbf{z}_j)(1 + \|\mathbf{z}_i - \mathbf{z}_j\|^2)^{-1}$ and the repulsion force from $\hat{w}_{ij}(\mathbf{z}_i - \mathbf{z}_j)(1 + \|\mathbf{z}_i - \mathbf{z}_j\|^2)^{-1}$.

6.4.3 Spherical stochastic neighbor embedding

sSNE [13] is a variation of the SNE that the lower dimensional maps are modeled by the Exit distribution. sSNE maps the high-dimensional data to unit l_2-norm nonlinear manifold; thus, it captures well important regularities of the data. The spherical manifold similarities are given as

$$\hat{w}_{ij} = \frac{\|\mathbf{z}_j - \rho\mathbf{z}_i\|^{-m}}{\sum_{k\neq i}\|\mathbf{z}_k - \rho\mathbf{z}_i\|^{-m}}, \tag{6.18}$$

where ρ is called a common concentration parameter. The Exit distribution is defined by a distribution on the surface of the unit norm sphere of dimension m, $\mathbb{S}^m$.

$$f(\mathbf{z}, \mathbf{z}_i, \rho) = \frac{1}{A_m}\frac{1-\rho^2}{\|\mathbf{z} - \rho\mathbf{z}_i\|^m}, \quad \mathbf{z} \in \mathbb{S}^m, \tag{6.19}$$

where A_m is the surface area of $\mathbb{S}^m$, that is, $A_m = 2\pi^{m/2}\Gamma(m/2)$ with $\Gamma(\dot{})$ being the Gamma function. $\mathbf{z}_i$ is associated with the mean direction of the distribution. As $\rho \to 0$, the density becomes the uniform distribution on the surface and as $\rho \to 1$, it becomes concentrated around $\mathbf{z}_i$. The inverse distance related distribution leads to a heavy-tailed characteristic; thus, separation is accentuated as the distance becomes smaller.

The gradient of the sum of Kullback–Leibler divergences is given as

$$\dot{\mathbf{z}}_i = -\frac{\partial U}{\partial \mathbf{z}_i} = 2m\rho\sum_j(\mathbf{z}_j - \rho\mathbf{z}_i)(w_{ij} - \hat{w}_{ij})\|\mathbf{z}_j - \rho\mathbf{z}_i\|^{-2}. \tag{6.20}$$

The attractive force comes from $w_{ij}(\mathbf{z}_j - \rho\mathbf{z}_i)\|\mathbf{z}_j - \rho\mathbf{z}_i\|^{-2}$ and the repulsion force from $\hat{w}_{ij}(\mathbf{z}_j - \rho\mathbf{z}_i)\|\mathbf{z}_j - \rho\mathbf{z}_i\|^{-2}$.

6.4.4 Large margin nearest neighbor embedding

LMNN learns a Mahalanobis distance metric for kNN classification by semidefinite programming [14]. The metric is trained with the goal that the kNNs always belong to the same class while examples from different classes are separated by a large margin.

Let $\{\mathbf{y}_i, x_i\}_{i=1}^n$ denote a training set of n labeled examples with inputs $\mathbf{y}_i \in \mathbb{R}^d$ and discrete class labels x_i. The binary matrix $x_{ij} \in \{0, 1\}$ indicates whether the labels x_i and x_j match. Our goal is to learn a linear transformation $\mathbf{L} : \mathbb{R}^d \to \mathbb{R}^m$ whose images are given as $\mathbf{z}_i = \mathbf{L}\mathbf{y}_i$, where $\mathbf{z}_i \in \mathbb{R}^m$ and $m < d$. We want to learn the linear transformation that optimizes kNN classification when distances are measure as follows:

$$d(\mathbf{z}_i, \mathbf{z}_j) = \|\mathbf{L}(\mathbf{y}_i - \mathbf{y}_j)\|^2. \tag{6.21}$$

Equivalently, we rewrite the above equation as

$$d(\mathbf{z}_i, \mathbf{z}_j) = (\mathbf{y}_i - \mathbf{y}_j)^T\mathbf{M}(\mathbf{y}_i - \mathbf{y}_j), \tag{6.22}$$

where the matrix $\mathbf{M} = \mathbf{L}^T\mathbf{L}$ parametrizes the Mahalanobis distance matrix induced by the linear transformation $\mathbf{L}$.

We need to specify k "target" neighbors, that is, k other inputs with the same label x_i that we wish to have minimal distance to $\mathbf{z}_i$ in (6.21). In the absence of prior knowledge, the target neighbors are identified as the kNNs, determined by Euclidean distance, that share the same label x_i. We use $\eta_{ij} \in \{0, 1\}$ to indicate whether input $\mathbf{y}_j$ is a target neighbor of input $\mathbf{y}_i$.

The cost function penalizes large distances between each input and its target neighbors while penalizing the small distances between each input and all other inputs that do not share the same class label. Specifically, the cost function is given by

$$U(\mathbf{L}) = \sum_{ij} \eta_{ij} \|\mathbf{L}(\mathbf{y}_i - \mathbf{y}_j)\|^2 + c \sum_{ijl} \eta_{ij}(1 - x_{il})[1 + \|\mathbf{L}(\mathbf{y}_i - \mathbf{y}_j)\|^2 - \|\mathbf{L}(\mathbf{y}_i - \mathbf{y}_l)\|^2]_+, \quad (6.23)$$

where $[z]_+ = \max(z, 0)$ denotes the standard hinge loss and c is a positive constant that is typically set by cross-validation. The first term acts as an attraction potential to pull close each other and the second term does as a repulsion potential that separates samples from other classes by a large margin [1].

6.5 Review of MAFE model

This section discuses an example of MAFE model that is compared to the other embedding methods. The contents are iterated from [1,25].

6.5.1 MAFE model by embedding artificial potentials

The superposed artificial field embedding was proposed from the idea that the pair-wise equilibrium distances is thought as a sink of a potential function [1]. The equilibrium distances are regarded as the minimum energy configuration between the points in the embedded map. The attractive potential function must be bounded from below. Otherwise, all the points merge at a single point. We choose the following form of the attractive potentials:

$$U_a^{ij}(\|\mathbf{z}_i - \mathbf{z}_j\|) = \xi_a w_{ij} \|\mathbf{z}_i - \mathbf{z}_j\|^p, \quad (6.24)$$

where ξ_a is an attraction magnitude parameter and w_{ij} represents the spectral signature similarity. Figure 6.1(a) shows the attractive potential created by the above equation for $p = 3$, $\xi_a = 0.00033$, and $w_{ij} = 1$.

A repulsive potential function must prevent the two points $\mathbf{z}_i$ and $\mathbf{z}_j$ from getting too close. Thus, there must be a potential barrier, when the distance is small. Obviously, the repulsive potential function must have a negligible effect on the points in the embedded map, when the distance between the pair of points becomes huge. So it must be a decreasing function of distance, $\|\mathbf{z}_i - \mathbf{z}_j\|$. Example functions include Gaussian, exponential, Cauchy, hyperbolic tangent, and inverse distance power functions. We choose the form of the unbounded repulsive potential function that is the inverse of the attractive potential function in (6.24).

$$U_r^{ij}(\|\mathbf{z}_i - \mathbf{z}_j\|) = \xi_r \frac{1}{\|\mathbf{z}_i - \mathbf{z}_j\|^q}, \quad (6.25)$$

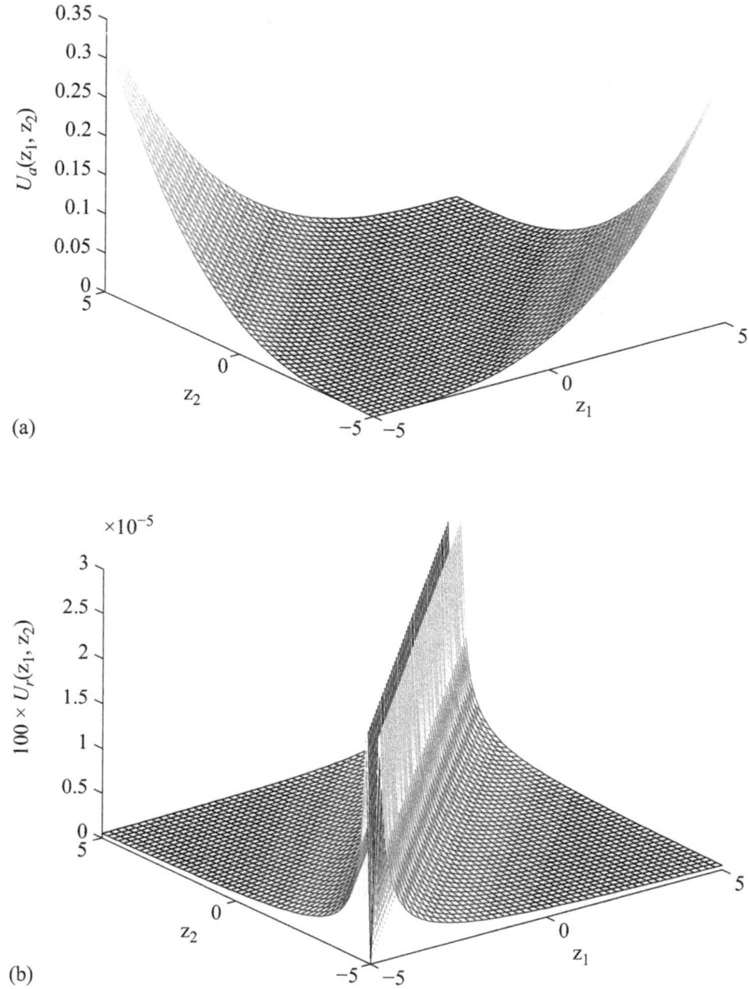

Figure 6.1 (a) Attraction potential for p = 3; (b) repulsive potential scaled up by 100 for q = 1

Figure 6.1(b) shows the scaled repulsive potential generated by (6.25) for $q = 1$ and $\xi_r = 5 \times 10^{-8}$. The superimposed field of the repulsive potential and the attractive potential in Figure 6.2(a) is given as

$$U(\mathbf{z}) = \sum_{i,j \neq i} \xi_a w_{ij} \|\mathbf{z}_i - \mathbf{z}_j\|^p + \xi_r \frac{1}{\|\mathbf{z}_i - \mathbf{z}_j\|^q}. \tag{6.26}$$

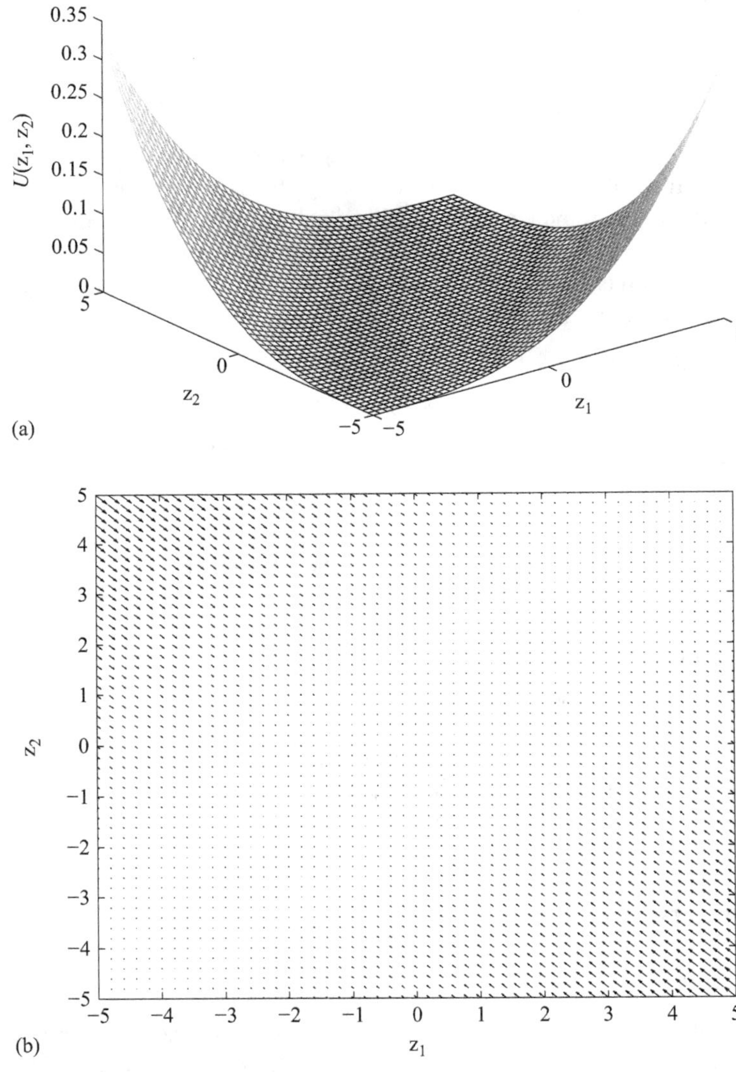

Figure 6.2 (a) Superimposed potential function for p = 3, q = 1, (b) arrows show negative gradient force field

The gradient of the potential gives the direction for which each pixel moves its position is given by

$$\dot{\mathbf{z}}_i = -\frac{\partial U(\mathbf{z})}{\partial \mathbf{z}_i} = -\sum_{j \neq i} (\mathbf{z}_i - \mathbf{z}_j)\xi_a w_{ij} p \|\mathbf{z}_i - \mathbf{z}_j\|^{p-2} - \xi_r q \frac{\mathbf{z}_i - \mathbf{z}_j}{\|\mathbf{z}_i - \mathbf{z}_j\|^{q+2}}. \quad (6.27)$$

Figure 6.2(b) shows the gradient field for a point where strong attraction force is present.

6.5.2 Bilateral kernel for spectral signature similarities

The bilateral filter was proposed to smooth images while preserving their edge information [26]. It combined the photometric similarity with the geometric closeness. Its application in the remote sensing applications includes [27,28]. Spatially sensitive kernel function was used for computing high-dimensional neighborhood graphs. The kernel function encodes spatial details that are essential for maintaining disjoint spectral classes in hyperspectral embedding:

$$K_s^{ij} = \exp\{-\|\mathbf{s_i} - \mathbf{s_j}\|^2\}, \tag{6.28}$$

where $\mathbf{s_i}$ denotes the spatial coordinates of pixel i. The above convolution kernel weighs image pixel values as a function of the spatial distance from the center pixel. There is also a photometric term:

$$K_y^{ij} = \exp\{-(\mathbf{y}_i - \mathbf{y}_j)^T \Sigma_y^{-1} (\mathbf{y}_i - \mathbf{y}_j)\}, \tag{6.29}$$

which weighs pixel values as a function of the photometric differences between the center pixel and its neighbor pixels [26,28]. $\mathbf{y}_i$ represents the photometric d-dimensional vector and Σ_y its covariance. Then, the unnormalized similarity of pixels i and j is defined as

$$w(\mathbf{s_i}, \mathbf{s_j}, \mathbf{y_i}, \mathbf{y_j}) = \mathbf{K_s^{ij} K_y^{ij}} \tag{6.30}$$

The inverse covariance matrix in the photometric similarity can be efficiently computed by the sparse matrix transform approach that solves the problem of eigen-decomposition of the covariance matrix [1,29].

6.6 Numerical analysis

6.6.1 Data sets

Hyperion data with eight identified classes of complex natural vegetation and one unidentified class were obtained over the Okavango Delta, Botswana [30]. Hyperspectral images are subsampled by three for efficient computations. The class groupings include seasonal swamps, occasional swamps, and woodlands in Table 6.1. Poor classification accuracies typically result from several classes whose signature spectra are overlapped. Embedding results are reported based on classification error rates for all nine classes.

Airborne hyperspectral data were acquired by the National Aeronautics and Space Administration over Kennedy Space Center (KSC). The data are subsampled by 11 in this work. A total of 176 features for 13 wetland and upland classes are present. Cabbage palm hammock (Class 3) and broad leaf/oak hammock (Class 6) are upland trees. Swamps (Classes 2 and 7) and marshes (Classes 8 and 9) are trees

Table 6.1 Experimental data: class labels and number of labeled samples

Botswana hyperion		**Kennedy Space Center**	
c1	Water (55)	c1	Scrub (70)
c2	Floodplain (77)	c2	Willow swamp (23)
c3	Riparian (73)	c3	Cabbage hamm (25)
c4	Firescar (60)	c4	Cabbage palm (20)
c5	Island interior (63)	c5	Slash pine (15)
c6	Woodlands (65)	c6	Oak (22)
c7	Savanna (56)	c7	Hardwood swamp (10)
c8	Short mopane (40)	c8	Graminoid marsh (37)
c9	Unclassified (38)	c9	Spartina marsh (48)
		c10	Cattail marsh (41)
		c11	Salt marsh (37)
		c12	Mud flats (41)
		c13	Water (85)

and grasses in wetlands. The spectral signatures are known to be mixed and their differences are subtle [1]. The embedding and classification results are reported for all 13 classes.

6.6.2 Method of experiments

For the comparison of the embedding algorithms, the dimensionality reduction tool box [31] is used. Visualization results are obtained by mapping to a 5D space, while classification results are created for a varying lower dimensional space. Land cover label information is not used for embedding but only used for classification purpose. The quality of the embedding methods is examined by the 1-nearest neighbor classification algorithm (1NN) that has many advantages in the following [1]. 1NN requires no a priori knowledge of class distribution and is an unbiased measure. It is also a fast method. For the iterative gradient-based approaches such as sSNE, tSNE, and MAFE, the trajectories of the embedded maps are also presented to examine the stability of the embedding map.

Both sSNE and tSNE terminates after 1,000 iterations of the gradient descent steps. The optimization of MAFE is stopped when the gradient of the cost function is less than 0.01 or after 1,000 iterations of the gradient descent steps. LMNN learns the linear transformation via the Mahalanobis distance metric for kNN classification, while LLE chooses the eigenvectors with the smallest m non-zero eigenvalues, where m represents the embedding dimension. Both LMNN and LLE assumes the neighborhood graph. The number of neighbors in LLE is chosen as 8 for Botswana data and 12 for KSC data. On the other hand, the number of neighbors in LMNN is set as 3 for Botswana data and 12 for KSC data.

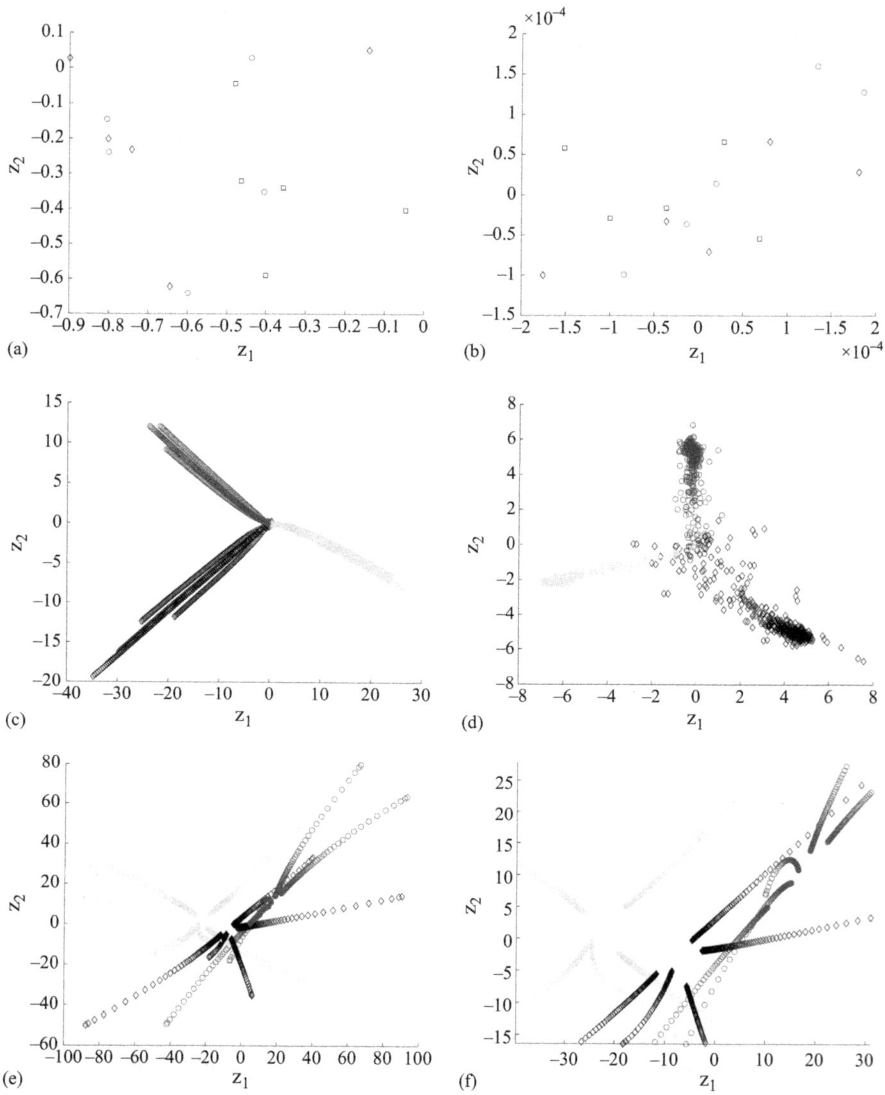

Figure 6.3 Illustrations of gradient-based optimization over 100 iterations for total 1,000 iterations of gradient descent steps. tSNE (d), MAFE (e), and MAFE zoom-in (f) are initialized from the same seed of 15 points from Botswana data of three different classes such as Firescar (diamond), Island Interior (circle), and Woodlands (square). sSNE (c) are initialized differently from the Exit distribution. Initial maps of sSNE and tSNE/MAFE are shown in (a) and (b), respectively

6.6.3 Trajectories in embedding space

Equation (6.2) formulates the model of the spectral manifolds. Travel of the embedded maps toward the minimum disposition of the graph helps us understand the embedding methods better. For example, the trajectories of the various mappings of 15 hyperspectral pixels are evaluated from three Botswana data classes. Five samples are randomly chosen from Woodlands, Firescar, and Island interior classes. Note that the initial map of sSNE follows the Exit distribution while those of tSNE/MAFE follows the Gaussian distribution. They are shown in Figure 6.3(a) and (b), respectively. Figure 6.3 shows tSNE optimization path against the trajectories of sSNE and MAFE. Figure 6.3(e) and (f) shows instabilities of MAFE such as many collisions. On the other hand, sSNE and tSNE produce stable paths in Figure 6.3(c) and (d), implying that the relation between the attraction force and the repulsion force is stable.

6.6.4 Visualization and embedding results

Let us start by elaborating on the typical "crowding problem" in the dimensional reduction field. Many classes of hyperspectral data tend to overlap each other because similar data are mapped to a single cluster in many embedding methods and similar classes of data can result in different manifolds due to their spacial location [1].

This section shows the results of many embedding methods including tSNE combined with two different similarity functions. Remarkable differences are found between the embedding results from tSNE combined with the bilateral similarity function (tSNE-BK) and "default" version of tSNE combined with Gaussian kernel function in (6.29). Thus, it is critical to have appropriate kernel functions that represent the geodesic structure of the data. For hyperspectral images, the bilateral similarity function looks competitive.

Figure 6.4(a) shows the ground truths of class sample pixels from the Botswana data. Figure 6.5 shows the experimental results from sSNE, tSNE-BK, tSNE, LMNN, LLE, and MAFE projections of the Botswana data. tSNE-BK and sSNE results show a superior embedding map compared to other methods. tSNE-BK does not have any overlap between the two classes that are known to be most difficult to

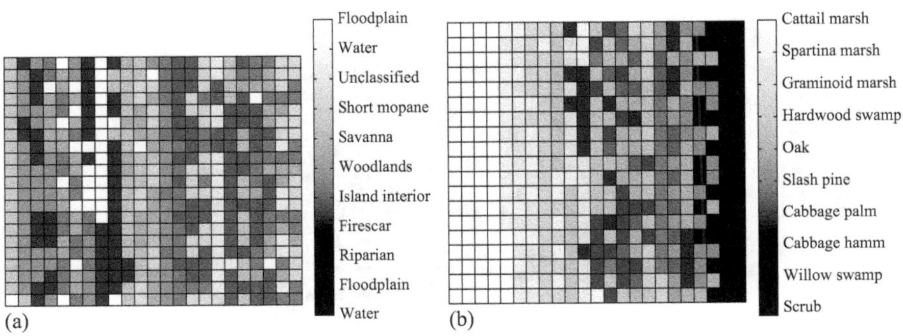

Figure 6.4 Class sample pixels for (a) Botswana data and (b) Kennedy Space Center data

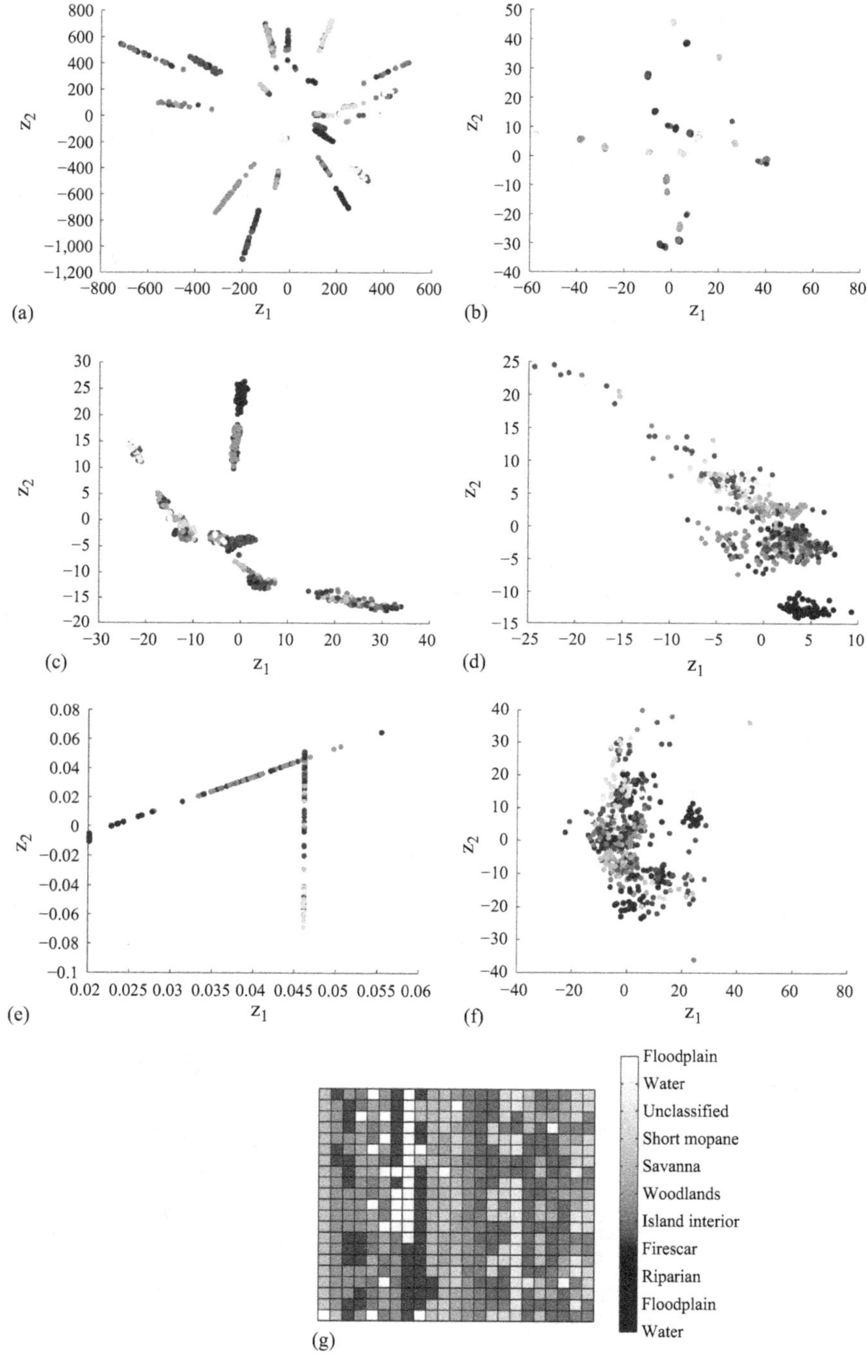

Figure 6.5 Embedding results of Botswana data from (a) sSNE, (b) tSNE-BK, (c) tSNE, (d) LMNN, (e) LLE, and (f) MAFE. Gray colored class labels are repeated in (g) from Figure 6.4(a)

separate [10]. The tSNE-BK result is the best in embedding the ground truth map. LMNN has a good clustering result but has a significant overlap between Riparian and Woodlands classes. tSNE also displays a similar trend as LMNN. Similarly, MAFE has a good clustering result but has a significant overlap between Short mopane and Unclassified classes. The hyperspectral manifold of the data sinks and classes are not well separated for LLE.

Figure 6.4(b) shows the ground references of the KSC data. Figure 6.6 displays the KSC embedding results from all the embedding techniques. LMNN and LLE show the crowding problem and overlap of classes. Almost all classes are not distinguishable for LMNN. The geometry of the embedded map collapses and classes except Water and Salt marsh are heavily overlapped for LLE. On the other hand, tSNE-BK and sSNE embedding map the high-dimensional hyperspectral data onto the disjoint clusters well. tSNE also has most classes overlapped except Water, Salt marsh, and Scrub. MAFE has classes overlapped fairly.

6.6.5 Classification results

Classification results are shown to compare further the representations of different embedding methods. Embedded pixel maps were randomly sampled to generate 70% training and 30% testing samples. The classification results were averaged over ten times. Standard deviations of the classification errors suggest that ten-time repetitions are indeed reliable as shown in Figures 6.7 and 6.8.

Tables 6.2 and 6.3 display the 1NN classification accuracy per class. They list the kappa statistics (KSs) [32], which is computed as follows:

$$KS = \frac{N \sum_{c=1}^{|C|} t_{cc} - \sum_{c=1}^{|C|} t_{c+}t_{+c}}{N^2 - \sum_{c=1}^{|C|} t_{c+}t_{+c}}, \tag{6.31}$$

where N is the number of testing samples, t_{cc} indicates the number of samples correctly classified in class c, t_{c+} denotes the number of testing samples labeled as class c, and t_{+c} does the number of samples predicted as belonging to class c. $|C|$ denotes the total number of classes in the data. The KS is known to be a more robust measure than simple percent accuracy calculation because it takes into the account the possibility of the accurate classification by chance. The kappa static shows that the classification results are reliable in spite of a small number of classification repetition in Tables 6.2 and 6.3.

tSNE-BK exhibits a clear separation of classes, so it accomplishes the best classification accuracy. On the other hand, tSNE with a Gaussian kernel neighborhood graph shows similar classification results as LMNN for Botswana data. The advantage of the bilateral kernel is proved by the difference between tSNE-BK and tSNE classification accuracy. LLE attained the lowest classification accuracy results for Botswana data. In addition, The Riparian (c3) and Woodlands (c6) classes have the lowest classification accuracy per class. The results are anticipated because the two classes are the most difficult to separate as shown in Figure 6.5. Note that tSNE and LMNN have similar results, although LMNN exploits the class label information.

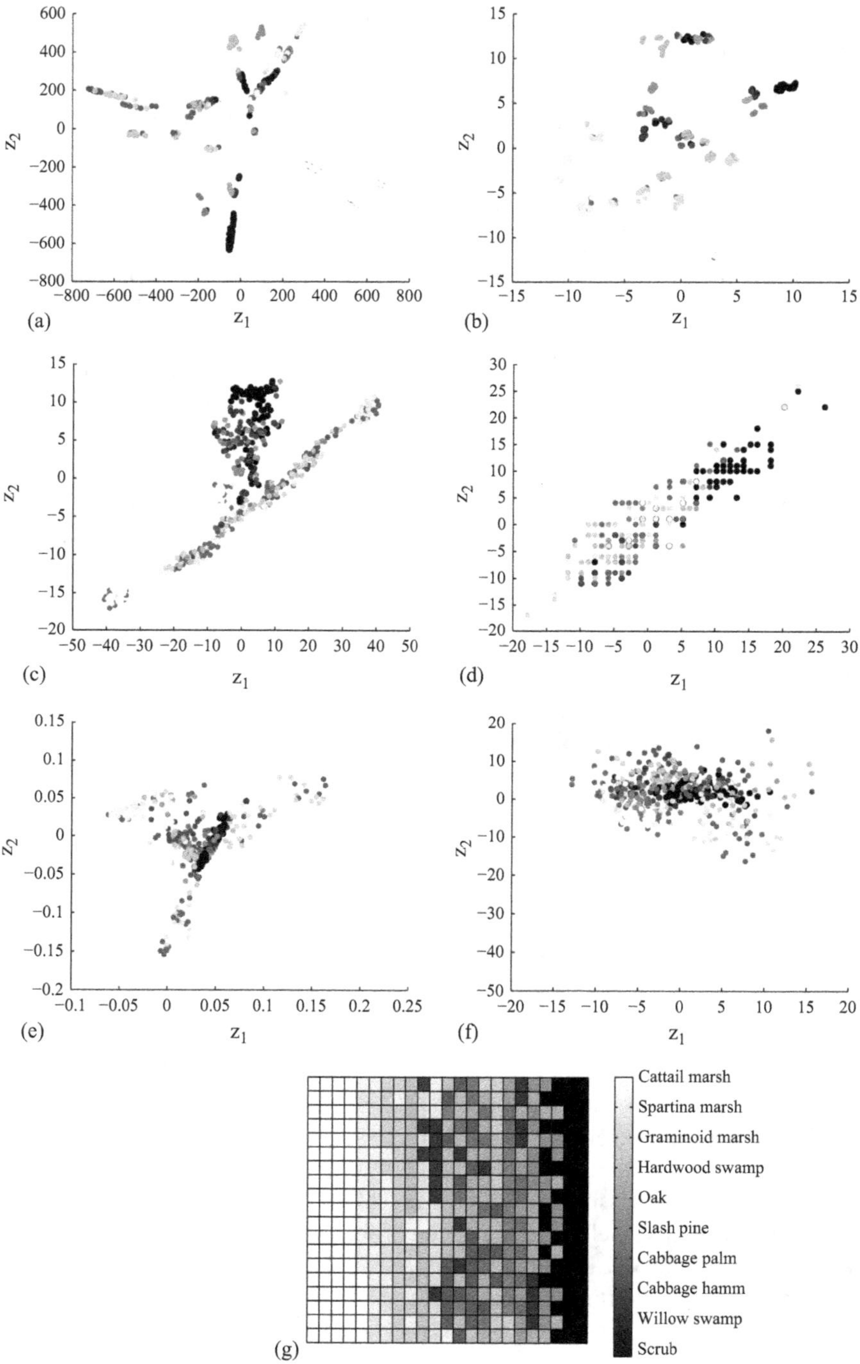

Figure 6.6 Embedding results of Kennedy Space Center data from (a) sSNE, (b) tSNE-BK, (c) tSNE, (d) LMNN, (e) LLE, and (f) MAFE. Gray colored class labels are repeated (g) from Figure 6.4(b)

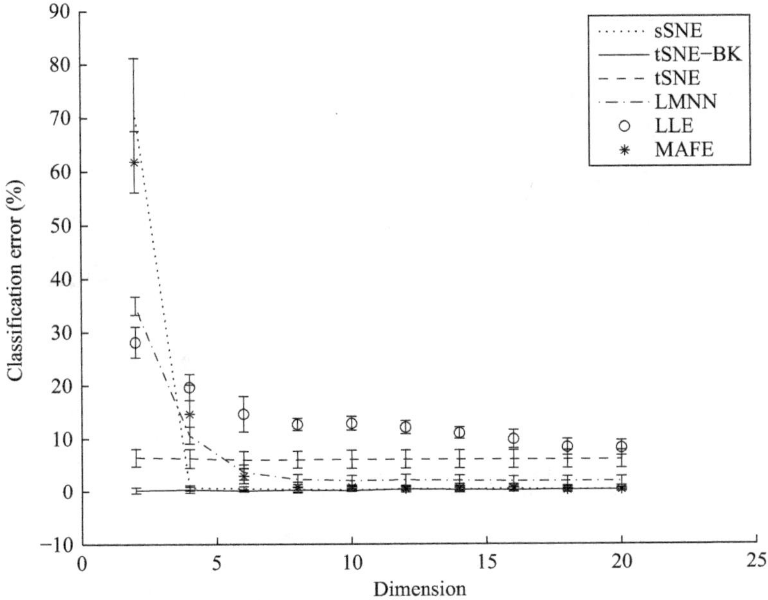

Figure 6.7 Comparison of misclassification errors in mean ±1 standard deviation for 1-nearest neighbor classifiers based on various embedding methods with varying dimension on Botswana data

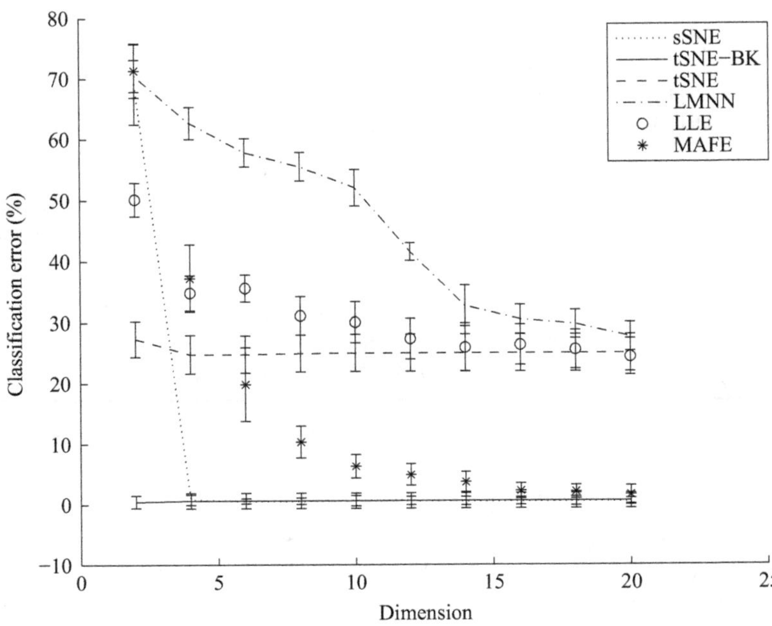

Figure 6.8 Comparison of misclassification errors in mean ±1 standard deviation for 1-nearest neighbor classifiers based on various embedding methods with varying dimension on KSC data

Table 6.2 Classification results from different embedding spaces for Botswana data

Class	Accuracy on single class					
	sSNE	tSNE-BK	tSNE	LMNN	LLE	MAFE
c1	0.9341	**1.0000**	1.0000	1.0000	0.9879	0.9275
c2	0.9345	**1.0000**	0.9836	0.9432	0.9426	0.9366
c3	**0.9210**	**1.0000**	**0.8444**	**0.8779**	**0.6675**	**0.9171**
c4	0.9281	**1.0000**	0.9730	0.9790	0.9678	0.9319
c5	0.9071	**1.0000**	0.9264	0.9328	0.8530	0.9296
c6	**0.9224**	**1.0000**	**0.8474**	**0.8856**	**0.6738**	**0.9245**
c7	0.9223	**1.0000**	0.9741	0.9228	0.8598	0.9186
c8	0.9270	**1.0000**	1.0000	0.9640	0.9312	0.8992
c9	0.9384	**0.9769**	0.9386	0.9596	0.9741	0.8278
KS	0.9159	**0.9981**	0.9308	0.9287	0.8445	0.9065

Table 6.3 Classification results from different embedding spaces for KSC data

Class	Accuracy on single class					
	sSNE	tSNE-BK	tSNE	LMNN	LLE	MAFE
c1	0.9488	**0.9906**	0.8901	0.8160	0.8038	0.9033
c2	**0.9000**	**1.0000**	**0.6123**	**0.1406**	**0.5161**	**0.6397**
c3	0.9189	**1.0000**	0.6097	0.5743	0.3909	0.6849
c4	**0.8516**	**0.9719**	**0.2111**	**0.2369**	**0.1815**	**0.7317**
c5	**0.9114**	**1.0000**	**0.4348**	**0.1886**	**0.2214**	**0.7442**
c6	**0.9095**	**1.0000**	**0.2719**	**0.2541**	**0.3443**	**0.8297**
c7	0.9118	**1.0000**	0.5606	0.4879	0.3857	0.7531
c8	0.9179	**1.0000**	0.5892	0.4616	0.5911	0.7812
c9	0.9231	**1.0000**	0.7610	0.5423	0.6730	0.8930
c10	0.9295	**1.0000**	0.8161	0.2976	0.6828	0.8198
c11	0.9230	**1.0000**	0.9469	0.6814	0.8723	0.8920
c12	0.9375	**0.9587**	0.7065	0.4740	0.7492	0.8121
c13	0.9389	**1.0000**	1.0000	0.7313	1.0000	0.9399
KS	0.9171	**0.9931**	0.7195	0.4857	0.6531	0.8220

tSNE-BK similarly displays the highest classification accuracy for the KSC data. The lowest accuracy per class comes from Classes 2, 4, 5, and 6. These classes correspond to Willow swamp, Cabbage palm, Slash pine, and Oak, respectively. Three of the classes are similar upland trees. Thus, they are difficult to distinguish due to their mixed spectral signatures. Notably, LMNN has the lowest accuracy for the KSC data. sSNE shows the second best performance here. tSNE with a Gaussian kernel neighborhood graph shows much lower accuracy than tSNE-BK for the KSC data, implying the spatial signature plays an important role in dimension reduction.

Figures 6.7 and 6.8 plot the mean ± 1 standard deviation of classification errors depending on the embedding dimension between 2 and 20. tSNE-BK clears shows the best classification results among the different embedding methods. Notably, sSNE works similar to tSNE-BK except when the embedding dimension is two. MAFE show very poor classification accuracies when the embedding dimension is less than six, although it works well for higher embedding dimensions. LMNN works better than LLE for Botswana data and the opposite happens for KSC data.

A priori determination of the number of neighbors in the graph is critical for manifold projection results for LLE and LMNN. If the size of the neighborhood is too small, then important data features will be lost from the hyperspectral manifold of the data as we see the result of LLE in Figures 6.5 and 6.6. On the other hand, the manifold projections would become noisy and unstable if the neighborhood is too large and a large intrinsic dimension is generated. MAFE, sSNE, and tSNE resolve such difficulties by automatically promoting a sparse neighborhood graph with the neighbors selected from the sparse nature of the data. Figures 6.7 and 6.8 suggest the optimal dimension as two for tSNE-BK mapping both the 145-dimensional Botswana spectral channels and the 176-dimensional KSC channels. On the other hand, the optimal dimension is eight except for LLE on Botswana data. For the KSC data, the dimension appears to be 16 except for LMNN.

6.7 Discussion

The main goal of this chapter is to review a nonlinear embedding framework that preserves the neighborhood relations of highly nonlinear and disjoint structures from the viewpoint of swarm intelligence combined with the potential field methods. Interpretation of neighbor embedding by the potential field methods provides insight to the understanding of swarm behavior in the embedding methods. An attractive potential pulls the objects in the lower dimensional space to the prior model, high-dimensional neighbor graph, while a repulsive potential pushes the object away from the other objects that are seen as its obstacles. Both attractive and repulsive potentials are dynamically determined by varying distances between the images of the data. So the dimensionality reduction problem is posed as finding minimum energy configuration or the graph topology that preserves the neighborhood.

However, the experimental results show that the flexible MAFE model proposed in the paper [1] does not have an advantage over tSNE and sSNE, which contradicts the main argument of the paper. A possible explanation is that comparison to sSNE and tSNE with the bilateral similarity function were missing from the paper. Still, MAFE provides an opportunity to derive new embedding models due to its very few assumptions.

References

[1] Dalton Lunga and Okan Ersoy. Multidimensional artificial field embedding with spatial sensitivity. *IEEE Transactions on Geoscience and Remote Sensing*, 52(2):1518–1532, 2014.

[2] John A Richards and Xiuping Jia. *Remote sensing digital image analysis*, volume 3. Springer, Berlin, Heidelberg, 1999.

[3] Charles M Bachmann, Thomas L Ainsworth, and Robert A Fusina. Exploiting manifold geometry in hyperspectral imagery. *IEEE transactions on Geoscience and Remote Sensing*, 43(3):441–454, 2005.

[4] Ian Jolliffe. *Principal component analysis*. Wiley Online Library, 2002.

[5] Joseph B Kruskal and Myron Wish. *Multidimensional scaling*, volume 11. Sage, 1978.

[6] Masashi Sugiyama. Dimensionality reduction of multimodal labeled data by local fisher discriminant analysis. *Journal of Machine Learning Research*, 8(May):1027–1061, 2007.

[7] Sam T Roweis and Lawrence K Saul. Nonlinear dimensionality reduction by locally linear embedding. *Science*, 290(5500):2323–2326, 2000.

[8] Mikhail Belkin and Partha Niyogi. Laplacian eigenmaps for dimensionality reduction and data representation. *Neural Computation*, 15(6):1373–1396, 2003.

[9] Joshua B Tenenbaum, Vin De Silva, and John C Langford. A global geometric framework for nonlinear dimensionality reduction. *Science*, 290(5500):2319–2323, 2000.

[10] Melba M Crawford, Li Ma, and Wonkook Kim. Exploring nonlinear manifold learning for classification of hyperspectral data. In *Optical Remote Sensing*, pages 207–234. Springer, Berlin, Heidelberg, 2011.

[11] Laurens van der Maaten and Geoffrey Hinton. Visualizing data using t-SNE. *Journal of Machine Learning Research*, 9(Nov):2579–2605, 2008.

[12] Geoffrey Hinton and Sam Roweis. Stochastic neighbor embedding. In *NIPS*, volume 15, pages 833–840, 2002.

[13] Dalton Lunga and Okan Ersoy. Spherical stochastic neighbor embedding of hyperspectral data. *IEEE Transactions on Geoscience and Remote Sensing*, 51(2):857–871, 2013.

[14] Kilian Q Weinberger and Lawrence K Saul. Distance metric learning for large margin nearest neighbor classification. *Journal of Machine Learning Research*, 10(Feb):207–244, 2009.

[15] Gerardo Beni and Jing Wang. Swarm intelligence in cellular robotic systems. In *Robots and Biological Systems: Towards a New Bionics?*, pages 703–712. Springer, Berlin, Heidelberg, 1993.

[16] Eric Bonabeau, Marco Dorigo, and Guy Theraulaz. *Swarm intelligence: from natural to artificial systems*. Oxford University Press, 1999.

[17] Yogeswaran Mohan and SG Ponnambalam. An extensive review of research in swarm robotics. In *Nature & Biologically Inspired Computing, 2009. NaBIC 2009. World Congress on*, pages 140–145. IEEE, 2009.

[18] Jean-Claude Latombe. *Robot motion planning*, volume 124. Springer Science & Business Media, 2012.

[19] Oussama Khatib. Real-time obstacle avoidance for manipulators and mobile robots. In *Autonomous robot vehicles*, pages 396–404. Springer, Berlin, Heidelberg, 1986.

[20] Laura Barnes, MaryAnne Fields, and Kimon Valavanis. Unmanned ground vehicle swarm formation control using potential fields. In *Control & Automation, 2007. MED'07. Mediterranean Conference on*, pages 1–8. IEEE, 2007.

[21] Laura E Barnes, MaryAnne Fields, and Kimon P Valavanis. Swarm formation control utilizing elliptical surfaces and limiting functions. *IEEE Transactions on Systems, Man, and Cybernetics, Part B (Cybernetics)*, 39(6):1434–1445, 2009.

[22] RK Sharma and D Ghose. Collision avoidance between UAV clusters using swarm intelligence techniques. *International Journal of Systems Science*, 40(5):521–538, 2009.

[23] Xiwang Dong, Bocheng Yu, Zongying Shi, and Yisheng Zhong. Time-varying formation control for unmanned aerial vehicles: theories and applications. *IEEE Transactions on Control Systems Technology*, 23(1):340–348, 2015.

[24] Veysel Gazi and Kevin M Passino. Stability analysis of swarms. *IEEE Transactions on Automatic Control*, 48(4):692–697, 2003.

[25] Wadzanai D Lunga. *Multidimensional nonlinear embedding models for image scene dimensionality reduction and visualization*. PhD thesis, Purdue University, 2012.

[26] Carlo Tomasi and Roberto Manduchi. Bilateral filtering for gray and color images. In *Computer Vision, 1998. Sixth International Conference on*, pages 839–846. IEEE, 1998.

[27] Peng Honghong and Raghuveer Rao. Hyperspectral image enhancement with vector bilateral filtering. In *Image Processing (ICIP), 2009 16th IEEE International Conference on*, pages 3713–3716. IEEE, 2009.

[28] Ketan Kotwal and Subhasis Chaudhuri. Visualization of hyperspectral images using bilateral filtering. *IEEE Transactions on Geoscience and Remote Sensing*, 48(5):2308–2316, 2010.

[29] James Theiler, Guangzhi Cao, Leonardo R Bachega, and Charles A Bouman. Sparse matrix transform for hyperspectral image processing. *IEEE Journal of Selected Topics in Signal Processing*, 5(3):424–437, 2011.

[30] Amy Lynn Neuenschwander. *Remote sensing of vegetation dynamics in response to flooding and fire in the Okavango Delta, Botswana*. ProQuest, 2007.

[31] Laurens JP van der Maaten, Eric O Postma, and Hendrik J van den Herik. *Matlab toolbox for dimensionality reduction*. MICC, Maastricht University, 2007.

[32] Jacob Cohen. A coefficient of agreement for nominal scales. *Educational and Psychological Measurement*, 20(1):37–46, 1960.

Chapter 7
Inertia weight control strategies for PSO algorithms

Ahmad Nickabadi[1], Reza Safabakhsh[1], and Mohammad Mehdi Ebadzadeh[1]

Abstract

Particle swarm optimization (PSO) is a stochastic population-based algorithm which was originally introduced by Kennedy and Eberhart [1]. This optimization algorithm is motivated by intelligent collective behavior of some animals such as flocks of birds or schools of fish. As in most of the metaheuristic optimization algorithms, in PSO, a population of individuals, known as particles, are evolved through successive iterations. The most important advantages of PSO, compared to other optimization strategies, are its easy implementation and few parameters requiring adjustment.

Since the initial development of PSO by Kennedy and Eberhart, several variants of this algorithm have been proposed by researchers. The first modification was introducing an inertia weight parameter in the velocity update equation of the initial PSO—a PSO model which is now accepted as the global best PSO algorithm [2]. The goal of the inertia weight parameter is to balance the exploration and exploitation characteristics of PSO. Generally, large inertia weight values are expected to increase the velocity of the particles and improve the long-range exploration of the PSO algorithm, while low inertia values increase the short-range exploration.

Due to the strong effect of the inertia weight on the performance of PSO, many researchers have investigated different inertia weight control approaches during the past decades, and many different strategies have been proposed. Various inertia weighting strategies can be categorized into three main classes (Figure 7.1): (1) constant or random, (2) time-varying, and (3) adaptive inertia weights. The first class contains strategies in which the value of the inertia weight is constant during the search or is determined randomly. In the second class, the inertia weight is defined as a function of time (iteration number), and hence these strategies are referred to as time-varying inertia weight strategies. These methods are not considered adaptive since they do not monitor the situation of the particles in the search space, and the value of the inertia weight in each iteration is known before the execution of the

[1]Computer Engineering Department, Amirkabir University of Technology, Iran

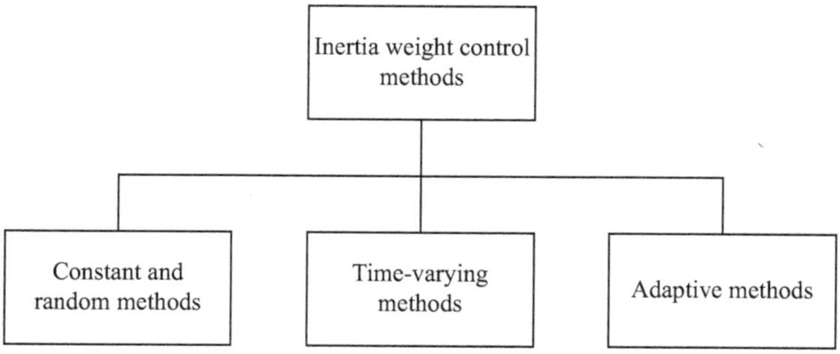

Figure 7.1 The three categories of the inertia weight control methods

Table 7.1 A short description of the symbols frequently used in this chapter

Symbol	Description
N	The number of particles
x_{ij}	Position of particle i along dimension j
v_{ij}	Velocity of particle i along dimension j
$pbest_i$	The best position found by particle i
$gbest$	The position of the best solution found by all particles
w	The inertia weight value
t	The iteration number
t_{max}	The maximum number of iterations
$w_{initial}$	The value of inertia at the first iteration ($t = 0$)
w_{final}	The value of inertia at the last iteration ($t = t_{max}$)
w_{min}	The minimum value of w
w_{max}	The maximum value of w
fit	The fitness function
D	The number of dimensions of the search space

algorithm. The third class of the inertia weight strategies consists of those methods which use a feedback parameter to monitor the state of the algorithm and adjust the value of the inertia weight based on the feedback parameter's value.

This chapter presents the major existing inertia weight control strategies in the above three categories and discusses examples of each category in detail. The remainder of this chapter is organized as follows. The next section gives a short review of the PSO algorithm, and the role of the inertia weight parameter in the velocity update equation. Sections 7.2–7.4 review the three groups of inertia weight strategies, constant or random, time-varying, and adaptive models. Section 7.5 reports some experimental evaluations of different inertia weight models. Finally, Section 7.6 concludes this chapter.

A short description of the symbols frequently used in this chapter is shown in Table 7.1.

7.1 PSO and the inertia weight parameter

PSO is an optimization algorithm in which each potential solution of the target problem is treated as a bird which is also called a *particle*. The set of all particles, also known as the *swarm of particles*, is flown through the D-dimensional search space of the problem. The position of each particle is changed in the algorithms iterations based on the best experiences of the particle itself as well as those of its neighbors. The position of the ith particle is presented as

$$x_i = (x_{i1}, x_{i2}, \ldots, x_{iD}), \tag{7.1}$$

where $x_{id} \in [l_d, u_d]$ is the dth dimension of the particle's position, and l_d and u_d are the lower and upper bounds of the dth dimension of the search space. Similar to the position, the velocity of each particle is represented with a vector. The ith particle's velocity is presented as

$$v_i = (v_{i1}, v_{i2}, \ldots, v_{iD}), \tag{7.2}$$

where v_{id} is the dth dimension of the particle's velocity vector.

At each time iteration, the positions and velocities of the particles are updated according to the following equations [1]:

$$v_{ij}(t+1) = v_{ij}(t) + R_{ij}^{(1)} c_1 (pbest_{ij} - x_{ij}(t)) + R_{ij}^{(2)} c_2 (gbest j - x_{ij}(t)) \tag{7.3}$$

$$x_{ij}(t+1) = x_{ij}(t) + v_{ij}(t+1) \tag{7.4}$$

where $R_{ij}^{(1)}$, $R_{ij}^{(2)}$ are two distinct random values in $[0, 1]$, c_1 and c_2 are acceleration constants, $pbest_i$ is the best previous position of particle i, and $gbest$ denotes the best previous position of all particles of the swarm (global best position). If c_1 is set to 0 and $c_2 \neq 0$, the PSO algorithm turns into the *social-only* model, and if c_2 is set to 0 and $c_1 \neq 0$, it becomes a *cognition-only* model.

The balance between the global and local search throughout the course of a run is critical to the success of an optimization algorithm [3]. Almost all of the evolutionary algorithms utilize some mechanisms to achieve this goal. The step size of the normal mutation in evolution strategies [4] and temperature parameter in simulated annealing [5] are two examples of such balance-controlling parameters. To bring about a balance between the exploration and exploitation characteristics of PSO, Shi and Eberhart proposed a PSO using inertia weight, w, in which the velocity of each particle is updated according to the following equation [2]:

$$v_{ij}(t+1) = wv_{ij}(t) + R_{ij}^{(1)} c_1 (pbest_{ij} - x_{ij}(t)) + R_{ij}^{(2)} c_2 (gbest_{gj} - x_{ij}(t)) \tag{7.5}$$

They claimed that a large inertia weight facilitates the global search, while a small inertia weight facilitates the local search. By changing the inertia weight dynamically, the search capability is dynamically adjusted. This is a general statement about the impact of w on PSO's search behavior shared by many other researchers, as well.

The initial inertia weight control mechanisms cannot be successfully employed in some of the new PSO models proposed in the last decade. For example, in *multiswarm* particle swarm optimizers, the main population is divided into a number of *subswarms* with the hope of exploring different areas of the search space [6,7]. While the inertia weight is only adjusted according to the iteration number of the overall algorithm, the subswarms may be created at any time during a run of the algorithm and hence require a w value compatible with their needs.

There are also some optimization problems that show the need for a new inertia weight adaptation mechanism. For example, in dynamic environments with moving peaks, the global best PSO fails to follow the extrema due to the small velocity values of the particles [8]. Adaptive inertia weight strategies, as shown in this chapter, can improve the performance of PSO in such environments.

There have been several proposals to modify the PSO algorithm by utilizing an adaptive value of the inertia weight in each iteration. This chapter will provide a comprehensive survey of these works, attempting to classify them based on the type of the feedback parameter used in them.

7.2 Constant or random strategies

The first category of inertia weight control strategies incorporate simple models which use a constant or random value as the inertia weight. In spite of their simplicity, these strategies have been shown to be very effective for some static and dynamic test functions [9]. In this section, a constant and a random inertia weight model are introduced. These models are summarized in Table 7.2.

7.2.1 Constant inertia weight

As mentioned earlier, the inertia weight parameter was originally introduced by Shi and Eberhart [2]. They used a range of constant w values and showed that by using large values of w, i.e., $w > 1.2$, PSO only performs a weak search by exploring the search space and with low values of this parameter, i.e., $w < 0.8$, PSO tends to trap in local optima. They suggest that with a w within the range of [0.8, 1.2], PSO finds the global optimum in a reasonable number of iterations. Shi and Eberhart also investigated the impact of the inertia weight and the maximum velocity on the performance of PSO [10].

Table 7.2 Constant or random inertia weight models

Label	Strategy	Reference
W1	$w = c$	[2,10]
W2	$w = 0.5 + \frac{uniform(0,1)}{2}$	[11]

7.2.2 Random inertia weight

To enable PSO to track the optima in a dynamic environment, Eberhart and Shi used a random value for the inertia weight [11] as

$$w = 0.5 + \frac{uniform(0, 1)}{2} \tag{7.6}$$

where *uniform*(0, 1) is a random number in [0, 1]; w is then a uniform random variable in the range [0.5, 1].

Generally, it is difficult to predict in dynamic environments whether at a given time exploration (large values of w) would be better or exploitation (small values of w). Hence, the above strategy selects a random value of w to address this problem. However, there are adaptive inertia weight control strategies specially developed for the dynamic environments which try to detect the situation of the particles in the search space and choose the inertia weight value accordingly.

7.3 Time-varying strategies

Most of the PSO variants use time-varying inertia weight strategies in which the value of the inertia weight is determined as a function of time (the iteration number). The strategies of this category are further divided into three subcategories (Figure 7.2): the increasing, decreasing, and oscillating models. Moreover, the increasing and decreasing models may use linear or nonlinear functions. In the following, the three main categories of time-varying inertia weight models are studied in more detail. These models are summarized in Table 7.3.

7.3.1 Decreasing inertia weights

The first and the most prevalent group of time-varying inertia weight models are the decreasing strategies. In these models, the inertia weight is set to an initial maximum value (w_{max}) at the start of the algorithm and is decreased toward a final minimum

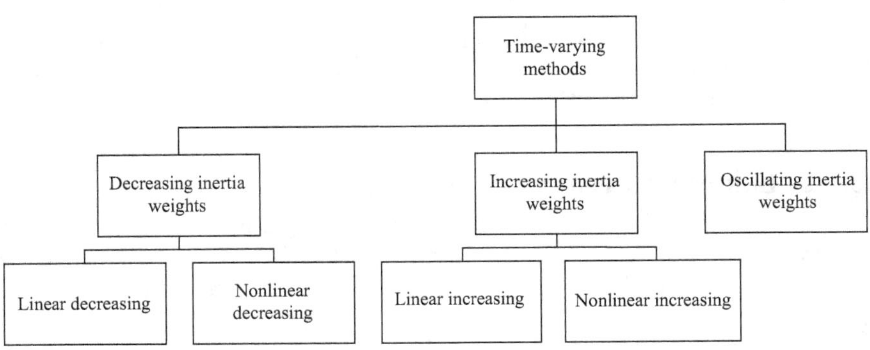

Figure 7.2 Time-varying inertia weight control methods

Table 7.3 Time-varying inertia weights (LD: linear decreasing, ND: nonlinear decreasing, LI: linear increasing, NI: nonlinear increasing, O: oscillating)

Label	Type	Strategy	Reference
T1	LD	$w(t) = \frac{t_{max}-t}{t_{max}}(w_{max} - w_{min}) + w_{min}$	[11,12]
T2	LD	$w(t) = \frac{t_{max}-t}{t_{max}}(w_{max} - w_{min}) + w_{min} \times z(t)$	[13,14]
T3	ND	$w(t) = \left\{\frac{(t_{max}-t)^n}{(t_{max})^n}(w_{max} - w_{min}) + w_{min}\right\}$	[15]
T4	ND	$w(t) = \frac{1-(\beta)}{1+s\beta}$	[16]
T5	ND	$w(t) = w_{initial} \times u^{-t}$	[17]
T6	ND	$w(t) = w_{max} - (w_{max} - w_{min})\left(\frac{t}{t_{max}}\right)^{\alpha}$	[18]
T7	ND	$w(t) = w_{min} + (w_{max} - w_{min}) \times e^{-\frac{Kt}{t_{max}}}$	[19]
T8	ND	$w(t) = w_{max} + (w_{max} - w_{min}) \times \log_{10}\left(\alpha + \frac{10t}{t_{max}}\right)$	[20]
T9	ND	$w(t) = \left(\frac{2}{t}\right)^{0.3}$	[21]
T10	LI	$w(t) = \frac{t_{max}-t}{t_{max}}(w_{min} - w_{max}) + w_{max}$	[22,23]
T11	NI	$w(t) = \frac{w_{min}-w_{max}}{1+e^{u(t-nt_{max})}} + w_{max}$	[24]
T12	O	$w(t) = \frac{w_{min}+w_{max}}{2} + \frac{w_{max}-w_{min}}{2}\cos\frac{2\pi t}{T}$	[25]
T13	O	$w(t) = \begin{cases} w_{max} & \text{if } \cos(\frac{2\pi t}{T}) \geq 0 \\ w_{min} & \text{otherwise.} \end{cases}$	[25]
T14	O	$w(t) = w_{max} - \lvert round(\alpha t) - \alpha t \rvert$	[26]

value (w_{min}) in the following steps of the algorithm. The inertia weight is defined as a linear or nonlinear function of time: $f : [0, t_{max}] \rightarrow [w_{min}, w_{max}]$. The models of this group follow the generally accepted idea that a large value of the inertia weight facilitates the exploration of the particles, while small inertia weights increase the convergence rate of PSO. As a result, they use large inertia weight values at the beginning to give the algorithm the opportunity to properly travel the search space and then employ small inertia weight values at the end to fine tune the best result found by the algorithm. Samples of the decreasing inertia weight models are shown in Figure 7.3.

7.3.1.1 Linear decreasing inertia weights

Eberhart and Shi introduce a linear decreasing inertia weight and show it to be effective in improving the fine-tuning characteristic of the PSO [12,27]. In this method, the value of w is linearly decreased from an initial value w_{max} to a final value w_{min} according to the following equation:

$$w(t) = \frac{t_{max} - t}{t_{max}}(w_{max} - w_{min}) + w_{min} \tag{7.7}$$

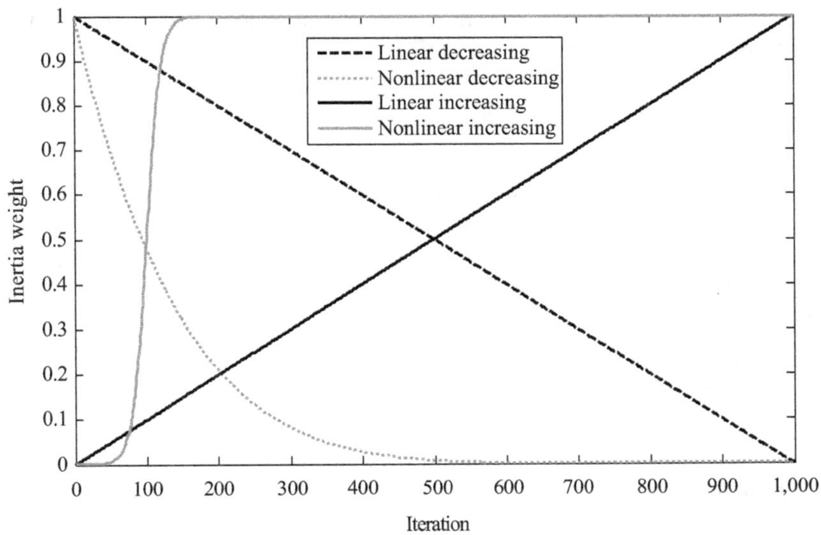

Figure 7.3 *The curves of increasing and decreasing time-varying inertia weight strategies: linear decreasing [27], nonlinear decreasing [15], linear increasing [22], and nonlinear increasing [24].* $w_{min} = 0$, $w_{max} = 1$, $t_{max} = 1{,}000$

where t is the current iteration of the algorithm and t_{max} is the maximum number of iterations the PSO is allowed to continue. This strategy is very common and most of the PSO algorithms follow this model for adjusting the value of the inertia weight.

Feng *et al.* [13,14] employ a modified version of the above model in which a chaotic term is added to the linear decreasing inertia weight. The proposed w is

$$w(t) = \frac{t_{max} - t}{t_{max}}(w_{max} - w_{min}) + w_{min} \times z(t) \tag{7.8}$$

where

$$z(t) = 4 \times z(t-1) \times (1 - z(t-1)) \tag{7.9}$$

is a chaotic sequence of values with $z(0)$ randomly generated from *Uniform*(0, 1).

The chaotic effect added to the base linear descending inertia weight model introduces some randomness into the inertia weight value at different iterations which is expected to improve the convergence velocity and global search ability of PSO [13].

7.3.1.2 Nonlinear decreasing inertia weights

Adopting the general idea of decreasing the inertia weight over time, many researchers have proposed nonlinear decreasing strategies. Chatterjee and Siarry [15] propose a nonlinear decreasing variant of the inertia weight in which at each time step, w is determined based on the following equation:

$$w(t) = \left\{ \frac{(t_{max} - t)^n}{(t_{max})^n}(w_{max} - w_{min}) + w_{min} \right\} \tag{7.10}$$

where n is the nonlinear modulation index. Different values of n result in different variations of the inertia weights all of which start from w_{max} and end at w_{min}. For $n = 1$, it is equal to the linear decreasing inertia weight studied before, and as n approaches infinity, $w(t)$ tends to constant value w_{min} for all $t > 0$.

Lei *et al.* use a Sugeno function for the inertia weight decline curve [16] as

$$w(t) = \frac{1 - (\beta)}{1 + s\beta} \tag{7.11}$$

where β is t/t_{max} and s is a constant larger than -1 that controls the shape of the inertia weight function. For $s = 0$, the inertia weight linearly decreases from 1 to 0 during a run of PSO. The inertia weight is a nonlinear convex or concave curve for $s < 0$ and $s > 0$, respectively.

Another nonlinear decreasing inertia weight has been proposed by Jiao *et al.* [17]. In this model, w is reduced over time based on the following equation:

$$w(t) = w_{initial} \times u^{-t} \tag{7.12}$$

where $w_{initial}$ is the initial inertia weight value selected in the range $[0, 1]$ and u is a constant value in the range $[1.0001, 1.005]$. In the experiments conducted by Jiao *et al.* [17], u is set to 1.0002, and the performance of the algorithm is evaluated for different values of $w_{initial}$.

Yang *et al.* [18] investigate the impact of some nonlinear time-varying variants of the inertia weight and acceleration coefficients on the PSO's performance. The goal of their proposed strategy is to have a faster reduction of the inertia weight at the earlier stages to increase the convergence rate and a slow reduction of the inertia weight at the later stages to precisely capture the solution. They utilize the following model.

$$w(t) = w_{max} - (w_{max} - w_{min}) \left(\frac{t}{t_{max}} \right)^{\alpha} \tag{7.13}$$

where α is a constant determining the shape of the inertia weight function. Again, for $\alpha = 1$, this model reduces to a linear function. Different values of α are examined, and the best value of this parameter is found as $\alpha = 1/\pi^2$.

Chen *et al.* propose a natural exponential inertia weight strategy as [19]

$$w(t) = w_{min} + (w_{max} - w_{min}) \times e^{-(Kt/t_{max})} \tag{7.14}$$

with $w_{min} = 0.4$ and $w_{max} = 0.9$. $K > 0$ is a constant value. Two different strategies of $e_1 - PSO$ and $e_2 - PSO$ are studied with $K = 10$ and $K = 4$, respectively. Larger values of K cause faster reduction of inertia weight over time.

Gao *et al.* [20] employ a logarithmic decreasing w as

$$w(t) = w_{max} + (w_{max} - w_{min}) \times \log_{10}\left(\alpha + \frac{10t}{t_{max}}\right) \tag{7.15}$$

where α is a constant with which the evolution speed is adjusted and is set to 1 [20].

There are also other similar approaches that use nonlinear decreasing inertia weights such as [21]

$$w(t) = \left(\frac{2}{t}\right)^{0.3} \tag{7.16}$$

7.3.2 *Increasing inertia weights*

While most of the time-varying inertia weight models have a decreasing form, some researchers have proposed increasing models. In the following, examples of linear and nonlinear increasing strategies are reviewed. Samples of these inertia weight models are shown in Figure 7.3.

7.3.2.1 Linear increasing inertia weights

Distinct from the widely used decreasing inertia weight, Zheng *et al.* proposed a PSO with an increasing inertia weight [22,23] and confirmed its validity in terms of convergence speed and solution precision by testing it with four standard test functions. In this model, the inertia weight is given by the following equation:

$$w(t) = \frac{t_{max} - t}{t_{max}}(w_{min} - w_{max}) + w_{max} \tag{7.17}$$

The experiments of Zheng *et al.* [22] show that the PSO with an increasing inertia weight (increasing from 0.4 to 0.9) outperforms the PSO with a decreasing inertia weight in all benchmarks used in their tests.

7.3.2.2 Nonlinear increasing inertia weights

Malik *et al.* [24] propose a sigmoid increasing model in which the inertia weight is given by the following function:

$$w(t) = \frac{w_{min} - w_{max}}{1 + e^{u(t - nt_{max})}} + w_{max} \tag{7.18}$$

where $u = 10^{\log(t_{max})-2}$ adjusts the sharpness and n is a constant to set partition of the sigmoid function.

7.3.3 *Oscillating inertia weights*

In an attempt to have exploration and exploitation steps periodically repeated after each other during a run of PSO, Kentzoglanakis [25] proposes an oscillating inertia weight strategy. In this model, the inertia weight value of PSO oscillates between the maximum (w_{max}) and the minimum (w_{min}) values. To this end, w is defined as a sinusoidal function:

$$w(t) = \frac{w_{min} + w_{max}}{2} + \frac{w_{max} - w_{min}}{2}\cos\frac{2\pi t}{T} \tag{7.19}$$

where T is the period of oscillation, i.e., the number of iterations within which PSO goes through a full cycle of w values defined as

$$T = \frac{2S_1}{3 + 2k} \tag{7.20}$$

where $k \in N$ controls the frequency of oscillation, and s_1 is the number of iterations the PSO is allowed to oscillate. After this period, the inertia weight remains constant at $w(t) = w_{min}$.

Two other versions of the above model are examined by Kentzoglanakis [25]. The first variant is a sinusoidal wave with decreasing amplitude resulting from substituting w_{max} with

$$w'_{max}(t) = \left(1 - \frac{t}{t_{max}}\right) w_{max}. \tag{7.21}$$

The second variant is a step function with two discrete values of w_{min} and w_{max} defined as

$$w(t) = \begin{cases} w_{max} & \text{if } \cos\left(\frac{2\pi t}{T}\right) \geq 0 \\ w_{min} & \text{otherwise.} \end{cases} \tag{7.22}$$

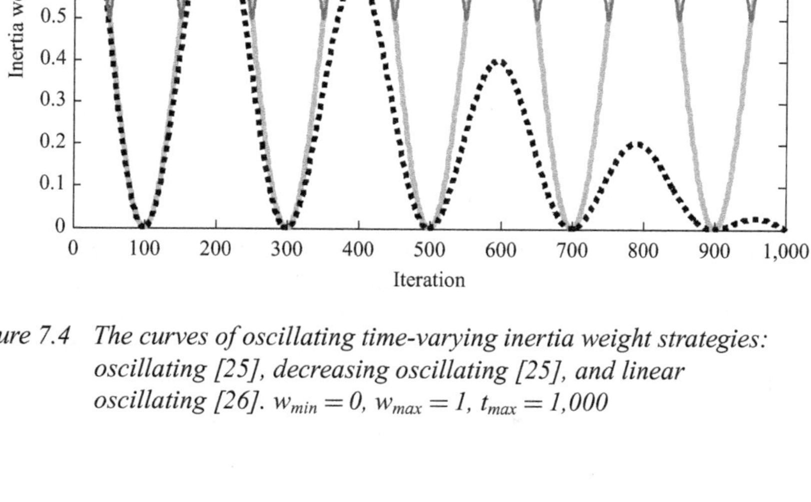

Figure 7.4 The curves of oscillating time-varying inertia weight strategies: oscillating [25], decreasing oscillating [25], and linear oscillating [26]. $w_{min} = 0$, $w_{max} = 1$, $t_{max} = 1{,}000$

Another oscillating, inertia weight is proposed by Kordestani *et al.* [26] in which the inertia weight is defined as a symmetric triangular wave function

$$w(t) = w_{max} - |round(\alpha t) - \alpha t| \tag{7.23}$$

where $\alpha \in [0.0, 0.5]$ determines the slope of the linear pieces of the resulting inertia function, and $round(x)$ is a function that rounds the real-valued number x to its nearest integer number.

Samples of oscillating inertia weight models are shown in Figure 7.4.

7.4 Adaptive strategies

In some optimization problems, e.g., dynamic functions, it is difficult to determine the best inertia weight value for each iteration of PSO before the execution of the algorithm. On the other hand, as mentioned earlier, most of the time-varying models use the total number of iterations as a parameter of their inertia weight function and hence are not applicable to an ongoing optimization problem. The last category of the inertia weight models, comprising the adaptive models, provides a solution to this type of problem. An adaptive PSO algorithm is one that changes its behavior at run time based on the information gathered from the optimization process and the available prior knowledge. The first requirement of an adaptive inertia weight control strategy is a set of feedback parameters through which the search space and the situation of the particles are monitored. The values of the feedback parameters can determine the state of the particles in the search space and provide information about the optimization problem which is going to be solved. The values of the feedback parameters are measured at periods of time, and the inertia weights of the particles are then adjusted using an adaptive mechanism.

So far, many adaptive inertia weight control strategies have been proposed. In this section, we categorize these models based on the feedback parameters utilized in each model (Figure 7.5). The positions and velocities of the particles in the search space, the fitness values, the ranks of the particles, and success rates of the particles or the whole swarm are examples of the feedback parameters utilized in the literature.

The adaptive inertia weight models reviewed in this section are summarized in Table 7.4.

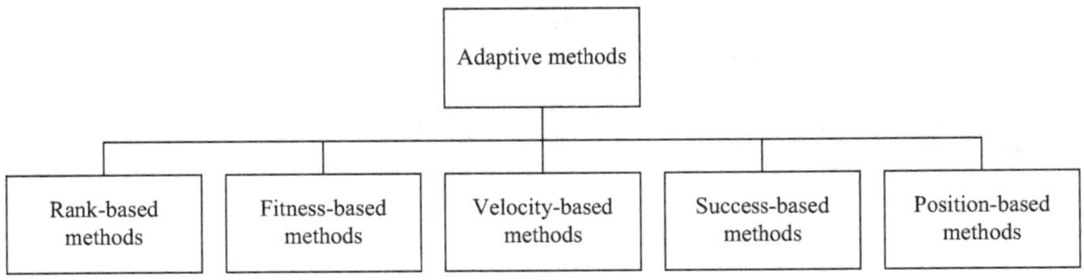

Figure 7.5 Adaptive inertia weight control methods

Table 7.4 Adaptive inertia weight strategies

Label	Strategy	Feedback	Reference
A1	$w_i = w_{min} + (w_{max} - w_{min})\frac{Rank_i(t)}{N}$	Rank	[28]
A2	$w_i(t+1) = \begin{cases} w_i(t) + \eta\Delta w & \text{for the best particle} \\ w_i(t) - \Delta w & \text{otherwise.} \end{cases}$	Rank	[29]
A3	A set of fuzzy rules	Fitness	[3,30]
A4	$w_i(t) = w_{initial} - \alpha(1 - h_i(t)) + \beta s$	Fitness	[31]
A5	$w(t) = 1.1 - \frac{fit(gbest)}{\frac{1}{N}\sum_{i=1}^{N} fit(pbest_i)}$	Fitness	[32]
A6	$w_i(t) = 1 - \left\lvert \frac{\alpha(1-c_i(t))}{(1+d_i(t))(1+\beta)} \right\rvert$	Fitness	[33]
A7	$w(t+1) = \begin{cases} w(t) - \Delta w & \text{if } v_{ave}(t) \geq v_{ideal}(t+1) \\ w(t) + \Delta w & \text{if } v_{ave}(t) < v_{ideal}(t+1) \end{cases}$	Velocity	[34]
A8	$w(t) = (w_{max} - w_{min})P_s(t) + w_{min}$	Success	[35]
A9	$w(t) = \frac{1}{1+1.5e^{-2.6f(t)}}$	Position	[36]
A10	$w_i(t+1) = \exp(-\exp(-R_i))$	Position	[37]

7.4.1 Particle rank-based inertia weight control strategies

7.4.1.1 Rank-based inertia weight

In the adaptive PSO algorithm proposed by Panigrahi *et al.*, different inertia weights are assigned to different particles based on the ranks of each particle [28]:

$$w_i = w_{min} + (w_{max} - w_{min})\frac{Rank_i(t)}{N} \tag{7.24}$$

where $Rank_i(t)$ is the rank of particle i at time step t when particles are sorted based on their particle-best fitness, and N is the total number of particles. The rational behind this model is that the positions of the particles are adjusted in a way that highly fitted particles move more slowly compared to the lowly fitted ones.

7.4.1.2 Self-regulating inertia weight

Tanweer *et al.* [29] propose to self-regulate the inertia weight of the best particle to accelerate its search toward the global optimum, while the inertia weights of the other particles are linearly decreasing. The self-regulating inertia function is given by

$$w_i(t+1) = \begin{cases} w_i(t) + \eta\Delta w & \text{for the best particle} \\ w_i(t) - \Delta w & \text{otherwise.} \end{cases} \tag{7.25}$$

where

$$\Delta w = \frac{w_{initial} - w_{final}}{t_{max}} = \frac{0.55}{t_{max}}, \tag{7.26}$$

and η is a constant controlling the acceleration.

7.4.2 Fitness-based inertia weight control strategies

7.4.2.1 Fuzzy adaptive particle swarm optimization

Some researchers have utilized a fuzzy system consisting of nine rules to dynamically adapt the inertia weight [3,30]. Each rule of this system has two input and one output variables. The NCBPE (Normalized Current Best Performance Evaluation), as a performance measure of the best candidate solution found so far, and the current inertia weight value are the input variables, and the weight change is the output of the system. NCBPE is calculated as follows:

$$NCBPE = \frac{CBPE - CBPE_{min}}{CBPE_{max} - CBPE_{min}} \tag{7.27}$$

All three fuzzy variables defined in this work have three fuzzy sets: LOW, MEDIUM and HIGH. An example of the rules of this system is

$$IF\ (NCBPE\ is\ MEDIUM)\ and\ (weight\ is\ LOW)\ THEN\ (weight\_change\ is\ HIGH). \tag{7.28}$$

7.4.2.2 Speed and aggregation degree factor method

Yang *et al.* adapt w based on two characteristic parameters of PSO, namely, the speed factor and the aggregation degree factor [31]. The speed factor of particle i at iteration t ($h_i(t)$) is defined as

$$h_i(t) = \left| \frac{\min(fit(pbest_i^{t-1}), fit(pbest_i^t))}{\max(fit(pbest_i^{t-1}), fit(pbest_i^t))} \right| \tag{7.29}$$

The aggregation degree is defined as:

$$s_i(t) = \left| \frac{\min(fit_{tbest}, \bar{fit}_t)}{\max(fit_{tbest}, \bar{fit}_t)} \right| \tag{7.30}$$

where $\bar{fit}_t$ is the mean fitness of all particles at iteration t and fit_{tbest} is the best fitness achieved by the particles at the same iteration. Using the speed and aggregation factors, the inertia weight is determined as follows:

$$w_i(t) = w_{initial} - \alpha(1 - h_i(t)) + \beta s \tag{7.31}$$

where α and β are two constants typically within the range [0, 1]. The performance of the PSO algorithm is evaluated for different values of α and β [31].

7.4.2.3 Fitness ratio based inertia weight

There are other researchers who have used the fitness of the particles as a characteristic of the particles to adapt the inertia weight. Arumugam and Rao [32] use the ratio of the global best fitness and the average local best fitness to determine the inertia weight in each iteration. Their inertia weight rule is

$$w(t) = 1.1 - \frac{fit(gbest)}{\frac{1}{N}\sum_{i=1}^{N} fit(pbest_i)} \tag{7.32}$$

7.4.2.4 Convergence and diffusion factors based inertia weight

The inertia weight control strategy of Li and Tan [33] is defined based on the convergence and diffusion factors of the particles. The convergence factor of particle i at iteration t is define as

$$c_i(t) = \frac{|fit(pbest_i(t-1)) - fit(pbest_i(t))|}{fit(pbest_i(t-1)) + fit(pbest_i(t))} \tag{7.33}$$

where $c_i(t) \in [0, 1]$ shows how successful the particle i was at iteration $t-1$. If a particle has performed well in the previous iteration (i.e., large $c_i(t)$), the inertia weight will be large, and for poor-performing particles (i.e., low $c_i(t)$), the inertia weight is decreased. On the other hand, the diffusion factor is defined as

$$d_i(t) = \frac{|fit(pbest_i(t)) - fit(gbest(t))|}{fit(pbest_i(t)) + fit(gbest(t))} \tag{7.34}$$

which compares the performance of particle i with the performance of the best particle of the swarm; more fitted particles with lower d_i are assigned lower inertia weights. The inertia weight of each particle is defined as

$$w_i(t) = 1 - \left| \frac{\alpha(1 - c_i(t))}{(1 + d_i(t))(1 + \beta)} \right| \tag{7.35}$$

where α and β are constants in the range of [0, 1], determining the shape of the inertia function.

7.4.3 Velocity-based inertia weight control strategies

7.4.3.1 Average velocity based inertia weight

Xu [34] divides the search process of PSO into two stages and believes that to have good exploration and exploitation characteristics, particles should have high velocities at the early stage and low velocities at the latter stage. According to this analysis, the ideal average velocity of the swarm of particles at time step t is defined by the following nonlinear function:

$$v_{ideal}(t) = v_{start}\frac{1 + \cos(t\pi / t_{max})}{2} \tag{7.36}$$

where $v_{start} = (x_{max} - x_{min})/2$ is the initial ideal average velocity. The average absolute velocity of the swarm of particles is then calculated as follows:

$$v_{ave}(t) = \frac{1}{m.d} \sum_{i=1}^{m} \sum_{j=1}^{d} |v_{ij}| \tag{7.37}$$

The average absolute velocity ($v_{ave}(t)$) is the feedback parameter is used to measure the size of the search space covered by the particles. This average velocity is then compared with the ideal velocity defined in (7.36). If the current average velocity is larger than the ideal value, the inertia weight will be decreased, and if the current absolute velocity is smaller than the ideal value, the inertia weight will be increased.

$$w(t+1) = \begin{cases} w(t) - \Delta w & \text{if } v_{ave}(t) \geq v_{ideal}(t+1) \\ w(t) + \Delta w & \text{if } v_{ave}(t) < v_{ideal}(t+1) \end{cases} \tag{7.38}$$

7.4.4 Success-based inertia weight control strategies

7.4.4.1 Success rate based inertia weight

Nickabadi *et al.* [35] propose an PSO model with an adaptive inertia weight (AI–WPSO) based on the success rate of the swarm of particles. For a minimization problem, the success of particle i at iteration t is defined as

$$S_i(t) = \begin{cases} 1 & \text{if } fit(pbest_i(t)) < fit(pbest_i(t-1)) \\ 0 & \text{if } fit(pbest_i(t)) = fit(pbest_i(t-1)) \end{cases} \tag{7.39}$$

Using the success values of the particles, the success percentage of the swarm is computed as

$$P_s(t) = \frac{\sum_{i=1}^{N} S_i(t)}{N} \tag{7.40}$$

where $P_s(t) \in [0, 1]$ is the percentage of the particles which have had an improvement in their fitness in the last iteration.

The success rate of the whole population of particles ($P_s(t)$) is then used to determine the situation of the particles in the search space. A high percentage of success usually indicates that the particles have converged to a point that is far from the optimum point and the entire swarm is slowly moving toward the optimum. On the other hand, a low percentage of success shows that the particles are oscillating around the optimum without much improvement. The same idea is utilized in evolutionary strategies [4,38]. So, it is proposed to choose the value of $w(t)$ proportional to $P_s(t)$. This way, the inertia weight can be any increasing function (f) of $P_s(t)$ defined as

$$w(t) = f(P_s(t)) \tag{7.41}$$

Nickabadi *et al.* chose a linear function is used to map the values of P_s to the possible range of inertia weights as follows [35].

$$w(t) = (w_{max} - w_{min})P_s(t) + w_{min} \tag{7.42}$$

Although P_s is in the range of [0, 1], w can be in any acceptable range. However, Nickabadi *et al.* chose the range of the inertia weight (w) to be [0, 1] [35].

The authors provide the motivation of the success rate based inertia weight model using the two possible situations of the particles in the search space. In the first state, the particles of the swarm are close to each other and far from the global optimum. Compared to their distance from the global optimum, the particles have a very small velocity resulting in a slow convergence. It is clear that in this situation, a large inertia weight helps the particles to accelerate toward the global optimum. In the proposed model of [35], this state of the swarm is identified by the high success rate of the swarm resulted from the movement of the particles toward the global optimum. As the inertia weight is defined as a monotonic increasing function of the success rate, the inertia weight is increased in this case.

The second state represents the situation where *gbest* is close to the global optimum, and the velocities of the particles is very large compared to the distance of the *gbest* from the global optimum. In this situation, the particles oscillate around *gbest* without any improvement in the fitness. The success rate of the particles properly distinguishes this situation from the previous example. In this case, the swarm achieves a very low success rate resulting in a small inertia weight value. Consequently, the velocity of the particles is decreased, and the particles rapidly converge to the global optimum.

7.4.4.2 Stability-based adaptive inertia weight

Taherkhani and Safabkhsh [39] propose a stability-based adaptive inertia weight method. In this model, the value of inertia weight is considered different for each dimension, in order to increase the convergence speed, especially in asymmetric environments. Displacement of the best position found by each particle is used as the feedback. To do so, the success of particle i at iteration t is defined as

$$\delta_i(t) = \begin{cases} 1 & \text{if } fit(x_i(t+1)) < fit(P_i^t) \\ 0 & \text{otherwise} \end{cases} \tag{7.43}$$

The successes of the particle in the two past iterations are then investigated and the inertia weight of the particle i for direction j is adjusted as

$$w_{ij}(t+1) = \begin{cases} w_{ij}(t) + (1 - w_{min}) \\ \qquad \times e^{(((x_{ij}(t+1) - Pbest_{ij}(t))^2)/-2\sigma^2)} + \varepsilon & \text{if } (\delta_i(t) = \delta_i(t-1) = 1) \\ w_{ij}(t) - w_{min} \\ \qquad \times (1 - e^{(((x_{ij}(t+1) - Pbest_{ij}(t))^2)/-2\sigma^2)}) - \varepsilon & \text{if } (\delta_i(t) = \delta_i(t-1) = 0) \\ w_{ij}(t) & \text{otherwise} \end{cases} \tag{7.44}$$

where σ is the Gaussian kernel's width adjusted in a way that covers the maximum particles' movement, and ε is a small positive number (e.g., $\varepsilon = 0.005$) used to ensure proper increase or decrease of the inertia weight.

The rationale behind the above strategy is that if a particle succeeds to improve its performance in two consecutive iterations, the particle is in the right direction and the inertia weight of the particle should be increased to speed up the convergence toward the local optimum. On the other hand, if a particle fails to find a better solution in two consecutive iterations, either the particle's direction or its step size is not appropriate and the inertia weight is decreased to redirect the particle toward the local optimum. In both cases, the difference between the current position of the particle and the best position found by the particle in each dimension is used to adjust the inertia weight of that dimension.

7.4.5 Position-based inertia weight control strategies

7.4.5.1 Evolutionary factor based method

The idea of the adaptive PSO proposed by Zhan *et al.* [36] is to automatically control the parameters of PSO via estimation of the search state. To this end, the mean distance of each particle i to all the other particles is calculated using a Euclidian metric.

$$d_i = \frac{1}{N-1} \sum_{j=1, j \neq i}^{N} \sqrt{\sum_{k=1}^{D} (x_i^k - x_j^k)^2} \tag{7.45}$$

The evolutionary factor (f) of the population is then defined based on the minimum and maximum mean distances (d_{min} and d_{max}) and the mean distance of the globally best particle (d_g)

$$f = \frac{d_g - d_{min}}{d_{max} - d_{min}} \in [0, 1] \tag{7.46}$$

Four evolutionary states, namely, convergence, exploitation, exploration, and jumping-out, are then defined based on the above evolutionary factor. It is claimed that the evolutionary factor shares some characteristics with the inertia weight in which both are relatively large at the exploration state and small in the convergence state. Hence, the inertia weight is defined based on the evolutionary factor using a sigmoid function:

$$w(t) = \frac{1}{1 + 1.5e^{-2.6f(t)}} \in [0.4, 0.9] \quad \forall f \in [0, 1] \tag{7.47}$$

In this model, the inertia weight is a monotonic increasing function of f and is adapted based on the state of the search environment.

7.4.5.2 The fine grained inertia weight

The adaptive PSO algorithm proposed by Deep *et al.* [37] assigns an individual inertia weight to each particle of the swarm based on the distance of the particle's best position from the global best position of the swarm and the iteration number by the following relation

$$W_i(t+1) = w_i(t)((w_i(t) - 0.4) \times e^{-(|gbest(t) - pbest_i(t)| \times (t/t_{max}))}) \tag{7.48}$$

In this model, the inertia weight is initialized to $w(0) = 0.9$. At each iteration, the inertia weight of the best particle of the swarm is set to 0.4. The inertia weight of the other particles is reduced by $c_i(t) \times (w_i(t) - 0.4)$, where $c_i(t)$ is a function of the iteration number and the Euclidean distance between $pbest_i(t)$ and $gbest(t)$.

A double exponential version of this strategy is also proposed by Chauhan *et al.* [40] to have a faster growing of inertia weight with respect to the particle's performance. As before, the performance index R_i is defined based on the distance between the personal and global best positions. Then, a Gompertz functions is applied to the performance index of each particle to produce its inertia weight.

$$R_i(t) = |gbest(t) - pbest_i(t)| \times \left(\frac{t_{max} - t}{t_{max}} \right) \quad (7.49)$$

$$w_i(t+1) = \exp(-\exp(-R_i)) \quad (7.50)$$

When particles are far away from the global best position (i.e., $R_i \gg 0$), the inertia weight tends to its maximum value ($w_i = 1$) which facilitates the rapid movement of the particles toward the global best positions by adding a high momentum to the velocity update rule of the particles. On the other hand, as particles get close to the global best position, the inertia weight approaches its minimum value ($W = \exp(-1)$) and increases the convergence speed of the particles.

7.5 Experimental evaluations

So far, we have studied several inertia weight control strategies in this chapter. A question may now arise: what is the best inertia weight strategy model? To answer this question, some researchers have performed several experiments on different test functions; see [41], [9] and [35] for example. In this section, two such studies are reviewed, and a number of experiments are conducted to evaluate and compare the performance of different inertia weight strategies.

7.5.1 Experimental setup

All of the experiments of this section are conducted using the same PSO algorithm with different inertia weight strategies. A swarm of 50 particles with a star topology is employed, and the position and velocity vectors of the particles are updated according to (7.4) and (7.3), respectively. The acceleration coefficients are set to $c_1 = c_2 = 2.0$. Thirty independent runs are executed for each inertia weight model on each test function, and each run continues for 5,000 iterations (250,000 fitness evaluations).

The inertia weight models used in this study and their corresponding parameters are listed in Table 7.5. One sample of each subcategory of inertia weight models introduced in the previous section is included in this list.

The suite of static test functions used in the experiments is summarized in Table 7.6. The set of test functions are selected from the well-known unimodal and multimodal benchmark functions providing a diverse set of fitness landscapes. The dimensionality of the search space is set to $D = 30$ for all test functions.

In addition to the above static test functions, the moving peaks benchmark (MPB) [42] is also used as a dynamic test environment to study the behavior of different inertia weight models in such problems. Here, the MPB function is defined as a three-dimensional sphere function whose optimum is repeatedly moved after T iterations.

In this study, three performance measures are considered: accuracy, convergence speed and success rate of the models. Accuracy is defined as the average fitness of the best solution found by each strategy, the convergence speed shows how fast PSO

Table 7.5 Inertia weight models used in the experiments

Label	**Reference**	**Description**	**Parameters**
CONS	Table 7.2 (W1)	Constant	$w = 0.7$
LD	Table 7.3 (T1)	Linear decreasing	$w_{min} = 0.4, w_{max} = 0.9$
NLD	Table 7.3 (T3)	Nonlinear decreasing	$w_{min} = 0.4, w_{max} = 0.9, n = 3$
LI	Table 7.3 (T10)	Linear increasing	$w_{min} = 0.4, w_{max} = 0.9$
OSC	Table 7.3 (T12)	Oscillating	$w_{min} = 0.3, w_{max} = 0.9, T = 500$
AV	Table 7.4 (A7)	Adaptive velocity-based	$w_{min} = 0.3, w_{max} = 0.9, \Delta w = 0.1$
AR	Table 7.4 (A1)	Adaptive rank-based	$w_{min} = 0.4, w_{max} = 0.9$
AS	Table 7.4 (A8)	Adaptive success-based	$w_{min} = 0.3, w_{max} = 1.0$
AP	Table 7.4 (A10)	Adaptive position-based	–

Table 7.6 The set of test functions

Name	**Formula**	**Search space**	**Modality**
Sphere	$f_1(x) = \sum_{i=1}^{D} x_i^2$	$[-100, 100]^D$	U
Absolute	$f_2(x) = \sum_{i=1}^{D} \lvert x_i \rvert$	$[-100, 100]^D$	U
Rosenbrock	$f_3(x) = \sum_{i=1}^{D-1} ((100x_{i+1} - x_i^2)^2 + (x_i - 1)^2)$	$[-10, 10]^D$	U–M
Quadric	$f_4(x) = \sum_{i=1}^{D} \left(\sum_{j=1}^{i} x_j\right)^2$	$[-100, 100]^D$	U
Step	$f_5(x) = \sum_{i=1}^{D} (\lfloor x_i + 0.5 \rfloor)^2$	$[-10, 10]^D$	U
Rastrigin	$f_6(x) = \sum_{i=1}^{D} (x_i^2 - 10\cos(2\pi x_i) + 10)$	$[-5.12, 5.12]^D$	M
Ackley	$f_7(x) = -20e^{-0.2\sqrt{\frac{1}{D}\sum_{i=1}^{D} x_i^2}} - e^{(1/D)\sum_{i=1}^{D}\cos(2\pi x_i)}$ $+ 20 + e$	$[-32, 32]^D$	M
Griewank	$f_8(x) = (1/4{,}000)\sum_{i=1}^{D} x_i^2$ $- \prod_{i=1}^{D} \cos((x_i/\sqrt{i})) + 1$	$[-600, 600]^D$	M
Levy	$f_9(x) = \sum_{i=1}^{D-1} (w_i - 1)^2[1 + 10\sin^2(\pi w_i + 1)]$ $+ (w_D - 1)^2[1 + \sin^2(2\pi w_D)]$ $+ \sin^2(\pi w_1) w_i = 1 + ((x_i - 1)/4)$	$[-10, 10]^D$	M
Salomon	$f_{10}(x) = 1 - \cos(2\pi\sqrt{\sum_{i=1}^{D} x_i^2}) + 0.1\sqrt{\sum_{i=1}^{D} x_i^2}$	$[-10, 10]^D$	M

converges to a local (global) optimum, and the success rate is defined as the number of times (out of 30 runs) that the algorithm properly achieves a fitness level close to the optimum fitness.

The summary of the results is given in Tables 7.7 and 7.8. The mean and standard deviation of the fitness values obtained from each inertia weight model for each test function is given in Table 7.7, and the number of successful runs of each strategy are given in Table 7.8. Here, a run is considered successful if $\lvert f(x^*) - f(x_{best}) \rvert < 0.01$, where x^* is the global optimum and x_{best} is the best solution found by the corresponding PSO algorithm.

Table 7.7 The mean and standard deviation of the best fitness values found by each strategy

	Sphere	**Absolute**	**Rosenbrock**	**Quadric**	**Step**
CONS	5.185E−04	3.421E−04	8.766E+01	1.466E+03	0.000E+00
	6.651E−04	2.713E−04	4.895E+01	2.748E+03	0.000E+00
LD	3.177E−31	1.358E−20	4.064E+01	1.005E+01	0.000E+00
	6.613E−31	1.611E−20	2.918E+01	1.201E+01	0.000E+00
LI	2.870E−29	1.674E−19	3.425E+01	1.818E+01	3.333E−02
	5.628E−29	2.339E−19	2.588E+01	2.246E+01	1.826E−01
NLD	3.213E−63	1.066E−40	3.188E+01	2.971E−02	0.000E+00
	8.539E−63	1.964E−40	2.885E+01	3.068E−02	0.000E+00
OSC	3.302E−45	4.179E−29	3.045E+01	4.355E−01	0.000E+00
	1.758E−44	1.294E−28	2.844E+01	4.783E−01	0.000E+00
AV	1.980E−02	9.378E−02	3.645E+01	7.560E−01	0.000E+00
	9.159E−03	1.864E−02	2.817E+01	5.287E−01	0.000E+00
AP	5.797E+03	2.763E+02	7.335E+01	2.876E+03	3.333E−02
	1.705E+03	4.361E+01	4.689E+01	2.363E+04	1.826E−01
AR	1.265E−51	5.162E−33	3.003E+01	2.534E+00	0.000E+00
	5.912E−51	8.890E−33	2.496E+01	9.268E−01	0.000E+00
AS	5.191E−66	1.653E−37	1.974E+01	7.169E−05	0.000E+00
	1.763E−65	1.918E−37	2.140E+01	6.186E−05	0.000E+00

	Rastrigin	**Ackley**	**Griewank**	**Levy**	**Salomon**
CONS	3.331E+01	8.498E−03	4.524E−02	9.373E−05	2.666E−01
	9.632E+00	1.061E−02	7.949E−02	1.263E−04	6.476E−02
LD	2.653E+01	1.492E−14	1.648E−02	4.644E−31	1.599E−01
	6.439E+00	3.415E−15	1.785E−02	1.098E−30	3.051E−02
LI	4.777E+01	7.582E−02	9.273E−03	1.263E+00	1.765E−01
	1.584E+01	2.927E−01	9.462E−03	2.538E+00	4.498E−02
NLD	2.996E+01	9.711E−15	1.557E−02	1.514E−02	1.665E−01
	7.848E+00	3.483E−15	1.564E−02	8.295E−02	3.790E−02
OSC	1.698E+01	1.089E−14	1.295E−02	1.514E−02	1.732E−01
	4.978E+00	3.605E−15	1.277E−02	8.295E−02	4.302E−02
AV	2.756E+01	6.355E−01	1.108E+00	4.696E−03	2.466E−01
	8.151E+00	7.325E−01	5.524E−02	1.609E−02	4.902E−02
AP	5.398E+01	4.815E+00	1.772E+02	7.140E−01	2.332E−01
	1.511E+01	1.178E+00	3.011E+01	7.952E−01	3.790E−02
AR	2.504E+01	1.007E−14	1.783E−02	1.514E−02	1.733E−01
	8.374E+00	3.501E−15	1.889E−02	8.295E−02	4.299E−02
AS	4.447E+01	9.711E−15	1.180E−02	1.288E+00	1.932E−01
	1.028E+01	3.483E−15	1.618E−02	1.862E+00	5.683E−02

7.5.2 Performance on unimodal test functions

As the reported results of Table 7.8 show, different inertia weights have relatively similar behavior in terms of the success rates. For simple test functions (Sphere, Absolute and Step), all strategies achieve a success rate of 100 percent, but for the harder example of Quadric functions, all fail to get close enough to the global

Table 7.8 The number of successful runs out of 30 independent runs for each test function

	Sphere	**Absolute**	**Rosenbrock**	**Quadric**	**Step**
CONS	30	30	0	0	30
LD	30	30	0	0	30
LI	30	30	0	0	29
NLD	30	30	0	8	30
OSC	30	30	0	0	30
AV	3	0	0	2	30
AP	0	0	0	0	29
AR	30	30	0	0	30
AS	30	30	0	30	30
	Rastrigin	**Ackley**	**Griewank**	**Levy**	**Salomon**
CONS	0	23	6	30	0
LD	0	30	15	30	0
LI	0	28	19	15	0
NLD	0	30	13	29	0
OSC	0	30	19	29	0
AV	0	0	0	29	0
AP	0	0	0	3	0
AR	0	30	16	29	0
AS	0	30	17	14	0

optimum (except for the success-rate-based adaptive strategy). However, a closer look at the accuracies of models reveals their difference. To have a better understanding of the impact of the inertia weigh strategy on the behavior of PSO on a single peak, the average fitness obtained from different strategies for two unimodal test functions (Sphere and Quadric) are shown in Figure 7.6. From these figures and the results of Table 7.7, it is clear that appropriate adjustment of inertia weight during a run of PSO can greatly improve the convergence speed and the accuracy of the final solution.

As the fitness curves on the Sphere function (Figure 7.6) show, adaptive strategies based on the position and velocity of the particles provides the worst results. This seems to occur due to the inappropriate feedback parameters used in these models. The low performance of the constant inertia weight model in this test is due to the inappropriate choice of the inertia weight in this experiment ($w = 0.7$). A value in the range of $[0.3, 0.4]$ considerably improves the performance of this model. However, the inertia weight needs to be properly set before the execution of the algorithm for each test function.

All other strategies properly locate the global optimum of the Sphere function, albeit with different accuracies and convergence speeds. The convergence speed of the linear increasing and linear decreasing strategies are half of the convergence speed of the success-rate-based adaptive model (in the logarithmic scale), as these models waste half of their search time due to the inappropriate values of the inertia weight at

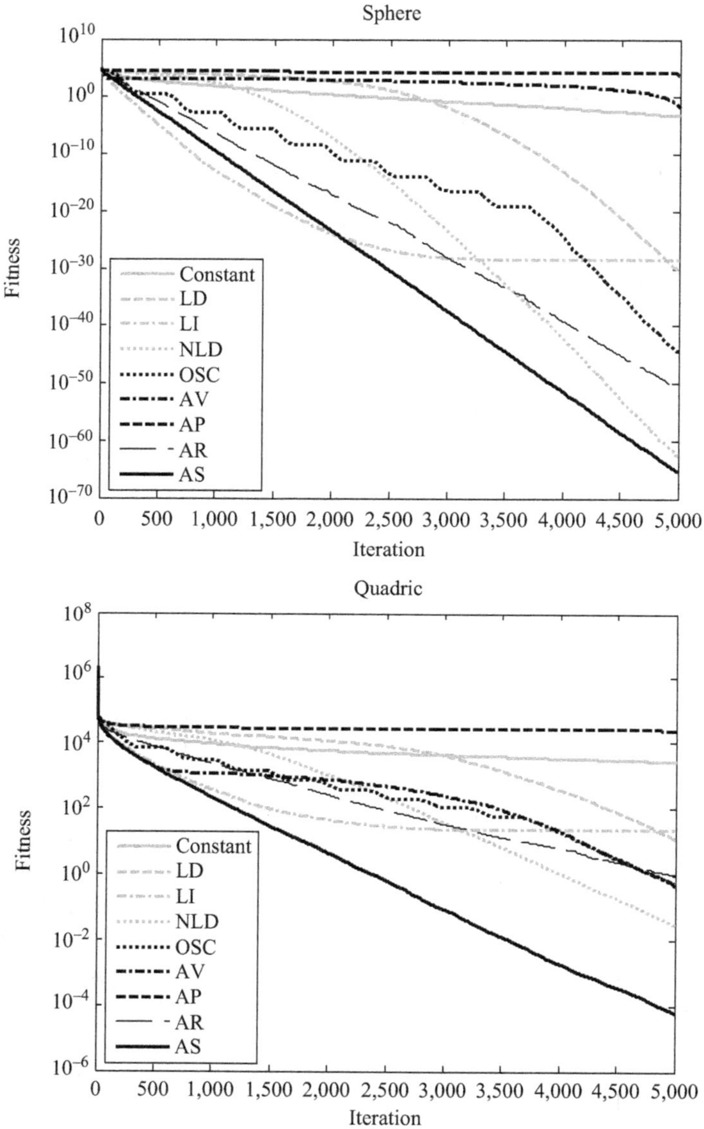

Figure 7.6 Performance of different inertia weight strategies on two unimodal test functions

some iterations. The oscillating method repeatedly switches between good and inappropriate values of the inertia weight resulting in high and low convergence speeds. The horizontal parts of the related curve indicate inappropriate inertia weight values. In the case of the nonlinear decreasing model, after the initial 1,500 iterations, the inertia weight value is close to w_{min} which is a good choice for this test function and hence results in a good performance.

Rank-based and success-rate-based adaptive models show a constant and fast convergence toward the optimum. The success of the rank-based adaptive inertia weight model is due to the use of a diverse set of inertia values at each iteration by different particles. The success-rate-based adaptive model is the best strategy in this case and tries to find the best inertia weight value for the whole swarm based on the success rate of the swarm.

The role of the convergence speed is more obvious for the Quadric test function (Figure 7.6). While the overall shape of fitness curves is similar to the previous example, most of the strategies fail to provide good results due to their low convergence speed.

7.5.3 Performance on multimodal functions

Regarding the multimodal test functions, the performance of inertia weight strategies mainly depend on the shape and difficulty of the test function. For example, despite its large number of local minima, the overall shape of Ackley function is similar to a unimodal function, and hence the behavior of the different inertia weight strategies on this test function is similar to the unimodal functions. As Table 7.8 shows, the success rate for this function is about 100 percent, and the fitness curves of strategies on Figure 7.7 are similar to the curves of Figure 7.6.

The Rastrigin function is a hard and highly multimodal test function. As Table 7.8 shows, PSO variants with all studied inertia weight strategies are trapped in a local optimum and cannot locate the global optimum. From Figure 7.7, it is clear that models with higher convergence speed get stuck in worse local optima. For example, the performance of the success-rate based models is lower than the performance of the constant inertia weight strategy in this case. This difference is more obvious for the Levy function. Levy is a multimodal test function with fewer local optima compared to Ackley and Rasrtigin functions, but the attraction basins of its local optima are larger. The large difference between the success rates of the constant and adaptive success-rate-based inertia weight models indicates that a long-term search is more preferred for this function.

Harrison *et al.* performed a similar experimental study to examine the performance of different inertia weight control methods [9]. In this study, a diverse set of inertia weight models were utilized in the same PSO algorithm. The resulting algorithm was applied to a number of test functions. A swarm of 30 particles with a global-best (star) topology was used, and each execution consisted of 5,000 iterations. A suite of 60 minimization problems proposed by Engelbrecht [43] were used in this study. This set of problems provide different optimization challenges including unimodal, multimodal, separable, nonseparable, rotated, shifted, noisy, and composition functions. Three performance measures were considered in the comparisons: *accuracy, consistency*, and *success rate*. Accuracy of each algorithm is defined as the fitness of the best solutions found by the corresponding PSO algorithm. Consistency shows the variance of different accuracies of an algorithm, and the success rate of an

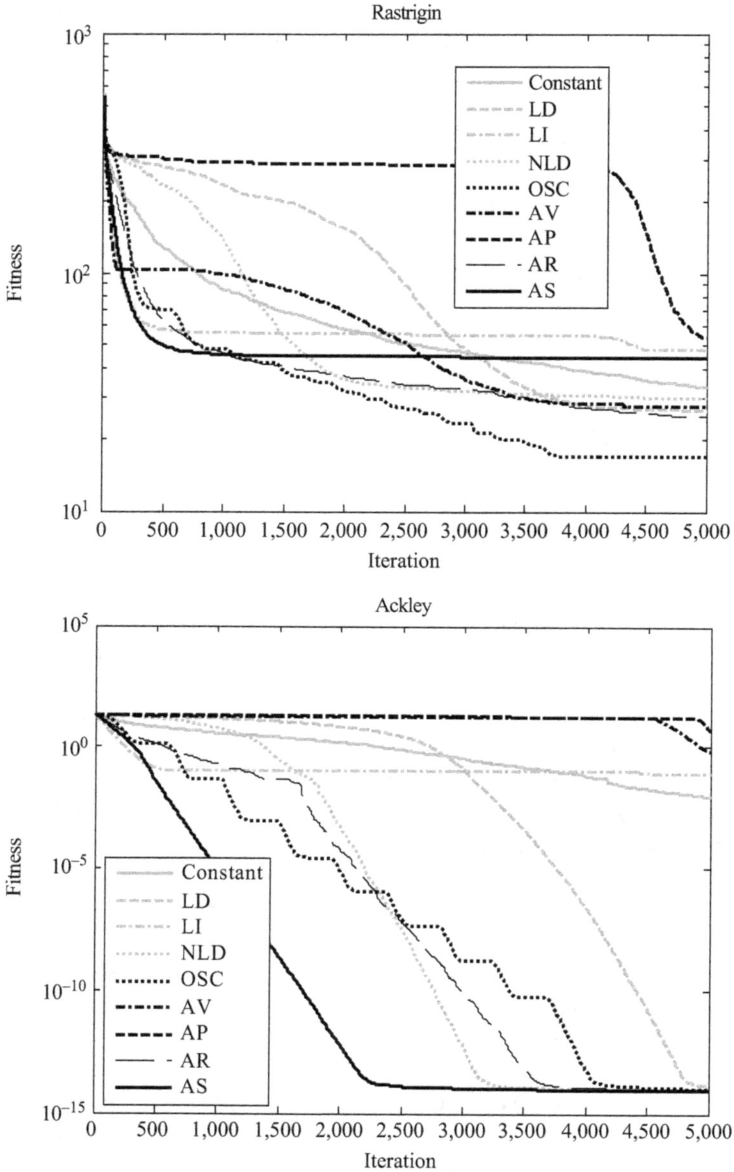

Figure 7.7 Performance of different inertia weight strategies on two multimodal test functions

algorithm refers to the percentage of the 50 independent runs within which the PSO algorithm reaches a specific accuracy level.

The overall conclusion of the Harrison *et al.* study [9] is that the constant and random inertia weight models achieve the best rank with respect to the three performance measures of accuracy, consistency, and success rate. Further investigation of this study shows a correlation between the particle movement and the performance of each PSO model. The best performing models are attributed with a moderate initial movement followed by a gradual decrease in the particles movement. However, it should be noted that the configurations of the PSO model may change the ranking of the inertia weight models. For example, linear decreasing [12] and success-rate-based inertia model [35] attain the highest ranks in terms of the accuracy for a PSO with a ring topology.

7.5.4 Performance in dynamic environments

As mentioned earlier, some of the inertia weight control strategies have been specially designed for dynamic optimization problems. For example, while it is very rare to see the situation where all particles are converged to a point far from a local or global optimum in static optimization problems, it is very common to see such situations in dynamic optimization problems with moving peaks. In such problems, after the convergence of the swarm of particles on a local optimum (peak), the local optimum may move, placing the particles in the above state. The optimization algorithms should be able to rapidly relocate the swarm of particles on the new position of the peak. To do so, the algorithm should first coarsely explore the vicinity of the current position of the swarm by increasing the velocities of the particles. After that, the velocities of the particles should be reduced to let the particles converge to the new position of the local (global) optimum.

To compare the inertia weight strategies in a dynamic environment, three inertia weight models are applied to a three-dimensional unimodal moving peak benchmark [42]. In this test function, the position of the global optimum is randomly changed after periods of 100 iterations. The parameters of the PSO algorithm are the same as before. The mean of the best fitness achieved by each strategy and the inertia weight values over 20 independent runs are shown in Figure 7.8.

As the fitness curves of Figure 7.8 show, the constant inertia weight is quite successful at the first period of optimization and rapidly converge toward the optimum. At the first iteration of this period, particles are moving through the search space with initial random velocities. However, after the first movement of the single peak of the objective function in the search space, the constant inertia weight completely fails to follow the peak due to the low velocities of the particles at the beginning of this period. As the inertia weight curves of Figure 7.8 show, the inertia weight is set to the constant value of 0.3 for all iterations in this strategy.

Linear decreasing inertia weight is a time-varying strategy. As the results show, it linearly decreases from the initial value of 1.0 to the final value of 0.0 during a course of run. The fitness curve of this model shows that the model is successful only at the periods where the inertia weight is set to a value at the range of [0.4, 0.6]. It should be noted that in real-world dynamic optimization problems, the optimization may continue forever and time-varying strategies cannot be applied easily.

Figure 7.8 indicates that after each change of the test function, the success-rate-based adaptive strategy rapidly converges to the new position of the peak. The inertia

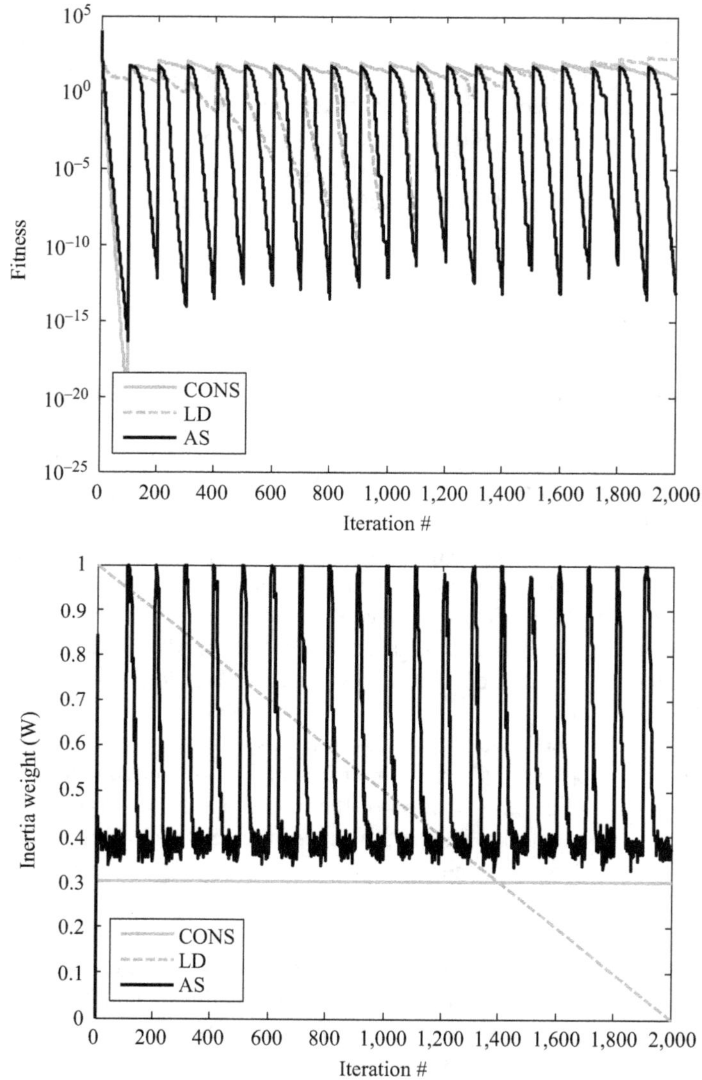

Figure 7.8 Performance of constant (CONS), linear decreasing (LD), and success-rate-based adaptive (AS) inertia weight strategies on a moving peak benchmark. Up: the fitness of the best solution found by the algorithm. Down: the inertia weight at each iteration

weight values of this model show that after each movement of the peak, the model instantly reacts to the environment change by increasing the inertia to its maximum value. After a few iterations, the swarm reaches the new position of the optimum, and the inertia weight is decreased to enable the algorithm to perform the extraction phase of its search.

7.6 Conclusion

PSO is an easy to implement optimization technique with a few parameters to adjust. Among the PSO parameters, the inertia weight has received the most attention during the past decades, and many parameter control strategies have been proposed for adjusting the value of this parameter during the execution of PSO. In this chapter, these models were introduced in the three categories of constant or random, time-varying, and adaptive inertia weight models. The time-varying inertia weight models were further divided to increasing, decreasing, and oscillating groups. The adaptive strategies were also categorized based on their feedback parameters. While every inertia weight model may be useful for some specific problems, the experimental studies show that despite the simplicity of constant and random inertia weight models, these models may be the best choice in static problems where no prior knowledge about the objective function is available. On the other hand, the constant or time-varying methods do not seem to be appropriate choices for dynamic optimization problems. In such problems, adaptive inertia strategies more rapidly react to the objective function's changes and provide better solutions during the execution of PSO.

References

[1] Kennedy J, Eberhart R. Particle swarm optimization. In: IEEE International Conference on Neural Networks. IEEE; 1995. p. 1942–1948.

[2] Shi Y, Eberhart R. A modified particle swarm optimizer. In: Evolutionary Computation Proceedings, 1998. IEEE World Congress on Computational Intelligence, The 1998 IEEE International Conference on. IEEE; 1998. p. 69–73.

[3] Shi Y, Eberhart RC. Fuzzy adaptive particle swarm optimization. In: Evolutionary Computation, 2001. Proceedings of the 2001 Congress on. vol. 1. IEEE; 2001. p. 101–106.

[4] Rechenberg I. Evolutions strategie: Optimierung technischer Systeme nach Prinzipien der biologischen Evolution. Frommann-Holzboog Verlag, Stuttgart. 1973.

[5] Kirkpatrick S, Gelatt CD, Vecchi MP, *et al.* Optimization by simulated annealing. Science. 1983;220(4598):671–680.

[6] Li X. Adaptively choosing neighbourhood bests using species in a particle swarm optimizer for multimodal function optimization. In: Genetic and Evolutionary Computation Conference. Springer; 2004. p. 105–116.

[7] Nickabadi A, Ebadzadeh MM, Safabakhsh R. DNPSO: a dynamic niching particle swarm optimizer for multi-modal optimization. In: Evolutionary Computation, 2008. CEC 2008 (IEEE World Congress on Computational Intelligence). IEEE Congress on. IEEE; 2008. p. 26–32.

[8] Nickabadi A, Ebadzadeh MM, Safabakhsh R. Evaluating the performance of DNPSO in dynamic environments. In: Systems, Man and Cybernetics, 2008. SMC 2008. IEEE International Conference on. IEEE; 2008. p. 2640–2645.

[9] Harrison KR, Engelbrecht AP, Ombuki-Berman BM. Inertia weight control strategies for particle swarm optimization. Swarm Intelligence. 2016;10(4):267–305.

[10] Shi Y, Eberhart RC. Parameter selection in particle swarm optimization. In: International Conference on Evolutionary Programming. Springer; 1998. p. 591–600.

[11] Eberhart RC, Shi Y. Tracking and optimizing dynamic systems with particle swarms. In: Evolutionary Computation, 2001. Proceedings of the 2001 Congress on. vol. 1. IEEE; 2001. p. 94–100.

[12] Shi Y, Eberhart RC. Empirical study of particle swarm optimization. In: Evolutionary Computation, 1999. CEC 99. Proceedings of the 1999 Congress on. vol. 3. IEEE; 1999. p. 1945–1950.

[13] Feng Y, Teng G, Wang A, *et al.* Chaotic inertia weight in particle swarm optimization. In: Innovative Computing, Information and Control, 2007. ICICIC'07. Second International Conference on. IEEE; 2007. p. 475.

[14] Feng Y, Yao Y, Wang A. Comparing with chaotic inertia weights in particle swarm optimization. In: Machine Learning and Cybernetics, 2007 International Conference on. vol. 1. IEEE; 2007. p. 329–333.

[15] Chatterjee A, Siarry P. Nonlinear inertia weight variation for dynamic adaptation in particle swarm optimization. Computers & Operations Research. 2006;33(3):859–871.

[16] Lei K, Qiu Y, He Y. A new adaptive well-chosen inertia weight strategy to automatically harmonize global and local search ability in particle swarm optimization. In: 1st International Symposium on Systems and Control in Aerospace and Astronautics. IEEE; 2006. p. 977–980.

[17] Jiao B, Lian Z, Gu X. A dynamic inertia weight particle swarm optimization algorithm. Chaos, Solitons & Fractals. 2008;37(3):698–705.

[18] Yang C, Gao W, Liu N, *et al.* Low-discrepancy sequence initialized particle swarm optimization algorithm with high-order nonlinear time-varying inertia weight. Applied Soft Computing. 2015;29:386–394.

[19] Chen G, Huang X, Jia J, *et al.* Natural exponential inertia weight strategy in particle swarm optimization. In: Intelligent Control and Automation, 2006. WCICA 2006. The Sixth World Congress on. vol. 1. IEEE; 2006. p. 3672–3675.

[20] Gao YL, An XH, Liu JM. A particle swarm optimization algorithm with logarithm decreasing inertia weight and chaos mutation. In: Computational Intelligence and Security, 2008. CIS'08. International Conference on. vol. 1. IEEE; 2008. p. 61–65.

[21] Fan SKS, Chiu YY. A decreasing inertia weight particle swarm optimizer. Engineering Optimization. 2007;39(2):203–228.

[22] Zheng Y, Ma L, Zhang L, *et al.* Empirical study of particle swarm optimizer with an increasing inertia weight. In: Evolutionary Computation, 2003. CEC'03. The 2003 Congress on. vol. 1. IEEE; 2003. p. 221–226.

[23] Zheng Y, Ma L, Zhang L, *et al.* On the convergence analysis and parameter selection in particle swarm optimization. In: Machine Learning and Cybernetics, 2003 International Conference on. vol. 3. IEEE; 2003. p. 1802–1807.

[24] Malik RF, Rahman TA, Hashim SZM, *et al.* New particle swarm optimizer with sigmoid increasing inertia weight. International Journal of Computer Science and Security. 2007;1(2):35–44.

[25] Kentzoglanakis K, Poole M. Particle swarm optimization with an oscillating inertia weight. In: Proceedings of the 11th Annual Conference on Genetic and Evolutionary Computation. ACM; 2009. p. 1749–1750.

[26] Kordestani JK, Rezvanian A, Meybodi MR. An efficient oscillating inertia weight of particle swarm optimisation for tracking optima in dynamic environments. Journal of Experimental & Theoretical Artificial Intelligence. 2016;28(1–2):137–149.

[27] Eberhart RC, Shi Y. Comparing inertia weights and constriction factors in particle swarm optimization. In: Evolutionary Computation, 2000. Proceedings of the 2000 Congress on. vol. 1. IEEE; 2000. p. 84–88.

[28] Panigrahi B, Pandi VR, Das S. Adaptive particle swarm optimization approach for static and dynamic economic load dispatch. Energy Conversion and Management. 2008;49(6):1407–1415.

[29] Tanweer MR, Suresh S, Sundararajan N. Self regulating particle swarm optimization algorithm. Information Sciences. 2015;294:182–202.

[30] Saber AY, Senjyu T, Urasaki N, *et al.* Unit commitment computation – A novel fuzzy adaptive particle swarm optimization approach. In: Power Systems Conference and Exposition, 2006. PSCE'06. 2006 IEEE PES. IEEE; 2006. p. 1820–1828.

[31] Yang X, Yuan J, Yuan J, *et al.* A modified particle swarm optimizer with dynamic adaptation. Applied Mathematics and Computation. 2007;189(2): 1205–1213.

[32] Arumugam MS, Rao M. On the improved performances of the particle swarm optimization algorithms with adaptive parameters, cross-over operators and root mean square (RMS) variants for computing optimal control of a class of hybrid systems. Applied Soft Computing. 2008;8(1):324–336.

[33] Li Z, Tan G. A self-adaptive mutation-particle swarm optimization algorithm. In: Natural Computation, 2008. ICNC'08. Fourth International Conference on. vol. 1. IEEE; 2008. p. 30–34.

[34] Xu G. An adaptive parameter tuning of particle swarm optimization algorithm. Applied Mathematics and Computation. 2013;219(9):4560–4569.

[35] Nickabadi A, Ebadzadeh MM, Safabakhsh R. A novel particle swarm optimization algorithm with adaptive inertia weight. Applied Soft Computing. 2011;11(4):3658–3670.

[36] Zhan ZH, Zhang J, Li Y, *et al.* Adaptive particle swarm optimization. IEEE Transactions on Systems, Man, and Cybernetics, Part B (Cybernetics). 2009;39(6):1362–1381.

[37] Deep K, Chauhan P, Pant M. A new fine grained inertia weight particle swarm optimization. In: Information and Communication Technologies (WICT), 2011 World Congress on. IEEE; 2011. p. 424–429.

[38] Beyer H, Schwefel H. Evolution strategies – A comprehensive introduction. Natural Computing. 2002;1(1):3–52.

[39] Taherkhani M, Safabakhsh R. A novel stability-based adaptive inertia weight for particle swarm optimization. Applied Soft Computing. 2016;38:281–295.
[40] Chauhan P, Deep K, Pant M. Novel inertia weight strategies for particle swarm optimization. Memetic Computing. 2013;5(3):229–251.
[41] Arasomwan MA, Adewumi AO. On the performance of linear decreasing inertia weight particle swarm optimization for global optimization. The Scientific World Journal. 2013;2013.
[42] Branke J. Memory enhanced evolutionary algorithms for changing optimization problems. In: Evolutionary Computation, 1999. CEC 99. Proceedings of the 1999 Congress on. vol. 3. IEEE; 1999. p. 1875–1882.
[43] Engelbrecht A. Particle swarm optimization: Global best or local best? In: Computational Intelligence and 11th Brazilian Congress on Computational Intelligence (BRICS-CCI & CBIC), 2013 BRICS Congress on. IEEE; 2013. p. 124–135.

Chapter 8
Robot path planning using swarms of active particles

C. Helbert Eduardo Espitia[1] *and E. Jorge Ivan Sofrony*[2]

Abstract

There exist a wide range of techniques that generate collision-free (optimal) trajectories, in an autonomous fashion, for mobile robotics. Probably, the most popular technique is that of artificial potential fields, where the robot is treated as a particle subject to a potential field that is generated by the obstacles and the goal position. The path generation problem is then treated as an optimization problem where gradient descent methods have been traditionally used. Particle swarm optimization has also been widely used to solve optimization problems, and although it has proven to be more efficient in the search of minima, it also suffers from early convergence, i.e. the swarm may get trapped in a local minima. Many of the modifications that cater for this weakness involve added complexity in the construction of the potential field and thus translates into ad-hoc algorithms. A particle swarm approach with the property of escaping local minima by forcing vortex-like dynamics when the gradient of the potential field is close to zero is proposed, proving to be more compact and intuitive than previously proposed algorithms.

8.1 Introduction

Path planning is of great interest in many robotics applications, especially in automated guided vehicles or unmanned aerial vehicles applications where an enhanced autonomy is desired. The path planning problem concentrates on determining a collision free path, from an initial location to a final destination, in a workspace under the presence of obstacles. The applications of trajectory generating algorithms go beyond the field of robotics and are being used in fields like computer animation, video games, assembly and disassembly problems and molecule folding. Path planning problems are a far more common concern in robotics, where a classical version

[1]Electronic Engineering, Universidad Distrital Francisco Jose de Caldas, Colombia
[2]Mechanical Engineering and Mechatronics, Universidad Nacional de Colombia, Colombia

of motion planning is sometimes referred to as the Piano Mover's Problem where the steps to move the piano from one room to another without hitting any of the walls must be determined. Robot motion planning usually ignores dynamics and other differential constraints and focuses primarily on the translations and rotations required to "move the piano."

Trajectory planning usually refers to the problem of taking the solution from a motion planning algorithm, and determining how to move along the solution in a continuous fashion, usually taking into account mechanical and dynamic limitations of the robot. The basic path planning problem involves computing a collision-free path between two configurations in a static environment of known obstacles. In this case, the constraints imposed on the solution arise exclusively from the geometric characteristics of both the obstacles and the "robot" (the object for which a path is being computed). If the robot can be represented as a single 2D rigid object that moves in a 2D environment, then the problem is referred to as a two-dimensional search space. In the case of a rigid object that rotates and translates in a 3D environment, the search space has six dimensions.

Path-planning algorithms have been studied extensively, and many have been developed. In general, these algorithms aim at finding a collision-free configuration space, within the search space, that joins the start and goal robot configurations. Early works constructed exact, or complete resolution, free-space representations and searched them systematically. The complexity of such strategies grows with the configuration space dimension, which is usually equal to the number joints or degrees-of-freedom; the worst case computation time grows in an exponential fashion with the robot's dimensions for planar and spatial systems. Practical algorithms have been developed for dimension three, notably for polygonal robots and obstacles. But the approach appears impractical for dimension six, which is the minimum for spatial planning or for multi-part planar planning.

Another approach is the development of probabilistic algorithms that construct "free-space" road maps by random sampling. Although extremely promising, probabilistic algorithms have some drawbacks: the probability of finding a path, when one exists, is a function of geometric constants that may be difficult to select and calculate; finding a solution depends on the type of sampling employed. In configuration spaces with a high number of obstacles and narrow passages, probabilistic methods require a large number of samples (or test nodes) in order to be able to find an appropriate path.

Regarding the difficulty of representing explicitly the configuration space, an alternative is to develop search algorithms that incrementally explore free space while searching for a path to the goal, for example potential field methods. The application of artificial potential fields (APFs) to obstacle avoidance was introduced by Khatib [1], where the use of repulsive fields around the obstacles (and forbidden regions) in order to repel the robot and an attractive potential field around goal to attract it (consequently, the robot experiences a generalized force equal to the negative of the total potential gradient), allowed an incremental exploration of the search space. This force drives the robot downhill towards its goal configuration until it reaches a minimum; thus, the algorithm may get stuck in local minima and consequently trap the robot before reaching its goal.

APFs (combined with different optimization techniques) have demonstrated themselves useful in mobile robotics; none the less, for high-dimensional systems and/or search spaces with a large number of obstacles, oscillations usually occur in between obstacles (or when the robot approaches an obstacle), and thus the robot must stay far from any obstacle in order to avoid this unwanted behavior. Another problem arises because local minima tend to trap the robot and thus the algorithm is rendered impractical under the conditions mentioned above. Particle swarms can model the behavior of living organisms [2] or can be used to solve unconventional optimization problems (i.e. particle swarm optimization or PSO), so it is not surprising that particle swarm models have been successfully applied to generate collision free trajectories [3]. The algorithm proposed in [4] uses a large model, where tuning of the parameters involved may be application dependent and difficult to choose appropriately.

A detrimental situation of many PSO algorithms consists of early convergence to local minima [5,6]; additionally, performance of these algorithms is highly susceptible to the selection of parameters as shown in [7,8]. Although many modifications that seek to avoid local minima by inducing an explosion of the swarm have been developed (see e.g. [9–15]), they tend to use complex procedures to induce this type of behavior and most of them exhibit great susceptibility to the choice of parameters with no real guidance on how to appropriately choose them. An analytical approach is presented in [8], where some analysis is provided in order to improve the selection of parameters, hence convergence of the algorithm.

Several PSO algorithms have built on the idea of incorporating a turbulence term. In [16], this operator is included in the traditional PSO algorithm as a stochastic variable of "madness." According to [17], this is done to maintain the diversity of the population. From an evolutionary algorithms viewpoint, the turbulence term has the same function as the mutation operator [18].

In [19] a logistic series is used to replace the random number in other PSO algorithms; in [20], a chaotic system based on disturbance of descent method for a virtual function associated with individual and group minimum is proposed. An appropriate selection of parameters is difficult to obtain for this algorithm is not easy, and no guide is provided. In [21], chaotic sequences to initialize the position of the particle are used, and a mutation operator was added to increase the diversity of the population and avoid premature convergence.

In traditional algorithms PSO algorithms, the inertia factor has been used to enhance performance. For example, in [22] a chaotic PSO algorithm based on improved adaptive inertia is proposed. The inertia value is adjusted adaptively by feedback of parameters, including the number of iterations, factor of aggregation and particle speed. Premature convergence is established by the variance of the population fitness and a chaotic disturbance is added to escape local optima.

The following sections will present a path planning algorithm for environments with local minima, built using known APF techniques [1,23–25] and a model of active particle swarm (APS) [26,27]. The algorithm is labeled *Vortex Particle swarm path planning* (VPSPP) [28], and its main novelty is that it presents vorticity in its dynamics, a fact that is exploited. Local minima evasion is achieved by exploiting the vortex-like behavior of the swarm, where each particle *explores closely* its surroundings once it is close to a minimum. Although a similar approach (in the sense that vortex-like behavior is used) has been proposed by McInnes *et al.*, the particle swarm model used in [3,4] is not self-propelled and uses Morse potentials in order to describe the interaction between particles. Although the approach proved to be effective, tuning of the parameters may become unintuitive; on the other hand, weakly nonlinear harmonic oscillators have been widely studied and have proven to effectively model (by a simple set of dynamic equations and few parameters) complex behavior in living organisms.

The concept of self-propelled particles was introduced in [29] where the collective movement of large groups of living organisms or chemical processes was studied. Self-propelled particle models consist of particles that interact locally with a form of propulsion intrinsic to the particle [30] and are common in modeling physical–chemical processes involving small particles with surface forces sensitive to chemical concentrations and temperature [31]. According to [30], the interaction between (randomly moving) neighbors may generate coordinated movement.

The chapter is organized as follows: it starts by giving a brief introduction to the well-known path planning problem; this is followed by a description and brief dynamic analysis of proposed model and its components; the next section makes use of the available parameters and introduces a method that forces the swarm to increase the dispersion of its vortex-like behavior; finally the effectiveness of the algorithm is validated in an environment with several well-known settings, including a U-type obstacle that generates a field potential with two local minima.

8.2 Path planning for mobile robots

The path-planning problem concentrates on determining a collision-free path, from an initial location to a final destination, in an environment that presents obstacle or *no-go* areas. The "basic path-planning problem" involves computing a collision-free path between two configurations in a static environment of known obstacles. In this case, the constraints imposed on the solution arise exclusively from the geometric characteristics of both the obstacles and the "robot" (the object for which a path is being computed). Path-planning algorithms can be divided into two categories depending on the environment: dynamic or static [32]. Only the static case will be considered, where full knowledge of the workspace is assumed and the obstacles are considered static. Trajectory planning refers to the problem of taking the output from a motion planning algorithm and determining how to move along the solution taking into account mechanical and dynamic limitations of the robot. A trajectory is optimal if it additionally meets a certain performance criterion [25]. A study of trajectory generation applications for mobile robots, for the case of static and dynamic obstacles, is made in [33].

A wide range of path-planning algorithms have been proposed, where initially exact (or complete) resolution representations where sought for the free space and a systematic search was performed on this representation [34]. Some of the most popular "conventional" techniques were based on Visibility Graphs, Voronoi diagrams, Modeling of Free Space and Cell Decomposition ([25,32,34]). More "modern" techniques determine optimal trajectories via stochastic search algorithms or optimization, where examples of such approaches use genetic algorithm [35], ant colonies [36] and rapidly exploring random trees. Additional work on path planning using bioinspired methods has made use of bacterial chemotaxis models [37] and other forms of PSO [38].

APFs is the most popular technique in the area of path planning for mobile robots, as it presents an intuitive and elegant way to solve the conventional path planning problem. Khatib [1] introduced this method in the 1980s, where he proposed that the obstacles (forbidden regions) and the robot(s) be considered as electrical charges of the same sign (generating a repulsion force between them), while the goal had an associated electric charge of opposite sign in order to attract the robot to the destination point. Conventionally attractive and repulsive potentials are generated separately, and the total workspace potential is taken as the sum of these. The goal is represented as a standard parabolic potential with center at the objective position (goal) and obstacles are represented by exponential potential barriers. The expressions for these potentials are

$$U_{obj}(r) = k_{obj}\|r - r_{obj}\|^2 \tag{8.1}$$

$$U_{obs,m}(r) = k_{obs}\|r - r_{obs,m}\|^{-1} \tag{8.2}$$

where r is the position vector of the particle, $U_{obj}(r)$ is the objective potential and $U_{obs,m}(r)$ is the potential of the mth obstacle. The vectors r_{obj} and $r_{obs,m}$ represent the goal position and the position of the mth obstacle, respectively. Considering m as the total number of obstacles, the total potential is given by

$$U_{esp}(r) = U_{obj}(r) + \sum_{m=1}^{M} U_{obs,m}(r) \tag{8.3}$$

Figure 8.1 depicts the APF for a U-trap type obstacle. The problem of path planning is now formulated as that of finding the trajectories that take a particle

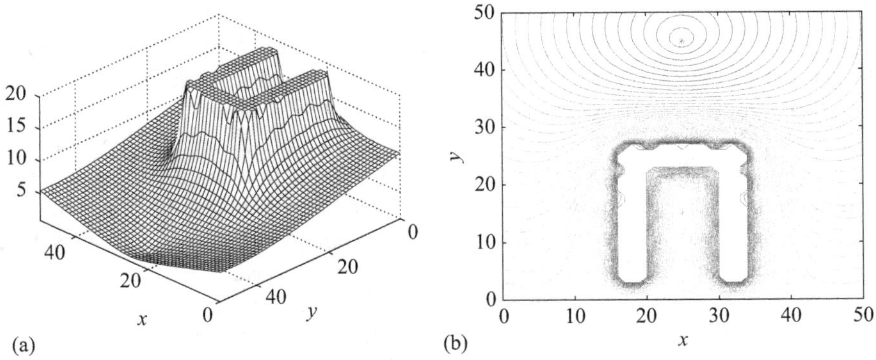

Figure 8.1 Artificial potential field generated by a U-type obstacle: (a) potential field, (b) contorno

immersed in a potential field, to the global minimum. The APF method has been traditionally solved using a gradient descent algorithm to obtain the evolution of the path. However, the search space is in general non-convex and presents local minima. Modifications to the original APF method have been made in order to address the mentioned limitations, such that a convex potential field is always generated: Volpe and Khosla [24] proposed Superquadric Artificial Potential Functions; an algorithm based on the solution of the Laplace equation for heat transfer is proposed separately by Connolly [39] and Kim and Khosla [40]. Additional methods that avoid local minima by generating convex APF can be found in [41–43]. A disadvantage of these methods is that the selection of the parameters that generate the vector field is not straightforward and may affect convergence properties. Another way of avoiding local minima in APF problems is to use recursive methods that tend to be more effective but also require more computational power (navigation function [44]).

The methods previously mentioned, due to their characteristic of adding relative potential fields relative to the obstacles, were baptized *Local Potential Fields*. In contrast, we have global potential fields that consider all the obstacles in order to build the potential field as a whole, and not as the addition of parts. Under this approach, navigation functions are proposed by Rimon and Koditchek [45], where the global APF is explicitly designed to present a single local minimum, i.e. the objective. The basic idea is to mask (fill) any additional equilibrium points; thus, it is necessary to fully understand the workspace in order to construct the navigation function; loss of simplicity and additional computational complexity is present in comparison with other methods [44].

In order to avoid the design of application-specific potential fields, strategies based on emergent behavior particle swarms have been proposed. PSO algorithms were initially introduced by James Kennedy and Russell Eberhart [8], and are believed to present a more efficient exploration of the search space. The proposed algorithm uses on the idea of vortex-like swarming to overcome local minima.

8.3 Active particle swarms

Social behavior of living organisms is based on the interaction amongst members (neighbors) of the same species. Some of these interactions have been shown to be optimal in tasks such as food search and predator evasion. Modeling of such interactions, and the behavior that emerges from this congregation of individuals, has been the focus of different research areas that have tried to explain the dynamics of living organisms such as bacteria, school of fish and zooplankton [27]. In addition to modeling the behavior of individuals within a community, these techniques have been sought to solve problems within the engineering context.

Within the family of APSs, we have self-propelled particles [46] that solve the Langevin equation using different dissipation functions. In [26], the interactions between individuals are represented as a parabolic potential, resulting in a simple and intuitive model. An interesting fact of APS models, as proposed in this work, is that vortex formations may be imprinted upon the system by incrementing a single parameter. An important difference of the proposed swarm model from the standard APS models with Brownian particles is that a stochastic component is not being used; it has been proven [47,48] that vortex-type motion may be induced even in the case when no stochastic component is present; hence, we will assume that the dynamics are totally deterministic.

8.3.1 Particle dynamics

In order to give more insight into how the swarm might evolve, an initial analysis of the dynamic properties of a single particle is performed. The initial analysis will concentrate on showing that a single particle has a stable limit cycle centered around a given optimal point, and not on finding the geometry of the swarm's trajectory. For ease of the presentation, all states will be assumed scalar, i.e. the particles move in a 1D space. Consider the scalar system described by

$$m\ddot{x} = -k\nabla U_{esp}(x) + (\alpha\dot{x} - \beta\dot{x}^3) \tag{8.4}$$

where m is the mass of the particle, $U_{esp}(x)$ is a (possibly non-convex) function of the position that represents a potential that produces a force $\nabla U_{esp}(x)$ on the system and k is a parameter associated with this force; ∇ is the partial derivative of the function with respect to the states. The parameter α is associated with the force of propulsion and β with the braking force. Due to the complexity of analyzing convergence properties for a general form of the potential $U_{esp}(x)$, the analysis is restricted to harmonic potentials, i.e. $U_{esp} = 1/2x^2$. Under this assumption, the system dynamics are described by the Raleigh equation (8.4) that describes a weakly nonlinear harmonic oscillator that can be analyzed using averaging of nonlinear systems (see [49]) or other Lyapunov methods. The latter approach is chosen since only existence of a stable limit cycle is sought.

Define $r = x$, $v = \dot{x}$ and rewrite (8.4) in state form as

$$\begin{aligned} \dot{r} &= v \\ \dot{v} &= -kr + g(v) \qquad\qquad g(v) = (\alpha v - \beta v^3) \end{aligned} \tag{8.5}$$

It is well known that this system has an unstable equilibrium point at the origin. Nonetheless, further analysis shows that the system is ultimately bounded. To observe this, consider a Lyapunov equation (or energy function)

$$V(\zeta) = \frac{1}{2}\zeta' P\zeta + \frac{\beta}{2}r^4$$

where $\zeta = [r,\ v]^T$. Setting

$$P = \begin{bmatrix} \alpha k & k \\ k & 2\alpha \end{bmatrix}$$

and taking its derivative with respect to time, and after some manipulations,

$$\begin{aligned} \dot{V}(\zeta) &= \zeta' P\dot{\zeta} + 2\beta v^3 \dot{v} \\ &< -k|r|^2 - 2\beta^2|v|^6 + (2\alpha^2 + k)|v|^2 + 3k\beta|r||v|^3 \end{aligned} \tag{8.6}$$

It may be concluded that $\dot{V}(\zeta) < 0$ for all $\zeta \in \mathbb{R}^2/\mathscr{X}$ where

$$\mathscr{X} = \left\{ \zeta \in \mathbb{R}^2 : |r| > \left(\frac{2\alpha^2 + k}{2\beta} \right)^{3/2}, \ |v| > \sqrt{\frac{2\alpha^2 + k}{2\beta}} \right\} \tag{8.7}$$

hence the system is bounded. Since the system dynamics have a unique unstable equilibrium with bounded dynamics, by Poincare–Bendixson's theorem, the system will exhibit a stable limit cycle.

Remark 1. *Observe that the state bounds depend on the parameters α and β; by increasing α the lower bound on the velocity is increased, hence the radius of rotation will grow. This fact will be used to enlarge the dispersion of the swarm.*

8.3.2 Particle swarm model

The particle swarm model selected is based on the locomotion technique of the *Daphnia* zooplankton (see [4]), which exhibits vortex-like behavior when escaping predators or searching for food. It is important to highlight that such approach has only been used to model the dynamics of living organisms; the APS approach has not been used in trajectory generation applications, thus the novelty in the proposed algorithm. Even more, APS models assume that both α and β are constant and analysis has concentrated on studying dynamic behavior for different values of these parameters; parameter variation has not been used to knowingly change the motion of the swarm. The model implemented for the generation of trajectories is based on [26] and consists of an N-particle swarm of position-coupled (slightly modified) Raleigh oscillators. The particle swarm model is as follows:

$$\dot{r}_i = v_i \tag{8.8}$$

$$m_i \dot{v}_i = \underbrace{(\alpha - \beta \, \|v_i\|^2) v_i}_{F_{pro,i}} - \underbrace{k(r_i - R)}_{F_{int,i}} - \underbrace{k_f \nabla_i U_{esp}(r_i)}_{F_{esp,i}} \tag{8.9}$$

where m_i, r_i and v_i represent the mass, position and velocity of the ith particle, respectively. The term F_{pro} represents the Raleigh–Helmholtz positive dissipation function, F_{int} represents the attractive forces amongst individuals (a parabolic well is considered as the potential field [26]), and F_{esp} represents forces exerted on the particles by some external potential field, in this case the APF generated by the objective position and obstacles [see (8.2)]; this force is weighted by k_f. The center of mass (c.o.m.) of the swarm is defined as

$$R = \frac{1}{N} \sum_{i=1}^{N} r_i \tag{8.10}$$

If a single particle and no external potential field (i.e. $U_{esp} = 0$) are considered, the equilibrium positions for the velocity would be given by $v = 0$ and $v = \sqrt{\alpha/\beta}$. Extensive numerical results with particle models have shown that the average magnitude for the velocity of the swarm converges to a common value [46]. Although no formal proof is provided, evidence suggests that, for small α and small particle misalignment (an ordered swarm close to rest), the magnitude of the velocities will converge to $\sqrt{\alpha/\beta}$. Although stating an explicit solution of the swarm trajectories is a cumbersome (almost impossible) task, in the next section some insight is provided on the type of formations that the swarm may take. It is important to highlight that the interest here lies in the exploration characteristics of the swarm; hence, no coordination or consensus of the particles is explicitly sought.

8.3.2.1 Energy analysis

Let K_i, U_i be the kinetic and potential energies for the ith particle, defined as

$$K_i = \frac{1}{2} m_i v_i^2, \qquad U_i = U_{int}(r_i) + k_f U_{esp}(r_i) \tag{8.11}$$

Define the energy of the ith particle as $E_i = K_i + U_i$, and the energy of the swarm as the sum for all particles.

Proposition 1. *Consider the N-particle swarm given in (8.9) with α, β, k and k_f all non-negative parameters and a harmonic external potential (i.e. $U_{esp} = (1/2)r^2$). The energy content of the swarm, i.e. $E = \sum_{i=1}^{N} E_i$, is constant if $\|v_i\| = 0$ or $\|v_i\|^2 = \alpha/\beta$.*

Proof. The statement is equivalent to proving that the time derivative of E is equal to zero. Taking the derivative of the particle's energy $E_i = K_i + U_i$, and noting that $U_{esp} = (1/2)r^2$ and $\sum_{i=1}^{N} (r_i - R) = 0$, it follows that

$$\dot{E}_i = (\alpha - \beta \, \|v_i\|^2) \, \|v_i\|^2 - k(r_i - R)\dot{R} \tag{8.12}$$

Adding the contributions of all particles, the following expression is obtained for the energy of the swarm

$$\dot{E} = \sum_{i=1}^{N} \dot{E}_i = \sum_{i=1}^{N} (\alpha - \beta \, \|v_i\|^2) \, \|v_i\|^2 . \tag{8.13}$$

It is observed that if $|v_i| = 0$ or $|v_i|^2 = \alpha/\beta$, then $\dot{E} = 0$ and the energy will remain constant. □

Unfortunately, this only shows boundedness of the velocity and no conclusions may be drawn as to the boundedness of the position of the particles. More insight may be obtained with a new representation of the swarm. The swarm can be represented as the dynamics of the c.o.m. R (and its velocity $V = \dot{R}$) and the dynamics of the relative position of each particle. Defining the relative position as $\delta r_i = r_i - R$ and after some algebraic manipulations, the following dynamic equations are obtained [50]

$$\begin{aligned} \dot{R} &= V \\ \dot{V} &= (\alpha - \beta V^2)V - R - 2\frac{\beta}{N-1}\sum_i \delta v_i^2 V \end{aligned} \tag{8.14}$$

$$\begin{aligned} \dot{\delta r_i} &= \delta v_i \\ \dot{\delta v_i} &= (\alpha - \beta \delta v_i^2)\delta v_i - \delta r_i - 2\frac{\beta}{N-1}\delta v_i V^2 \end{aligned} \tag{8.15}$$

If we assume a swarm that is almost at rest, two extreme conditions may be obtained: the center of gravity is zero with zero velocity; the particles' relative position is zero with zero relative velocity. In the latter case, (8.15) is equal to zero and the dynamics of the c.o.m. (8.14) are reduced those of a Raleigh oscillator; hence, the swarm will implode into a single super-particle that presents a stable limit cycle.

For the prior case, (8.14) is now reduced to zero and the relative position dynamics are, again, a Raleigh oscillator. This means that all particles will exhibit a stable limit cycle around the c.o.m. Even more, the particles must be symmetrically located within the limit cycle trajectory (i.e. $\sum_i \delta r_i = 0$ and $\sum_i \delta v_i = 0$). Such extreme dynamics are difficult to obtain naturally if initial conditions are not close to this rest states.

Consider the c.o.m. dynamics given by (8.14) and assume that the relative velocities are bounded. Then the dynamics of the c.o.m. may be averaged around δv_i to obtain

$$\begin{aligned} \dot{R} &= V \\ \dot{V} &= (\alpha_1 - \beta V^2)V - R \end{aligned} \tag{8.16}$$

where $\alpha_1 = \alpha - \mu$, $\mu = 2(\beta^2/(N-1))\sum_i \delta v_i^2$ is the mean quadratic relative velocity of the particles. It is interesting to observe that if μ is larger than α, then $\alpha_1 < 0$ and the c.o.m. will present asymptotically stable dynamics. If $0 < \alpha_1$, then the c.o.m. will exhibit a limit cycle but with a smaller velocity than that of the particles (since $\alpha_1 < \alpha$). The previous analysis suggests that the particles may move about the c.o.m. with a velocity less than $\sqrt{\alpha/\beta}$ if the c.o.m. rotates around a minimum (i.e. $\nabla U_{esp}(r^*) = 0$) with a small velocity.

8.3.2.2 Analysis of angular momentum

The objective of this section is to show that if the swarm is at rest (has reached a stable limit cycle), the particles will start describing a rotational movement around R and the angular momentum of the swarm will remain constant. For analysis purposes, it is assumed (again) that $\nabla U_{esp}(r) = (1/2)r$. The angular momentum of the ith particle about the origin is given by

$$L_i = r_i \times m_i v_i, \tag{8.17}$$

which changes in time as

$$\dot{L}_i = v_i \times m_i v_i + r_i \times m_i \dot{v}_i \tag{8.18}$$

$$= r_i \times m_i \dot{v}_i \tag{8.19}$$

$$= r_i \times \left[(\alpha - \beta \, \|v_i\|^2) v_i - k(r_i - R) - k_f \nabla_i U_{esp}(r_i) \right] \tag{8.20}$$

Adding for all particles, the variation of the angular momentum of the swarm can be written as

$$\begin{aligned} \dot{L} &= \sum_{i=1}^{N} \dot{L}_i \\ &= \sum_{i=1}^{N} r_i \times \left[(\alpha - \beta \, \|v_i\|^2) v_i - k(r_i - R) - k_f \nabla_i U_{esp}(r_i) \right] \\ &= \sum_{i=1}^{N} r_i \times (\alpha - \beta \, \|v_i\|^2) v_i - \sum_{i=1}^{N} r_i \times k_f \nabla_i U_{esp}(r_i). \end{aligned} \tag{8.21}$$

where the fact $r_i \times r_j + r_j \times r_i = 0$ has been used. Assuming the swarm is close to a (local) minimum, i.e. the potential field $\nabla U_{esp}(R) \approx 0$, and that $\nabla U_{esp}(r) = (1/2)r$, (8.21) is reduced to

$$\dot{L} \approx \sum_{i=1}^{N} r_i \times (\alpha - \beta \, \|v_i\|^2) v_i \tag{8.22}$$

Note that the swarm is at rest if it does not change its angular momentum; hence, $\sqrt{\alpha/\beta}$ achieves this goal. If the particle swarm is at a local minimum, i.e. the gradient of the potential field on the c.o.m. is exactly equal to zero, and the potential is harmonic, then the formation will evolve to a circular motion around the (local) minima with constant rotation speed.

8.3.3 *Qualitative behavior of the swarm*

A complete dynamic analysis of the position coupled Raleigh oscillators is out of the scope of this work, so instead evidence to back up the previous statements is collected via numerical simulation. A *qualitative* analysis of the swarm is presented, and based on the intrinsic and emergent behavior of the swarm, an algorithm to solve the APF path-planning problems is proposed.

In general, the swarm exchanges energy between motion of the c.o.m. and motion of the particles. Figure 8.3 depicts how the magnitude of the velocities converge to the common value $\sqrt{\alpha/\beta}$ for evolution of swarm presented in Figure 8.2. The distinct panels show the different stages previously described. This exchange is responsible for the vortex-like behavior of the swarm, which exhibits the following dynamic steps:

- **Initialization**: Starting from some initial conditions ($R \neq 0$), the swarm will first invest its energy into aligning velocities such that all the particles point towards the goal position $\nabla U_{esp}(r^*) = 0$ [see panel (a) of Figure 8.2].

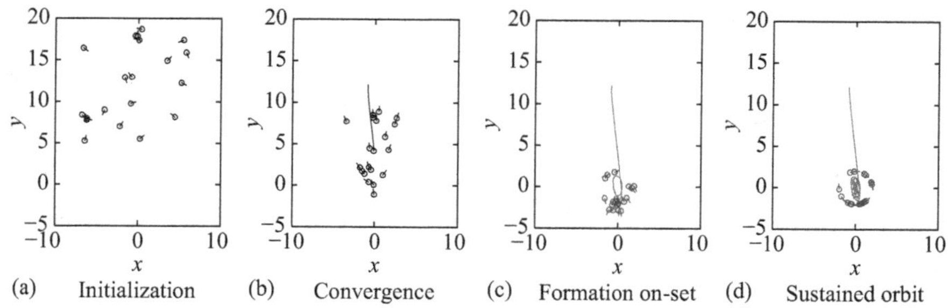

Figure 8.2 (a) Typical behavior of the swarm model with goal position is set at the origin and a harmonic external potential $U_{esp}(R)$, $N = 20$, $\alpha = 1$, $\beta = 0.25$, $k = k_f = 1$. (b) Time step 130: the swarm starts moving towards the origin; (c) time step 150: the swarm is close to the origin and a formation starts to emerge; (d) time step 600: the swarm describes a sustained circular motion about the c.o.m.

- **Convergence to local minimum**: The swarm moves towards the goal position r^* and once it reaches a *suitable* vicinity, the particle starts to organize around the optimal point r^* [see panel (b) of Figure 8.2].
- **Formation on-set**: Once the mean position is within a certain vicinity of r^*, the swarm is expected to have one of two types of behavior [see panel (c) of Figure 8.2]:
  - All the particle velocities are aligned and converged such that the entire swarm moves consistently as a single super-particle orbiting r^* ($\sum v_i \neq 0$).
  - The mean velocity converges to zero and the particles seek a symmetric configuration around the mean position $R \approx r^*$ and $\sum v_i \approx 0$.
- **Sustained orbit**: Once the swarm has reached a state of minimum energy [i.e. (8.13) is zero], the swarm will invest its energy in maintaining formation [i.e. (8.22) is zero] and the angular momentum remains constant [see panel (d) of Figure 8.2].

It is important to mention that these steps are presented to facilitate understanding and sketching of the interaction of the swarm as a whole entity. The steps described above are ideal; thus, the "real" swarm exhibits a more complex behavior, which reflects the choices of the different parameters and initial conditions. Figure 8.3 shows how the magnitude of the velocities converge into the common value $\sqrt{\alpha/\beta}$ for the swarm presented in Figure 8.2. Note how the distinct panels show the different stages previously described.

8.4 Vortex particle swarm path planning algorithm for APF

So far, we have described the approximate behavior of the swarm when it reaches a rest configuration. Here, we present a method to perturb the system in such a way

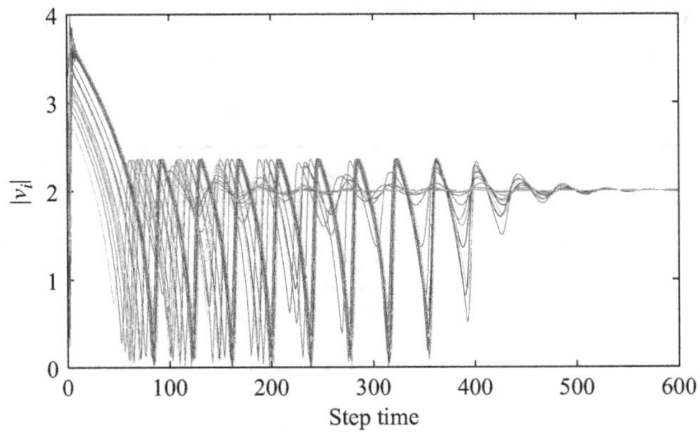

Figure 8.3 Evolution of the speed for each particle in the swarm. Magnitudes of the particle velocities converge to (or very close to) a constant value

it will escape from local minima configurations for non-convex eternal potentials. The main idea relies on the fact that the particles' velocity is increased by increasing the parameter α, which in turn will increase the radius of exploration. Then, by forcing dispersion (increase α) and convergence[1] (decreasing α), it is possible force the swarm out of local minima. The pseudo-code for the VPSPP algorithm is presented in Algorithm 1.

Algorithm 1: Vortex particle swarm path planning algorithm pseudo-code.

Initialize the swarm in the robot APF;
Initialize U_{min};
begin
    while *Maximum number of iterations or the convergence to the goal* **do**
        Establish R and determine U_{min} (8.23);
        Compute $\dot{\alpha}_e$, equation (8.24);
        Compute new α_e;
        for $i = 1$ *until* N **do**
            Compute new v_i;
            Compute new r_i;
        end
    end
    Establish the path for robot displacement;
end

---

[1]With convergence we refer to having a swarm that presents a stable limit cycle around a minimum, with a small radius of rotation.

Table 8.1 Experimental setup for different configurations of the systems parameters; $\alpha_0 = 1$, $m_i = 1$, $k = 1$, $k_f = 1$, $\tau_d = 5$ *and* $\alpha_{\max} = 10$

	C.1	**C.2**	**C.3**	**C.4**
N	10	10	20	20
β	0.2	0.5	0.2	0.5

The underlying idea is that while the swarm presents convergence dynamics (α small), it is mildly disperse and travels towards a minimum of APF. Once the APF evaluated at the c.o.m. does not decrease any further, α is increased and a dispersion process is induced. It is assumed that the dispersion process helps the swarm to find a new lower value. If this happens, α is set back to a small value and the swarm will now travel towards a new minimum value and regroup. Finally, by repeating this sequence, the swarm will travel from minimum to minimum until arrive to goal.

In this way two distinct stages of the algorithm may be observed:

- ***Convergence***: This phase is present after the swarm finds a better value of the APF; therefore, the swarm travels to this point. In this step, the algorithm has a similar behavior to the method of gradient descent.
- ***Dispersion***: After the swarm converges to a (local) minimum of the APF, circular motion of the particles with increasing radius of rotation is induced by increasing α. At this stage, the algorithm is expected to find a new minimum of the APF; otherwise, we assume that the swarm has found a global minimum.

For implementation purposes, first define a variable U_{min}, which is continuously updated with the minimum value of the potential field $U_{esp}(R)$, as

$$U_{min} = \begin{cases} U_{esp}(R), & \text{if } U_{min} \geq U_{esp}(R); \\ U_{min}, & \text{if } U_{min} < U_{esp}(R). \end{cases} \tag{8.23}$$

Note that the only purpose of the parameter α is assumed to have constant and variable parts, i.e. $\alpha = \alpha_0 + \alpha_e$, where the constant α_0 is the nominal propulsion factor and the term $0 \leq \alpha_e \leq \alpha_{max}$ represents the variable part. The term α_e is updated via the following rule

$$\dot{\alpha_e} = \begin{cases} \tau_c & \text{if } U_{min} \leq U_{esp}(R); \\ -\tau_d & \text{if } U_{min} > U_{esp}(R) \\ 0 & \text{otherwise.} \end{cases} \tag{8.24}$$

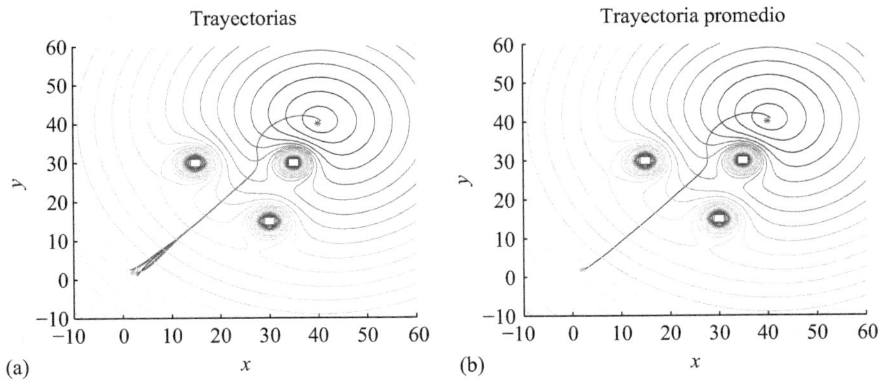

Figure 8.4 Parameter selection C.1. (a) Particle trajectories, (b) c.o.m. trajectory

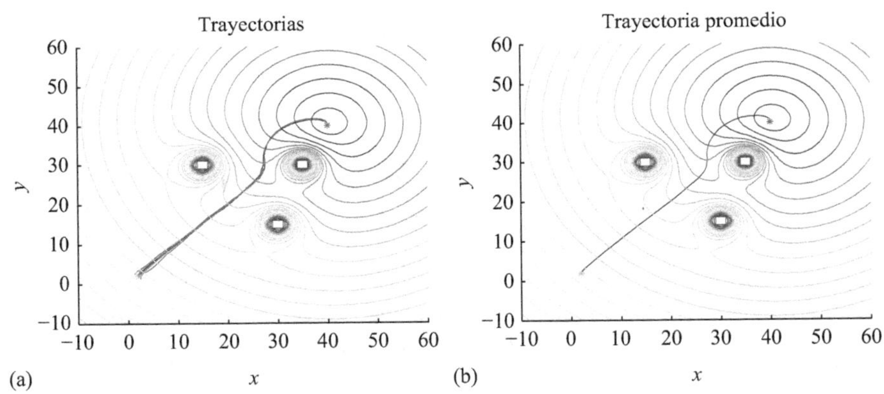

Figure 8.5 Parameter selection C.4. (a) Particle trajectories, (b) c.o.m. trajectory

Finally, as the criterion of completion, we have the maximum number of iterations or the convergence to the goal.

8.5 Experimental results

This section presents simulation results of the proposed VPSPP algorithm using different configurations of the workspace. Performance and dynamics of the swarm are discussed and compared for different values of system parameters.

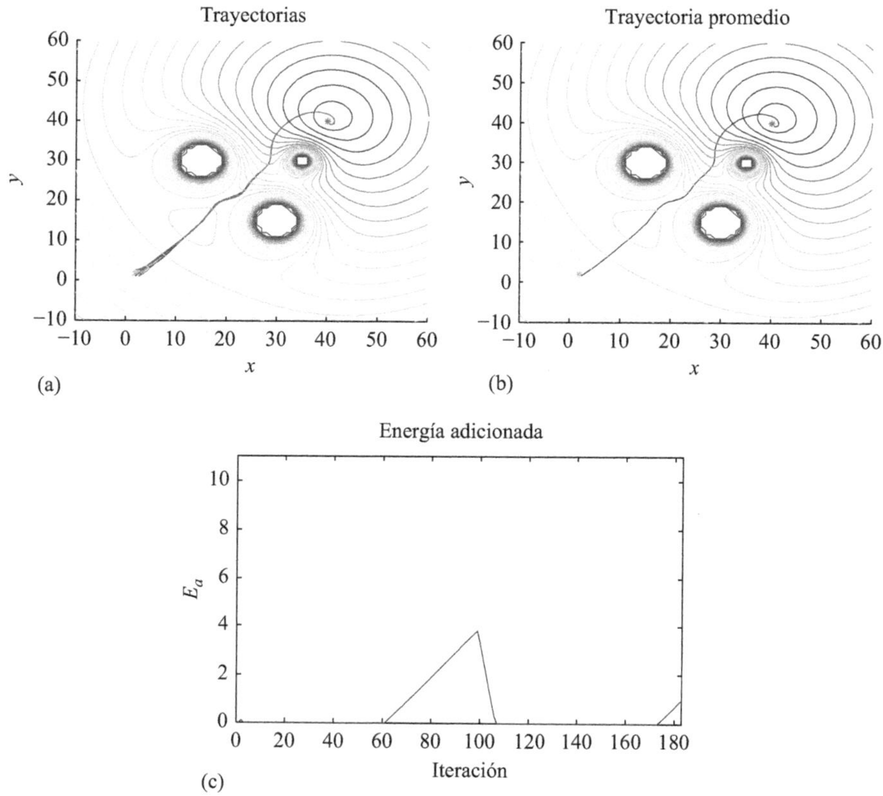

Figure 8.6 Evolution of the swarm with parameter configuration C.1. (a) Trajectory for each particle of the swarm, (b) trajectory for R, (c) variation of α_e, induced to escape the local minima

For compactness, α_0, m_i, k and k_f are assumed constant and only the number of particles N and the breaking factor β are varied in the former; the following table summarizes the different parameter configurations studied in different environment setting. Note that this chapter is not intended as a comparison study but rather presents a novel idea that we believe is more intuitive previous work. Further work must include such comparisons.

A detailed description of the simulation results is presented. The following analysis is intended as the evidence that supports the claims of the qualitative description of the swarm (see Section 8.3). Different workspace setups are selected using previously studied, problematic obstacle, configurations and are intended as a way to compare the behavior of the swarm. Some of the most problematic types of obstacles that

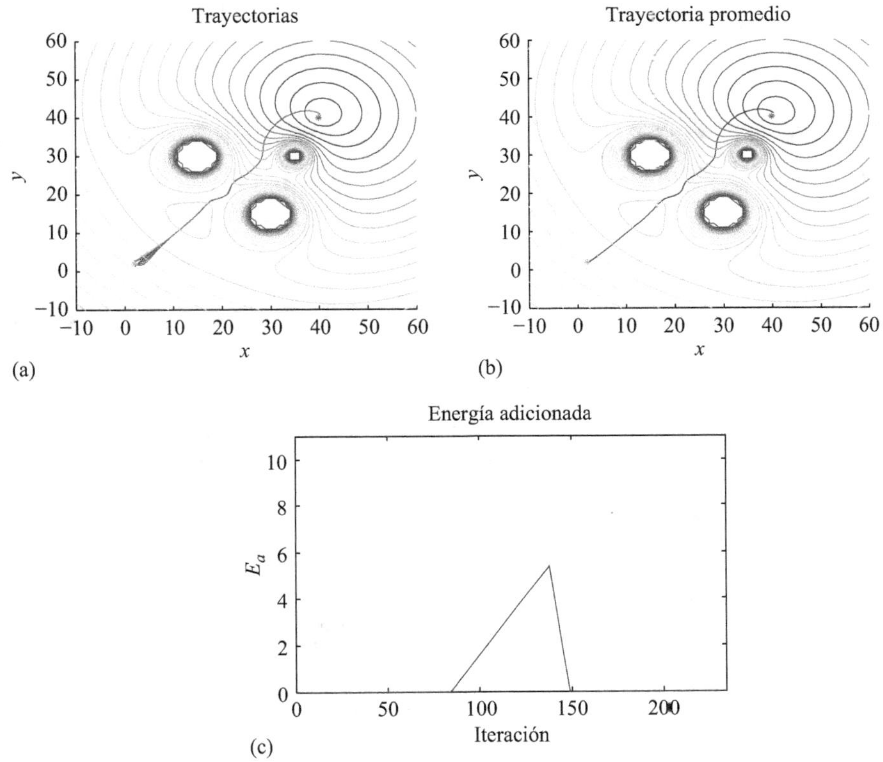

Figure 8.7 Evolution of the swarm with parameter configuration C.2. (a) Trajectory for each particle of the swarm, (b) trajectory for R, (c) Variation of α_e, induced to escape the local minima

have been observed in practice are narrow passages [32] and U-type obstacles [4,51]. In both of these situations, a marked local minima is present and it has proven to be problematic. Simulations are presented for different types of obstacles and parameter selection (see Table 8.1).

8.5.1 Convex external potential fields

Figures 8.4 and 8.5 depict the trajectory of the particles and the trajectory of the c.o.m.; the value of the propulsion parameter α is not shown since it remained constant throughout. In general, the trajectories generated for the different parameter configurations are very similar with only small differences being present. The results obtained coincide to a great extent to those of a gradient descent algorithm; hence,

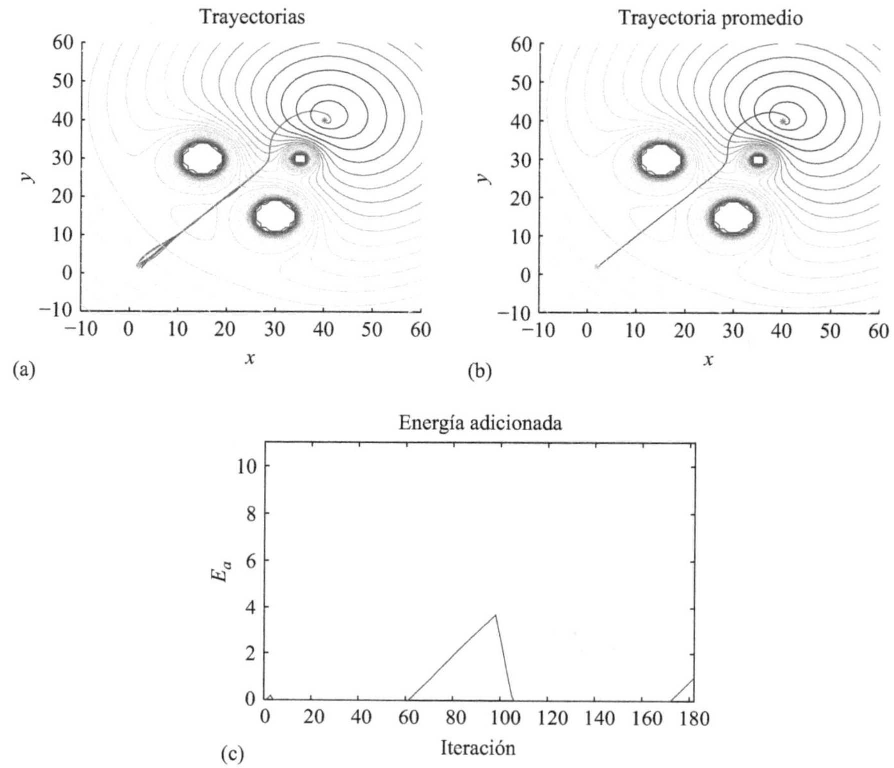

Figure 8.8 Evolution of the swarm with parameter configuration C.3. (a) Trajectory for each particle of the swarm, (b) trajectory for R, (c) variation of α_e, induced to escape the local minima

VPSPP algorithm does not present fundamental differences with other traditional algorithms in this particular case.

8.5.2 External potential fields with local minima

An external potential is constructed with a similar object arrangement as the previous example, but this time the object size is increased in such a way that a local minimum is present. Observe (see Figures 8.6–8.9) that the swarm converges to the local minimum that is in front of it, time at which the propulsion term is increased. Notice that the swarm does not linger around the local minimum and passes over it almost instantly. As a consequence, no real rotational movement is exhibited. The propulsion parameter

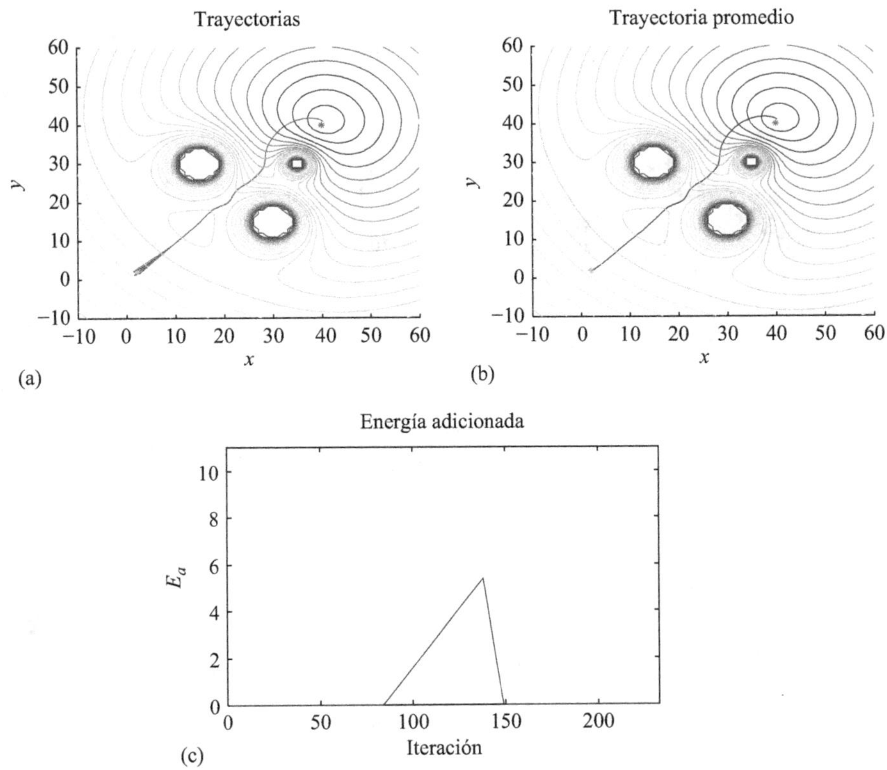

Figure 8.9 Evolution of the swarm with parameter configuration C.4. (a) Trajectory for each particle of the swarm, (b) trajectory for R, (c) variation of α_e, induced to escape the local minima

α does not reach its maximum and it is possible to observe that a new best solution U_{min} is found after approximately 50 iterations. Figures 8.6–8.9 show the evolution of the particles and the c.o.m.

8.5.3 External potential fields for barrier-type obstacles

Barrier-type obstacles generate an external field presenting a local minima with the added restriction that the particle must go around the obstacle and cannot just pass over the local minimum. Figures 8.10–8.13 depict the trajectories of the particles and the c.o.m. Note how the swarm approaches the minimum, but when it meets the obstacle, it has to find its way around it before finding the global solution.

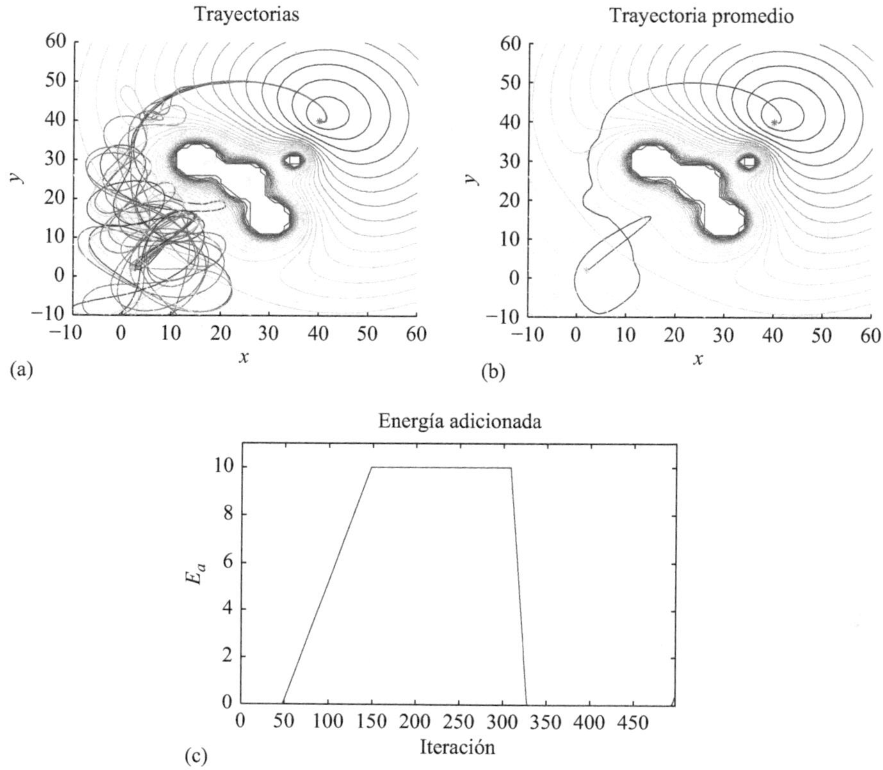

Figure 8.10 Evolution of the swarm with parameter configuration C.1. (a) Trajectory for each particle of the swarm, (b) trajectory for R, (c) variation of α_e, induced to escape the local minima

In this example, it is possible to better appreciate the dispersion stage of the algorithm and the influence of the braking parameter β on the dynamics of the swarm. Observe how for lower β, the swarm presents larger dispersion; this is expected as the rest velocity $\sqrt{\alpha/\beta}$ increases. For all cases, the maximum propulsion is reached and it may be appreciated how for parameter configurations C.1 and C.3 maximum propulsion is retained for a longer number of iterations (α_e is large).

Although in all cases the swarm reaches its goal position, some of the trajectories present un-optimal trajectories; hence, some smoothing post-processing is suggested, which eliminates redundant section of the trajectory.

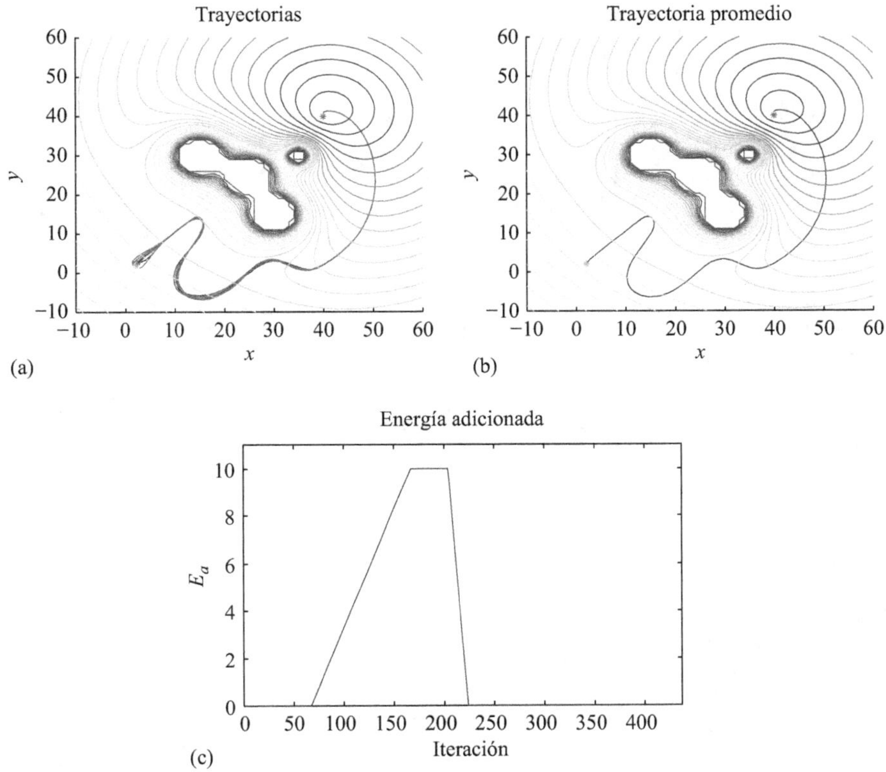

Figure 8.11 Evolution of the swarm with parameter configuration C.2. (a) Trajectory for each particle of the swarm, (b) trajectory for R, (c) variation of α_e, induced to escape the local minima

8.5.4 External potential fields for narrow passage obstacles

When two obstacles are close together, a narrow passage is generated due to the opposing fields. This type of configuration has proven to be difficult in path planning, especially when the field of the individual objects is large, inducing oscillations in the trajectory (Figure 8.14). Figures 8.15–8.18 show the evolution of the trajectories when considering an obstacle with a local minimum (located at the start point) and a narrow passage. Observe how the swarm exhibits a rotational motion around the local minimum until it manages to escape and move through the narrow passage.

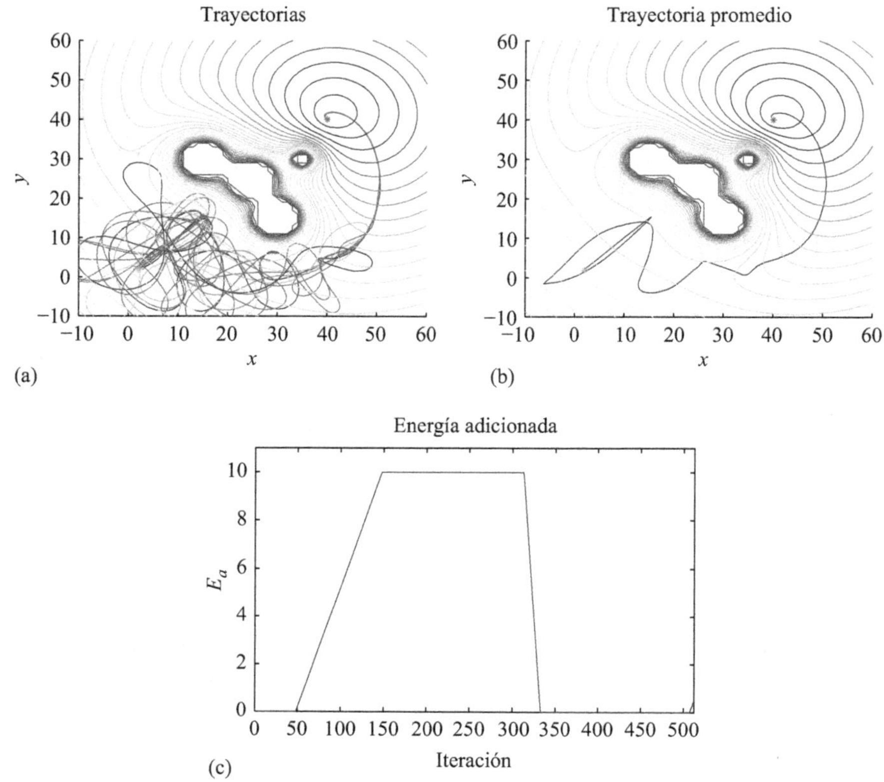

Figure 8.12 *Evolution of the swarm with parameter configuration C.3. (a) Trajectory for each particle of the swarm, (b) trajectory for R, (c) variation of α_e, induced to escape the local minima*

Once again, parameter configurations with a lower breaking force have larger dispersion but escape the local minimum in less time.

8.5.5 External potential fields for U-type obstacles

U-type obstacles have been widely studied [4,51] and present at least one local minimum. The obstacle configuration is depicted in Figure 8.19, where the existence of two local minima can be appreciated.

The swarm presents an initial rotational motion due to the fact that it starts close to a local minimum; once the propulsion parameter is increased, it escapes and moves towards the second local minima, point at which it increases the propulsion parameter again to escape this new local minimum. The oscillating trajectory observed when

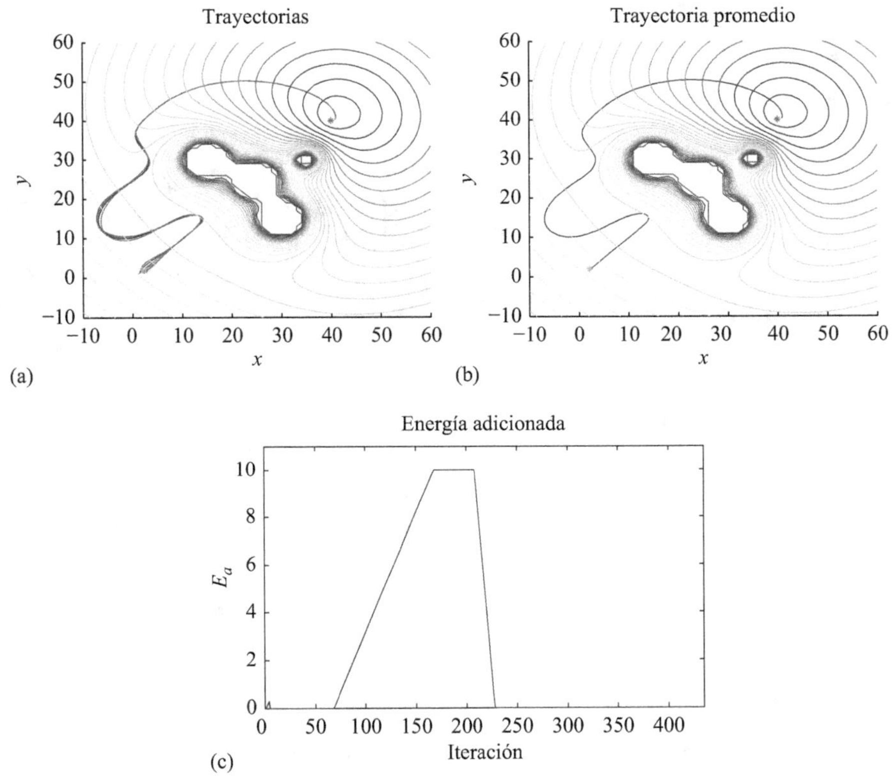

Figure 8.13 *Evolution of the swarm with parameter configuration C.4. (a) Trajectory for each particle of the swarm, (b) trajectory for R, (c) variation of* α_e*, induced to escape the local minima*

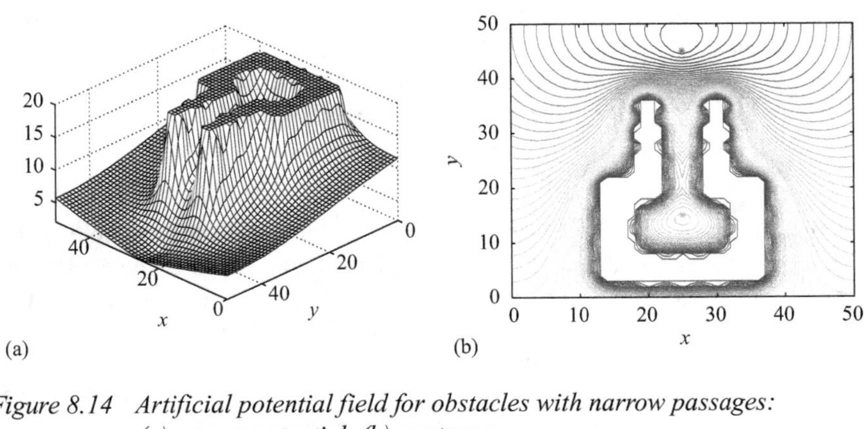

Figure 8.14 *Artificial potential field for obstacles with narrow passages: (a) campo potential, (b) contorno*

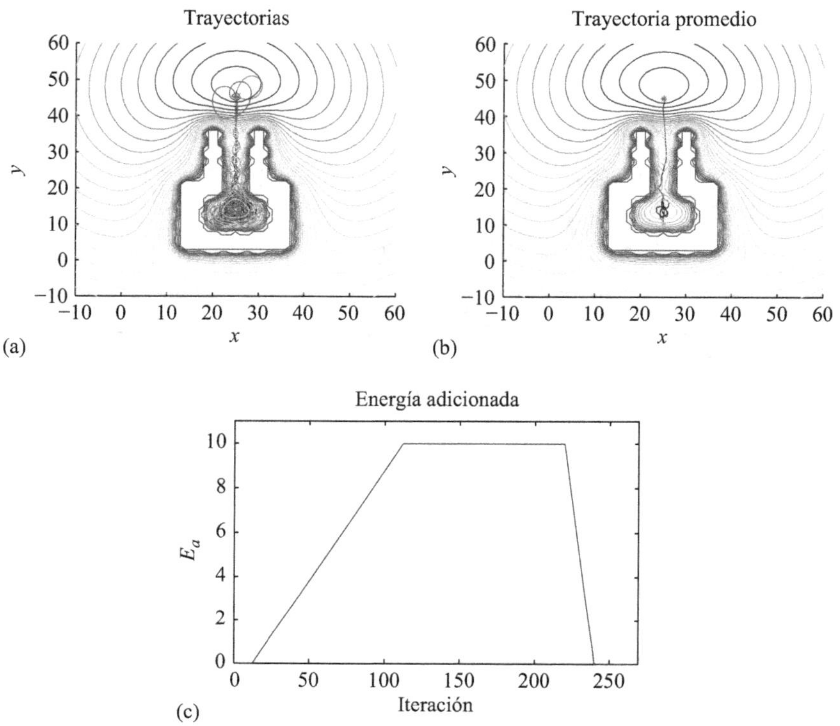

Figure 8.15 Evolution of the swarm with parameter configuration C.1. (a) Trajectory for each particle of the swarm, (b) trajectory for R, (c) variation of α_e, induced to escape the local minima

escaping the U-type obstacle is due to the similarity in configuration to a narrow passage obstacle. It is interesting to observe that for this type of potential, parameter configuration C.4 obtains a more compact swarm that does not need to achieve large dispersions in order to escape the minima (Figures 8.20–8.23).

8.5.6 Statistical analysis

A statistical analysis is conducted in order to better understand the role of the breaking parameter and the number of individuals in the swarm in the exploration behavior. The methodology proposed is standard and starts by performing normality and homoscedasticity tests, where it was revealed that the data did not generally meet the conditions; hence, nonparametric methods are applied for the analysis ([52,53]). Normality and homoscedasticity were determined with a significance level 0.05, which means that only a 5% error is accepted in the tests.

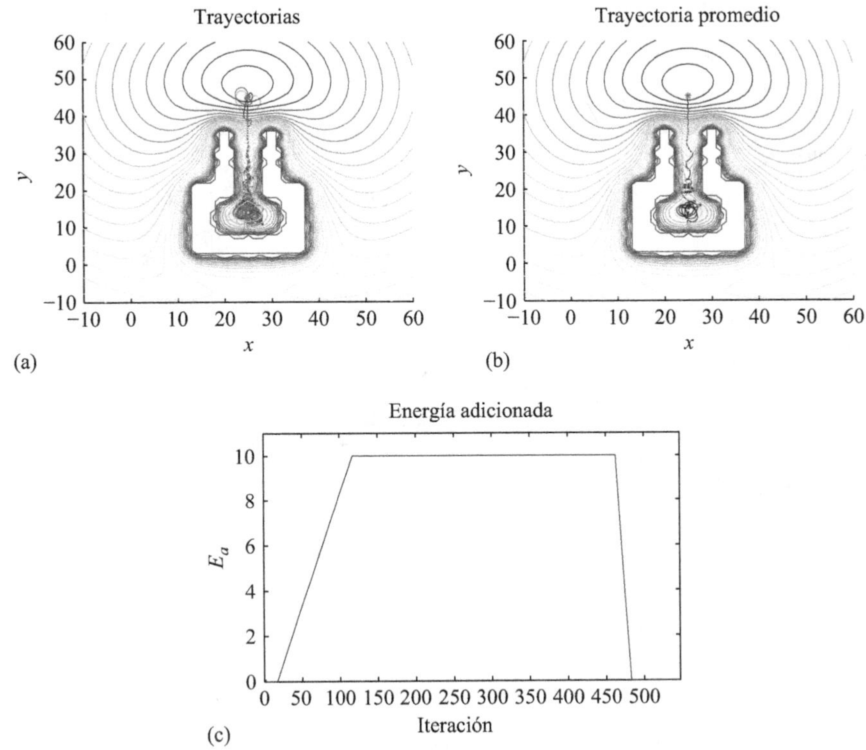

Figure 8.16 Evolution of the swarm with parameter configuration C.2. (a) Trajectory for each particle of the swarm, (b) trajectory for R, (c) variation of α_e, induced to escape the local minima

The most common nonparametric analysis tool, and the one used in this work, is the Kruskal–Wallis test (other tests like the Bonferroni–Dunn [54,55] can be applied). After the statistical test has been performed, different contrast methods like Duncan, Newman–Keuls, Bonferroni, Scheffe or Tukey's HSD [56] may be applied in order to better understand the data set that presents the differences.

As a result of the nonparametric Bonferroni test, intervals of reliability are obtained and can be used to conclude statistical difference between the results. A graphical interpretation of this results is presented where the mean ranking and its variations are presented for each parameter configuration. According to [57], sets are statistically different if their rankings do not overlap.

In order to define the number if runs that must be performed in order to obtain statistical significance, the recommendations of [8,58] are used; a total of 50 runs per parameter configuration are performed. The metrics used for comparison are the

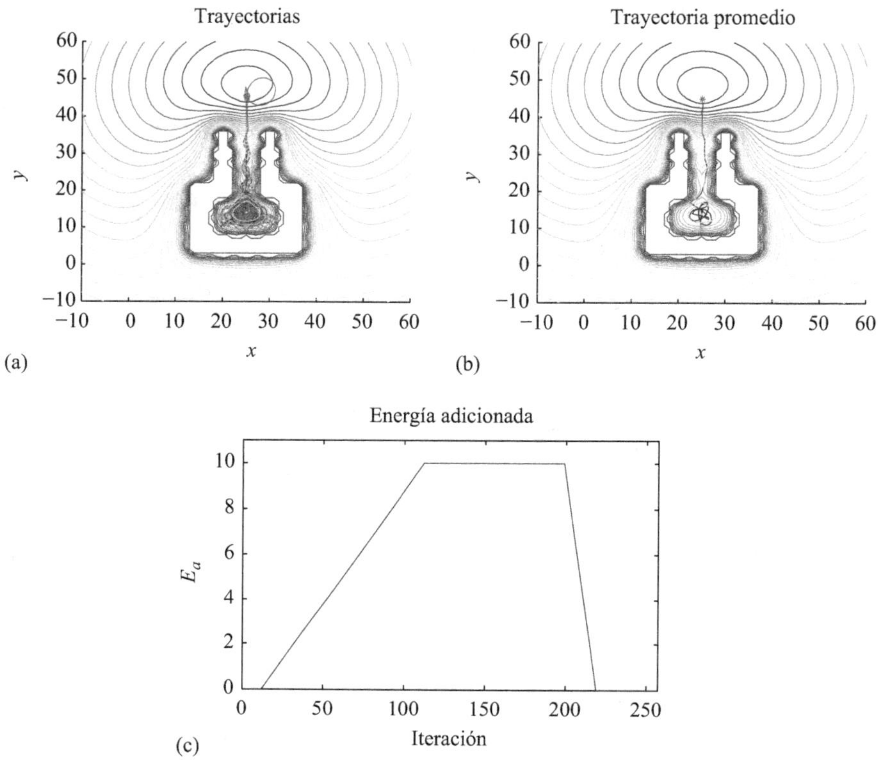

Figure 8.17 Evolution of the swarm with parameter configuration C.3. (a) Trajectory for each particle of the swarm, (b) trajectory for R, (c) variation of α_e, induced to escape the local minima

variance (VAR), the standard deviation (STD) and the mean range (MR), which are the most common in the Kruskal–Wallis and Bonferroni tests.

The summary of the results for a convex external potential field is shown in Table 8.2. Figure 8.24 shows that changes in the breaking parameter β produce significant changes on the swarm's behavior. This is not the case for the number of particles N, where no difference is observed when increasing it from $N = 10$ to $N = 20$. From the results obtained, it can be concluded that C.1 and C.4 have the best performance.

Table 8.3 shows the results for 50 runs of this type of non-convex external field (with local minimum). Performing the Kruskal–Wallis, it is concluded that there are significant differences between different parameter configurations. The graphical results of the Bonferroni test are depicted in Figure 8.25. As in the previous case, the change in breaking force has a significant change in the swarm's behavior, while this is not true for the number of particles.

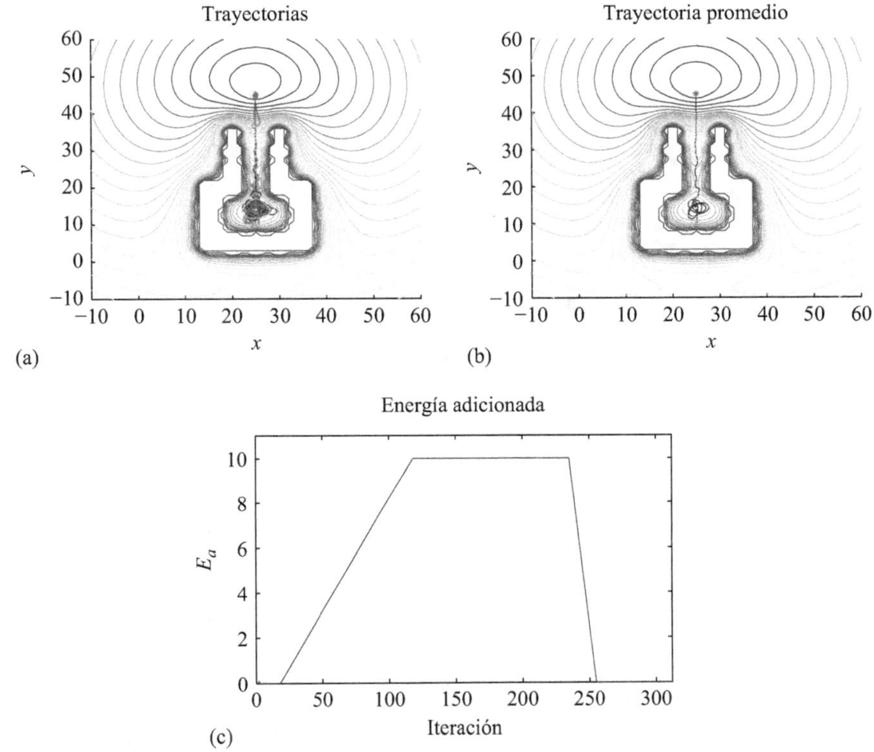

Figure 8.18 Evolution of the swarm with parameter configuration C.4. (a) Trajectory for each particle of the swarm, (b) trajectory for R, (c) variation of α_e, induced to escape the local minima

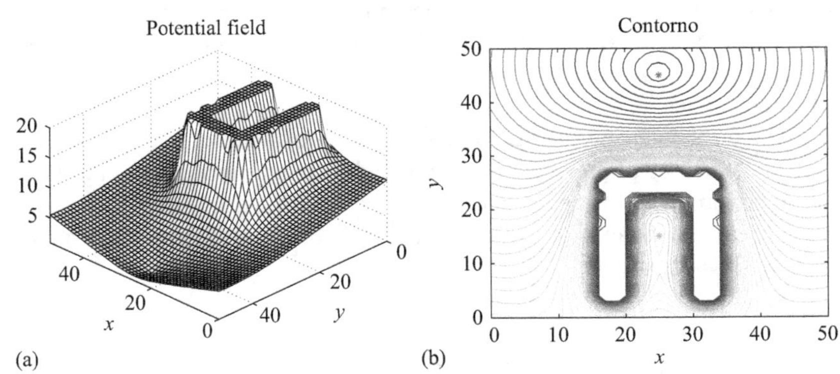

Figure 8.19 (a) Artificial potential field of a U-type obstacle, (b) contour of the potential field

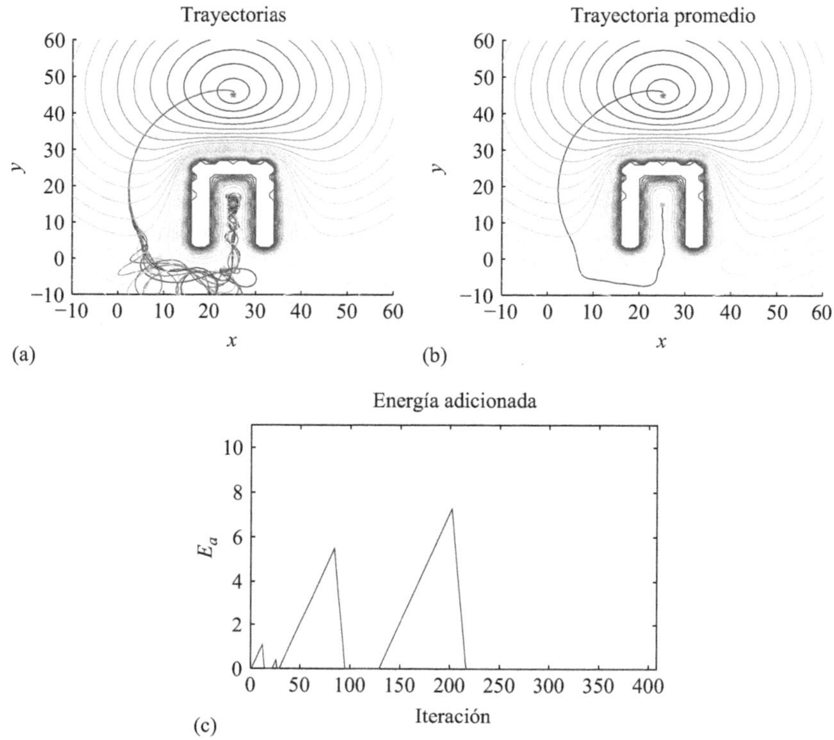

Figure 8.20 Evolution of the swarm with parameter configuration C.1. (a) Trajectory for each particle of the swarm, (b) trajectory for R, (c) variation of α_e, induced to escape the local minima

Results for the 50 runs of an external potential field for barrier-type obstacles are shown in Table 8.4; the Bonferroni test results are shown in Figure 8.26. Statistical difference is observed between C.3 and C.2, C.3 and C.4. No difference is found for any other cases; hence, no conclusion can be made.

Results for the narrow passage case are present in Table 8.5 and Figure 8.27. It can be concluded, as in previous cases, that only changes in the breaking force β induce significant changes on the behavior of the swarm.

Finally, the results for a for U-type obstacle in Table 8.6 and Figure 8.28 reinforce previous observations.

8.6 Conclusions and future work

By studying a swarm of self-propelled particles, a novel path-planning algorithm is proposed. The main problem has been posed as an optimization problem, where

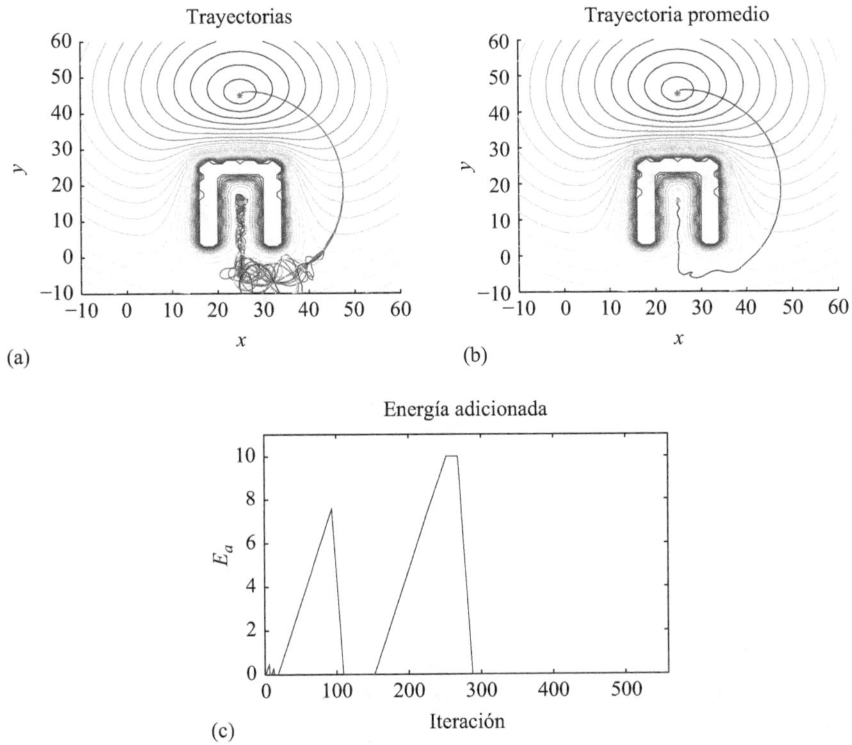

Figure 8.21 Evolution of the swarm with parameter configuration C.2. (a) Trajectory for each particle of the swarm, (b) trajectory for R, (c) variation of α_e, induced to escape the local minima

efforts have been focused on providing local-minima escape capabilities to a swarm of self-propelled particles. A Lyapunov-based analysis was performed for a single particle, providing sufficient conditions for boundedness of the state, hence a stable limit cycle. This is not surprising since the dynamic model of the particle resembles a Raleigh oscillator and hence it inherits its well-studied dynamical behavior. When looking at the multi-particle case, and based on numerical findings for similar models, it is shown that the relative velocity of the particles with respect to the c.o.m. will all converge to a similar (almost constant) value. It was also shown how the c.o.m. will approximately have Raleigh oscillator dynamics for small mean quadratic relative velocity and will converge to a minimum in the opposite case.

Under the assumption of a harmonic external potential, sufficient conditions are given such that it may be concluded that the states reach a configuration of minimum energy. It was also proven that when the swarm converges to a local minima of the potential function (i.e. where the potential for the c.o.m. R is locally minimized), then

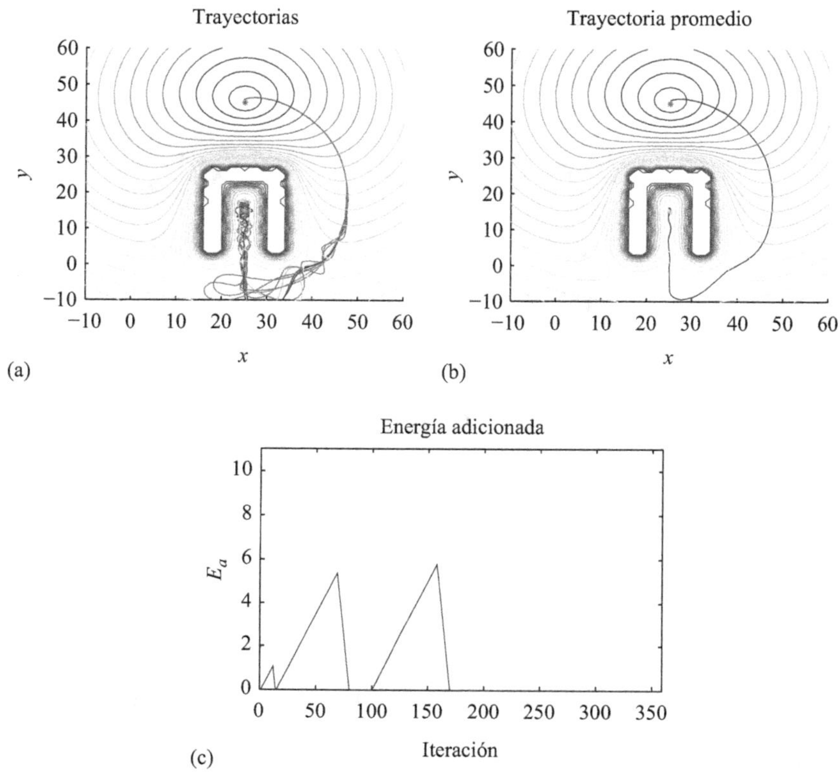

Figure 8.22 Evolution of the swarm with parameter configuration C.3. (a) Trajectory for each particle of the swarm, (b) trajectory for R, (c) variation of α_e, induced to escape the local minima

the formation describes a rotational (possibly circular) motion, at constant speed, about R. It was found that the velocity and radius of rotation are dependent on the breaking and propulsion parameters, fact that is exploited to propose a technique to allow the swarm to escape from local minima configurations by increasing its radius of rotation until the symmetry around R gets broken.

There are some interesting open questions derived from this presentation: analytical evidence to establish the convergence of the speeds to a constant value, independent of the initial conditions of the swarm, is required; similarly, and more related to the path-planning problem, a better description of the trajectory's shape and its relation to the characteristics of the potential field is of interest. Recall that discussions on convergence, energy conservation and conservation of the angular momentum around a local minima for the potential of R were derived assuming a harmonic potential field.

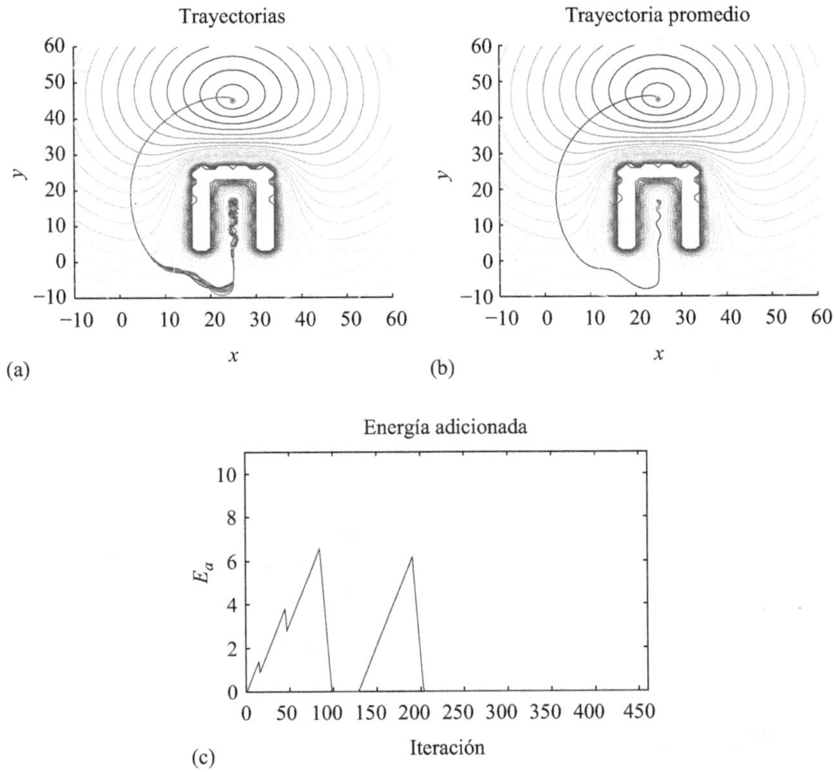

Figure 8.23 Evolution of the swarm with parameter configuration C.4. (a) Trajectory for each particle of the swarm, (b) trajectory for R, (c) variation of α_e, induced to escape the local minima

This paper showed that the proposed algorithm manages to evade local minima in APF path planning, thus proving to be efficient. The main feature of the proposed swarm technique is that it uses a more compact and intuitive particle model, and thus less complex than previous techniques that use vorticity as its strategy to escape local minima. Nonetheless, the choice of parameters to obtain enhanced exploration and good optimization results is still an open question.

Future work may propose another function for the propulsion factor, thus escaping local minima more efficiently. In order to expand our exploration of more efficient algorithms, it is possible to consider other models of self-propelled particles, including Brownian particles. It is also possible to focus future work to the improvement of the calculation of the estimate of the final trajectory, considering that it is possible to present circular trajectories before evading local minima.

Table 8.2 Summary of the results: convex external potential fields

Conf.	Conf. 1	Conf. 2	Conf. 3	Conf. 4
Media	174.06	236.34	173.94	236.14
VAR	0.30	0.56	0.26	0.45
STD	0.55	0.75	0.51	0.67
MR	83.47	285.23	73.99	271.73

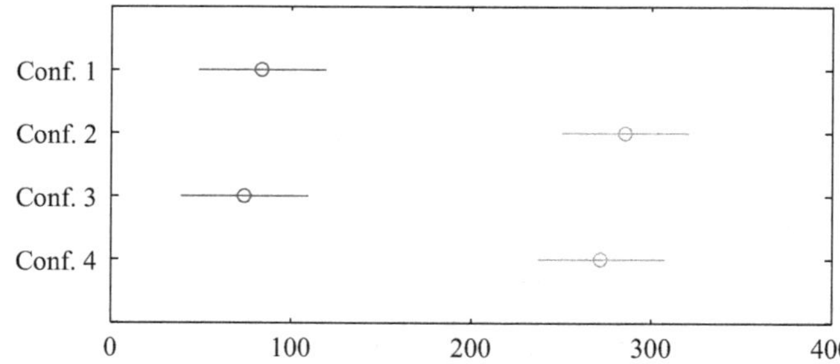

Figure 8.24 Graphical representation of ratings: convex external potential fields

Table 8.3 Summary of the results: external potential fields with local minimum

Conf.	Conf. 1	Conf. 2	Conf. 3	Conf. 4
Media	182.06	234.40	181.92	234.16
VAR	0.59	3.47	0.40	0.95
STD	0.77	1.86	0.63	0.98
MR	83.57	251.05	72.10	250.13

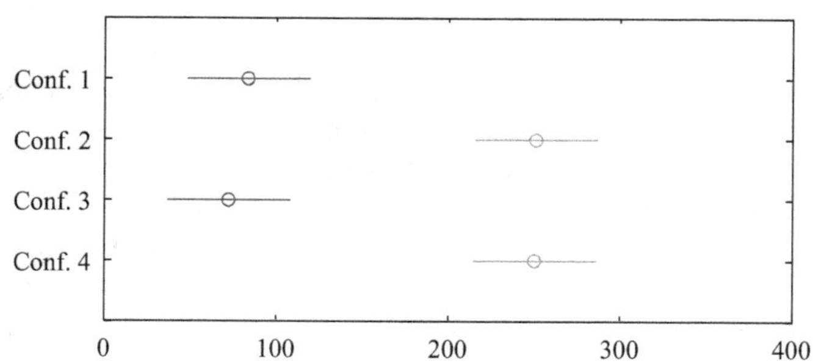

Figure 8.25 Graphical representation of ratings: external potential fields with local minimum

Table 8.4 Summary of the results: external potential fields for barrier-type obstacles

Conf.	Conf. 1	Conf. 2	Conf. 3	Conf. 4
Media	481.18	461.12	536.50	442.10
VAR	17769.3	11811.7	27216.4	1036.3
STD	133.30	108.68	164.97	32.19
MR	163.80	130.09	207.90	129.24

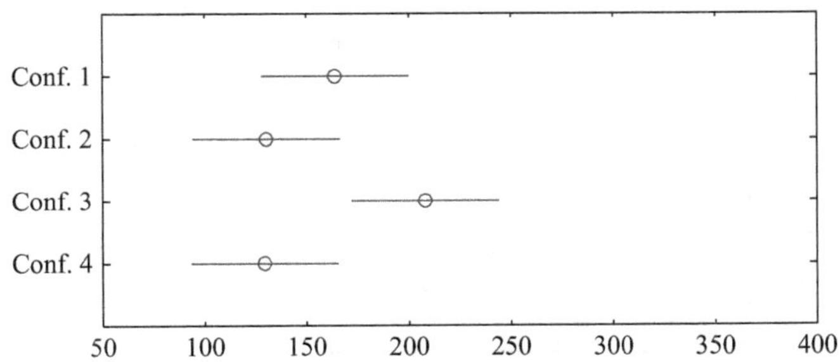

Figure 8.26 Graphical representation of ratings: external potential fields for barrier-type obstacles

Table 8.5 Summary of the results: external potential fields for narrow passage obstacles

Conf.	Conf. 1	Conf. 2	Conf. 3	Conf. 4
Media	279.98	942.70	295.38	911.16
VAR	1,733.2	299,316.8	2,463.7	341,174.7
STD	41.63	547.10	49.64	584.10
MR	112.57	290.69	128.48	283.26

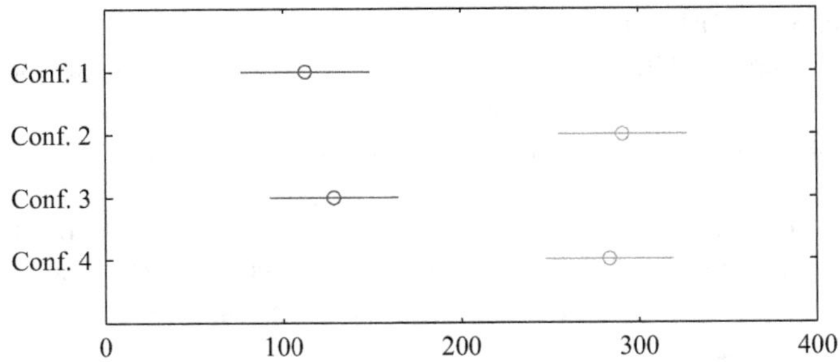

Figure 8.27 Graphical representation of ratings: external potential fields for narrow passage obstacles

Table 8.6 Summary of the results: external potential fields for U-type obstacles

Conf.	Conf. 1	Conf. 2	Conf. 3	Conf. 4
Media	409.44	598.78	423.40	643.74
VAR	8,608.1	19,042.0	8,105.3	24,169.0
STD	92.78	137.99	90.03	155.46
MR	94.38	256.37	112.00	283.64

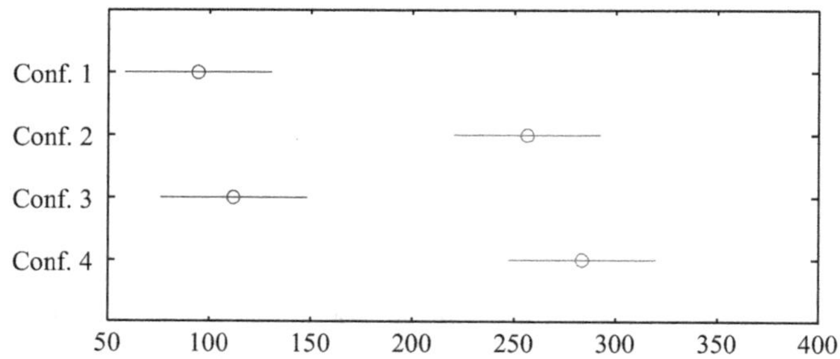

Figure 8.28 Graphical representation of ratings: external potential fields for U-type obstacles

References

[1] Khatib O. Real-time obstacle avoidance for manipulators and mobile robots. International Journal of Robotic Research. 1986;5(1):500–505.

[2] Passino KM. Biomimicry for optimization, control, and automation. London, UK: Springer-Verlag; 2005.

[3] Mabrouk MH, McInnes CR. Solving the potential field local minimum problem using internal agent states. Robotics and Autonomous Systems. 2008;56(1–10):1050–1060.

[4] Abdel Wahid MHM, McInnes CR. Wall following to escape local minima for swarms of agents using internal states and emergent behavior. In: International Conference of Computational Intelligence and Intelligent Systems; 2008. p. 24–31.

[5] Evers GI. An automatic regrouping mechanism to deal with stagnation in particle swarm optimization. University of Texas–Pan American; 2009.

[6] Schutte JF. Particle swarms in sizing and global optimization. University of Pretoria; 2002.

[7] Hvass ME. Tuning and simplifying heuristical optimization. University of Southampton; 2010.

[8] van den Bergh F. An analysis of particle swarm optimizers Pretoria. University of Pretoria; 2001.

[9] Feng CS, Cong S, Feng XY. A new adaptive inertia weight strategy in particle swarm optimization. In: IEEE Congress on Evolutionary Computation; 2007. p. 4186–4190.

[10] Yin L, Liu X. A PSO algorithm based on biologe population multiplication (PMPSO). In: Proceedings of the Second Symposium International Computer Science and Computational Technology; 2009. p. 1–6.

[11] Bratton D, Kennedy J. Defining a standard for particle swarm optimization. In: Proceedings of IEEE Swarm Intelligence Symposium SIS; 2007.

[12] Clerc M, Kennedy J. The particle swarm-explosion, stability, and convergence in a multidimensional complex space. IEEE Transactions on Evolutionary Computation. 2002;6:58–73.

[13] Hendtlass T. A particle swarm algorithm for high dimensional, multi-optima problem spaces. In: IEEE Swarm Intelligence Symposium; 2005. p. 149–154.

[14] Liang JJ, Qin AK, Suganthan PN, *et al.* Comprehensive learning particle swarm optimizer for global optimization of multimodal functions. IEEE Transactions on Evolutionary Computation. 2006;10:281–295.

[15] Hatanaka T, Takeshi K, Nobuhiko K, *et al.* Search performance improvement for PSO in high dimensional space. InTech Particle Swarm Optimization; 2009.

[16] Fieldsend J, Singh S. A multiobjective algorithm based upon particle swarm optimisation, an efficient data structure and turbulence. Workshop on Computational Intelligence. 2002;27(23).

[17] Cervantes A. Clasificacion mediante enjambre de prototipos. Leganes: Universidad Carlos III de Madrid, Departamento de Informatica; 2009.

[18] Deb K, Padhye N. Development of efficient particle swarm optimizers by using concepts from evolutionary algorithms. In: Proceedings of the 12th Annual Conference on Genetic and Evolutionary Computation; 2010. p. 55–62.

[19] Yuemei Y. The chaotic particle swarm optimization based on sub-energy tunneling. In: Fifth International Joint Conference on Computational Sciences and Optimization; 2012. p. 150–154.

[20] Tatsumi K, Ibuki T, Nakashima S, *et al.* Improved chaotic particle swarm optimization with a perturbation-based chaotic system for a virtual quartic function. In: IEEE International Conference on Systems Man and Cybernetics; 2013. p. 208–213.

[21] Yu H, Tiantian X, Pu H. An improved quantum particle swarm optimization and its application in system identification. In: The 26th Chinese Control and Decision Conference (CCDC); 2014. p. 1132–1134.

[22] Li J, Cheng Y, Chen K. Chaotic particle swarm optimization algorithm based on adaptive inertia weight. In: The 26th Chinese Control and Decision Conference (CCDC); 2014. p. 1310–1315.

[23] Ge SS, Cui YJ. New potential functions for mobile robot path planning. IEEE Transactions on Robotics and Automation. 2000;16(5):615.

[24] Khosla P, Volpe R. Superquadric artificial potentials for obstacle avoidance and approach. In: IEEE International Conference on Robotics and Automation; 1988. p. 1–6.

[25] Ollero Baturone A. Robotica: Manipuladores y robots moviles. Espana: Marcombo; 2001.

[26] Erdmann U, Ebeling W, Mikhailov A. Noise induced transition from translational to rotational motion of swarms. Physical Review E. 2005;71(5):051904.

[27] Erdmann U, Ebeling W, Schimansky-Geierand L, *et al.* Active Brownian particle and random walk theories of the motions of zooplankton: application to experiments with swarms of *Daphnia*. Journal of Theoretical Biology. 2008;9.

[28] Espitia HE, Sofrony J. Path planning of mobile robots using potential fields and swarms of Brownian particles. In: IEEE Congress on Evolutionary Computation (CEC); 2011. p. 123–129.

[29] Vicsek T. Universal patterns of collective motion from minimal models of flocking. In: Second IEEE International Conference on Self-Adaptive and Self-Organizing Systems; 2008. p. 3–11.

[30] Czirok A, Vicsek T. Collective behavior of interacting self-propelled particles. Physica A. 2000;281:17–29.

[31] Mikhailov AS, Zanette DH. Noise-induced breakdown of coherent collective motion in swarms. Physical Review E. 1999;60(44):4571–4575.

[32] De Los Santos EA. Heuristica para la generacion de configuraciones en pasajes estrechos aplicada al problema de los clavos. Universidad de las Americas Puebla; 2004.

[33] Poty A, Melchior P, Oustaloup A. Dynamic path planning for mobile robots using fractional potential field. In: First International Symposium on Control, Communications and Signal Processing; 2004. p. 557–561.

[34] LaValle S. Planning algorithms. Cambridge University Press; 2006.

[35] Gemeinder M, Gerke M. GA-based path planning for mobile robot systems employing an active search algorithm. Applied Soft Computing. 2003;3(2):149–158.

[36] Porta Garcia MA. Planeacion De Trayectorias Para Robotica Movil Mediante Optimizacion Por Colonia De Hormigas. Tijuana, BC, Mexico: Instituto Politecnico Nacional; 2007.

[37] dos Santos CL, Sierakowski C. Bacteria colony approaches with variable velocity applied to path optimization of mobile robots. ABCM Symposium Series in Mechatronics. 2006;2.

[38] Sierakowski CA, Coelho L. Study of two swarm intelligence techniques for path planning of mobile robots. In: 16th IFAC World Congress; 2005. p. 1–6.

[39] Connolly CI, Burns JB. Path Planning using Laplace's equation. In: IEEE Conference on Robotics and Automation; 1994. p. 2102–2106.

[40] Kim JO, Khosla PK. Real-time obstacle avoidance using harmonic potential functions. In: IEEE Conference on Robotics and Automation; 1991. p. 790–796.

[41] Chengqing L, Ang MH, Krishnan H, Yong LS. Virtual obstacle concept for local-minimum-recovery in potential-field based navigation. In: IEEE International Conference on Robotics and Automation; 2000. p. 983–988.
[42] Prahlad V, Kay CT, Wang ML. Evolutionary artificial potential fields and their application in real time robot path. In: Proceedings of the Congress on Evolutionary Computation; 2000. p. 256–263.
[43] Zou XY, Zhu J. Virtual local target method for avoiding local minimum in potential field based robot navigation. Journal of Zhejiang University – Science A. 2003;4(3):264–269.
[44] Leng-Feng L. Decentralized motion planning within an artificial potential framework (APF) for cooperative payload transport by multi-robot collectives. Buffalo: University of New York at Buffalo; 2004.
[45] Rimon E, Koditchek DE. Exact robot navigation using artificial potential functions. IEEE Transactions on Robotics and Automation. 1992;8(5): 501–518.
[46] Levine H, Rappel WJ, Cohen I. Self-organization in systems of self-propelled particles. Physical Review E. 2000;63:017101.
[47] Ebeling W. Active Brownian motion of pairs and swarms of particles. Acta Physica Polonica B. 2007;38:1657–1671.
[48] Erdmann U, Ebeling W, Anishchenko VS. Excitation of rotational modes in two-dimensional systems of driven Brownian particles. Physical Review E. 2002;65:061106.
[49] Khalil HK. Nonlinear systems. 3rd ed. Prentice Hall; 2002.
[50] Ebeling W, Schimansky-Geier L. Swarm dynamics, attractors and bifurcations of active Brownian motion. The European Physical Journal, Special Topics. 2008;157(1):17–31.
[51] Esmaeili O, Sai T, Ismail N. Development of a new minimum avoidance system for a behavior-based mobile robot. Fuzzy Sets and Systems. 2009;160(13):1929–1946.
[52] Arriaza Gomez AJ, Fernandez Palacin FM, Lopez Sanchez A, *et al.* Estadistica basica con R y R-Commander. Servicio de Publicaciones de la Universidad de Cadiz; 2008.
[53] Cecilia CL. Optimizacion mono y multiobjetivo a traves de una heuristica de inteligencia colectiva. Argentina: Universidad Nacional de San Luis; 2010.
[54] Demsar J. Statistical comparisons of classifiers over multiple data sets. Journal of Machine Learning Research. 2006;7:1–30.
[55] Garcia S, Herrera F. An extension on statistical comparisons of classifiers over multiple data sets for all pairwise comparisons. Journal of Machine Learning Research. 2008;7:2677–2694.
[56] Douglas M. Diseno y analisis de experimentos. Limusa Wiley; 2003.
[57] Hochberg Y, Tamhane AC. Multiple comparison procedures. New York: Wiley; 1987.
[58] Zhang XH, Meng HY, Jiao LC. Intelligent particle swarm optimization in multiobjective optimization. IEEE Congress on Evolutionary Computation. 2005;1:714–719.

Chapter 9

MAHM: a PSO-based multiagent architecture for hybridisation of metaheuristics

Givanaldo R. de Souza[1], Elizabeth F. G. Goldbarg[2], Anne M. de P. Canuto[2], Marco C. Goldbarg[2], and Iloneide C. de O. Ramos[3]

Abstract

Hybridisation of metaheuristics is an important subject that has been explored by several researchers. Multiagent systems have been important tools to accomplish the task of hybridising metaheuristics. In those multiagent approaches, however, each metaheuristic is performed separately, and the potential of the hybridisation is not fully explored. In order to bridge this gap, this chapter presents MAHM (multiagent architecture for hybridisation of metaheuristics), a multiagent approach for metaheuristics hybridisation inspired on particle swarm optimisation (PSO). Introduced as a metaheuristic, PSO can be viewed as a multiagent system once particles can be thought as agents that interact and work together to achieve specific goals. In this context, particles are identified to autonomous virtual entities with social ability, reactivity, and pro-activeness. Each agent in MAHM may use different sets of predefined metaheuristics in different moments of the search to look for high-quality solutions of an optimisation problem. Thus, several elements of different metaheuristics can be found running in the swarm every MAHM iteration. In order to show the potentiality of the proposed architecture, computational experiments were carried out with the travelling salesman problem and the quadratic assignment problem, two important test grounds for algorithmic ideas.

[1]Diretoria Acadêmica do Campus Parnamirim (DIAC/PAR), Instituto Federal de Educação, Ciência e Tecnologia do Rio Grande do Norte, Brazil
[2]Departamento de Informática e Matemática Aplicada (DIMAp), Universidade Federal do Rio Grande do Norte (UFRN), Brazil
[3]Departamento de Estatística, Universidade Federal do Rio Grande do Norte (UFRN), Brazil

9.1 Introduction

A multiagent system consists of a number of agents which interact with one another through communication [1]. Multiagent systems provide an interesting framework to optimisation algorithms which have been developed to solve several optimisation problems, such as reactive power design [2], staff scheduling [3], ship design [4], economic power dispatch [5], k-graph partition [6], travelling salesman [7,8] and vehicle routing [9].

Optimisation problems model a number of real world applications and have a fundamental role in important fields. Most of them are non-deterministic polynomial (NP)-hard and can be tackled, for example, by approximate algorithms such as metaheuristics. In the last decades, it has become evident that an efficient combination of metaheuristics can provide a more efficient behaviour and a higher flexibility for optimisation systems, emerging to a field called hybridisation of metaheuristics. The promising behaviour of hybrid metaheuristics is, possibly, due to the fact that the hybridisation combines good characteristics of each metaheuristic, strengthening the strategy as a whole. The main motivation behind the hybridisation of different algorithms is to exploit the complementary character of different optimisation strategies, i.e. hybrids are believed to benefit from synergy [10,11]. Hybridisation of metaheuristics is also an important application for multiagent systems, and various researches addressing this issue have emerged in the last years such as the works presented in [7,8,12–14]. In the existing multiagent approaches for metaheuristic hybridisation, each metaheuristic is usually performed separately. In order to bridge this gap, this chapter proposes MAHM (multiagent architecture for hybridisation of metaheuristics), a multiagent approach inspired on particle swarm optimisation (PSO). Introduced in [15], the PSO can be viewed as a multiagent system since particles can be thought as agents that interact and work together to achieve specific goals as pointed out in [2]. In this context, particles are identified as autonomous virtual entities with social ability, reactivity and pro-activeness. According to [5], the decision-making process of agents matches the basic nature of a particle from the PSO framework. Each agent in MAHM can use a set of predefined metaheuristics in different moments. Thus, several elements from different metaheuristics can be found in the swarm every iteration.

In order to investigate the potential of the proposed architecture, an empirical investigation was conducted with two combinatorial optimisation problems that have been test grounds for algorithmic ideas: Travelling salesman problem (TSP) and quadratic assignment problem (QAP).

This chapter is divided into seven sections and organised as follows. Section 9.2 presents a brief review of related work. Section 9.3 presents the general description of the proposed architecture. Section 9.4 describes the combinatorial optimisation problem used in the experiments. The computational experiments are presented in section 9.6. Finally, conclusions and future works are addressed in Section 9.7.

9.2 Related works

This section presents multiagent systems that have been proposed for metaheuristic hybridisation.

MAGMA (multiagent metaheuristic architecture) was proposed in [12], which addresses a conceptual and practical framework where hybrid metaheuristic algorithms were accommodated with the interaction among different kinds of agents distributed in different levels. In the basic architecture, there are agents that provide feasible solutions for upper levels, agents that deal with solutions improvements and agents that have a global view of the search space. The authors describe the MAGMA architecture of GRASP (greedy randomised adaptive search procedure), ACO (ant colony optimisation), ILS (iterated local search), MA (memetic algorithm) and a cooperative search using GRASP and MA.

MAOS (multiagent optimisation system) was proposed in [8], which is a nature-inspired method similar to genetic algorithms (GAs). Each agent is considered as an autonomous entity with its own memory and behavioural components. The agent's memory allows the acquisition of knowledge by using past experiences of the agents. Also, agents explore the environment in parallel, based on socially and individual learning, and interact indirectly with other agents by sharing public information organised in the environment. MAOS was also applied to the TSP, including two improving heuristics based on a generalised edge assembly recombination and 3-opt local search.

MAM (multiagent architecture for metaheuristics) was proposed in [13] proposed, whose objective is to combine metaheuristics for solving combinatorial optimisation problems through the multiagent approach. Each metaheuristic is developed as an autonomous agent, interacting cooperatively with an environment and among themselves in a cooperative and competitive way, in an attempt to find the best solution to combinatorial optimisation problems. The general structure of MAM is composed of different types of agents: constructor, local search, coordinator, solution analyser and metaheuristic. The authors describe an application to the vehicle routing problem with time windows with implementations of GA, ILS and VNS (variable neighbourhood search).

An agent-based hyper-heuristic framework was proposed in [14], which combines different metaheuristics. There are four types of agents with specific functions: problem, solution pool, algorithm and adviser. Among other problems, it was applied to the TSP with hybridisation of GA, TS (tabu search), SA (simulated annealing) and ACO.

ABGLS (agent-based guided local search) was proposed in [16], which is a hybrid approach for solving combinatorial optimisation problems in which guided local search (GLS) metaheuristics [17] are incorporated into a cooperative multiagent framework. It is based on the concept of asynchronous teams (A-Teams) [18], where each agent performs a GLS. This approach is composed of a set of agents working asynchronously and in parallel within an A-Team, a common shareable memory and a flow of messages between agents. Two classes of agents compose the ABGLS, the solution manager agent, responsible for managing the entire process of finding the best solution and manages the population of individuals in the memory. In addition, it is composed of a set of GLS agents, where each one is an implementation of the GLS algorithm. Computational experiments were applied to 26 instances of vehicle routing problem, and four GLS algorithms were used as GLS agents.

In the aforementioned approaches, each metaheuristic is usually performed separately, performing a type of static combination of the different metaheuristics. In these cases, the full potential of the hybridisation of metaheuristics is not fully explored. In the system proposed in this chapter, a set of metaheuristic is used simultaneously, using the full potential of a hybridisation approach. The idea is to have different metaheuristics in the swarm, in the same iteration, cooperating among each other.

9.3 The proposed approach

The MAHM was proposed to address the modelling of autonomous entities which achieve the specific goal of searching the space of solutions of hard combinatorial optimisation problems. The particles are autonomous agents with memory, learning and decisions methods. The memory of these agents contains proprietary and acquired information. Proprietary information is constituted of metaheuristics and enables the agent to explore the space of solutions on its own. Acquired information comes from other agents, for example the attractiveness of positions occupied by those particles, search methods and configuration parameters.

Once the agents are in a one-to-one correspondence in the PSO framework proposed here, hereafter, the term *particles* will be taken as referring to *agents*.

The particles defined for MAHM can be classified as *reactive agents with memory* [1], since their organisation is thought to be in a swarm, and they act and react in the environment. Furthermore, they keep some former states in order to use it in their learning methods. According to a hierarchical classification for hybridisation of metaheuristics proposed in [19], the hybridisation process used in MAHM is classified as *heterogeneous high-level teamwork*, since several different metaheuristics are performed in parallel by different particles towards a common goal.

Section 9.3.1 describes briefly PSO and the general architecture is presented in Section 9.3.2.

9.3.1 *Particle swarm optimisation*

PSO was proposed in [15], originally introduced for continuous optimisation problems. It can be considered as a stochastic search method inspired by the behaviour of a birds flock. The captured behaviour of a particle swarm is defined as follows: birds are randomly arranged in a flock and they should look for food and for places in which a nest can be built. In the beginning, as there exists no information about food and nest, birds should fly with no guidance.

It is important to emphasise that they fly in flocks until food or nest is found when the birds who are close to food or nest are attracted by this. Then, all birds can find food or nest. Intelligence in the behaviour of birds is basically social [15]. Therefore, it is possible for a bird to learn from the knowledge of a different one since the chance of other birds to find food or nest increases drastically when one bird finds them.

In an analogy, the search space of a problem may be defined as the area flown by the birds (particles). In addition, an optimum solution is the one in which a bird finds food or nest. As a result, PSO is an algorithm formed by a set of particles, and each one defines a possible solution to a problem. During its functioning, the search for an improved position in the search space is done by each particle.

This is done by updating a velocity parameter, and it is based on some rules inspired by behavioural models of flocks of birds.

PSO can be efficiently applied for optimisation problems. Therefore, it is necessary to define how the basic parameters will work, for example the interaction between position and its velocity. The proposed approach was inspired by the previous research of the authors on the application of PSO in discrete optimisation problems by applying new concepts to velocity operators, achieving effective results for TSP [20,21] and for bi-objective degree constrained minimum spanning tree [22].

Other attempts to apply PSO for TSP were limited to adapt the formulas for continuous optimisation problems [23,24] to calculate the distance between particles [25] and to use space transformation from continuous to discrete [26]. However, computational results for these cited approaches regarding solution quality were not effective considering the state-of-the-art algorithms for TSP [21].

9.3.2 Multiagent architecture for hybridisation of metaheuristics

For modelling MAHM as a framework for hybridisation of metaheuristics based on PSO, it is necessary to plan the implementation of each component of PSO (particle, position, velocity, communication between particles and neighbourhoods). In MAHM, the position of a particle is the representation of a solution. The particle can move from one position to another using a *movement operator* which builds a trajectory from a current position to a target one. During that trajectory, new positions can be found, and the particle can decide between interrupting its trajectory on an intermediate position or continue until reaching the target. This movement operator is thought as the velocity defined in the PSO for discrete problems as described in [21].

The *particle* is the central component of MAHM. Each particle knows the value (objective function) associated with its position in the search space and has a memory. The particle can also access information from another. The way they do it depends on the way MAHM was implemented. In this study, every particle knows the information of the best one. The best particle is the one associated with the best position. However, in distributed systems implementations, mechanisms for agent's interaction are commonly found in the literature of multiagent systems, such as KQML, ACL and Blackboard [1]. For example, Blackboard can be used as a central memory in which the stored information is shared between particles.

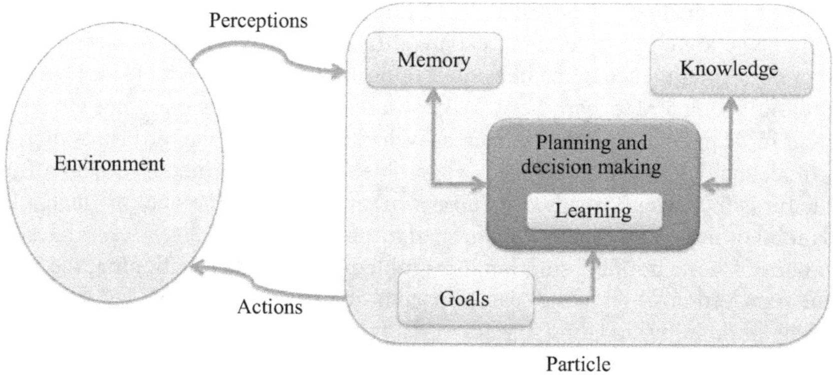

Figure 9.1 The structural composition of a particle in MAHM

Figure 9.1 illustrates the structural composition of a particle in MAHM and its interaction with the environment through actions and perceptions. The key components are knowledge, memory, goals, learning, planning and decision-making. The knowledge module contains information about the particle's position, the evaluation of that position usually given by the value of the objective function, the best particle (or the best neighbour particle, if applicable) of the swarm. In the memory module, information about the particle is stored, such as decision and learning methods and the best position ever reached by the particle. Also, the goals module explores the search space to find near-optimum solutions. The learning module contains the decision method and the learning method. Each particle has behavioural characteristics, based on its learning and decision strategies, to achieve its goal. In the particle composition, the other components are *perceptions* which are the updating of position and velocity; *actions* which comprehend the execution of decision and learning methods, and the updating of the objective function value; and the *environment* which is the search space of combinatorial optimisation problem.

The memory module is a central element, and its structure is shown in Figure 9.2. The memory module contains information about the particle and its search trajectory. Some examples of such type of information are the position, objective function value, the best position ever reached, a list of decision and learning strategies already used and the current decision and learning strategies adopted by the particle.

The *decision method* is used to select a strategy to modify the particle's position. The strategy is a metaheuristic (or the element of a metaheuristic). There is a set of metaheuristics available in a centralised repository (Figure 9.2) to be chosen by the decision method. For this work, only trajectory metaheuristics were used. Also, recombination elements can be utilised, as in population-based metaheuristics. For example, we can define a set of particles (population) for the population metaheuristics to be performed, hybridising different operators as decision strategies.

Initially, the decision method selects a strategy randomly from the repository and associates it to the particle. During the execution of the algorithm, if the adopted

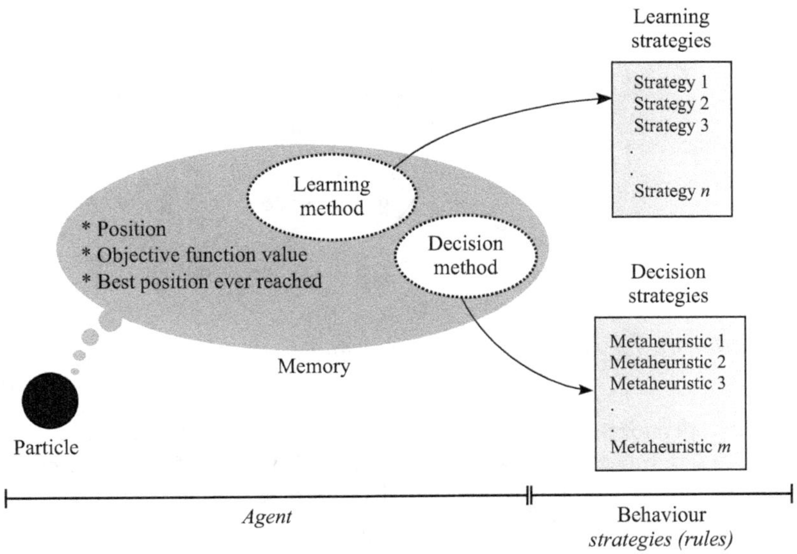

Figure 9.2 The memory module of a partition (agent) in MAHM

strategy is failing in finding good solutions, the particle can decide to select another strategy based on a successful history. The more successful a strategy, the more likely is to be chosen by the decision method. The most promising strategy can be chosen or can be selected from among the n most promising strategies. The success and failure rules and how the strategy is selected are configurations to be defined in the implementation of MAHM.

Particle's memory maintains a history of the search performed by that particle. This history contains information about the search, such as positions considered good (elite solutions), metaheuristics parameters and success and failure statistics. The *learning method* is used to update the search history of the corresponding particle. The learning method accesses a centralised repository of learning strategies (Figure 9.2). The learning strategies guide the decision method in their choice of the decision strategies. Examples of learning strategies are the method to use some of the best positions ever reached by a particle, the method to account for success and failure, the method for changing parameters of a metaheuristic, etc. The learning method starts by randomly selecting a strategy and associating it with its corresponding particle. Subsequently, if the selected learning strategy does not succeed in aiding to find a better position, the learning method changes it. Each learning strategy has the function of updating the particle's memory. They are applied as the decision strategies are performed, collecting necessary information for the memory.

When a metaheuristic is applied to a particle's position, a new position is found, and the particle assumes this new position. Particles move in the search space from one position towards another (p_{best} or g_{best}) using a velocity operator. This movement can be defined in several ways, including the particle simply assuming the target position.

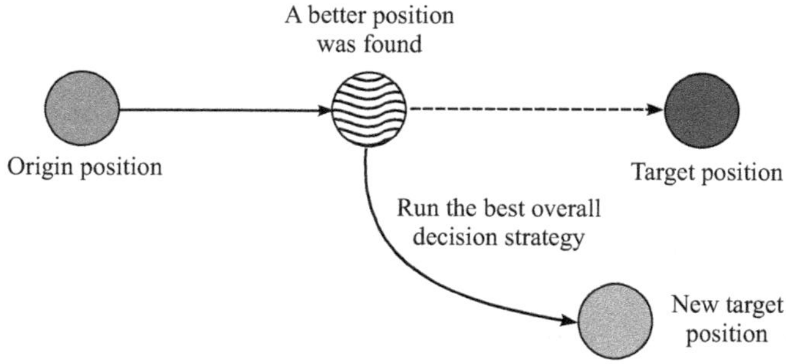

Figure 9.3 Velocity operator with an interrupted trajectory

The velocity operator used in MAHM is the *path-relinking* proposed in [27], which allows building a trajectory between two positions, through intermediate positions. The origin and target positions (x_o and x_t) can be defined according to three strategies as follows. (i) The origin position is the current particle's position, and the target position is the best position already visited by that particle (p_{best}) or the best position already visited by the swarm (g_{best}) [21]. The target position is chosen by setting probabilities for p_{best} and g_{best}. These probabilities change during execution. (ii) The origin position is selected randomly from p_{best} or g_{best}, and the target position is the current position of the particle. In this case, it follows the observation in [28], which suggests that the trajectory should be done from the best solution to the worst one. (iii) Combine the two aforementioned strategies.

Once the origin and target positions are defined, the particle starts its trajectory which can be complete or truncated (interrupted in some intermediate position). The result of a whole trajectory is the best position found among the intermediate positions. In the case of an interrupted trajectory, i.e. stopping when a position better than the initial is found, this new position is set to the current particle's position. In any case, a post-optimisation can be applied. This can be done by applying the most effective decision strategy. The goal is to find an even better position. Figure 9.3 illustrates an interrupted trajectory. When a position better than the origin position is found, the most efficient decision strategy is performed at that position, resulting in an even better position.

Considering minimisation problems, the pseudo-code with the MAHM general idea is presented in Algorithm 1.

The algorithm has as input parameters the swarm size, the stopping criteria, the initial probability, and the decreasing factor of the probability. Some examples of stopping criteria are the number of iterations without improvement, the maximum number of iterations and the maximum runtime. Particle i is represented by p_i and its position is represented by x_i. Following the PSO idea, the best position ever reached by particle p_i and the best position ever reached by any particle are represented respectively by p_{best_i} and g_{best}. In the first lines of the algorithm, the heuristic methods to be hybridised are loaded in *DS*, the set of decision strategies. Then, the learning strategies are set to *LS*.

Algorithm 1: Pseudo-code of MAHM

```
   Input: swarm_size, stopping criteria
1  DS ← load_decision_strategies();
2  LS ← load_learning_strategies();
3  value(g_best) ← ∞;
4  for i ← 1 to swarm_size do
5      x_i ← generate_position(p_i);
6      value(p_best_i) ← ∞;
7      select_decision_strategy(p_i, DS);
8      select_learning_strategy(p_i, LS);
9  end
10 repeat
11     for i ← 1 to swarm_size do
12         value(x_i) ← evaluate(x_i);
13         if value(x_i) < value(p_best_i) then
14             p_best_i ← x_i;
15         end
16         if value(x_i) < value(g_best) then
17             g_best ← x_i;
18         end
19     end
20     for i ← 1 to swarm_size do
21         start_learning_method(p_i);
22         x_i ← apply_decision_method(p_i);
23         x_t ← choose(p_best_i, g_best);
24         run_velocity_operator(x_i, x_t);
25     end
26 until stopping criteria satisfied;
   Result: g_best position.
```

In the loop from line 4 to 9, an initial position, decision and learning strategies from DS and LS, respectively, are associated with each particle. Each particle maintains information used by the learning and decision methods, its position, value of the objective function related to the position and the value of p_{best}.

The main loop between lines 10 and 26 contains two independent inner loops. In the first inner loop (lines 11–19), the position x_i of each particle p_i is evaluated (line 12). If the value of the position associated with p_i is better than the best value ever associated with it, the best position achieved by p_i is set to p_{best_i}. The best value ever reached by any particle, g_{best}, is updated, if necessary.

The learning method, in the second inner loop, collects new information and updates the memory of each particle (line 21). Results from the learning method are important for the decision method since the strategies adopted by each particle can change from one iteration to another. The information collected during the execution of the decision method and the velocity operator is used to improve the efficiency of the learning method. The type of the collected information is defined according to the learning strategies.

Let the position of the *i*th particle be x_i. A selected strategy is executed in the decision method (line 22), and the *i*th particle searches for a better position in the solution space. After applying the strategy selected from the list of metaheuristics of the *i*th particle, a new solution, x_i', is obtained. The best position among x_i and x_i' is set to x_i.

For the path-relinking, the target position, x_t, is selected among p_{best_i} and g_{best} (line 23). x_t is chosen at random with a probability that varies during the execution of MAHM. The initial probability to choose g_{best}, *prob_ini*, is a user-defined parameter (1-*prob_ini* to choose p_{best_i}). The *prob_ini* value is updated as follows. If g_{best} is selected as the target position and the search reaches a solution better than x_i, *prob_ini* is increased by a factor (*prob_ini* = *prob_ini* × *factor*). Otherwise, *prob_ini* is decreased by the same factor. The *prob_ini* value is limited to the range from 0 to 100 (0%–100%), and the *factor* value is limited to the range from 0.1 to 0.9.

Positions x_i and x_t are submitted to the velocity operator (line 24). If a new position, x_{new}, reached in the trajectory between x_i and x_t, is better than x_i, the trajectory is interrupted and x_{new} is submitted to the most efficient decision strategy. Then, if the new position reached by the particle is better than x_{new}, the particle moves to this new position. Otherwise, x_{new} becomes the current position of the particle.

The stopping criterion of the main loop can be any of the usual ones, such as the number of iterations without improvement, the maximum number of iterations or a limit for the maximum processing time.

9.4 The travelling salesman problem

The TSP is defined on a graph $G = (V, E)$, where $V = \{1, \ldots, n\}$ is the set of nodes and $E = \{1, \ldots, m\}$ is the set of edges. A cost, c_{ij}, is associated with each edge linking vertices *i* and *j*, where $i,j \in V$. The problem consists in finding the Hamiltonian cycle, *C*, in graph *G*, such that the cost of *C*, calculated by the sum of the costs of the edges in *C*, is minimum over all Hamiltonian cycles in *G*. The TSP is symmetric if $c_{ij} = c_{ji}$ for all pairs $i,j \in V$ such that $(i,j) \in E$. The TSP is said to be asymmetric if $\exists i,j \in V$ and $(i,j), (j,i) \in E$ such that $c_{ij} \neq c_{ji}$.

TSP is certainly one of the most studied combinatorial optimisation problems found in the literature. Due to its wide applicability and difficulty to be solved by exact methods, a number of heuristics have been proposed to it by engineers and computer scientists. A review of the TSP is presented in [29], and a history of computation and applications is presented in [30].

Among recent PSO implementations, a discrete PSO model was proposed in [21] and applied to the TSP. In that work, the velocity equation was replaced by two techniques: local search and path-relinking. The Lin & Kernighan's method [31] was used. An extension of that approach that includes methods to combine distinct velocity operators was also presented. Computational experiments with a large set of benchmark instances showed that those algorithms produced high-quality solutions when compared with efficient heuristics for the TSP. The first roots of the research that led to the proposal of the multiagent system for heuristics hybridisation presented here were presented in [21].

The application of MAHM to the TSP is described as follows. Given a TSP instance and S_n, the set of permutations of $\{1, \ldots, n\}$, the particle's position is a permutation $\pi \in S_n$, $\pi = (\pi(1), \ldots, \pi(n))$, where $\pi(i)$ is the ith city visited in the cycle represented by π. The particle's position is evaluated using the TSP objective function, calculated by (9.1).

$$f(x) = \sum_{i=1}^{n-1} c_{\pi(i),\pi(i+1)} + c_{\pi(n),\pi(1)} \tag{9.1}$$

Initial positions are generated with a greedy randomised version of the nearest neighbour heuristic presented in [32]. The procedure is based on the construction phase of a GRASP algorithm. GRASP is a metaheuristic proposed in [33] and in its first phase, called construction phase, a solution is iteratively built with a greedy randomised adaptive algorithm. To generate a position, $\pi(1)$ is selected at random and marked as 'visited'. $\pi(2)$ is randomly selected among the 5% unvisited closest neighbours of $\pi(1)$ and marked as 'visited'. This process continues until all vertices are marked as 'visited', forming a TSP solution.

Strategies for decision and learning methods are organised in arrays whose indices indicate which strategy to use. Regarding the learning strategy, only one is used in the experiments: the analysis of the statistics of success and failure in the application of decision strategies. Each particle has counters of success and failure for each decision strategy. For each time a strategy improves the position of the particle in question, its success counter is incremented. Otherwise, its failure counter is incremented. Thus, in the final iterations, the decision strategies with the best success rate will be more likely to be selected by the decision method.

The *path-relinking* procedure makes a permutation of adjacent elements of the origin position, considering these permutations as circular lists. First, the origin position is rotated until its first element is equal to the first element of the target position, then it goes to the analysis of the second element of the target position. The corresponding element in the origin solution is swapped to the left until it is the second element in the corresponding sequence. This process continues until the original position has reached the target position (they are the same). This procedure is illustrated in Figure 9.4, whose position is represented by five elements.

9.5 Quadratic assignment problem

The QAP has been introduced in [34]. The QAP can be stated as in (9.2) where $F = (f_{kl})$ and $D = (d_{ij})$ are $n \times n$ matrices, and S_n is the set of permutations π of $\{1, 2, \ldots, n\}$.

$$min_{\pi \in S_n} \sum_{i,j} f_{\pi(i)\pi(j)} d_{ij} \tag{9.2}$$

One of the major applications of the QAP is in the location theory where F is a flow matrix, i.e. f_{kl} is the flow of materials between facilities k and l and D is a distance matrix, i.e. d_{ij} represents the distance from location i to j. The objective is

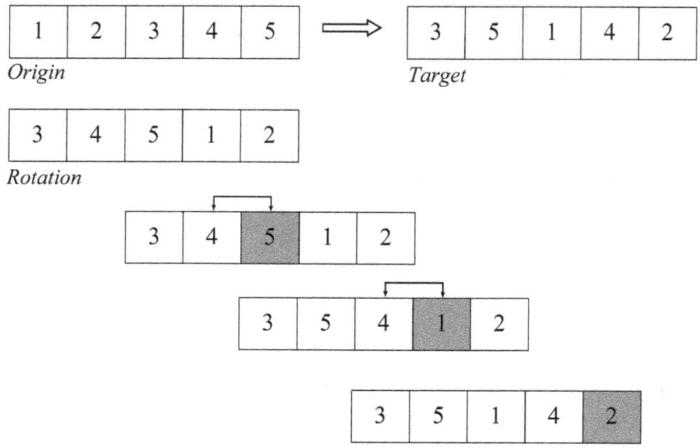

Figure 9.4 Path-relinking for the TSP

to find an assignment of facilities to locations (a permutation $\pi \in S_n$), such that the assignment cost is minimised. In [35], it was proved the QAP is strong NP-hard. A survey presenting a number of real world applications, exact and heuristic algorithms for this problem, was presented in [36].

To apply MAHM to the QAP, positions are defined similarly to the TSP. Given a QAP instance and S_n, the set of permutations of $\{1, \ldots, n\}$, the particle's position is a permutation $\pi \in S_n$, $\pi = (\pi(1), \ldots, \pi(n))$, where $\pi(i)$ stands for the facility assigned to the ith location. The particle's position is evaluated with (9.2). Initial positions are generated with a random association of facilities to locations since effective constructive heuristics for QAP are not known [37].

The MAHM architecture was applied to the QAP, named MAHM-QAP, with the hybridisation of four metaheuristics: Simulating Annealing [38], Robust TS [39], GRASP with Path-relinking [40] and variable depth search [41]. The learning strategy used in the experiments was the same used for the TSP, i.e. the analysis of the success and failure statistics. The *path-relinking* procedure is the same described in Section 9.4 for the TSP. MAHM-QAP used the backward path-relinking strategy.

9.6 Computational experiments

The algorithms were implemented in C++ in a Xeon QuadCore 2.8 GHz, 8 Gb of RAM with Scientific Linux. Thirty independent executions of each heuristic algorithm were performed for each instance. We present the relative percentage deviation (RPD) of the solution produced by the heuristic algorithm from the best solution known for the specific instance. The RPD was calculated using (9.3), where S_{alg} and S_{alg} stand for the heuristic solution and the best-known solution, respectively.

$$RPD(\%) = \frac{S_{alg} - S_{opt}}{S_{opt}} \times 100 \tag{9.3}$$

Table 9.1 Groups of instances

Group 1	Group 2	Group 3	Group 4	Group 5
dsj100	vm1084	nrw1379	u2319	rl11849
pr1002	pcb1173	u1817	fnl4461	usa13509
u1060	fl1400	rl1889	pla7397	brd14051
d1291	fl1577	u2152		d15112
rl1304	d2103	pcb3038		d18512
rl1323	fl3795	rl5915		
u1432		rl5934		
d1655				
vm1748				
pr2392				

The tables shown in this section present the average percentage deviation (RPD_{avg}), i.e. the value set to S_{alg} in (9.3) was the average of the best solutions from the 30 independent executions. The tables also show the average processing time in seconds, T_{avg}.

The Mann–Whitney U-test [42] at a significance level of 0.05 was applied to verify whether significant differences existed between the results. The test was implemented in the R software [43].

The input parameters for the MAHM variants were $swarm\_size = 10, prob\_ini = 90$ and $factor = 0.1$. These parameters are the same used in [20].

The stopping criteria for the TSP were to reach the optimal solution, 10 iterations without improvement of the best solution found so far and a maximum limit of 20 iterations. The stopping criteria for the QAP were to reach the best-known solution; 10 iterations without improvement of the best solution found so far and the maximum limit of 20 iterations.

9.6.1 Experiments with the TSP

A set of 31 symmetric TSPLIB instances [44] with n ranging from 1,000 to 18,512 was used in the experiments. The instances were divided into five groups to facilitate the analysis.

The first four groups consisted of instances with n ranging from 1,000 to 7,397 and were created concerning their RPD values using the Ward's method [45]. The fifth group consists of larger instances with n ranging from 11,849 to 18,512. Table 9.1 presents the groups of instances.

The graphic in Figure 9.5 shows the cluster analysis for grouping similar instances, obtained by Ward's method, configuring the *joint level* parameter for generating four groups. In Ward's minimum-variance method, the distance between two clusters is the ANOVA sum of squares between the two clusters added up over all the variables [46].

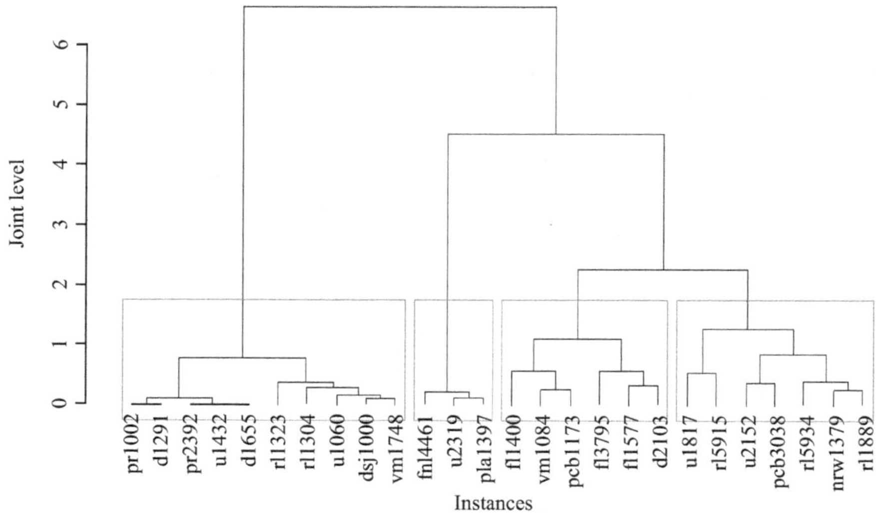

Figure 9.5 Clusters obtained by the Ward's method

Three MAHM variations, named *MAHM-MLS*, *MAHM-LS-BF* and *MAHM-LS-B*, were tested. The metaheuristics hybridised in *MAHM-MLS* were *SA* [47], *TS* [48], GRASP [33,49], 2 and 3-Opt [50] and Lin–Kerninghan local search [51,52]. Two Lin-Kerninghan implementations were used in the experiments: one from the Concorde software presented in [51], named LKC, and the Helsgaun's algorithm [52], named LHK. We used the 2 and 3-Opt implementations made available in [51] in the Concorde TSP Solver[1]. The parameters of the *SA* were an initial temperature of $T_0 = 100,000$, and a temperature reduction factor of 0.9997. The *TS* parameters were tabu list size, $|T| = 10$, and neighbourhood size, $|V| = 100$. The improvement phase of the GRASP was divided into two parts. First, the solution built in the constructive phase was submitted to the 2-Opt local search. Then the solution was submitted to the United States procedure from the GENIUS heuristic [53]. There are several strategies for the path-relinking heuristic. They are described in the study presented in [49]. The back-and-forward path-relinking strategy was implemented in the *MAHM-MLS*.

The *MAHM-LS-BF* and *MAHM-LS-B* were implemented with the 2-Opt, 3-Opt, LKC and LKH local searches. The back-and-forward path-relinking strategy was implemented in the *MAHM-LS-BF* and the backward strategy in the *MAHM-LS-B*. These variants were used to investigate the MAHM behaviour concerning different strategies.

A previous work presented in [54] showed that the MAHM-MLS outperformed the *SA*, *TS*, GRASP, 2 and 3-Opt and the two Lin-Kerninghan implementations.

[1] http://www.math.uwaterloo.ca/tsp/concorde/index.html

Table 9.2 Comparison between MAHM-LS-BF and MAHM-MLS

Group	**Instance**	**MAHM-LS-BF**		**MAHM-MLS**		***p-Value***	
		RPD_{avg}	T_{avg}	RPD_{avg}	T_{avg}	RPD_{avg}	T_{avg}
1	dsj1000	0.0027	8.46	0.0027	9.72	1.0000	0.0716
	pr1002	0	1.20	0	1.88	1.0000	**0.0484**
	u1060	0	1.29	0	2.22	1.0000	**0.0086**
	d1291	0	2.05	0	3.10	1.0000	0.2247
	rl1304	0	1.82	0	3.06	1.0000	0.0941
	rl1323	0	1.92	0	3.45	1.0000	**0.0023**
	u1432	0	2.47	0	4.17	1.0000	**<0.0001**
	d1655	0	53.41	0	53.57	1.0000	0.6070
	vm1748	0	49.66	0	53.84	1.0000	0.1209
	pr2392	0	10.94	0	16.02	1.0000	**<0.0001**
2	vm1084	0	1.18	0.0010	2.55	0.1669	0.0660
	pcb1173	0	2.50	0.0006	6.81	**0.0407**	**0.0004**
	fl1400	0	2.20	0	3.04	1.0000	0.5530
	fl1577	0	3.38	0	7.12	1.0000	**<0.0001**
	d2103	0	6.41	0.0002	24.69	0.1669	**<0.0001**
	fl3795	0	28.68	0	56.54	1.0000	**<0.0001**
3	nrw1379	0	2.64	0.0008	10.54	0.0804	**<0.0001**
	u1817	0.0223	36.77	0.0221	51.03	0.4208	**0.0125**
	rl1889	0	4.83	0.0001	12.26	0.1669	**<0.0001**
	u2152	0.0021	9.42	0.0020	23.95	0.5190	**<0.0001**
	pcb3038	0	20.98	0.0003	56.46	0.0804	**<0.0001**
	rl5915	0.0008	806.62	0.0040	759.30	0.6603	**0.0011**
	rl5934	0	99.29	0	299.65	1.0000	**<0.0001**
4	u2319	0	8.37	0	15.85	1.0000	**<0.0001**
	fnl4461	0	58.49	0.0006	186.00	0.0804	**<0.0001**
	pla7397	0	287.85	0	396.54	1.0000	**<0.0001**

That work also showed that the MAHM-MLS outperformed some effective algorithms for the TSP: Tourmerge [51], PHGA [55], MAOS [8] and HBMO [56].

For this chapter, we first analyse the potential of the three MAHM variations: MAHM-MLS, MAHM-LS-BF and MAHM-LS-B. Then, the best variation is compared to two algorithms from the literature: PSO-LK-C1 [21] and MAOS [8]. PSO-LK-C1 is an implementation of PSO for discrete problems based on local search and path-relinking. MAOS is an effective multiagent-based approach.

Table 9.2 shows the results of the MAHM-LS-BF and MAHM-MLS for four groups of instances. The Mann–Whitney *U*-test [42] was applied to detect significant differences of solution quality and processing time. The *p-values* are shown in the last two columns of Table 9.2, and better results are in bold.

The results show that, regarding solution quality, the null hypothesis can be rejected only for the pcb1173 instance. The statistical test for the processing time,

Table 9.3 Comparison between MAHM-LS-BF and MAHM-LS-B

Group	Instance	MAHM-LS-BF		MAHM-LS-B		*p-Value*	
		RPD_{avg}	T_{avg}	RPD_{avg}	T_{avg}	RPD_{avg}	T_{avg}
1	dsj1000	0.0027	8.46	0.0027	1.70	1.0000	**<0.0001**
	pr1002	0	1.20	0	1.64	1.0000	1.0000
	u1060	0	1.29	0	1.80	1.0000	1.0000
	d1291	0	2.05	0	2.74	1.0000	1.0000
	rl1304	0	1.82	0	2.57	1.0000	1.0000
	rl1323	0	1.92	0	2.69	1.0000	1.0000
	u1432	0	2.47	0	3.39	1.0000	1.0000
	d1655	0	53.41	0	9.82	1.0000	**<0.0001**
	vm1748	0	49.66	0	6.18	1.0000	**<0.0001**
	pr2392	0	10.94	0	15.05	1.0000	1.0000
2	vm1084	0	1.18	0	1.67	1.0000	1.0000
	pcb1173	0	2.50	0.0004	3.66	0.9626	1.0000
	fl1400	0	2.20	0	2.97	1.0000	1.0000
	fl1577	0	3.38	0	4.64	1.0000	1.0000
	d2103	0	6.41	0	9.16	1.0000	1.0000
	fl3795	0	28.68	0	29.43	1.0000	0.5442
3	nrw1379	0	2.64	0	3.75	1.0000	1.0000
	u1817	0.0223	36.77	0.0175	8.08	0.1601	**0.0081**
	rl1889	0	4.83	0	6.52	1.0000	1.0000
	u2152	0.0021	9.42	0	10.79	0.1669	1.0000
	pcb3038	0	20.98	0	20.35	1.0000	0.4267
	rl5915	0.0008	806.62	0.0015	142.54	0.5871	0.9965
	rl5934	0	99.29	0	133.68	1.0000	1.0000
4	u2319	0	8.37	0	11.89	1.0000	1.0000
	fnl4461	0	58.49	0.0001	58.14	0.8493	0.9488
	pla7397	0	287.85	0	245.61	1.0000	1.0000
5	rl11849	0.0001	2743.66	0.0002	935.01	0.7281	0.9962
	usa13509	0.0009	22125.41	0.0010	1860.99	0.9692	**<0.0001**
	brd14051	0.0005	18655.26	0.0015	1430.64	1.0000	**0.0179**
	d15112	0.0003	50567.28	0.0006	1623.22	0.8626	**<0.0001**
	d18512	0.0002	87841.61	0.0007	3269.91	0.9989	**<0.0001**

Note: Bold values represent that the results are statistically significant.

however, indicated that the null hypothesis was rejected for 19 of the 26 instances. We can conclude that the MAHM-LS-BF is more efficient than the MAHM-MLS regarding processing time. The processing time differences are more accentuated for groups 2, 3 and 4. These results indicate that the successful search effort was accomplished by the local search methods, with less influence of the SA, TS and GRASP metaheuristics.

Table 9.3 shows the results of the MAHM-LS-BF and MAHM-LS-B for the five groups of instances. The objective of the comparison between these two methods is to verify if the backward path-relinking strategy is as effective as the back-and-forward strategy.

Table 9.4 Comparison between PSO-LK-C1 and MAHM-LS-B

Group	Instance	PSO-LK-C1		MAHM-LS-B		p-Value RPD_{avg}
		RPD_{avg}	T_{avg}	RPD_{avg}	T_{avg}	
1	dsj1000	0.0027	22.21	0.0027	**1.70**	0.1182
	pr1002	0	**0.90**	0	1.64	1.0000
	u1060	0.0008	6.17	0	**1.80**	**0.0426**
	d1291	0	**2.38**	0	2.74	1.0000
	rl1304	0	**1.50**	0	2.57	1.0000
	rl1323	0.0030	17.98	0	**2.69**	**0.0008**
	u1432	0	5.99	0	**3.39**	1.0000
	d1655	0	99.44	0	**9.82**	1.0000
	vm1748	0	103.22	0	**6.18**	1.0000
	pr2392	0	23.34	0	**15.05**	1.0000
2	vm1084	0.0016	6.76	0	**1.67**	**0.0427**
	pcb1173	0.0003	8.82	0.0004	**3.66**	0.1828
	fl1400	0	**1.53**	0	2.97	1.0000
	fl1577	0.0200	67.19	0	**4.64**	**<0.0001**
	d2103	0.0123	112.73	0	**9.16**	**0.0001**
	fl3795	0.0214	35.40	0	**29.43**	**0.0161**
3	nrw1379	0.0058	49.07	0	**3.75**	**<0.0001**
	u1817	0.1068	110.74	0.0175	**8.08**	**<0.0001**
	rl1889	0.0037	94.14	0	**6.52**	**<0.0001**
	u2152	0.0711	138.27	0	**10.79**	**<0.0001**
	pcb3038	0.0343	478.66	0	**20.35**	**<0.0001**
	rl5915	0.0435	277.89	0.0015	**142.54**	**<0.0001**
	rl5934	0.0797	389.75	0	**133.68**	**<0.0001**
4	u2319	0.0319	336.05	0	**11.89**	**<0.0001**
	fnl4461	0.0421	257.20	0.0001	**58.14**	**<0.0001**
	pla7397	0.0348	**223.12**	0	245.61	**<0.0001**

Notes: Bold values represent that the results are statistically significant. Scale coefficient for runtime = 0.27.

Regarding solution quality, there is no evidence that the MAHM-LS-B improves the results of the MAHM-LS-BF. This conclusion comes from the observation of the RPD_{avg} column related to the *p*-values where there is no value smaller than 0.05. Therefore, the null hypothesis cannot be rejected for any instance. Regarding processing time, the MAHM-LS-B presented statistically better results than the MAHM-LS-BF for three of ten instances in the group 1 and for one of seven instances in the group 3. For the five largest instances in the group 5, the MAHM-LS-B presented better results for four of them. Thus, the null hypothesis can be rejected for eight instances of 31.

We can conclude from the results presented in Tables 9.2 and 9.3 that the MAHM-LS-B presented the best trade-off between solution quality and processing time, regarding the three MAHM variants. Thus, the MAHM-LS-B was compared to the PSO-LK-C1 and MAOS.

Table 9.4 shows the results of the PSO-LK-C1 and MAHM-LS-B. The PSO-LK-C1 was executed on a Pentium IV 3.0 GHz (single core), 1 GB of RAM with

Table 9.5 Comparison between MAOS and MAHM

Group	Instance	MAOS			MAHM-LS-B			*p*-Value	
		RPD_{avg}	T_{avg}	N_{opt}	RPD_{avg}	T_{avg}	N_{opt}	RPD_{avg}	T_{avg}
1	dsj1000	0.0027	68.23	0/30	0.0027	1.70	0/30	1.0000	**<0.0001**
	pr1002	0	54.40	30/30	0	1.64	30/30	1.0000	**<0.0001**
	u1060	0	58.63	30/30	0	1.80	30/30	1.0000	**<0.0001**
	d1291	0	41.63	30/30	0	2.74	30/30	1.0000	**<0.0001**
	rl1304	0	40.06	30/30	0	2.57	30/30	1.0000	**<0.0001**
	rl1323	0	46.21	30/30	0	2.69	30/30	1.0000	**<0.0001**
	u1432	0	106.87	30/30	0	3.39	30/30	1.0000	**<0.0001**
	d1655	0	97.74	30/30	0	9.82	30/30	1.0000	**<0.0001**
	vm1748	0	130.20	30/30	0	6.18	30/30	1.0000	**<0.0001**
	pr2392	0	235.75	30/30	0	15.05	30/30	1.0000	**<0.0001**
2	vm1084	0	48.27	30/30	0	1.67	30/30	1.0000	**<0.0001**
	pcb1173	0	63.05	30/30	0.0004	3.66	27/30	0.9626	**<0.0001**
	fl1400	0	76.82	30/30	0	2.97	30/30	1.0000	**<0.0001**
	fl1577	0.0069	78.79	16/30	0	4.64	30/30	**<0.0001**	**<0.0001**
	d2103	0.0025	121.93	19/30	0	9.16	30/30	**0.0002**	**<0.0001**
	fl3795	0.0626	421.08	16/30	0	29.43	30/30	**<0.0001**	**<0.0001**
3	nrw1379	0	140.11	30/30	0	3.75	30/30	1.0000	**<0.0001**
	u1817	0	141.80	30/30	0.0175	8.08	21/30	0.9994	**<0.0001**
	rl1889	0	106.50	30/30	0	6.52	30/30	1.0000	**<0.0001**
	u2152	0	172.70	30/30	0	10.79	30/30	1.0000	**<0.0001**
	pcb3038	0	482.51	30/30	0	20.35	30/30	1.0000	**<0.0001**
	rl5915	0	797.15	30/30	0.0015	142.54	23/30	0.9974	**<0.0001**
	rl5934	0	809.88	27/30	0	133.68	30/30	**0.0407**	**<0.0001**
4	u2319	0.0117	336.80	1/30	0	11.89	30/30	**<0.0001**	**<0.0001**
	fnl4461	0.0009	1521.19	0/30	0.0001	58.14	29/30	**<0.0001**	**<0.0001**
	pla7397	0.0124	1020.86	0/30	0	245.61	30/30	**<0.0001**	**<0.0001**

Note: Bold values represent that the results are statistically significant.

32 bits Ubuntu Linux. Therefore, there is a scale coefficient for the equalisation of the processing times. Based only on the processors clocks, the coefficient should be $3.0/2.8 \approx 1.07$. However, since the platform where the MAHM was executed had 8 Gb of RAM and a quadcore processor, the result must be divided by four, resulting in the scale coefficient of 0.27.

MAHM-LS-B outperformed the PSO-LK-C1 for 16 of 26 instances regarding solution quality. The null hypothesis was rejected for five of ten instances in the group 1, for four of six instances in the group 2 and for all instances of the groups 3 and 4. The MAHM-LS-B spent less processing times for 21 instances (considering the scale coefficient). The MAHM-LS-B presented better results for seven of ten instances in the group 1, for five of six instances in the group 2, for all seven instances in the group 3 and for two of three instances in the group 4.

Table 9.5 shows the comparison between MAOS and MAHM-LS-B. MAOS source code is available on http://www.adaptivebox.net/main/maos-tsp-algorithm, and it was performed in the same computational environment that MAHM-LS-B.

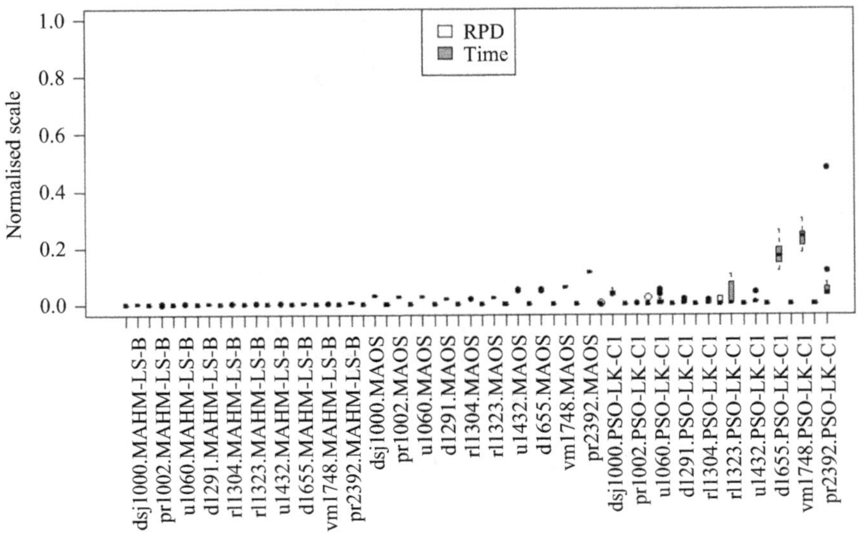

Figure 9.6 Boxplot for instances of group 1

Analysing the results in the Table 9.5, with regard to solution quality, the *p*-values show that the null hypothesis is rejected for seven instances (three of six in the group 2, one of seven in the group 3 and all three of them in the group 4), for the most part in the larger instances of the experiment. This means that the MAHM-LS-B presents better results as the size of the instance increases, a result already seen in previous experiments.

Regarding runtime, MAHM-LS-B presented better results for all instances in all four groups. The *p*-values less than 0.05 show that the null hypothesis is rejected for all instances in terms of runtime. This means that the MAHM-LS-B outperforms the MAOS with regard to runtime.

Regarding a number of optimal solutions found by the algorithms (N_{opt} column), in a total of 780 executions, MAHM-LS-B found 730 optimal solutions (about 94% success), while MAOS found 619 optimal solutions (about 79% success).

The graphics in Figures 9.6–9.9 show the boxplots for the four groups obtained by Ward's method for the MAHM-LS-B, MAOS and PSO-LK-C1 approaches. Analysing the graphics, we can confirm that, in general, MAHM-LS-B outperforms both MAOS and PSO-LK-C1.

9.6.2 Experiments with QAP

The experiments were executed for a set of 31 QAPLIB instances [57] with *n* from 10 to 100. The instances were divided into the four classes presented in [58]: uniform, grid, real-life and real-like. The entries of the matrixes of the uniform instances were randomly generated from a uniform distribution. The entries of the distance matrix of grid instances were based on the Manhattan distance on a grid. The entries of the flow matrix of grid instances were randomly generated. Real-life instances arose

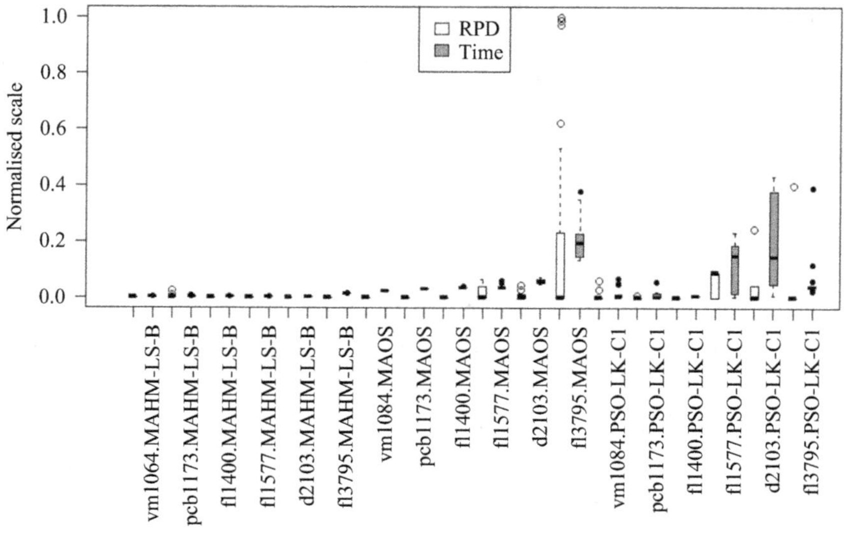

Figure 9.7 Boxplot for instances of group 2

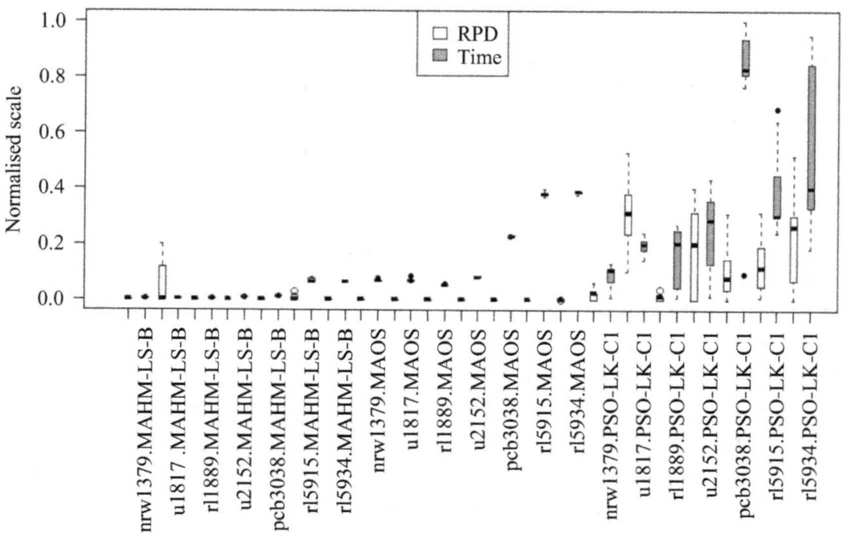

Figure 9.8 Boxplot for instances of group 3

from practical applications. Finally, real-like instances have the same characteristics of real-life problems. The entries of their distance matrixes are the Euclidean distances between n points in the plane. Table 9.6 presents the instances used in the computational experiments.

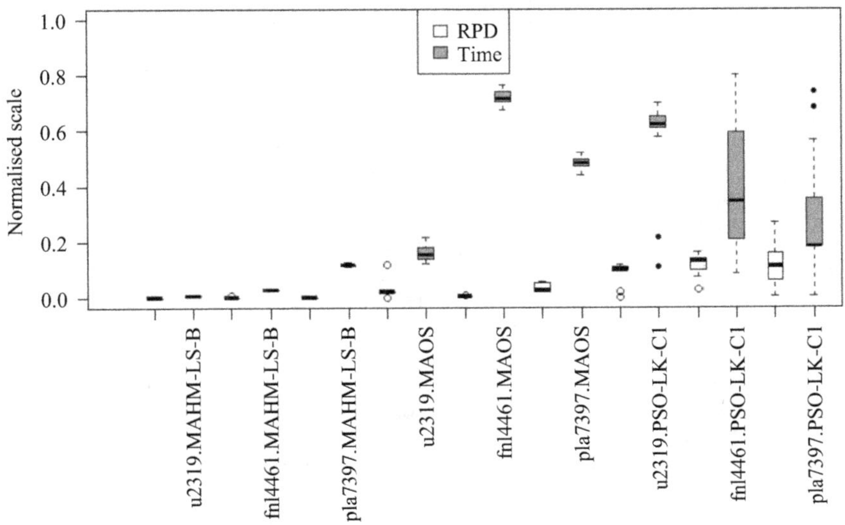

Figure 9.9 Boxplot for instances of group 4

Table 9.6 QAP instances

Class	Instances
Uniform	tai10a, tai12a, tai15a, tai17a, rou20, tai20a, tai30a, tai35a, tai40a, tai50a, tai60a, tai80a, tai100a
Grid	nug30, tho30, sko72, sko81, sko100a
Real-life	chr25a, kra30a, kra30b, ste36a, ste36b
Real-like	tai20b, tai25b, tai30b, tai35b, tai40b, tai50b, tai60b, tai80b

The experiments compared the MAHM-QAP with three hybrid approaches from the literature: ANGEL [59], GRASP+SA-TS [60] and HGA [61]. The results reported in this section were taken from the original papers were the approaches were published.

The ANGEL, presented in [59], hybridises ant system, GA and local search (2-opt variant). Table 9.7 shows the results of the MAHM-QAP and ANGEL. The ANGEL processing time was normalised with the scale coefficient shown in Table 9.7. This was made to allow processing time comparison between two different platforms.

Regarding solution quality, the MAHM-QAP presented the best RPD_{avg} results for nine instances. For the remaining 13 instances, there were draws. Regarding the average processing time, the MAHM-QAP presented the best result for all instances.

The GRASP+SA–TS, presented in [60], hybridises GRASP, SA and TS. Table 9.8 shows the results of the GRASP+SA–TS and MAM-QAP. Processing times were not reported in [60]. Thus, we do not present processing times for the MAHM-QAP in Table 9.8. The MAHM-QAP presented the best results for seven

Table 9.7 Results of MAHM-QAP and ANGEL

Instance	ANGEL		MAHM-QAP	
	RPD_{avg}	T_{avg}	RPD_{avg}	T_{avg}
nug30	0	3.39	0	**<0.01**
tho30	0	1.27	0	**<0.01**
kra30a	0	1.31	0	**<0.01**
kra30b	0	7.24	0	**<0.01**
tai30a	0.34	15.25	**0**	**<0.01**
tai30b	0	2.71	0	**<0.01**
tai35a	0.49	38.18	**0**	**<0.01**
tai35b	0	3.44	0	**<0.01**
ste36a	0	6.75	0	**<0.01**
ste36b	0	1.15	0	**<0.01**
tai40a	0.59	140.97	**0.19**	**0.03**
tai40b	0	6.35	0	**<0.01**
tai50a	0.85	320.22	**0.62**	**0.04**
tai50b	0	31.23	0	**<0.01**
tai60a	1.03	631.74	**0.79**	**0.04**
tai60b	0	30.56	0	**<0.01**
sko72	0.02	764.94	0.02	**0.06**
tai80a	0.86	1,643.77	**0.64**	**0.11**
tai80b	0	651.13	0	**0.04**
sko81	0.03	1,213.14	**0.02**	**0.09**
sko100a	0.02	1,842.46	**0**	**0.17**
tai100a	0.80	3,624.68	**0.60**	**0.20**

Notes: Bold values represent that the results are statistically significant.
Scale coefficient (ANGEL) = 0.18.

and twelve instances, regarding RPD_{min} (minimum deviation) and RPD_{avg}. The remaining results were draws.

The HGA, presented in [61], hybridises GA and TS. Table 9.9 presents the results of the HGA and the MAHM-QAP for 17 instances. The processing times presented in Table 9.9 come from similar computational environments. Thus, normalisation was not necessary.

Regarding solution quality, the MAHM-QAP presented the best RPD_{avg} results for two instances, and the HGA for five instances. For the remaining nine instances, there were draws. Regarding processing time, the MAHM-QAP presented the best result for all instances.

From the computational experiments presented in this section, we can conclude that the MAHM-QAP is an efficient approach for the four classes of QAP instances. Regarding solution quality, the MAHM-QAP outperformed the ANGEL and the GRASP+SA−TS. However, the approach proposed in this study did not outperform the HGA. Regarding processing time, the MAHM-QAP was able to overcome the three hybrid algorithms.

Table 9.8 Results of MAHM-QAP and GRASP+SA−TS

Instance	**GRASP+SA−TS**		**MAHM-QAP**	
	RPD_{min}	RPD_{avg}	RPD_{min}	RPD_{avg}
tai10a	0	0	0	0
tai12a	0	0	0	0
tai15a	0	0	0	0
tai17a	0	0	0	0
tai20a	0	0.16	0	**0**
rou20	0	0.03	0	**0**
chr25a	0	0.40	0	**0**
tai30a	0.82	1.51	**0**	**0**
nug30	0	0.01	0	**0**
kra30a	0	0.74	0	**0**
kra30b	0	0	0	0
tai35a	2.27	2.57	**0**	**0**
tai40a	2.18	2.83	**0.07**	**0.19**
tai50a	3.58	3.72	**0.31**	**0.62**
tai60a	2.93	3.57	**0.50**	**0.79**
tai80a	3.13	3.57	**0.62**	**0.64**
tai100a	3.02	3.49	**0.49**	**0.60**

Note: Bold values represent that the results are statistically significant.

Table 9.9 Results of MAHM-QAP and HGA

Instance	**HGA**		**MAHM-QAP**	
	RPD_{avg}	T_{avg}	RPD_{avg}	T_{avg}
tai20a	0	0.46	0	**<0.01**
tai20b	0	0.05	0	**<0.01**
tai25b	0	0.09	0	**<0.01**
tai30a	0	1.93	0	**<0.01**
tai30b	0	0.13	0	**<0.01**
tai35a	0	3.40	0	**<0.01**
tai35b	0	0.21	0	**<0.01**
tai40a	**0.090**	5.70	0.192	**0.03**
tai40b	0	0.28	0	**<0.01**
tai50a	**0.250**	13.73	0.618	**0.04**
tai50b	0	0.60	0	**<0.01**
tai60a	**0.270**	28.67	0.791	**0.04**
tai60b	0.003	1.03	**0**	**<0.01**
tai80a	**0.404**	91.67	0.643	**0.11**
tai80b	0.102	2.25	**0**	**0.04**
tai100a	**0.275**	223.33	0.599	**0.20**

Note: Bold values represent that the results are statistically significant.

9.7 Conclusions

This study proposed a multiagent approach for hybridisation of metaheuristics inspired in the PSO framework. The motivation for developing it is that PSO is inherently multiagent, so a MAHM based on the PSO could be straightforward and efficient, comprising the full potential of multiagent systems. In the proposed approach, each particle is a reactive agent with memory, decision and learning methods.

We applied three variations of the proposed approach to the symmetric TSP, and the results were compared to efficient approaches proposed previously in the literature. The results from the experiments were promising, since, although the potential of the proposed approach was not fully explored, the algorithm presented a satisfactory performance obtaining high-quality solutions. The results showed that, for the tested instances, the proposed approach is useful for solving TSP.

We also applied one implementation of MAHM to QAP, and the results were compared to efficient hybrid approaches proposed previously in the literature. The results from the experiments were promising, and the algorithm presented a satisfactory performance obtaining high-quality solutions, mainly with regard to the runtime. The results showed that, for the tested instances, the proposed approach is also useful for solving QAP.

Future works will investigate other learning methods and the hybridisation of other metaheuristics. The potential of the proposed approach on other hard combinatorial optimisation problems such as multi-objective problems will also be investigated.

References

[1] Wooldridge M. An Introduction to Multiagent Systems. 2nd ed. Chichester, UK: Wiley; 2009.

[2] Zhao B, Guo CX, Cao YJ. A multiagent-based particle swarm optimization approach for optimal reactive power dispatch. IEEE Transactions on Power Systems. 2005;20(2):1070–1078.

[3] Günther M, Nissen V. Particle Swarm Optimization and an Agent-Based Algorithm for a Problem of Staff Scheduling. In: Conference on the Applications of Evolutionary Computation. Springer, Berlin, Heidelberg, 2010. p. 451–461.

[4] Cui H, Turan O. Application of a new multi-agent hybrid co-evolution based particle swarm optimisation methodology in ship design. Computer-Aided Design. 2010;42(11):1013–1027.

[5] Kumar R, Sharma D, Sadu A. A hybrid multi-agent based particle swarm optimization algorithm for economic power dispatch. International Journal of Electrical Power & Energy Systems. 2011;33(1):115–123.

[6] Hammami M, Ghédira K. COSATS, X-COSATS: two multi-agent systems cooperating simulated annealing, tabu search and X-over operator for the K-graph partitioning problem. In: Knowledge-Based Intelligent Information and Engineering Systems. Springer; 2005. p. 647–653.

[7] Tiejun Z, Yihong T, Lining X. A multi-agent approach for solving traveling salesman problem. Wuhan University Journal of Natural Sciences. 2006;11(5):1104–1108.

[8] Xie XF, Liu J. Multiagent optimization system for solving the traveling salesman problem (TSP). IEEE Transactions on Systems, Man, and Cybernetics, Part B: Cybernetics. 2009;39(2):489–502.

[9] Meignan D, Koukam A, Créput JC. Coalition-based metaheuristic: a self-adaptive metaheuristic using reinforcement learning and mimetism. Journal of Heuristics. 2010;16(6):859–879.

[10] Raidl G. A Unified View on Hybrid Metaheuristics. In: Almeida F, Blesa Aguilera M, Blum C, *et al.*, editors. Hybrid Metaheuristics. vol. 4030 of Lecture Notes in Computer Science. Springer Berlin/Heidelberg; 2006. p. 1–12.

[11] Blum C, Puchinger J, Raidl GR, *et al.* Hybrid metaheuristics in combinatorial optimization: a survey. Applied Soft Computing. 2011;11(6):4135–4151.

[12] Milano M, Roli A. MAGMA: a multiagent architecture for metaheuristics. IEEE Transactions on Systems, Man, and Cybernetics, Part B: Cybernetics. 2004;34(2):925–941.

[13] Fernandes FC, Souza SRd, Silva M, *et al.* A multiagent architecture for solving combinatorial optimization problems through metaheuristics. In: IEEE International Conference on Systems, Man and Cybernetics, 2009; 2009. p. 3071–3076.

[14] Malek R. An agent-based hyper-heuristic approach to combinatorial optimization problems. In: IEEE International Conference on Intelligent Computing and Intelligent Systems (ICIS). vol. 3; 2010. p. 428–434.

[15] Kennedy J, Eberhart R. Particle swarm optimization. In: Proceedings of the IEEE International Conference on Neural Networks. vol. 4; 1995. p. 1942–1948.

[16] Barbucha D. Agent-based guided local search. Expert Systems with Applications. 2012;39(15):12032–12045.

[17] Voudouris C, Tsang E. Guided local search. University of Essex, UK; 1995.

[18] Talukdar S, Baerentzen L, Gove A, *et al.* Asynchronous teams: cooperation schemes for autonomous agents. Journal of Heuristics. 1998;4(4): 295–321.

[19] Talbi EG. Metaheuristics: From Design to Implementation. Wiley Publishing; 2009.

[20] Goldbarg EFG, Souza GR, Goldbarg MC. Particle Swarm for the Traveling Salesman Problem. In: Gottlieb J, Raidl G, editors. Evolutionary Computation in Combinatorial Optimization. vol. 3906 of Lecture Notes in Computer Science. Springer Berlin/Heidelberg; 2006. p. 99–110.

[21] Goldbarg EFG, Souza GR, Goldbarg MC. Particle Swarm Optimization Algorithm for the Traveling Salesman Problem. In: Greco F, editor. Traveling Salesman Problem. Vienna: InTech; 2008. p. 75–96.

[22] Goldbarg EFG, Souza GR, Goldbarg MC. Particle swarm optimization for the bi-objective degree constrained minimum spanning tree. In: IEEE Congress on Evolutionary Computation, 2006. CEC 2006; 2006. p. 420–427.

[23] Clerc M. Discrete particle swarm optimization, illustrated by the traveling salesman problem. In: New Optimization Techniques in Engineering. Springer; 2004. p. 219–239.

[24] Wang KP, Huang L, Zhou CG, *et al.* Particle swarm optimization for traveling salesman problem. In: Machine Learning and Cybernetics, 2003 International Conference on. vol. 3. IEEE; 2003. p. 1583–1585.

[25] Machado T, Lopes H. A hybrid particle swarm optimization model for the traveling salesman problem. In: Adaptive and Natural Computing Algorithms. 2005; p. 255–258.

[26] Pang W, Wang KP, Zhou CG, *et al.* Modified particle swarm optimization based on space transformation for solving traveling salesman problem. In: Machine Learning and Cybernetics, 2004. Proceedings of 2004 International Conference on. vol. 4. IEEE; 2004. p. 2342–2346.

[27] Glover F, Laguna M, Marti R. Fundamentals of scatter search and path relinking. Control and Cybernetics. 2000;29(3):653–684.

[28] Ribeiro CC, Uchoa E, Werneck RF. A hybrid GRASP with perturbations for the Steiner problem in graphs. Journal on Computing. 2002;14(3):228–246.

[29] Gutin G, Punnen AP. The Traveling Salesman Problem and Its Variations. vol. 12. Kluwer Academic Publishers; 2002.

[30] Applegate DL, Bixby RE, Chvatal V, *et al.* The Traveling Salesman Problem: A Computational Study. Princeton University Press; 2006.

[31] Lin S, Kernighan BW. An effective heuristic algorithm for the traveling-salesman problem. Operations Research. 1973;21:498–516.

[32] Bellmore M, Nemhauser GL. The traveling salesman problem: a survey. Operations Research. 1968;16(3):538–558.

[33] Feo TA, Resende MGC. A probabilistic heuristic for a computationally difficult set covering problem. Operations Research Letters. 1989;8(2):67–71.

[34] Koopmans TC, Beckmann MJ. Assignment problems and the location of economic activities. Econometrica. 1957;25:53–76.

[35] Sahni S, Gonzalez T. P-complete approximation problems. Journal of the Association of Computing Machinery. 1976;23:555–565.

[36] Loiola EM, Abreu NMM, Boaventura-Netto PO, *et al.* A survey for the quadratic assignment problem. European Journal of Operational Research. 2007;176:657–690.

[37] Stützle T. Iterated local search for the quadratic assignment problem. European Journal of Operational Research. 2006;174(3):1519–1539.

[38] Connolly DT. An improved annealing scheme for the QAP. European Journal of Operational Research. 1990;46(1):93–100.

[39] Taillard E. Robust taboo search for the quadratic assignment problem. Parallel Computing. 1991;17(4):443–455.

[40] Oliveira CA, Pardalos PM, Resende MG. GRASP with path-relinking for the quadratic assignment problem. In: Experimental and Efficient Algorithms. Springer; 2004. p. 356–368.

[41] Goldbarg EFG, Goldbarg MC. An experimental study of variable depth search algorithms for the quadratic assignment problem. Pesquisa Operacional. 2012;32(1):165–196.
[42] Conover WJ. Practical Nonparametric Statistics. New York: Wiley; 1999.
[43] R Core Team. R: A Language and Environment for Statistical Computing. Vienna, Austria; 2012. ISBN 3-900051-07-0. Available from: http://www.R-project.org/.
[44] Reinelt G. TSPLIB – A traveling salesman problem library. ORSA Journal on Computing. 1991;3(4):376–384.
[45] Ward Jr JH. Hierarchical grouping to optimize an objective function. Journal of the American Statistical Association. 1963;58(301):236–244.
[46] Hair JF, Black WC, Babin BJ, *et al.* Multivariate Data Analysis. 5th ed. New Jersey: Prentice Hall; 1998.
[47] Kirkpatrick S, Gelatt CD, Vecchi MP. Optimization by simulated annealing. Science. 1983;220(4598):671–680.
[48] Glover F, Laguna M. Tabu Search. vol. 1. Springer, Boston, MA; 1998.
[49] Resende MGC, Ribeiro CC. GRASP with path-relinking: recent advances and applications. In: Metaheuristics: Progress As Real Problem Solvers. 2005; p. 29–63.
[50] Flood MM. The traveling-salesman problem. Operations Research. 1956;4(1):61–75.
[51] Applegate D, Cook W, Rohe A. Chained Lin–Kernighan for large traveling salesman problems. INFORMS Journal on Computing. 2003;15(1):82–92.
[52] Helsgaun K. An effective implementation of the Lin–Kernighan traveling salesman heuristic. European Journal of Operational Research. 2000;126(1):106–130.
[53] Gendreau M, Hertz A, Laporte G. New insertion and postoptimization procedures for the traveling salesman problem. Operations Research. 1992;40(6):1086–1094.
[54] Souza GR, Goldbarg EFG, Goldbarg MC, *et al.* A multiagent approach for metaheuristics hybridization applied to the traveling salesman problem. In: Neural Networks (SBRN), 2012 Brazilian Symposium on. Washington: IEEE Computer Society; 2012. p. 208–213.
[55] Nguyen HD, Yoshihara I, Yamamori K, *et al.* Implementation of an effective hybrid GA for large-scale traveling salesman problems. IEEE Transactions on Systems, Man, and Cybernetics, Part B: Cybernetics. 2007;37(1):92–99.
[56] Marinakis Y, Marinaki M, Dounias G. Honey bees mating optimization algorithm for the Euclidean traveling salesman problem. Information Sciences. 2011;181(20):4684–4698.
[57] Burkard RE, Karisch SE, Rendl F. QAPLIB – a quadratic assignment problem library. Journal of Global Optimization. 1997;10(4):391–403.
[58] Taillard ED. Comparison of iterative searches for the quadratic assignment problem. Location Science. 1995;3(2):87–105.

[59] Tseng LY, Liang SC. A hybrid metaheuristic for the quadratic assignment problem. Computational Optimization and Applications. 2006;34:85–113.

[60] Gunawan A, Ng KM, Poh KL. Solving the quadratic assignment problem by a hybrid algorithm. International Journal on Advanced Science, Engineering and Information Technology. 2011;1(1):36–41.

[61] Drezner Z, Misevičlus A. Enhancing the performance of hybrid genetic algorithms by differential improvement. Computers & Operations Research. 2013;40:1038–1046.

Chapter 10

The critical state in particle swarm optimisation

Adam Erskine[1], Thomas Joyce[1], and J. Michael Herrmann[1]

Abstract

Particle swarm optimisation (PSO) is a widely used meta-heuristic algorithm, which has been successfully applied to a large variety of problems in optimisation, prediction, classification, visualisation and robotics. However, the underlying mechanism that leads to good solutions in these domains is not well understood. A number of parameters influence the algorithm mainly by determining the balance between exploration and exploitation. In this chapter, we describe the behaviour of PSO as a product of random matrices, and use a Lyapunov exponent to precisely characterise the critical parameters for PSO. The theoretical results are discussed based on numerical experiments for standard benchmark problems.

10.1 Introduction

Particle swarm optimisation (PSO) and its variants enjoy widespread application in search and optimisation problems. The quality of solutions found depends upon our ability to overcome the algorithm's drawbacks: the selection of suitable parameter values and the avoidance of the swarm's premature convergence. For a given of problem function, one can empirically determine the correct parameter values or the best restart strategy, yet these solutions may fail to be optimal if presented with a different arbitrary function to solve.

PSO took its original inspiration from the flocking behaviour of bird flocks. Such biological swarms often demonstrate near simultaneous movements across their populations such as sudden coordinated direction changes in the response to an approaching threat. Those individuals closest to an oncoming predator adjust their heading to avoid the encounter. These changed directions are communicated through the swarm instantly. Such a loss of length scales is reminiscent of state changes in physical systems. Ferromagnetic materials if held at a specific temperature may show wide

[1]School of Informatics, Institute for Perception, Action and Behaviour, University of Edinburgh, United Kingdom

responses than small local magnetic perturbations. Such behaviour is described as being critical or showing criticality in its response.

Consideration of the properties of the product of the stochastic dynamic matrix of PSO's update rules shows that the optimum capabilities of the algorithm occur for a locus of PSO parameter values. These result in the algorithm behaving critically. The set of parameter values that deliver these optimal results differ from those suggested in earlier studies. To confirm our results, we present empirical results.

We will first set out backgrounds to PSO, and to criticality. Next, we discuss some previous attempts to explore the theoretical dynamics of PSO. The earliest of these offer some insight into the behaviours that may be expected from the algorithm, but they have typically ignored the multiplicative role of the random element of the update rules. These have often been replaced by their expectation values or simply viewed as a *drunken walk* about the predicted dynamics of the system, enhancing the local search capabilities. We show that in fact they contribute to the dynamics of the swarm's capabilities in a manner that alters its expected behaviour. More recent analyses consider the random elements by considering expectation and variance values of the update matrices. These predict a locus of parameter values that yield convergent swarms but we believe are still too conservative. In Section 10.2, we consider the products of random matrices, in particular those that represent the PSO algorithm. we derive the location of the locus of α and ω values that give rise to swarms that will neither diverge nor collapse over time. This curve matches the observed behaviour of an unconstrained PSO swarm i.e. one where neither positional or velocity constraints are imposed on the dynamic of the swarm.

10.1.1 Particle swarm optimisation

PSO is a meta-heuristic method for obtaining solutions to search or optimisation type problems. The aim is to locate the best solution to problems that may have high dimensionality or parametrisation. The *best* solution is at a global minimum (or depending upon problem construction a global maximum). How best to find the solution? If time is no object, a random search will assuredly find it. This is usually not the case! For simple problem manifolds e.g. a sphere function, gradient descent may be all that is required to find good solutions. However, real world problems are rarely as conducive to simple methods: they may exhibit discontinuities or other potentially deceptive features.

PSO has enjoyed much attention and widespread usage. Around 700 PSO papers are grouped into 26 discernible application areas in [1]. Google Scholar reveals over 150,000 results for "PSO" in total and 24,000 for the year 2014. It was inspired by the cooperative behaviours of flocks of birds or schools of fish to locate a problem's solution [2]. The swarm is constructed from a number of particles, each of which occupies a position within the problem space being evaluated. The problem is evaluated at each particle location: essentially sampling the problem space. Additionally, particles have velocity states. New locations are computed by applying the particle velocities as an update to their positions. Their velocities are then updated based on a sum of an inertia term and terms that represent the sharing of local and global knowledge that the particles have gleaned about the problem space derived from their sampling procedure. Their future dynamics are thus modified, in part, by the knowledge of the problem surface that they are uncovering.

The position and velocity of particles, after random initialisation, are iteratively updated using the following linear dynamic of the following form.

$$\mathbf{v}_{i,t+1} = \omega \mathbf{v}_{i,t} + \alpha_2 \mathbf{R}_1(\mathbf{p}_i - \mathbf{x}_{i,t}) + \alpha_2 \mathbf{R}_2(\mathbf{g} - \mathbf{x}_{i,t}) \tag{10.1}$$

$$\mathbf{x}_{i,t+1} = \mathbf{x}_{i,t} + \mathbf{v}_{i,t+1} \tag{10.2}$$

Here $\mathbf{x}_{i,t}$ and $\mathbf{v}_{i,t}$, $i = 1, \ldots, N$, $t = 0, 1, 2, \ldots$, represent, respectively, the d-dimensional position and velocity vectors of the ith particle in the swarm at time t. The velocity update contains an inertial term parametrised by ω and includes attractive forces towards the personal best location $\mathbf{p}_i$ and the global best location $\mathbf{g}$, which are parametrised by α_1 and α_2, respectively. The symbols $\mathbf{R}_1$ and $\mathbf{R}_2$ denote diagonal matrices whose non-zero entries are uniformly distributed in the unit interval.[1] The number of particles N is quite low in most applications, usually amounting to a few dozens.

In order to function as an optimiser, problems are cast as a non-negative cost function $F : \mathbb{R}^d \to \mathbb{R}$. $F(\mathbf{x}) = 0$ defines $\mathbf{x}$ as an optimal solution of the problem. In many problems, where PSO is applied, states with near-zero costs can also be considered as good solutions. Indeed, for a given problem function, one can compare the success of different algorithms by considering which obtains the smallest value solution for a given criteria such as the number of function evaluations or computational cost etc. The cost function is evaluated at the state of each particle for each time step. If $F(\mathbf{x}_{i,t})$ is better than $F(\mathbf{p}_i)$, the personal best $\mathbf{p}_i$ is replaced by $\mathbf{x}_{i,t}$. Similarly, if one of the particles arrives at a state with a cost less than $F(\mathbf{g})$, the $\mathbf{g}$ term is replaced in all particles by the position of the particle that has discovered the new solution. If its velocity is non-zero, a particle will depart from the current best location, but it may still have a chance to return guided by the force terms in the dynamics.

The history of PSO has explored many variants. Instead of this fully connected swarm, different connectivities have been shown to yield improved performances [3,4]. However, in all cases, the success of the algorithm for a given problem will often require careful selection of the parameters ω, α_1 and α_2. Poorly chosen parameters can result in the stagnation of the algorithm in a suboptimal location. Here we are using a simple PSO with its original fully connected topology.

10.1.1.1 PSO parameter selection and resultant behaviours

In order to perform optimally, the algorithm has to ensure an appropriate mix of local search, i.e. the exploitation of existing knowledge of the problem space, and exploration of areas not yet adequately sampled. The optimal balance of exploration and exploitation will depend on the nature of the problem space and to some extent

[1]This implements a component-wise multiplication of the difference vectors with a random vector and is known to introduce a bias into the algorithm as particles prefer to stay near the axes of the problem space [6,7]. We have retained this form of stochasticity in our algorithm for the purpose of comparability.

how long the algorithm is allowed to run for. In PSO, this mix is achieved via the parametrisation of the algorithm. PSO is often applied to problems with little formal specification, and trial-and-error search remains the only general option for parameter tuning. Small ω and large α_1 and α_2 values (within limits) will encourage the particles to converge towards the currently known good locations. Overdone, the swarm may prematurely converge to a local minima and achieve no further improvements. This is termed stagnation and is often deemed a failing of the algorithm. However, this is not always the case. If the algorithm is given a time limit, collapse to a local minima at the end of this time may be the best action it could take. For example, [5] utilises both failure to locate better solutions and loss of mean particle velocity to calculate a trigger signal. Once the signal crosses a threshold, a diversity increasing procedure is initiated for all except the best particle. This is successfully applied to a typical PSO algorithm. Sometimes, the detection of stagnation can be used to trigger a change in the algorithm either by simply restarting the particles in freshly initialised locations or used in a hybrid algorithm to switch behaviour to one more attuned to exploitation.

Typically, PSO variants are tested against a number of test functions which have known optima. A number of criteria for success are measured (including average error, average number of iterations of obtaining an optimum solution and minimum error). Other approaches have looked to use information from all particles in the update mechanisms. A comprehensive learning PSO [8] stochastically selects from the best location of one particle and another random particle on a dimension by dimension basis. All particles can thus influence the evolution of the swarm. They claim to be good on multi-modal functions but poor on unimodal ones. This highlights the difficulty in finding a single strategy that succeeds against all criteria.

The algorithm's parameters play a major role in the dynamics of the swarm. One set of parameters may lead to a collapse in swarm diversity, another set to an exploding swarm expanding beyond the confines of the defined problem space. Often this can be controlled for. Particle velocities may be simply truncated to limit the motion of the particles and reduce the likelihood of divergence [4]. Similarly, the particle position states outside the problem space may be repositioned on its boundary. Velocity multipliers, derived from the algorithm's parameters, provide another means to constrict velocities [9]. The mathematical derivation of this constriction factor essentially limits the parameter values so that divergence is avoided. Other parameter controls may be applied for instance via linear decreases in the ω value as the algorithm progresses [10]. Larger ω values tend to encourage greater exploration by the swarm of the problem space. By reducing this parameter's value, the swarm is guided to spend more time exploiting the locations that seem promising. Other approaches to varying PSO's parameters have been explored including, random, increasing, decreasing and chaotic [11]. These mechanisms will generally restrict a diverging swarm. However, this risks encouraging premature convergence. Other mechanism have thus been used to avoid stagnation: by using particle repulsion [12,13] or random velocities [14]. Other means to affect the balance between exploratory and exploitative behaviours have been derived from other biological inspirations. For instance, animals such as bat [15], cuckoo [16] and fish [17] use, respectively, echolocation, brood parasitism or leaping motion as inspirations. The cuckoo search algorithm also co-opts an earlier idea of using Lévy flights to drive local searches. Lévy flights are random walks where the step-lengths are drawn from a heavy tail distribution. This is suggestive of potential benefits arising from adding power-law distributed movements to PSO.

Rather than modifying the algorithm's parameters, one can alter how the search near known good locations is conducted. Indeed the standard PSO algorithm adds stochastic noise to the linear sum of the various knowledge terms. This has the effect allowing the particles to have a somewhat randomised search that is drawn towards the known good locations. However, this has the tendency to focus searches along the axes of the search space. A chaotic search was explored [18]. Here they combined a parameter update strategy with a chaotic local search mechanism. Particles with good results had their ω parameter linearly reduced, whereas particles with poor results had their ω values increased to a maximum value. On its own, this assured a mix of diverse behaviours. The logistic function, $x_{n+1} = \mu x_n(1 - x_n)$, was employed to create a chaotic search pattern about the location of the best performing particle. Results compare favourably with many PSO variants, but it is not clear whether both of these components are needed to achieve the results reported.

10.1.1.2 Alternative meta-heuristics

PSO is not the only meta-heuristic to provide solutions to optimisation problems. Algorithms may use an adaptive probability model to choose from a number of strategies in a success-dependent fashion. Self-adaptive algorithms have been proposed for differential evolution [19] and also for PSO [20,21]. These hyperheuristic methods measure the behaviour of the swarm and use this to determine which algorithm is most appropriate to run at that particular time. A stagnating swarm may be switched to follow a rule set that favours increasing diversity, or a divergent swarm may be changed to encourage future convergence. The TRIBES mechanism removes the velocity update component of PSO and the inherent need to set parameter values [22]. Multiple sub-swarms grow and shrink (removing the need to set a swarm size) depending on performance. The swarms exchange information on good locations and, new positions are created by weighted sums of the positions of particles representing good solutions. Recent successful approaches have been derived from the Covariance Matrix Analysis Evolutionary Strategy (CMA-ES) [23]. This technique has performed well in recent benchmark tests, capturing top spots [24,25]. This is an (μ, λ) evolutionary strategy. A multivariate normal distribution derived from the μ best parents are used to generate the next generation of solutions (λ in number). After each iteration, the distribution's covariance matrix is adapted to direct the future selection towards new and better solutions. The technique may still suffer from premature convergence. However, detection of such stagnation may be used as a signal to trigger algorithmic restarts or other procedures to increase the diversity of the swarm to obtain good results [24].

10.1.1.3 The problem of making comparisons between algorithms

PSO is therefore only one of many potential algorithms proposed to solve a given problem. To differentiate between them, one needs to find a means to make comparisons: to determine which is better. Of course better can mean different things, such as best solution, best solution after N iterations or least computationally intensive solution. Competition formats have proved popular method to establish the capabilities of multiple algorithms. These typically consist of a suite of test functions that provide a broad range of challenges, be it high dimensional spaces, multi-modalities, deceptive landscape features etc. Competitors are expected to locate the best locations within each function given a given budget of function calls. Statistical comparisons can then be computed to provide a ranking of solution finding capabilities or of computation load. One weakness of the competition format may be that the winning algorithms are, by definition, good at the functions included, but that other function suites may yield different results. It has been shown using genetic programming techniques that objective functions may be evolved that favour different algorithms [26]. Here PSO, differential evolution, CMA-ES and simple hill climbing algorithms were used. In each pair of algorithms, the authors were able to find problems that favoured either of the algorithms. This does not negate the value of the competition format but does mean that it is necessary to show that the functions used in the format map to range of real world problems, although intellectually wider function types retain an interest. In this chapter, most of our empirical results are obtained using the 28 functions defined for the 2013 International Conference on Evolutionary Computation (CEC2013).

An interesting approach, certainly augmenting knowledge of the algorithm's dynamics, is to seek to use the sampling of the problem space to characterise the nature of the function being explored. This can be used to modify the future operation of any algorithm. Geometries of problem surfaces have been examined in this context [27], whereas [28] utilises information-like measures to capture problem space features that can similarly be used to guide selection of appropriate algorithms.

10.1.2 On criticality

Criticality in equilibrium thermodynamics is used to refer to the properties of a system at a transition point between phases. At this point, small perturbations can propagate throughout the whole system [29]. Ferromagnetic materials, for example, gain their magnetism ultimately via the alignment of magnetic dipole moments arising from electron spin states. The state of any individual dipole may be influenced by the magnetic field it finds itself in. If this material is at a high temperature, the thermal noise will be greater than the influence of small perturbations and no effect is seen. At low temperatures, the dipole moments are effectively frozen in whichever state they find themselves in. Small external magnetic perturbations will have little or no effect in either case. When the system is tuned by heating it to point between these two states, any small external magnetic perturbations are able to flip the state of nearby dipoles. Further, these changes can propagate throughout the material resulting in a chain of dipole flips. A characteristic of such systems is that changes in system states can exhibit power-law distributions. The small external magnetic perturbation will frequently cause small localised shifts in magnetic states and domain structures in the material but can also result in occasional large shifts in the magnetic state of the material. In order to show this type of behaviour, this system needs to be very finely tuned to a critical temperature.

More widely, the term criticality may refer to any dynamical system which behaves in a manner like this [30]. Bak notes that there are many systems in nature that exhibit similar power-law type distributions. We see such heavy-tail distributions in the earthquakes described by the Gutenberg–Richter law. Similarly, power laws appear in Zipf's law describing ranked distributions of word usage in English, fractal geometries in nature and $1/f$ type noise in physical systems. The ferromagnetic example above requires tuning, but it seems infeasible for all the occurrences of power laws to require such fine tuning. It is proposed that systems consisting of many interacting units may evolve automatically into a critical state: effectively self-organising to a point poised between order and chaos [31]. No outside agent is required to tune the system. If perturbed, such a system will respond with events that follow a power-law distribution but will result in the system returning to its critical state. Models such as this are referred to as exhibiting self-organised criticality (SOC).

One means to explore the properties of such systems may be explored by looking at simple models such as the proposed sandpile model [30]. This numerical model imagines a square lattice grid: at each locale, sand grains may be deposited. As the sand pile grows due to the slow deposition of grains, the gradient between grid locations can grow as each grain added adds $+1$ to the local gradient. If the gradient at any location increases beyond some specified threshold (say 3), the grains at that location collapse. Four grains are redistributed to the adjacent squares, and the location of collapse in reduced by the corresponding amount (-4). Early in the sandpile's evolution, the collapses are minor; however, as time passes on, grains continue to be added, more and more locations increase towards the point of collapse. Now a single added grain may lead to a collapse, and as the grains are reallocated to neighbours, those adjacent cells are found to also lead to collapse. The avalanche triggered by a single grain may be small or large. Avalanche sizes in this model are shown to follow a power-law distribution. Similar dynamics have noted earthquake magnitudes [32], punctuated evolution [33], neuronal avalanches [34–36] and even rainfall [37]. The presence of criticality in a system can be shown to optimise the system in some way. [38] showed the criticality of cortical neuronal avalanches results in optimal information capacity and transmission. The functional benefits of critical dynamics for information processing in the brain have been discussed, see the overview by [39]. These studies gave reason to the present study on the question whether criticality can be beneficial as well in dynamical systems for optimisation and search.

In such systems, the presence of power laws in system event sizes imply that events of any size are possible. Practical limits arise from the finite size of systems. If we were to engineer critical dynamics into the swarm size of a PSO algorithm, we would be assured that the swarm would sooner or later extend throughout the whole problem space and thus avoid stagnation. A random search strategy would likewise avoid stagnation, but our approach assures that the algorithm can still favour exploitation of the problem space over exploration.

Previously, there have been approaches to adding criticality to PSO. A sandpile-like approach was employed by adding an additional counter to each particle [40]. If a particle came close to another, it incremented its counter. Once over a threshold the particle relocated within the problem space and redistributed its accrued value to other particles allowing *avalanches* to occur. It is somehow elegant that feedback from the swarm's own behaviour is used to modify the future behaviour of the swarm. There was limited evidence that the swarm behaved in a critical manner, and the problem of setting parameter values remained. [41] implemented a PSO algorithm inspired by the Lévy flight random walk behaviour of many foraging animals. They modified a Gaussian PSO algorithm, which performs velocity updates by drawing from Gaussian distributions scaled by distance of particle from local and global best locations, to use Lévy distributions instead. The nature of this power-law approach produced a greater number of outliers in a given problem space, resulting in a more powerfully exploring swarm. Their results showed that this Lévy swarm outperformed both standard PSO and Gaussian PSO approaches. The Bak–Sneppen algorithm [33] provides a SOC model of punctuated evolution seen historically in the number of species in the fossil record. This has been used to create a criticality augmented PSO variant [42]. The generated power-law distributed random numbers are used directly to modify the velocity update parameters in a statistically balanced way. Additionally, local search about the particles position was driven by random numbers from the same distribution. This power–law distribution may be created offline adding little or no overhead to the algorithm. Favourable results are demonstrated in comparison to standard PSO.

10.1.3 The dynamics of PSO

As it is known that parameter selection can make a large difference to the performance of the PSO algorithm, a great deal of effort has been expended on locating the *best* values for a given problem. As we have discussed, speed of solution convergence or the need to deal with different problems often requires that we search anew for values to use for new challenges. Previous studies have also examined the dynamics that arise from consideration of the update rules but reach a different conclusion. We will present our approach and provide empirical support.

An early exploration of the PSO dynamics [4] considered a single particle in a one-dimension space where the personal and global best locations were taken to be the same. The random vectors in (10.1) were replaced by their averages such that apart from random initialisation, the algorithm was deterministic. As each dimension is updated separately, the system is cast as essentially one-dimensional. Also, as particle interactions only arise from the communication of updated global best values, one can considered the case where the global best is genuinely the optimum result. No updates occur, and the dynamics of a single particle can be studied alone. Varying the parameters was shown to result in a range of periodic motions and divergent behaviour for the case of $\alpha_1 + \alpha_2 \geq 4$, see (10.1). The addition of the random vectors was seen as beneficial as it adds noise to the deterministic search.

Control of velocity, not requiring the enforcement of an arbitrary maximum value as in [4], is derived in an analytical manner by [9]. Here eigenvalues derived from the dynamic matrix of a simplified version of the PSO algorithm are used to imply various search behaviours. Thus, again the $\alpha_1 + \alpha_2 \geq 4$ case is expected to diverge. For $\alpha_1 + \alpha_2 < 4$, various cyclic and quasi-cyclic motions are shown to exist for a non-random version of the algorithm.

In [43], the analysis again considered the dynamics of a single particle in a one-dimensional problem space, using a deterministic version of PSO, setting $\mathbf{R}_1 = \mathbf{R}_2 = 0.5$. The eigenvalues of the system were determined as functions of ω and a combined $\alpha = \alpha_1 + \alpha_2$ (and $\alpha_1 = \alpha_2$). This led to the derivation of three conditions for, respectively, convergence, harmonic oscillations and 'zigzagging' dynamics. The conditions specify regions in the (ω, α)-parameter space which are not fully disjoint. Where parameter values lie in more than one region, we will see behaviours that combine these differing dynamics. For example, oscillatory motions whose amplitudes decrease as the particle converges to a solution etc. As with the preceding papers, the discussion of the random numbers in the algorithm views them purely as enhancing the search capabilities by adding a *drunken walk* to the particle motions. Their replacement by expectation values was thus believed to simplify the analysis with no loss of generality.

Empirical evidence for the role of parameter selection in the capability of the PSO algorithm to location solutions can be carried out. Whilst the algorithm is shaped by three parameters, it is often the case that α_1 and α_2 are assigned the same values. Thus, we can explore the parameter space by considering different pairs of ω and α. Clearly, if the ω value is greater than unity, each particle's velocity will tend to increase on each update. The twin attractions of personal and global best locations will tend to be overcome by the inertial term. The swarm will expand outwards. We can also consider negative ω values. Again values less than negative one will also result in diverging swarms. Typical implementations of PSO usually set α values between 0 and 4. Larger or negative values found to yield divergent swarms. For values within these broad regions, it is not apparent which values yield the best results. Locating good parameter pairs for specific problems often involves a degree of trial and error. We can explore PSO capabilities by repeatedly executing a simple variant of the algorithm for each pair of parameters in the (α, ω) space. With many repetitions and a suitably large iterations budget for each run, results' can be averaged to even out variations in performance seen in individual executions. Figure 10.1 shows the performance landscape for a typical problem function. The (x, y)-plane represents the position in parameter space, and the vertical position shows the average value solution for the algorithm. For positive ω values between zero and one, there appears to be a broadly banana-curve shaped portion of the parameter space that results in the best performances. For negative ω values, the best pairs appear to have a linear relationship. Parameter pairs that perform well occupy the regions in the valley of this surface plot. This pattern consistently emerges over all functions. With different objective functions, the speed with which this picture emerges is variable. Not all locations in the valley are equally good, nor is the variation between the foot of the valley and locations up its slopes equivalent across all functions. Nevertheless, the topology of the performance landscapes for all 28 functions explored has a very similar form, with the notable exception of the 'sphere' function where the algorithm finds the trivial solution also for suboptimal parameters. It is, however, interesting

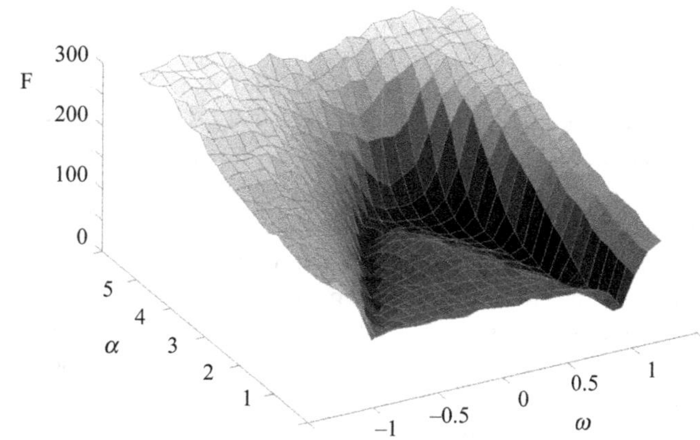

Figure 10.1 Typical PSO performance as a function of its ω and α parameters. Here a 25 particle swarm was run for pairs of ω and α values ($\alpha_1 = \alpha_2$). Cost function here was the d = 10 non-continuous rotated Rastrigin function [48]. Each pair was repeated 100 times and the minimal costs after 2,000 iterations were averaged

that the theoretical analyses mentioned above do not appear to suggest that this curved relationship (for positive ω) should exist between parameters.

The failing of these early papers stems from the treatment of the stochastic elements. Rather than replacing the R_1 and R_2 vectors by 0.5, the dynamic behaviour can be explored by considering their expectation values and variances. An early work doing this produced a predicted best performance region in parameter space similar to the curved valley of best values that is seen empirically [44]. The authors explicitly consider the convergence of means and variances of the stochastic update matrix. The curves they predict marks the locus of (ω, α) pairs they believed guaranteed swarm convergence i.e. parameter values between this line and the origin will result in convergence. Similar approaches yielded matching lines [45] and utilising a weaker stagnation assumption [46]. We will refer to this locus line as the *Jiang* line. This line is displayed for comparison in Figure 10.2.

Experimentally, one can run PSO against a variety of functions and explore the convergence or otherwise of the swarms. This can then be compared with these theoretical approaches. In [45], the comparison is made with very short runs (30 iterations). Their intention being to perform a huge number of repeats in order to confirm the statistics of the expectation and variance beliefs underlying their approach. The algorithm was initialised already in a state of stagnation in order to confirm that convergence occurs as predicted. In [46], iteration budgets were also limited. Their protocol uses 3,100 function evaluations per run (about 155 iterations). They studied a range of parameter pairs both inside and outside the Jiang line. Whilst those within the line do converge, even those outside occasionally achieved good results. It is

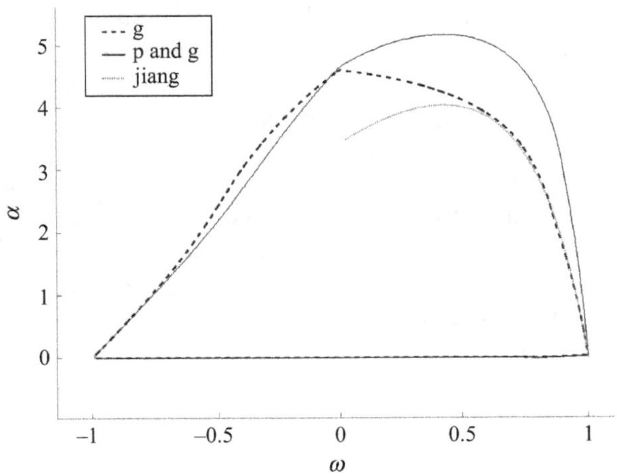

Figure 10.2 *Solution of (10.14) representing a single particle in one dimension with a fixed best value at* $\mathbf{g}=\mathbf{p}=0$. *The curve that has higher* α*-values on the right (solid line) is for* $\alpha_1 = \alpha_2$, *the other curve (dotted line) is for* $\alpha = \alpha_2$, $\alpha_1 = 0$. *Except for the regions near* $\omega = \pm 1$, *where numerical instabilities can occur, a simulation produces an indistinguishable curve. In the simulation, we tracked the probability of a particle to either reach a small region* (10^{-12}) *near the origin or to escape beyond a radius of* 10^{12} *after starting from a random location on the unit circle. Along the curve, both probabilities are equal. The dotted line (inside the dashed curve) is the result predicted by [44] as given in [53] for comparison*

shown in this paper that this is due to the region of convergence extending beyond the Jiang line. As one approaches the locus of parameter values proposed in this paper, one needs exponentially greater iteration counts to ensure convergence. Low numbers of function evaluations will likely result in poor results. In [47], PSO is run with experimental budgets of 2,000 iterations (and 64 particles). Good agreement with the Jiang line is shown. Their problem spaces are 50 dimensional making improved solutions even harder to find. They also (I believe) cap diverging swarms in some manner which will impact the dynamics. With much greater iteration budgets, initially diverging swarms may eventually return to a converging state. This paper shows that with iteration budgets well beyond 2,000, the best performing parameter pairs often lie beyond the Jiang line.

The Jiang line does not mark the division between convergent and divergent swarms. The iterated use of the random factors $\mathbf{R}_1$ and $\mathbf{R}_2$ adds a further level of complexity to the dynamics of the swarm which affects the behaviour of the algorithm in a non-trivial way. Empirical results are shown later than demonstrate the existence of best parameter values outside the Jiang line. For this purpose, we will formulate the PSO algorithm as a random dynamical system and present an analytical solution for the swarm dynamics in a simplified but representative case (Section 10.2) and compare the theoretical predictions to results of simulation for a number of benchmark functions (Section 10.3). Finally, in Section 10.4, we discuss the relation of our model to the original PSO algorithm.

We present an analysis of the quasi-linear swarm dynamics that arise from consideration of the iterated contributions of the random components of the algorithm's dynamic matrix. Our study shows that a region of the (ω,α)-parameter plane results in a division line between convergent and divergent behaviours i.e. the location where a PSO swarm is poised between infinite exploration and collapse to stagnation. The form of the relationship between the parameters that generate this dividing line differs from those suggested in previous analyses. Using an unconstrained swarm we show that empirically the algorithm performs best when approaching this margin.

Deviations from stability represent the ability of the swarm to sample the problem space. Such deviations are possible when poised on the line, yet on average the swarm will return from such deviations. Thus the swarm avoids the two extremes of behaviour: convergence (and its stagnation) and divergence (and the inability to locate good solutions). Near this line, the best performance of PSO is empirically confirmed (for positive ω) to follow a curved relationship between ω and $\alpha = \alpha_1 + \alpha_2$, with $\omega \approx 1$ at small α, and $\alpha \gtrapprox 4$ at small ω. Large values of both α and ω are found to cause the particles to diverge leading to results far from optimality, while at small values for both parameters, the particles converge to a nearby solution which sometimes is acceptable.

For other cost functions, similar relationships are observed in numerical tests (see Section 10.3) unless no good solutions are found due to problem complexity or runtime limits. For trivial cost functions, such as a single well potential in low dimensions, parameter combinations with small ω and small α lead to optimal results.

10.1.4 Matrix formulation

In order to analyse the behaviour of the algorithm, it is convenient to use a matrix formulation by inserting the velocity explicitly in (10.2). If we consider particle states $\mathbf{z} = (\mathbf{v}, \mathbf{x})^T$, we can rearrange the update rules. In (10.3), we stack the updates for a single particle.

$$\begin{pmatrix} \mathbf{v_{t+1}} \\ \mathbf{x_{t+1}} \end{pmatrix} = \begin{pmatrix} \omega \mathbf{v_t} + \alpha_1 R_1 (p - \mathbf{x_t}) + \alpha_2 R_2 (g - \mathbf{x_t}) \\ x_t + \omega \mathbf{v_t} + \alpha_1 R_1 (p - \mathbf{x_t}) + \alpha_2 R_2 (g - \mathbf{x_t}) \end{pmatrix} \tag{10.3}$$

In (10.4), we separate the state dependant parts from those elements that are independent of velocity or position.

$$\begin{pmatrix} \mathbf{v_{t+1}} \\ \mathbf{x_{t+1}} \end{pmatrix} = \begin{pmatrix} \omega & -\alpha_1 R_1 \mathbf{x_t} - \alpha_2 R_2 \mathbf{x_t} \\ \omega & 1 - \alpha_1 R_1 \mathbf{x_t} - \alpha_2 R_2 \mathbf{x_t} \end{pmatrix} \begin{pmatrix} \mathbf{v_t} \\ \mathbf{x_t} \end{pmatrix} + \alpha_1 R_1 \begin{pmatrix} p \\ p \end{pmatrix} + \alpha_2 R_2 \begin{pmatrix} g \\ g \end{pmatrix} \tag{10.4}$$

In terms of our state variable $\mathbf{z}$, this is more generally as shown in (10.5)

$$\mathbf{z}_{t+1} = M\mathbf{z}_t + \alpha_1 \mathbf{R}_1 (\mathbf{p}, \mathbf{p})^\top + \alpha_2 \mathbf{R}_2 (\mathbf{g}, \mathbf{g})^\top \tag{10.5}$$

with

$$M = \begin{pmatrix} \omega \mathbf{I}_d & -\alpha_1 \mathbf{R}_1 - \alpha_2 \mathbf{R}_2 \\ \omega \mathbf{I}_d & \mathbf{I}_d - \alpha_1 \mathbf{R}_1 - \alpha_2 \mathbf{R}_2 \end{pmatrix}, \tag{10.6}$$

where $\mathbf{I}_d$ is the unit matrix in d dimensions. Since the second and third term on the right in (10.5) are constant most of the time, the analysis of the algorithm can focus on the properties of the matrix M. PSO's wide applicability has led to the analyses discussed earlier. These focused either on simplifying the algorithm to make it deterministic or on the expectation and variance values of the update matrix. Here we recognise that the long-term behaviour of the swarm depends upon both the probability distribution of the particles within the z state space and the properties of the infinite product of the stochastic matrix.

10.2 Critical swarm conditions for a single particle

10.2.1 PSO as a random dynamical system

We study the dynamics of the particle swarm by looking at the behaviour of a single particle as done previously in earlier analyses [4,43]. This can be justified because the particles interact only via the shared knowledge of the global best position such that, while $\mathbf{g}$ is unchanged, single particles exhibit qualitatively the same dynamics as in the swarm. For the one-particle case, we have necessarily $\mathbf{p} = \mathbf{g}$, such that shift invariance allows us to set both to zero in (10.5). This gives us the following stochastic-map formulation of the PSO dynamics.

$$\mathbf{z}_{t+1} = M \mathbf{z}_t \tag{10.7}$$

Extending earlier approaches, we will explicitly consider the randomness of the dynamics, i.e. instead of averages over $\mathbf{R}_1$ and $\mathbf{R}_2$, we consider a random dynamical system with dynamical matrices M chosen from the set

$$\mathcal{M}_{\alpha,\omega} = \left\{ \begin{pmatrix} \omega \mathbf{I}_d & -\alpha \mathbf{R} \\ \omega \mathbf{I}_d & \mathbf{I}_d - \alpha \mathbf{R} \end{pmatrix}, \quad \mathbf{R}_{ij} = 0 \text{ for } i \neq j \text{ and } \mathbf{R}_{ii} \in [0,1], \right\} \tag{10.8}$$

with $\mathbf{R}$ being in both rows the same realisation of a random diagonal matrix that combines the effects of $\mathbf{R}_1$ and $\mathbf{R}_2$ (10.2). The parameter α is the sum $\alpha_1 + \alpha_2$ with $\alpha_1, \alpha_2 \geq 0$ and $\alpha > 0$. As the diagonal elements of $\mathbf{R}_1$ and $\mathbf{R}_2$ are uniformly distributed in $[0, 1]$, the distribution of the random variable $\mathbf{R}_{ii} = (\alpha_1/\alpha)\mathbf{R}_{1,ii} + (\alpha_2/\alpha)\mathbf{R}_{2,ii}$ in (10.7) is given by a convolution of two uniform random variables, namely

$$P_{\alpha_1,\alpha_2}(s) = \begin{cases} \frac{\alpha^2 s}{\alpha_1 \alpha_2} & \text{if } 0 \leq s \leq \min\{\frac{\alpha_1}{\alpha}, \frac{\alpha_2}{\alpha}\} \\ \frac{\alpha}{\max\{\alpha_1,\alpha_2\}} & \text{if } \min\{\frac{\alpha_1}{\alpha}, \frac{\alpha_2}{\alpha}\} < s \leq \max\{\frac{\alpha_1}{\alpha}, \frac{\alpha_2}{\alpha}\} \\ \frac{\alpha^2(1-s)}{\alpha_1 \alpha_2} & \text{if } \max\left\{\frac{\alpha_1}{\alpha}, \frac{\alpha_2}{\alpha}\right\} < s \leq 1 \end{cases} \tag{10.9}$$

if the variable $s \in [0, 1]$ and $P_{\alpha_1,\alpha_2}(s) = 0$ otherwise. $P_{\alpha_1,\alpha_2}(s)$ has a tent shape for $\alpha_1 = \alpha_2$ and a box shape in the limits $\alpha_1 \to 0$ or $\alpha_2 \to 0$. Thus, the selection of particular α_1 and α_2 parameters will determine the distribution for the random multiplier $\mathbf{R}_{ii}$ in (10.8).

We expect that the multi-particle PSO is well represented by the simplified version for $\alpha_2 \gg \alpha_1$ or $\alpha_1 \gg \alpha_2$, the latter case being irrelevant in practice. For $\alpha_1 \approx \alpha_2$ deviations from the theory may occur because in the multi-particle case $\mathbf{p}$ and $\mathbf{g}$ will be different for most particles. We will discuss this as well as the effects of the switching of the dynamics at discovery of better solutions in Section 10.4.3.

10.2.2 Stability

Lyapunov exponents provide a means to describe the dynamics of a system. If one considers two particles initially positioned arbitrarily close together and subject to the dynamics of the PSO update rules, we may observe whether the particles move apart or move together as the algorithm runs. Equation (10.10) provides a model of our particles' motions. The λ value captures the nature of the dynamics. For two points at time $t = 0$ a distance δZ_0 apart, we can relate the future separation of these points δZ_t. For swarms where particles are converging, the calculated λ is less than zero. For divergent swarms, λ is greater than zero.

$$\delta Z_t = e^{\lambda t} \delta Z_0 \tag{10.10}$$

Clearly, with each iteration of the PSO algorithm, we will generate a different set of random numbers. These update to our particle states, given by (10.7). The iterated behaviour of our swarm is thus determined by the product of stochastic matrices drawn from set $\mathcal{M}_{\alpha,\omega}$. We need to consider how such products behave on average. Such products have been studied for several decades [49] and have found applications in physics, biology and economics. In terms of stability of the swarm, we wish to determine the α and ω pairs that result in swarms that are neither convergent nor divergent i.e. those that give rise to a Lyapunov exponent $\lambda = 0$. The analysis below derives this, alternatively one may use a numerical approach such as the resampled Monte Carlo method [50].

Whilst a stable swarm does not discover any new solutions, its dynamical properties are determined by an infinite product of matrices from the set $\mathcal{M}_{\alpha,\omega}$ given by (10.8). This provides a convenient way to explicitly model the stochasticity of the swarm dynamics such that we can claim that the performance of PSO is determined by the stability properties of the random dynamical system described by (10.7).

Since (10.7) is linear, the analysis can be restricted to vectors on the unit sphere in the $(\mathbf{v}, \mathbf{x})$ space, i.e. to unit vectors

$$\mathbf{a} = \frac{(\mathbf{x}, \mathbf{v})^\top}{\| (\mathbf{x}, \mathbf{v})^\top \|}, \tag{10.11}$$

where $\| \cdot \|$ denotes the Euclidean norm. In order to determine whether the swarm is on average expanding, contracting or stable, we assess on each iteration whether particles move from the unit circle: outward for divergence, inward for convergence. Unless the set of matrices shares the same eigenvectors (which is not the case here), standard stability analysis in terms of eigenvalues is not applicable. Instead we will use

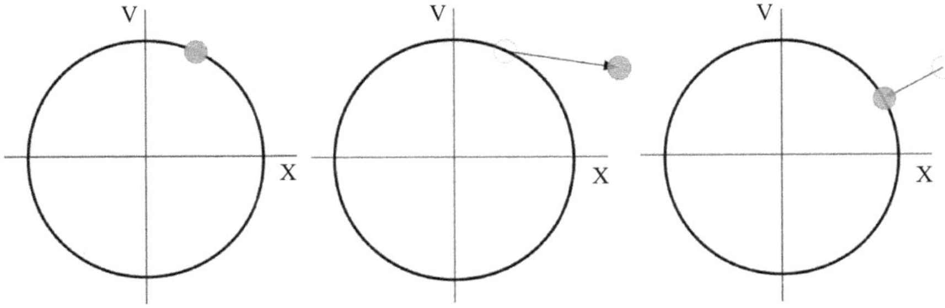

Figure 10.3 *PSO update and projection onto unit circle process. Here we visualise the process of a single particle in one dimension. Left panel shows a single particle sitting on the unit circle in this state space. Centre panel applying PSO update rules results in the particle moving to a new state which is not on the unit circle. If the particle moves out from the circle it is expanding, if inward then contracting. Right panel the particle is re-projected back to the unit circle ready for the next update. The process thus results in the particle remaining a single unit away from the state space origin, but its position on the unit circle may change on each update*

means from the theory of random matrix products in order to decide whether the set of matrices is stochastically contractile. Figure 10.3 visualises the process. A particle in the PSO swarm may be shown on a unit circle in the system's velocity: position state space. On each iteration, the algorithm's update rules move the particle elsewhere in this space. If the particle moves outward from the unit circle, it is contributing to expansion of the swarm (as shown in the figure). If moving within the circle, it is contracting. To assess the behaviour of the whole swarm, we will wish to sum these individual moves. At the end of the iteration, we project the particles back to the unit circle. The process is repeated.

The resultant effect of the update shown in Figure 10.3 was to shift the position of the particle on the unit circle. Other particles, that started elsewhere, may have finished in other locations. Any given update of particles in a swarm's state space will result in some particles moving outward, or inward from the unit circle. The likelihood of these moves is dependent upon where on the unit sphere a particle lies prior to the update being applied. In order to assess the whole effect, we therefore need to determine the distribution of particles on the unit circle under the dynamics of the update rules and the specific parameter values. Whilst updates are stochastic, some portions of the state space act somewhat like attractor states i.e. particles in these portions of the unit circle are likely to remain their perhaps only rare matrix updates will appreciably shift them elsewhere. Figure 10.4 visualises this for a single iteration of a small swarm. The particles in the top right are moved outward during the shown update. Their re-projection will return them to a similar location. The remaining particles move inward, by a larger amount. On this iteration, the sum of these moves may suggest the swarm is perhaps contracting a little. However, the particle in the

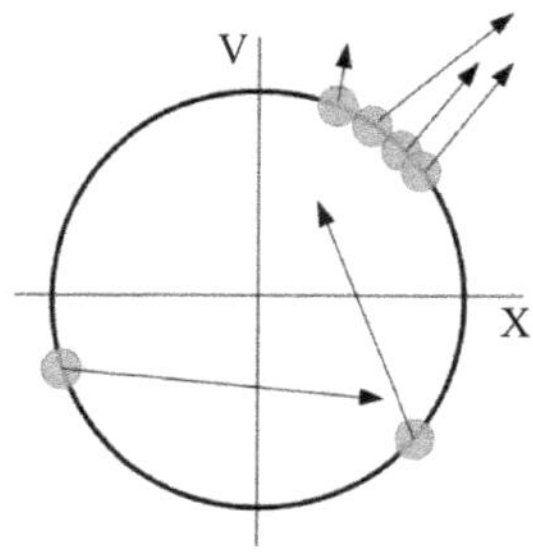

Figure 10.4 *PSO update and projection onto unit circle process. Here we visualise a number of particles on the unit circle. To assess whether the whole swarm is expanding or contracting, we need to consider the effects of both the location on the unit circle that each particles lies, and the dynamics of the PSO algorithm as applied to a particle at that location. Here the PSO update rules are moving the lower particles closer to the origin, so they are contractile for those particles. However, the upper particles are moving outward representing a divergent move. Summing the sizes and directions of these moves give an indication of the effect on the swarm for this iteration. However, we can see that when the moved particles are re-projected onto the unit circle, there will now be five particles in the top right area*

bottom right will be re-projected to join the main group of particles. If this upper left location has an overall tendency to be expansive, on subsequent iterations more and more particles may be contributing more divergent behaviour to the overall swarm.

We can estimate the stationary distribution of particles on the unit circle, $\nu_{\alpha,\omega}(\mathbf{a})$, by the following process. A number of particles, $\mathscr{N}_p$, are placed around our unit circle. For each particle, we create a set of new particle locations by applying the update (10.7) $\mathscr{N}_u$ times. Each update is re-projected onto the unit circle [as per (10.11)]. This gives us $\mathscr{N}_{new} = \mathscr{N}_p \mathscr{N}_u$ new particles. We can then randomly sample $\mathscr{N}_p$ of these and repeat the process. Different PSO parameter values result in different stochastic update matrices ($\mathscr{M}_{\alpha,\omega}$). These yield different stationary distributions of the particles. Figure 10.5 shows a number of these. Equation (10.12) expresses this process for state space of dimension d.

$$\nu_{\alpha,\omega}(\mathbf{a}) = \int d\nu_{\alpha,\omega}(\mathbf{b}) \int dP_{\alpha,\omega}(M)\, \delta\left(\frac{\mathbf{a}, M\mathbf{b}}{\|M\mathbf{b}\|}\right), \quad \mathbf{a}, \mathbf{b} \in S^{2d-1}. \tag{10.12}$$

The existence of the invariant measure requires the dynamics to be ergodic which is ensured if at least some of elements of $\mathscr{M}$ have complex eigenvalues, such as being the case for $\omega^2 + \alpha^2/4 - \omega\alpha - 2\omega - \alpha + 1 < 0$ (see above, [43]). This condition excludes a small region in the parameters space at small values of ω, such that there we have to take all ergodic components into account. There are not more than two components which due to symmetry have the same stability properties. It depends on the parameters α and ω and differs strongly from a homogeneous distribution, see Figure 10.5 for a few examples in the case $d = 1$.

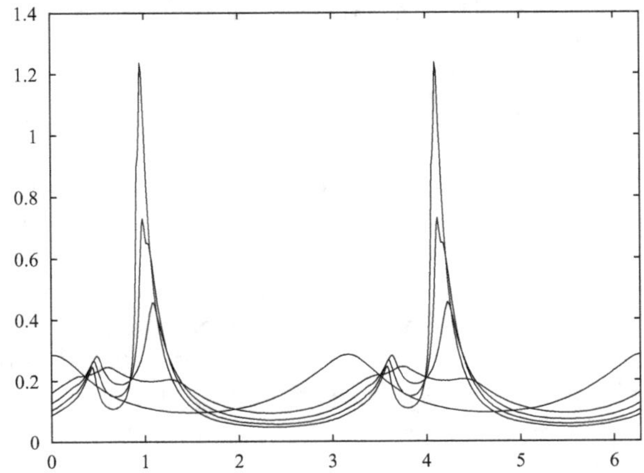

Figure 10.5 *Stationary distribution $\nu_{\alpha,\omega}(\mathbf{a})$ on the unit circle ($\mathbf{a} \in [0,2\pi)$) in the (x, v) plane for a one-particle system (10.7) for $\omega = 0.7$ and $\alpha = \alpha_1 + \alpha_2 = 0.5,\ 1.5,\ 2.5,\ 3.5,\ 4.5$ (the distribution with peak near π is for $\alpha = 0.5$, otherwise main peaks are highest for largest α)*

The properties of the asymptotic dynamics of the PSO swarm can be described based on a double Lebesgue integral over the unit sphere S^{2d-1} and the set $\mathcal{M}$ [51,52]. As with Lyapunov exponents, the effect of the dynamics is measured in logarithmic units in order to account for multiplicative action.

$$\lambda\,(\alpha,\omega) = \int d\nu_{\alpha,\omega}\,(\mathbf{a}) \int dP_{\alpha,\omega}\,(M)\ \log \|M\mathbf{a}\| \tag{10.13}$$

If $\lambda\,(\alpha,\omega)$ is negative, the algorithm will converge with probability 1, while for positive λ, arbitrarily large fluctuations are possible. While the measure for the inner integral in (10.13) is the distribution of the random multiplier as given earlier in (10.9), the stationary distribution ν on the unit sphere for the outer integral was given in (10.12).

Critical parameters are obtained from (10.13) by the relation

$$\lambda\,(\alpha,\omega) = 0. \tag{10.14}$$

Solving (10.14) is difficult in higher dimensions, so we rely on the linearity of the system when considering the ($d = 1$)-case as representative. Figure 10.2 represents solutions of(10.14). The figure shows three lines, representing three distinct solutions to this problem. Parameter pairs that lie inside a given line given rise to swarms whose largest Lyapunov exponent is less than zero. These swarms are therefore stable in the sense will tend to converge or become static. For parameter pairs outside each line, the swarm generated will have a largest Lyapunov exponent greater than zero and will tend to explode. PSO experiments use finite iteration counts, so any individual trial may yield a set of random matrices whose product may generate a behaviour that for the limited nature of a single experiment differs from this theoretical approach. Averaged, though, we would expect this picture to emerge. The set of parameter pairs on the line result in a critical swarm, whose deviations are not described by either convergence or divergence.

The solid black curve in Figure 10.2 represents the solution for $d = 1$, $\alpha = \alpha_1 + \alpha_2$ and $\alpha_1 = \alpha_2$. The black dotted line is the solution for $d = 1$, $\alpha = \alpha_2$ and $\alpha_1 = 0$. This results in a larger variance rendering the dynamics less stable. This results in the locus of stability moving inwards, towards the origin somewhat. The dotted line touching the dashed in from below shows the predicted curve of stability [44], and given as a polynomial in [53]. Our approach generates the upper two lines (dashed and solid) depending on whether the personal and global best locations are coincident or not. Where they differ the swarm remains stable for more of the parameter space. Experimentally, it is shown below that the outer lines, rather than the Jiang line, represent the limit parameter pairs that lead to convergent swarms.

Inside the contour $\lambda\,(\alpha, \omega)$ is negative, meaning that the state will converge with probability 1. Along the contour and in the outside region, large state fluctuations are possible. Interesting parameter values are expected near the curve where due to the coexistence of stable and unstable dynamics (induced by different sequences of random matrices) a theoretically optimal combination of exploration and exploitation is possible. For specific problems, however, deviations from the critical curve can be expected to be beneficial. In order to solve a practical problem, there is usually a finite iteration budget for the algorithm. In that case, it is beneficial to allow the swarm to converge at some point. Thus, the swarm being somewhat sub-critical, to allow such a convergence is desirable. The degree to which we should allow this is a function of the iteration budget and is also related to the nature of the problem space being explored.

10.2.3 Personal best vs. global best

Due to linearity, the particle swarm update rule (10.2) is subject to a scaling invariance which was already used in (10.11). We now consider the consequences of linearity for the case where personal best and global best differ, i.e. $\mathbf{p} \neq \mathbf{g}$. For an interval where $\mathbf{p}_i$ and $\mathbf{g}$ remain unchanged, the particle i with personal best $\mathbf{p}_i$ will behave like a particle in a swarm where together with $\mathbf{x}$ and $\mathbf{v}$, $\mathbf{p}_i$ is also scaled by a factor $\kappa > 0$. The finite-time approximation of the Lyapunov exponent [see (10.13)]

$$\lambda(t) = \frac{1}{t} \log \langle \| (\mathbf{x}_t, \mathbf{v}_t) \| \rangle \tag{10.15}$$

will be changed by an amount of $(1/t) \log \kappa$ by the scaling. Although this has no effect on the asymptotic behaviour, we will have to expect an effect on the stability of the swarm for finite times which may be relevant for practical applications. For the same parameters, the swarm will be more stable if $\kappa < 1$ and less stable for $\kappa > 1$, provided that the initial conditions are scaled in the same way. Likewise, if $\|\mathbf{p}\|$ is increased, then the critical contour will move inwards, see Figure 10.6. Note that in this figure, the low number of iterations leads to a few erroneous trials at parameter pairs outside the outer contour which have been omitted here. We also do not consider the behaviour near $\alpha = 0$ which is complex but irrelevant for PSO. The contour (10.14) can be seen as the limit $\kappa \to 0$ such that only an increase of $\|\mathbf{p}\|$ is relevant for comparison with

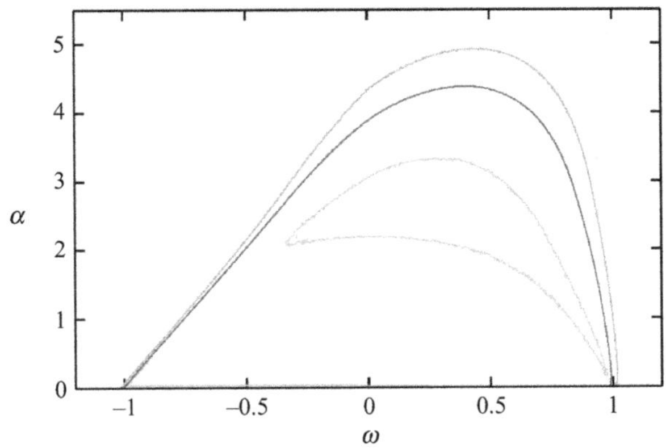

Figure 10.6 *For* $\mathbf{p} \neq \mathbf{g}$*, we define neutral stability as the equilibrium between divergence and convergence. Convergence means here that the particle approaches the line connecting* **p** *and* **g***. Curves are for a one-dimensional problem with* $\mathbf{p} = 0.1$ *and* $\mathbf{g} = 0$ *scaled (see Section 10.2.3) by* $\kappa = 1$ *(outer curve),* $\kappa = 0.1$ *and* $\kappa = 0.04$ *(inner curve). For small values of* κ *the algorithm tends to lose stability (inner curve). Results are for 200 iterations and averaged over 100,000 repetitions*

the theoretical stability result. When comparing the stability results with numerical simulations for real optimisation problems, we will need to take into account the effects caused by differences between **p** and **g** in a multi-particle swarm with finite runtimes.

10.3 Simulations

Meta-heuristic algorithms are often tested in competition against benchmark functions designed to present different problem space characteristics. To confirm the stability results above, testing against the 28 functions defined for the 2013 International Conference on Evolutionary Computation [48] was performed. This contains a mix of unimodal, basic multimodal and composite functions designed to provide a range of challenges to various meta-heuristic algorithms. In the competition, the domain of the functions in this test set are all defined to be $[-100, 100]^d$ where d is the dimensionality of the problem. We retain this only as the spatial region within which the particles are initialised. We use ten-dimensional problems throughout. Our implementation of PSO performed no spatial or velocity clamping if the swarm is reasonably well behaved. Particles are free to travel where they wish so long as the they are not clearly diverging. We define this as the case where absolute particle positions or velocities are greater than 10^{12}. In this way, we ensure minimal interference with the dynamics of the particles in motion. In all trials, a swarm of 25 particles was used. We

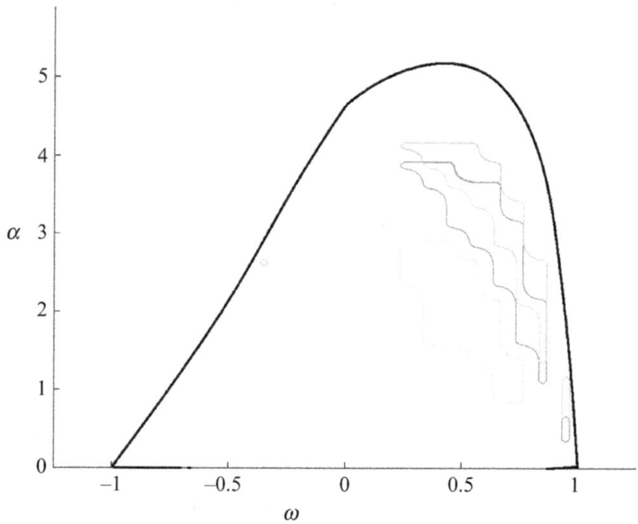

Figure 10.7 The 5% best parameter regions for 20 (light grey), 200 (dark grey) and 2,000 (medium grey) iterations. As we increase the number of iterations performed, the best performing parameter pairs shifts towards the critical line. Functions costs were averaged 100 runs and all 28 CEC benchmark functions. The line of zero Lyapunov exponent for $N = 1$, $d = 1$, $\alpha = \alpha_1 + \alpha_2$, and $\alpha_1 = \alpha_2$ is shown in black

repeated the algorithm 100 times, on each occasion allowing 20, 200, 2,000 iterations to pass before recording the best solution found by the swarm. For the competition, 50,000 fitness evaluation were allowed which corresponds to 2,000 iterations with 25 particles. Other iteration numbers were included for comparison. This protocol was carried out for pairs of $\omega \in [-1.1, 1.1]$ and $\alpha \in [0, 6]$. This was repeated for all 28 functions. The averaged solution costs as a function of the two parameters showed curved valleys similar to that in Figure 10.1 for all problems. For each function, we obtain different best values along (or near) the theoretical curve defined by the stability condition (10.14). There appears to be no preferable location within the valley. Some individual functions yield best performance near $\omega = 1$. This is not the case near $\omega = 0$, although the global average performance overall test functions is better in the valley near $\omega = 0$ than near $\omega = 1$, see Figure 10.7.

10.3.1 Combining the results from multiple cost functions

Whilst the stochastic map, (10.7) defined the essential dynamics of the system, a real swarm with many particles and many shifts in their p_i and g values will cause small deviations that may make interpretation of results less clear. These effects will differ for different problems. Exploration of a sphere function will result in many early shifts in the particle knowledge terms leading to quick agreement of a

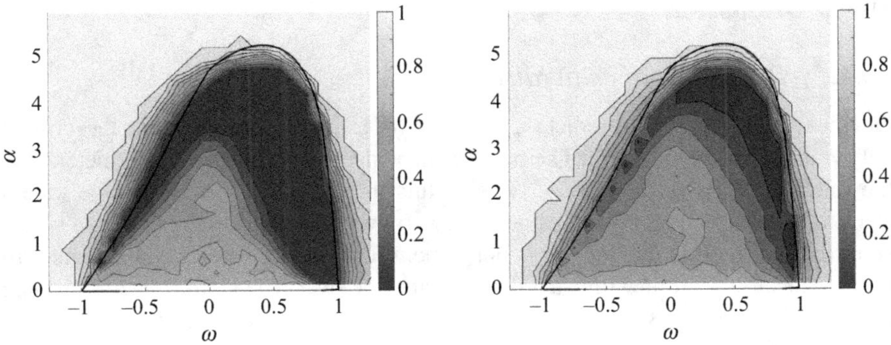

Figure 10.8 Cost function values rescaled using estimate of average values. Left hand image is for a Sphere function. This cost function returns values that range over many decades of scale. Right hand image is of non-continuous rotated Rastrigin's function. This cost function returns a much more restricted range of values. Here, each function's average has been estimated and used to scale the results. Both now show results in approximately in the range [0,1]. For each function, our PSO algorithm is run 100 times at each (ω, α) pair. Each run lasted 2,000 iterations. Both retain the characteristic shape showing the best performing parameter pairs

solution. It is less easy to envisage the effects of more complex function surfaces. Therefore, it is useful to average these effects by combining the 28 functions used. The potential range of cost values returned by each function varies widely. The Sphere function may return values whose magnitudes may vary over many decades. Other functions have differences between minimum and maximum values of only one or two decades. Simply averaging the results from these will result in the values from poorly performing swarms acting on a Sphere problem swamping the result.

To more fairly combine the results, an estimate of the average value of each function is used to normalise the results obtained for each problem function. The following procedure was used: each function was randomly sampled within the $[-100, 100]^d$ spatial domain. These data were used to estimate an average value of each function. These average values, V_{ave}, are then used to scale the results. This is not quite normalising into the range [0, 1] as poorly performing swarms may in fact do worse that the average of random sampling. In practise, results are broadly scaled to similar ranges, see Figure 10.8.

$$V' = \frac{(V - V_{min})}{(V_{ave} - V_{min})} \tag{10.16}$$

For each α, ω pair, we have an average of 100 executions of the PSO algorithm. This forms the set of results V, the smallest of which is V_{min}. V' is the normalised set of results.

10.4 Discussion

10.4.1 Relevance of criticality

Our analytical approach predicts a locus of (α, ω)-pairs that maintain the critical behaviour of the PSO swarm. The precise shape depends upon these parameters and the relative values of α_1 and α_2. Outside this line, the swarm will diverge unless steps are taken to constrain it. Inside, the swarm will eventually converge to a single solution. On the line, the swarm is posed between divergence and convergence. In order to locate a solution within a problem space, the swarm needs to converge at some point, so the line represents an upper bound on the exploration–exploitation mix that a swarm manifests. For parameters on the critical line, fluctuations are still arbitrarily large. Therefore, sub-critical parameter values can be preferable if the settling time is of the same order as the scheduled runtime of the algorithm. Additionally, if a typical length scale of the problem is known, then the finite standard deviation of the particles in the stable parameter region can be used to decide about the distance of the parameter values from the critical curve. These dynamical quantities can be approximately set, based on the theory presented here, such that a precise control of the behaviour of the algorithm is in principle possible.

The observation of the distribution of empirically optimal parameter values along the critical curve confirms the expectation that critical or near-critical behaviour is the main reason for success of the algorithm. Critical fluctuations are a plausible means to perform well in a search problem if little is known about the cost landscape apart from perhaps certain smoothness assumptions. The majority of excursions will exploit the smoothness of the cost function by local search, whereas the fat tails of the distribution allow the particles to escape from local minima.

10.4.2 Comparison with earlier explanations

In Figure 10.2, critical loci of parameter values are plotted. These, as discussed, are derived from the treatment of PSO as a random dynamical system. Two lines are plotted in black: one (inner) derived for the case of $\alpha = \alpha_1 + \alpha_2$, and $\alpha_1 = \alpha_2$; the other (outer) for the case of $\alpha = \alpha_2$, and $\alpha_1 = 0$. We also plot a line (dotted) showing an oft cited earlier analysis of swarm stability [44]. Empirically, we can show that this latter solution must be wrong.

10.4.2.1 Lines of stability when viewed against average cost function results

Following the procedure in Section 10.3.1 allows us to average the solutions achieved for the 28 functions used. Figure 10.9 shows where in parameter space the best results can be obtained (on average) across all functions. With lower iteration budgets available to the algorithm, better results occur where the swarm is able to converge early and exploit whatever knowledge is gained early. As we allow greater iterations, the algorithm, as expected, benefits from slower convergence. This figure also shows the stability lines predicted in this paper and that previously presented by [44]. For the lower iteration counts, all the good locations occur within both lines. With 2,000

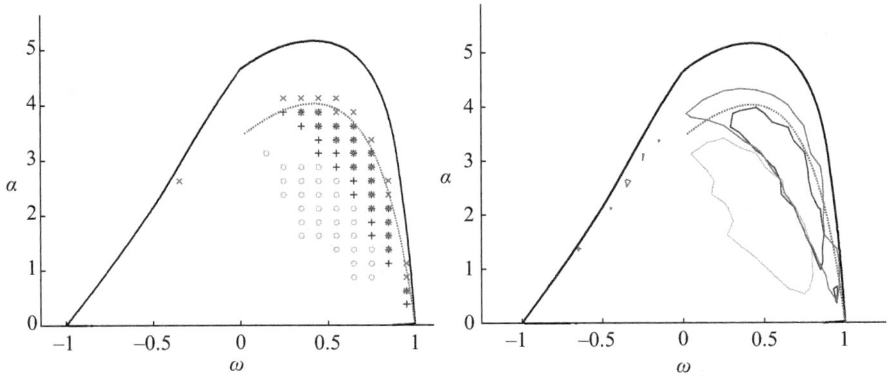

Figure 10.9 Average cost function value as function of iteration budget. Iteration budgets of 20 (light gray circles), 200 (dark gray pluses) and 2,000 (medium gray crosses) are shown. The points in the left hand figure and the contours in the right hand image represent the 5% best parameter pairs. The locus of stable parameter pairs as predicted by this paper (black solid line) and Jiang (gray dotted line) are shown

iterations, some of the best locations sit outside the Jiang line. These lines represent the point beyond which we would expect divergence in the PSO swarm. We would expect that good solutions should not occur beyond these lines. Of course as the algorithm is stochastic, it is feasible that some runs with parameters taken from outside the line may result in good solutions. However, this result is suggestive that the Jiang line may not be a good solution for stability prediction.

10.4.2.2 Lines of stability when viewed against individual cost function results

Average results for PSO run against 28 cost functions for 2,000 iteration suggest that the Jiang stability line does not contain all the best parameter pairs. Looking at the best locations for individual functions demonstrates that variability exists between cost functions. Certain problems yield good solutions for parameter pairs further within the stable region than others. The effect of averaging over all cost functions will thus tend to move the best locations further within the stable region. The Jiang line will tend to look like a better solution when viewed on average rather than against more extreme results.

Yet the upper limit of stability should be the same for all functions. We can look at specific functions to highlight this. Figure 10.10 shows that for some functions, the location of best results in the parameter space is outside the Jiang line.

10.4.2.3 Lines of stability when viewed against detailed cost function results

Jiang's result appears at odds with these empirical results. The critical locus predicted by treating PSO as a random dynamical system appears to be consistent with the

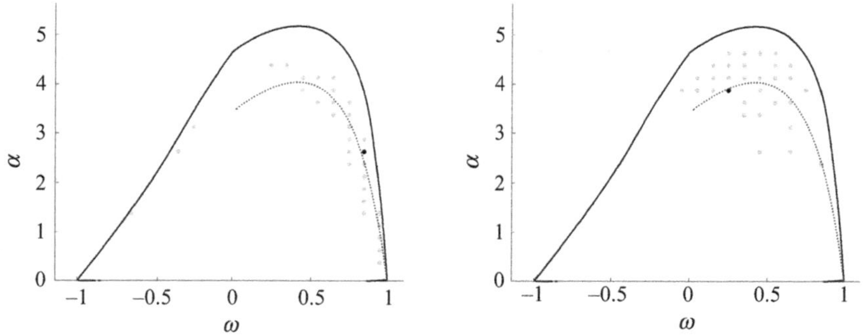

Figure 10.10 Specific cost function value as function of iteration budget. PSO is run for 2,000 iterations against CEC2013 functions 7 (left) and 21 (right). 5% best parameter pairs are shown (gray dots), best location (black dot). As elsewhere, this paper's stability line is shown as a black solid line and Jiang's as a gray dotted line

results so far seen. All best results lie on parameter pairs within the locus of stability predicted. We would expect that the as iteration budget is increased, we would continue to move away from the Jiang line but remain constrained by the line our treatment predicts.

To explore this, we chose two functions (7 and 21 in the CEC2013 set) and ran the PSO algorithm for parameter pairs along a slice of the parameter space lying on $\omega = 0.55$. The α values were varied with a much finer granularity from 2 to 6. Again 2,000 repetitions were performed and iteration budgets of 20, 200, 2,000 and 20,000 were used.

Figure 10.11 shows the result. With increasing iteration budget, we appear to get closer to the stability limit and achieve better results. It is also clear that results are better outside Jiang's predicted line.

10.4.3 Switching dynamics at discovery of better solutions

Equation (10.5) shows that the discovery of a better solution affects only the constant terms of the linear dynamics of a particle, whereas its dynamical properties are governed by the linear coefficient matrices. However, in the time step after a particle has found a new solution, the corresponding force term in the dynamics is zero [see (10.2)] such that the particle dynamics slows down compared to the theoretical solution which assumes a finite distance from the best position at all (finite) times. As this affects usually only one particle at a time and because new discoveries tend to become rarer over time, this effect will be small in the asymptotic dynamics, although it could justify the empirical optimality of parameters in the unstable region for some test cases.

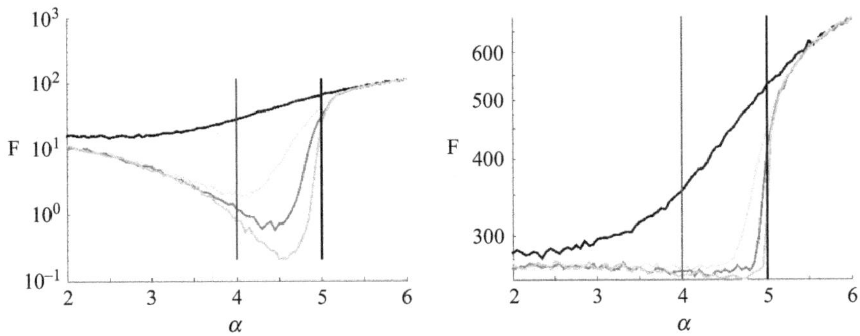

Figure 10.11 Detailed cost function value as function of iteration budget. PSO is run for (top to bottom) 20 (black), 200 (light gray), 2000 (dark gray) and 20,000 (medium gray) iterations against CEC2013 functions 7 (left) and 21 (right). The parameter $\omega = 0.55$ throughout, α varies from 2 to 6. As elsewhere, this papers' stability line is shown as a black solid lines and Jiang's as a gray dotted line. Note that a logarithmic scale is used

The question is nevertheless, how often these changes occur. A weakly converging swarm can still produce good results if it often discovers better solutions by means of the fluctuations it performs before settling into the current best position. For cost functions that are not 'deceptive', i.e. where local optima tend to be near better optima, parameter values far inside the critical contour (see Figure 10.2) may give good results, while in other cases, more exploration is needed. In solving a particular problem, knowledge of the iterative budget to be allowed should assist to gauge to what degree we should wish the swarm should behave sub-critically.

10.4.4 The role of personal best, present best and best ever

A numerical scan of the (α_1, α_2) plane shows a valley of low-fitness values, which, at small fixed ω, is roughly linear and described by the relation $\alpha_1 + \alpha_2 = \text{const}$, i.e. only the joint parameter $\alpha = \alpha_1 + \alpha_2$ matters. For large ω, and accordingly small predicted optimal α values, the valley is less straight. This may be because the effect of the known solutions is relatively weak, so the interaction of the two components becomes more important. In other words, if the movement of the particles is mainly due to inertia, the relation between the global and local best is non-trivial, while at low inertia, the particles can adjust their $\mathbf{p}$ vectors quickly towards the $\mathbf{g}$ vector such that both terms become interchangeable.

At medium values of ω, the difference between the analytical solutions for $\alpha_1 = \alpha_2$ and for $\alpha_1 = 0$ is strongest, see Figure 10.7. In simulations, this shows to a lesser extent revealing a shortcoming of the one-particle approximation. Because in the multi-particle case, $\mathbf{p}$ and $\mathbf{g}$ are often different, the resulting vector will have a smaller norm than in the one-particle case, where $\mathbf{p} = \mathbf{g}$. The case $\mathbf{p} \neq \mathbf{g}$ violates

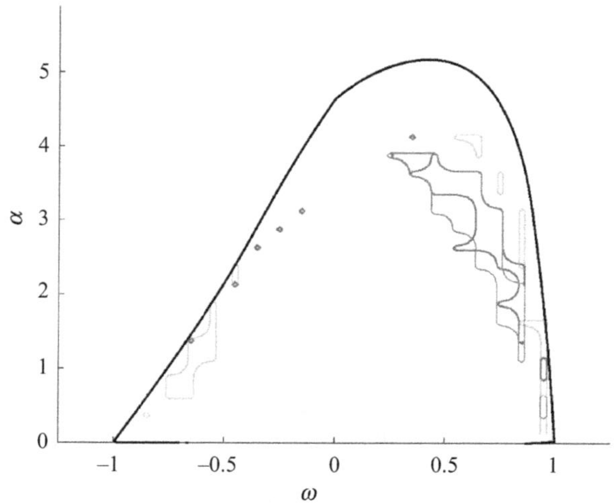

Figure 10.12 The 5% (left) best parameter regions for 2D (light gray), 5D (dark gray) and 10D (medium gray) problem spaces: for lower dimensional problem spaces, the region shifts towards the critical line. Cost averaged over 100 runs and 28 CEC benchmark functions. Zero Lyapunov exponent for $N = 1$, $d = 1$, $\alpha_1 = \alpha_2$ (black)

the assumption of that the dynamics can be described based unit vectors. While a particle far away from both **p** and **g** will behave as predicted from the one-particle case, at length scales smaller than $\|\mathbf{p} - \mathbf{g}\|$, the contractile forces will tend to be reduced such that the inertia becomes more effective and the particle is locally less stable which shows numerically in optimal parameters that are smaller than predicted.

Using the best present position in the swarm as **g** position instead of the best position found so far is sometimes suggested by pointing to the fact that at least for one particle, the personal best position **p** and the global best coincide. If $\alpha_1 \approx \alpha_2$, this particle will be subject to fluctuations due to constantly changing **g** and will not affect other particles unless the dynamics is strongly convergent. Therefore, the dynamics are either not sufficiently exploratory, or the search does not benefit from collaboration among the particles. The drawbacks of setting **g** to the present best are also confirmed in simulations.

Finally, we should mention that more particles, longer runtime (as shown earlier) as well as lower search space dimension increase the potential for exploration. They all lead to the empirically determined optimal parameters being closer to the critical curve than in the opposite cases where search in a relatively larger space benefits from a local search such that smaller parameters are favoured. The effect of lower spatial dimension can be seen in Figure 10.12. The best parameter pairs move closer to critical line as the problem space is reduced.

10.5 Conclusion

PSO is a widely used optimisation scheme which is theoretically not well understood. Historically, its inherent stochasticity is handled via deterministic simplifications or by consideration of expectation values and variance of the random elements. It is known that the standard PSO algorithm requires parameters to be set to ensure good performance for a specific problem. However, correctly choosing these parameter values for a given problem can be difficult. It is shown here that the system is more correctly understood by casting PSO as a random dynamical system. Analysis of this system shows that there exists a locus of (ω,α) pairs that result in the swarm behaving in a critical manner. These may be plotted as a line in parameter space. The PSO update rules are rewritten in matrix form. The behaviour of the swarm as runtime tends to infinity are then studied considering the properties of the infinite product of these stochastic matrices. Outside this line, the swarm will diverge unless steps are taken to constrain it. Inside, the swarm will eventually converge to a single solution.

This predicted locus of stability differs from the previous approaches. Empirical evidence is presented to support the correctness of the random dynamical system approach.

For parameters on the critical line, fluctuations are still arbitrary large. Therefore, sub-critical parameter values can be preferable for practical solutions to be located. PSO is usually run with time or iteration constraints, so it makes sense to use parameters that lie within the critical line. This ensures convergence to some solution before the execution finishes. The shorter the run or the fewer the iterations, the more sub-critical the swarm needs to be to find good solutions. Empirically, this is confirmed.

The location of the critical line differs for particles which have personal best locations equal to the global best. This lies within the critical line of the more general case. Swarms necessarily contain particles of both types (at least early in the swarms evolution), so optimum behaviour would be somewhere between the two. The nature of the problem space will also play a role here. Functions with many local minima, distantly separated, will tend in swarms that have particles with different personal and global bests for a longer time.

Real world problems tend to present as a black box function. Thus, we can't know if we are achieving good solutions. It would be useful to be able to be confident in our selection of parameter values such that we know that the solutions we achieve are at least the best the algorithm and this function will yield. Given problem dimensionality, swarm size and the number of iterations allowed this may be achievable. This could be combined with techniques such as exploratory landscape analysis attempt to gauge the nature of the problem space's topology given the values already sampled from its landscape.

It would be interesting to explore whether other meta-heuristics that employ stochasticity can be similarly analysed as random dynamical systems. Does such analysis predict similar critical regions in their parameter space? If so, do empirical results in such regions show optimal behaviour?

To summarise, PSO and its variants enjoy widespread application in locating optimal or near optimal solutions to complex problems. However, the quality of solutions found depend upon our ability to overcome the algorithm's drawbacks: the selection of suitable parameter values and the avoidance of the swarm's premature convergence. Previous stability analysis suggestion regions of parameter space that guarantee convergence. It was shown here that these regions are too conservative. We consider the quasi-linear swarm dynamics that arise from consideration of the random components of the algorithm's dynamic matrix. We find a relationship between the algorithm's parameters at the edge between convergent and divergent swarm behaviours. Empirically, it is shown that as the iteration budget of the algorithm is increased, the best parameter values exist outside the previously assumed stable region but remain within our region. The algorithm can be seen experimentally to operate at its best towards this margin.

Acknowledgement

This work was supported by the Engineering and Physical Sciences Research Council, grant number EP/K503034/1.

References

[1] Riccardo Poli. Analysis of the publications on the applications of particle swarm optimisation. *Journal of Artificial Evolution and Applications*, 2008:3, 2008.

[2] James Kennedy and Russell Eberhart. Particle swarm optimization. In *Neural Networks, 1995. Proceedings., IEEE International Conference on*, volume 4, pages 1942–1948. IEEE, 1995.

[3] Daniel Bratton and James Kennedy. Defining a standard for particle swarm optimization. In *Swarm Intelligence Symposium, 2007. SIS 2007. IEEE*, pages 120–127. IEEE, 2007.

[4] James Kennedy. The behavior of particles. In V. W. Porto, N. Saravanan, D. Waagen, and A. E. Eiben, editors, *Evolutionary programming VII*, pages 579–589. Springer, 1998.

[5] Chukiat Worasucheep. A particle swarm optimization with stagnation detection and dispersion. In *Evolutionary Computation, 2008. CEC 2008. (IEEE World Congress on Computational Intelligence). IEEE Congress on*, pages 424–429. IEEE, 2008.

[6] Stefan Janson and Martin Middendorf. On trajectories of particles in PSO. In *Swarm Intelligence Symposium (SIS 2007)*, pages 150–155. IEEE, 2007.

[7] William M. Spears, Derek T. Green, and Diana F. Spears. Biases in particle swarm optimization. *International Journal of Swarm Intelligence Research (IJSIR)*, 1(2):34–57, 2010.

[8] Jing J. Liang, A. Kai Qin, Ponnuthurai N. Suganthan, and S. Baskar. Comprehensive learning particle swarm optimizer for global optimization of multimodal functions. *IEEE Transactions on Evolutionary Computation*, 10(3):281–295, 2006.

[9] Maurice Clerc and James Kennedy. The particle swarm-explosion, stability, and convergence in a multidimensional complex space. *IEEE Transactions on Evolutionary Computation*, 6(1):58–73, 2002.

[10] Yuhui Shi and Russell C. Eberhart. Empirical study of particle swarm optimization. In *Evolutionary Computation, 1999. CEC 99. Proceedings of the 1999 Congress on*, volume 3. IEEE, 1999.

[11] Jagdish Chand Bansal, Praveen Kumar Singh, Mukesh Saraswat, Abhishek Verma, Shimpi Singh Jadon, and Ajith Abraham. Inertia weight strategies in particle swarm optimization. In *Nature and Biologically Inspired Computing (NaBIC), 2011 Third World Congress on*, pages 633–640. IEEE, 2011.

[12] Jacques Riget and Jakob S. Vesterstrøm. A diversity-guided particle swarm optimizer—the ARPSO. *EVALife Technical Report no. 2002-02*, 2002.

[13] Souma Chowdhury, Weiyang Tong, Achille Messac, and Jie Zhang. A mixed-discrete particle swarm optimization algorithm with explicit diversity-preservation. *Structural and Multidisciplinary Optimization*, pages 1–22, 2013.

[14] Alberto García-Villoria and Rafael Pastor. Introducing dynamic diversity into a discrete particle swarm optimization. *Computers & Operations Research*, 36(3):951–966, 2009.

[15] Xin-She Yang. A new metaheuristic bat-inspired algorithm. In *Nature Inspired Cooperative Strategies for Optimization (NICSO 2010)*, pages 65–74. Springer, Berlin, Heidelberg, 2010.

[16] Xin-She Yang and Suash Deb. Cuckoo search via Lévy flights. In *Nature & Biologically Inspired Computing, 2009. NaBIC 2009. World Congress on*, pages 210–214. IEEE, 2009.

[17] Cui-Ru Wang, Chun-Lei Zhou, and Jian-Wei Ma. An improved artificial fish-swarm algorithm and its application in feed-forward neural networks. In *Machine Learning and Cybernetics, 2005. Proceedings of 2005 International Conference on*, volume 5, pages 2890–2894. IEEE, 2005.

[18] Bo Liu, Ling Wang, Yi-Hui Jin, Fang Tang, and De-Xian Huang. Improved particle swarm optimization combined with chaos. *Chaos, Solitons & Fractals*, 25(5):1261–1271, 2005.

[19] A. Kai Qin and Ponnuthurai N. Suganthan. Self-adaptive differential evolution algorithm for numerical optimization. In *IEEE Congress on Evolutionary Computation*, volume 2, pages 1785–1791. IEEE, 2005.

[20] Yu Wang, Bin Li, Thomas Weise, Jianyu Wang, Bo Yuan, and Qiongjie Tian. Self-adaptive learning based particle swarm optimization. *Information Sciences*, 181(20):4515–4538, 2011.

[21] Zhi-Hui Zhan, Jun Zhang, Yun Li, and HS-H Chung. Adaptive particle swarm optimization. *IEEE Transactions on Systems, Man, and Cybernetics, Part B: Cybernetics*, 39(6):1362–1381, 2009.

[22] Maurice Clerc. *Particle swarm optimization*, volume 93. John Wiley & Sons, 2010.

[23] Nikolaus Hansen and Andreas Ostermeier. Completely derandomized self-adaptation in evolution strategies. *Evolutionary Computation*, 9(2):159–195, 2001.

[24] Ilya Loshchilov. CMA-ES with restarts for solving CEC 2013 benchmark problems. In *Evolutionary Computation (CEC), 2013 IEEE Congress on*, pages 369–376. IEEE, 2013.

[25] Tianjun Liao and Thomas Stutzle. Benchmark results for a simple hybrid algorithm on the CEC 2013 benchmark set for real-parameter optimization. In *Evolutionary Computation (CEC), 2013 IEEE Congress on*, pages 1938–1944. IEEE, 2013.

[26] William B. Langdon and Riccardo Poli. Evolving problems to learn about particle swarm optimizers and other search algorithms. *IEEE Transactions on Evolutionary Computation*, 11(5):561–578, 2007.

[27] Olaf Mersmann, Bernd Bischl, Heike Trautmann, Mike Preuss, Claus Weihs, and Günter Rudolph. Exploratory landscape analysis. In *Proceedings of the 13th Annual Conference on Genetic and Evolutionary Computation*, pages 829–836. ACM, 2011.

[28] Katherine Mary Malan and Andries Petrus Engelbrecht. Characterising the searchability of continuous optimisation problems for PSO. *Swarm Intelligence*, 8(4):275–302, 2014.

[29] Henrik Jeldtoft Jensen. *Self-organized criticality: emergent complex behavior in physical and biological systems*, volume 10. Cambridge University Press, 1998.

[30] Per Bak, Chao Tang, Kurt Wiesenfeld, *et al.* Self-organized criticality. *Physical Review A*, 38(1):364–374, 1988.

[31] Per Bak. *How nature works*. Oxford University Press, Oxford, 1997.

[32] Zeev Olami, Hans Jacob S. Feder, and Kim Christensen. Self-organized criticality in a continuous, nonconservative cellular automaton modeling earthquakes. *Physical Review Letters*, 68:1244–1247, Feb 1992.

[33] Per Bak and Kim Sneppen. Punctuated equilibrium and criticality in a simple model of evolution. *Physical Review Letters*, 71(24):4083–4086, 1993.

[34] Christian W. Eurich, J. Michael Herrmann, and Udo A. Ernst. Finite-size effects of avalanche dynamics. *Physical Review E*, 66(6):066137, 2002.

[35] John M. Beggs and Dietmar Plenz. Neuronal avalanches in neocortical circuits. *The Journal of Neuroscience*, 23(35):11167–11177, 2003.

[36] Anna Levina, J. Michael Herrmann, and Theo Geisel. Dynamical synapses causing self-organized criticality in neural networks. *Nature Physics*, 3(12):857–860, 2007.

[37] Anna Deluca, Nicholas R. Moloney, and Álvaro Corral. Data-driven prediction of thresholded time series of rainfall and self-organized criticality models. *Physical Review E*, 91(5):052808, 2015.

[38] Woodrow L. Shew, Hongdian Yang, Shan Yu, Rajarshi Roy, and Dietmar Plenz. Information capacity and transmission are maximized in balanced cortical

networks with neuronal avalanches. *The Journal of Neuroscience*, 31(1):55–63, 2011.

[39] Woodrow L. Shew and Dietmar Plenz. The functional benefits of criticality in the cortex. *The Neuroscientist*, 19(1):88–100, 2013.

[40] Morten Lovbjerg and Thiemo Krink. Extending particle swarm optimisers with self-organized criticality. In *Evolutionary Computation, 2002. CEC'02. Proceedings of the 2002 Congress on*, volume 2, pages 1588–1593. IEEE, 2002.

[41] Toby J. Richer and Tim M. Blackwell. The Lévy particle swarm. In *Evolutionary Computation, 2006. CEC 2006. IEEE Congress on*, pages 808–815. IEEE, 2006.

[42] Carlos M. Fernandes, Juan J. Merelo, and Agostinho C. Rosa. Controlling the parameters of the particle swarm optimization with a self-organized criticality model. In *Parallel Problem Solving from Nature – PPSN XII*, pages 153–163. Springer, Berlin, Heidelberg, 2012.

[43] Ioan Cristian Trelea. The particle swarm optimization algorithm: convergence analysis and parameter selection. *Information Processing Letters*, 85(6): 317–325, 2003.

[44] Ming Jiang, Yupin Luo, and Shiyuan Yang. Stagnation analysis in particle swarm optimization. In *Swarm Intelligence Symposium, 2007. SIS 2007. IEEE*, pages 92–99. IEEE, 2007.

[45] Riccardo Poli. Mean and variance of the sampling distribution of particle swarm optimizers during stagnation. *IEEE Transactions on Evolutionary Computation*, 13(4):712–721, 2009.

[46] Qunfeng Liu. Order-2 stability analysis of particle swarm optimization. *Evolutionary Computation*, 23(2):187–216, 2015.

[47] Christopher W. Cleghorn and Andries P. Engelbrecht. Particle swarm convergence: an empirical investigation. In *Evolutionary Computation (CEC), 2014 IEEE Congress on*, pages 2524–2530. IEEE, 2014.

[48] CEC2013, 2013. URL http://www.ntu.edu.sg/home/EPNSugan/index_files/CEC2013/CEC2013.htm.

[49] Harry Furstenberg and Harry Kesten. Products of random matrices. *Annals of Mathematical Statistics*, 31(2):457–469, 1960.

[50] Jacques Vanneste. Estimating generalized Lyapunov exponents for products of random matrices. *Physical Review E*, 81(3):036701, 2010.

[51] Valerii Nikolaevich Tutubalin. On limit theorems for the product of random matrices. *Theory of Probability & Its Applications*, 10(1):15–27, 1965.

[52] Rafail Zalmanovich Khas'minskii. Necessary and sufficient conditions for the asymptotic stability of linear stochastic systems. *Theory of Probability & Its Applications*, 12(1):144–147, 1967.

[53] Christopher W. Cleghorn and Andries P. Engelbrecht. A generalized theoretical deterministic particle swarm model. *Swarm Intelligence*, 8(1):35–59, 2014.

Chapter 11

Bounded distributed flocking control of nonholonomic mobile robots

Thang Nguyen[1], Hung M. La[2], Vahid Azimi[3], and Thanh-Trung Han[1]

Abstract

There have been numerous studies on the problem of flocking control for multiagent systems whose simplified models are presented in terms of point-mass elements. Meanwhile, full dynamic models pose some challenging problems in addressing the flocking control problem of mobile robots due to their nonholonomic dynamic properties. Taking practical constraints into consideration, we propose a novel approach to distributed flocking control of nonholonomic mobile robots by bounded feedback. The flocking control objectives consist of velocity consensus, collision avoidance, and cohesion maintenance among mobile robots. A flocking control protocol that is based on the information of neighbor mobile robots is constructed. The theoretical analysis is conducted with the help of a Lyapunov-like function and graph theory. Simulation results are shown to demonstrate the efficacy of the proposed distributed flocking control scheme.

11.1 Introduction

It is well known that the collective behavior of self-propelled organisms constitutes flocking [1]. The coherent motion of the flock inspires various researches on flocking control of multiagent systems. A typical objective is to achieve a desired collective motion that can be produced by a constructive flocking control procedure. For numerous models, which are described from simplest models such as point-mass models to actual physical models, design protocols have been systematically proposed for multiagent systems [2,3]. Several control strategies were also addressed in noisy environments where the agent's position is affected by noise [4–6]. With point-mass models, the problem of flocking control of multiple agents has been addressed

[1]Faculty of Electrical and Electronics Engineering, Ton Duc Thang University, Vietnam
[2]Department of Computer Science and Engineering, University of Nevada, USA
[3]School of Electrical and Computer Engineering, Georgia Institute of Technology, USA

with typical results reported in [7–11]. For a wide range of engineering applications, extensive studies in flocking control of mobile robots have been done in various scenarios [12–14].

In this chapter, we study the problem of distributed flocking control of mobile robots by bounded feedback, which takes into consideration nonholonomic nature of mobile robots as well as the implementation issue posed by the physical limit of the motor speed. Our flocking control problem employs the full dynamic model of the mobile robot derived in [15]. Similar to [16,17], due to the nonholonomic property of the dynamics of mobile robots, our proposed design framework constructed to achieve velocity consensus is modular. In other words, the consensuses on the linear speed and orientation angles are obtained separately.

In this chapter, we are interested in agents with nonholonomic dynamics and boundedness constraints. Specifically, a coordination function is proposed to ensure that the induced attractive and repulsive forces are bounded and hence can be incorporated in the bounded velocity control. Using the results of Barbalat's lemma and graph theory, the theoretical analysis is conducted, which shows that the maximal value of the coordination function determines the basin of attraction for the flocking convergence.

In this chapter, graph theory will be employed as in the case of nearest neighbor communication [7,18]. We will employ the velocity control law reported in [16,17] in a decentralized sense, which helps one to avoid collision and maintain a linear speed consensus. In addition, the orientation consensus will be achieved using a modified approach, which is inspired by the one in [19], where the input constraint is taken into account.

The organization of the chapter is as follows. Section 11.2 summarizes some research work in the literature related to the topic in this chapter. In Section 11.3, the multiple-goal control problem for flocking of nonholonomic mobile robots and preliminaries are introduced. Section 11.4 describes main results where a modular design framework is proposed for bounded velocity control and bounded orientation control and the theoretical analyses are introduced. In Section 11.5, a description of an obstacle avoidance scheme is presented. Section 11.6 shows some simulation results. Section 11.7 concludes the chapter by some conclusions.

Notations: $\mathbb{R}$ and $\mathbb{R}^+$ are the sets of real numbers and nonnegative real numbers, respectively; for $q = [q_1, \ldots, q_n]^T$, $\nabla_q = [\partial/\partial q_1, \ldots, \partial/\partial q_n]^T$ is the del operator [20]; for two vectors a and b, $a \cdot b$ is their scalar product; $(a_1, \ldots, a_n)$ is $[a_1^T, \ldots, a_n^T]^T$; $|\cdot|$ is the absolute value of scalars; and $\|\cdot\|$ is the Euclidean norm of vectors.

11.2 Related work

In many applications, the mission carried out by a single complicated robotic system can be equivalently completed by a coordination of a mobile robotic system with much simpler configurations, whose advantages lie in scalability, flexible deployment, cheaper costs, reliability, etc. Therefore, more sophisticated tasks can be fulfilled using a group of small mobile robots with lower cost and higher efficiency than a complex unit; see [2,14,21–39] and references therein.

Flocking control of mobile robots was widely addressed with different control schemes; see [9,18,21,40–42] and references therein. Recently, a new measure-theoretic approach that systematically provides a framework for obtaining flocking protocols for mobile robots was reported in [13].

The common assumption in many papers is the availability of information of all agents or all-to-all communication. Numerous control protocols for mobile robots have been constructed based on this assumption. This centralized communication control architecture yields inflexibility and large computation costs for the controller of each agent, especially when the number of agents is large. Meanwhile, a distributed control protocol can offer an ease of implementation and less computational burden as each element of the system needs only the information of neighbor agents. In this direction, a range of decentralized control schemes for mobile robots have been proposed [8,10,18,43,44].

For a wide range of engineering applications, cohesion maintenance and collision avoidance (CMCA) properties of a mobile robotic system are of importance. As reported in [21,45], the attractive and repulsive forces cannot be included in the control for CMCA of mobile robots, as it is possible for point-mass agents [7]. In [13], desired attractive and repulsive forces for CMCA of mobile robots were achieved using a new rearrangement strategy. In [7,10,18], the graph theory was employed to generate control protocols that maintain CMCA of multiagent systems with double integrator models.

In [19], a distributed flocking control approach was proposed but no constraints on the control inputs are imposed. The work in [16,17] considers the bounded feed-back flocking control problem for nonholonomic mobile robots without a flocking desired heading angle. The problem of interest in this chapter is to address bounded control of nonholonomic dynamic mobile vehicles, which also achieves CMCA and obstacle avoidance. We also consider a flocking desired heading angle, which reveals a collective flocking behavior.

11.3 Problem formulation

Similarly to [16,17], we investigate a collective system of N identical autonomous mobile robots whose respective equations of motion are [15]

$$\begin{aligned}\dot{q}_i &= v_i e(\theta_i)\\ \dot{\theta}_i &= w_i\\ \dot{v}_i &= u_i\\ \dot{w}_i &= \tau_i\end{aligned} \tag{11.1}$$

where $i = 1, \ldots, N$, $q_i = [x_i, y_i]^T \in \mathbb{R}^2$, and $\theta_i \in \mathbb{R}$ are, respectively, the position and the heading angle of the ith robot in the inertial frame Oxy; $v_i \in \mathbb{R}$ is the linear speed, and $e(\theta_i)$ is the unit vector $[\cos\theta_i, \sin\theta_i]^T$; $w_i \in \mathbb{R}$ is the angular speed, and $u_i, \tau_i \in \mathbb{R}$ are control inputs.

Following the same vein as in [16], we define $0 < r_0 < R_0$. Then, the flocking control problem for (11.1) is to construct the control inputs u_i, τ_i as bounded functions of the collective state $(q_1, \ldots, q_N, \theta_1, \ldots, \theta_N, v_1, \ldots, v_N, w_1, \ldots, w_N)$ in a distributed fashion to satisfy the following multiple goals

(G1) *Velocity consensus:*

$$\lim_{t\to\infty} (\dot{q}_i(t) - \dot{q}_j(t)) = 0, \quad \forall i,j = 1, \ldots, N \tag{11.2}$$

(G2) *Collision avoidance:* $r_{ij}(t) = \|q_i(t) - q_j(t)\| \geq r_0, \qquad \forall t \geq 0, \quad \forall i \neq j$

(G3) *Cohesion maintenance:* $r_{ij}(t) \leq R_0, \qquad \forall t \geq 0, \quad \forall i \neq j.$

Similarly to [16,17], we have the following definition.

Definition 11.1. *A control* $\dot{\zeta} = g(\zeta, y), u = c(\zeta, y), (\zeta, y) \in \mathbb{R}^d \times \mathbb{R}^m$ *of a system* $\dot{x} = f(x, u), y = h(x, u)$ *is said to be bounded if there is a finite constant* $M > 0$ *such that* $\|c(\zeta, y)\| \leq M, \forall(\zeta, y) \in \mathbb{R}^d \times \mathbb{R}^m$.

To achieve the goals (G2) and (G3), we consider the coordination function U : $\mathbb{R}^+ \to \mathbb{R}^+$ which satisfies the following properties:

(P1) there is a constant $U_M > 0$ such that

$$0 \leq U(r) \leq U_M, \qquad \forall r \in \mathbb{R} \tag{11.3}$$

(P2) $U(r)$ is continuously differentiable on $[r_0, R_0]$;

(P3) $\lim_{r\to r_0^+} U(r) = U_M$; and

(P4) $\lim_{r\to R_0^-} U(r) = U_M$.

For a link between agents *i* and *j* of the flock, we aim to maintain $r_{ij}(t) \in [r_0, R_0]$. Without loss of generality, we assume that $U(0) = 0$ and hence $U(r)$ is well defined for $r_{ii} = 0$ [16].

We are interested in the function U with the dead zone $[a, A]$ since even distribution of agents may not be achievable by a common coordination function U. Accordingly, we use the zone $[a, A]$ for free alignment. A function U satisfying the above requirements is shown in Figure 11.1.

For bounded control, we shall use the linear saturation functions σ_1 and σ_2, which are continuous and nondecreasing functions and satisfy, for given positive constants $L_i \leq M_i, i = 1, 2$ [16].

1. $\sigma_i(-s) = -\sigma_i(s)$ for all s;
2. $\sigma_i(s) = s$ for $s \leq L_i$; and
3. $|\sigma_i(s)| \leq M_i, \forall s \in \mathbb{R}$.

Similarly to other works on distributed for multiagent systems [7,10,17,18], the graph theory will be utilized to address our problem. A digraph associated with

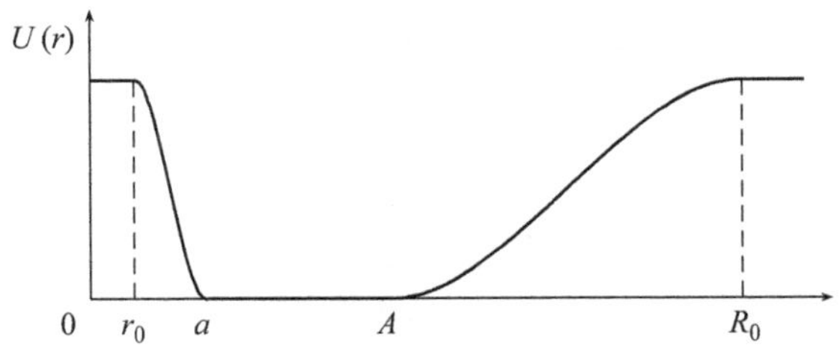

Figure 11.1 Coordination function (extracted from [16])

(11.1) is called $\mathscr{G}(t) = (\mathscr{V}, \mathscr{E}(t))$ where $\mathscr{V} = 1, \ldots, N$ and $\mathscr{E} \subseteq \mathscr{V} \times \mathscr{V}$. The set $\mathscr{V}$ is denoted as the node set of $\mathscr{G}(t)$ and the set $\mathscr{E}(t)$ is defined as the edge set of $\mathscr{G}(t)$. In addition, $\mathscr{N}_i(t)$ denotes the neighbor set of the node i for $i = 1, \ldots, N$.

As in [10], the description of the edge $\mathscr{E}(t)$ is presented as follows.

Given any $R > 0$, $\varepsilon_2 \in (0, R)$, and $\varepsilon_1 \in (0, R - \varepsilon_2)$, for any $t \geq 0$, $\mathscr{E}(t) = \{(i,j)|i,j \in \mathscr{V}\}$ is defined such that

1. $\mathscr{E}(0) = \{(i,j)|\varepsilon_1 < \|q_i(0) - q_j(0)\| < (R - \varepsilon_2)\}$;
2. if $\|q_i(0) - q_j(0)\| \geq R$, then $(i,j) \notin \mathscr{E}(t)$;
3. for $i = 1, \ldots, N$, $j = 1, \ldots, N$, if $(i,j) \notin \mathscr{E}(t^-)$ and $\|q_i(t) - q_j(t)\| < R - \varepsilon_2$, then $(i,j) \in \mathscr{E}(t)$;
4. for $i = 1, \ldots, N$, $j = 1, \ldots, N$, if $(i,j) \in \mathscr{E}(t^-)$ and $\|q_i(t) - q_j(t)\| < R$, then $(i,j) \in \mathscr{E}(t)$.

As in [17], the following results will be employed for the main results.

Lemma 11.1. *Let $\sigma : \mathbb{R} \to \mathbb{R}$ be a function satisfying $\sigma(-s) = -\sigma(s), \forall s \in \mathbb{R}$. Then, for all a_i, b_i, it holds true that*

$$\frac{1}{2}\sum_{i=1}^{N}\sum_{j\in\mathscr{N}_i(t)}(a_i - a_j)\sigma(b_i - b_j) = \sum_{i=1}^{N}\sum_{j\in\mathscr{N}_i(t)} a_i\sigma(b_i - b_j). \tag{11.4}$$

Proof: Since $\sigma(-s) = -\sigma(s)$ and $\mathscr{G}(t)$ is an undirected graph, we have

$$\begin{aligned}\sum_{i=1}^{N}\sum_{j\in\mathscr{N}_i(t)} a_j\sigma(b_i - b_j) &= -\sum_{i=1}^{N}\sum_{j\in\mathscr{N}_i(t)} a_j\sigma(b_j - b_i)\\ &= -\sum_{i=1}^{N}\sum_{j\in\mathscr{N}_i(t)} a_i\sigma(b_i - b_j).\end{aligned} \tag{11.5}$$

Hence,

$$\begin{aligned}
&\sum_{i=1}^{N}\sum_{j\in\mathcal{N}_i(t)}(a_i-a_j)\sigma(b_i-b_j)\\
&\quad=\sum_{i=1}^{N}\sum_{j\in\mathcal{N}_i(t)}a_i\sigma(b_i-b_j)-\sum_{i=1}^{N}\sum_{j\in\mathcal{N}_i(t)}a_j\sigma(b_i-b_j)\\
&\quad=2\sum_{i=1}^{N}\sum_{j\in\mathcal{N}_i(t)}a_i\sigma(b_i-b_j)
\end{aligned}\tag{11.6}$$

which implies (11.4). ■

Remark 11.1. *Lemma 11.1 plays an important role in the theoretical analysis of the main results. Here, the lemma is similar to the one in [46]. However, [46] considers all-to-all communication in the multiagent system. Our problem in this chapter is focused on the distributed fashion, which requires the employment of the neighbor set $\mathcal{N}_i(t)$ of robot i.*

Lemma 11.2. *The linear saturation functions $\sigma_i, i = 1, 2, 3$ satisfy*

$$(\sigma_i(\theta_1)-\sigma_i(\theta_2))\sigma_i(\theta_1-\theta_2)\geq 0,\quad \forall\theta_1,\theta_2. \tag{11.7}$$

Proof: Without loss of generality, suppose that $\theta_1 \geq \theta_2$. Since σ_i are nondecreasing functions, this implies that

$$\sigma_i(\theta_1)-\sigma_i(\theta_2)\geq 0. \tag{11.8}$$

Furthermore, as $\sigma_i(0) = 0$, $\theta_1 \geq \theta_2$ and the nondecreasing property of σ_i imply that

$$\sigma_i(\theta_1-\theta_2)\geq 0. \tag{11.9}$$

Multiplying (11.8) and (11.9) side-by-side, we obtain (11.7). ■

11.4 Main results

Our constructive strategy is to design u_i to achieve consensus on v_i, and τ_i to achieve consensus on θ_i. The design for u_i is derived from [17], while the construction for τ_i is built based on the approach in [19].

Note that $U(r_{ij}) = U(\|q_i - q_j\|)$, which is the symmetric function of q_i and q_j. As a result, we write $U(q_i, q_j)$ with the understanding that $U(q_i, q_j) = U(q_j, q_i)$. The control protocols u_i and τ_i are constructed based on Lyapunov theory. Specifically, a positive definite function V is presented such that the time derivative of V is a negative definite function. Regarding the distribution control problem, the graph theory will be employed to show the connectivity preservation for our multiagent system. Then, we apply the LaSalle's invariance principle [47] to conclude the desired consensuses.

Similarly to [17], the initial state of the collective system of agents (11.1) is chosen such that the graph $\mathscr{G}(0)$ is connected. The parameters of the graph $\mathscr{G}(0)$ are chosen as follows:

$$R = R_0, \tag{11.10}$$

$$r_0 \leq \varepsilon_1 < a, \tag{11.11}$$

$$0 < \varepsilon_2 \leq R_0 - a. \tag{11.12}$$

11.4.1 Speed consensus and connectivity preservation

The derivation of this subsection is essentially similar to the control design for the linear speed in [17]; hence, it is presented here for completeness. Consider the energy function for system (11.1)

$$V_1 = \frac{1}{2}\sum_{i=1}^{N}\sum_{j\in\mathscr{N}_i(t)} U(q_i, q_j) + \frac{1}{2}\sum_{i=1}^{N} v_i^2. \tag{11.13}$$

We assume that $U(r)$ is designed such that

$$U(R_0) = U_M > V_{1max}, \tag{11.14}$$

where

$$V_{1max} \triangleq \frac{1}{2}\sum_{i=1}^{N} v^2(0) + \frac{N(N-1)}{2}U(R_0 - \varepsilon_2). \tag{11.15}$$

Let m_0 be the number of the links of the initial graph. The simplest connected graph of N agents is a tree whose number of links is $n-1$. Hence, $m_0 \geq n-1$. Let

$$V_1(0) \leq V_{1max} - \frac{(N-1)(N-2)}{2}U(R_0 - \varepsilon_2). \tag{11.16}$$

Note that $U(q_i, q_j)$ is a symmetric function of q_i and q_j. We compute the derivative of V_1 with respect to (11.1)

$$\begin{aligned}\dot{V}_1 &= \sum_{i=1}^{N}\sum_{j\in\mathscr{N}_i(t)} \nabla_{q_i}U(q_i, q_j)\cdot \dot{q}_i + \sum_{i=1}^{N} v_i u_i \\ &= \sum_{i=1}^{N} v_i\left(\sum_{j\in\mathscr{N}_i(t)} \nabla_{q_i}U(q_i, q_j)\cdot e(\theta_i) + u_i\right).\end{aligned} \tag{11.17}$$

From (11.17), a control law for the speed consensus protocol is chosen as

$$u_i = -\sum_{j\in\mathscr{N}_i(t)} \nabla_{q_i}U(q_i, q_j)\cdot e(\theta_i) - \sum_{j\in\mathscr{N}_i(t)} \sigma_1(v_i - v_j) \tag{11.18}$$

where σ_1 is the linear saturation function introduced in Section 11.3.

Substituting (11.18) into (11.17), we obtain

$$\dot{V}_1 = -\sum_{i=1}^{N} \sum_{j \in \mathcal{N}_i(t)} v_i \sigma_1(v_i - v_j). \tag{11.19}$$

We have the following speed consensus theorem [17].

Theorem 11.1. *Suppose that the collective system (11.1) subject to the protocol (11.18) is initiated such that $V_1(0) < V_{1max}$. Then, the following properties hold:*

1. *$\mathcal{G}(t)$ is connected for all $t \geq 0$ and there exists t_k such that for $t \geq t_k$, $\mathcal{G}(t) = \mathcal{G}(t_k)$*
2. *Collision avoidance is guaranteed, i.e. $\|q_i - q_j\| > r_0$ for all $i,j \in N$ and $i \neq j$*
3. $\lim_{t\to\infty} (v_i(t) - v_j(t)) = 0$

Proof: Assume that $\mathcal{G}(t)$ switches at time t_k $(k = 1, 2, \ldots)$. Hence, $G(t) = G(0)$ for all $t \in [0, t_1)$. In other words,

$$\begin{aligned} \mathcal{G}(t) &= G(0), \ t \in [0, t_1) \\ \mathcal{G}(t_1) &\neq G(0). \end{aligned} \tag{11.20}$$

We prove that $G(0) \subset G(t_1)$. Using control law (11.18), we have

$$\dot{V}_1 = -\sum_{i=1}^{N} v_i \sum_{j \in \mathcal{N}_i(t)} \sigma_1(v_i - v_j). \tag{11.21}$$

According to Lemma 11.1,

$$\dot{V}_1 = -\frac{1}{2}\sum_{i=1}^{N} \sum_{j \in \mathcal{N}_i(t)} (v_i - v_j)\sigma_1(v_i - v_j). \tag{11.22}$$

Because $\sigma_1(s)$ is an odd function, $(v_i - v_j)\sigma_1(v_i - v_j) \geq 0$. Thus, $\dot{V}_1 \leq 0$, which deduces

$$V_1(t) \leq V_1(0) < V_{1max} < U_M \text{ for } [0, t_1). \tag{11.23}$$

From the definition of $U(r)$, $U(R_0) > V_{1max} \geq V_1(0)$. Hence for any $(i,j) \in \mathcal{G}(t)$ for $t \in [0, t_1)$

$$U(q_i, q_j) \leq V_1(t) < U_M = U(r_0) = U(R_0). \tag{11.24}$$

By the continuity of $U(r)$, (11.24) shows that $r_0 < \|q_i - q_j\| < R_0$. This implies that no existing links are deleted at time t_1 and collision avoidance is achieved. As a result, new links must be added to the current graph at the switching time t_1. We assume that there are m_1 new links being added to the network at time t_1. Since the number of the current links before switching is $m_0 \geq N - 1$ and the complete graph possesses $(N(N-1))/2$ edges, $m_1 \leq ((N(N-1))/2) - (N-1) = (((N-1)(N-2))/2)$. Hence, we have

$$V_1(t_1) = V_1(t_1^-) + m_1 U(R_0 - \varepsilon_2). \tag{11.25}$$

Due to (11.16),

$$V_1(t_1^-) \leq V_1(0) < V_{1max} - \frac{(N-1)(N-2)}{2} U(R_0 - \varepsilon_2). \tag{11.26}$$

Thus,

$$V_1(t_1) < V_{1max}. \tag{11.27}$$

By induction, for $t \in [t_{k-1}, t_k)$,

$$\dot{V}_1 = -\frac{1}{2} \sum_{i=1}^{N} \sum_{j \in \mathcal{N}_i(t)} (v_i - v_j)\sigma_1(v_i - v_j) \leq 0, \tag{11.28}$$

and therefore $V_1(t) \leq V_1(t_{k-1}) \leq V_{1max}$. This shows that no edges are lost at time t_k and $V_1(t_k) \leq V_{1max}$. As a result, the size of the set of the links of $\mathcal{G}(t)$ forms an increasing sequence, bounded above by $(N(N-1))/2$, which is the number of the links of a complete graph. Thus, there exists a finite integer $k > 0$ such that

$$\mathcal{G}(t) = \mathcal{G}(t_k), \ t \in [t_k, \infty). \tag{11.29}$$

Hence, for $t \geq t_k$, we have

$$\dot{V}_1 = -\frac{1}{2} \sum_{i=1}^{N} \sum_{j \in \mathcal{N}_i(t_k)} (v_i - v_j)\sigma_1(v_i - v_j) \leq 0. \tag{11.30}$$

Next, we will show that the linear velocities of all agents converge to the same value. Using the fact that $U(q_i, q_j) \leq V_1(t) \leq V_{1max} < U_M$ and the properties of U, we deduce $\|q_i - d_j\| > r_0$. This shows that no collision takes place among agents. Since $0 \leq V_1(t) \leq V_{1max}$ and $\dot{V}_1 \leq 0$, by Barbalat's lemma, $\lim_{t\to\infty} \dot{V}_1(t) = 0$. Because the graph $\mathcal{G}(t)$ is connected for all t and $s\sigma_1(s) \geq 0$ for all s, from (11.30),

$$\lim_{t\to\infty} (v_i - v_j) = 0, \ \text{for all } i, j = 1, 2, \ldots, N. \tag{11.31}$$

■

Remark 11.2. *The proof of the theorem follows similar approaches as in [10,48], where graph theory was employed as a means for proving the connectivity of mobile networks. The potential in this chapter is similar to the one in [10] in the sense that it is bounded. In contrast, the potential function used in [48] goes to infinity at singularities. Note that the mobile robots in this work are nonholonomic while [10,48] addressed double integrator systems.*

Remark 11.3. *The first sum in (11.18) consists of the gradients of $U(q_i, q_j)$ and the unit vector $e(\theta_i)$ that are bounded by definition. The second sum in (11.18) comprises $\sigma_1(.)$, which is a linear saturation function defined in Section 11.3. Hence, as a whole, the control law (11.18) for each agent is bounded. This satisfies our objective on the boundedness of the control input u_i.*

Theorem 11.1 shows that the design (11.18) achieves speed consensus and the goals (G2) and (G3). In the next subsection, we will design τ_i for orientation consensus completing the goal (G1).

11.4.2 Orientation consensus

Motivated by the orientation consensus design method presented in [19], we shall develop a bounded control approach that employs a saturation function in Section 11.3.

Define the orientation trajectory error for agent i as

$$e_i = \theta_i - \theta_r, \tag{11.32}$$

where θ_r is the desired orientation of the flock. Thus, the angle difference between two agents i and j is

$$\theta_i - \theta_j = \theta_i - \theta_r - (\theta_j - \theta_r) = e_i - e_j. \tag{11.33}$$

Similarly to [19], the following lemma is employed for our convergence analysis.

Lemma 11.3. *Suppose that the flock possesses a graph $\mathcal{G}(t)$, then the trajectory error signals of the group have the following property:*

$$\frac{1}{2}\sum_{i=1}^{N}\sum_{j\in\mathcal{N}_i(t)}(e_i - e_j)(\dot{e}_i - \dot{e}_j) = \sum_{i=1}^{N}\sum_{j\in\mathcal{N}_i(t)} e_i(\dot{e}_i - \dot{e}_j). \tag{11.34}$$

Proof: The proof is similar to the one employed in Lemma 11.1. ■

We have the following orientation consensus theorem.

Theorem 11.2. *Assume that the desired orientation θ_r and its first and second derivation are bounded, and the collective system (11.1) is subject to the following protocol:*

$$\tau_i = \ddot{\theta}_r - \sigma_2(\dot{\theta}_i - \dot{\theta}_r) - \frac{k_\theta}{n_i + 1}\left[(n_i + 1)\theta_i - \sum_{j\in\mathcal{N}_i(t)}\theta_j - \theta_r\right], \tag{11.35}$$

where n_i is the number of the neighbors of robot i and k_θ is a positive parameter. Then, all the mobile robots eventually reach consensus on the heading angles θ_i in the sense that

$$\lim_{t\to\infty}(\theta_i(t) - \theta_j(t)) = 0, \quad \forall i,j. \tag{11.36}$$

Proof: Consider the following Lyapunov function candidate

$$V_2(t) = \frac{1}{2}\sum_{i=1}^{N}\frac{k_\theta}{n_i + 1}e_i^2 + \frac{1}{2}\sum_{i=1}^{N}\dot{e}_i^2 + \frac{1}{4}\sum_{i=1}^{N}\sum_{j\in\mathcal{N}_i(t)}\frac{k_\theta}{n_i + 1}(\theta_i - \theta_j)^2. \tag{11.37}$$

According to Theorem 11.1, there exists t_k such that for $t \geq t_k$, $\mathcal{G}(t) = \mathcal{G}(t_k)$. The derivative of $V_2(t)$ with respect to t for $t \geq t_k$ is given as

$$\begin{aligned}
\dot{V}_2(t) &= \sum_{i=1}^{N} \frac{k_\theta}{n_i+1} e_i \dot{e}_i + \sum_{i=1}^{N} \dot{e}_i(\tau_i - \ddot{\theta}_r) + \sum_{i=1}^{N} \sum_{j \in \mathcal{N}_i(t)} \frac{k_\theta}{n_i+1} (\theta_i - \theta_j)(\dot{\theta}_i - \dot{\theta}_j) \\
&= \sum_{i=1}^{N} \frac{k_\theta}{n_i+1} e_i \dot{e}_i - \sum_{i=1}^{N} \frac{k_\theta}{n_i+1} \dot{e}_i e_i - \sum_{i=1}^{N} \dot{e}_i \sigma_2(\dot{e}_i) \\
&\quad - \sum_{i=1}^{N} \sum_{j \in \mathcal{N}_i(t)} \frac{k_\theta}{n_i+1} \dot{e}_i (\theta_i - \theta_j) + \frac{1}{2} \sum_{i=1}^{N} \sum_{j \in \mathcal{N}_i(t)} \frac{k_\theta}{n_i+1} (\theta_i - \theta_j)(\dot{\theta}_i - \dot{\theta}_j) \\
&= -\sum_{i=1}^{N} \dot{e}_i \sigma_2(\dot{e}_i) - \sum_{i=1}^{N} \sum_{j \in \mathcal{N}_i(t)} \frac{k_\theta}{n_i+1} \dot{e}_i (e_i - e_j) \\
&\quad + \frac{1}{2} \sum_{i=1}^{N} \sum_{j \in \mathcal{N}_i(t)} \frac{k_\theta}{n_i+1} (e_i - e_j)(\dot{e}_i - \dot{e}_j).
\end{aligned} \tag{11.38}$$

Using Lemma 11.3, we obtain

$$\dot{V}_2(t) = -\sum_{i=1}^{N} \dot{e}_i \sigma_2(\dot{e}_i) \leq 0 \tag{11.39}$$

since $\sigma_2(.)$ is a nondecreasing function defined in Section 11.3. Since $\theta_i, \theta_r \in [-pi, pi]$ and $\ddot{\theta}_r$ is bounded, the control law (11.35) implies $\ddot{e}_i$ is bounded. By the Barbalat's lemma, from (11.39), $\dot{e}_i \to 0$. Also, since $\dddot{\theta}_r$ is bounded, $\ddot{e}_i \to 0$. Therefore, the control law (11.35) implies that $\theta_i \to \theta_j$, which proves the theorem. ■

Remark 11.4. *The boundedness of the control law (11.35) is guaranteed by the properties of linear saturation function σ_2 and the fact that $\theta_i, \theta_j, \theta_r \in [-\pi, \pi]$. This demonstrates that the proposed control scheme meets the requirement on the physical limits of the control inputs.*

Remark 11.5. *It should be noted that our control law for orientation consensus is similar to the one in [19] but here the boundedness of the control input is taken into account. The scheme in this chapter also shares the same objective as the one in [17] but offers a more simple form and implementation.*

Combining Theorems 11.1 and 11.2, we have the following bounded flocking theorem.

Theorem 11.3. *Suppose that the collective system (11.1) is subject to the bounded protocols (11.18) and (11.35). Suppose further that the initial configuration of the collective system (11.1) is such that $\mathcal{N}(0)$ is connected. Then, all the multiple flocking goals of velocity consensus, CMCA are achieved.*

Proof: The proof is straightforward from the results of Theorems 11.1 and 11.2. ■

11.5 Avoidance of obstacles

The problem of obstacle avoidance has been extensively studied in the literature [17,19,22]. In this section, we employ the idea from [19] to derive our control algorithm in which the agents are able to pass obstacles. It is shown that a convex obstacle can be extrapolated by a round shape or a rectangle [17,19]. In [19], a convex obstacle is presented by a circle, which is used in our work. Let $q_r = [x_r, y_r]^T$ be the coordinate of the robot, $q_{jobs} = [x_{obs}, y_{obs}]^T$ be the projection point of the robot onto obstacle $j = 1, 2, \ldots, n_{obs}$ where n_{obs} is the number of obstacles, and $\bar{O}_k = [x_k, y_k]^T$ be the center of the obstacle. From [19],

$$q_{jobs} = \frac{r}{\|q_r - \bar{O}_k\|} q_r + \left(1 - \frac{r}{\|q_r - \bar{O}_k\|}\right) \bar{O}_r \tag{11.40}$$

where r is the radius of the obstacle. The projection point has the following velocity

$$v_{jobs} = \frac{vr \sin \alpha}{\|q_r - q_{jobs}\|} \tag{11.41}$$

where v is the velocity of the agent, α is the angle between the heading of the robot and the straight line that connects the robot to the center of the obstacle. The projection point moves in the direction [19]

$$\theta_{jobs} = -\frac{\pi}{2} + \alpha + \theta, \quad \alpha > 0 \tag{11.42}$$

$$\theta_{jobs} = \frac{\pi}{2} + \alpha + \theta, \quad \text{otherwise.} \tag{11.43}$$

The fact that the projection point possesses a position, velocity, and orientation enables it to be an agent. The above descriptions of the robot and obstacle are demonstrated in Figure 11.2.

We have the following orientation consensus theorem.

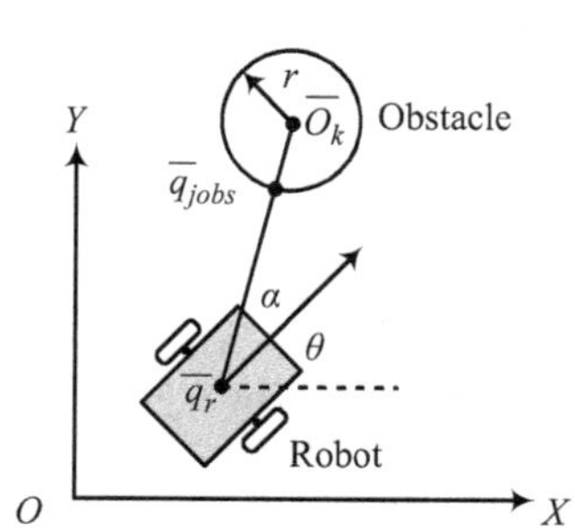

Figure 11.2 Illustration of a robot and a convex obstacle

Theorem 11.4. *The following control protocol guarantees that the robot avoids obstacles with arbitrary boundary shapes:*

$$u_i = -\sum_{j\in\mathcal{N}_{obs}} \nabla_{q_i} U(q_i, q_j)\cdot e(\theta_i) - \sum_{j\in\mathcal{N}_{obs}} \sigma_1(v_i - v_j) \tag{11.44}$$

$$\tau_i = \ddot{\theta}_r - \sigma_2(\dot{\theta}_i - \dot{\theta}_r) - \frac{k_\theta}{n_{obs}+1}\left[(n_{obs}+1)\theta_i - \sum_{j\in\mathcal{N}_i(t)} \theta_j - \theta_r\right], \tag{11.45}$$

where $\mathcal{N}_{obs}$ is the set of obstacles, n_{obs} is the number of obstacles and

$$\theta_r = \frac{1}{n_{obs}} \sum_{j\in\mathcal{N}_{obs}} \theta_{jobs}. \tag{11.46}$$

Proof: The proof can be derived using the same approach as in [19]. ■

Remark 11.6. *It should be noted that the speed control law in (11.44) enjoys the boundedness due to the saturation function $\sigma_1(.)$, which is different from the one in [19]. Since θ_i, $\theta_r \in (-\pi, \pi]$ and the saturation function $\sigma_2(.)$ is bounded, (11.45) reveals that the heading control law is also bounded.*

11.6 Simulation

We conducted simulation for a multiagent system of 15 mobile robots of the model (11.1). A bump function is used to generate the smooth coordination function U. As the control (11.18) invokes the gradient forces $\nabla_{q_i} U$, we designed the coordination function in the form

$$U(r) = \int_0^r \varphi(s)ds \tag{11.47}$$

where φ is a compact support function given by

$$\varphi(s) = \begin{cases} p_1 \exp\left(\frac{-(s-s_0)^2}{((a-r_0)/2)^2-(s-s_0)^2}\right) & \text{if } s \in (r_0, a) \\ p_2 \exp\left(\frac{-(s-s_1)^2}{((R_0-A)/2)^2-(s-s_1)^2}\right) & \text{if } s \in (A, R_0) \\ 0 & \text{otherwise} \end{cases}$$

where $s_0 = (r_0 + a)/2$, $s_1 = (R_0 + A)/2$, and p_1, p_2, a, A, r_0, and R_0 are design parameters.

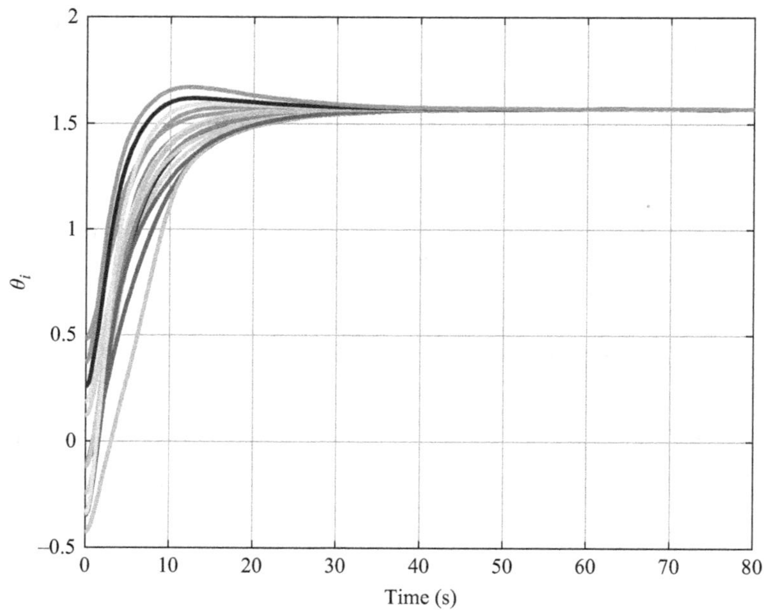

Figure 11.3 Orientation consensus

The parameters of the coordinate function are $r_0 = 1, a = 3, A = 6, R_0 = 8$, and $U_M = 15$. The parameter for the control law (11.35) is $k_\theta = 1.5$. The initial positions of 15 mobile robots are randomly distributed on three circles. Their coordinates are

$$x(i) = \Gamma \ \sin(z\pi(i-1)/\Gamma + \pi)$$
$$y(i) = \Gamma \ \cos(z\pi(i-1)/\Gamma + \pi)$$

where $z = 1$, and

$$\Gamma = \begin{cases} 4 & \text{if } i < 6 \\ 8 & \text{otherwise} \end{cases} \tag{11.48}$$

The initial values of θ_i and v_i are randomly chosen. The initial value of w_i is 0. The desired orientation of the flock is $\theta_r = \pi/2$.

We obtained the simulation results shown in Figures 11.7–11.9. It is shown in Figures 11.3 and 11.5, the heading angles converge to θ_r and angular speeds of all agents converge to 0 after $t = 40$ s. The linear speeds are depicted in Figure 11.4, where the convergence of all agents takes place after $t = 45$ s. In Figure 11.5, the angular speeds converge after 40 s. Hence, Figures 11.3 and 11.4 demonstrate that consensuses on orientation and linear speed of the mobile robots have been obtained. The speed control and steering control signals are shown to be bounded in Figures 11.6 and 11.7. The minimum distance among agents is described in Figure 11.8, which

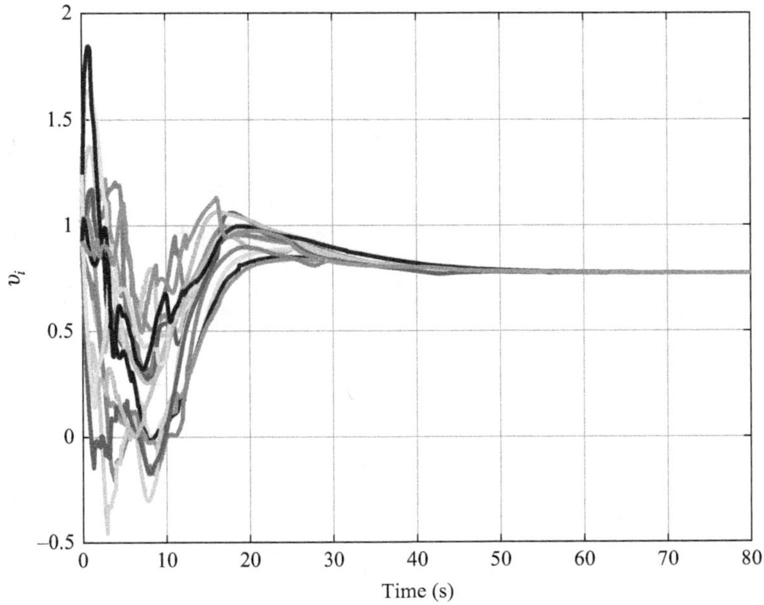

Figure 11.4 Linear speed consensus

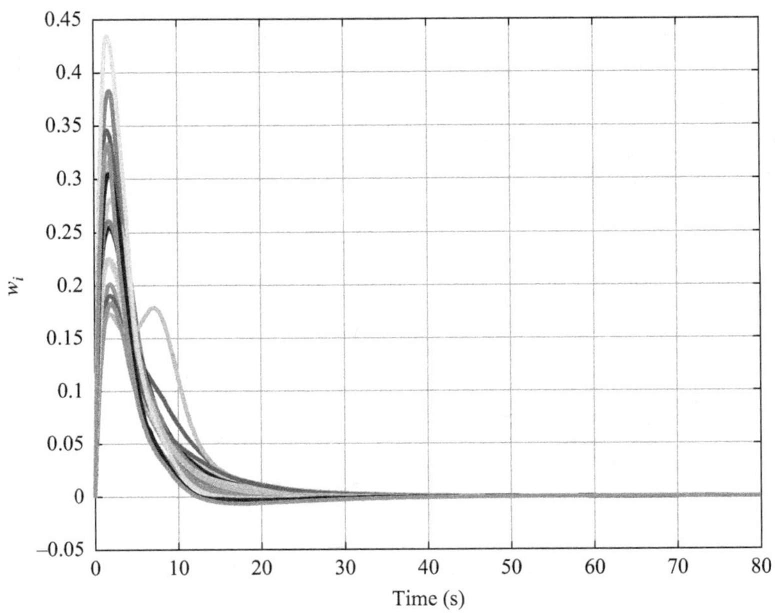

Figure 11.5 Angular speed

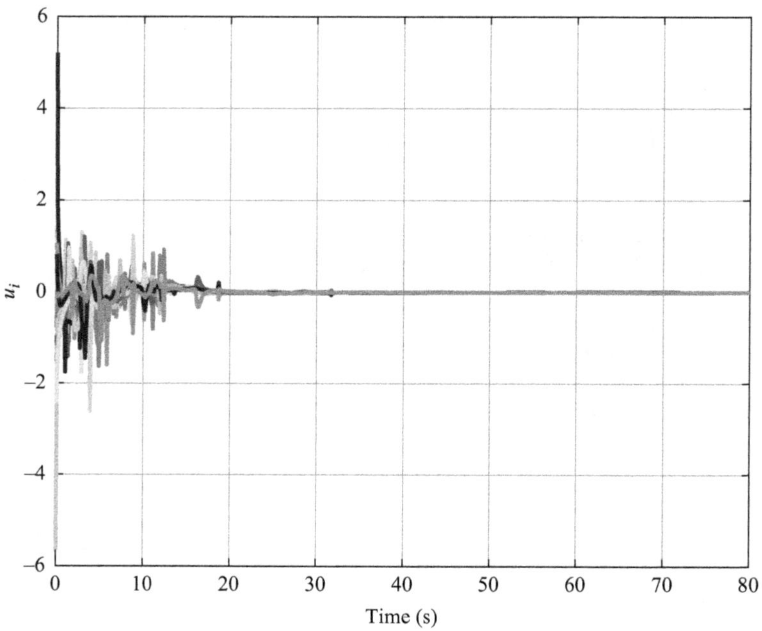

Figure 11.6 Translational force

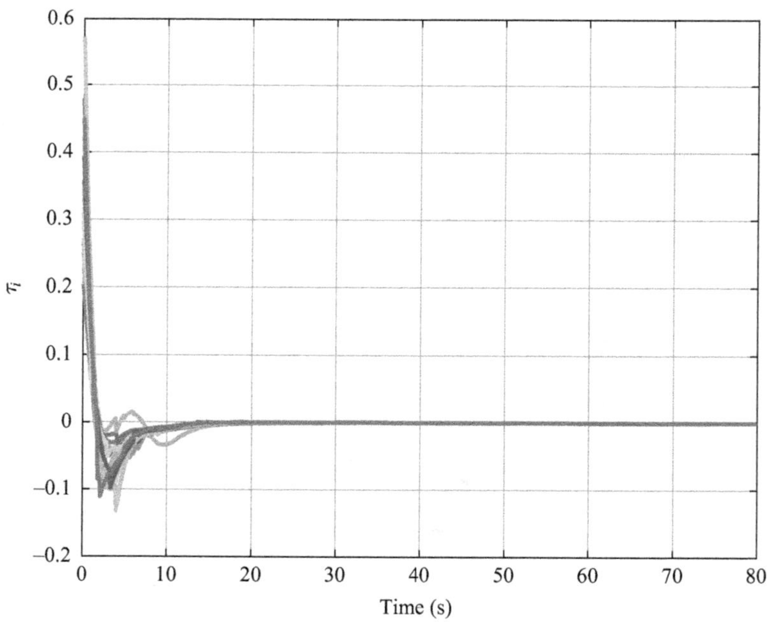

Figure 11.7 Steering control

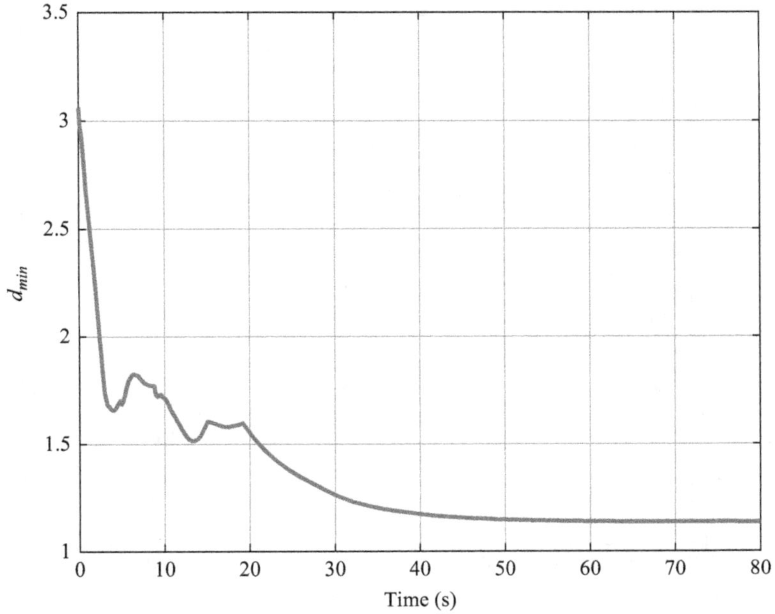

Figure 11.8 The minimum distance among agents

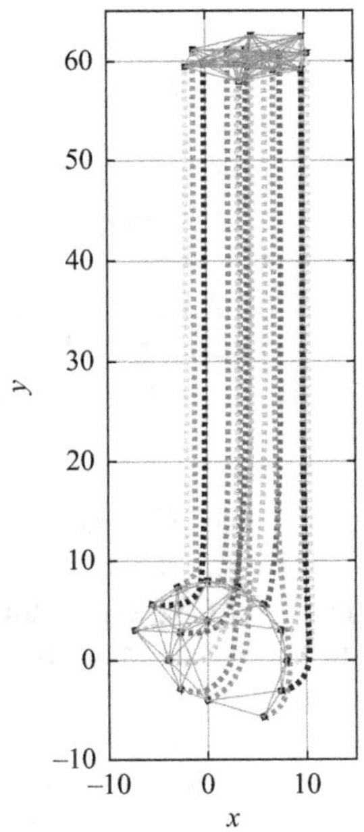

Figure 11.9 Distributed flocking of 15 mobile robots

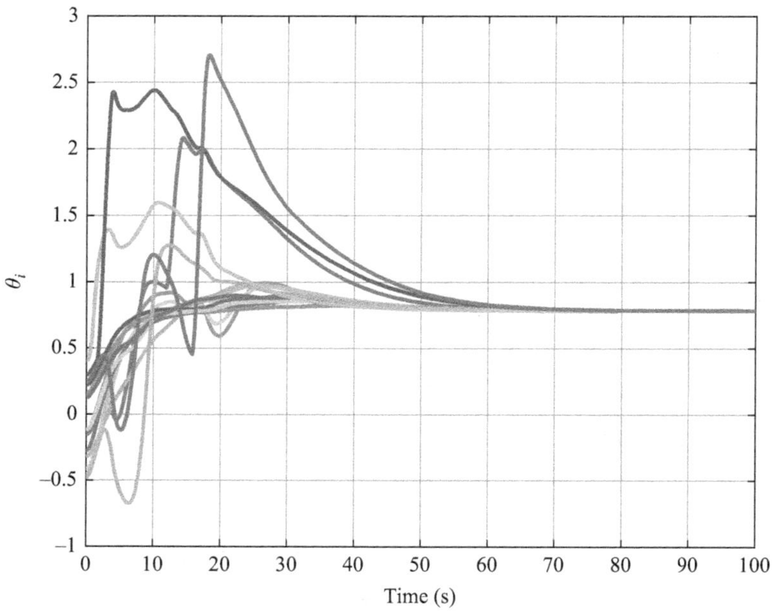

Figure 11.10 Orientation consensus for the obstacle avoidance case

shows that collision avoidance is guaranteed. The flocking behavior is shown in Figure 11.9, where no collision occurred.

Next, we consider a simulation for a multiagent system with an obstacle. The obstacle is a circle whose coordinate is $[12, -1]$ with a radius of 1. The configuration of the multiagent system is the same as in the obstacle-free case above. The desired group heading is chosen as $\theta_r = \pi/4$. An agent is chosen as a leader of the group. When the collision avoidance mechanism of the leader is inactive, a speed control law is designed to drive the group at a constant speed, that is,

$$u_l = -\sigma_1(v_l - v_r) \tag{11.49}$$

where u_l is the speed control of the leader, v_l is its linear speed, and $v_r = 0.2$. During the first 50 s of their evolution, the mobile robots encounter the obstacle. The control laws (11.44) and (11.45) enable them to avoid potential collisions with the obstacle and with other neighboring agents. The orientations of the robots in Figure 11.10 converge to $\pi/4$ after 60 s. Similarly, in Figure 11.11, the linear speeds of the agents converge after 50 s. In Figure 11.12, the angular speeds converge faster to 0 after 50 s. Figures 11.13 and 11.14 demonstrate that the control inputs both are bounded. Finally, Figure 11.15 shows that no collision takes place during the evolution of the robots. The evolution of the multiagent system is shown in Figure 11.16 in which the robots cooperate to form a flocking and avoid the obstacle.

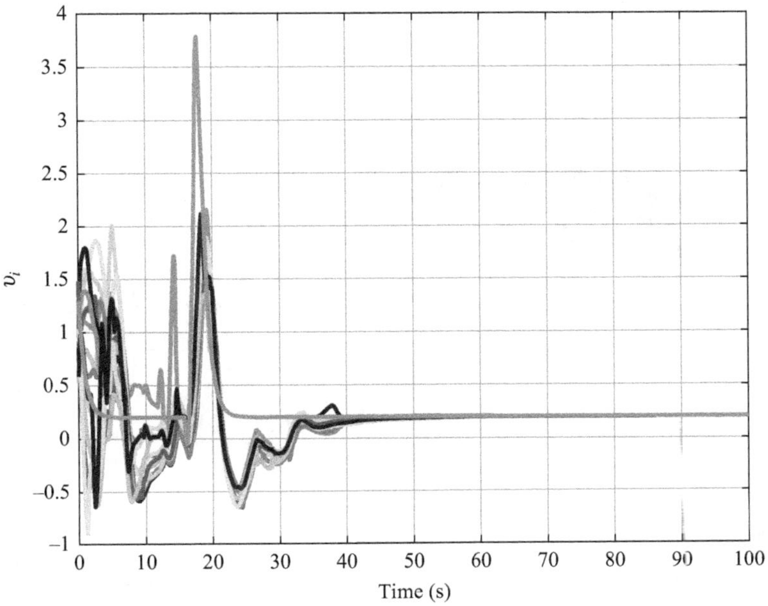

Figure 11.11 Linear speed consensus for the obstacle avoidance case

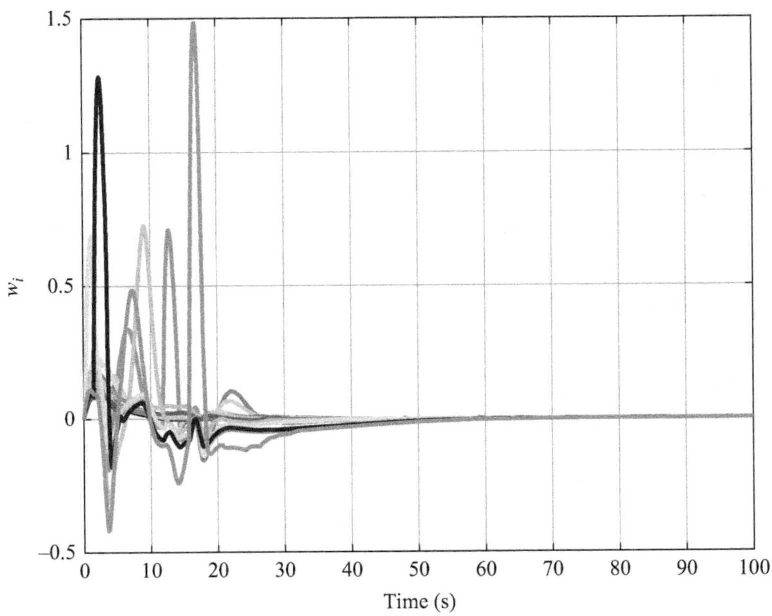

Figure 11.12 Angular speed for the obstacle avoidance case

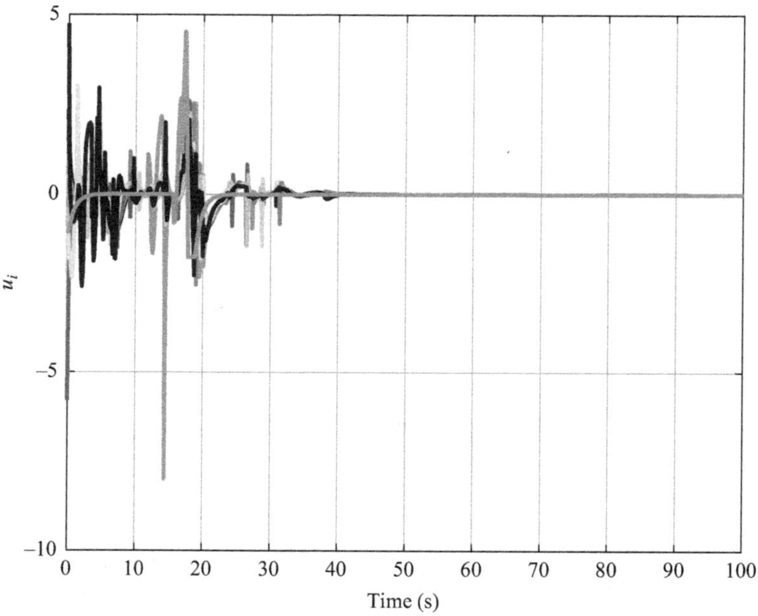

Figure 11.13 Speed control for the obstacle avoidance case

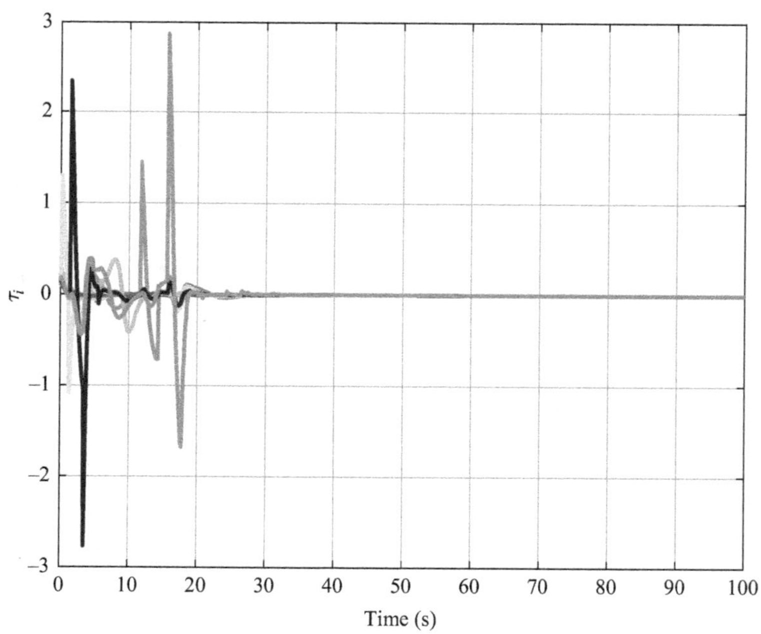

Figure 11.14 Steering control for the obstacle avoidance case

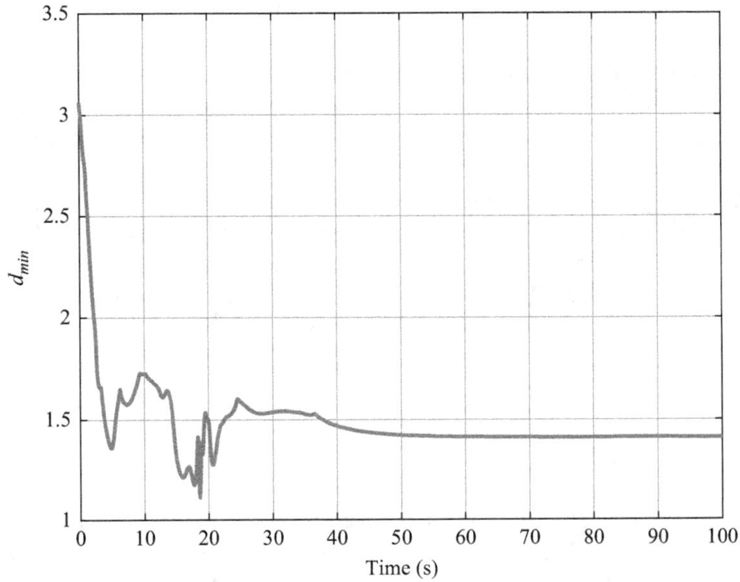

Figure 11.15 The minimum distance among agents for the obstacle avoidance case

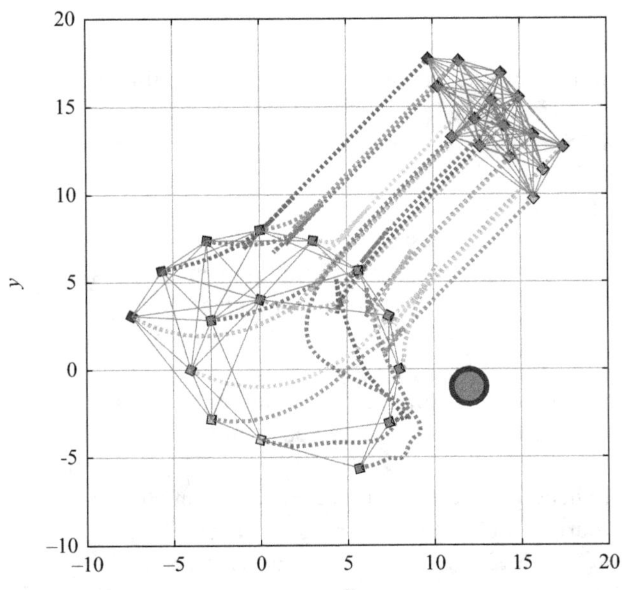

Figure 11.16 Distributed flocking of 15 mobile robots for the obstacle avoidance case

11.7 Conclusions

This chapter has presented a bounded decentralized control protocol for the flocking problem of mobile robots by a systematic fashion, where the control laws only require the information of neighbor agents. The proposed scheme is modular in designing the speed control and steering control separately. Theoretical and numerical results have shown that using our proposed method, a collective system of mobile robots achieves all the multiple objectives of the flocking control: velocity consensus, CMCA.

Future work would consider the shape and size of each mobile robot and obstacle avoidance for which a similar context was studied in [49]. Noisy and uncertain environments can affect the performance of the proposed scheme. Robustness analysis and improved methods can be proposed to address this issue.

Acknowledgments

This work was partially supported by the National Science Foundation under grant no. NSF-NRI1426828, the National Aeronautics and Space Administration (NASA) under grant no. NNX15AI02H issued through the Nevada NASA Research Infrastructure Development Seed Grant, and the University of Nevada, Reno.

References

[1] Toner J, Tu Y. Flocks, herds, and schools: A qualitative theory of flocking. Physical Review E. 1998;58:4828–4858.

[2] La HM, Sheng W. Distributed sensor fusion for scalar field mapping using mobile sensor networks. IEEE Transactions on Cybernetics. 2013 April;43(2):766–778.

[3] Muñoz F, Espinoza Quesada ES, La HM, *et al.* Adaptive consensus algorithms for real-time operation of multi-agent systems affected by switching network events. International Journal of Robust and Nonlinear Control. 2017;27(9):1566–1588. Rnc.3687. Available from: http://dx.doi.org/10.1002/rnc.3687.

[4] La HM, Sheng W. Flocking control of multiple agents in noisy environments. In: 2010 IEEE International Conference on Robotics and Automation; 2010. p. 4964–4969.

[5] La HM, Sheng W. Multi-agent motion control in cluttered and noisy environments. Journal of Communications. 2013;8(1):32–46.

[6] Dang AD, La HM, Horn J. Distributed formation control for autonomous robots following desired shapes in noisy environment. In: 2016 IEEE International Conference on Multisensor Fusion and Integration for Intelligent Systems (MFI); 2016. p. 285–290.

[7] Olfati-Saber R. Flocking for multi-agent dynamic systems: algorithms and theory. IEEE Transactions on Automatic Control. 2006 March;51(3):401–420.

[8] Tanner HG, Jadbabaie A, Pappas GJ. Flocking in fixed and switching networks. IEEE Transactions on Automatic Control. 2007;52(5):863–868.

[9] Cucker F, Smale S. Emergent behavior in flocks. IEEE Transactions on Automatic Control. 2007 May;52(5):852–862.

[10] Dong Y, Huang J. Flocking with connectivity preservation of multiple double integrator systems subject to external disturbances by a distributed control law. Automatica. 2015;55:197–203.

[11] Ghapani S, Mei J, Ren W, *et al.* Fully distributed flocking with a moving leader for Lagrange networks with parametric uncertainties. Automatica. 2016;67:67–76.

[12] Moshtagh N, Jadbabaie A. Distributed geodesic control laws for flocking of nonholonomic agents. IEEE Transactions on Automatic Control. 2007;52(4):681–686.

[13] Han TT, Ge SS. Styled-velocity flocking of autonomous vehicles: a systematic design. IEEE Transactions on Automatic Control. 2015 August;60(8):2015–2030.

[14] La HM, Sheng W, Chen J. Cooperative and active sensing in mobile sensor networks for scalar field mapping. IEEE Transactions on Systems, Man, and Cybernetics: Systems. 2015 January;45(1):1–12.

[15] Amar K, Mohamed S. Stabilized feedback control of unicycle mobile robots. International Journal of Advanced Robotic Systems. 2013;10(187):1–8.

[16] Han TT, La HM, Dinh BH. Flocking of mobile robots by bounded feedback. In: 2016 IEEE International Conference on Automation Science and Engineering (CASE); 2016. p. 689–694.

[17] Nguyen T, Han TT, La HM. Distributed flocking control of mobile robots by bounded feedback. In: 2016 54th Annual Allerton Conference on Communication, Control, and Computing (Allerton); 2016. p. 563–568.

[18] Jadbabaie A, Lin J, Morse AS. Coordination of groups of mobile autonomous agents using nearest neighbor rules. IEEE Transactions on Automatic Control. 2003 June;48(6):988–1001.

[19] Liang Y, Lee HH. Avoidance of multiple obstacles for a mobile robot with nonholonomic constraints. In: ASME Proceedings of Dynamic Systems and Control; 2005. p. 1657–1663.

[20] Riley KF, Hobson MP, Bence SJ. Mathematical Methods for Physics and Engineering. 3rd ed. Cambridge: Cambridge University Press; 2006.

[21] Tanner HG, Jadbabaie A, Pappas GJ. Flocking in teams of nonholonomic agents. In: Kumar V, Leonard N, Morse S, editors. Cooperative Control. vol. 309 of Lecture Notes in Control and Information Sciences. Springer, Berlin, Heidelberg; 2004. p. 229–239.

[22] Liang Y, Lee HH. Decentralized formation control and obstacle avoidance for multiple robots with nonholonomic constraints. In: 2006 American Control Conference; 2006. p. 5596–5601.

[23] Nguyen T, La HM. Formation control of multiple rectangular agents with limited communication ranges. In: 10th International Symposium on Advances in Visual Computing. Springer International Publishing; 2014. p. 915–924.

[24] Nguyen T, La HM, Jafari M. On the formation control of a multi vehicle system. In: ISSAT International Conference on Modeling of Complex Systems and Environments; 2015.

[25] Cruz D, McClintock J, Perteet B, *et al.* Decentralized cooperative control—A multivehicle platform for research in networked embedded systems. IEEE Control Systems. 2007;27(3):58–78.

[26] Nguyen T, La HM. Distributed formation control of nonholonomic mobile robots by bounded feedback in the presence of obstacles. In: 2017 IEEE International Conference on Real-time Computing and Robotics (RCAR); 2017. p. 206–211.

[27] Pham HX, La HM, Feil-Seifer D, *et al.* A distributed control framework for a team of unmanned aerial vehicles for dynamic wildfire tracking. In: 2017 IEEE/RSJ International Conference on Intelligent Robots and Systems (IROS); 2017. p. 6648–6653.

[28] La HM, Sheng W. Flocking control of a mobile sensor network to track and observe a moving target. In: 2009 IEEE International Conference on Robotics and Automation; 2009. p. 3129–3134.

[29] La HM, Sheng W. Moving targets tracking and observing in a distributed mobile sensor network. In: 2009 American Control Conference; 2009. p. 3319–3324.

[30] La HM, Sheng W. Adaptive flocking control for dynamic target tracking in mobile sensor networks. In: 2009 IEEE/RSJ International Conference on Intelligent Robots and Systems; 2009. p. 4843–4848.

[31] Nguyen MT, La HM, Teague KA. Compressive and collaborative mobile sensing for scalar field mapping in robotic networks. In: 2015 53rd Annual Allerton Conference on Communication, Control, and Computing (Allerton); 2015. p. 873–880.

[32] La HM, Sheng W, Chen J. Cooperative and active sensing in mobile sensor networks for scalar field mapping. In: 2013 IEEE International Conference on Automation Science and Engineering (CASE); 2013. p. 831–836.

[33] La HM, Sheng W. Cooperative sensing in mobile sensor networks based on distributed consensus. In: SPIE Optical Engineering + Applications. International Society for Optics and Photonics; 2011. p. 81370Y.

[34] La HM, Nguyen TH, Nguyen CH, *et al.* Optimal flocking control for a mobile sensor network based a moving target tracking. In: 2009 IEEE International Conference on Systems, Man and Cybernetics; 2009. p. 4801–4806.

[35] La HM, Lim RS, Sheng W. Hybrid system of reinforcement learning and flocking control in multi-robot domain. In: Proceedings Conference Applied Computing Science; 2010. p. 7–13.

[36] La HM, Lim RS, Sheng W, *et al.* Cooperative flocking and learning in multi-robot systems for predator avoidance. In: 2013 IEEE International Conference on Cyber Technology in Automation, Control and Intelligent Systems; 2013. p. 337–342.

[37] Jafari M, Sengupta S, La HM. In: Bebis G, Boyle R, Parvin B, *et al.*, editors. Adaptive Flocking Control of Multiple Unmanned Ground Vehicles by Using a UAV. Cham: Springer International Publishing; 2015. p. 628–637. Available from: http://dx.doi.org/10.1007/978-3-319-27863-6_58.

[38] La HM. Multi-robot swarm for cooperative scalar field mapping. In: Handbook of Research on Design, Control, and Modeling of Swarm Robotics. 2015;p. 383–395.

[39] La HM, Sheng W. Flocking control algorithms for multiple agents in cluttered and noisy environments. In: Bio-Inspired Self-Organizing Robotic Systems. Springer Berlin Heidelberg; 2011. p. 53–79.

[40] Carrillo JA, Fornasier M, Rosado J, *et al.* Asymptotic flocking dynamics for the kinetic Cucker-Smale model. SIAM Journal on Mathematical Analysis. 2010;42(1):218–236.

[41] La HM, Lim R, Sheng W. Multirobot cooperative learning for predator avoidance. IEEE Transactions on Control Systems Technology. 2015 January;23(1):52–63.

[42] Sepulchre R, Paley DA, Leonard NE. Stabilization of planar collective motion: all-to-all communication. IEEE Transactions on Automatic Control. 2007 May;52(5):811–824.

[43] Zavlanos MM, Tanner HG, Jadbabaie A, *et al.* Hybrid control for connectivity preserving flocking. IEEE Transactions on Automatic Control. 2009;54(12):2869–2875.

[44] Dong W. Flocking of multiple mobile robots based on backstepping. IEEE Transactions on Systems, Man, and Cybernetics, Part B (Cybernetics). 2011;41(2):414–424.

[45] Moshtagh N, Michael N, Jadbabaie A, *et al.* Vision-based, distributed control laws for motion coordination of nonholonomic robots. IEEE Trans Robotics. 2009 August;25(4):851–860.

[46] Han TT, La HM, Dinh BH. Flocking of mobile robots by bounded feedback. In: 2016 IEEE International Conference on Automation Science and Engineering (CASE); 2016. p. 689–694.

[47] Khalil HK. Nonlinear Systems. 3rd ed. Upper Saddle River, NJ: Prentice Hall; 2002.

[48] Zavlanos MM, Jadbabaie A, Pappas GJ. Flocking while preserving network connectivity. In: 2007 46th IEEE Conference on Decision and Control; 2007. p. 2919–2924.

[49] Nguyen T, La H, Le TD, *et al.* Formation control and obstacle avoidance of multiple rectangular agents with limited communication ranges. IEEE Transactions on Control of Network Systems. 2017 December;4(4):680–691.

Chapter 12

Swarming in forestry environments: collective exploration and network deployment

Micael S. Couceiro[1] and David Portugal[1]

Abstract

Landscape maintenance is a pivotal factor to reduce wildfire hazard potential. Being a difficult, time-consuming and generally expensive task, humans often face great difficulties to address the problem effectively. Considering its spatial distribution and exploration capabilities, swarm robotics presents itself as a natural solution to assist humans or other robotic agents in the reconnaissance of forestry environments and identification of fuel accumulation, such as forest debris. In this work, we propose an optimization framework for the exploration problem of forestry environments with a team of swarming robots, which also entails the deployment of a network infrastructure formed by the robots to assist agents during fire-prevention tasks. Experiments show the benefits of employing the multi-robot coordination algorithm described, which provides an effective solution for the exploration task and guarantees a final formation that maximizes network coverage for mission information exchange, considering obstacles density and other interfering factors. Thus, the system proposed might assist human operators with relevant forestry information acquired during exploration and maintain a fail-safe network for communication while agents operate in the field.

12.1 Introduction

According to statistics from the European Commission, about 65,000 fires occur in Europe every year, devastating around half a million hectares of forest areas [1]. Within these, more than 85% of the area burned corresponds to the Mediterranean region, in which Portugal stands out with an average of 18,000 annual fires over the last 25 years, followed by Spain, Italy, and Greece, with an average of 15, 9, and 1.5 thousand annual fires, respectively. This situation has a significant economic impact, whereas only in wood, each hectare of burned forest represents a loss of about 3,000€,

[1]Ingeniarius, Ltd., Portugal

which in turn corresponds to a direct loss of more than 20 million euros only during the year of 2014 in Portugal. Additionally, forest fires have an even greater impact on the economy, in particular associated with a lack of forestry regeneration capacity, damage to the tourism sector, the abandonment and depopulation of rural areas, rising unemployment, progressive lack of interest in forest management and immeasurable environmental losses [2]. Given the above, forest fires are considered as the third natural disaster with the greatest economic impacts in Mediterranean Europe [3].

In fact, the wildfire issue has been one of the most disturbing issues in the scientific community and public entities [4]. The number of occurrences and their burning area have been acknowledged for some decades, and the media attention involved has given a rise to several studies and methodologies with the purpose of carrying out a reflection and raising awareness to forest fires and to follow preventive measures that mitigate the current numbers [1,5–7].

According to Ribeiro *et al.* [2], one of the most efficient measures for the prevention of forest fires has been landscape maintenance. The accumulation of fuel, i.e., forest debris, such as pine needles, pine cones, twigs, bushes, lumber debris, firewood, hardwood leaves, cork, etc. is identified as a key factor to the spreading of wildfires. Clearing and managing forests is a costly and time-consuming task, which is not available to all. The cleaning of 1,000 m^2 of forest costs on average around 300€, making it difficult for landowners to comply with the existing legislation, which imposes fines for misconduct infringement, needing therefore the effective support from forestry associations, municipalities and/or boards of parish. The measures adopted until now have not solved the crisis of forest fires and their impact on society. Existing human resources for forest clearing are scarce and the technological means are reduced or practically nonexistent. The need to keep these forestry environments clean, particularly through weed control and forest clearing, is reflected in a high investment need, with a special focus on human resources [2].

Maintaining forestry landscape mainly via weed control and forest clearance is an application domain with unquestionable economical, environmental, and societal benefits, which significantly reduces the potential of wildfires. Thus, it deserves greater attention from the scientific community. Regular control and intervention is essential to ensure high plant productivity and vegetation growth rate [8]. This is commonly done using man-operated machinery. However, due to the lack of control of some species, erosion problems, high financial costs and other factors, it is necessary to invest in more economical and potentially more efficient alternatives, guaranteeing an ecological cleaning of forested areas.

In this chapter, we propose to develop a supporting swarm robotics system to assist human teams in the reduction of fuel accumulation to assist in landscape maintenance tasks. The focus of robotics assistance considered is twofold: (i) collective exploration and (ii) network deployment and connectivity. We describe work in progress toward the ultimate goal of providing a supporting system, which combines local positioning, navigation, and multi-robot coordination in forests, assisting humans not only in the identification of forest debris, but also in the deployment of a network infrastructure to allow for communications covering a large area of the forest.

In the next section, a literature review is conducted and the contributions of this chapter are described. Afterward, the collective exploration and network deployment problems are defined. Section 12.4 describes the swarm optimization strategy applied to solve the aforementioned problems using swarm robotics. Section 12.5 presents the preliminary results obtained through simulation experiments, as well as an in-depth analysis of the evidence obtained. Additionally, in Section 12.6 we discuss the potential of employing our approach in the real world, and finally, the chapter ends with conclusions and open issues for future research.

12.2 Related work

The literature in landscape maintenance technology focuses almost exclusively in human-in-the-loop solutions, based on three paradigms: (i) man-operated equipment [9] (e.g., brush cutters, rotary harrows, ploughs, mowers, saws); (ii) forestry machines driven by humans [10,11] (e.g., mulchers, tractors, harvesters, forwarders, skidders, etc.); and (iii) remotely controlled machines, typically manned systems that are converted for teleoperation, such as feller bunchers, tree shear excavators, or modified bobcat systems [12].

Semi-autonomous and remotely controlled systems allow the user to maneuver heavy forestry equipment in outdoor terrain with steep slopes and up to a mile away. The user can monitor the progress via cameras mounted on the equipment or at a safe distance, driving around obstacles while taking advantage of human perception. Based on this, several different designs have been presented for vegetation clearing robots [13] and harvesting robots [14]. Despite the presentation of a variety of forestry robot designs with different forms and sizes (see Figure 12.1), the lack of fully autonomous robots for vegetation removal and brush clearing for fire prevention is an evident gap in Robotics research. Regardless of the considerable number of studies and proofs-of-concept, the development of fully autonomous robotic solutions for wildfire prevention is still in a very early stage. This stems from the huge challenges imposed by rough terrain traversability [15] (e.g., steep slopes), autonomous outdoor navigation and locomotion systems [16], limited perception capabilities [17], and reasoning and planning under a high-level of uncertainty [18]. Hence, related work still presents evident limitations.

In fact, the work in robotics addressing vegetation deals almost exclusively with thick vegetation identification for outdoor traversability, so as to safely drive in challenging outdoor environments, and overcome vegetation similarly to obstacles [19–21]. However, advances in related areas such as field and service robots, firefighting robots and agricultural robots provide good prospects for robotic applications in landscape maintenance and forest fuel reduction. For instance, Ohkawa *et al.* [22] described a modular robotic brush-cutter with a center articulated body, a manipulator, two laser imaging detection and ranging devices, global positioning system (GPS), and an innovative sensor steering mechanism for the articulated vehicle. Preliminary tests achieved interesting autonomous behavior and performance in brush cutting within fields with grass height up to 50 cm without leaving strips of unmowed

Figure 12.1 *Examples of forestry robots found in the literature: (a) Ponsse's Harvester Scorpion 8 Wheel; (b) TreeRover, the tree-planting robot prototype from Iota Enterprises; (c) 3D printed robot platform designed for debris retrieval by RobotMissions.org; (d) forest monitor robot developed by the Oregon State University*

grass. Nevertheless, the wheeled locomotion system and cutting mechanism would face high difficulties in terrains with rougher relief and denser vegetation. Furthermore, GPS does not provide the intended accuracy in the presence of thick vegetation, which makes it unfeasible for localization and mapping in this context [23]. Presently, there are several autonomous lawn mowing commercial solutions with proven performance, like the solar-powered Vitirover, the Husqvarna Automower, and others [24]. Most of these products are able to maintain lawns in any weather, with rough terrain and slopes up to 35%. Typically, when they meet obstacles, they stop and redirect themselves, and they also charge automatically when needed.

Hellström *et al.* [25] presented a study that points to the limited degree of automation and autonomy for forest vehicles. The authors pinpoint the importance of support for autonomous decision-making and the vehicle navigation system. They also present a robotic prototype developed at Umeå University (Sweden), which learns from records of feasible paths driven by a human operator and drives autonomously along the demonstrated paths. Achieving precise localization with available multimodal sensory solutions, such as GPS, odometry, and inertial measurements, is discussed,

Figure 12.2 *Examples of multi-robot systems: (a) an arena with S1R robots from the Department of Cybernetics, Slovak University of Technology in Bratislava [38]; (b) diverse robots from the Distributed Intelligence Laboratory at the University of Tennessee [39]; (c) the DARPA Centibots Project (SRI International, Stanford University, University of Washington and ActivMedia) [40]; (d) a team of Turtlebots from the center for Robotics and Intelligent Machines at Georgia Tech Research Institute [41]*

as well as using laser scanners and radars for obstacle detection, and autonomous driving in predetermined paths.

Another evident gap in the robotics literature for vegetation removal and fire prevention is the lack of multi-robot and swarm solutions. Yet, the advantages of using multiple robots instead of single robot solutions for forest coverage [26–29] and several other applications [30,31] have been vastly documented. Examples of multi-robot systems are presented in Figure 12.2. Besides distributed control, increased autonomy, communicative agents and greater fault-tolerance [32], multi-robot, and swarm systems can benefit from the assistance of nearby teammates during emergencies, failures, or malfunctions. Robotic teams allow having many robots in numerous places, carrying out diverse tasks at the same time, i.e., space distribution. In fact, most missions are solved much quicker if robots operate in parallel [33–35]. Advantages and limitations of team operation, such as communication requirements and scalability, were discussed in [36], as well as the effects of distributed group configurations in collective exploration and inspection tasks. In general, multi-robot systems and swarm robotic systems have the potential to increase robustness and reliability of solutions by introducing redundancies in the capabilities across robot team member and graceful performance degradation, remaining functional in case some of the robots fail. From the standpoint of cost-effectiveness, it can actually be cheaper and more practical to build a set of less capable and simpler robots that cooperate, instead of one single robot to perform the entire mission. However, one of the main difficulties when approaching these systems is to coordinate many robots to perform a complex, global task in an efficient manner, maximizing group performance under a wide range of conditions, with the flexibility to take advantage of the resources available, embrace the requirements and constraints imposed and resolve issues like action selection, coherence, conflict resolution, and communication. This cannot be done by just increasing the number of robots assigned to a task. A coordination mechanism must exist to establish relationships between agents so that they can accomplish the mission effectively [37].

In this chapter, we propose a coordination framework for swarm robotic systems in collective forest exploration and network deployment tasks, with the ultimate goal to provide assistance to human teams in landscape maintenance and fire prevention tasks by identifying forest debris and deploying a communication infrastructure in large forestry areas. To this end, we will make use of a swarm robotics coordination algorithm for exploration named robotic Darwinian particle swarm optimization (RDPSO) proposed by the authors previously [33]. Despite the increase of the use of machinery for clearing forests during the last decades, such equipment do not provide autonomous ways to clean and cut bushes effectively, and they must be exclusively manipulated by humans, requiring the constant presence of an operator. This entails high risks for human operators and high costs for commissioners. We envision to innovate and enable people to carry on with their jobs without risking their safety and physical integrity. To this end, the design of teams of autonomous robots, supervised by teams of human agents, allows one to carry out a vast set of tasks of prevention of fires and cleaning of lands in hostile environments, which are extremely challenging to human beings, such as terrains with steep slopes, near forest shores, rocky relief, and other natural obstacles, where the fires are commonly at an early stage of deflagration.

12.3 Problem definition

In this chapter, we consider an autonomous swarm of robots with variable size which operates in a bounded outdoor environment, represented by a 2D occupancy grid map. For simplicity, the team of robots is assumed to be composed by N homogeneous differential drive mobile robots that can rotate in place and translate continuously at bounded speed. Robots are equipped with a distance sensor that has a maximum range $r_s > 0$ and an angular field-of-view $\phi \in [0, \pi]$. The sensor is modeled after a light detection and ranging sensor, which cannot penetrate obstacles but is capable of detecting the distances to obstacles visible to it within the sensor's field-of-view. The problem is then divided into two phases: (i) collective exploration and (ii) network deployment.

12.3.1 Collective exploration

The mission begins with a swarm of robots, which coordinates itself to explore and cooperatively map an unknown area. This phase implies the spatial distribution of robots in the environment, while maintaining the connectivity of a mobile ad-hoc network (MANET) with occasional disconnections, so as to maximize the explored area. The main goal is for the fleet of cooperative mobile robots to explore thoroughly the forestry scenario, performing a cooperative mapping of the environment that can be used to signalize fuel accumulation (e.g., forest debris). Hence, the objective function of the team is defined as a cost function in which robots need to minimize the map's entropy (cf., [42]). Therefore, during exploration, each robot n computes its best frontier cell as

$$m_i^s = \underset{m_i \in \mathcal{N}(x_n[t], r_s)}{\arg\max} \; \Psi(x_t, m_i) \cdot ||\vec{\nabla} H(m_i)||, \tag{12.1}$$

wherein $\mathcal{N}(x_n[t], r_s)$ represents the set of frontier cells located in the neighborhood of robot n with sensing radius r_s. The coefficient $\Psi(x_t, m_i) \in [0, 1]$ measures whether the cell m_i is in line-of-sight from a position $x_n[t]$, which also implies that cell m_i is likely to be obstacle-free. Moreover, the entropy of the cell m_i is represented by $H(m_i)$ and may be calculated as

$$H(m_i) = -p(m_i) \cdot \log_2 (p(m_i)) - (1 - p(m_i)) \cdot \log_2 (1 - p(m_i)), \tag{12.2}$$

being $p(m_i)$ the probability that a grid cell is occupied.

12.3.2 Network deployment

After a certain degree of confidence in the built representation of the scenario, the team of robots begins the network deployment stage. Consider a group of N robots where each robot is both an exploring agent of the environment and a mobile node of a MANET that performs packet forwarding, according to a paradigm of multi-hop communication [43]. The goal is to ensure that robots maximize the network, while ensuring that the MANET remains connected throughout the mission. Robots are endowed with wireless communication capabilities that allow multi-hop messaging within a maximum one-hop range of $d_{max} > 0$. Trying to maintain the network connectivity by only taking into account the communication range d_{max} does not match reality since the propagation model is more complex, i.e., the signal depends not only on the distance but also on the multiple paths from trees and other obstacles. However, in simulation, the communication range is a good approach, being simpler to implement.

In order to overcome the non-connectivity between robots, the desired position of each robot $x_n[t + 1]$ must be controlled, and the connectivity between robots can be monitored by means of a link matrix $L = [l_{ij}] \in \mathbb{R}^{N \times N}$ for an N robot network, wherein each entry represents the link between robot i and j.

In the next section, we propose a distributed swarm optimization solution for the collective exploration and network deployment problems in a forestry environment.

12.4 A swarm optimization solution for the collective exploration and network deployment problems

This section briefly presents the RDPSO algorithm previously proposed [44] and extended in subsequent works [33,36]. The RDPSO approach is an adaptation of the Darwinian PSO (DPSO), presented by Tillett *et al.* [45] to real mobile robots, wherein the entire population of robots is divided into multiple smaller coopetitive swarms (or subgroups). As such, several features have been added: (i) a novel "punish-reward" mechanism to emulate the deletion (i.e., swarm exclusion) and creation of robots (i.e., acceptance of excludes); (ii) an improved inertial influence based on the fractional calculus concept, taking into account convergence dynamics; (iii) an obstacle avoidance behavior to avoid collisions; (iv) an algorithm to ensure that the MANET remains connected throughout the mission; and (v) a novel methodology to establish the initial planar deployment of robots preserving the connectivity of the MANET, while spreading out the robots as most as possible. This will be the strategy adopted in this work. Thus, the behavior of robot n can be described by the following discrete equations at each discrete time, or iteration, $t \in \mathbb{N}_0$:

$$\begin{aligned} v_n[t+1] &= \alpha v_n[t] + \frac{1}{2}\alpha(1-\alpha)v_n[t-1] + \frac{1}{6}\alpha(1-\alpha)(2-\alpha)v_n[t-2] \\ &\quad + \frac{1}{24}\alpha(1-\alpha)(2-\alpha)(3-\alpha)v_n[t-3] + \sum_{i=1}^{4} \rho_i r_i(\chi_i[t] - x_n[t]), \quad (12.3) \end{aligned}$$

$$x_n[t+1] = x_n[t] + v_n[t+1]. \quad (12.4)$$

The parameters α, $0 < \alpha \leq 1$ and ρ_i, $\rho_i > 0$ and $i = 1,2,3,4$ assign weights to the inertial influence, the local best (cognitive component), the global best (social component), the obstacle avoidance component, and the communication connectivity component when determining the new velocity. Coefficients r_i, $i = 1,2,3,4$, are random matrices where each component is generally a uniform random number between 0 and 1. The variables $v_n[t]$ and $x_n[t]$ represent the velocity and position vector of robot n, respectively, and $\chi_i[t]$ denotes the best position of the cognitive, social, obstacle, and MANET matrix components. The cognitive $\chi_1[t]$ and social components $\chi_2[t]$ are common in the PSO algorithm, where $\chi_1[t]$ represents the local best position, and $\chi_2[t]$ represents the global best position of robot n. The obstacle avoidance component $\chi_3[t]$ is represented by the position of each robot that optimizes a monotonically decreasing or increasing function $g(x_n[t])$ that describes the distance to a sensed obstacle. The MANET component $\chi_4[t]$ is represented by the position of the nearest neighbor increased by a vector, whose intensity corresponds to the maximum communication range d_{max} and whose direction leads toward the robot's nearest neighbor position.

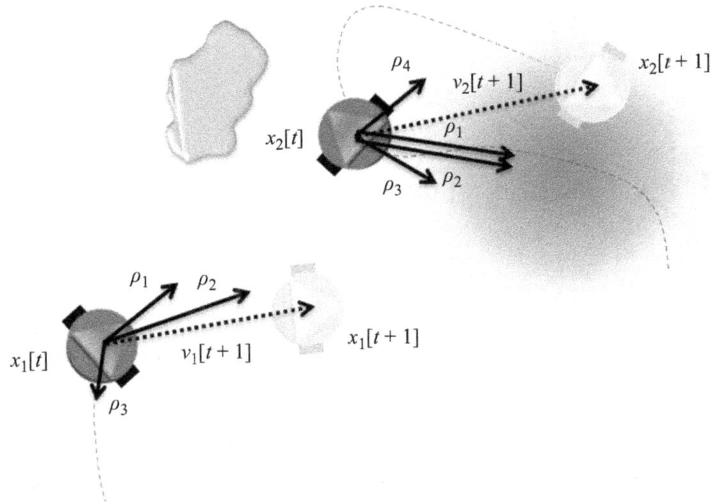

Figure 12.3 Geometrical illustration of the RDPSO using a subgroup of two robots. The cognitive coefficient ρ_1 influences robots to improve their own individual solutions. For robot 2, both cognitive ρ_1 and social coefficients ρ_2 influence motion toward the same position since this is the best performing robot of the subgroup. On the other hand, the social coefficient ρ_2 of robot 1 attracts it to the global best position found so far by robot 2. The obstacle susceptibility weight ρ_3 influences robots to move to a previous position, in which obstacles were not detected within robots range. As for the communication component ρ_4, since robot 2 was the first to choose robot 1 as is nearest neighbor, being at a distance inferior to d_{max}, it is slightly repelled by it. Note that the fractional coefficient α influences the next position of both robots $x_n[t+1]$, $n = 1,2$, with an inertial factor that considers their trajectory

A higher ρ_4 will enhance the ability to maintain the network connected ensuring a specific communication range between robots. Figure 12.3 presents an illustration of the RDPSO approach on a small team of two robots.

Considering the core of this chapter, the RDPSO is applied in the two phases previously defined in Section 12.3.

12.4.1 Collective exploration with RDPSO

The RDPSO is a meta-heuristic that consists of multiple dynamic swarms, thus enabling a distributed approach since the population of robots that may had been previously defined by the whole group is divided into multiple smaller populations (one for each swarm). In large scenarios, the computation of the global exploration map can be significantly reduced, as robots are not so constrained by their teammates,

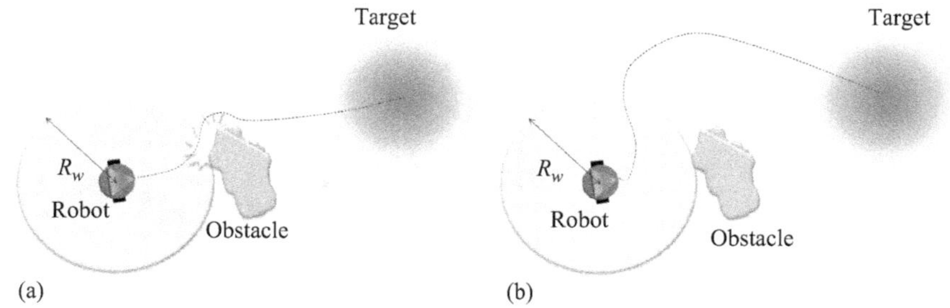

Figure 12.4 Illustrative example of the obstacle avoidance behavior of a robot. The relation between ρ_3 and the other weights, namely ρ_1, ρ_2, and ρ_4, depends on the susceptibility of each robot to the main objective and the obstacle avoidance behavior. For instance, in (a) $\rho_3 < min(\rho_1, \rho_2)$ may lead to a faster convergence to the final solution, but it may also lead to obstacle collisions, while in (b) $\rho_3 > min(\rho_1, \rho_2)$ may increase the performance of obstacle avoidance but may lead to a slower convergence to the main objective

and each swarm acts independently and in parallel. Therefore, by itself the strategy of dividing the whole population into distinct swarms may allow a faster exploration of the environment.

In order to utilize the RDPSO approach for swarm exploration, besides assigning a large weight for the cognitive component ($\rho_1 = 1.8$) in (12.2), the social component weight is minimal, $\rho_2 = 0.1$, allowing robots to spread out in the environment. A large obstacle susceptibility weight, $\rho_3 = 5.0$, is necessary to avoid collisions not only with the obstacles, but also between robots (cf., Figure 12.4). Finally, the MANET communication constraint weight is kept low, $\rho_4 = 0.5$, so as not to overly restrict the exploration ability of robots.

Since we are considering a bounded outdoor environment, we are able to calculate the expected environment area prior to the mission, and we compute the exploration ratio η using

$$\eta(\%) = \frac{explored\_area}{total\_area} \times 100. \tag{12.5}$$

When $\eta > 75\%$, the exploration task finishes and the network deployment stage may begin.

12.4.2 Network deployment with RDPSO

When the behavior is switched from the exploration task to the network deployment, all robots become now part of a unified swarm, as they all need to strictly coordinate to maintain the MANET, as mentioned in Section 12.3.2.

Assuming that the network supports multi-hop connectivity, the communication between two end nodes (i.e., robots) is carried out through a number of intermediate

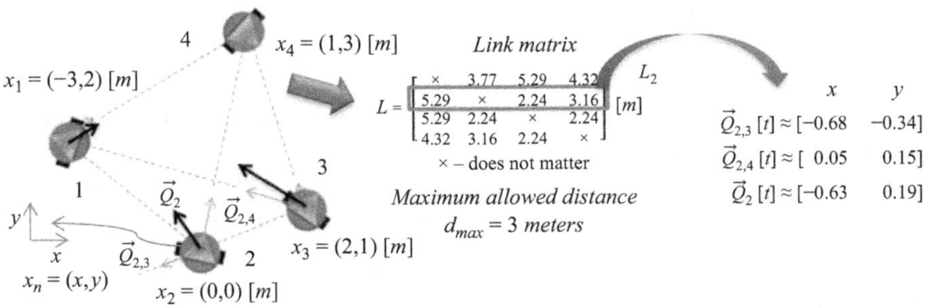

Figure 12.5 Example of a MANET topology. The dashed lines represent the vector $\vec{Q}_n[t]$ for each robot extracted from ith line of the Link Matrix L; thinner arrows represent the force vectors $\vec{Q}_{n,i}[t]$ regarding the chosen neighbor i; and larger arrows represent the resulting MANET component $\chi_4[t]$

nodes whose function is to relay information from one point to another. Considering that nodes are mobile, it is necessary to guarantee the communication between all nodes. One way to ensure the full connectivity of the MANET is to "force" the connectivity between a robot and its nearest neighbor, which in turn has not chosen it as its nearest neighbor. Since connectivity depends on distance, the connection between robots may be ensured by computing the minimum value of each line of the link matrix L, after excluding zeros and $\langle i, j \rangle$ pairs previously chosen. This is depicted in Figure 12.5.

The exchanged data concerning to the robots' position allows the implicit processing of the RDPSO algorithm by the team in a distributed way. However, as a limitation, every robot needs to be aware (or at least estimate) of the position of all other robots in order to proceed to the next iteration of the algorithm. Each robot identifies its nearest neighbor in order to compute $\chi_4[t]$, and a high value of ρ_4 should be chosen to ensure the MANET connectivity.

Once again, to adapt the RDPSO for the network deployment task we only need to readjust the RDPSO coefficient weights, when the transition from the exploration task to the network deployment task occurs. During network deployment, there is no need to limit the robot's motion through the cognitive and social components. Thus, we set $\rho_1 = 0$ and $\rho_2 = 0$. We maintain a large obstacle susceptibility weight as before, $\rho_3 = 5.0$ and we increase the MANET communication constraint weight to $\rho_3 = 2.0$.

As a stopping condition, we evaluate the mobility of the whole population N of robots (in the last s discrete steps) using a centroid displacement variable κ, given by

$$\overline{\mathbf{X}}_{(1\times N)} = \left[\sum_{i=t-s}^{t} \frac{x_1[i]}{s+1} \quad \cdots \quad \sum_{i=t-s}^{t} \frac{x_N[i]}{s+1} \right], \tag{12.6}$$

$$\mathbf{K}_{(1\times N)} = \left[|x_1[t] - \overline{\mathbf{x}}_{1,1}| \quad \cdots \quad |x_N[t] - \overline{\mathbf{x}}_{1,N}|\right], \tag{12.7}$$

$$\kappa = \widetilde{\mathbf{K}}. \tag{12.8}$$

We expect robots to converge into a final position, guaranteeing that the network is maintained. Thus, when $\kappa < \kappa_T$, the network deployment mission finishes, and the robots remain in their final position. In our simulation experiments, we have used $s = 100$ steps, and $\kappa_T = 2.5$, as the centroid displacement threshold.

12.5 Results

In order to test different distributed group configurations for the swarm robotics system in collective exploration and network deployment missions in a forestry environment, several simulations using the multi-robot simulator (MRSim) were run. This is an evolution of the Autonomous mobile robotics toolbox SIMROBOT (SIMulated ROBOTs) previously developed for an obsolete version of MATLAB®. A beta version of this toolbox may be found online.[1]

We ran experiments with different robot population sizes $N = \{6, 12, 24\}$ initially positioned randomly in a polygonal scenario that resembles a forest environment with trees of different sizes spread within the area, as illustrated in Figure 12.6. We have set the initial subgroup/swarm sizes of $N_S = 6$, and a communication range for each robot of $d_{max} = 15$ m, and we have conducted 30 trials for each test configuration, allowing for a comprehensive analysis of the results.

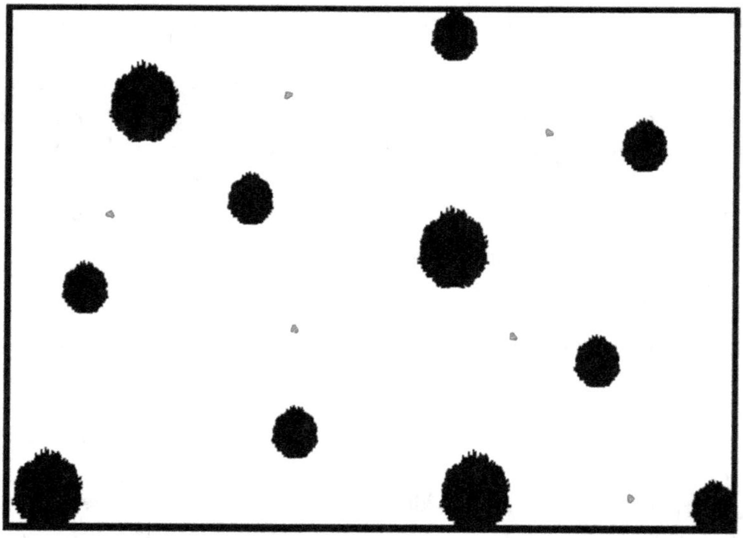

Figure 12.6 Illustrative example of the final robot formation for network deployment with six mobile units

[1] http://www.mathworks.com/matlabcentral/fileexchange/38409-mrsim-multi-robot-simulator–v1-0-

In Figures 12.6–12.8, we present the typical final configuration of the swarm robotic system positioning at the end of the mission with different team sizes. As one can see, the robots tend to keep a constant distance between them and their teammates so as to maximize network coverage in the network deployment stage. Clearly, as more

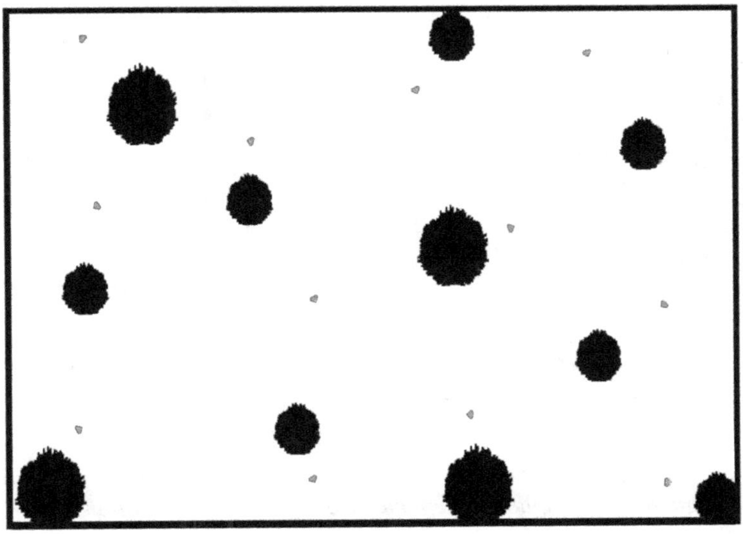

Figure 12.7 Illustrative example of the final robot formation for network deployment with 12 mobile units

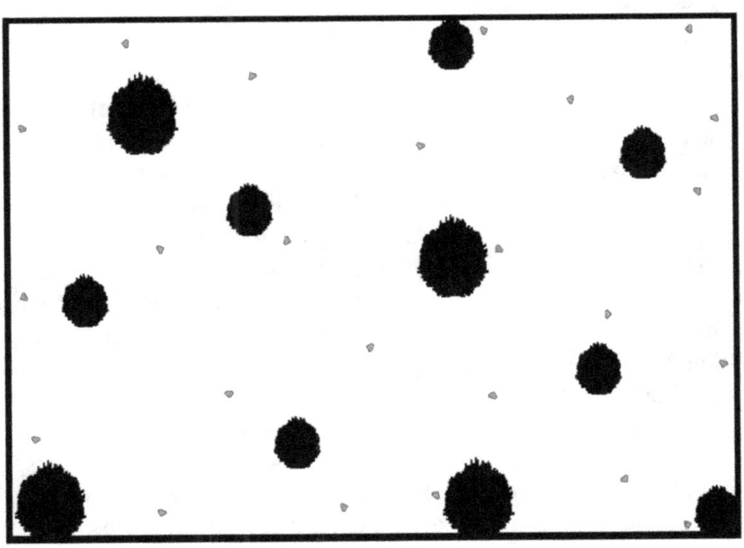

Figure 12.8 Illustrative example of the final robot formation for network deployment with 24 mobile units

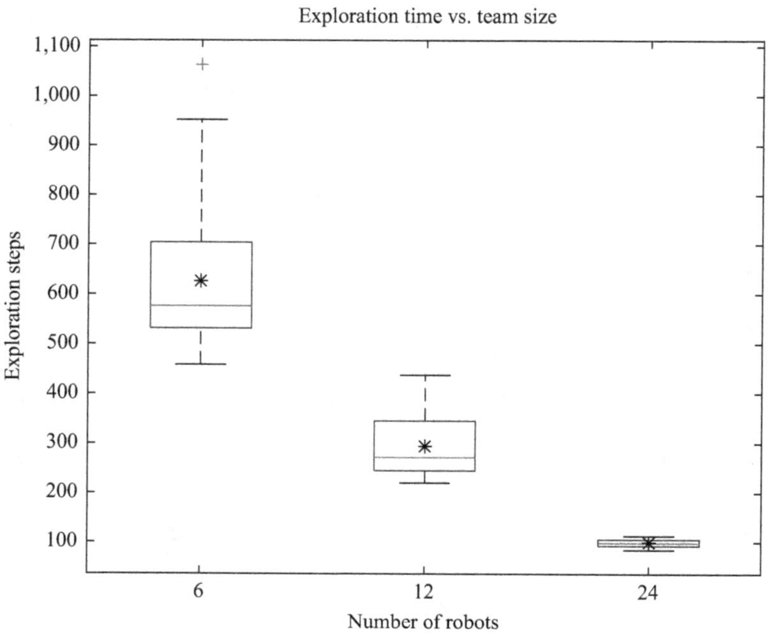

Figure 12.9 Impact of the number of robots on exploration time

robots are deployed in the environment, the larger the network coverage is expected to be. However, this comes at the cost of a much slower network deployment (cf., Figure 12.10).

The box plots in Figures 12.9–12.11 represent different mission stages' duration (exploration time, network deployment time, and final mission time) for each tested number of robots. The average value along 30 trials is represented by an asterisk. The ends of the boxes and the horizontal line in between correspond to the first and third quartiles and the median values of the mission stage duration, respectively. An analysis of these box plots show that the exploration stage of the mission tends to occur much faster with a high number of robots, as opposed to the network deployment stage, which is quicker with a low number of robots. This results in similar mission duration for 6 and 12 robots, while on the 24 robots' case, robots spend too much time coordinating themselves to appropriately deploy a communication network, which has an important impact on the final mission duration.

It is also interesting to note that the average value of exploration time is a bit superior to the median value for all team sizes. This means that the exploration time values are positively skewed, i.e., most values are below the average, and as a consequence, adjacent values are above the third quartile (see the cross in Figure 12.9 for the six robots case). Contrarily, the trend seems to be the opposite for the case

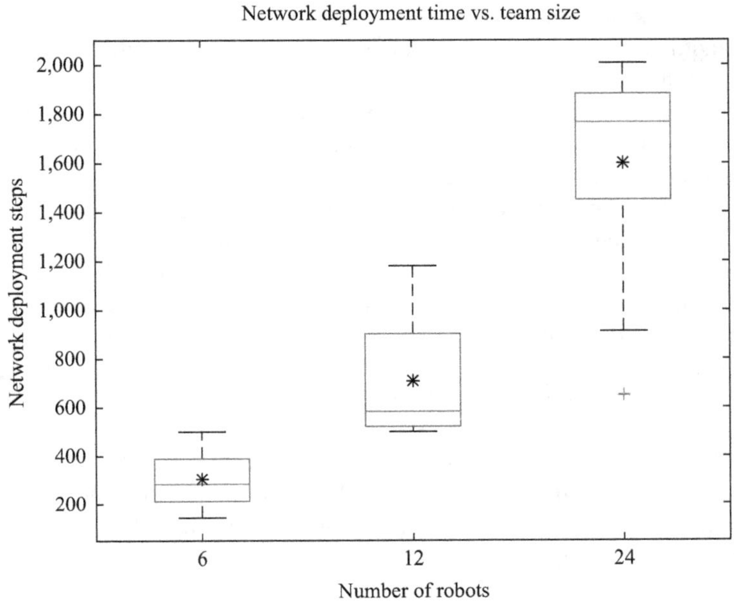

Figure 12.10 Impact of the number of robots on network deployment time

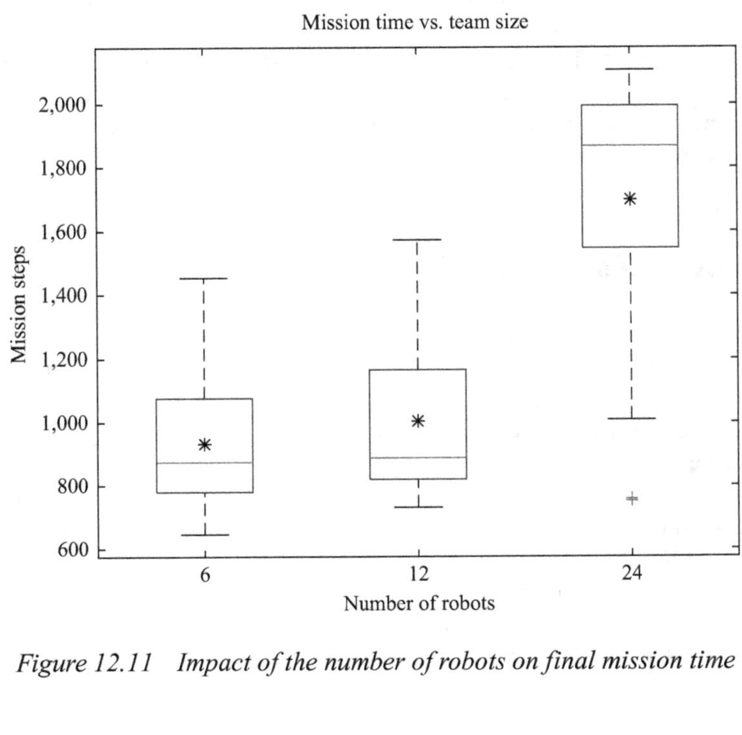

Figure 12.11 Impact of the number of robots on final mission time

of the network deployment time with 24 robots, where the average time is below the median and therefore the adjacent value is below the first quartile (see the cross in Figure 12.10 for the 24 robots case).

Regarding the collective exploration, besides taking longer to explore the forestry scenario with six robots, it also becomes evident that with this team size, the swarm of robots usually cannot explore the full environment timely within the mission. Considering that all robots are equipped with a distance sensor with maximum range $r_s = 3\,\text{m}$, a scan angle of 180°, and a resolution of 180 points, in our experiments the team of six robots only reached an exploration ratio of $\eta = 78.03\% \pm 2.42\%$, with maximum and minimum values of 86.34% and 75.47%, respectively. This is illustrated by the curve in Figure 12.12(a). In the 12 robots case, the exploration ratio in 30 trials was $\eta = 84.65\% \pm 3.42\%$ with maximum and minimum values of 91.86% and 77.71%, respectively. An illustration of the cooperative map building with 12 robots in the trial with maximum exploration ratio is shown in Figure 12.13. With 24 robots, the swarm was able to explore the full scenario in all the 30 trials ($\eta = 100\%$). The dashed line in Figure 12.12(a) represents the threshold of 75% for the exploration ratio, which was defined as the transition from the collective exploration to the network deployment task.

On the other hand, we show in Figure 12.12(b) the evolution of the largest subnetwork coverage along missions with different numbers of robots. The largest subnetwork represents the maximum network coverage area that derives from the subset of any N_S connected robots, i.e., within communication range, considering multi-hop packet forwarding. As expected, the network coverage is much higher with larger teams of robots. In general, an increasing network coverage trend can be seen along the missions. Still, the 24-robot case is somehow a special case, as the number of mobile nodes is too high considering the communication range and the limited area of the scenario, which results in the 24 units being connected to each other during most of the mission time.

12.6 Discussion

The preliminary results extracted in this chapter provide interesting hints for the deployment of real-world swarm robotic systems in forestry environments. Due to their spatial distribution and mobility, parallel exploration capabilities, network infrastructure deployment and coverage, high energetic autonomy, greater fault-tolerance, and robustness, swarm robotics presents itself as a natural solution to assist humans or other robotic agents in the reconnaissance of forestry environments and identification of fuel accumulation, such as forest debris. Our proposed optimization framework based on the RDPSO approach has shown to be extremely effective in solving the exploration problem of a forestry environment, as well as the deployment of a network infrastructure in simulations, simply by tuning the algorithm parameters. Furthermore, RDPSO is particularly suited to solve real-world problems, due to ability

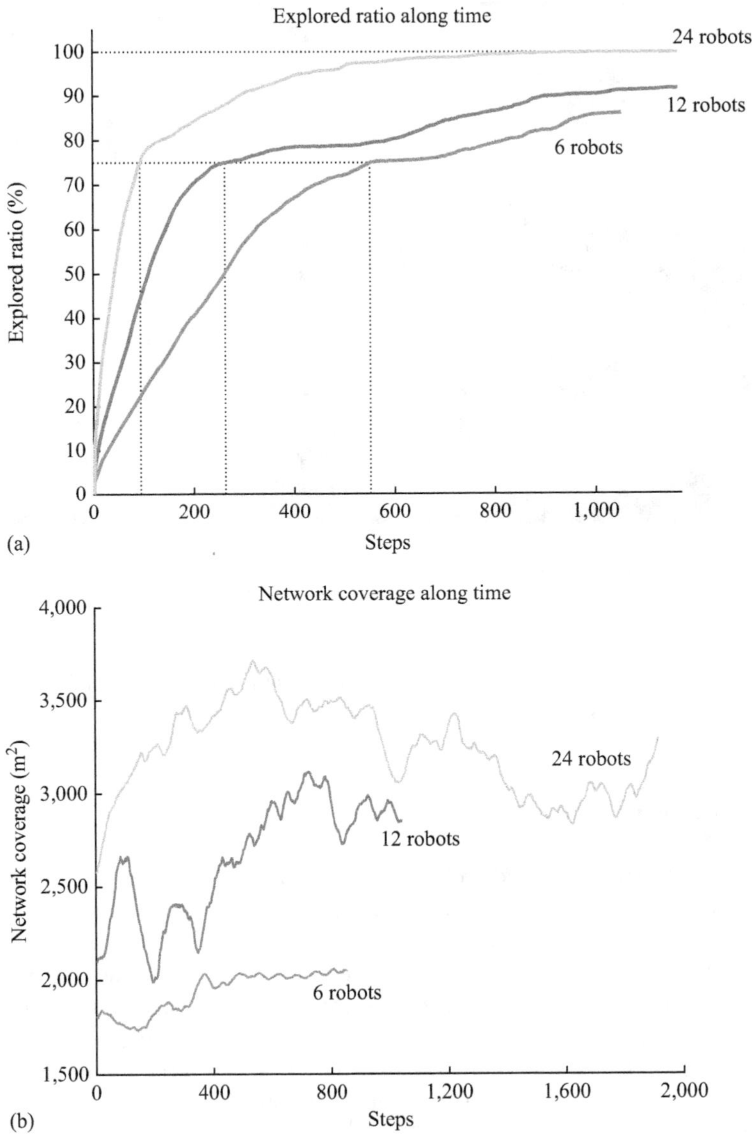

Figure 12.12 Exploration ratio and network coverage with teams of 6, 12, and 24 robots: (a) evolution of the exploration ratio η along a mission with distinct N and (b) evolution of the largest subnetwork coverage along a mission with distinct N

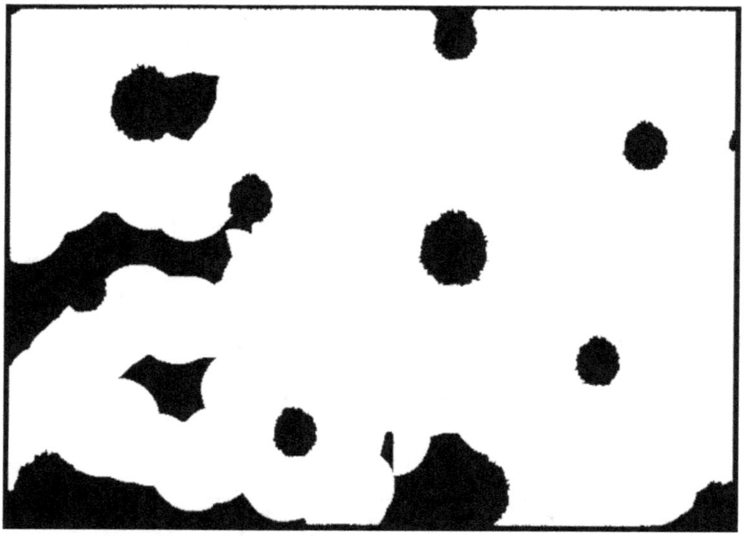

Figure 12.13 Example of cooperative map building using 12 robots ($\eta = 91.86\%$)

of the algorithm to turn a computational particle (present in all PSO-based approaches) into a swarm robotic agent.

12.7 Conclusions and future work

In this work, we have described preliminary concepts and work in progress toward a swarm robotics framework for collective forest exploration and network deployment, with the ultimate goal to assist human operators in cleaning forest debris for landscape maintenance and fire prevention. Simulation results with different configurations have shown that it is crucial to dimension the team, having into account the area where the mission occurs and the communication constraints. These aspects are fundamental to guarantee high exploration ratio and network coverage in the minimum possible time, while at the same time not overloading each robot.

Experiments have also shown the benefits of employing the multi-robot coordination algorithm described, which provides an effective solution for the exploration task and guarantees a final formation that maximizes network coverage for mission information exchange, considering obstacles density and other interfering factors. Thus, the system proposed has the potential to assist human operators with relevant forestry information acquired during exploration and maintain a fail-safe network for communication while agents operate in the field.

By validating this approach in simulation experiments, the foundations were laid for a whole series of more thorough experiments to be carried out on different test and real-world scenarios. In the future, we intend to address the needs for physical deployment of swarm robots, such as hardware prototypes, multi-robot localization and mapping, and testing the approach in the physical world. We also aim at addressing the cooperation between teams of robots and humans, in order to improve the landscape maintenance missions effectiveness, mainly focusing on the efficient sharing of forestry information, e.g., sending map information to human agents, automatic detection and flagging of forestry debris, etc. through information utility assessment.

Acknowledgment

This work was supported by the **STOP**: **S**eguranças robó**T**icos co**OP**erativos (*Cooperative Robotic Securities*) research project funded by "Agência Nacional de Inovação" (*National Innovation Agency*) under the *Portugal 2020 Programme* with reference CENTRO-01-0247-FEDER-017562.

References

[1] San-Miguel-Ayanz, J., Schulte, E., Schmuck, G., *et al.* (2012). *Comprehensive monitoring of wildfires in Europe: the European Forest Fire Information System (EFFIS)*, European Commission, Joint Research Centre Italy.

[2] Ribeiro, C., Valente, S., Coelho, C. & Figueiredo, E. (2015). *A look at forest fires in Portugal: technical, institutional, and social perceptions*. Scandinavian Journal of Forest Research, 30(4), pp. 317–325.

[3] Ribeiro, M. I. (2014). *Prevenção e Deteção de Incêndios Florestais: Análise Holística e Sistemas Tecnológicos* (In Portuguese). Master Dissertation on Environmental Engineering, Faculty of Engineering, University of Porto, Portugal.

[4] Fernandes, S. (2015). *Incêndios Florestais em Portugal Continental fora do "período crítico" – Contributos para o seu conhecimento* (In Portuguese). Master Dissertation on Physiography, Department of Geography of the Language and Literature Faculty, University of Coimbra, Portugal.

[5] De Dios, V. R., Fischer, C., & Colinas, C. (2007). *Climate change effects on Mediterranean forests and preventive measures*. New Forests, 33(1), pp. 29–40.

[6] Zhang, J., Li, W., Yin, Z., Liu, S. & Guo, X. (2009). *Forest fire detection system based on wireless sensor network*. In 4th IEEE Conference on Industrial Electronics and Applications, ICIEA 2009, pp. 520–523, IEEE.

[7] Kolström, M., Lindner, M., Vilén, T., *et al.* (2011). *Reviewing the science and implementation of climate change adaptation measures in European forestry*. Forests, 2(4), pp. 961–982.

[8] Veroustraete, F., Sabbe, H. & Eerens, H. (2002). *Estimation of carbon mass fluxes over Europe using the C-Fix model and Euroflux data*. Remote Sensing of Environment, 83, pp. 376–399.

[9] González-García, S., Bonnesoeur, V., Pizzi, A., Feijoo, G. & Moreira, M. T. (2013). *The influence of forest management systems on the environmental impacts for Douglas-fir production in France*. Science of the Total Environment, 461–462, pp. 681–692.

[10] Recchia, L., Boncinelli, P., Cini, E., Vieri, M., Pegna, F. G. & Sarri, D. (2011). *Agricultural and Forestry Mechanization*. In Multicriteria Analysis and LCA Techniques, pp. 57–97, Springer, London.

[11] Bolding, M. (2006). *An Integrated Study of Mechanical Forest Fuel Reduction: Quantifying Multiple Factors at the Stand Level*. Ph.D. Thesis, Department of Forest Engineering and College of Forestry, Oregon State University, USA.

[12] Skibba, B. (2011). *Range Rovers: Robots Roam Camp Guernsey to Prove UXO Range Clearance*, Unmanned Systems, 29(10), pp. 30–32.

[13] *McConnel's All-Terrain Robocut* (2017), Available Online: http://www.mcconnel.com/remote-control-technology/_product/20/robocut/.

[14] Hannan, M. W. & Burks, T. F. (2004). *Current developments in automated citrus harvesting*. In 2004 ASAE Annual Meeting (p. 1). American Society of Agricultural and Biological Engineers.

[15] Suger, B., Steder, B. & Burgard, W. (2015). *Traversability analysis for mobile robots in outdoor environments: a semi-supervised learning approach based on 3D-lidar data*. In Proc. of the IEEE International Conference on Robotics and Automation (ICRA 2015), Seattle, Washington, May 26–30.

[16] Siegwart, R., Lamon, P., Estier, T., Lauria, M. & Piguet, R. (2002). *Innovative design for wheeled locomotion in rough terrain*. Robotics and Autonomous Systems, 40, pp. 151–162.

[17] Habib, M. K. & Baudoin, Y. (2010). *Robot-assisted risky intervention, search, rescue and environmental surveillance*. International Journal of Advanced Robotic Systems, 7(1), pp. 1–8.

[18] Panzieri, S., Pascucci, F. & Ulivi, G. (2012). *An outdoor navigation system using GPS and inertial platform.*, IEEE/ASME Transactions on Mechatronics, 7(2), pp. 134–142.

[19] Wurm, K. M., Kmmerle, R. & Stachniss, K. (2009). *Improving robot navigation in structured outdoor environments by identifying vegetation from laser data*. In Proc. of the IEEE/RSJ International Conference on Intelligent Robots and Systems, pp. 1217–1222.

[20] Bradley, D. M. & Unnikrishnan, R. (2007). *Vegetation detection for driving in complex environments*. In Proc. of the IEEE International Conference on Robotics and Automation, pp. 503–508.

[21] Nguyen, D. V., Kuhnert, L. & Thamke, S. (2012). *A novel approach for a double-check of passable vegetation detection in autonomous ground vehicles*. In Proc. of the 15th International IEEE Conference on Intelligent Transportation Systems, pp. 230–236.

[22] Ohkawa, S., Takita, Y. & Date, H. (2014). *Development of the autonomous brush-cutting robot using center articulated steering vehicle*. Transactions of the Japan Society of Mechanical Engineers, 80(812), pp. 1–6.

[23] D'Eon, R. G., Serrouya, R., Smith, G. & Kochanny, C. O. (2002). *GPS radiotelemetry error and bias in mountainous terrain*. Wildlife Society Bulletin, 30(2), pp. 430–439.

[24] Hicks, R. W. & Hall, E. L. (2000). *Survey of robot lawn mowers*. In Intelligent Systems and Smart Manufacturing (pp. 262–269). International Society for Optics and Photonics.

[25] Hellström, T., Lärkeryd, P., Nordfjell, T. & Ringdahl, O. (2009). *Autonomous forest vehicles: historic, envisioned, and state-of-the-art*. International Journal of Forest Engineering, 20(1), pp. 31–38.

[26] Zheng, X., Koenig, S., Kempe, D. & Jain, S. (2010). *Multirobot forest coverage for weighted and unweighted terrain*. IEEE Transactions on Robotics, 26(6), pp. 1018–1031.

[27] Miettinen, M., Ohman, M., Visala, A. & Forsman, P. (2007). *Simultaneous localization and mapping for forest harvesters*. In Proc. of the IEEE International Conference on Robotics and Automation, pp. 503–508.

[28] Brust, M. R. & Strimbu, B. M. (2015). *A networked swarm model for UAV deployment in the assessment of forest environments*. In 2015 IEEE 10th International Conference on Intelligent Sensors, Sensor Networks and Information Processing (ISSNIP), pp. 1–6, IEEE.

[29] Garzón, M., Valente, J., Roldán, J. J., Cancar, L., Barrientos, A. & Del Cerro, J. (2015). *A multirobot system for distributed area coverage and signal searching in large outdoor scenarios*. Journal of Field Robotics, 33(8), pp. 1087–1106.

[30] Blender, T., Buchner, T., Fernandez, B., Pichlmaier, B., & Schlegel, C. (2016). *Managing a mobile agricultural robot swarm for a seeding task*. In IECON 2016-42nd Annual Conference of the IEEE Industrial Electronics Society, pp. 6879–6886, October, IEEE.

[31] Tan, Y. & Zheng, Z. Y. (2013). *Research advance in swarm robotics*. Defence Technology, 9(1), pp. 18–39.

[32] Dorigo, M., Floreano, D., Gambardella, L. M., *et al.* (2013). *Swarmanoid: a novel concept for the study of heterogeneous robotic swarms*. IEEE Robotics & Automation Magazine, 20(4), pp. 60–71.

[33] Couceiro, M. S., Figueiredo, C. M., Rocha, R. P. & Ferreira, N. M. (2014). *Darwinian swarm exploration under communication constraints: initial deployment and fault-tolerance assessment*. Robotics and Autonomous Systems, 62(4), pp. 528–544, April.

[34] Portugal, D., Couceiro, M. S. & Rocha, R. P. (2013). *Applying Bayesian learning to multi-robot patrol*. In Proceedings of the 2013 International Symposium on Safety, Security and Rescue Robotics (SSRR 2013), Linköping, Sweden, October 21–26.

[35] Portugal, D. & Rocha, R. P. (2016). *Multi-robot patrol with Bayesian learning*. Autonomous Robots Journal, 40(5), pp. 929–953, June.

[36] Couceiro, M. S., Portugal, D. & Rocha, R. P. (2013). *A collective robotic architecture in search and rescue scenarios*. In Proceedings of the 28th ACM Symposium on Applied Computing (SAC 2013), Special Track on Cooperative

Multi-Agent Systems and Applications, pp. 64–69, Coimbra, Portugal, March 18–22.

[37] Bicho, E., Louro, L. & Erlhagen, W. (2004). *Coordinated transportation with minimal explicit communication between robots*. In Proceedings of the 5th IFAC Symposium on Intelligent Autonomous Vehicles (IAV 2004), Lisbon, Portugal, July 5–7.

[38] Chudoba, J., Faigl, J., Kulich, M., Krajník, T., Kosnar, K. & Preucil, L. (2011). *A technical solution of a robotic e-learning system in the SyRoTek project*. In Proceedings of the 3rd International Conference on Computer Supported Education (CSEDU 2011), Noordwijkerhout, The Netherlands, May 6–8.

[39] Parker, L. E., Kannan, B., Tang, F. & Bailey, M. (2004). *Tightly-coupled navigation assistance in heterogeneous multi-robot teams*. In Proceedings of the IEEE International Conference on Intelligent Robots and Systems (IROS 2004), Sendai, Japan, 28 September–2 October.

[40] Fox, D., Ko, J., Konolige, K., Limketkai, B., Schulzz, D. & Stewart, B. (2006). *Distributed multi-robot exploration and mapping*. In Proceedings of the IEEE, 94(7), pp. 1325–1339, July.

[41] Pippin, C., Christensen, H. & Weiss, L. (2013). *Performance based task assignment in multi-robot patrolling*. In Proceedings of the 2013 ACM Symposium on Applied Computing (SAC 2013), pp. 70–76, Coimbra, Portugal, March 18–22.

[42] Rocha, R., Dias, J. & Carvalho, A. (2005). *Cooperative multi-robot systems: a study of vision-based 3-D mapping using information theory*. Robotics and Autonomous Systems, 53(3), pp. 282–311.

[43] Fernandes, A., Couceiro, M. S., Portugal, D., Machado Santos, J. & Rocha, R. P. (2015). *Ad hoc communication in teams of mobile robots using ZigBee technology*. Computer Applications in Engineering Education, 23(5), pp. 733–745.

[44] Couceiro, M. S., Rocha, R. P. & Ferreira, N. M. (2011). *A novel multi-robot exploration approach based on particle swarm optimization algorithms*. In IEEE International Symposium on Safety, Security, and Rescue Robotics, SSRR 2011. Kyoto, Japan.

[45] Tillett, J., Rao, T. M., Sahin, F., Rao, R. & Brockport, S. (2005). *Darwinian Particle Swarm Optimization*. Em B. Prasad (Ed.), Proceedings of the 2nd Indian International Conference on Artificial Intelligence, (pp. 1474–1487). Pune, India.

Chapter 13

Guiding swarm behavior by soft control

Jing Han[1,2] and Caiyun Wang[1,2]

Abstract

Guiding swarm behavior is one of important problems in the area of swarm intelligence. While in many multiagent systems, the local rules for each agent is difficult or impossible to be redesigned. Even for artificial systems, it is not easy to design simple local rules for each agent which can lead the group emerge desired collective behavior. In this chapter, we introduce a new method called "Soft Control" which does not need to change the underlying local rules of agents, but to add some special agents, called "shills" which can be controlled by us, into the system. It is a nondestructive intervention method because shills are still treated as normal agents by normal ones. The soft-control method is effectively applied to three different multiagent system models to guide the swarm behavior: (1) guide consensus of a Vicsek-like flocking model by adding one intelligent shill, (2) promote cooperation for a group of players that play the end-unknown repeated prisoner's dilemma game by adding a cooperating team of shill plays and (3) add one or some shills to guide the collective opinion of the DeGroot model system.

13.1 Introduction

Collective behavior is the macroscopic property of a self-organized system which consists of many microscopic agents. This microscopic property usually cannot be found from the single individual. For examples, aggregation, synchronization, pattern formation, phase transition, fashion, swarm intelligence, etc. People found this kind of phenomena in many systems, such as flocking of birds [1,2], school of fish [1], cooperation in ant colonies [3], panic of crowds [4], locust collective motion [5], norms in economic systems [6], etc.

[1]Key Laboratory of Systems and Control, Institute of Systems Science, Academy of Mathematics and Systems Science, Chinese Academy of Sciences, China
[2]School of Mathematical Sciences, University of Chinese Academy of Sciences, China

Agents usually interact locally in a distributed multiagent system. If the local rule is carefully designed, the system will emerge expected and useful function, such as swarm intelligence, which is a kind of collective behavior of decentralized, self-organized natural or artificial multiagent systems.

However, for some systems under some circumstances, the emergent collective behavior might not be what people expect. Then what is the solution? There are at least two ways:

1. Destructive intervention. For example, redesign the local rule of the agents or add some special agents into the system, such as leaders [7,8], mediators [9], etc. These both require additional attributes of the already existing normal agents to recognize and interact with the special ones. So these intervention methods are destructive to the system.
2. Nondestructive intervention keep the local rules of the already existing agents in the current system. In reality, some multiagent systems acutely cannot be redesigned, nor can the local rules of the existing agents be changed. In this case, how do we intervene in the system and guide the collective behavior? We proposed a novel method called "**soft control**" [10]: add one or a few special agents, called the **shill**, to locally interact with their neighbors and spread their influence to the whole system. The shill is treated as a normal agent by normal ones. It is not like the leader or mediator which is treated differently. So in this case, normal agents do not change their way when interacting with others even after one or some shills are added.

The features of soft control are listed below:

1. Nondestructive intervention: it is a way to control the system without changing the local rules of the existing agents (also called normal agents).
2. Add shills: to add the control interface into the system, one (or a few) special agent, called "shills," which can be designed and controlled. Shill's behavior does not necessarily obey the local rules as the normal agents do. They are redesigned for the control purpose.
3. Soft power: because the interaction local rules are kept for normal agents, normal agents still treat shills as normal ones. A shill is not like the stronger influential leader [7], virtue leader [8] or mediator [9] which should be treated differently by normal agents. The influence of a shill and that of a normal agent are equal in strength, while stronger influential leaders or mediator have more influence on others than normal agents. Thus, by adding a shill, influence is "softly" put on the system.

13.2 Related works

All related works and our approach can be classified into three categories, as mentioned in our early paper [11]:

1. ***Analysis*: Given the local rules of agents, what is the swarm behavior of the overall system?** This is more on analytical work, by using mathematics tools,

mean-field theory in statistical physics [12,13], distributed control which studies the performance of the overall system and subsystems, etc.

2. ***Design*: Given the desired swarm behavior, what are the local rules for agents?** This is a design problem. For example, design a practical distributed system for resource sharing and cooperation. Typical work is the ant colony optimization algorithm [14]. The self-organized collective behavior can be used as a function for optimization, and the system shows intelligence. Another example is formation control [15] which is to design a team of locally interacting robots to maintain a geometric shape.
3. ***Intervention*: Given the local rules of agents, how we intervene and guide behavior?** In some real applications, changing the local rules of the already existing agents (e.g., the behavioral rules of people in panic and the flying strategies of birds) is very difficult or even impossible. However, for the purpose of avoiding danger or improving efficiency, controlling the system is very necessary. Then what is the feasible way that can utilize the property of swarm behavior to intervene in the swarm behavior? Our invented method, called soft control, is the way that can guide the swarm behavior while keeping the basic local rule of the existing agents.

There are some related literatures which are guiding swarming behavior by adding special agents. Jadbabaie *et al.* proposed a leader–follower model. In their model, a leader agent is added into a linearized Vicsek model defined system. And they found that when the neighborhood graph satisfies the joint connectivity condition, the leader can lead the group converge to its heading. While this condition is pretty difficult to be satisfied, no algorithm for the implementation is given in their paper.

In the work of Couzin *et al.* [16], a few informed individuals are added into the system for the purpose of influencing the moving decision of animal groups. These informed individuals have pertinent information, such as knowledge about the location of a food source, or of a migration route. And they update their direction by balancing the influence of their preferred direction and their social interactions with a weighting term which is different from the updating rule of normal agents. Couzin *et al.* found that as the proportion of informed individuals increases the accuracy of group motion increases asymptotically and they revealed the larger the group the smaller the proportion of individuals needed to guide the group. Later Liu *et al.* [17] proved that under the assumption of the initial density of the group is big enough, all agents can be let to converge to the desired heading with probability of 1 as long as the proportion of fixed-heading shills (informed agents) $\geq \sqrt{15/2 \cdot \sqrt{3/\pi}}\sqrt[4]{\log n/(nr_n^2)}$, where r_n is the interaction radius and it satisfies $\sqrt{\log n/n} \ll r_n \ll 1$, v_n is the velocity of agents and it satisfies $v_n \leq r_n/8 \cdot 961$. That means these informed individuals cannot lead the group to the preferred direction when the proportion of informed individuals is small, while we can use our soft-control method to guide a moving system (the Vicsek model defined system) to our desired direction by just adding one special agent (shill).

Based on the property that the degree distribution of many real social networks is the power-law distribution, Wang *et al.* [18] studied how to stabilize the whole network by just controlling a small number of nodes, called "pinning control." They found that the whole network can be stabilized on a homogeneous stationary state by imposing local linear feedback information to a small number of nodes. Furthermore, they found that it need less local controllers if the controlled node is specifically selected. The pinning control is to impose control to some already existing nodes (agents) but not to add some special agent into the system.

For the purpose of investing the role of prominent agents, such as politicians and officers of the state who are able to manipulate the opinion of individuals, Acemoglu *et al.* [19] add some prominent agents into the opinion dynamic system. They found that the opinions of prominent agents might spread but can not be persistent. These prominent agents influence others without being influenced to the same degree. And different neighbors of one prominent agents rely on the prominent agent in different degree, while our shills can be treated as equal degree as normal agent.

Mobilia *et al.* [20,21] studied how a special agent, called "zealot," affect a group of voters. The zealot favors one opinion which allows the zealot to change its state from its unfavored state to its favored state with a rate without regard to his neighbors. That means the zealot behaves different from the normal agents. Through algebraically method, they found that in one and two-dimensional lattice the voter model with zealot toward the unanimity state chosen by the zealot, while in higher dimensions the system cannot reach the zealot's favored state by adding one zealot. Our soft-control method can be applied to all kinds of networks but not limited to lattice structure in the opinion dynamic DeGroot model.

By using the soft-control method, Sartoretti *et al.* [22] added one shill into the group to successfully guide the whole group to avoid an obstacle. Later, in [23], they used the soft-control method to successfully pilot the whole group to a given goal. Furthermore, Sartoretti [24] give four example for apparent/hidden leaders and apparent/hidden shills guide swarm behavior based on different multiagent systems.

Genter *et al.* [25,26] used a subset of ad hoc agents, named influencing agents, to lead a flock of agents to a desired orientation. These special agents can be controlled by the their algorithms, and they can not be recognized by normal agents. Later, in [27], they designed three algorithms for these influencing agents to influence the flock. In this case, these special ad hoc agents behave like a shill and similar to our soft-control method, but it came latter after our approach.

Albi *et al.* [28] studied how to use special agents, leaders, to guide the crowd who have no information about the exit toward to the exit. Through simulations, they found that leaders with "go-to-target" strategy can lead the crowd congest around the exist in a circular shape. Furthermore, they found that leaders whose strategy is obtained by the compass search algorithm can guide the whole crowd to the exit without congestion. In their works, these special agents have their own strategy and cannot be distinguished by normal ones. This idea follows our soft-control method.

Used the soft-control idea, Su *et al.* [29] introduced a fixed-opinion shill into the Hegselmann–Krause opinion dynamic model for the purpose of guiding the group. They demonstrated that by the influencing of the fixed-opinion shill, the group can converge to the opinion value of the shill within finite time.

There are also some approaches for guiding real artificial systems by special agents. In 2007, Halloy *et al.* [30] added robots into groups of cockroaches to control self-organized choices. In their experiment, there are two shelters in a circular arena. These two shelters have different colors. Without guidance, cockroaches randomly explore and choose one shelter, but they prefer the darker one. Some chemically dressed robots, which seems to be accepted by real cockroaches, are added into the group of cockroaches to guide the behavior of cockroaches. Through experiences, authors found that more cockroaches will stay under the lighter shelter if the robots stay there.

In the work of Rubenstein *et al.* [31], the initial group consists of more than 1,000 autonomous robots. Authors developed a collective algorithm for shape formation by placing four specially programmed seed robots, which define the origin and orientation of the coordinate system, next to the initial group. Real experimental results show these seed robots can guide the robotics system form a desired shape.

13.3 Guiding flocking swarm by adding one intelligent shill

In 1987, Reynolds proposed a multiagent systems (with three local rules: alignment, separation and cohesion) which can show flocking phenomena [1] by computer simulation. Later Vicsek *et al.* [32] proposed a simplified model, the Vicsek model, which only has the alignment rule. This simplified system can emerge the self-order motion and a flocking-like cooperative behavior and many other models [33–35] more or less base on this model despite of the complexity in reality of the physical and biological systems.

This section, based on the Vicsek model, introduces the method of adding an intelligent shill to guide the heading of a group of mobile agents.

13.3.1 Model—the Vicsek model

The Vicsek model is a discrete time system composed of n agents $(\boldsymbol{X}_i(\cdot), \theta_i(\cdot))$, labelled from 1 to n. Each agent moving with the same absolute velocity v in the plane and heading of the agent is updated by the vector average of all its neighboring agents. Those agents whose distance from agent i's current position is no larger than a constant radius r are neighbors of agent i, i.e., the neighborhood of agent i at time t is defined as $N_i(t) = \{j \,|\, \|\boldsymbol{X}_j(t) - \boldsymbol{X}_i(t)\| \leq r, j = 1, 2, \ldots, n\}$ where $\boldsymbol{X}_i$ is the position of agent i at time t. Obviously, each agent is the neighbor of itself. Because each agent moving with the same absolute velocity v in the plane, it updates its position according to (13.1) below:

$$\begin{cases} x_i(t+1) = x_i(t) + v\cos\theta_i(t) \\ y_i(t+1) = y_i(t) + v\sin\theta_i(t) \end{cases} \quad i = 1, 2, \ldots, n, \tag{13.1}$$

where $\theta_i(t)$ is the heading of agent i at time t, and it is updated as follows:

$$\theta_i(t+1) = \arctan\left(\frac{\sum_{j\in N_i(t)} \sin(\theta_j(t))}{\sum_{j\in N_i(t)} \cos(\theta_j(t))}\right) + \Delta\theta \qquad i = 1, 2, \ldots, n, \tag{13.2}$$

with some necessary regulations reflecting angles of $[0, 2\pi)$, where $\Delta\theta$ is the noise and it is a random number chosen with a uniform probability from the interval $[-\vartheta/2, \vartheta/2]$. Without special instruction, we assume $\vartheta = 0$.

Definition 13.1. *The Vicsek model is called reach consensus if headings of all agents satisfy:*

$$\lim_{t\to\infty} \theta_i(t) = \theta, \quad i = 1, 2, \ldots, n,$$

where θ is dependent on the initial state $\boldsymbol{X}_i(0)$, $\{\theta_i(0), i = 1, 2, \ldots, n\}$, r and v.

Thought simulations, Vicsek *et al.* found that the Vicsek model presents a kinetic phase transition. Given different initial conditions, the system will present different phenomenon: all agents will reach consensus; all agents will separate into several subgroups and never get back again. So under what circumstances, the Vicsek model can reach consensus? The nonlinear coupling of the Vicsek model (13.2) makes the theoretical analysis of the problem very difficult. Jadbabaie *et al.* [36] showed that under some connectivity conditions, the linearized Vicsek model will be consensus. But the jointly connected condition of [36] is very difficult to verify in general. So Tang *et al.* [37] proposed a probabilistic framework to this problem and found that the overall multiagent system can reach consensus with large probability as long as the number of agents is large enough. Based on their results, Liu *et al.* [38] gave a sufficient condition to guarantee the consensus of the Vicsek model, which is only imposed on the model parameters. Later, Chen *et al.* [39] gave a smallest possible interaction radius condition of the linearized Vicsek model that can ensure the system be consensus.

However, the self-organized swarm sometimes does not reach consensus, or the consensus heading is not what we expect. In these cases, how does soft control help?

13.3.2 Model with soft control

In this subsection, we consider the case of guiding the swarm behavior of the Vicsek model by soft control. We propose the first shill strategy which is the more general one than the shill strategy in [10,11], and we give the corresponding theoretical analysis. Then an improved strategy [40,41] is given as well as theoretical analysis and simulations (Figure 13.1).

13.3.2.1 The first soft-control strategy

Han *et al.* [10,11] considered adding one controlled special, called shill, into the Vicsek model to control the system converge to the expected heading θ^*, where $\theta^* = \pi$. In this chapter, we study a more general case where the expected heading $\theta^* \in [0, \pi]$. Initial headings of normal agents are $\theta_i(0) \in [0, \pi)$, $1 \le i \le n$.

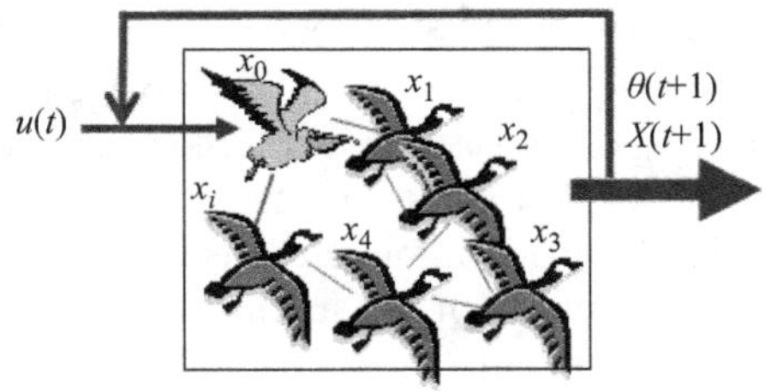

Figure 13.1 The soft control for the bird flocking model, by adding a shill. Two birds are linked if they are neighbors of each other at time t [11]. Reproduced by permission of Springer

The position and the heading of the shill are, respectively, denoted as $\boldsymbol{X}_0(t)$ and $\theta_0(t)$. The shill can affect a normal agent i if it is the neighbor of agent i, i.e., $\|\boldsymbol{X}_0 - \boldsymbol{X}_i\| \leq r$, which means the neighborhood of each agent may include the shill as well. So the neighborhood of agent i ($i = 1, 2, \ldots, n$) is redefined as follow:

$$N_i(t) = \{j \mid \|\boldsymbol{X}_j(t) - \boldsymbol{X}_i(t)\| \leq r, \quad j = 0, 1, 2, \ldots, n\}. \tag{13.3}$$

Then normal agents still keep their update rules defined in (13.1) and (13.2) because normal agents treat a shill as a normal one.

To effective intervene the system, the shill should know the local rule and the state information (headings and positions) of all normal agents at any time. Suppose the shill can be controlled by us, then we propose the first effective control law u_γ of the shill which is defined as:

$$\boldsymbol{X}_0(t) = \boldsymbol{X}_{s(t)}(t) \tag{13.4}$$

$$\theta_0(t) = \begin{cases} \theta_{s(t)}(t) + \gamma, & \text{if } \theta_{s(t)} \leq \theta^* - \gamma \\ \theta_{s(t)}(t) - \gamma, & \text{if } \theta_{s(t)} \geq \theta^* + \gamma \\ \theta^*, & \text{if } \left|\theta_{s(t)} - \theta^*\right| < \gamma \end{cases} \tag{13.5}$$

where $\gamma \in (0, \pi)$ is a constant, and $s(t) = \text{argmax}_{1 \leq i \leq n}\{|\theta_i(t) - \theta^*|\}$ which means $s(t)$ is the "worst" agent (or one of the "worst" agents).

Suppose, the distance between the angel of the "worst" agent and the objective angel θ^* at time t is $\delta(t) = \left|\theta_{s(t)}(t) - \theta^*\right|$, then we have the following main technical result of the first strategy:

Theorem 13.1. *For any $n \geq 2$, $r \geq 0$, $v \geq 0$, and any initial headings $\theta_i(0) \in [0, \pi)$ and any initial positions $x_i(0) \in R^2$, $(1 \leq i \leq n)$, the update rule of normal agents (13.1)–(13.3) and the soft-control law u_γ defined by (13.4)–(13.5) will lead the group to be consensus, i.e.,* $\lim_{t\to\infty} \delta(t) = 0$.

The proof line of Theorem 13.1 is the similar as that of Theorem 1 of [11], so we do not give the proof Theorem 13.1.

13.3.2.2 The second soft-control strategy—"*consistent moving*"

In the first soft-control strategy, the shill is arbitrarily placed in any different target positions without considering the consistency of the moving direction and the heading of shill. Although the shill exploit the state information (headings and positions) of all agents, these online feedback information is only used to select the target position rather than making decision of the schematic movements of the shill. The key problem is how to avoid putting "bad" effects (leading neighboring agents to direction other than the purpose one) while the shill moves from one location to another location without putting negative effects (i.e., drawing heading of agents to a nondesired direction). This is not solved by the first soft-control strategy. The heading of the shill applied in the first soft-control strategy is not consistent with its moving direction, and the shill speed has no limit. Therefore, Han *et al.* [40,41] later proposed a comprehensive algorithm, called "consistent moving," for shill which can solve this problem completely (see Figure 13.2).

Considering the Vicsek model consisting of n normal agents whose initial headings are uniformly randomly picked up from $(-\pi/2, \pi/2)$. The initial positions of normal agents are randomly and uniformly distributed inside a square with length L. We add one shill into the system to guide all normal agents converge toward a desired heading $\theta^* = 0$, i.e., $\max_{i\in\{1,\ldots,n\}} \cos\theta_i(t) \to 1$ when $t \to \infty$. The shill has its own speed v_s, and it is allowed to speedup and slowdown during the evolution but the shill's speed is limited, i.e., $v_s(t) \leq v_0$ for all $t \geq 0$, where v_0 denotes the *maximal shill speed*. That means, the speed of the shill is not a constant. Under the above assumptions, to complete the task, the shill should satisfy the following two rules, denoted by $\mathbf{\Lambda}$:

1. **The shill always keeps the desired heading θ^* when it has neighboring agents. That is to say, the heading of the shill sensed by normal agents is always θ^*;**
2. **All agents will be directly or indirectly affected periodically by the shill inside a time period H.**

Figure 13.3(a) gives an example of the simplest idea for a shill to move between two locations. In order to avoid put negative effect on normal agents, the shill need to hold the desired heading θ^* when it can influence any agent. So the shill simply moves forward with the desired heading and far outside the group and then makes a big U-turn and gets back to the group from behind, finally changes its heading to *zero* and hits the target. We can prove that the shill with a limited speed is able to affect all agents in a period without meeting any normal agent during the U-turn path and eventually guide the group to consensus.

However, this route is not efficient. As showed in Figure 13.3(b), the shill can find a more efficient short-cut route by moving inside the group. It will plan a more subtle path by following the rule of "keep shill's heading zero when there are neighboring agents around the shill." We design an algorithm for this refined strategy of the shill, the "consistent moving" strategy, which is described by a finite-state machine (see the state transition diagram in Figure 13.4). By using this diagram, the shill will always take heading zero when it is the neighbor of normal agents, so the shill will not put

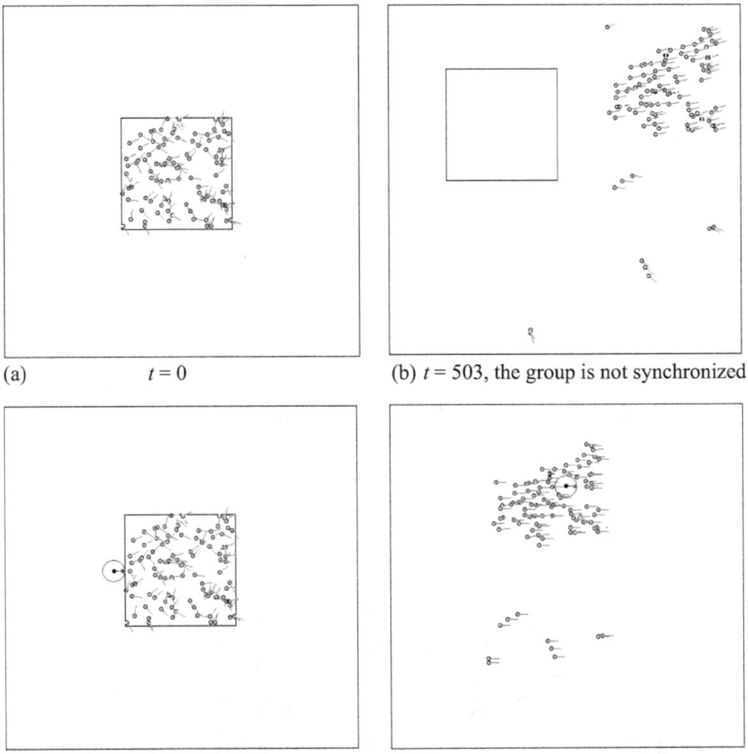

(a) $t = 0$ (b) $t = 503$, the group is not synchronized

(c) $t = 0$, a shill is added (d) $t = 1{,}999$, the group is synchronized with the help of the shill

Figure 13.2 Snapshots of computer simulation. Panels (a) and (b) are for the evolution of a self-organized group which does not reach consensus. Panels (c) and (d) are for the group with a shill guiding the group to consensus. Each case starts from the same configuration which has 100 agents, $v = 0.03$ and $r = 1$ (indicated by the little circle). The shill's speed is $v_0 = 0.3$. Little circle represents the normal agents. The small line represents its heading, while the little dark dot with large circle (which indicated the neighborhood) represents the shill [41]

negative effect on normal agents although it will move inside the group. The simple U-turn route is a special case that can be produced by this state transition diagram, if the speed of the shill is very large. Otherwise, with a low and median speed, its moving route will be much more subtle like the ones shown in Figure 13.3(b).

Explanations of Boolean functions for each transition of Figure 13.4 are listed below:

Affected: it returns *true* after the target is affected by the shill.

TargetAhead: it returns *true* if keeps moving in heading zero the shill will hit the target agent.

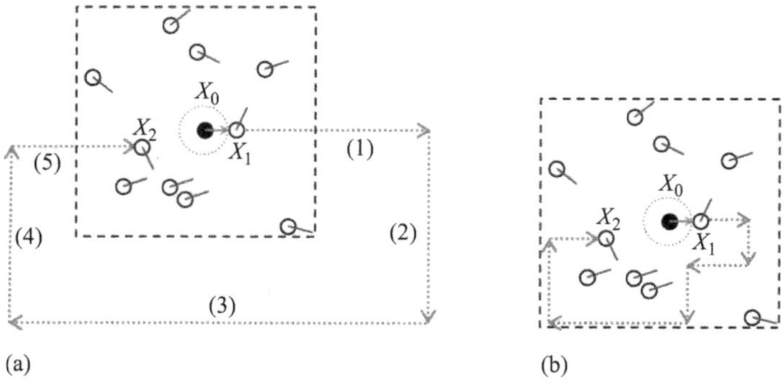

Figure 13.3 Two moving routes of shill from location of X_1 to location of X_2. (a) a simple U-turn route. (b) A more efficient route [41]

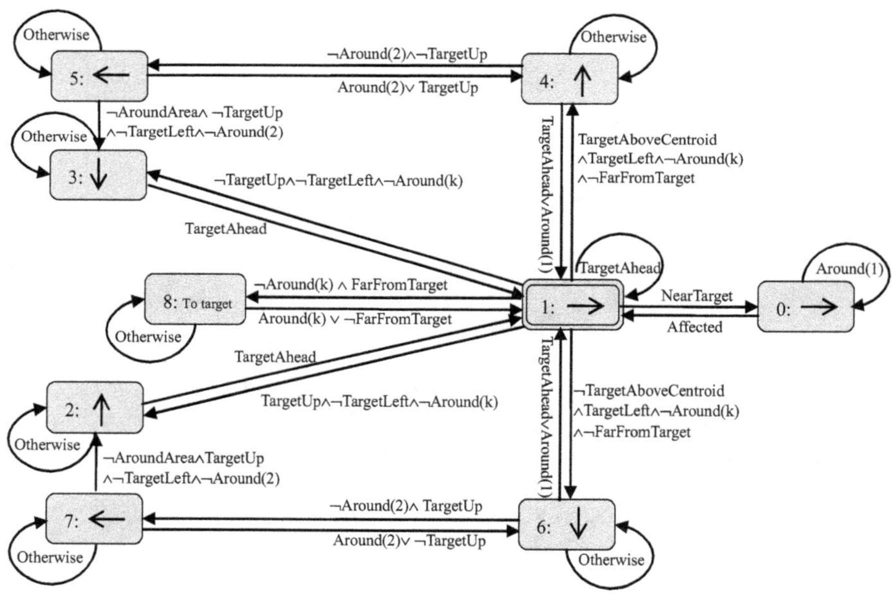

Figure 13.4 State transition diagram of how the shill decides its heading in "consistent moving." The heading in each state is indicated by the arrow inside the state box. For example, heading of state 0 and state 1 are zero, while headings of state 3 and state 6 are $-\pi/2$. In particular, heading of state 8 is pointing to the current location of the target. State 1 is the beginning and ending state of a loop. Once a target agent is affected, the new loop begins for the next target agent. Loop of state $1 \leftrightarrow 6 \leftrightarrow 7 \rightarrow 2 \leftrightarrow 1 \leftrightarrow 0$ is for the route bypassing from underneath. Loop of state $1 \leftrightarrow 4 \leftrightarrow 5 \rightarrow 3 \leftrightarrow 1 \leftrightarrow 0$ is for the route bypassing from above [41]

Around(k): it returns *true* if during the next k steps, the shill might have one or more neighboring agents (i.e., the shill might affect one or some normal agents).

TargetAboveCentroid: it returns *true* if $target.y > avg.y$, where $target.y$ is the y-coordinate of the target agent, and $avg.y = \left(\sum_{i=1}^{n} y_i\right)/n$ is the y-coordinate of the centroid of n normal agents.

TargetUp: it returns *true* if the location of the target agent is above the shill, i.e., $target.y > shill.y$.

TargetLeft: it returns *true* if the target agent locates in the shill's left side, i.e., $target.x < shill.x$.

AroundArea: it returns *true* if it will not have any neighboring normal agent during its movements upward/downward to the horizontal target position ($target.y$).

NearTarget: it returns *true* if the shill is neighboring to the target.

FarFromTarget: it returns *true* if the distance between the shill and the target is bigger than $G * r$, where G is a heuristic value ($G = 5$ in the simulations).

otherwise: This is not a specified condition function. It is *true* if conditions on other transitions are *false*. For example, if from state 8, it does not satisfy conditions to transit to state 1, it will keep state 8.

For example, at time t, if the shill is in state 8, its heading is pointing to the current location of target agent at time t. It checks the conditions: if *Around(k)* is *false* and *FarFromTarget* is *true*, at time $t + 1$ it will change to state 1 and its heading is updated to be zero. Otherwise, it will keep in state 8 at time $t + 1$ with its heading pointing to the location of target agent at time $t + 1$.

The time computational complexity of all above Boolean functions are not high: $O(1)$ for *Affected*, *TargetAhead*, *TargetAboveCentroid*, *TargetUp*, *TargetLeft*, *NearTarget* and *FarFromTarget*. To reduce the computation, *Around(k)* and *AroundArea* are considered in the worst case, since exact prediction is not required in these heuristic functions. For example, when *Around(k)* is checked, it is not necessary to simulate k steps evolution of the whole system to see whether during k steps the shill will have any neighboring agent. Instead, we check whether there are agents inside the circle of radius kr centered at the shill. So the computational complexity for *Around(k)* and *AroundArea* are $O(n)$.

The current strategy needs global information which is demanding. However, as we can see from the case of adding simple fixed-heading shills that do not require global information, the system will fail in synchronization starting from some configurations. So in this sense global information is necessary. But it does not mean that we need to know location information of all normal agents at each state. Actually, only local information is needed to compute Boolean functions of *Affected*, *Around(k)* and *NearTarget*. While *TargetAhead*, *TargetUp*, *TargetLeft* and *FarFromTarget* need no information of normal agents but only the location information of the target agent, *AroundArea* needs local sense of locations of nearby agents and the location information of the target agent as well. The only one that needs location information of all agents are *TargetAboveCentroid*. Actually, this function is not necessary. Indeed, the shill can always take the loop from underneath or above (see Figure 13.4), the only problem is that it might be less efficient than using the centroid information to decide going underneath or above. As another solution, we might try to find some way of determining the approximate centroid without complete knowledge of the normal agent positions.

13.3.2.3 Algorithm

The overall system after adding one shill evolves in this way: first, n normal agents and a shill are randomly and uniformly distributed inside a square with length L, and initial headings of normal agents are uniformly randomly picked up from $(-\pi/2, \pi/2)$; then, at each step, normal agents and the shill simultaneously update their headings, respectively, according to (13.2) and the finite-state machine (Figure 13.4), and their positions using (13.1).

The following pseudocode roughly shows the complete algorithm about how the system with a shill evolves. After given the initial information of normal agents and the shill, the system evolves as long as it has not reached consensus, i.e., $\max_{i\in\{1,2,\ldots,n\}} \cos\theta_i(t) \geq 1 - \varepsilon$ with $\varepsilon > 0$. During evolution, normal agents update their positions and headings according to (13.1) and (13.2) and the shill is moving according to the "consistent moving" strategy (Figure 13.4). Once a target agent is affected, the shill will choose one agent from the set of unaffected normal agents within current period according to the heuristic mentioned above and affect this normal agent, and repeat the above processes until all agents are affected. Then a new period begins. The program evolves until $\max_{i\in\{1,2,\ldots,n\}} \cos\theta_i(t) < 1 - \varepsilon$.

But if the number of steps in each period is limited, suppose the upper bound is H, can the shill periodically affect all agents? The shill should try to use small v_0 in most steps as possible as it can until the number of left steps is no more than five times of the number of unaffected agents n' in the current period, then the shill should speed up to a big value (which has a upper bound v_b). Through five steps, shill with such speed set can surely affect one unaffected agent by using the big U-turn shown in Figure 13.3(a) to find route to the next target normal agent (see Appendix S2 of [41] for detailed proof). Thus, the mission of affect n' unaffected agents in the remained

Algorithm 1: Procedure *main*

$t = 0$; initialization:
  $X_i(\theta)$ = randomly and uniformly picked up from a square of length L, for $i = 0, 1, \ldots, n$.
  $\theta_i(0)$ = randomly and uniformly picked up from $(-\pi/2, \pi/2)$, for $i = 1, \ldots, n$.
  state = 0;   *//this refers to the state number in the state transition diagram*
  $\theta_0(0) = 0$;   *//heading of the shill*
  j = *target_agent_selection*;
while $\max_i \cos\theta_i(t) > 1 - \varepsilon$ **do**   *// check whether the group reach consensus.*
  $t = t + 1$;
  if $j = 0$ **then** j = *target_agent_selection*;
  updates $\theta_0(t)$ and *state* according to the state transition diagram (Figure 13.4)
  if *state= 0* **then** $j = 0$.   *//if the targer agent is been affected, it will find another target*
  updates positions and headings of n agents according to (13.1) and (13.2).
  updates position $X_0(t)$ of the shill according to (13.1).
end while.

time $5n'$ steps can be accomplished by the shill, i.e., the shill can affect all agents in one period of H steps.

13.3.2.4 Analysis

We need to prove that the shill with a limited speed can affect all agents periodically. For this purpose, below we prove that all normal agents can be covered by a limited circle during the evolution at any time. Therefore, the shill does not need an unlimited speed to accomplish its task.

A dynamic reference point $\bar{X}(t) = (\bar{x}(t), \bar{y}(t))$ for the group of normal agents is defined as below:

$$\begin{cases} \bar{x}(t+1) = \bar{x}(t) + v\cos\theta^* \\ \bar{y}(t+1) = \bar{y}(t) + v\sin\theta^*, \end{cases} \tag{13.6}$$

where $(\bar{x}(0), \bar{y}(0))$ is the center of the minimal circle covering all normal agents in the initial state. It is worth noting that the reference point is just a virtual point which is constructed by us, and it will not affect other agents. The following Theorem 13.2 gives the theoretical support of the "consistent moving" strategy.

Theorem 13.2. *If the shill strategy satisfies condition* $\mathbf{\Lambda}$, *it can lead [41]*

1. headings of all normal agents exponentially converge to the desired value θ^*, *and*

$$\max_{i,j\in\{0,1,2,\ldots,n\}} |\tan\theta_i(t) - \tan\theta_j(t)| \le b\lambda^{t-1}\Delta_0,$$

where

$$b = \sigma^{-1}, \quad \lambda = \sigma^{1/\mu}, \quad \sigma = 1 - \left(\frac{\cos\bar{\theta}}{0}\right)^{\mu}, \quad \mu = nH,$$

$$\bar{\theta} = \max_{i\in\{0,1,2,\ldots,n\}} |\theta_i(0)|, \quad \Delta_0 = \max_{i,j\in\{0,1,2,\ldots,n\}} \{\tan\theta_i(0) - \tan\theta_j(0)\},$$

2. locations of all normal agents can be covered by a circle with a fixed radius R^* *at every step, and*

$$\bar{d}_i(t) \doteq \|\bar{X}(t) - X_i(t)\| \le R^*, \qquad \forall t \ge 0, \quad \forall i \in \{1,\ldots,n\},$$

where

$$R^* = D + vb\Delta_0\frac{2-\lambda}{1-\lambda},$$

D is the radius of the minimal circle enclosing all the normal agents at time 0.

To give the detailed proof of Theorem 13.2, we first introduce some preparation knowledge. We use *directed* graph $D_t = (V, E_t)$ to model the interaction topology among agents at time t. Here, the vertex set $V = \{0, 1, 2, \ldots, n\}$ consist of a shill (denoted by agent 0) and n normal agents. $E_t \subset V \times V$ is the arc set, and an arc

$(i,j) \in E_t$ means that agent j can receive information from agent i at time t, i.e., i is a neighbor of agent j. D_t is called neighbor graph.

For a directed graph (digraph) D, a *path* from vertex i to j is a sequence of distinct vertexes $i_0, i_1, i_2, \ldots, i_m$, where $i_0 = i, i_m = j$ and arc $(i_l, i_{l+1}) \in E, 0 \leq l \leq m-1$. A digraph is called *strongly connected* if for any ordered pair of distinct vertices, there is a path in D connecting them. A digraph is said to have a *spanning tree* if and only if there exists a vertex $i \in V$, called *root*, such that there is a path from i to any other vertex. The *union* of a collection of digraphs $\{D_1, D_2, \ldots, D_k\}$ with the same vertex set V, is a digraph D with vertex set V and arc set $E = E_1 \bigcup E_2 \cdots \bigcup E_k$.

By using the above terminologies, the second condition of $\mathbf{\Lambda}$ means that the union of interaction graphs $\{D_{kH+1}, \ldots, D_{(k+1)H}\}$ has a spanning tree rooted at agent 0 for any $k \geq 0$.

Now, we will analyze the dynamic of all agents. For convenience, we introduce a transformation $\tan\theta_k(t)$, then for agent $1 \leq k \leq n$, the heading rule (13.2) without noise can be rewritten as

$$\tan\theta_k(t+1) = \sum_{i \in N_k(t)} \frac{\cos\theta_i(t)}{\sum_{j \in N_k(t)} \cos\theta_j(t)} \tan\theta_i(t), \tag{13.7}$$

where $N_k(t)$ contains the shill if it is in agent k's neighborhood.

For the shill, its heading updates complicatedly according to the state transition diagram. So it is difficult to describe its dynamic process by an equation. However, our goal is to synchronize the normal agents, and we can just focus on their headings. From the normal agents' view, the shill always keeps the desired heading θ^* because of the first condition of $\mathbf{\Lambda}$. Thus, the normal agents can assume

$$\theta_0(t) = \theta^*, \quad \forall t \geq 0.$$

Let $\tan\theta(t) = [\tan\theta_0(t), \tan\theta_1(t), \ldots, \tan\theta_n(t)]^\tau$. From the normal agents' view, the dynamic of the whole group can be written as the following matrix form:

$$\tan\theta(t+1) = P(t)\tan\theta(t), \tag{13.8}$$

where

$$P(t) \doteq \begin{bmatrix} 1 & 0 \\ \gamma(t) & A(t) \end{bmatrix}, \tag{13.9}$$

$\gamma(t) = [\gamma_1(t), \ldots, \gamma_n(t)]^\tau$ with

$$\gamma_i(t) = \begin{cases} \dfrac{\cos(\theta_0(t))}{\sum_{k \in N_i(t)} \cos(\theta_k(t))}, & \text{if the shill is a neighbor of agent } i \text{ at time } t; \\ 0, & \text{otherwise.} \end{cases}$$

$A(t) = [a_{ij}(t)]_{i,j \in \{1,\ldots,n\}}$ with

$$a_{ij} = \begin{cases} \dfrac{\cos\theta_j(t)}{\sum_{k \in N_i(t)} \cos\theta_k(t)}, & \text{if agent } j \text{ is a neighbor of agent } i \text{ at time } t; \\ 0, & \text{otherwise.} \end{cases}$$

From the problem definition, we know the initial headings of all agents belong to $(-\pi/2, \pi/2)$. Thus, all elements of $P(0)$ are nonnegative. It is also easy to check that the row sum of $P(t)$ is 1 independent of t. Recursively, we can get for any agent $i \in \{0, 1, \ldots, n\}$,

$$\min_{j\in\{0,1,\ldots,n\}} \tan\theta_j(t) \leq \tan\theta_i(t+1) \leq \max_{j\in\{0,1,\ldots,n\}} \tan\theta_j(t), \quad \forall t \geq 0. \tag{13.10}$$

Furthermore, from the monotonicity of function $\tan(\cdot)$, we know that the maximal heading of all agents is nonincreasing, and the minimal heading is nondecreasing over time. So, at any time, the headings of agents belong to $(-\pi/2, \pi/2)$ and the elements of $P(t)$ are nonnegative. Thus, $\{P(t), t \geq 0\}$ is a sequence of stochastic matrix[1] [42].

From (13.8), we know

$$\tan\theta(t+1) = P(t)\cdots P(0)\tan\theta(0), \quad \forall t \geq 0. \tag{13.11}$$

Thus, in order to prove the synchronization, we need to study the product of stochastic matrices. Here, synchronization means the difference of headings of all agents are zero. Because the shill always affects normal agents with heading θ^*, the synchronization of system (13.8) means that headings of all agents converge to θ^*.

From (13.11), we know that if the product $P(t)\cdots P(0)$ tends to a matrix with the same row as $t \to \infty$, the difference among entries of $\tan\theta(t+1)$ tends to be zero. A characterization of consensus is defined for the stochastic matrix $B = [B_{ij}]_{(m)\times(m)}$:

$$\tau(B) = \frac{1}{2}\max_{i,j}\sum_{s=1}^{m}|B_{is} - B_{js}|. \tag{13.12}$$

From [43], we know that for any stochastic matrices $B(1)$ and $B(2)$,

$$\tau(B(1)B(2)) \leq \tau(B(1))\tau(B(2)). \tag{13.13}$$

Furthermore, the function $\tau(\cdot)$ also has the following property.

Lemma 13.1. *Let $y = [y_1, \ldots, y_n]^\tau \in \mathbf{R}^m$ be an arbitrary vector and denote $\Delta y = \max_{i,j}|y_i - y_j|$. For a stochastic matrix $B = [B_{ij}]_{m\times m}$, if $z = Py$, $z = [z_1, \ldots, z_m]^\tau$, then we have [43]*

$$\Delta z \leq \tau(P)\Delta y.$$

Note that, Δy represents the difference among the entries of vector y. By Lemma 13.1 and formula (13.11), if $\tau(P(t)\cdots P(0))$ tends to zero, we have $\lim_{t\to\infty}\Delta\tan\theta(t) = 0$ means the system converge to synchronization. Define

$$\lambda(B) = \min_{i,j}\sum_{s=1}^{m}\min(B_{is}, B_{js}). \tag{13.14}$$

If $\lambda(B) > 0$, then B is called *scrambling matrix* [43]. It is also easy to check that

$$\tau(B) = 1 - \lambda(B).$$

Thus, if B is a scrambling matrix, then $\tau(B) < 1$.

[1]A matrix is called stochastic if its elements are all nonnegative and the sum of each row is 1.

Based on the second condition of $\mathbf{\Lambda}$, we will see that the product of matrices $\{P(t)\}$ becomes scrambling matrix periodically, which leads to the exponential decrease of $\tau(P(t)\cdots P(t))$. To account for this, we need to study the relationship between the product of stochastic matrices $\{P(t)\}$ and the union of neighbor graphs $\{D_t\}$. Before elaborating this, we introduce some concepts and well-known results.

For a matrix $B \in R^{m\times m}$, its associated directed graph, denoted by $\Gamma(B)$, is a digraph on m nodes $\{1, 2, \ldots, m\}$ such that there is a directed arc in $\Gamma(B)$ from i to j if and only if $b_{ji} \neq 0$ [42]. From the definition of $P(t)$, we know the digraph associated with the matrix $P(t)$ is the neighbor graph D_t.

The composition of a directed graph G_1 with a directed graph G_2 [44] is the directed graph with vertex set $\{1, 2, \ldots, m\}$, and arc set defined as follows: (i,j) is an arc of the composition just in case there is a vertex k such that (i, k) is an arc of G_1 and (k,j) is an arc of G_2.

According to the concept of associated directed graph $\Gamma(B)$ for a matrix B, we know that for any two matrices, B_1, B_2, the composition of $\Gamma(B_1)$ with $\Gamma(B_2)$ is in fact the digraph $\Gamma(B_2B_1)$.

A neighbor-shared graph is a graph that each pair of two distinct vertices share a common neighbor. From the definition of scrambling matrix, it is easy to see that the following two statements are equivalent: stochastic matrix B is scrambling, the associated graph $\Gamma(B)$ is neighbor-shared.

Based on the above concepts, we now introduce a result from which the periodic scrambling matrix hidden in the shill's strategy will be presented.

Lemma 13.2. *For the rooted graphs with m vertices that each one has a self-arc, the composition of any set of $N \geq m-1$ such rooted graphs is neighbor-shared [44].*

From the second condition of $\mathbf{\Lambda}$, we know that the union of neighbor graph sequence $\{D_{kH+1}, \ldots, D_{(k+1)H}\}$ is a rooted graph for any $k \geq 0$. The union graph has $n+1$ vertices. Thus, the composition of n union graphs is neighbor-shared by Lemma 13.2. Combining the neighbor-shared graph with the scrambling matrix and defining $\mu = nH$, we know

$$\mathbf{P((k+1)\mu)\cdots P(k\mu+1)\ is\ a\ scrambling\ matrix,}\ \forall \mathbf{k} \geq \mathbf{0}.$$

Thus, $\tau(P((k+1)\mu)\cdots P(k\mu+1)) < 1$, the exponential stable of $\Delta \tan\theta(t)$ becomes possible. The detailed analysis will be given in the proof of the theorem.

Up to now, we only investigate the headings of all agents. In fact, positions of agents affect the neighborhood which further affect their headings of the next step. And the headings will influence their future positions. That is, the headings and positions are coupled together. The reason why the above heading analysis can implement independently is that the motion of the shill is designed by us. We program the shill to affect every agent at least once in a fixed period. This guarantees the normal agents are affected by shill periodically, which further makes $\Delta \tan \theta(t)$ decrease. Combining with the above analysis, next we will give a complete proof of the main result.

Proof of Theorem 13.2

Proof. Let $\Delta_t \doteq \Delta \tan \theta(t) = \max_{i,j\in\{0,1,2,\ldots,n\}} |\tan \theta_i(t) - \tan \theta_j(t)|$. Based on the previous analysis (13.10), we know that the function $\max_i \tan \theta_i(t)$ is nonincreasing, and the function $\min_i \tan \theta_i(t)$ is nondecreasing. So, Δ_t is a monotonous sequence.

For the system (13.8), we define

$$\Phi(k+1,i) = P(k)\Phi(k,i), \quad \Phi(i,i) = I, \ \forall k \geq i \geq 0,$$

then,

$$\tan \theta(t+1) = \Phi(t+1,0) \tan \theta(0), \tag{13.15}$$

and by applying Lemma 13.1, we have

$$\Delta_t \leq \tau(\Phi(t,0))\Delta_0. \tag{13.16}$$

From the strategy designed for the shill, we know that for any $k \geq 0$

$\Phi((k+1)\mu + 1, k\mu + 1)$ is a scrambling matrix.

Let $\bar{\theta} = \max_{i\in\{0,1,2,\ldots,n\}} |\theta_i(0)|$. Since the maximal heading of all agents is nonincrease and the minimal heading is nondecrease, for any $t \geq 0$, all the nonzero entries of $P(t)$ are large than or equal to $(\cos\bar{\theta})/(n+1)$. Thus,

$$\lambda(\Phi((k+1)\mu + 1, k\mu + 1)) \geq \left(\frac{\cos\bar{\theta}}{n+1}\right)^{\mu}, \quad \forall k \geq 0.$$

Hence,

$$\begin{aligned} \tau(\Phi((k+1)\mu + 1, k\mu + 1)) &= 1 - \lambda(\Phi((k+1)\mu + 1, k\mu + 1)) \\ &\leq 1 - \left(\frac{\cos\bar{\theta}}{n+1}\right)^{\mu} \doteq \sigma, \quad \forall k \geq 0. \end{aligned} \tag{13.17}$$

For any $t \geq 1$, there exists an integer $k_0 \geq 0$ such that $k_0\mu \leq t-1 < (k_0+1)\mu$. By (13.13), we have

$$\begin{aligned} \tau(\Phi(t,0)) &\leq \tau(\Phi(t,k_0\mu+1))\tau(\Phi(k_0\mu+1,(k_0-1)\mu+1))\cdots(\Phi(\mu+1,1))\tau(P(0)) \\ &\leq \sigma^{k_0} \leq \sigma^{(t-1/\mu)-1} = \sigma^{-1}(\sigma^{(1/\mu)})^{t-1}. \end{aligned}$$

Now, let $b \doteq \sigma^{-1}$, $\lambda \doteq \sigma^{(1/\mu)}$. Combining with (13.16), we have

$$\Delta_t \leq b\lambda^{t-1}\Delta_0, \tag{13.18}$$

which proves the first part of Theorem 13.2.

According to the update rules (13.1) and (13.6), with the fact that heading of the shill felt by normal agents is always θ^*, we know that at time t the distance between agent i, $i \in \{1, \ldots, n\}$ and the reference point $\bar{X}$ is

$$\begin{aligned} \bar{d}_i(t+1) &\leq \bar{d}_i(t) + 2v \left| \sin \frac{\theta_i(t) - \theta^*}{2} \right| \\ &\leq \bar{d}_i(t) + v|\theta_i(t) - \theta^*| \\ &\leq \bar{d}_i(t) + v \max_{i,j \in \{0,1,2,\ldots,n\}} |\theta_i(t) - \theta_j(t)|. \end{aligned} \tag{13.19}$$

Because $\tan\theta - \theta$ is an increasing function when $\theta \in (-\pi/2, \pi/2)$, we have

$$\max_{i,j \in \{0,1,2,\ldots,n\}} |\theta_i(t) - \theta_j(t)| \leq \max_{i,j \in \{0,1,2,\ldots,n\}} |\tan\theta_i(t) - \tan\theta_j(t)|. \tag{13.20}$$

Thus

$$\bar{d}_i(t+1) \leq \bar{d}_i(t) + v\Delta_t. \tag{13.21}$$

Next, we will use mathematical induction to prove

$$\bar{d}_i(t) \leq R^*, \qquad \forall t \geq 0, \quad \forall i \in \{1, \ldots, n\}. \tag{13.22}$$

It is obvious at $t = 0$. By using $\sigma < 1$ and the nonincrease of Δ_t, it is easy to check that (13.22) is true at $t = 1$. Thus, the shill with a limited speed can finish its task at $t = 1$.

Now, suppose (13.22) is true when $t \leq T$. Thus the shill can accomplish its task with a limited speed in time interval $[1, T]$.

Next, we will prove formula (13.22) is true at time $T + 1$.

Case 1: if $T \leq \mu$
By using the monotonicity of Δ_t, we have for any $i \in \{1, \ldots, n\}$

$$\begin{aligned} \bar{d}_i(T+1) &\leq \bar{d}_i(0) + v\sum_{t=0}^{T} \Delta_t \leq \bar{d}_i(0) + v(T+1)\Delta_0 \\ &\leq \bar{d}_i(0) + v(\mu+1)\Delta_0 \leq R^*. \end{aligned} \tag{13.23}$$

Case 2: if $T > \mu$
By applying (13.16), we have

$$\begin{aligned} \bar{d}_i(T+1) &\leq \bar{d}_i(0) + v\sum_{t=0}^{T} \Delta_t \leq \bar{d}_i(0) + v\sum_{t=1}^{T} b\lambda^{t-1}\Delta_0 + v\Delta_0 \\ &\leq \bar{d}_i(0) + vb\Delta_0 \frac{2-\lambda}{1-\lambda} \leq R^*. \end{aligned} \tag{13.24}$$

Combining the above two cases, we know that locations of all normal agents can be covered by a circle with radius R^* at time $T + 1$. So, the shill with a limited speed can accomplish its task at time $T + 1$. By using mathematical induction, we know (13.22) is true for all $t \geq 0$. So the speed of the shill can be bounded by a constant. ■

13.3.2.5 Computer simulations

Here are settings of simulations: there are $n = 10, 30, 50, 100, 200, 300, 400$ normal agents with speed $v = 0.03$ initialized in a square with length $L = 10$. Then a shill is added into the system to test the effectiveness of the "consistent moving" strategy under different set of maximal speed of the shill $v_0 = 1.5v, 2v, 5v, 10v, \infty$, where $v_0 = \infty$ means that once the shill affects the target agent, it will slow down a bit if there are neighbors around, otherwise it will get to the position of the next target agent in the next step, no matter how big the distance between them. The interaction radius $r = 1$. For each realization, when $1 - \max_i \cos\theta_i < 0.0001$, the system is regarded as reaching consensus, then the simulation is terminated and the total number of steps (denoted as *totalStep*) can be measured.

Figure 13.5 shows the effectiveness of the "consistent moving" strategy. It is also shows the effect of the group size n of the performance. All results showed in Figure 13.5 satisfy "v_0 is smaller than the theoretical bound v_b." We find in Figure 13.5 that one intelligent shill with "consistent moving" can guide the group converge to the desired heading within limited steps.

Figure 13.6 shows comparisons of adding one intelligent "consistent moving" shill with adding n_f fixed-heading simple shills. In [16,17,36], a number of fixed-heading simple shills (leaders) are introduced into the system to guide the heading of a group with dens population. Figure 13.6 shows that adding an intelligent "consistent moving" shill can significantly improve the consensus level of the system which is much better than a number of fixed-heading simple shills when there is no noise in the model.

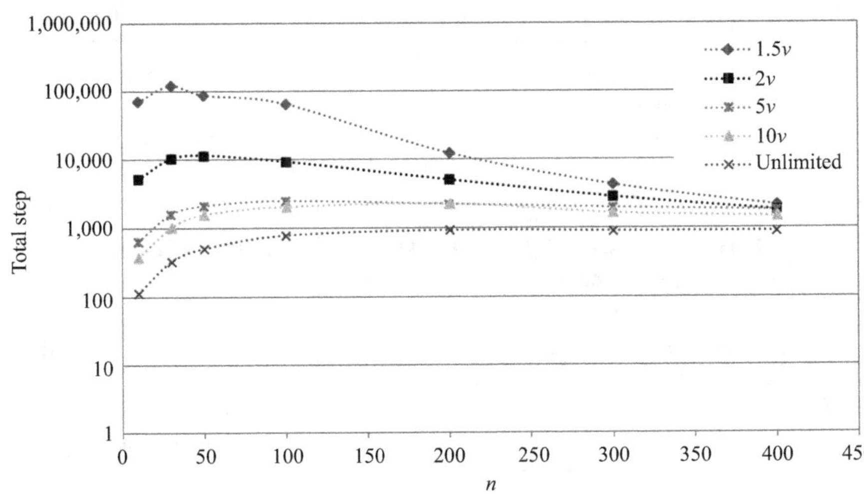

Figure 13.5 Mean totalStep (over 500 realizations) for consensus ($\max_i \cos\theta_i > 0.9999$) when one "consistent moving" strategy shill is added [41]

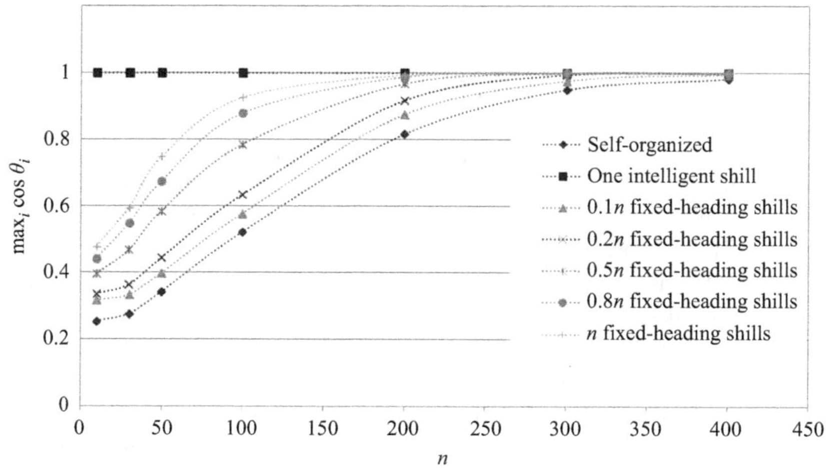

Figure 13.6 Consensus level (mean $\max_i \cos\theta_i$*) respectively for self-organized evolution, adding one intelligent "consistent moving" shill with* $v_0 = 0.5$*, and adding different number of fixed-heading simple shills [41]*

And then we compare the intelligent shill methods with the fixed-heading shills (informed individuals) in the case of model with noise in Figure 13.7. Since there is noise, the system would not really reach consensus. We use order parameter $v_a \doteq (1/nv)|\sum_{i=1}^{n} V_i|$ to measure the performance of soft control. v_a can be regarded as the *consensus level* to measure the state of the system. In Figure 13.7, we can see that one intelligent shill outperforms $0.1n, 0.2n, 0.5n, 0.8n$ or n fixed-heading simple shills. And the intelligent shill works very well when noise is very small ($\eta = 0.01$), and it still achieves nice performance when noise is not big ($\eta \leq 0.05$).

13.4 Guiding a group of game players by adding shills to promote cooperation

Except application of soft control on the flocking model, the soft-control method can also be applied in the game system to improve the cooperation level.

Since 1970s, cooperative behavior has been one of the hot topics. Cooperation can be widely found in human society and animal groups. Most of these systems are composed of self-interested agents. It raises a question: how does cooperation emerge in a group of self-interested agents without centralized control? The game theory is used to understand the choice of strategy, the equilibrium among agents with conflicting interests and the cooperation among self-interested agents. Prisoner's dilemma is one of the most famous model in game theory. It reveals the contradiction

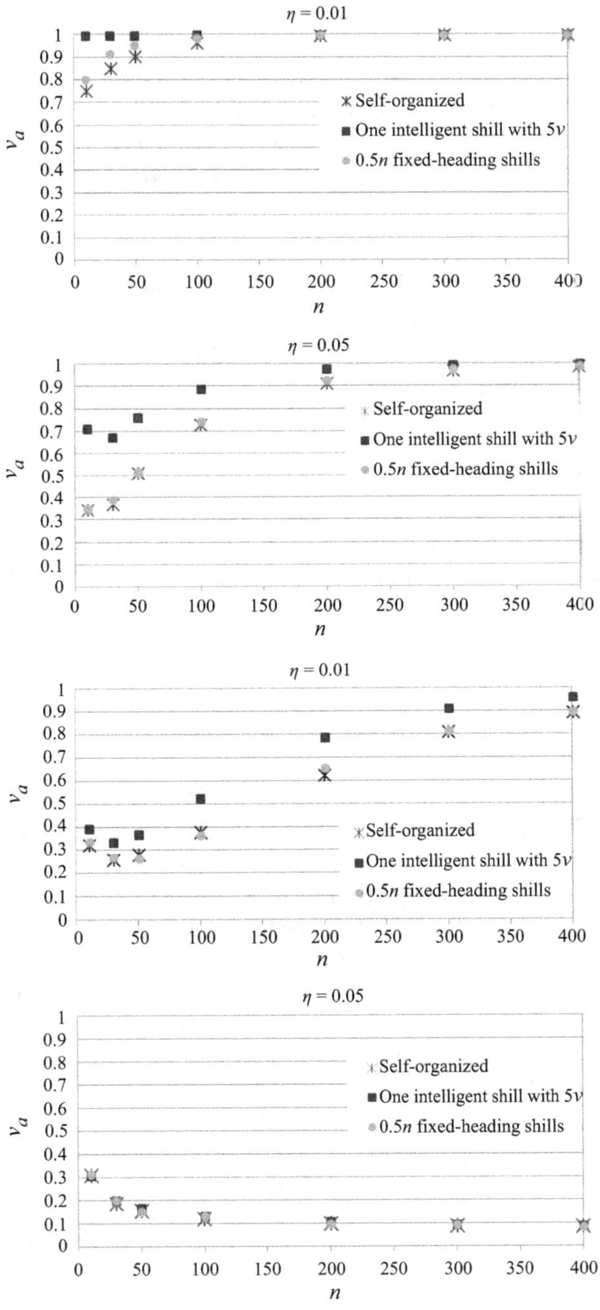

Figure 13.7 The average value of v_a over 500 realizations at $t = 10{,}000$ for different noise η in three scenarios (self-organized system, adding one intelligent shill with $v_0 = 5v$ and $0.5n$ fixed-heading shills) [41]

between individual and group rationalities. It is widely used to explore the origin of cooperation. For the multiperson repeated prisoner's dilemma (RPD) model, if the spontaneous evolution of the population can not reach a high level of cooperation, how can we intervene the system to improve the level of cooperation? Wang and Han [45] applied the soft-control method to the multiperson RPD. By adding some shills to the system, the level of cooperation of the system is greatly promoted.

13.4.1 Model—the multiperson repeated prisoner's dilemma game

In the prisoner's dilemma game, both two players simultaneously decide their actions, cooperation (C) or defection (D). Actions players decide their payoffs. The payoff matrix considered in this section has the following form:

$$\begin{array}{cc} & \begin{array}{cc} C & D \end{array} \\ \begin{array}{c} C \\ D \end{array} & \begin{pmatrix} (R,R) & (S,T) \\ (T,S) & (P,P) \end{pmatrix} \end{array}$$

where parameters in the matrix satisfy $T > R > P > S$ and $R > (T+S)/2$.

According to the above matrix, (D, D) is the Nash equilibrium point [46] of the system. But they can receive higher payoffs, R points, if they choose cooperate with each other. This is the dilemma between individual and collective rationalities.

Suppose there are N_A normal agents in the system and let $\mathcal{A} = \{1, 2, \ldots, N_A\}$. The total number of agents in each generation during evolution is constant. For each generation, pairs of players are randomly selected and play a β round PD game (β-stage RPD). The pairing can be a complete interaction, i.e., every two agents will play a β-stage RPD; or incomplete interaction, i.e., players interact with a proportion α ($\alpha \in \mathbb{R}[0, 1]$) of the group. α is called the interaction locality. If $\alpha = 1$, the interaction is a complete interaction. Each normal agent takes a mixed reactive strategy [47], which is described as $(y, p, q) \in \mathbb{R}[0, 1]^3$, where y, p and q, respectively, represent cooperation probability on the first stage, on the stage when the opponent's last move is defection or cooperation.

At the end of each generation, agent i ($i \in \mathcal{A}$) updates its payoff. $f_{ij}(t)$ denotes the total payoff agent i gains in the game playing with agent j in generation t. Then $f_i(t) = \sum_{j \in \mathcal{A} \setminus \{i\}} f_{ij}(t)$ denotes agent i's total payoff. And then it is the reproduction process at the end of each generation. Let $E\{\#_i(t+1)\}$ denote the expected number of agent i's offsprings in $t+1$ generation, and it can be calculated as follows:

$$E\{\#_i(t+1)\} = \frac{f_i(t)}{\sum_{k \in \mathcal{A}} f_k(t)} N_A \quad \forall i \in \mathcal{A} \tag{13.25}$$

In the following sections, if there is no special instruction, the discussion is based on the case of complete population interaction.

13.4.2 Model with shills

The cooperation level of the swarm, which is measured by the fraction of cooperation actions taken by all players in a generation, also called the frequency of cooperation, is denoted as f_c. If f_c is low, how do we use the soft-control method to increase it? N_s shills with the purpose of promoting cooperation are added into the group. Let $N = N_A + N_S$ denote total number of the population. Let $\mathcal{P} = \{1, 2, \ldots, N\}$.

13.4.2.1 The shill strategy

Shills have no preliminary knowledge of the normal agents before they play in the beginning of each generation. But shills can store the action history of their opponents and share with other shills. For example, when a shill plays with agent i, (m_i, n_i) will be updated, where m_i is how many stages they play, and n_i is how many cooperation actions agent i takes so far. (m_i, n_i) are initialed as (0, 0) at the beginning of each generation, and it is shared and accessed by all shills. When a shill begins to play with a normal agent, it will calculate the agent's cooperation frequency according to the shared history and make decision based on estimation of its cooperation tendency, i.e., the shill cooperates with the probability n_i/m_i. This is called Frequency-based Tit for Tat strategy.

Therefore, the role of shills is to reward cooperative behavior and punish defective ones by playing games with normal agents.

13.4.2.2 Computer simulations

Some numerical experiments are used to study the performance of soft control. Here are settings of simulations: $N_A = 500$ normal agents whose strategies are uniformly generated in $\mathbb{R}[0, 1]^3$. Each simulation result is the average of 100 randomly runs on independent realizations. Parameters of the payoff matrix are set as follows: $R = 3$, $T = 5$, $S = 0$ and $P = 1$ [48].

Complete population interaction vs. incomplete population interaction

The pattern of the performance of soft control with different N_S for the complete population interaction case is demonstrated in Figure 13.8(a). Without shills ($N_S = 0$), the self-organized group of normal agents emerge low-cooperation level, while the

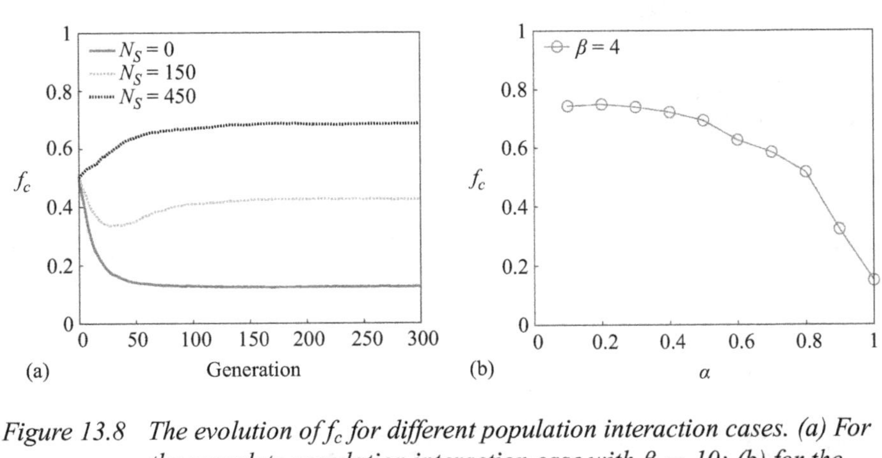

Figure 13.8 The evolution of f_c for different population interaction cases. (a) For the complete population interaction case with $\beta = 10$; (b) for the incomplete population interaction case with $\beta = 4$ and $N_S = 50$ [45]

cooperation is promoted significantly after adding shills ($N_S = 150$ and $N_S = 450$). Figure 13.8(b) shows that in the incomplete interaction cases, adding shills will promote cooperation in a more significant way comparing to the complete interaction cases. That is because with the increase of interaction locality α, shills lose the advantage on selection.

Sharing knowledge vs. nonsharing knowledge among shills

Figure 13.9 shows action history sharing will enhance the soft-control performance, even though only sharing with a small number of other shills.

The robustness of soft control to noise

Figure 13.10 illustrates the robustness of soft control to noise. So we suppose that any player can take the other action with the probability p_n because of noise in each

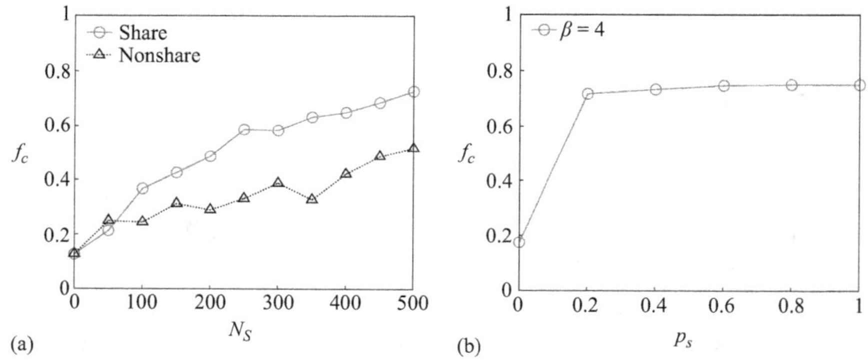

Figure 13.9 The importance of sharing knowledge under different settings. (a) For the complete population interaction case with $\beta = 10$; (b) for the incomplete population interaction case with $N_S = 50$ and $\beta = 4$ [45]

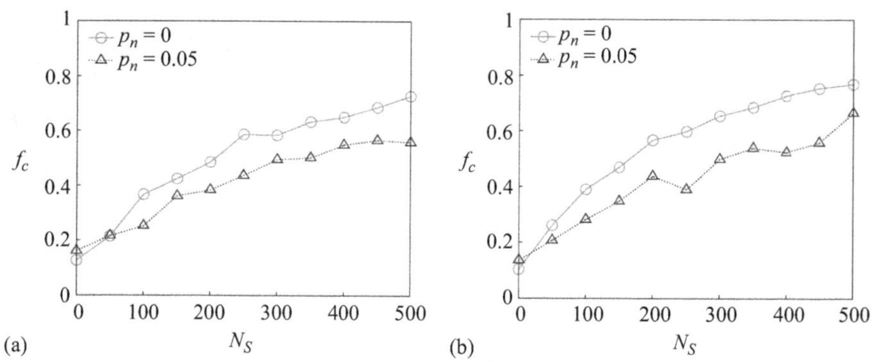

Figure 13.10 The sensitivity of soft control to noise. (a) For $\beta = 10$; (b) for $\beta = 100$ [45]

stage. We find that in the system with noise, soft control still works in the short-term ($\beta = 10$) and long-term ($\beta = 100$) RPD. Therefore, it is robust to noise.

13.5 Guiding collective opinion by adding shills

Humans do not live alone. They are social animals. They interact with each other and affect each others opinion. Opinions are evolving during the process. This is what the field of opinion dynamics studies. It uses multiagent models with real numbers representing their opinions and show how opinion values change over time.

The DeGroot model has been widely studied over a long time [49] in the field of opinion dynamics. The DeGroot model is a basic model and the update rule of it is simple. This model has beautiful mathematical results. Especially, it has some basic features of social learning [50,51], and it can be used to study some problems of information and learning [52].

The collective opinion is actually a kind of swarm behavior. Most studies in this area focus on the mechanism that can lead to opinion consensus or cleavage. In fact, intervention of collective opinion of the swarm is also important, i.e., guiding opinions evolve toward to an expected direction. In this section, we will discuss how to use the soft-control method to intervene opinion dynamics of a group without changing the underlying mechanism. The related literatures have been reviewed in the second section of this chapter.

We add one or a few shills into the DeGroot model to change the collective opinion. Shills behave as same as normal ones. If the underline neighborhood network is strongly connected, the DeGroot model defined system can achieve consensus [43]. So the connected group can also reach consensus, but with a different value, after adding shills. We proved that the initial opinion and influential value of the shill as well as how the shill connects to normal agents decide the change of the convergent opinion value varies. Through theoretical analysis and computer simulations, we discover an interesting and counterintuitive phenomenon: adding a small initial opinion value shill might increase the convergent collective opinion value, and vice versa. We also found that the in the undirected interaction neighborhood scale-free network, the convergent speed is positively related to the degree of the normal agent that connects to the shill.

13.5.1 Model—the DeGroot model

Suppose there are n agents $(a_1, a_2, \ldots, a_n)$ in the system. $x_i(t)$, a real number, represents agent i's opinion at time t. Therefore, vector $\boldsymbol{X}(t) = (x_1(t), x_2(t), \ldots, x_n(t))$ represents all agents' opinions in the group, where $t = 0, 1, 2, \ldots$. Especially, the initial opinion of all agents is represented by $\boldsymbol{X}(0)$. The neighborhood structure of all agents are represented as a graph $G = (V, E)$, where V is the set of nodes which represent agents and E is the set of edges which represent neighborhood relationships among agents, i.e., $e_{ij} \in E$ means that agent i and j are neighbors of each other.

In standard DeGroot model, every agent updates its opinion by the following formula:

$$x_i(t) = \sum_{e_{ji} \in E} w_{ji} x_j(t-1),$$

where w_{ji} denotes the weight that agent j puts on agent i's opinion. For the standard DeGroot model, for every agent i, it is assumed that

$$\sum_{e_{ji} \in E} w_{ji} = 1, \qquad i = 1, 2, \ldots, n. \tag{13.26}$$

Here we put some constraints on the weights. Because in reality, the influential agent will have big impact on all its neighbors. Therefore, we modified the DeGroot model: every agent a_i has a intrinsic influential value u_i $(i = 1, 2, \ldots, n)$, which indicates its influential power on its neighbors. With this change, the new updated formula for all agents at time t is:

$$x_i(t) = \frac{\sum_{e_{ji} \in E} u_j x_j(t-1)}{\sum_{e_{ji} \in E} u_j} \qquad i = 1, 2, \ldots, n \tag{13.27}$$

The weight that agent i puts on agent j in this modified DeGroot model is:

$$w_{ji} = \frac{u_j}{\sum_{e_{ki} \in E} u_k} \qquad i = 1, 2, \ldots, n \tag{13.28}$$

It still satisfies (13.26), so our modified model is a special case of the standard DeGroot model. All agents update their opinions as follows:

$$\boldsymbol{X}(t) = A\boldsymbol{X}(t-1). \tag{13.29}$$

Matrix A is the updating matrix of the system, and it has the following form:

$$A = \begin{bmatrix} \frac{b_{11}u_1}{\sum_{j=1}^{n} b_{j1}u_j} & \frac{b_{21}u_2}{\sum_{j=1}^{n} b_{j1}u_j} & \cdots & \frac{b_{n1}u_n}{\sum_{j=1}^{n} b_{j1}u_j} \\ \frac{b_{12}u_1}{\sum_{j=1}^{n} b_{j2}u_j} & \frac{b_{22}u_2}{\sum_{j=1}^{n} b_{j2}u_j} & \cdots & \frac{b_{n2}u_n}{\sum_{j=1}^{n} b_{j2}u_j} \\ \vdots & \vdots & & \vdots \\ \frac{b_{1n}u_1}{\sum_{j=1}^{n} b_{jn}u_j} & \frac{b_{2n}u_2}{\sum_{j=1}^{n} b_{jn}u_j} & \cdots & \frac{b_{nn}u_n}{\sum_{j=1}^{n} b_{jn}u_j} \end{bmatrix} \tag{13.30}$$

where $b_{ij} = 1$ if $e_{ij} \in E$; $b_{ij} = 0$ if $e_{ij} \notin E$, i.e., $B = (b_{ij})$ is the adjacent matrix of graph G. Because G is a undirected graph, so we have $b_{ij} = b_{ji}$ for $i = 1, 2, \ldots, n$ and $j = 1, 2, \ldots, n$. If G is strongly connected and each node has its loop, then the updating matrix A is a nonnegative stochastic primitive matrix.

Theorem 13.3. *Suppose T is an $n \times n$ nonnegative primitive matrix [43]. Then there exists an eigenvalue ρ such that:*

1. ρ is a real number and $\rho > 0$;
2. with ρ can be associated strictly positive left and right eigenvectors;

3. $\rho > |\lambda|$ *for any eigenvalue* $\lambda \neq \rho$*;*
4. the eigenvectors associated with ρ *are unique to constant multiples.*

Corollary 13.3.1. $\min_i \sum_{j=1}^{n} t_{ij} \leq \rho \leq \max_i \sum_{j=1}^{n} t_{ij}$ *[43].*

Theorem 13.4. *For a primitive matrix* T*, let* λ_2 *represent the eigenvalue which has the second largest absolute value of eigenvalues of* T*, then we have [43]:*
(a) if $\lambda_2 \neq 0$*, then as* $k \to \infty$

$$T^k = \rho^k \boldsymbol{w}\boldsymbol{v}' + O(k^s|\lambda_2|^k),$$

where $s = m_2 - 1$ *and* m_2 *is the multiplicity of* λ_2*;*
(b) if $\lambda_2 = 0$*, then for* $k \leq n - 1$

$$T^k = \rho^k \boldsymbol{w}\boldsymbol{v}'.$$

In both above cases, $\boldsymbol{w}$*,* $\boldsymbol{v}'$ *are any positive right and left eigenvectors corresponding to* ρ *guaranteed by Theorem 13.3, providing only they satisfy* $\boldsymbol{v}'\boldsymbol{w} = 1$.

Suppose G is strongly connected and each node has its loop, then the updating matrix A is a nonnegative stochastic primitive matrix. By Theorem 13.3 and Corollary 13.3.1, we know that the largest eigenvalue of the updating matrix A is $\rho = 1$. Then, by Theorem 13.4, we know that the system will reach consensus, i.e., $\lim_{t\to\infty} A^t\boldsymbol{X}(0) = (\bar{x}, \bar{x}, \ldots, \bar{x})'$. Furthermore, $\bar{x}$ can be calculated as $\bar{x} = \boldsymbol{Y}\boldsymbol{X}(0)$, where $\boldsymbol{Y}$ is the normalized left eigenvector of A with eigenvalue 1.

How can we change the convergent value without changing settings and update rules of the already existing agents? We use the soft-control idea to solve this problem. One or several shills are added into the system by connecting to some selected normal agents following some strategies. We will show how to add one or some shills into the group and select normal agents to connect with shills to change the convergent value $\bar{x}$ in the following subsection.

13.5.2 Model with evolvable-opinion shills

l ($l \geq 1$) *evolvable-opinion* shills are added into the system to increase (or decrease) the convergent opinion value. Shills connect to some agents directly, but they do not connect with each other. We use $x_{n+1}(0), x_{n+2}(0), \ldots, x_{n+l}(0)$ and $u_{n+1}, u_{n+2}, \ldots, u_{n+l}$ denote shills' initial opinions and shills' influential values. Shills evolve following the same update rule. All agents' (including normal agents' and shills') opinions are updated as follows:

$$\boldsymbol{X}(t) = A_1\boldsymbol{X}(t-1) \quad t = 1, 2, 3, \ldots \tag{13.31}$$

where $\boldsymbol{X}(t) = (x_1(t),\ x_2(t),\ \ldots,\ x_{n+l}(t))'$ and

$$A_1 = \begin{bmatrix} \frac{b_{11}u_1}{\sum_{j=1}^{n+l} b_{j1}u_j} & \cdots & \frac{b_{n1}u_n}{\sum_{j=1}^{n+l} b_{j1}u_j} & \cdots & \frac{b_{n+1,1}u_{n+l}}{\sum_{j=1}^{n+l} b_{j1}u_j} \\ & & \vdots & & \\ \frac{b_{1n}u_1}{\sum_{j=1}^{n+l} b_{jn}u_j} & \cdots & \frac{b_{nn}u_n}{\sum_{j=1}^{n+l} b_{jn}u_j} & \cdots & \frac{b_{n+l,n}u_{n+l}}{\sum_{j=1}^{n+l} b_{jn}u_j} \\ & & \vdots & & \\ \frac{b_{1,n+l}u_1}{\sum_{j=1}^{n+l} b_{j,n+l}u_j} & \cdots & \frac{b_{n,n+l}u_n}{\sum_{j=1}^{n+l} b_{j,n+l}u_j} & \cdots & \frac{b_{n+l,n+l}u_{n+l}}{\sum_{j=1}^{n+l} b_{j,n+l}u_j} \end{bmatrix}, \tag{13.32}$$

where $b_{ij} = 1$ if $e_{ij} \in E_1$; $b_{ij} = 0$ if $e_{ij} \notin E_1$. $G_1 = (V_1, E_1)$ is the neighborhood graph of the system with *evolvable-opinion* shills.

13.5.2.1 Undirected neighborhood graph

We first consider the case of the undirected neighborhood graph G, i.e., $b_{ij} = b_{ji}$ for $i = 1, 2, \ldots, n+l$ and $j = 1, 2, \ldots, n+l$. In this case, G is connected and each agent has its loop, then the normalized left eigenvector of A with eigenvalue 1 can be calculated as follows:

$$\boldsymbol{Y} = \left(\frac{u_1 \sum_{j=1}^{n} b_{1j}u_j}{\sum_{i=1}^{n}\sum_{j=1}^{n} b_{ij}u_iu_j}, \frac{u_2 \sum_{j=1}^{n} b_{2j}u_j}{\sum_{i=1}^{n}\sum_{j=1}^{n} b_{ij}u_iu_j}, \ldots, \frac{u_n \sum_{j=1}^{n} b_{nj}u_j}{\sum_{i=1}^{n}\sum_{j=1}^{n} b_{ij}u_iu_j} \right) \tag{13.33}$$

So the convergent opinion value of the original system $\bar{x}$ can be calculated as follow:

$$\bar{x} = \frac{\sum_{i=1}^{n}\sum_{j=1}^{n} b_{ij}u_iu_jx_i(0)}{\sum_{i=1}^{n}\sum_{j=1}^{n} b_{ij}u_iu_j} \tag{13.34}$$

Similarly, let $\bar{x}$ represent the convergent opinion value of the new system with *evolvable-opinion* shills. It has the following form:

$$\bar{x} = \frac{\sum_{i=1}^{n+l}\sum_{j=1}^{n+l} b_{ij}u_iu_jx_i(0)}{\sum_{i=1}^{n+l}\sum_{j=1}^{n+l} b_{ij}u_iu_j} \tag{13.35}$$

Then, with (13.34) and (13.35), we have

$$\Delta\bar{x} = \bar{x}' - \bar{x} = \sum_{k=n+1}^{n+l} \frac{u_k^2(x_k(0) - \bar{x}) + U_ku_k(M_k + x_k(0) - 2\bar{x})}{\sum_{i=1}^{n+l}\sum_{j=1}^{n+l} b_{ij}u_iu_j} \tag{13.36}$$

where

$$M_k = \frac{\sum_{j=1}^{n} b_{jk}u_jx_j(0)}{\sum_{i=1}^{n} b_{ik}u_i} \qquad k = n+1, \ldots, n+l. \tag{13.37}$$

$$U_k = \sum_{j=1}^{n} b_{jk}u_j \qquad k = n+1, \ldots, n+l. \tag{13.38}$$

(M_k, U_k) are the initial opinion and the influential value of a "virtual agent" which represents all agents that connect to shill k. This is because the effect of "shill k connects to a number of normal agents" can be regarded as the effect of "shill k connects of one virtual agent (M_k, U_k)."

Equation (13.36) shows that the effect of l shills is a linear sum of each shill's independent effect. So the basic problem is to study the effect of adding one shill, say shill (x_{n+1}, u_{n+1}). For convenience, (x_{n+1}, u_{n+1}) is denoted as (x_s, u_s) in the following. The related virtual agent that connects to (x_s, u_s) is also denoted as (M, U), where M and U are correspondingly denoted as the initial opinion and the influential value of the virtual agent.

With (13.36), we have

$$\Delta\bar{x} = \frac{u_s^2(x_s(0) - \bar{x}) + Uu_s(M + x_s(0) - 2\bar{x})}{\sum_{i=1}^{n+1}\sum_{j=1}^{n+1} b_{ij}u_iu_j} \tag{13.39}$$

where $M = (\sum_{j=1}^{n} b_{j,n+1}u_jx_j(0))/(\sum_{i=1}^{n} b_{i,n+1}u_i)$ and $U = \sum_{j=1}^{n} b_{j,n+1}u_j$.

From (13.39), we know that when $M < \bar{x}$, we have

$$\Delta\bar{x} \begin{cases} <0 & x_s(0) \le \bar{x} \\ <0 & x_s(0) \in (\bar{x}, 2\bar{x} - M), u_s \in (0, u_c) \\ \ge 0 & \text{others} \end{cases} \tag{13.40}$$

where

$$u_c = \frac{(x_s(0) + M - 2\bar{x})U}{\bar{x} - x_s(0)} \tag{13.41}$$

When $M > \bar{x}$, we have

$$\Delta\bar{x} \begin{cases} >0 & x_s(0) \ge \bar{x} \\ >0 & x_s(0) \in (2\bar{x} - M, \bar{x}), u_s \in (0, u_c) \\ \le 0 & \text{others} \end{cases} \tag{13.42}$$

When $M = \bar{x}$, we have

$$\Delta\bar{x} \begin{cases} >0 & x_s(0) > \bar{x} \\ <0 & x_s(0) < \bar{x} \\ =0 & x_s(0) = \bar{x} \end{cases} \tag{13.43}$$

From (13.40), we can see that even if $x_s(0)$ is larger than $\bar{x}$, the convergent value does not necessarily increase: when M is less than $\bar{x}$, the convergent value decreases under certain conditions of $x_s(0)$ and u_s. Similarly, from (13.42), when $x_s(0)$ is less than $\bar{x}$, the convergent value does not necessarily decreases. When M is equal to $\bar{x}$, increase or decrease of the convergent value depends on the difference between $x_s(0)$ and $\bar{x}$ [see (13.43)].

In order to increase the convergent value of the system, we can connect the shill to a set of normal agents whose related virtual agent initial opinion value M is large. Similarly, the shill can decrease the convergent value by connecting with those agents whose related virtual initial opinion M is small. Again, it might be able to increase (or decrease) the convergent value by adding a shill who has a smaller (or larger) initial opinion value under some conditions.

The convergent speed and the network topology
As we studied above [Equations (13.40), (13.42) and (13.43)], the initial opinion value and the influence value of the normal agent which is connected to the shill will affect the change of convergent value $\Delta\bar{x}$, while $\Delta\bar{x}$ is not affected by the topological property of the connected agent, e.g., the degree. However, we find that the convergent speed of the system is related to the degree of the connected agent through the following simulations. According to Theorem 13.4, we know that in our modified DeGroot model with *evolvable-opinion* shills, the convergent speed is related to the second largest eigenvalue λ_2 of the update matrix.

In the following simulations on a number of scale-free networks, we find that in most cases, the larger the degree of the connected agent leads to smaller λ_2, and that means faster convergent speed.

In the following simulations settings of influence values of normal agents $\boldsymbol{U} = (u_1, u_2, \ldots, u_n)$ are different. Initial opinion values do not need to set because λ_2 has no relationship with $\boldsymbol{X}(0)$. We use the Barabási–Albert method [54] to create these networks with 400 nodes. The weight of the shill $u_s = 200$. Two scenarios are considered: (1) the shill connects to the smallest degree agent m; (2) the shill connects to the largest degree agent l. We let $u_m = u_l$ for agents a_m and a_l. Figure 13.11 shows

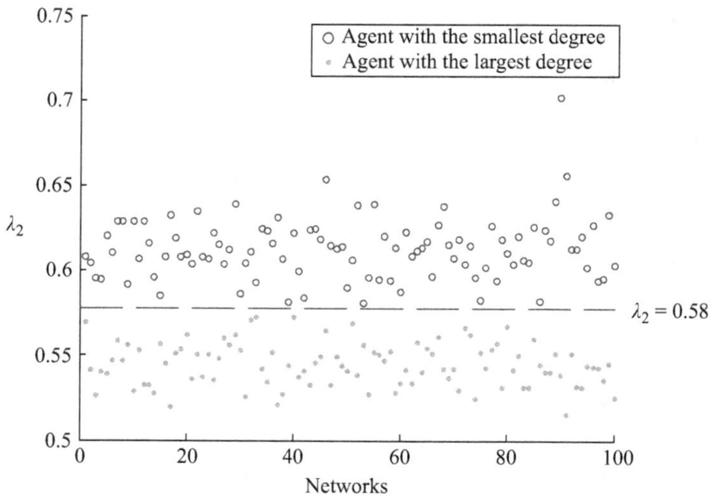

Figure 13.11 The second largest eigenvalue (avg. over 100 scale-free networks) of the system λ_2 (x-axis) based on an undirected neighborhood graph. The hollow circles represent λ_2 in the case of adding the shill to the agent with the smallest degree; the solid dot means λ_2 in the case of adding a shill connecting to the agent with the largest degree. The hollow circles are all set above [53]. Reproduced by permission of Springer

that connecting the shill to smallest degree agent will give a larger λ_2 than connecting the shill to largest degree agent. This implies that connecting the shill to a large degree agent will probably lead to a fast convergence.

13.5.2.2 Directed neighborhood graph

What we considered above is the undirected neighborhood graph, now we consider directed graph. If the neighborhood graph of the original system is a strongly connected graph and each node has its loop, the updating matrix is a nonnegative stochastic primitive matrix. Then the original system will reach consensus. In this case, the soft-control method can be used to intervene the system.

Suppose one evolvable-opinion shill a_{n+1} whose initial opinion value is $x_s(0)$ and influential value is u_s is added into the original system. Shill a_{n+1} is connected with one agent a_n with bidirectional edges. Apparently, if the updating matrix of the original system is a nonnegative stochastic primitive matrix, the updating matrix of the new system is also a nonnegative stochastic primitive matrix. So under the assumption of the strongly connected neighborhood graph, the new system with an evolvable-opinion shill can reach consensus. Because the neighborhood graph is a directed graph, b_{ij} may not equal to b_{ji}. Therefore, (13.33)–(13.35) are no longer valid. In this case, how does add one *evolvable-opinion* shill affect the convergent opinion value of the system? Will we get similar results? Will we also find the counterintuitive phenomenon that adding a small (or large) initial opinion value shill might increase (or decrease) the convergent collective opinion value $\bar{x}$? Below we do some simulations give answers to these questions.

In the simulations, two neighborhood graphs are used: one is a strongly connected directed random network and the other one is a strongly connected directed scale-free network. The directed random network constructed from 400 isolated nodes and then a directed edge for every pair of nodes is added with probability of 0.4. The directed scale-free networks is created by the method which is similar to the Barabási–Albert method [54]: each network is initialized as a complete graph of 20 nodes, then 10 direct edges from each new added node to the already existing nodes are added with a probability which is proportional to the in-degree of these already existing nodes, and at the same time 10 direct edges from already existing nodes to each new node are added with probabilities which is proportional to the in-degree of these already existing nodes until all 400 nodes are added into the system. The values of $x_i(0)$ and u_i for $i = 1, 2, \ldots, n$ are set as a random number over 1 to 400. An *evolvable-opinion* shill is added to the original system to change the convergent opinion value of the system. From Figure 13.12, we can see that when the neighborhood graph is a directed graph, adding one shill does change the convergent opinion value of the system. And a similar counterintuitive phenomena also exists [shadow areas in Figure 13.12(a)–(d)], similar to the undirected neighborhood graph cases. Furthermore, the line of $\Delta\bar{x} = 0$ surprisingly coincides with the analytical $\Delta\bar{x} = 0$ [defined by (13.41)] of the undirected neighborhood graph cases. This phenomena is surprising, and it is needed to testify in our future work.

We also investigate what affects the convergent speed in the directed neighborhood cases. By Theorem 13.4, if the neighborhood graph of the original system is

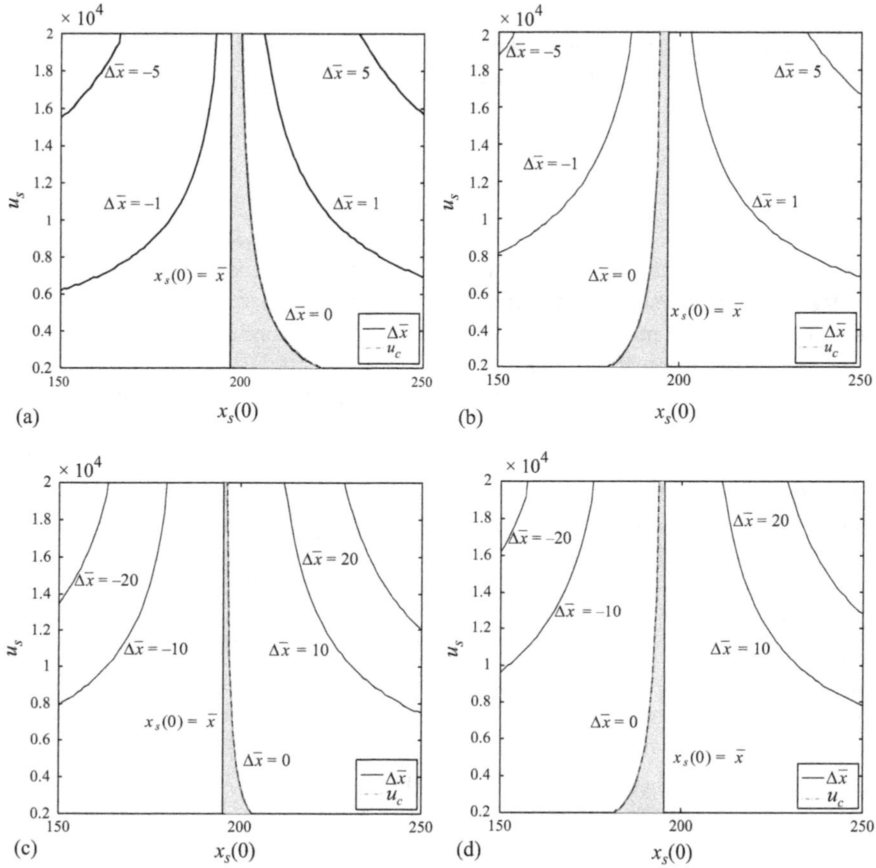

Figure 13.12 Contours represent different values of $\Delta\bar{x}$. Shadow area of (a) [or (c)] represents $\Delta\bar{x} < 0$ but $x_s(0) > \bar{x}$ based on a directed random network (or a directed scale-free network). Shadow area of (b) (or (d)) represents $\Delta\bar{x} > 0$ but $x_s(0) < \bar{x}$ based on a random network work (or a scale-free network). Shadow areas in (a)–(d) indicate the counterintuitive phenomena

strongly connected and each node has its loop, the smaller the second largest absolute value of eigenvalues λ_2 of the updating matrix of the system reach consensus.

Similar to the undirected neighborhood graph case, the following simulations are based on 100 directed scale-free networks. We add one evolvable-opinion shill with $u_s = 200$ into the system. We also consider two scenarios which is similar to the setting of Figure 13.12: (1) a double direction edge is added between the shill and the smallest out-degree agent m; (2) a double direction edge is added between the shill and the largest out-degree agent l. Similarly, we set $u_m = u_l$. Figure 13.13 shows

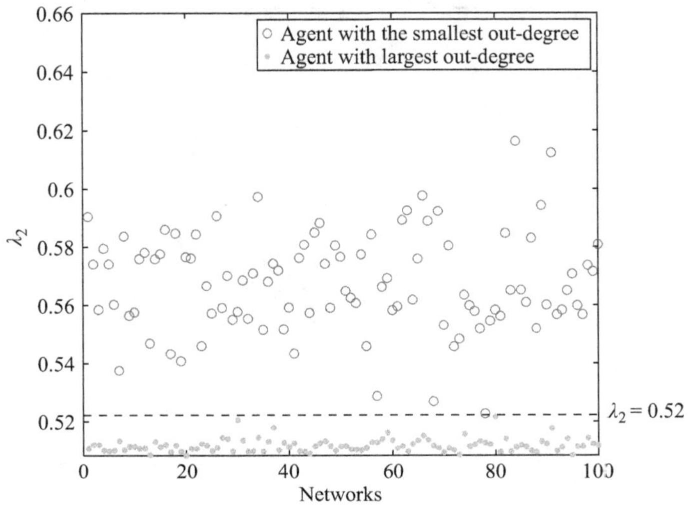

Figure 13.13 The second largest eigenvalue (avg. over 100 scale-free networks) of the system λ_2 based on a directed neighborhood graph. The hollow circles represent λ_2 in the case of adding the shill to the agent with the smallest degree; the solid dot means λ_2 in the case of adding a shill connecting to the agent with the largest degree. The hollow circles are all set above

that connecting the shill to the smallest out-degree agent will give a larger λ_2 than connecting the shill to the largest out-degree agent. This implies that connecting the shill to a large out-degree agent will probably lead to a fast convergence.

To further investigate how does the out-degree of the connected normal agent affect the convergent speed, some simulations are done. We select two agents (a_k and a_h) with different out-degree (suppose $d_k > d_h$) from one of 1,000 directed scale-free networks, and let $u_k = u_h$, $\Delta_d = d_k - d_h$. Let $\lambda_{2,h}$ and $\lambda_{2,k}$, respectively, denote the absolute value of the second largest eigenvalue of the updating matrix of the new system when the *evolvable-opinion* shill is connected with a_h or a_k. From Figure 13.14, we can see that with the increase of Δd, $\Delta\lambda_2$ has the tendency of increasing. That means a faster convergent speed can be ensured by letting the evolvable-opinion shill connect with the larger out-degree agent.

13.5.3 Model with fixed-opinion shills

The above study is based on adding several evolvable-opinion shills. In this section, we consider adding several *fixed-opinion* shills. Its opinion value is always set to be θ_s, and it does not evolve with time. Therefore, the system will eventually converge to θ_s if the neighborhood graph is strongly connected.

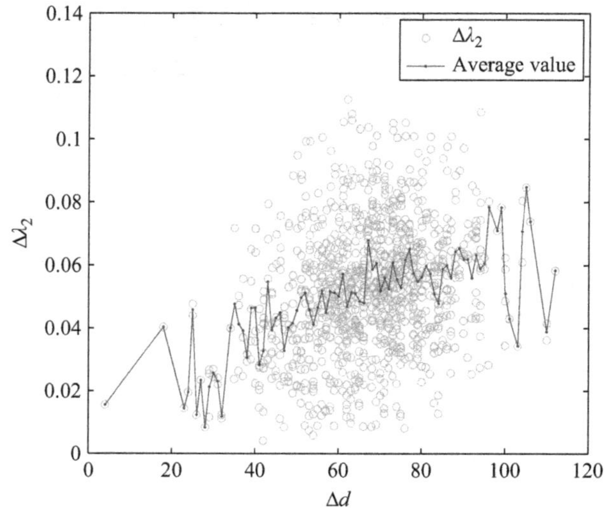

Figure 13.14 The relationship between $\Delta\lambda_2$ and Δd. Hollow circles represent test results. The solid dots represent the average $\Delta\lambda_2$ for each Δd

For the purpose of guiding the group to our desired opinion value θ_s, l $(l \leq 1)$ *fixed-opinion* shills $(a_{n+1}, a_{n+2}, \ldots, a_{n+l})$ are added into the group. Shills connect to some normal agents directly, but they do not connect with each other. Let $G_2 = (V_2, E_2)$ represent the neighborhood graph of the system with *fixed-opinion* shills. Opinion values of these *fixed-opinion* shills are all set to be θ_s, and they do not evolve with time. We use $u_{n+1}, u_{n+2}, \ldots, u_{n+l}$ denote influential values of these *fixed-opinion* shills. For convenience, we let $E_2^1 = \{e_{ij} \| e_{ij} \in E_2 \text{ for } i = 1, 2, \ldots, n \text{ and } j = 1, 2, \ldots, n\}$ and $E_2^2 = \{e_{ij} \| e_{ij} \in E_2 \text{ for } i = n+1, n+2, \ldots, n+l \text{ and } j = 1, 2, \ldots, n\}$. That means E_2^1 is the set of edges among normal agents, and E_2^2 is the set of edges whose starting node is one shill and ending node is a normal agent. Then for each normal agent i, its opinion is updated as follows:

$$x_i(t) = \frac{\sum_{e_{ji} \in E_2^1} u_j x_j(t-1) + \sum_{e_{ji} \in E_2^2} u_j x_s}{\sum_{e_{ji} \in E_2^1} u_j + \sum_{e_{ji} \in E_2^2} u_j} \qquad i = 1, 2, \ldots, n \tag{13.44}$$

Let $C = (c_{ij})_{n \times n}$ where $c_{ii} = \sum_{e_{ji} \in E_2^1} u_j$ for $i = 1, 2, \ldots, n$ and $c_{ij} = 0$ for $i \neq j$; $H = (h_{ij})_{n \times n}$ where $h_{ii} = \sum_{e_{ji} \in E_2^2} u_s$ for $i = 1, 2, \ldots, n$ and $h_{ij} = 0$ for $i \neq j$; $F = (f_{ij})_{n \times n}$ where $f_{ij} = u_j$ for $e_{ji} \in E_2^1$ and $f_{ij} = 0$ for $e_{ji} \notin E_2^1$; $\boldsymbol{X}(t) = (x_1(t), x_2(t), \ldots, x_n(t))'$. Then we have

$$\boldsymbol{X}(t) = (C + H)^{-1}(F\boldsymbol{X}(t-1) + H\mathbf{1}x_s). \tag{13.45}$$

Let $\boldsymbol{\varepsilon}(t) = \boldsymbol{X}(t) - \mathbf{1}x_s$, then

$$\begin{aligned}\boldsymbol{\varepsilon}(t+1) &= \boldsymbol{X}(t+1) - \mathbf{1}x_s \\ &= (C+H)^{-1}(F\boldsymbol{X}(t) + H\mathbf{1}x_s) - \mathbf{1}x_s \\ &= (C+H)^{-1}(F\boldsymbol{X}(t) + H\mathbf{1}x_s - (C+H)\mathbf{1}x_s) \\ &= (C+H)^{-1}(F\boldsymbol{X}(t) - C\mathbf{1}x_s) \\ &= (C+H)^{-1}(F\boldsymbol{X}(t) - F\mathbf{1}x_s), \quad \text{because } F\mathbf{1} = C\mathbf{1} \\ &= (C+H)^{-1}F(\boldsymbol{X}(t) - \mathbf{1}x_s) \\ &= (C+H)^{-1}F\boldsymbol{\varepsilon}(t) \end{aligned} \tag{13.46}$$

Let $Q = (C+H)^{-1}F$. If the neighborhood graph of the original system is strongly connected and each node has its loop, Q is a nonnegative stochastic primitive. Let λ_1 denote the biggest eigenvalue of Q, then we have λ_1 as real and $\lambda_1 \leq 1$ by Theorem 13.3 and Corollary 13.3.1. Therefore, the system with fixed-heading shills will eventually converge to x_s. In this case, we use the convergent speed to measure the performance of fixed-opinion shills. By Theorem 13.4, we have

$$Q^k = \lambda_1^k wv' + O(k^s|\lambda_2|_k) \quad (\text{as} \quad k \to \infty), \tag{13.47}$$

where w, v' are any positive right and left eigenvectors corresponding to λ_1 when $v'w = 1$. λ_2 is the second largest eigenvalue, and s is the index of λ_2 minus 1.

Because $((k^s|\lambda_2|^k)/\lambda_1^k) \to 0 \quad (k \to \infty)$, infer form (13.47) becomes

$$Q^k = \lambda_1^k wv' + o(\lambda_1^k) \quad (\text{as} \quad k \to \infty). \tag{13.48}$$

Therefore, the convergent speed is mainly decided by λ_1. Smaller λ_1 means better performance of *fixed-opinion* shills. Then, what affects the value of λ_1?

We try to use simulation to find some clues. Again, one *fixed-opinion* shill is connected to one normal agent in 100 direct scale-free networks. The setting of simulations is the same as that of Section 13.5.2.1. Figure 13.15 shows the relationship between λ_1 and the degree of agent that the shill connects to: λ_1 does not have apparently correlation with the degree of the connected agent. The factors that decided the convergent speed of the system with a *fixed-opinion* shill are needed to study in our future work.

13.6 Conclusions

The swarm intelligence of multiagent systems is the macroscopic feature that we concern and study in this chapter. If the self-organized swarm intelligence, such as the collective behavior, is not desirable, how can we control or intervene in multiagent systems without changing the local rule of the existing agents? It is an important issue but not well recognized yet. We use a novel framework, soft control, to solve this problem. Without changing the local rule of the existing agents, we add one or several

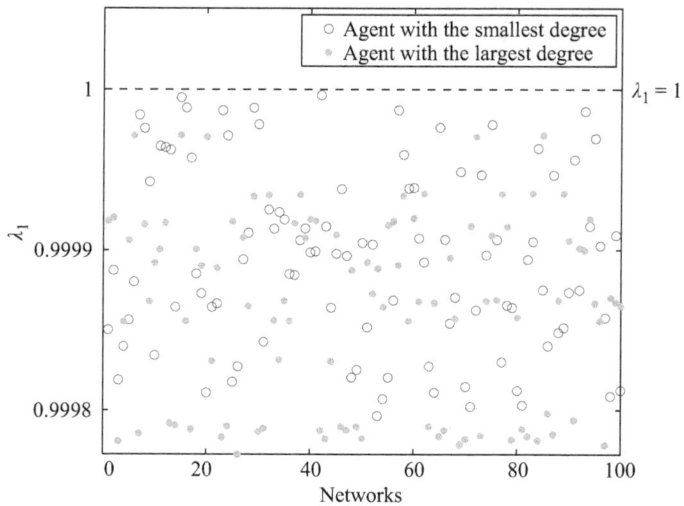

Figure 13.15 The largest eigenvalue of the system λ_1 in two scenarios in 100 scale-free networks (x-axis). The hollow circles represent λ_1 in the scenario of connecting shills with the smallest degree agent; the solid dots represent λ_1 in the scenario of connecting shill to the largest degree agent

shills into the systems and designed the control law for the shill carefully considering the control purpose. Shills do not necessarily behave as normal agents, and they are recognized as normal agents by normal ones.

We apply the soft-control method to three different swarm systems and show its feasibility:

1. Guiding the flocking swarm system to reach consensus by adding one intelligent shill.
2. Guiding cooperation of a group of game players by adding a number of cooperative shills.
3. Guiding collective opinion of a group by adding one or some shills.

The intervention purpose is very different in (1), (2) and (3)—(1) is to guide consensus, (2) is to promote cooperation and (3) is to change to the convergent value. But they are all properties of the collective swarm behavior, i.e., behavior at the macroscopic level of the system.

Shills are different. In (1), the shill added into the group is intelligent. It can lead the group reach consensus in a deterministic framework while fixed-heading "leader" can only work in a stochastic framework. And the intelligent shill beats fixed-heading "leader" even with noise. Here intelligent shill is a finite-state machine which uses feedback in formation. In (2), we focus on the feedback information and teamwork of a number of shills. They share or partially share history of game playing with normal agents, which significantly promote the cooperation level comparing to the case of no history sharing among shills. In (3), the shill is simple, actually like a normal agent except that the initial opinion and influential value are set artificially. Here, because the property of the DeGroot model, the effect of adding some unconnected shills are a linear sum of each shill's effect. The key point here is which normal agents the shill should connect to.

The interaction neighborhood networks of (1), (2) and (3) are different. In (1), it is a dynamic network because agents are moving. So the shill should also move in order to influence other agents. Therefore, the key part of the shill's strategy is its moving strategy. In (2), it is actually a random network underline. So we can get results averaged over the space. In (3), it is a static network. We find that the convergent speed positively correlates to the degree of the normal agent which connects to the shill in the scale-free networks.

To further investigate the interaction neighborhood network effect, we also studied the effect of interaction radius r on the performance of soft control in [55]. This work is based on the simplified Vicsek model (the velocities of all agents are set to be *zeros*) with nonperiodic boundary. Through simulations, we found that cluster together *fixed-heading* shills favor large r, while cluster together *evolvable-heading* shills favor small r, and there is no significant effect of r on the soft-control performance (for both *fixed-heading* and *evolvable-heading* shills). We will further investigate the relationship between the interaction network features and the performance of soft control.

In (1), we study how the interaction radius affects the soft-control performance and obtain some interesting phenomena. It also reveals teamwork (staying together) are crucial for evolvable shills which will be affected by their neighbors. So cooperation of shills is an important issue in soft control, and more research should be done in the future.

From the above three applications of soft control in different types of multiagent swarm systems, we can see that the soft-control method is flexible, and it can be applied to a wide range of swarm systems to guide their swarm behavior and help emergence of swarm intelligence. There are still a lot to do inside the soft-control framework. For example, can we define and analyze the controllability of soft control in general framework? What are the conditions when applying soft control? Are there any general strategy when distribute shills inside the swarm? On the other hand, we want to point out that soft control is not only an intervention method but also a design method. In many cases, it is difficult to achieve desired swarm intelligence only by designing the local interaction rules among agents. Soft control may provide an auxiliary method to the design of multiagent systems. Adding one or more shills to coordinate the swarm behavior may release designers from the hard work of subtle design.

Acknowledgments

This work is supported by National Natural Science Foundation of China (No. 61374168) and the Frontier Science Key Programs of the Chinese Academy of Sciences (No. QYZDJ-SSW-JSC003).

References

[1] Reynolds CW. Flocks, herds, and schools: a distributed behavioral model. ACM SIGGRAPH Computer Graphics. 1987;21(4):25–34.
[2] Morse CW. Ecological aspects of some mixed-species foraging flocks of birds. Ecological Monographs. 1970;40(1):119–68.
[3] Hölldobler B, Wilson EO, editors. The Ants. Cambridge, MA: Harvard University Press; 1990.
[4] Helbing D, Farkas I, Vicsek T. Simulating dynamical features of escape panic. Nature. 2000;407(6803):487–90.
[5] Buhl J, Sumpter DJ, Couzin ID, *et al.* From disorder to order in marching locusts. Science. 2006;312(5778):1402–06.
[6] Axtell RL, Chakravarty S. Radicals, revolutionaries and reactionaries in a multi-agent model of class norms. Private Communication.
[7] Hong Y, Chen G, Bushnell L. Distributed observers design for leader-following control of multi-agent networks. Automatica. 2008;44(3):846–50.
[8] Mu S, Chu T, Wang L. Coordinated collective motion in a motile particle group with a leader. Physica A: Statistical Mechanics and Its Applications. 2005;351(2):211–26.
[9] Monderer D, Tennenholtz M. Strong mediated equilibrium. Artificial Intelligence. 2009;173(1):180–95.
[10] Han J, Guo L, Li M. Guiding a group of locally interacting autonomous mobile agents. In: Proceedings of 24th Chinese Control Conference; 2005 July 15–18; Guangzhou, China. Piscataway, NJ: IEEE; 2005. p. 184–87.
[11] Han J, Li M, Guo L. Soft control on collective behavior of a group of autonomous agents by a shill agent. Journal of Systems Science and Complexity. 2006;19(1):54–62.
[12] Kyriakopoulos N, Ginelli F, Toner J. Leading birds by their beaks: the response of flocks to external perturbations. New Journal of Physics. 2016;18(7):073039.
[13] Pearce DJG, Giomi L. Linear response to leadership, effective temperature, and decision making in flocks. Physical Review E. 2016;94(2):022612.
[14] Bonabeau E, Dorigo M, Theraulaz G, editors. Swarm Intelligence: From Natural to Artificial Systems. New York: Oxford University Press; 1999.
[15] Desai JP, Ostrowski JP, Kumar V. Modeling and control of formations of nonholonomic robots. IEEE Transactions on Robotics and Automation. 2001;17(6):905–8.
[16] Couzin ID, Krause J, Franks NR, *et al.* Effective leadership and decision-making in animal groups on the move. Nature. 2005;433(7025):513–6.
[17] Liu Z, Han J, Hu X. The proportion of leaders needed for the expected consensus. Automatica. 2011;47(12):2697–2703.
[18] Wang XF, Chen G. Pinning control of scale-free dynamical networks. Physica A: Statistical Mechanics and Its Applications. 2002;310(3):521–31.
[19] Acemoglu D, Ozdaglar A. Opinion dynamics and learning in social networks. Dynamic Games and Applications. 2011;1(1):3–49.

[20] Mobilia M. Does a single zealot affect an infinite group of voters. Physical Review Letters. 2003;92(2):028701.

[21] Mobilia M, Petersen A, Redner S. On the role of zealotry in the voter model. Journal of Statistical Mechanics. 2007;2007(08):P08029.

[22] Sartoretti G, Hongler MO. Soft control of swarms: analytical approach. In: Proceedings of the 5th International Conference on Agents and Artificial Intelligence. 2013;1:147–153.

[23] Sartoretti G, Hongler MO. Soft control of self-organized locally interacting brownian planar agents. EUROCAST. 2013;8111:45–52.

[24] Sartoretti G. Leader-based versus soft control of multi-agent swarms. Artificial Life and Robotics. 2016;21(3):302–7.

[25] Genter K, Agmon N, Stone P. Ad hoc teamwork for leading a flock. In: Proceedings of the 2013 International Conference on Autonomous Agents and Multi-agent Systems. International Foundation for Autonomous Agents and Multiagent Systems; 2013. p. 531–38.

[26] Genter K, Stone P. Influencing a flock via ad hoc teamwork. Lecture Notes in Computer Science; 2014;8667:110–21.

[27] Genter K, Stone P. Ad hoc teamwork behavior for influencing a flock. Acta Polytechnica. 2016;56(1):18–26.

[28] Albi G, Bongini M, Cristiani E, *et al.* Invisible control of self-organizing agents leaving unknown environments. SIAM Journal on Applied Mathematics. 2016;76(4):1683–1710.

[29] Su W, Chen G, Yu Y, *et al.* Finite-time elimination of disagreement of opinion dynamics via covert noise. IET Control Theory & Applications. 2018;12(4):563–570.

[30] Halloy J, Sempo G, Caprari G. Social integration of robots into groups of cockroaches to control self-organized choices. Science. 2007;318(5853): 1155–58.

[31] Rubenstein M, Cornejo A, Nagpal R. Programmable self-assembly in a thousand-robot swarm. Science. 2014;345(6198):795–99.

[32] Vicsek T, Czirók A, Ben-Jacob E, *et al.* Novel type of phase transitions in a system of self-driven particles. Physical Review Letters. 1995;75(6):1226–29.

[33] Vicsek T, Zafeiris A. Collective motion. Physics Reports. 2012;517(3):71–140.

[34] Kunz H, Hemelrijk CK. Artificial fish schools: collective effects of school size, body size, and body form. Artificial Life. 2003;9(3):237–53.

[35] Ballerini M, Cabibbo N. Interaction ruling animal collective behavior depends on topological rather than metric distance: evidence from a field study. Proceedings of the National Academy of Sciences. 2008;105(4):1232–37.

[36] Jadbabaie A, Lin J, Morse AS. Coordination of groups of mobile autonomous agents using nearest neighbor rules. IEEE Transactions on Automatic Control. 2003;48(6):988–1001.

[37] Tang G, Guo L. Convergent of a class of multi-agent systems in probabilistic framework. Journal of Systems Science and Complexity. 2007;20(2):173–97.

[38] Liu Z, Guo L. Connectivity and synchronization of Vicsek model. Science in China Series F: Information Sciences. 2008;51(7):848–58.

[39] Chen G, Liu Z, Guo L. The smallest possible interaction radius for flock synchronization. SIAM Journal on Control and Optimization. 2012;50(4): 1950–70.
[40] Han J, Wang L. New strategy of the shill: 'consistent moving'. In: Proceedings of 29th Chinese Control Conference; 2010 July 29–31; Beijing, China. Piscataway, NJ: IEEE; 2010. p. 4459–64.
[41] Han J, Wang L. Nondestructive intervention to multi-agent systems through an intelligent agent. PLoS One. 2013;8(5):e61542.
[42] Horn RA, Johnson CR, editors. Matrix Analysis. New York: Cambridge University Press; 1985.
[43] Seneta E, editor. Non-negative Matrices and Markov Chains. Berlin: Springer; 1981.
[44] Cao M, Morse AS. Reaching a consensus in a dynamically changing environment: a graphical approach. SIAM Journal on Control and Optimization. 2008;47(2):575–600.
[45] Wang X, Han J, Han H. Special agents can promote cooperation in the population. PLoS One. 2011;6(12):e29182.
[46] Nash J. Non-cooperative games. Annals of Mathematics. 1951;54(2):286–95.
[47] Nowak M, Sigmund K. Oscillations in the evolution of reciprocity. Journal of Theoretical Biology. 1989;137(1):21–26.
[48] Axelrod RM, editor. The Evolution of Cooperation. New York: Basic Books; 2006.
[49] Hegselmann R, Krause U. Truth and cognitive division of labour first steps towards a computer aided social epistemology. Journal of Artificial Societies and Social Simulation. 2006;9(3):10–38.
[50] Golub B, Jackson MO. Naive learning in social networks and the wisdom of crowds. American Economic Journal: Microeconomics. 2010;2(1):112–49.
[51] Chandrasekhar AG, Larreguy H, Xandri JP. Testing models of social learning on networks-evidence from a lab experiment in the field. NBER Working Paper No. 21468. Cambridge, MA: National Bureau of Economic Research; 2015.
[52] Jackson MO, editor. Social and Economic Networks. Princeton, NJ: Princeton University Press; 2008.
[53] Han H, Qiang C, Wang C, *et al.* Intervention of DeGroot model by soft control. Journal of Systems Science and Complexity. 2017;30(3):550–67.
[54] Barabási AL, Albert R. Emergence of scaling in random networks. Science. 1999;286(5439):509–12.
[55] Wang C, Han J. The impact of interaction radius on soft control performance based on the Vicsek Model. In: Proceedings of 35th Chinese Control Conference; 2016 July 27–29; Guangzhou, China. Piscataway, NJ: IEEE; 2016. p. 1175–80.

Chapter 14

Agreeing to disagree: synergies between particle swarm optimisation and complex networks

Mihai Udrescu[1] and Andrei Lihu[1]

Abstract

Due to its numerous applications in fields like systems biology, medicine, technology, engineering, or social sciences, the new science of complex networks (CNs) has become extremely popular over the last couple of decades. Particle swarm optimisation (PSO) and CN science have significant common ground, as they both deal with a set of agents (particles for PSOs, vertices for CNs) which interact according to an underlying topology. Moreover, one of the most important branches of CN science is represented by social networks, while PSO was originally conceived for optimisation through the simulation of social behaviour, thus further emphasising the strong tie between the two scientific fields. A prominent problem in CNs is the opinion dynamics in social networks. To this end, opinion interaction models are tested in order to verify if they can recreate a realistic social behaviour; a ubiquitous feature of such real-life social behaviours is persistent opinion disagreement. One of the most important characteristics of a social network which fosters disagreement is that it contains stubborn agents, namely vertices which never change their opinion. We discuss models of disagreement, such as that with stubborn agents and tolerance-based models, then offer a new perspective in exploring the frontiers between network science and PSO.

14.1 Introduction

The science of complex networks (CNs) has emerged as a promising, stand-alone body of science which combines statistical physics, graph theory, computer algorithms, and big data technologies in order to undertake hard problems pertaining to complex systems, such as biological, technological, social, or semantic [1,2]. The ability of network science to represent and analyse real-world complex systems has made it very attractive for a wide array of scientists; therefore, it comes as no surprise the fact that researchers from the field of particle swarm optimisation (PSO) get inspiration

[1]Department of Computer and Information Technology, Politehnica University of Timisoara, Romania

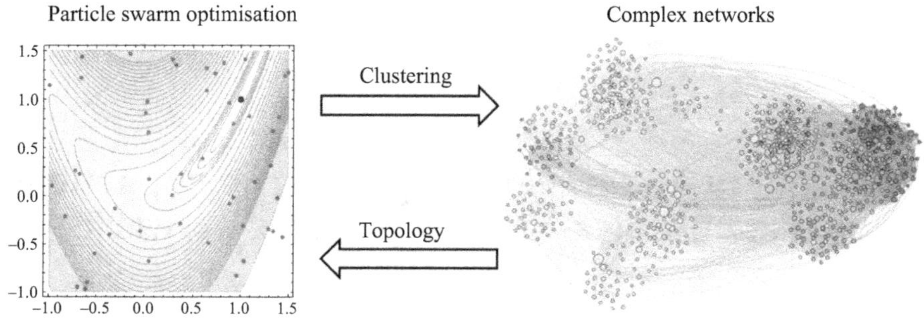

Figure 14.1 Graphical description of the way PSO and network science influence each other: PSOs use complex topologies from complex networks to optimise search, whereas PSOs are used to build clustering tools in complex networks. The example figures are taken from [7] (PSO for 2D problems example in Wolfram Demonstrations Project – left panel) and [8] (a drug–drug interaction complex network – right panel)

from CN. Indeed, even at first glance, there seems to be significant resemblance: both fields deal with a set of nodes or agents that are connected according to a topology.

Over time, the gap between PSOs and CNs was narrowed by the following two directions:

1. Finding inspiration in CN topologies to improve the efficiency of PSO search [3,4].
2. Using the PSO algorithmic framework to cluster vertices/nodes in CNs [5,6].

Figure 14.1 presents the current state of the art of the interplay between several characteristics of CN and PSO, by pointing the main source of inspiration on the corresponding arrow.

This paper presents a new perspective on the CN–PSO fusion: the fertile ground at the frontiers between CNs and PSOs is represented by opinion spread in complex social networks. The way opinions and beliefs spread and oscillate in social networks is an important topic of interest in science and engineering [9]. One of the most influential works in opinion spread in social networks is represented by the model in [10,11]; the novelty of the model consists in proving that when the social network contains *stubborn agents* (i.e. influencers, or agents who never change their opinion), the general opinion in the social network will continuously oscillate because regular (i.e. non-stubborn) social agents will find themselves – exactly as in real life – in a continuous disagreement. Moreover, the concept of disagreements from [10,11] has fuelled new research in PSOs, as it seemed that disagreements in PSO could bring additional search efficiency [12].

In this paper, we argue that there could be more to explore in-between PSOs and CNs. We believe that CN elements which PSO can benefit from are more than complex graph topologies; specifically, we identify the CN spread models, which could be used for PSO particle interactions in order to induce disagreement and make the optimisation process better.

The remainder of this chapter is organised as follows: Section 14.2 introduces the fundamental concepts in network science, including opinion spread social network models along with Acemoglu's work on continuous disagreement between stubborn agents. Section 14.3 reviews current efforts to leverage CNs concepts in improving the performance of PSO. In Section 14.4, we advocate for the research of new PSO topologies and update formulas based on social agent interaction models, then we lay the concluding remarks.

14.2 Background

14.2.1 Complex network science

Although the field of CNs is multidisciplinary, it is mostly related to physics – specifically to statistical physics and complex systems. The practical applications of such theoretical insights are extensive: biology, medicine, economy, social sciences, engineering (including electrical, computer, and software engineering), computer science, and physics. Therefore, according to the field of application, the vast domain of CNs can be divided into

- **Biological networks:** disease networks, food networks in given environments, gene networks, pathway networks, metabolic networks, protein interaction networks, drug interaction networks, metabolome networks etc.
- **Technological networks:** computer networks, world wide web, road and transportation networks, power distribution networks, electronic components networks, software classes networks etc.
- **Social networks:** online social networks, political networks, economic networks, friendship networks, collaboration networks etc.
- **Semantic networks:** LISP semantic networks, natural language word networks etc.

Technological CN techniques and methodologies can be used in computer engineering applications which entail a big amount of complexity [13]. Meanwhile, the overarching field of information technology includes various approaches in which computer algorithms and applications are used for achieving advancements in biology, medicine, pharmacology, or social physics. Indeed, the last decade has witnessed significant progress in precision medicine based on Big Data algorithms and computer technologies such as CN analysis and machine learning [14]. Moreover, the advance in social system physics has gained a lot of momentum since the global dissemination of online social networks began [15].

Herein, we present some basic theoretical and taxonomic elements used for CN analysis. Nonetheless, a thorough review on this vast field is beyond the scope of this chapter; therefore, we will indicate further references for a more detailed analysis [1,2,16,17].

14.2.1.1 Basic parameters

Definition 14.1. *A CN is a graph $G=(V,E)$ with V being the set of vertices (or nodes) $V=(v_i,w_i^v)$ and E the set of edges (or links) $E=(e_j,w_j^e)$ which connect some of the vertices in V, where v_i and e_j are individual vertices and edges, respectively, along with their corresponding vertex and edge weights w_i^v and w_j^e for $i=\overline{1,N}$ and $j=\overline{1,|E|}$.*

Usually, N is also referred as the *network size*, while E (i.e. all edges in G) are referred as the *network topology*. As opposed to conventional graph theory, the total number of vertices N and edges $|E|$ is large (up to millions or billions in the case of social networks). A big difference between conventional graphs and CNs is that graphs have regularly placed edges (i.e. regular topologies), whereas CNs are characterised by irregular, complex topologies.

Definition 14.2. *An* unweighted complex network *is a CN G where all vertex and edge weights are equal; therefore, $w_i^v=w_j^e$ for all i and j. As such, we do not take into account weights, such that $G=(V,E)$, with $v_i\in V$, and $e_j\in E$.*

In many applications, it is useful to graphically depict CNs. As such, in order to describe how the CNs and their topologies are graphically represented, we introduce the following definition:

Definition 14.3. *Given a CN $G=(V,E)$ and an Euclidean d-dimensional space $\mathbb{R}^d$, a layout algorithm maps each vertex $v\in V$ to a position $x_v\in\mathbb{R}^d$ and assigns an Euclidean distance $\|x_v-x_w\|$ to each edge $[v,w]\in E$ (v and w are vertices in V).*

In order to analyse CN topologies and behaviour, we use the parameters that are described by the following definitions.

The distance between two vertices v and w pertaining to CN G is given by the edge number of the shortest path between v and $w\in G$ (see Figure 14.2 for a graphical example).

Definition 14.4. *The average path length L is the average or – in other words – typical distance, obtained by averaging the distances between all possible pairs of vertices $(v,w)\in G$. Also, the diameter Dmt of graph G is the biggest distance between any two vertices from G.*

Definition 14.5. *The degree of vertex v is the number of its direct connections, k_v. For network G, the average degree is expressed as $\langle k\rangle=(1/N)\sum_{v\in V}k_v$, representing the average value for the degrees of all vertices in graph G.*

Definition 14.6. *The clustering coefficient C_v of a vertex/node v in graph G is given by the fraction of existing edges between the $n(v)=k_v$ neighbours of node v, relative to the number of all possible edges between the neighbours $(1/2)k_v(k_v-1)$. $C_v=(2E[n(v)]/n(v)[n(v)-1])$ (see Figure 14.3). The average clustering coefficient of the entire graph G is given by $C=(1/N)\sum_{v\in V}C_v$.*

Definition 14.7. *The network density Dst is defined as the ratio of all edges in the network relative to the total number of possible edges $(2|E|/N(N-1))$.*

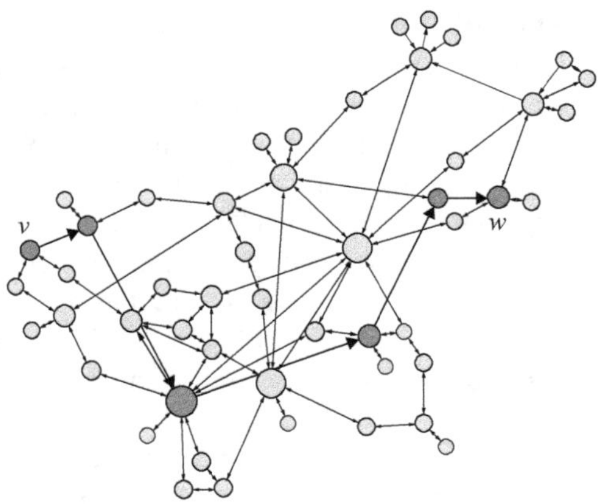

Figure 14.2 A graph with a distance between nodes v and w of five

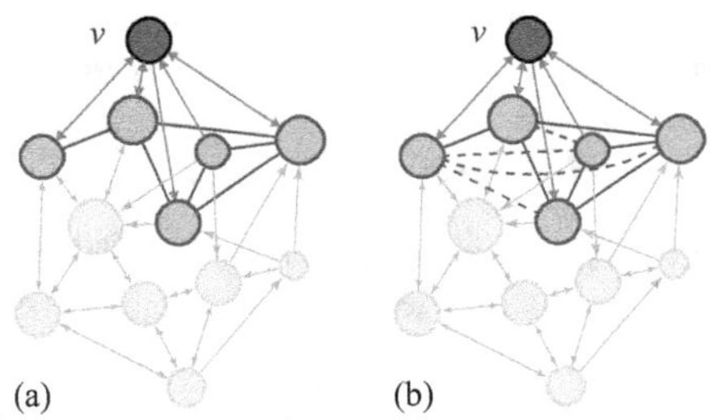

Figure 14.3 The clustering coefficient of the highlighted network node v is the ratio between the E[n(v)] = 6 existing edges – undirected in panel (a) – which connect the n(v) = 5 neighbours of node v, and the number (1/2)n(v)[n(v) − 1] = (1/2) · 5 · 4 = 10 of all possible edges in panel (b); in our example, C_v = (6/10) = (3/5)

Network clustering consists of classifying all vertices $v \in V$ in one of the disjoint vertex subsets (i.e. clusters) C_i pertaining to the set of disjoint subsets $C = \{C_1, C_2, \ldots, C_m\}$, with $\cup_{i=1}^{m} C_i = V$ [18,19]. Several network parameters and metrics are used for clustering in CNs [20–22], but one of the most accredited is the *modularity* introduced by Newman and Girvan [23].

Definition 14.8. *For an unweighted network, the modularity of clustering C is* $Mod_C = \sum_{C_i \in C} ((|E_{C_i}|/|E|) - (((1/2)k_{C_i}{}^2)/((1/2)k^2)))$, *where* $|E_{C_i}|$ *is the number of edges in cluster* C_i, $|E|$ *is the total number of edges in the network,* k_{C_i} *is the total degree of nodes in cluster* C_i, *and k is the total degree of nodes in the entire network. Assuming uniform network density,* $|E_{C_i}|/|E|$ *represents the fraction of intra-cluster edge density relative to the density of the entire network, while* $((1/2)k_{C_i}{}^2)/((1/2)k^2)$ *represents the expected fraction value. Modularity grows when the clustering process produces clusters with larger than expected edge densities.*

14.2.1.2 Network centralities

Network centralities represent network parameters or metrics which characterise the importance of vertices (or nodes) in a CN. Although centralities represent individual nodes, computing the average centrality values seems not to be the most appropriate way of characterising the network; the *distribution* of these centrality values is paramount for classifying and characterising CNs.

One of the most important centrality distributions – which are used to analyse complex graphs – is the *degree distribution*, namely the probability $P(k)$ that the degree k has a certain value. The degree distribution is a histogram that depicts the number nodes in the graph for each degree value k (ranging from one to the highest degree in the graph).

Another important centrality that renders insightful distributions is *node betweenness*. The concept of betweenness, characterising any vertex's ability of being a mediator or *in-between*, can be extended to edges (i.e. links) as well, thus obtaining the *link* or *edge betweenness*.

Definition 14.9. *The betweenness $b(v)$ of vertex $v \in V$ in complex graph $G = (V, E)$ is defined as $\sum_{i,j \in V} ((\sigma_{ij}(v))/\sigma_{ij})$, where $i \neq j \neq v$, σ_{ij} is the total number of shortest paths from node i to node j and $\sigma_{ij}(v)$ is the number of shortest paths from node i to node j which includes node v in their path sequence.*

Other centralities are also used for the analysis of CNs by capitalising on their distributions: *closeness*, *eigenvector*, and *page rank*. Computing the inverse of the sum of shortest path lengths between the reference node and all other vertices gives the *node closeness*. The *eigenvector* centrality computes relative scores for all nodes by considering that the connections to high-influence vertices are more important than the connections to low-influence vertices; as an observation, *page rank* is merely a variant of eigenvector that is employed by Google Search to rank websites [2,24].

For measuring the similarity between two CNs in terms of network parameters and centralities, the network statistical fidelity metric φ was introduced by [25,26]:

$$\varphi^j = \begin{cases} \frac{1}{n}\sum_i \frac{\overline{m_i}}{2\overline{m_i} - \overline{m_i^j}} & \text{if } \overline{m_i^j} < \overline{m_i} \\ \frac{1}{n}\sum_i \frac{\overline{m_i}}{\overline{m_i^j}} & \text{if } \overline{m_i^j} \geq \overline{m_i} \end{cases} \tag{14.1}$$

In (14.1), j represents the index of the network that is compared against a reference network. The index of the metric which describes the two compared models (e.g. average path length, average degree etc.) is denoted by $i = \overline{1 \ldots n}$, where n is the total number of common metrics taken into consideration to assess similarity. Statistical fidelity takes values between 0 and 1 (but it can also be represented as percentiles), with 1 representing perfect similarity. The metric measurements on the reference model are m_i, while measurements on the model being compared are m_i^j.

14.2.1.3 Fundamental complex network models

One of the main purposes of CN theory is to provide models that can accurately characterise real-world systems. To this end, the analysis of complex systems, which

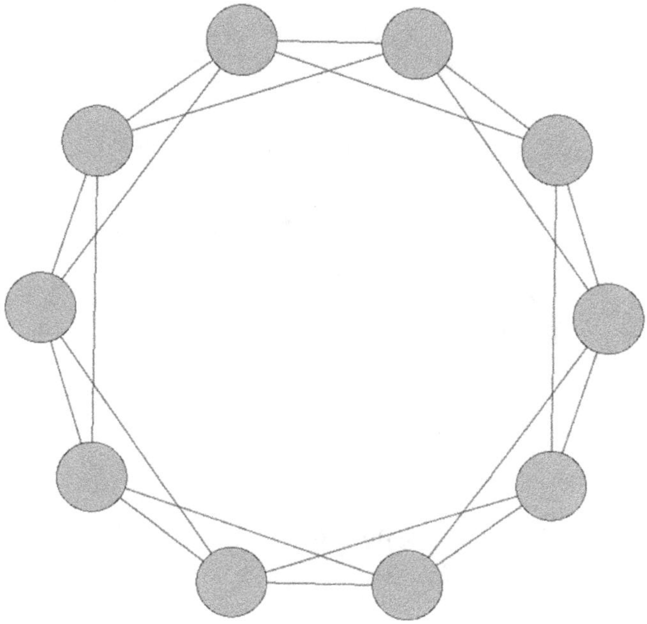

Figure 14.4 Regular graph example, with N = 10 vertices

can be modelled as graphs, has revealed that real-world networks (technological, biological, social, or semantic) are generally sparse (i.e. with low *Dst*), characterised by a low-average path length, high clustering, and complex degree distribution, such as power-law or Gaussian [17]. Also, most real-world CNs also exhibit complex distributions of other centralities: betweenness, closeness, or eigenvector [2].

Generally, in traditional graph theory, analysis can only be performed on regular graphs such as meshes or Euclidian lattices, like the one presented in Figure 14.4.

However, it is obvious that real-world systems do not exhibit the regular structure of Euclidean lattices. As opposed to real-world, regular networks have big average path lengths (because L grows linearly with N) and only simple sequences for degree distributions. However, regular networks have high clustering coefficients C similar to real-world networks [17], e.g. in Figure 14.4, $C = (3/6) = (1/2)$.

The first step in describing and modelling complex graphs was made at the end of the 1950s, by Hungarian mathematicians Paul Erdös and Alfréd Rényi, who came out with the idea of **Random Networks (RNs)**. According to the RN model, when a network consists of $|V| = N$ vertices, there are $(N(N-1)/2)$ potential edges between these vertices; however, each edge is actually placed in the network with probability p (see Figure 14.5 for an illustrative example). Accordingly, the average degree in RNs is $\langle k \rangle = pN$, $L \sim (\ln N/\langle k \rangle)$, and $C = p = (\langle k \rangle/N) \ll 1$; this means that in RNs, we have low average path lengths (similar to real-world networks), but also low clustering coefficients (unlike real-world networks). As opposed to regular networks

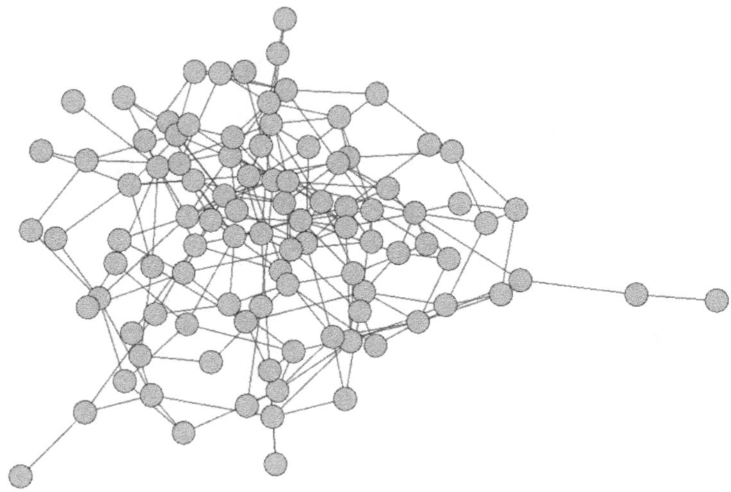

Figure 14.5 Random network example with $N = 100$ vertices

where degree distributions are simple sequences, RNs exhibit a complex structure because they have Poisson degree distributions with the peaks corresponding to $\langle k \rangle$.

From a fundamental standpoint, RNs are particularly significant because they narrow the gap between graph theory and probability theory [17]. However, the main contribution of the RN model is the so-called network effect, which characterises the phase-transition from a disconnected to a connected RN when probability p becomes bigger than threshold $p_c \sim (\ln N/N)$.

As both regular and RN models fail to accurately describe real-world systems, better models were sought. Mathematicians Duncan Watts and Steven Strogatz noted that a more realistic model will lie somewhere between regular and random models by taking the best of both worlds: low L from RNs and high C from regular networks [27].

The objective of exploring the space between completely regular and completely RNs is attained by performing a simple algorithm: we consider a regular network such as the lattice from Figure 14.4 as the starting point, then randomly pick – with probability p – a vertex pertaining to an edge that will be 'rewired'. The new destination of the rewired edge will be a randomly picked vertex from the pool of all other vertices. The only constraint is that there are no connections from a node to itself, nor there are double connections or links between two nodes. This simple procedure is executed for all N nodes from the initial lattice. As illustrative example, Figure 14.6 presents, the network obtained with the algorithm of Watts and Strogatz applied over Figure 14.4 lattice for $p = 0.05$.

When increasing p from 0 to 1, it can be noticed that after a relatively small p, the average path length L decreases significantly, whereas clustering coefficient C decreases only when p becomes significantly bigger. Therefore, for a certain range of p values, we can create a network which is somewhere in-between regular and

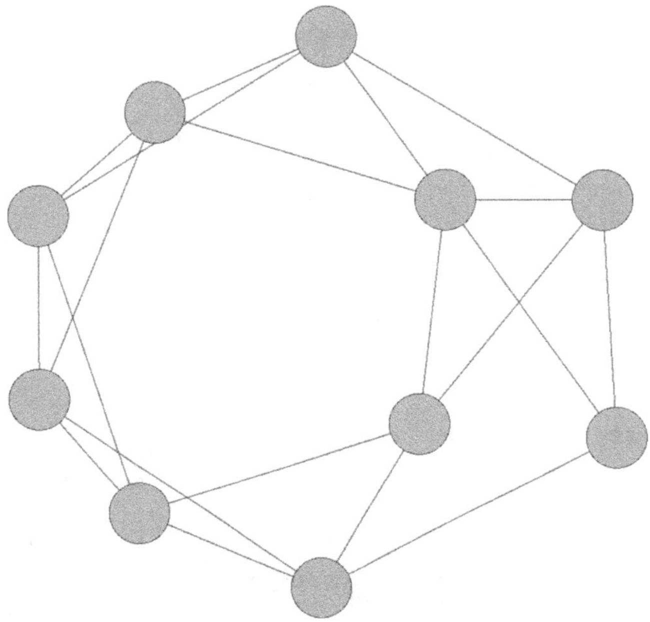

Figure 14.6 Small world network with $N = 10$ nodes, obtained by using Watts and Strogatz algorithm, which starts with the lattice from Figure 14.4, with a rewiring probability of $p = 0.05$

RNs, such that the newly created complex graph has low L and big C – similar to the values that real-world CNs exhibit. In this particular region of low L and high C, the so-called small-world (SW) effect appears, namely the property of having a relatively short distance between any two nodes, despite the fact that the network is significantly clustered.

Figure 14.7 presents this exploration methodology of finding the SW effect zone when the starting regular network is a lattice with $N = 1{,}000$ nodes; the log-normal plot represents both normalised clustering coefficient and average path length's $((C(p)/C(0))$ and $(L(p)/L(0)))$ evolution with p. $C(0)$ and $L(0)$ represent the clustering coefficient and average path length of the regular network because these values correspond to $p = 0$ as rewiring probability. Conversely, when $p = 1$, the network becomes fully random; therefore, the $C(p)$ and $L(p)$ values correspond to an RN.

However, in most real-life networks, degree distribution $P(k)$ abides by a power law; therefore, it is not normally distributed. Still, the SW networks that are built with Watts and Strogatz algorithm have a normally distributed degree (see Figure 14.8); this situation is explained by the fact that each rewiring used for the SW network algorithm is randomly constructed with the same probability, p.

The latest fundamental network model in CN science is the scale-free network introduced by Albert-László Barabási and Réka Albert in 1999 [28]. The main feature

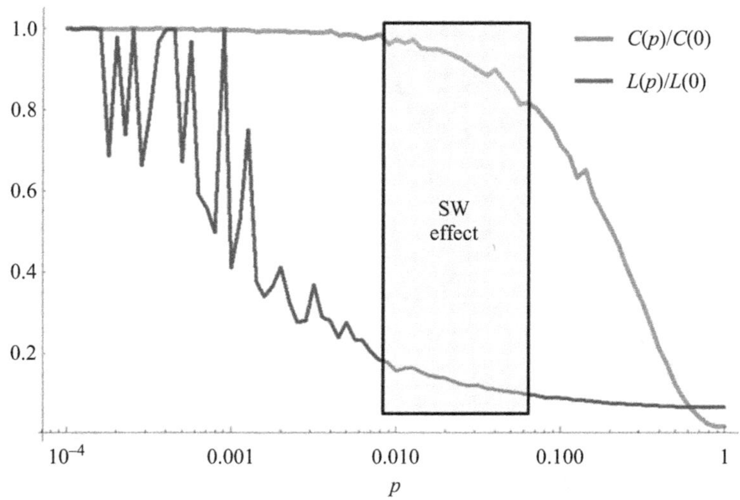

Figure 14.7 Representation in log-normal coordinates for the evolution of a normalised clustering coefficient and average path length values (C(p)/C(0)) and (L(p)/L(0)), when p grows from 0 to 1. The figure also highlights the window where the SW effect is generated because L decreases significantly and C remains high

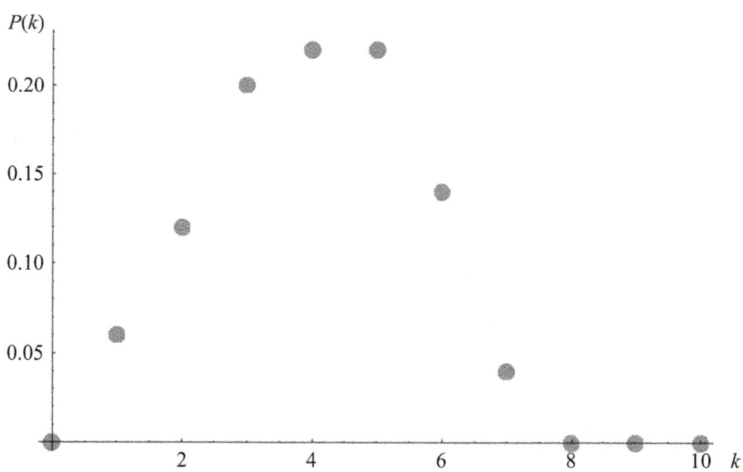

Figure 14.8 Normal degree distribution in a small-world network with 100 nodes

of this model is that it is dynamic; therefore, the network grows in each step of the algorithm. A brief description of the algorithm is given as follows:

1. Start with an arbitrary network with M nodes; this M-node network can be a SW or a random graph.

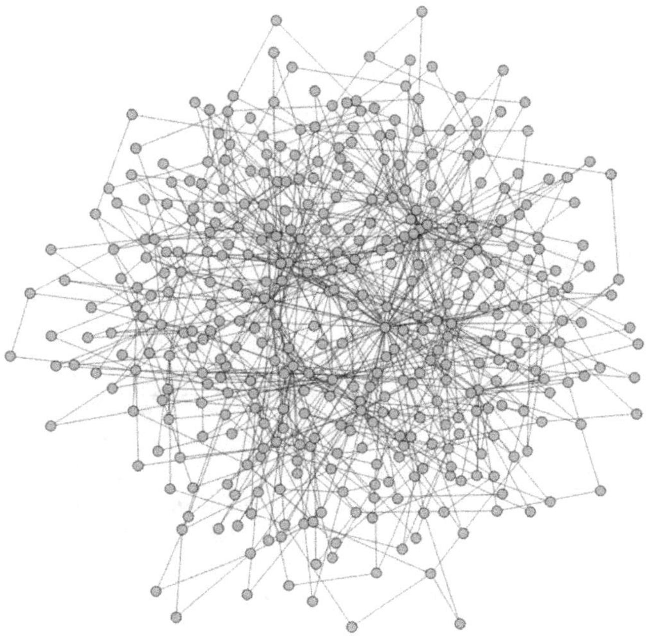

Figure 14.9 Scale-free network created with Barabási–Albert algorithm, N = 1,000 nodes when 2 nodes are added at each step

2. A number of m new nodes are added. The probability of linking each of the m new nodes to node v from the M existing nodes is $p_v = \sum_{i=1}^{M} (k_v/k_i)$ (k_v and k_i are the degrees of nodes v and i).
3. The new M is updated with $M + m$. If the new $M \neq M_{max}$, then go to the previous step, else exit.

An example of a scale-free network generated with Barabási–Albert algorithm for $M_{max} = 1{,}000$ and $m = 2$ is presented in Figure 14.9. In this scale-free network, the degree distribution is a power-law $P(k) = k^{-\gamma}$, as presented in Figure 14.10.

The main ingredient that produces the power-law distribution of degree in scale-free networks is the preferential attachment mechanism which assures that the nodes with already high degree in the network will have an even higher degree because the probability of attachment $p_v = \sum_{i=1}^{M} (k_v/k_i)$ and the degree k_v are directly proportional. Oftentimes it is said that in scale-free networks, the degree is distributed according to the 'rich gets richer principle'.

In conclusion, scale-free networks are realistic in terms of degree distribution and low average path length L. However, scale-free networks tend not to be clustered enough to sufficiently resemble real-life networks. Therefore, a lot of research has recently been done in order to combine the fundamental network models by adding randomness to scale-free networks or by adding the SW property to scale-free models [29]. Moreover, one of the weaknesses of the scale-free network model is that, even

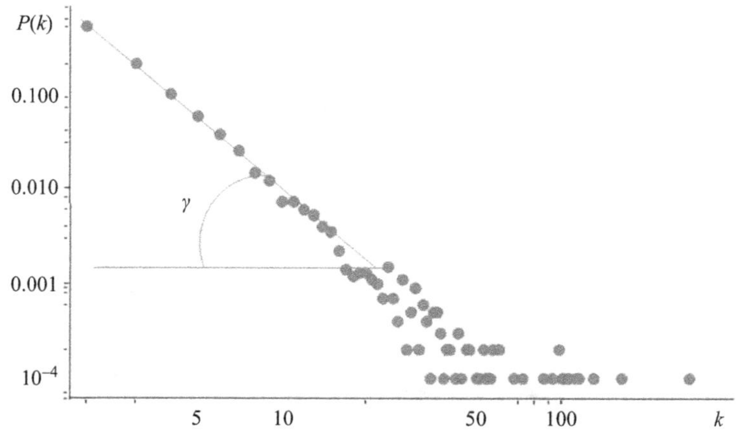

Figure 14.10 Log–log representation of the power-law degree distribution in a $N = 10{,}000$ nodes scale-free network with two nodes added at each step of the Barabási–Albert algorithm. Up to the cutting point indicated by the vertex of the added angle, the distribution abides by the power-law $P(k) = k^{-\gamma}$, where γ is the slope, namely the angle suggested in the figure

if it is very robust to random attacks, it is very vulnerable to targeted attacks. Specifically, if nodes are randomly removed from a scale-free network, the network remains connected (i.e. with one connected component) even if many nodes are removed. If the process of removal specifically targets high-degree nodes, then the network becomes disconnected (i.e. with several disconnected components) even after a small number of removed nodes.

14.2.2 Persistent opinion disagreement in social networks

Social networks are particular cases of CNs that specifically describe human collective behaviour and emerging phenomena. In a social network $G = (V, E)$, vertices or nodes $v \in V$ represent social actors or individuals, whereas edges or links $e \in E$ represent social relationships (such as friendship or collaboration) and interactions.

The advent of online social networks not only facilitates social interaction but also offer an environment for recording huge amounts of data on human activity and interaction, to be analysed and interpreted. Following this trend, the field of social networks has been intensely researched during the last decades [30,31]; the main topics of investigation are social network structure (i.e. topology) and dynamic behaviour (e.g. interaction, emergent phenomena etc.). Nonetheless, the structural and dynamical aspects are closely interrelated, as current structure often mirrors the dynamic processes from the past. Thus, most social network studies encompass both structure and dynamics in order to analyse friendship, collaboration, economic exchange, or opinion spread networks. Opinion and information spread in social networks is one of the most investigated topics in CNs, as researchers are continuously refining and testing new and sophisticated social interaction models.

In terms of network *structure*, social network analysis (SNA) also assumes basic topologies such as mesh, random [32], SW [27], and scale-free [28]. On the other hand, based on the last decade of research on realistic social network topology generation, which either adds the SW property to scale-free models [33–35], or adds a power-law degree distribution to the SWs [29,36–38], the Watts–Strogatz with degree distribution topology is considered [36] in many such social network studies.

In terms of *opinion dynamics*, SNA often relies on opinion interaction models that can be classified as graph and threshold based [39]; the threshold determines opinion phase transition. Most models use fixed values for such thresholds [40,41]. Some models actually use dynamic thresholds [42,43], but threshold value evolution is not determined internally by social agents. On the other hand, empirical dataset examples such as those from Facebook, Twitter, MemeTracker, and Yelp indicate that opinion oscillates continuously and consensus is a rare case in the real world [44]. Recent research using either discrete [45] or continuous [11] opinion representation indicate that, if the social network considers the presence of *stubborn agents* who never change their opinion, the society experiences continuous disagreement – a situation that is consistent with most empirical social behaviour.

In accordance with [11,45], given a social network $G = \{V, E\}$, where vertices/nodes are social agents $V = \{1, 2, \ldots, N\}$ and edges E represent social links or relations, the neighbourhood of agent $i \in V$ is defined as $N_i = \{j \mid (i, j) \in E\}$. The disjoint sets of stubborn agents, V_0, $V_1 \in V$ (representing agents that never change their binary opinion – 0 or 1) and *null agents* $V_{null} \in V$ (representing non-participant agents that hold no opinion) never change their opinion state, while all other (i.e. regular) agents $V \backslash \{V_0 \cup V_1 \cup V_{null}\}$ change their opinion states according to interactions with the opinion of their neighbours.

We denote the real-time opinion of agent i at time t as $o_i(t)$. Regular agents start with a predefined opinion value $o_i(0) \in [0, 1]$, which is generally considered as a random value. The process of changing the opinion states of regular agents consists of either being influenced by the opinion of a randomly chosen direct neighbour or by the average opinion of all direct neighbours, N_i.

We represent with $s_i(t)$ the discrete opinion state of agent i at moment t, which holds a continuous opinion $o_i(t)$. When the discrete opinion representation is adopted, $o_i(t) = s_i(t)$; when the continuous opinion representation is adopted, $s_i(t)$ is given by (14.2).

$$s_i(t) = \begin{cases} 0 & \textit{if } \ 0 \leq o_i(t) < 0.5 \\ NONE & \textit{if } \ o_i(t) = 0.5 \\ 1 & \textit{if } \ 0.5 < o_i(t) \leq 1 \end{cases} \tag{14.2}$$

According to the approach in [11,45], $s(t)$ denotes the average state of the population at time t, obtained by averaging the opinion of all individual agents $i \in V$ in the social network.

$$s(t) = \frac{1}{|V|} \sum_{i \in V} s_i(t) \tag{14.3}$$

At the same time, the concept of *opinion change* ω is introduced as the ratio of agents which have changed their current state at discrete time step t with respect to the previous moment (i.e. time step $t-1$):

$$\omega(t) = \frac{1}{|V|}\sum_{i \in V} |s_i(t) - s_i(t-1)| \tag{14.4}$$

If an agent changes its state from one discrete opinion to a complementary one, the absolute difference $|s_i(t) - s_i(t-1)|$ will be 1; it will be 0 if the agent's discrete opinion state does not change. These differences, averaged over all agents in the social network at (discrete) interaction moment t, define the opinion change of the population $\omega(t)$; this parameter is used to observe the current tolerance level across the entire social network.

In the interaction model proposed in [11,45], the social agent a updates its current opinion $o_a(t)$ when interacting with a randomly chosen neighbour n (which holds opinion $o_n(t)$) at the next time step $t+1$. In the given social network, which is characterises a specific topology, agents a and n are neighbour nodes/agents if there is a link/edge that directly connects them. Some agents do not have any opinion, so interacting with these NULL agents does not result in an opinion update. If agent a interacts with an agent n which holds an opinion, the updated opinion of node a at time $t+1$ is given by

$$o_a(t+1) = \theta_a o_n(t) + (1-\theta_a) o_a(t). \tag{14.5}$$

Most existing social interaction models [40,42,46–50] do not assign to nodes (representing individuals or social agents) basic individual humans traits, although it is well known that humans evolve, learn, react, and adapt over time. The explanation for the relative simplicity of the existing models is dual; on one hand, state-of-the-art models are mostly build for theoretical contexts; therefore, additional complexity in the interaction model would generate significant difficulty of mathematical analysis. On the other hand, any attempt to gauge human personality (e.g. quantifying an individual's trust, credibility, or emotional state) is a cumbersome task. Thus, (14.5) assumes that the individual's tolerance θ_a (of node a) towards other agents' opinions does not evolve over time.

14.2.3 Tolerance-based opinion interaction models in social networks

In reference [44], it is argued that tolerance should not be a fixed value. Indeed, in real-life, the predisposition for accepting other people's ideas evolves according to each individual's experiences. As such, an agent that is exposed to a high diversity of opinions tends to be more tolerant; conversely, an agent that is exposed to biased opinions that do not change becomes intolerant. Moreover, the evolution towards both tolerance and intolerance is not linear because, in time, a person under constant influence becomes intolerant at an increased rate. If that person will eventually face an opposing opinion, he/she will start to slowly and progressively build confidence in that other opinion. Therefore, the modified tolerance-based model from [44,51] employs a non-linear tolerance fluctuation function, unlike most other models in literature [52,53]. By taking into consideration several socio-psychological characteristics in the dynamical interaction model, the tolerance evolution for each node or social agent a is modelled as

$$\theta_a(t) = \begin{cases} \max\left(\theta_a(t-1) - \alpha_0\varepsilon_0,\ 0\right) & \text{if } s_a(t-1) = s_n(t) \\ \min\left(\theta_a(t-1) + \alpha_1\varepsilon_1,\ 1\right) & \text{otherwise} \end{cases} \tag{14.6}$$

In (14.6), tolerance decreases by a factor of $\alpha_0\varepsilon_0$ if the agent's state prior social interaction, $s_a(t-1)$ is equal to the state of the neighbour that it interacts with (randomly chosen from all directly connected neighbours), $s_n(t)$. However, if the states are not equal, and the agent comes in contact with an opposite opinion, the tolerance will be adjusted with the dynamically computed product $\alpha_1\varepsilon_1$. Variable t represents the time step when an opinion updating via social interaction is triggered; these time steps are considered as being randomly distributed. The two scaling factors, α_0 and α_1, are both initially set to 1 and act as weights that are increased by counting in order to keep track and take into consideration all the events in which the initiating agent maintains its previous opinion (i.e. tolerance decreasing) or changes its previous opinion (i.e. tolerance increasing). The scaling factors are

$$\alpha_0 = \begin{cases} \alpha_0 + 1 & \text{if } s_a(t-1) = s_a(t) \\ 1 & \text{otherwise} \end{cases} \tag{14.7}$$

$$\alpha_1 = \begin{cases} 1 & \text{if } s_a(t-1) = s_a(t) \\ \alpha_1 + 1 & \text{otherwise} \end{cases} \tag{14.8}$$

The scaling factors α_0 and α_1 are used to model the social bias which is taken into consideration by the tolerance-based opinion spread model: whenever an event (i.e. social interaction resulting in either confirming or contradicting the agent's previous opinion) occurs, the α corresponding to the other type of event is reset. The α factors are used to increase the magnitude of the two tolerance modification ratios: ε_0 (intolerance modifier weight) and ε_1 (tolerance modifier weight); the two ratios are chosen in [44] as fixed values: $\varepsilon_0 = 0.002$ and $\varepsilon_1 = 0.01$. These parameter values are determined such that if ε_0 is increased, most nodes will quickly become intolerant, as opinion will cease to percolate; if ε_0 is decreased, the society will become tolerance-inclined, with random opinion fluctuations. The ratio ε_0:ε_1 is chosen to be 1:5 through consistent experimentation to provide a good balance between the deviations towards tolerance and intolerance, respectively [44].

By simulating the social network opinion evolution with SocialSim[1] [44], using a tolerance-based opinion interaction network, it was confirmed that opinion disagreement is constant and never ceases, even when simulations assume CN topologies that drastically degrade tolerance (e.g. the scale-free model) (Figure 14.11).

[1] https://sites.google.com/site/alexandrutopirceanu/projects/socialsim.

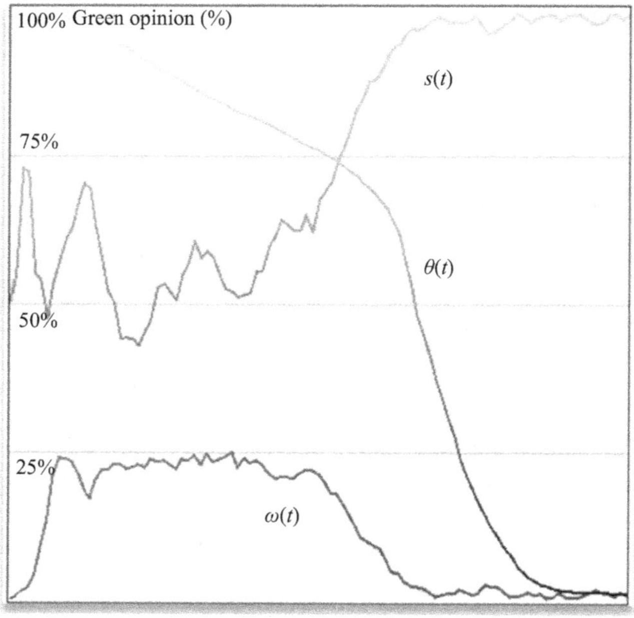

Figure 14.11 Simulation of opinion dynamics with SocialSim in a scale-free social network with 100,000 agents (32 stubborn agents holding opinion '0' (or 'red' opinion), 32 stubborn agents holding opinion '1' ('green' opinion), and no null agents, assuming the tolerance model of opinion interaction [44]. Opinion state $s(t) =$ '1' is represented by higher values, and opinion $s(t) =$ '0' is represented by lower values. Because the underlying topology of the social network is scale-free, thus leading to a high influence from the high degree agents, the tolerance θ drastically degrades. Even if the social network is inherently intolerant, there is constant disagreement between social agents, as opinion never ceases to change (see the evolution of $\omega(t)$)

14.3 Narrowing the gap between PSO and complex networks

PSO is a socially inspired meta-heuristic because it simulates natural swarm behaviours such as bird flocking or fish schooling. The algorithm consists of a population of particles represented as vectors that roams around the search space. Each particle's position depends on its previous best and its neighbourhood.

The algorithm does not need to make too many assumptions about the problem to be optimised. Its execution steps are simple: after the particles are initially randomly positioned across the search space, until termination, during each iteration, a particle's position, x_i, which is part of a candidate solution, is updated as follows:

$$v_{ij}(t+1) = wv_{ij}(t) + c_1 r_{1j}(t)[y_{ij}(t) - x_{ij}(t)] + c_2 r_{2j}(t)[\hat{y}_j(t) - x_{ij}(t)], \quad (14.9)$$

$$x_i(t+1) = x_i(t) + v_i(t+1), \quad (14.10)$$

Figure 14.12 Graphical representation of the gbest model with 25 particles: a fully connected graph of 25 vertices

where c_1 is the personal coefficient, c_2 is the social coefficient, $c_1, c_2 \in (0, 2]$ and $r_1, r_2 \sim \mathcal{U}(0, 1)$ are random vectors. The first term of (14.9) is the previous velocity influenced by an inertial weight w, the second term determines the particle to move around its best position (y), while the third term is providing the movement towards its neighbourhood's best position ($\hat{y}$) (see [54,55]).

Topologies determine how particles exchange information during every iteration. For example, the fully connected topology seen in Figure 14.12, also known as the *global best (gbest)* model, implies that all updates take into consideration the best position so-far of the entire swarm. This configuration is anecdotally known to provide fast convergence. On the other hand, the *local best (lbest)* model of the algorithm takes into consideration only several neighbours when updating each particle. In this case, the convergence is slower; however, it can find some solutions that could have been otherwise unexplored by the *gbest* model. One of the many *lbest* topologies is the ring topology, as seen in Figure 14.13, in which each particle is updated using the best position of its two neighbours ($k = 2$). Other topologies worth mentioning are the star topology (Figure 14.14(b)), in which all particles are connected to a central one, the grid topology (flattened Von Neumann topology with N, E, S, W in Figure 14.14(a)) and the hypercube topology (as in Figure 14.14(c)) [56]. The concepts of random and dynamic topologies have also been explored in the scientific literature

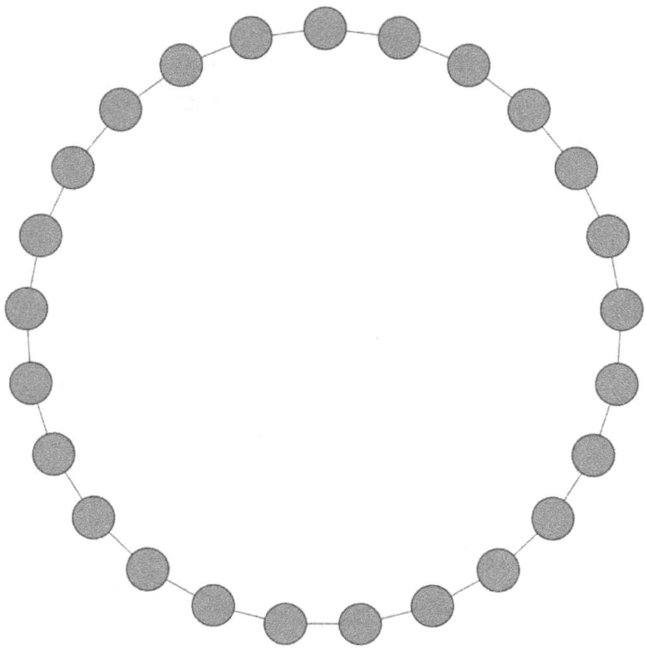

Figure 14.13 Graphical representation of the lbest ($k = 2$) model with 25 particles: a ring topology graph of 25 vertices

(e.g. [57,58]). Regarding the link between social networks and PSO, an example of a SW topology can be found in [59].

In connection to CNs, but irrespective of the underlying topology, inspired from Acemoglu's stubborn agents in [10], Lihu and Holban proposed a new method for increasing PSO's exploration capabilities and for escaping potential local minima: several particles are selected at each iteration to disagree with the updating operation [12]. The procedure increases randomness and the diversity across candidate solutions; a particle can partially or can extremely disagree with its next position; therefore, its final position will vary accordingly, as shown in Figure 14.15.

And finally, taking into consideration the above concepts, we arrive at the central idea of this chapter: there is no PSO model yet that includes both a CN-network topology and a specific CN agent behaviour such as the stubborn agents/disagreements model. We strongly believe that it might turn out a promising research direction.

14.4 Conclusion

This chapter presented some fundamental elements from CN science and SNA, in order to explore the synergies between PSO and CN science. The efforts to bring

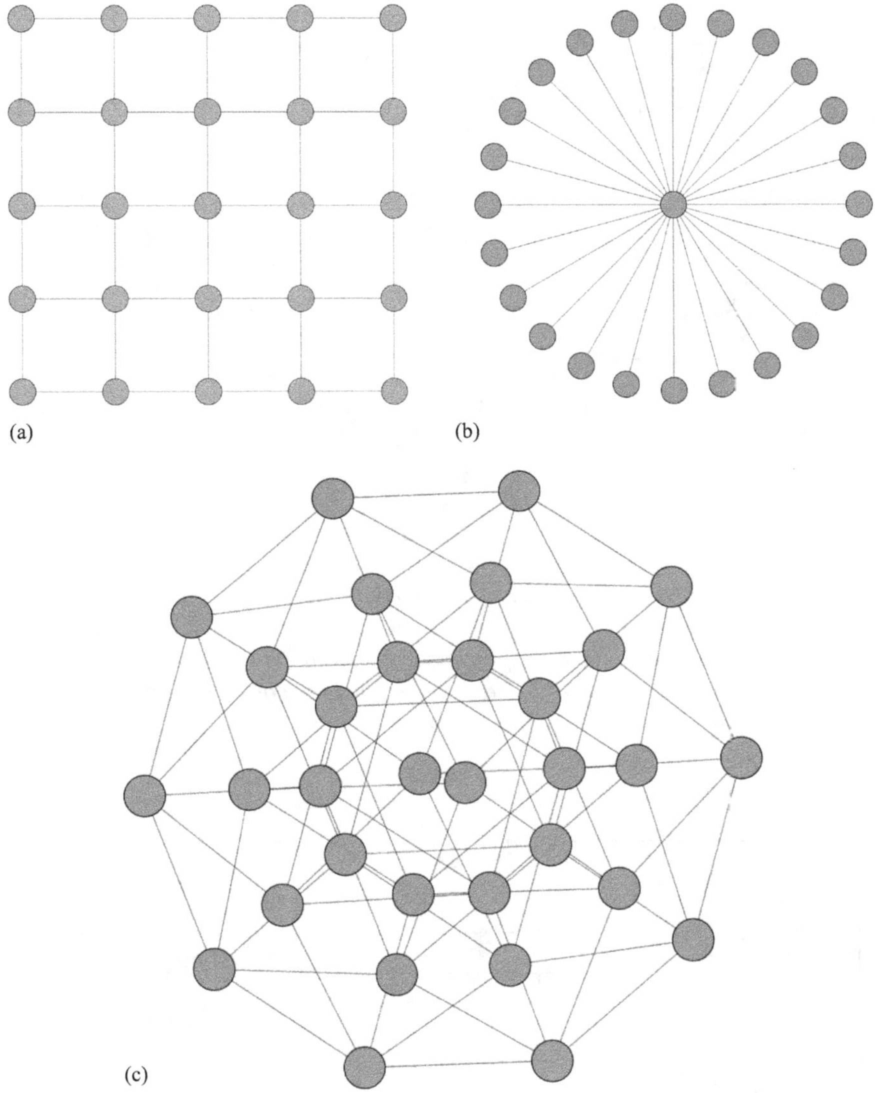

Figure 14.14 Other topologies: (a) Von Neumann topology (neighbours to the N, E, S, W), (b) star topology; all other particles are connected to a central particle, (c) hypercube topology with 32 agents

together these fertile scientific fields is not new and inspiration is constantly drawn from both directions. We noticed that in the field of CNs, the concepts of stubborn agents and the opinion spread models are continuously improved and developed.

We mostly focused on the role of opinion spread models in social networks for the development of new PSO algorithms. In particular, we presented the new

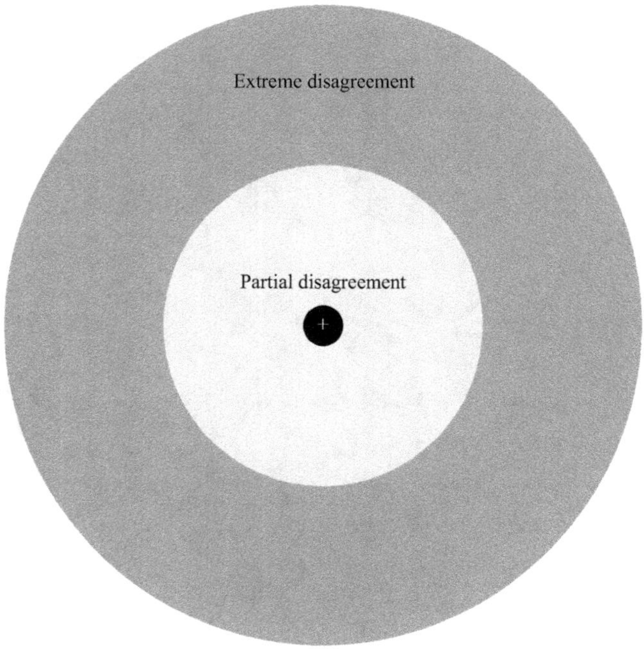

Figure 14.15 *Two types of disagreements: partial and extreme. The particle is positioned in the black dot at the centre of the figure. If it has chosen to disagree, then the particle will end up either in the inner region – if there is a partial disagreement, or in the outer region – if there is an extreme disagreement*

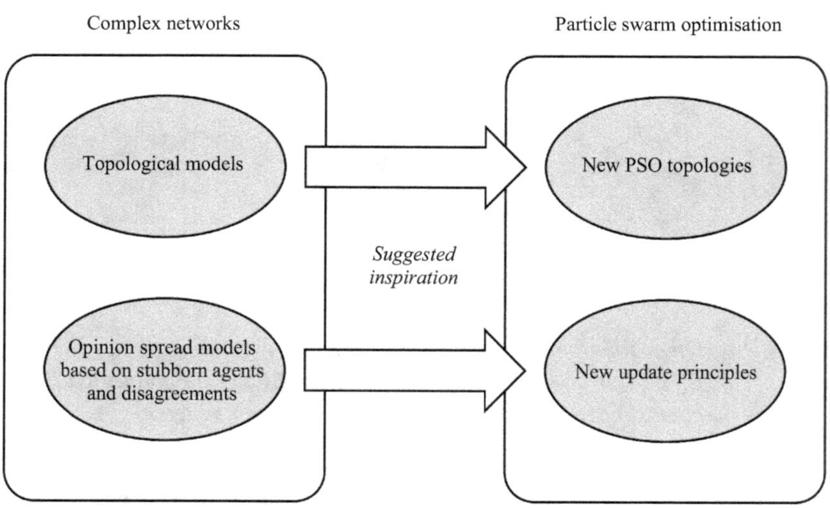

Figure 14.16 *A new perspective on finding inspiration in complex network science for improving particle swarm optimisation. Previous similar approaches have identified complex network topologies as possible solutions for PSO; however, this paper advocates the use of opinion diffusion models from social networks (based on stubborn agents and continuous disagreements) as efficient particle position updating principles*

tolerance-based opinion spread model in social networks and argued that it could be utilised to bring a more accurate social picture to PSO populations and therefore optimise the search.

We have pointed out that the original concept of constant disagreement and stubborn agents for [10,11,45] was already used for the creation of disagreements in PSOs to further improve the search efficacy.

Subsequently, as presented in Figure 14.16, inspired by the recent improvements in opinion spread models based on stubborn agents and disagreements presented in [44], we identified a new research perspective that could bring together both the topologies and the opinion spread features from CNs.

References

[1] A.L. Barabási, *Network Science*, Cambridge: Cambridge University Press, 2016.

[2] E. Estrada, *The Structure of Complex Networks*, New York: Oxford University Press, 2011.

[3] M. Li, W. Du, and F. Nian, *An adaptive particle swarm optimization algorithm based on directed weighted complex network*, Mathematical Problems in Engineering, 2014:434972, 2014.

[4] F. Li, J. Guo, *Topology optimization of particle swarm optimization*, Advances in Swarm Intelligence, Volume 8794, Lecture Notes in Computer Science, Springer, Cham, pp. 142–149, 2014.

[5] Z. Shi, Y. Liu, and J. Liang, *PSO-based community detection in complex networks*, Second International Symposium on Knowledge Acquisition and Modeling, pp. 114–119, 2009.

[6] M. Gong, Q. Cai, X. Chen, and L. Ma, *Complex network clustering by multiobjective discrete particle swarm optimization based on decomposition*, IEEE Transactions on Evolutionary Computation, 18(1):82–97, 2014.

[7] H. Binous, A. Bellagi, A.K. Al-Matar, and B.G. Higgins, *Particle Swarm Optimization for 2D Problems*, http://demonstrations.wolfram.com/ParticleSwarmOptimizationFor2DProblems/ Wolfram Demonstrations Project, 2014.

[8] L. Udrescu, L. Sbârcea, A. Topîrceanu, *et al.*, *Clustering drug–drug interaction networks with energy model layouts: community analysis and drug repurposing*, Scientific Reports 6:32745, 2016.

[9] C. Chamley, A. Christophe, and L. Li, *Models for the diffusion of beliefs in social networks: an overview*, IEEE Signal Processing Magazine, 30(3):16–39, 2013.

[10] D. Acemoglu, A. Ozdaglar, *Opinion dynamics and learning in social networks*, Dynamic Games and Applications, 1(1):3–49, 2011.

[11] D. Acemoglu, G. Como, F. Fagnani, and A. Ozdaglar, *Opinion fluctuations and disagreement in social networks*, Mathematics of Operations Research, 38(1):1–27, 2013.

[12] A. Lihu, S. Holban, *Particle swarm optimization with disagreements*, Advances in Swarm Intelligence, Volume 6728, Lecture Notes in Computer Science, Springer, Berlin, Heidelberg, pp. 46–55, 2011.

[13] R. Milo, S. Shen-Orr, S. Itzkovitz, N. Kashtan, and U. Alon, *Network motifs: simple building blocks of complex networks*, Science, 298(5594):824–827, 2002.

[14] F.S. Collins, H. Varmus, *A new initiative on precision medicine*, New England Journal of Medicine, 372(9):793–795, 2015.

[15] F. Alkemade, C. Castaldi, *Strategies for the diffusion of innovations on social networks*, Computational Economics, 25(1–2):3–23, 2005.

[16] M.E.J. Newman, *The structure and function of complex networks*, SIAM Review, 45:167–256, 2003.

[17] X.F. Wang, G. Chen, *Complex networks: scale-free and beyond*, IEEE Circuits and Systems Magazine, 3(2):6–20, 2003.

[18] M. Gaertler, Clustering. In: Brandes U., Erlebach T. (eds) Network Analysis. Lecture Notes in Computer Science, vol 3418. Springer, Berlin, Heidelberg, 2005.

[19] A. Noack, *Modularity clustering is force-directed layout*, Physical Review E, 79:026102, 2009.

[20] D. Diez, A. Augusti, and C.E. Wheelock, *Network analysis in the investigation of chronic respiratory diseases. From basics to application*, American Journal on Respiratory and Critical Care Medicine, 190:981–988, 2014.

[21] R. Faner, T. Cruz, A. Lopez-Giraldo, and A. Agustí, *Network medicine, multimorbidity and the lung in the elderly*, European Respiratory Journal, 44(3):775–788, 2014.

[22] R. Faner, A. Agustí, *Multilevel, dynamic chronic obstructive pulmonary disease heterogeneity. A challenge for personalized medicine*, Annals of the American Thoracic Society, 13(5):S466–S470, 2016.

[23] M. Girvan, M.E.J. Newman, *Community structure in social and biological networks*, Proceedings of the National Academy of Sciences, 99(12): 7821–7826, 2002.

[24] M.E.J. Newman, *Networks: An Introduction*, New York: Oxford University Press, 2010.

[25] A. Topîrceanu, M. Udrescu, and M. Vlăduţiu, *Network fidelity: a metric to quantify the similarity and realism of complex networks*, Third IEEE International Conference on Cloud and Green Computing (CGC), pp. 289–296, 2013.

[26] A. Topîrceanu, M. Udrescu, *Statistical fidelity: a tool to quantify the similarity between multi-variable entities with application in complex networks*, International Journal of Computer Mathematics, 94(9):1787–1805, 2016.

[27] D.J. Watts, S.B. Strogatz, *Collective dynamics of small-world networks*, Nature, 393(6684):440–442, 1998.

[28] A.L. Barabási, R. Albert, *Emergence of scaling in random networks*, Science, 286(5439):509–512, 1999.

[29] L. Jian-Guo, D. Yan-Zhong, and W. Zhong-Tuo, *Multistage random growing small-world networks with power-law degree distribution*, Chinese Physics Letters, 23(3):746, 2006.

[30] D. Lazer, A. Pentland, L. Adamic, *et al.*, *Life in the network: the coming age of computational social science*, Science, 323(5915):721–723, 2009.

[31] T. Tuna, E. Akbas, A. Aksoy, *et al.*, *User characterization for online social networks*, Social Network Analysis and Mining, 6:104, 2016.

[32] P. Erdös, A. Rényi, *On the evolution of random graphs*, Publications de l'Institut des Mathematiques Appliquees de l'Academie des Sciences de Hongrie, 5:17–61, 1960.

[33] P. Holme, B.J. Kim, *Growing scale-free networks with tunable clustering*, Physical Review E, 65(2):026107, 2002.

[34] P. Fu, K. Liao, *An evolving scale-free network with large clustering coefficient*, 9th IEEE International Conference on Control, Automation, Robotics and Vision, ICARCV'06, pp. 1–4, 2006.

[35] Y. Li, X. Qian, and D. Wang, *Extended HK evolving network model*, 24th IEEE Chinese Control and Decision Conference (CCDC), pp. 4095–4097, 2012.

[36] Y.W. Chen, L.F. Zhang, and J.P. Huang, *The Watts–Strogatz network model developed by including degree distribution: theory and computer simulation*, Journal of Physics A: Mathematical and Theoretical, 40(29):8237, 2007.

[37] J. Wang, L. Rong, *Evolving small-world networks based on the modified BA model*, IEEE International Conference on Computer Science and Information Technology ICCSIT'08, pp. 143–146, 2008.

[38] F. Zaidi, *Small world networks and clustered small world networks with random connectivity*, Social Network Analysis and Mining, 3(1):51–63, 2013.

[39] A. Guille, H. Hacid, C. Favre, and D.A. Zighed, *Information diffusion in online social networks: a survey*, ACM SIGMOD Record, 42(2):17–28, 2013.

[40] A. Das, S. Gollapudi, and K. Munagala, *Modeling opinion dynamics in social networks*, Proceedings of the 7th ACM International Conference on Web Search and Data Mining, pp. 403–412, 2014.

[41] Y. Li, W. Chen, Y. Wang, and Z.L. Zhang, *Influence diffusion dynamics and influence maximization in social networks with friend and foe relationships*, Proceedings of the Sixth ACM International Conference on Web Search and Data Mining, pp. 657–666, 2013.

[42] H. Fang, J. Zhang, and N.M Thalmann, *A trust model stemmed from the diffusion theory for opinion evaluation*, Proceedings of the 2013 International Conference on Autonomous Agents and Multi-Agent Systems, pp. 805–812, 2013.

[43] L. Deng, Y. Liu, and F. Xiong, *An opinion diffusion model with clustered early adopters*, Physica A: Statistical Mechanics and Its Applications, 392(17):3546–3554, 2013.

[44] A. Topîrceanu, M. Udrescu, M. Vlăduţiu, and R. Mărculescu, *Tolerance-based interaction: a new model targeting opinion formation and diffusion in social networks*, PeerJ Computer Science, 2:e42, 2016.

[45] E. Yildiz, A. Ozdaglar, D. Acemoglu, A. Saberi, and A. Scaglione, *Binary opinion dynamics with stubborn agents*, ACM Transactions on Economics and Computation, 1(19):4, 2013.

[46] G. Deffuant, D. Neau, F. Amblard, and G. Weisbuch, *Mixing beliefs among interacting agents*, Advances in Complex Systems, 3(1–4):87–98, 2000.

[47] M.A. Javarone, T. Squartini, *Conformism-driven phases of opinion formation on heterogeneous networks: the q-voter model case*, Journal of Statistical Mechanics: Theory and Experiment, 2015(10):P10002, 2015.

[48] L. Li, A. Scaglione, A. Swami, and Q. Zhao, *Phase transition in opinion diffusion in social networks*, IEEE International Conference on Acoustics, Speech and Signal Processing (ICASSP 2012), pp. 3073–3076, 2012.

[49] H.F. Chau, C.Y. Wong, F.K. Chow, and C-H.F. Fung, *Social judgment theory based model on opinion formation, polarization and evolution*, Physica A: Statistical Mechanics and its Applications, 415:133–140, 2014.

[50] L. Li, A. Scaglione, A. Swami, and Q. Zhao, *Trust, opinion diffusion and radicalization in social networks*, Conference Record of the Forty Fifth IEEE Asilomar Conference on Signals, Systems and Computers (ASILOMAR), pp. 691–695, 2011.

[51] M. Udrescu, A. Topîrceanu, *Probabilistic modeling of tolerance-based social network interaction*, 3rd IEEE European Conference on Network Intelligence (ENIC), pp. 48–54, 2016.

[52] R. Hegselmann, U. Krause, *Opinion dynamics and bounded confidence models, analysis, and simulation*, Journal of Artificial Societies and Social Simulation, 5(3), pp. 1–2, 2002.

[53] W. Weidlich, *Sociodynamics—a systematic approach to mathematical modelling in the social sciences*, Nonlinear Phenomena in Complex Systems, 5(4):479–487, 2012.

[54] J. Kennedy, R. Eberhart, *Particle swarm optimization*, Proceedings of IEEE International Conference on Neural Networks, pp. 1942–1948, 1995.

[55] Y. Shi, R.C. Eberhart, *A modified particle swarm optimizer*, Proceedings of IEEE International Conference on Evolutionary Computation, pp. 69–73, 1998.

[56] J. Kennedy, R. Mendes, *Population structure and particle swarm performance*, Proceedings of the Congress on Evolutionary Computation, Honolulu, pp. 1671–1676, 2002.

[57] Q. Ni, J. Deng, *A new logistic dynamic particle swarm optimization algorithm based on random topology*, The Scientific World Journal, vol. 2013, Article ID 409167, 8 pages, 2013. doi:10.1155/2013/409167.

[58] W.H. Lim, N.A.M. Isa, *Particle swarm optimization with increasing topology connectivity*, Engineering Applications of Artificial Intelligence, 27:80–102, 2014.

[59] Y. Gong, J. Zhang, *Small-world particle swarm optimization with topology adaptation*, In Proceedings of the 15th Annual Conference on Genetic and Evolutionary Computation, pp. 25–32, ACM, 2013.

Chapter 15

Ant colony algorithms for the travelling salesman problem and the quadratic assignment problem

Nikola Ivković[1]

Abstract

The ant colony optimization (ACO) is a metaheuristic, which has been successfully used to solve computationally difficult optimization problems, especially combinatorial optimization problems which belong to the class of non-deterministic polynomial (NP)-hard problems. This chapter explains ACO algorithms, their most important variants, and hybridization with local optimization methods. Practical considerations for successful ACO implementation and parameter setting are given. The working of the algorithm is explained in detail by using simple examples. The chapter ends with overview of research trends in ACO.

15.1 Introduction

Ant colony optimization (ACO) is a swarm-based metaheuristic that is successful in solving hard optimization problems. It was inspired by the foraging behaviour of biological ants. These ants leave pheromone trails (chemical) in their environment to influence the behaviour of their colony. In ACO, metaheuristic (artificial) ants iteratively construct solutions. This process of solution construction is guided by the pheromone trails associated with solution components. By using pheromone trails the colony of artificial ants progressively learn about the problem instance, they are trying to solve. The collective knowledge and experience from the previous iterations is passed on to future attempts by using pheromone trails to build a probabilistic model of the problem instance. At the beginning of the ACO algorithm, all the possible solution components seem equally appealing, but after some iteration of the algorithm, the components that are often part of good solutions become more appealing than other.

[1]Department of Computing and Technology, Faculty of Organization and Informatics, University of Zagreb, Croatia

Like other optimization heuristics, ACO cannot guarantee to find an optimal solution in reasonable time. But there are very useful for solving NP-hard optimization problems for which there are no available exact algorithms that can solve the problems in reasonable (polynomial) time. ACO algorithms are naturally suitable for combinatorial optimization problems, but there are also variants that can deal with continuous or mixed variable optimization problems. It is also possible to use ACO to solve multiobjective [1–4], dynamic [5–9], and stochastic [10–12] optimization problems, but in this chapter the focus is set on single-objective combinatorial optimization problems.

This chapter is intended for researchers who want to start research in ACO, but also for students and practitioners who are interested in solving specific optimization problems. Survey papers like [13] can also be useful for this purpose. Here is an updated overview of advances in the area of ACO, but unlike survey papers, detailed explanations of the algorithm with simple examples and recommendations for successful implementation of algorithms are also given. The structure of this chapter is as follows: Section 15.2 explains the general ideas of ACO class of metaheuristics whose most important variants are explained in Section 15.3. Since ACO metaheuristic is a rather abstract set of ideas and recipes that require a lot of effort and experience to produce successful ACO algorithm for some particular optimization problem, Section 15.4 gives practical examples for travelling salesmen problem (TSP) and quadratic assignment problem (QAP). Section 15.5 gives some suggestions for successful application of ACO algorithms, and Section 15.6 presents an overview of research trends.

15.2 Ant colony optimization metaheuristic

The first ant-inspired algorithms were proposed by Marco Dorigo in his Ph.D. dissertation [14]. One variant that proved to be successful was later entitled ant system (AS). After the publication of AS, Dorigo and other researchers developed various variants of ant-inspired algorithms which were later collectively described by ACO framework [15]. In general, on the high level of abstraction, the ACO algorithms can be described with the simple pseudocode shown in Algorithm 1.

Algorithm 1: The general metaheuristic of ant colony optimization

```
INITIALISE()
while stopping condition is not met do
    COLONYCONSTRUCTSOLUTIONS()
    DAEMONACTIONS()                          ▷ optional
    UPDATEPHEROMONETRAILS()
end while
```

The ACO algorithms are constructive methods that in every iteration of the algorithm construct a population of solutions. Before the algorithm can start generating solutions inside the loop, some preparing steps have to be performed in INITIALISE()

procedure. The algorithm needs to read the input problem instance, prepare appropriate data structures, initialize pheromone trails and heuristic values, and set up the algorithmic parameters.

After the initialization is complete, the algorithm repeats COLONYCONSTRUCTSOLUTIONS() and UPDATEPHEROMONETRAILS() procedures. The procedure DAEMONACTIONS() is optional. In the general framework, the order of execution inside the loop is not strictly defined and also these procedures might be executed in parallel. In particular, variants of the algorithms here presented order are usually preserved. The stopping conditions might be something like a fixed number of iterations, allowed execution time, reaching some prescribed solution quality, some combination of previous conditions, or something else.

15.2.1 Constructing the solutions

Normally multiple solutions are constructed inside COLONYCONSTRUCTSOLUTIONS() procedure. The solutions are constructed from the beginning, in each iteration of the algorithm, by selecting solution components one after another. This job is done by the colony of artificial ants. The number of ants in the colony is regularly denoted with m and normally stays constant during entire execution of the algorithm. Each ant constructs its own solution independently from other ants, in ANTCONSTRUCTSOLUTION(), but in the process of solution construction the ant uses colony-shared pheromone trails and heuristic values. In one iteration of the algorithm, m solutions are generated by m ants. Based on this, and possibly some solutions in previous iterations, the pheromone trails are updated inside UPDATEPHEROMONETRAILS().

The COLONYCONSTRUCTSOLUTIONS() procedure can be implemented in sequential or parallel manner. In sequential implementation, one ant constructs solution after another ant until each ant in the colony has constructed its own solution. This procedure is described by the following pseudocode:

```
procedure SequentialColonyConstructSolutions()
    for k = 1 to m do
        Ant_kConstructSolution()
    end for
end procedure
```

The natural parallelization of the COLONYCONSTRUCTSOLUTIONS() procedure is achieved simply by letting ants to construct solutions concurrently. This implementation can greatly improve algorithmic run time on modern multicore machines with hardware multithreading.

```
procedure ParallelColonyConstructSolutions()
    for all ants in the colony do in parallel
        AntConstructSolution()
    end for
end procedure
```

In the beginning of the ANTCONSTRUCTSOLUTION() procedure, the ant starts with partial solution s^P that is empty. In each following step, the ant chooses one solution component c from the (multi)set of possible components C^P by using some

Algorithm 2: The procedure of constructing a solution

function ANTCONSTRUCTSOLUTION(τ, η, *other parameters*)
    $s^P = ()$
    while $s^P \neq s$ **do**
        Add one solution component $c \in C^P$ into s^P by using some pheromone trail and heuristic value probability rule
    end while
     return s
end function

pheromone trail and heuristic information probability rule. The components with higher pheromone trail and heuristic information values have higher probabilities of being selected in partial solution s^P, but the exact rule depends on the variant of the ACO algorithm. The (multi)set of components C^P depends not only on problem instance but also on current partial solution s^P. After a component is chosen, this component is removed from the (multi)set C^P together with other components that would now cause the current partial solution to violate constraints of the problem. Adding component by component is repeated until the ant has constructed complete solution s as it is shown on the description of this procedure in Algorithm 2. The size of solution $|s|$ can be fixed sized for some optimization problems, like TSP or QAP, but also might be variable size in some problems where the optimization objective is to minimize or maximize the number of components in the solution, e.g., in the case of maximum clique problem (MCP) or minimum feedback arc set. In some cases, s^P and s are ordered list (tuple) but in other the order is not important and then s^P and s can be considered multisets or sets. The growth of empty partial solution s^P towards complete solution s is displayed in the following table.

Step	Partial solution
1	$s^P = (c_1)$
2	$s^P = (c_1, c_2)$
3	$s^P = (c_1, c_2, c_3)$
4	$s^P = (c_1, c_2, c_3, c_4)$
$\vdots$	$\vdots$
$\lvert s\rvert$	$s = (c_1, c_2, c_3, c_4, \ldots, c_{\lvert s\rvert})$

15.2.2 Updating pheromone trails

The pheromone trails are the most important component of any ACO algorithm. Their values have impact on solution construction procedure COLONYCONSTRUCTSOLUTIONS(). By using pheromone trails, the algorithm can learn about the problem instance it is trying to solve. The main goal of UPDATEPHEROMONETRAILS() procedure is to identify solution components that build good solutions, i.e., make pheromone trails of good solution components high and pheromone trails of bad solution components low. In a way, the artificial ants use pheromone trails to gather collective knowledge and experience about the problem instance that they are solving. Every variant of ACO algorithm has its own way of doing pheromone update and many variants of this metaheuristic differ essentially only in the UPDATEPHEROMONETRAILS() procedure. This procedure often has the following subtasks:

- Identify the best solution that was constructed
- Decrease pheromone trails, e.g., use pheromone evaporation
- Reinforce the pheromone trails for 'good' components

Earlier variants of ACO algorithms reinforce pheromone trails of the components from all solution constructed in the current iteration. The amounts by which the pheromone trails are increased generally depend on the solution quality and thus the components of the better solutions are reinforced more. Modern variants usually reinforce only components of one solution, which is in some way the 'best' solution. This solution can be the best solution found in the current iteration, the best solution found from the beginning of the algorithm, the best solution constructed in previous κ iterations – so-called κ-best strategy, or its approximation max-κ-best strategy [16].

15.2.3 Optional daemon actions

The procedure DAEMONACTIONS() from Algorithm 1 can contain some activities specific to particular optimization problem or variant of the algorithm and perform some centralized actions, but the most often this is some sort of local optimization. After the solutions are constructed by ants in COLONYCONSTRUCTSOLUTIONS() procedure, it is often beneficial for the overall performance of the algorithm to perform local optimization technique on all or some solutions. Local optimization technique usually has low computational complexity and can efficiently search neighbourhood of the original solution in order to find even a better solution. Local optimization and other Daemon Actions are optional and can be omitted from the algorithm.

15.3 Important variants of ant colony optimization algorithms

With AS, Dorigo proved that algorithm inspired by the foraging behaviour of ants can be successful in solving hard combinatorial problems. Subsequently, many different ACO algorithms were proposed to improve some aspects of the algorithm. Some algorithm described in this section have mainly historical value like AS, elitist ant system (EAS), rank-based ant system (AS_{RANK}). Other algorithms like approximate non-deterministic tree search (ANTS) and best–worst ant system (BWAS) introduced some unique techniques but did not gain wide popularity.

Modern ACO algorithms that might offer the best performance on a wide range of hard optimization problems are ant colony system (ACS), MAX–MIN ant system (MMAS), three bound ant system (TBAS), and population-based ant colony optimization (P-ACO). For standard combinatorial problems, the most popular algorithm is probably MMAS which have a slower start, but given enough time, often can outperform ACS which is greedier. Three bound AS is the newest of all mentioned algorithms. It has lower computational complexity than MMAS and depending on arbitrary chosen parameter ω can behave very similarly as MMAS or rather differently when this is preferred. P-ACO is specially designed for dynamic problems, although specially tuned ACS and MMAS can give competitive results for this type of problems.

15.3.1 Probabilistic rules for choosing solution components

Choosing solution components in ACO algorithms is determined by the pheromone trail and heuristic information probability rule which differs between variants of ACO algorithms, but most of them use the random-proportional rule or pseudorandom-proportional rule. In one step of ANTCONSTRUCTSOLUTION(), it is necessary to pick one solution component from the multiset C^P of all components that are possible to add into partial solution s^P. In the case of random-proportional rule, this probability is calculated by using expression (15.1), where $\tau_{c(l)}$ stands for the pheromone trail associated with the component c_l and $\eta_{c(l)}$ stands for the heuristic information associated with c_l,

$$p\left(c_l \mid s^P\right) = \frac{\tau_{c(l)}^{\alpha} \cdot \eta_{c(l)}^{\beta}}{\sum_{c \in C^P}\left(\tau_c^{\alpha} \cdot \eta_c^{\beta}\right)}, \quad \forall c_l \in C^P. \tag{15.1}$$

Based on some insights about nature of the optimization problem, for each component c_l it is possible to associate heuristic information $\eta_{c(l)}$ that reflects the desirability of that component. For the TSP, where the goal is to find the shortest tour that visits all the cities and to come back to the starting city, the desirability of particular road between two cities is inversely proportional to its distance. For some problems, heuristic information is not used, i.e., its value for all the solution components is equal to 1. The sum in the denominator of the fraction in expression (15.1) involves all the solution components that can be added into partial solution s^P in the current step of the solution construction procedure.

The relative importance or influence of pheromone trails on the probability of choosing a solution component is controlled by the parameter α and the influence of heuristic information is adjusted with the parameter β. The appropriate values for these parameters depend on the type of the problem for which particular ACO algorithm is designed, the number of allowed iterations and other parameters of the algorithm. The optimal values for parameters depend even on problem instance and thus the parameter tuning is necessary to achieve the best results. Often the parameter α is set to 1 and parameter β to values greater than 1.

The components with higher pheromone trails have higher probabilities of being chosen into partial solution and similarly apply to the heuristic information. Very often solution components, and consequently pheromone trails and heuristic information, can be organized in the form of matrix. Then the components, pheromone trails, and heuristic values are labelled with c_{ij}, τ_{ij}, and η_{ij} respectfully, where i and j stand for indices of the matrix. When only components from the same row are considered in C^P then the random-proportional rule from expression (15.1) can be rewritten in the following form

$$p\left(c_{ij} \mid s^P\right) = \frac{\tau_{ij}^{\alpha} \cdot \eta_{ij}^{\beta}}{\sum_{c_{il} \in C^P} \left(\tau_{il}^{\alpha} \cdot \eta_{il}^{\beta}\right)}. \tag{15.2}$$

Some ACO algorithms use pseudorandom-proportional rule where the random-proportional rule is used with some probability $1 - q_0$, and with the remaining probability q_0 the ant selects the component that has the highest probability of being chosen from all solution components in C^P. This means that random-proportional rule is a special case of pseudorandom-proportional rule where the pseudorandom-proportional rule can be a greedier variant.

Let q be the random variable which is uniformly distributed in interval [0, 1]. Then according to pseudorandom-proportional rule (15.3), the component c_l which will be added to partial solution s^P is

$$c_l = \begin{cases} \arg\max_{c \in C^P} \left(\tau_c^{\alpha} \cdot \eta_c^{\beta}\right), & q < q_0, \\ \text{use random-proportional rule,} & q \geq q_0. \end{cases} \tag{15.3}$$

Therefore, if the variable q is less than the parameter q_0, then the ant selects the component for which the product of the pheromone trail raised to the power of α and the heuristic information raised to the power of β is the largest. Else if $q \geq q_0$, then the ant selects component by using random-proportional rule (15.1). The parameter α is usually fixed to 1.

The third type of probabilistic rule is proposed with ANTS [17], where multiplication is replaced with addition and powering with multiplication to simplify the calculation. This ANTS-style random-proportional rule is defined with expression (15.4). In this rule, balancing between influence of pheromone trail and heuristic value is achieved by parameter $\zeta \in [0, 1]$

$$p(c_l \mid s^P) = \frac{\zeta \cdot \tau_{c(l)} + (1 - \zeta) \cdot \eta_{c(l)}}{\sum_{c \in C^P} (\zeta \cdot \tau_c) + (1 - \zeta) \cdot \eta_c}, \quad \forall c_l \in C^P. \tag{15.4}$$

15.3.2 Ant system

AS is the first published variant of ACO algorithm [18,19] and was created as part of the doctoral thesis of Marco Dorigo [14]. In fact, it was the first ant algorithm, and thus also the first ACO algorithm, although the class of ACO algorithms was defined later.

The probability of choosing component c_l for adding into partial solution s^P is calculated based on random-proportional rule (15.1). Updating of pheromone trails is performed after all ants have constructed the solutions. In the first phase of pheromone update, the pheromone trails of all components are evaporated by multiplying with $(1 - \rho)$ as defined with expression (15.5). The parameter ρ is evaporation rate that can be arbitrary set to value from interval $[0, 1\rangle$, and value $(1 - \rho)$ is known as pheromone persistency rate.*

$$\tau_c = (1 - \rho) \cdot \tau_c, \quad \forall c \in C. \tag{15.5}$$

After evaporation of pheromone trails, the components of all the solutions constructed by the colony of m ants are reinforced by additional value by using expression (15.6)

$$\tau_c = \tau_c + \sum_{k=1}^{m} \Delta\tau_c^{(k)}, \tag{15.6}$$

where $\Delta\tau_c^{(k)}$ is defined by

$$\Delta\tau_c^{(k)} = \begin{cases} \frac{Q}{f(s^k)} & \text{if } c \in s^k, \\ 0 & \text{else if } c \notin s^k, \end{cases} \tag{15.7}$$

and Q is the parameter that is usually set to 1. The solution that was constructed by kth ant in the current iteration is denoted as s^k and $f(s^k)$ stands for quality or cost of the solution s^k. The processes of pheromone evaporation and pheromone reinforcement, defined by expressions (15.5) and (15.6), can be combined in one step by applying the following expression:

$$\tau_c = (1 - \rho) \cdot \tau_c + \sum_{k=1}^{m} \Delta\tau_c^{(k)}. \tag{15.8}$$

15.3.3 Elitist ant system

AS algorithm showed encouraging results, but the performance for this algorithm was competitive only for smaller problem instances. For larger problem instances, this algorithm searches a solution space too widely. To enable more focused search for good solutions and promote faster convergence towards an optimal solution, it was proposed that the components of the best solution from the beginning of the algorithm s^{gb} receive additional pheromone trail reinforcement. The new variant of the algorithm was named EAS. It has identical ColonyConstructSolutions() procedure as AS algorithm, but for the pheromone update the following expression should be used.

$$\tau_c = (1 - \rho) \cdot \tau_c + \sum_{k=1}^{m} \Delta\tau_c^{(k)} + e \cdot \Delta\tau_c^{(gb)}. \tag{15.9}$$

where $\Delta\tau_c^{(k)}$ is defined by expression (15.7) and $\Delta\tau_c^{(gb)}$ by the following expression:

$$\Delta\tau_c^{(gb)} = \begin{cases} \frac{1}{f(s^{\mathrm{gb}})} & \text{if } c \in s^{\mathrm{gb}}, \\ 0 & \text{else if } c \notin s^{\mathrm{gb}}. \end{cases} \tag{15.10}$$

*Originally, in older works the symbol ρ was used for pheromone persistency rate, but the same symbol is later used only for pheromone evaporation rate.

In EAS, e is the new parameter of the algorithm. Experimental results presented in [19] by Dorigo *et al.* suggest that EAS finds better solutions than AS and that these better solutions are obtained after fewer iterations.

15.3.4 Ant colony system

Dorigo and Gambardella proposed ACS in [20] where solutions are constructed by using pseudorandom-proportional rule (15.3) inside COLONYCONSTRUCTSOLUTIONS() procedure. The parameter α is by default fixed to 1. The ACS algorithm is actually an improved variant of ANT-Q algorithm† that was previously proposed by the same authors in [21].

Unlike many other ACO algorithms, ACS uses global pheromone evaporation and local pheromone evaporation. Each ant that is constructing a new solution, after adding a solution component c_l into partial solution s^P performs local evaporation for pheromone trail associated with that component according to following expression:

$$\tau_{c(l)} = (1 - \xi) \cdot \tau_{c(l)}, \tag{15.11}$$

where parameter $\xi \in \langle 0, 1 \rangle$ is local pheromone evaporation rate.

The global pheromone evaporation and the pheromone reinforcement are performed at the end of every iteration of the algorithm only for the best found solution from the beginning of the algorithm s^{gb}, according to the following expression:

$$\tau_c = (1 - \rho) \cdot \tau_c + \Delta\tau_c^{(gb)}, \quad \forall c \in s^{\mathrm{gb}}. \tag{15.12}$$

For all others pheromone trails, the pheromone update is not performed. It is also worth noting that often parameter ξ is set equal to the parameter ρ.

15.3.5 Rank-based ant system

A variant ACO algorithm that was also devised by modifying the AS algorithm is AS_{RANK}, first and published in [22]. For this algorithm, all the solutions constructed by the colony in the current iteration are sorted and ranked according to their goodness. The ranking is done in a way that the best solution has rank 1, followed by the larger ranks and the possible ties are solved either randomly or in some other way. Solution construction techniques performed inside COLONYCONSTRUCTSOLUTIONS() are identical to the one used in AS and EAS, i.e., ants use random-proportional rule (15.1). Inside UPDATEPHEROMONETRAILS() procedure, pheromone evaporation is the same as in AS and uses expression (15.5). The pheromone trails reinforcement is performed for the first $w - 1$ solutions with smallest (best) ranks and for the best solution found from the beginning of the algorithm $s^{(gb)}$. The parameter w controls how many solutions are used for reinforcing pheromone trails since for the ants with solutions whose ranks are greater than or equal to w are not allowed to deposed

†This algorithm is not used anymore.

additional pheromone values. The pheromone reinforcement is performed by applying the following expression:

$$\tau_c = \tau_c + \sum_{r=1}^{w-1} (w - r) \cdot \Delta\tau_c^{(r)} + w \cdot \Delta\tau_c^{(gb)}. \tag{15.13}$$

where $\Delta\tau_c^{(r)}$ is defined by expression (15.7) and $\Delta\tau_c^{(gb)}$ by expression (15.10). This was the ants with lesser (better) rank deposit higher pheromone trails with greater weight than the ants with greater (worse) ranks.

15.3.6 Approximate nondeterministic tree search

Maniezzo proposed in [17] ANTS that uses a different rule for constructing solutions than the most other ACO algorithms. This ANTS-style random-proportional rule is defined by expression (15.4). Heuristic information $\eta_{c(l)}$ for ANTS is calculated dynamically when the component c_l is considered for adding into partial solution s^P.

In the process of solution construction, the lower bound (for minimization problems) of complete solution s is estimated. In the case that this optimistic estimation of the complete solution s for the considered partial solution s^P is worse than the best found solution from the beginning of the algorithm, then the current solution can be discarded as in the case of the well-known method of branch and bound.

This algorithm does not have explicit pheromone evaporation inside UPDATEPHEROMONETRAILS() procedure and the entire pheromone update is defined by expression (15.14)

$$\tau_c = \tau_c + \sum_{k=1}^{m} \Delta T_c^{(k)}, \tag{15.14}$$

where $\Delta T_c^{(k)}$ is defined by

$$\Delta T_c^{(k)} = \begin{cases} \vartheta \cdot \left(1 - \frac{f(s^k) - B_L}{L_{avg} - B_L}\right) & \text{if } c \in s^k, \\ 0 & \text{else if } c \notin s^k, \end{cases} \tag{15.15}$$

and ϑ is the parameter of the algorithm. The symbol B_L denotes lower bound for the optimal solution for which it is always true that $B_L \leq f(s^*)$, where $f(s^*)$ is the cost of the optimal solution. The lower bound B_L is used only in the case of minimization problems and in the maximization problems the upper bound B_U should be used instead. The moving average L_{avg} is calculated for last t constructed solutions, where t is another parameter of the algorithm.

15.3.7 MAX–MIN ant system

Stützle and Hoos proposed MMAS in [23], which is an ACO algorithm that keeps all the pheromone trails between minimal and maximal pheromone value. The main difference between AS, EAS, and MMAS is in the way that pheromone values are updated, while the method of constructing solutions is the same in all these algorithms – the random-proportional rule defined by expression (15.1) is used in

COLONYCONSTRUCTSOLUTIONS(). Unlike AS, EAS, AS_{RANK}, and ANTS, MMAS uses only one, in some sense the best solution, for pheromone reinforcement. In the procedure UPDATEPHEROMONETRAILS(), all the pheromone values have to stay inside interval $[\tau_{\min}, \tau_{\max}]$. This means that when the pheromone evaporation is performed, it is necessary to ensure that pheromone values do not decrease under $\tau_{\min}$ bound. This is accomplished by pheromone evaporation defined by expression (15.16)

$$\tau_c = \max\{\tau_{\min}, (1 - \rho) \cdot \tau_c\}, \quad \forall c \in C, \tag{15.16}$$

and also the reinforcement of pheromone value must not increase beyond $\tau_{\max}$ bound. This is ensured when reinforcement is performed according to expression (15.17), where the symbol s^{best} stands for the best solution from the beginning of the algorithm s^{gb} or the best constructed solution in the current iteration s^{ib}:

$$\tau_c = \min\{\tau_{\max}, \tau_c + \Delta\tau_c^{(best)}\}, \quad \forall c \in s^{best}. \tag{15.17}$$

The upper pheromone bound $\tau_{\max}$ is set to asymptotic value that some pheromone trail could obtain if it is reinforced in every iteration of the algorithm like it is part of the optimal solution, which is specified by expression (15.18), where $f(s^*)$ is cost of the optimal solution s^*

$$\tau_{\max} = \frac{1}{\rho \cdot f(s^*)}. \tag{15.18}$$

Since the cost of the optimal solution $f(s^*)$ is usually unknown, at the beginning of the algorithm the parameter $\tau_{\max}$ is calculated by using some simple estimation for $f(s^*)$. For example, in the case of the TSP, the cost of the solution obtained by the nearest neighbour method $f(s^{nn})$ is used instead. Initial value for all pheromone trails τ_0 in MMAS is set equal to $\tau_{\max}$

$$\tau_0 = \frac{1}{\rho \cdot f(s^*)}. \tag{15.19}$$

Setting the initial pheromone trails to upper pheromone bound $\tau_{\max}$ is due to an expectation of the authors of MMAS that this will enable better exploration of the search space at the beginning of the algorithm. Another novelty in this algorithm is the option to reset the pheromone trails to initial value τ_0 if the algorithmic stagnation is detected. For some optimization problems, like TSP, asymmetric travelling salesman problem (ATSP), and QAP, there are mathematical expressions [24] that can help in choosing appropriate lower pheromone bound $\tau_{\min}$, but for other problems this is done by experimentally tuning parameter $\vartheta = \tau_{\min}/\tau_{\max}$.

15.3.8 Best–worst ant system

BWAS uses the idea of pheromone mutation, the technique borrowed from the evolutionary algorithms, and in the pheromone update both the best and the worst solutions are used, hence the name of the algorithm [25–27]. The construction of new solutions inside COLONYCONSTRUCTSOLUTIONS() procedure is done by applying the random-proportional rule defined by expression (15.1) in the same way as in AS, EAS, AS_{RANK}, and MMAS algorithms. The pheromone update is done by applying pheromone evaporation first, subsequently followed by the pheromone reinforcement, and at the end, another pheromone evaporation for some components of the worst solution. The combined pheromone evaporation and pheromone reinforcement is done by applying expression (15.20) and expression (15.21), where s^{gb} is the global best solution (the best solution from the beginning of the algorithm).

$$\tau_c = (1-\rho)\cdot\tau_c + \Delta\tau_c^{(gb)}\,, \tag{15.20}$$

$$\Delta\tau_c^{(gb)} = \begin{cases} \frac{1}{f(s^{\mathrm{gb}})} & \text{if } c \in s^{\mathrm{gb}}, \\ 0 & \text{else if } c \notin s^{\mathrm{gb}}. \end{cases} \tag{15.21}$$

After combined pheromone evaporation and pheromone reinforcement, the components of the worst solution in the current iteration s^{iw} that are not contained in the best solution s^{gb} are additionally evaporated as defined by the following expression:

$$\tau_c = (1-\rho)\cdot\tau_c, \qquad \forall c \in s^{\mathrm{iw}}, \quad c \notin s^{\mathrm{gb}}. \tag{15.22}$$

The BWAS can use the optional technique of resetting the pheromone values to the initial value if the algorithmic stagnation is detected just the same as MMAS algorithm.

15.3.9 Population-based ant colony optimization

Guntsch and Middendorf devised P-ACO for application on dynamic optimization problems [5,28], although this algorithm can be applied to classical (static) optimization problems also. Taking into consideration that dynamical optimization problem is changing while it is tackled by the ACO algorithm, it is necessary for successful algorithm to abandon the information in pheromone trails that are outdated. Because of special design of the algorithm, it is possible in P-ACO to completely cancel the influence on pheromone trails for solutions older than the chosen number of iterations. Every solution that is used for pheromone reinforcement is saved into a population of solutions. From this population, the solution can be removed, and when the solution is removed, its influence on pheromone trails is revoked. It is usual that entering the population and leaving the population of solutions is organized by the First In, First Out principle.

The construction of solutions inside ColonyConstructSolutions() is conducted by pseudorandom-proportional rule defined by expression (15.3), the same way as in ACS. The pheromone trails are modified by adding or subtracting $\Delta\tau$ defined by expression (15.23). The upper bound of pheromone trails $\tau_{\max}$ is the parameter of the algorithm, and so is the initial value of pheromone trails τ_0. The parameter τ_0 is set to $1/(n-1)$ for TSP, and to $1/n$ in QAP. The size of the population of solutions is denoted by k.

$$\Delta\tau = \frac{(\tau_{\max} - \tau_0)}{k}\,. \tag{15.23}$$

In the original version of the algorithm, there was a parameter $w_e \in [0, 1]$ which was multiplied with $\Delta\tau$, but in later versions it was removed, i.e., it is fixed to $w_e = 1$.

Inside UPDATEPHEROMONETRAILS(), there is no pheromone evaporation in its usual form. The influence of old solution on pheromone trails is cancelled by subtracting $\Delta\tau$ and the reinforcement for a new solution that is entering the population is increased by adding $\Delta\tau$. Therefore, when the solution s^{in} is entering the population its components are updated by the following expression:

$$\tau_c = \tau_c + \Delta\tau, \quad \forall c \in s^{in}, \tag{15.24}$$

and when the solution s^{out} is leaving the population its components are decreased by the following expression:

$$\tau_c = \tau_c - \Delta\tau, \quad \forall c \in s^{out}. \tag{15.25}$$

15.3.10 Three bound ant system

TBAS started with the attempt to maintain excellent characteristics of MMAS that was attested on many optimization problems and at the same time to lower computational burden of updating pheromone trails [29,30]. The finial algorithm has indeed decreased the time complexity of pheromone update procedure that in practical implementation can result from unnoticeable up to significant speedup. The main differences between MMAS and TBAS is that TBAS

- has three pheromone bounds (the lower bound τ_{LB}, the upper bound τ_{UB}, and the contraction bound $\tau_{CB} = \omega \cdot \tau_{\text{UB}}$),
- uses occasional pheromone contractions instead of regular evaporations,
- has different pheromone reinforcement procedure,
- initial pheromone value τ_0 is equal to lower pheromone bound τ_{LB}.

TBAS uses the random-proportional rule (15.1) for constructing solutions. The pheromone update procedure of TBAS starts with pheromone reinforcement where different strategies like iteration best, κ-best, or max-κ-best [16] can be used to select the solution whose components will receive additional pheromone values. If some pheromone trail outgrows the upper pheromone bound τ_{UB}, then the pheromone contraction procedure is performed to shrink pheromones from $[\tau_{\text{LB}}, \tau_{\text{UB}}]$ interval into $[\tau_{\text{LB}}, \tau_{CB}]$ interval. All the components of solution s^{best} selected for pheromone reinforcement are modified according to expression (15.26)

$$\tau_c = \tau_c + \frac{Q_i}{f(s^{best})}, \quad \forall c \in s^{best}, \tag{15.26}$$

where $Q_0 = 1$ is initial value of the changing parameter that grows after i subsequent iterations into

$$Q_{i+1} = \frac{Q_i}{1-\rho}. \tag{15.27}$$

The upper pheromone bound τ_{UB} and the lower pheromone bounds τ_{UB} are set to

$$\tau_{\text{UB}} = \frac{1}{\rho \cdot f(s^*)}, \tag{15.28}$$

$$\tau_{\text{LB}} = \vartheta \cdot \tau_{\text{LB}}, \tag{15.29}$$

where ϑ can be selected either with the help of mathematical expressions, e.g., TSP, ATSP, and QAP [24], or can be experimentally determined if the mathematical expressions are not available. The initial pheromone value τ_0 is equal to lower pheromone bound τ_{LB}. Like in MMAS, there is an option to reset the pheromone trails to initial value τ_0 if the algorithmic stagnation is detected.

15.3.11 Other variants of ACO and ant-inspired algorithms

Previously described ACO algorithms are important because of good features and proved success in solving many hard optimization problems or because of some unique techniques or historical value. There are many other variants of ACO algorithms, some specifically designed for particular optimization problem or specific domain of application. ACO for huge graphs (ACO_{hg}) is specially designed for dealing with huge graphs whose size is unknown when the algorithm starts or the graph is too big to fit into memory [31]. Also, it is common to see a hybridization of ACO algorithms with other optimization techniques, e.g., beam-ACO is hybridizing ACO with beam search [32].

Although ACO algorithms are originally developed for discrete optimization problems and have the most success in that area, researchers proposed variants of ACO for optimization problems with continuous variables and mixed variables (where both discrete and continuous variables exit). Socha and Dorigo proposed ACO for continuous domains ($ACO_{\mathbb{R}}$) which is a direct extension of classical ACO algorithm for continuous variable problems. In one step of the solution construction procedure, the ant chooses a value for one of n variables in n-dimensional continuous space problem. The selection is based on probability density function made up of weighted sums of different Gaussian functions, which together form a Gaussian kernel. For each dimension there is a special Gaussian kernel that is generated on the basis of solutions kept in an archive of solutions – similar to the technique used in P-ACO. Like in classical ACO algorithms (for discrete problems), $ACO_{\mathbb{R}}$ uses random-proportional rule defined by expression (15.1) to choose solution components for partial solution s^P. Leguizamón and Coello Coello proposed an alternative for algorithm $ACO_{\mathbb{R}}$, named by the authors $DACO_{\mathbb{R}}$ which is designed with the intent to improve the properties of the algorithm for problems of large dimensions [33]. An important requirement for $DACO_{\mathbb{R}}$ was to retain the diversity in search for a longer run than $ACO_{\mathbb{R}}$. There are also other ant-inspired algorithms for solving continuous variable optimization problems like continuous ACO [34], the API algorithm [35], and continuous interacting ant colony [36], but they have significantly departed from principles used in ACO framework specified by Algorithm 1.

Besides ACO algorithms, there exist other algorithms inspired by the behaviour of biological ants. In order for ant algorithm or ant-based algorithm to belong to the class of ACO algorithms, they have to use pheromone trails to construct solutions and obey the logic of general ACO framework described by Algorithm 1. For example, hybrid ant system (HAS) proposed by Gambardella in [37] is used for solving discrete optimization problems, but it does not belong to ACO class. The algorithm HAS uses the pheromone trails, but not for solution construction. Instead, the pheromone trails are used for guiding solution modification procedure. Some ant algorithms are proposed for routing in networks. The ant-based control is intended for circuit switching networks [38]. For the networks with packet switching, Caro and Dorigo proposed AntNet [39]. AntHocNet proposed by Caro, Ducatelle, and Gambardella in [40] is specially designed for ad-hoc networks.

15.4 Practical examples of ant colony optimization

ACO metaheuristic and its more specific variants like MMAS, TBAS, ACS, P-ACO describe on very high level of abstraction only generic recipes on how to devise an ACO algorithm. For particular optimization problem, the designer of the algorithm has many choices to make that significantly influence the behaviour and finial performance of the algorithm. The designer of the algorithm has to choose what are the components of the solution, how the solution will be constructed, will the algorithm use heuristic information and what will it be, decide about potential local optimization and numerous parameters of the algorithm. In order to give a more detailed algorithm, it is necessary to consider particular optimization problem that should be tackled. In the following subsections, TSP and QAP are explained since they are suitable examples for the detailed description of algorithms that will be given.

15.4.1 Travelling salesman problem

The TSP may be the most researched NP-hard problem of combinatorial optimization. It is easy to define, but very difficult to solve TSP in the general case. Informally, the problem can be defined as follows. The travelling salesmen must find the shortest tour that will allow him to visit n given cities and return to the starting city by visiting each city exactly once. Distances between cities are known in advance. More formal definition of the problem is usually given with the help of weighted graph $G = (V, E)$, where nodes or vertexes $v \in V$ represent cities, and the weight on the edges $e \in E$ represents distances between cities. The problem instance of size n (with n cities) can be stored in the matrix $D_{n,n}$ where matrix elements specify distance between cites, i.e., the weight of the edges.

$$D_{n,n} = \begin{bmatrix} d_{1,1} & d_{1,2} & \cdots & d_{1,n} \\ d_{2,1} & d_{2,2} & \cdots & d_{2,n} \\ \vdots & \vdots & \ddots & \vdots \\ d_{n,1} & d_{n,2} & \cdots & d_{n,n} \end{bmatrix}. \tag{15.30}$$

One solution of this problem could be represented by the ordered list (tuple) of vertexes or by the set of edges, where the edge $e_{i,j}$ connects vertexes v_i and v_j. For example, a tour represented by the tuple $\pi_1 = (v_1, v_2, v_3, \dots, v_n)$ is also represented by the set

Figure 15.1 An example of Euclidean TSP with 100 nodes

$\{e_{1,2}, e_{2,3}, e_{3,4}, \ldots, e_{n-1,n}, e_{n,1}\}$. The total cost of this tour (π_1) would be $d_{1,2} + d_{2,3} + d_{3,4} + \cdots + d_{n-1,n} + d_{n,1}$.

An example of TSP with 100 randomly generated 'cities' inside Euclidean plane is shown in Figure 15.1. A suboptimal solution of that problem with cost 892 is shown in Figure 15.2, and an optimal solution with cost 773 is shown in Figure 15.3. For comparison, the random solutions for this problem instance have much worst cost, often around 5,000.

Despite the simple formulation, TSP is NP-hard problem, and decidable formulation is NP-complete [41]. In general, formulation TSP is NP-complete [42], which means that it belongs to a class of NP optimization problems that are hardest to approximate regarding the size of the approximation error. Metric TSP, which is a restricted type of TSP where the distance between any three nodes must obey the rule of the triangle inequality, is approximable (APX)-complete [43]. There exists approximation algorithm that ensures that the ratio of obtained and optimal solution for metric TSP is not greater than $R = 3/2$ [44]. Euclidean TSP is a special version of the metric TSP whose nodes are located in Euclidean area. This restricted version is also NP-hard, but also there is a polynomial time approximation schemes algorithm, meaning that for any selected ratio $R > 1$ the algorithm can find an approximate solution in polynomial time regarding the size of the problem [45].

Symmetric travelling salesman problem (STSP) is a special form of TSP, where $d_{i,j} = d_{j,i}$ always holds. If this property is not met, then this fact is often emphasized by the name ATSP. In literature, it is usual to refer to STSP simply as TSP. For symmetric metric TSP, especially for Euclidean TSP, there are specialized methods which effectively can solve large instances of the problem with small deviations from optimum and often even find optimal solutions. The available methods for ATSP are

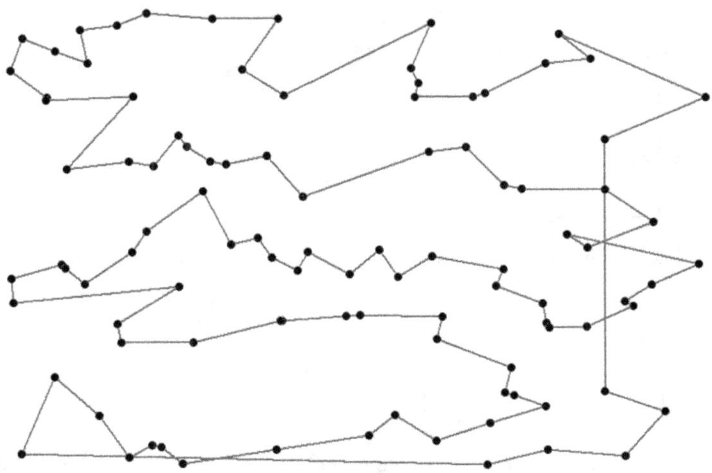

Figure 15.2 A rather good but suboptimal solution of 100-node TSP with solution cost equal to 892

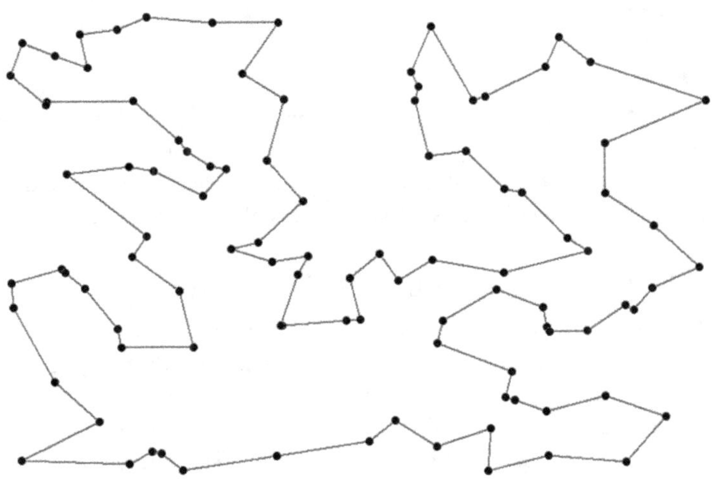

Figure 15.3 An optimal solution of 100-node TSP with solution cost equal to 773

not as nearly that successful. Besides great theoretical importance, TSP has many practical applications (for more information about the applications, consult [46–49]).

15.4.2 Quadratic assignment problem

QAP can be defined as a problem of placing facilities at the locations so that the sum of the products of distances between the locations and flows between the facilities is minimized. The number of facilities n equals the number of locations. Distances between location is usually stored in distance matrix

$$D_{n,n} = \begin{bmatrix} d_{1,1} & d_{1,2} & \cdots & d_{1,n} \\ d_{2,1} & d_{2,2} & \cdots & d_{2,n} \\ \vdots & \vdots & \ddots & \vdots \\ d_{n,1} & d_{n,2} & \cdots & d_{n,n} \end{bmatrix}, \tag{15.31}$$

and flows between facilities in flow matrix

$$F_{n,n} = \begin{bmatrix} f_{1,1} & f_{1,2} & \cdots & f_{1,n} \\ f_{2,1} & f_{2,2} & \cdots & f_{2,n} \\ \vdots & \vdots & \ddots & \vdots \\ f_{n,1} & f_{n,2} & \cdots & f_{n,n} \end{bmatrix}. \tag{15.32}$$

The solution of QAP can be stored in the set of n assignments (pairs) where each facility is assigned to exactly one location. More compact notation allows that a solution can be represented by the ordered list π, where facility i is assigned to location $\pi(i)$. The goal of the problem is to find an order that minimizes the following sum:

$$\sum_{i=1}^{n}\sum_{j=1}^{n} f_{i,j} \cdot d_{\pi(i),\pi(j)}. \tag{15.33}$$

The problem is NP-hard and is considered as one of the more difficult combinatorial problems. Combinatorial optimization problems, such as travelling salesman problem, graph partitioning problem, MCP, linear order problem, and subgraph isomorphism may be formulated as a special cases of QAP [50,51].

15.4.3 A variant of MMAS algorithm for TSP

An example of a simple MMAS algorithm is described in its usual variant for solving TSP. Further on, this sequential algorithm without local optimization will be labelled with MMAS-TSP acronym. The solution components for MMAS-TSP will be the edges $e_{i,j} \in E$. The pheromone trails are therefore associated with edges, i.e., for every edge $e_{i,j}$ there is associate pheromone trail $\tau_{i,j}$. The pheromone trails can be conveniently stored into matrix:

$$T_{n,n} = \begin{bmatrix} \tau_{1,1} & \tau_{1,2} & \cdots & \tau_{1,n} \\ \tau_{2,1} & \tau_{2,2} & \cdots & \tau_{2,n} \\ \vdots & \vdots & \ddots & \vdots \\ \tau_{n,1} & \tau_{n,2} & \cdots & \tau_{n,n} \end{bmatrix}. \tag{15.34}$$

The heuristic information $\eta_{i,j}$ can be defined for each edge $e_{i,j}$ by the following expression

$$\eta_{i,j} = \frac{1}{d_{i,j}}, \tag{15.35}$$

Algorithm 3: The pseudocode of MMAS-TSP algorithm

1: Load problem matrix $D_{n,n}$
2: Set parameters $\alpha, \beta, \rho, \tau_0, \tau_{\max}, \tau_{\min}, \ldots$ ▷ Start with initialization
3: **for** $i = 1$ to n **do**
4:     **for** $j = 1$ to n **do**
5:         $\tau_{i,j} \leftarrow \tau_0$
6:         $\eta_{i,j} \leftarrow \frac{1}{d_{i,j}}$
7:     **end for**
8: **end for**
9: **for** $t = 1$ to $t_{\max}$ **do** ▷ Main loop
10:     **for** $k = 1$ to m **do**
11:         $s_k, f(s_k) \leftarrow$ ANTCONSTRUCTTSPSOLUTION($D_{n,n}, T_{n,n}, H_{n,n}, \alpha, \beta$)
12:     **end for**
13:     $s^{\mathrm{ib}} \leftarrow s_1$
14:     **for** $k = 2$ to m **do**
15:         **if** $f(s_k) < f(s^{\mathrm{ib}})$ **then**
16:             $s^{\mathrm{ib}} \leftarrow s_k$
17:         **end if**
18:     **end for**
19:     **if** $t = 1$ OR $f(s^{\mathrm{ib}}) < f(s^{\mathrm{gb}})$ **then**
20:         $s^{\mathrm{gb}} \leftarrow s^{\mathrm{ib}}$
21:     **end if**
22:     UPDATETSPPHEROMONETRAILS($T_{n,n}, \tau_{\min}, \tau_{\max}, \rho, s^{best}, f(s^{best})$)
23:     **if** algorithmic stagnation is detected **then**
24:         reset all pheromone trails to τ_0
25:     **end if**
26: **end for**

and can also be stored into matrix

$$H_{n,n} = \begin{bmatrix} \eta_{1,1} & \eta_{1,2} & \cdots & \eta_{1,n} \\ \eta_{2,1} & \eta_{2,2} & \cdots & \eta_{2,n} \\ \vdots & \vdots & \ddots & \vdots \\ \eta_{n,1} & \eta_{n,2} & \cdots & \eta_{n,n} \end{bmatrix}. \tag{15.36}$$

The partial solution s^P can use the tuple representation, where tuple grows from 0-tuple at the beginning up to n-tuple when the solution is finished. Algorithm 3 describes the necessary steps and procedures. The first eight rows implement INITIALIZATION() procedure from the ACO framework. The main loop between rows 9 and 26 has a preselected number of iterations $t_{\max}$. Inside the inner loop between rows 10 to 12, for each ant in the colony of m ants, the algorithm calls ANTCONSTRUCTTSP-SOLUTION() described in Algorithm 4. The best solution from the current iteration is stored in s^{ib} and the best solution from the beginning of the algorithm in s^{gb}. When

Algorithm 4: Functions and procedures used in MMAS-TSP

```
function AntConstructTspSolution(D_{n,n}, T_{n,n}, H_{n,n}, α, β)
    cost ← 0
    Randomly chose starting city v_f
    v_a ← v_f
    s^P ← (v_a)
    for i = 2 to n do
        Chose one edge e_{a,b} that goes from v_a to any unvisited node v_b using random-proportional rule (15.1)
        s^P ← s^P&(v_b)
        cost ← cost + d_{a,b}
        Mark v_b as visited node
        a ← b
    end for
    cost ← cost + d_{b,f}
    return s, cost
end function

procedure UpdateTspPheromoneTrails(T_{n,n}, τ_min, τ_max, ρ, s^best, cost_best)
    for i = 1 to n do
        for j = 1 to n do
            τ_{i,j} ← (1 − ρ) · τ_{i,j}
            if τ_{i,j} < τ_min then
                τ_{i,j} ← τ_min
            end if
        end for
    end for
    for all components c_{r,s} of best solution s^best do
        τ_{r,s} ← τ_{r,s} + 1/(ρ·cost_best)
        if τ_{r,s} > τ_max then
            τ_{r,s} ← τ_max
        end if
        if problem is symmetrical then
            τ_{s,r} ← τ_{r,s}
        end if
    end for
end procedure
```

the main loop ends, the solution s^{gb} is reported as solution that was obtained by the algorithm. For the process of pheromone reinforcement any strategy for choosing s^{best} can be implemented like κ-best, max-κ-best, or simple just s^{ib} of s^{gb}. Both pheromone evaporation and pheromone reinforcement are described in UpdateTspPheromoneTrails() procedure described in Algorithm 4. Here the construction of one solution is such that the artificial ant is walking through graph, i.e., in two consecutive steps the ant has to choose edges where common node v_b is the end of one edge $e_{a,b}$ and the start of another edge $e_{b,c}$. An alternative possibility, which is not used in practical implementations, would be that ant can choose any possible edge that can be added into partial solution. In that case, the partial solution would be represented with set of edges and the computational complexity of solution construction would be higher.

To achieve good performance, it is very important to set up appropriate parameters of the algorithm. For example in the presented algorithm, $\alpha = 1$, $\beta = 4$, $\rho = 0.02$ would yield good solutions for many TSP instances (assuming that other parameters are also appropriately set), but the best performance is archived if the parameters are tuned for particular problem instance and particular stopping criteria. More details about parameter setting are provided in Section 15.5.2.

15.4.4 A sketch of MMAS for QAP

MMAS algorithms for QAP, further denoted with MMAS-QAP, have many similarities with MMAS-TSP explained in the previous subsection. To avoid much duplication between subsections, only parts that are significantly different are explained in detail. The main point in producing successful ACO algorithm is to define appropriate pheromone model and efficient solution construction procedure. In the case of QAP, the pheromone trails are associated with assignments of particular facilities to particular locations. This is schematically shown in Figure 15.4. As the left side of the figure shows, each of n facility can be assigned to each of n location, meaning that pheromone can be conveniently stored in $T_{n,n}$ matrix. The right side of the figure represents one solution, where only assignments with full lines are part of the solution. If this is the optimal solution, then it would be preferred if these full lines also represent pheromone trails with higher values and dotted lines the pheromone trails with lower values.

There is no known way of defining heuristic information for QAP, which would improve the long-term performance of MMAS algorithm, thus the heuristic information is not used, i.e., $\eta_c^\beta = 1$ for all components c. For this reason, it is more important to use some form of local optimization like 2-opt, 3-opt, or short runs of tabu search that can greatly improve the performance of the algorithm. It would be possible in one step of the solution construction procedure to choose between all possible assignments (for which both location and facility are yet unused). In common variant of MMAS-QAP, at first, one available facility (or location) is chosen completely random and then by using random-proportional rule (15.1) and presented pheromone model, ant chooses assignment, i.e., location (or facility).

15.4.5 A detailed example of solution construction

In this subsection, random-proportional rule is explained on simple example as it is used in MMAS-TSP algorithm presented in Section 15.4.3. The identical process would be in the case of TBAS or other algorithm that uses random-proportional rule. Let us assume a very simple TSP instance with five nodes visualized by the graph in Figure 15.5. All the edges are drawn with the lines of the same thickness considering that all pheromone trails are equal at the beginning of the algorithm. For the sake of

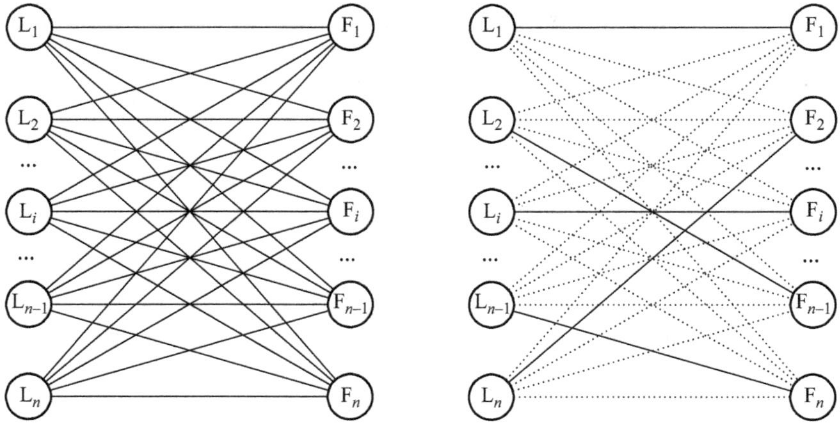

Figure 15.4 Schematic description of pheromone trails model for QAP

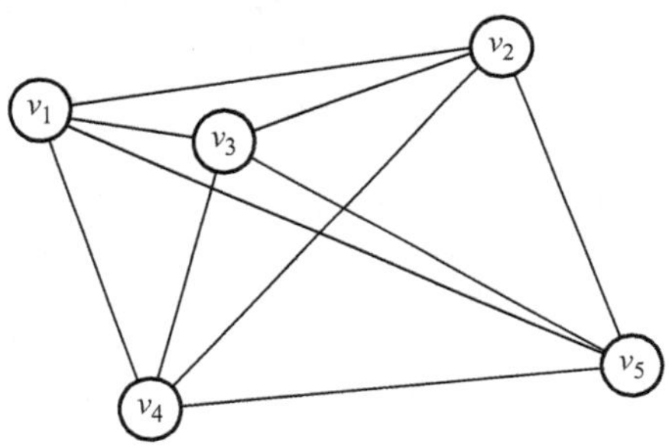

Figure 15.5 An example of TSP with 5 nodes and 10 edges

simplicity, let all heuristic values be equal to 1, such that $\eta_{i,j}^{\beta} = 1$ for every $e_{i,j} \in E$. Let parameter $\alpha = 1$. After some iterations the values of pheromone trails might change to

$$T_{5,5} = \begin{bmatrix} \text{d.c.} & 0.1 & 2 & 8 & 0.2 \\ 0.1 & \text{d.c.} & 2 & 0.3 & 4 \\ 2 & 2 & \text{d.c.} & 0.2 & 0.4 \\ 8 & 0.3 & 0.2 & \text{d.c.} & 6 \\ 0.2 & 4 & 0.4 & 6 & \text{d.c.} \end{bmatrix}. \tag{15.37}$$

The values on diagonal of matrix $T_{5,5}$ showed on expression (15.37) labelled with 'd.c.' stands for 'don't care' since edges on diagonal are not relevant to TSP. This is visualized in Figure 15.6 where ticker lines represent edges with higher pheromone

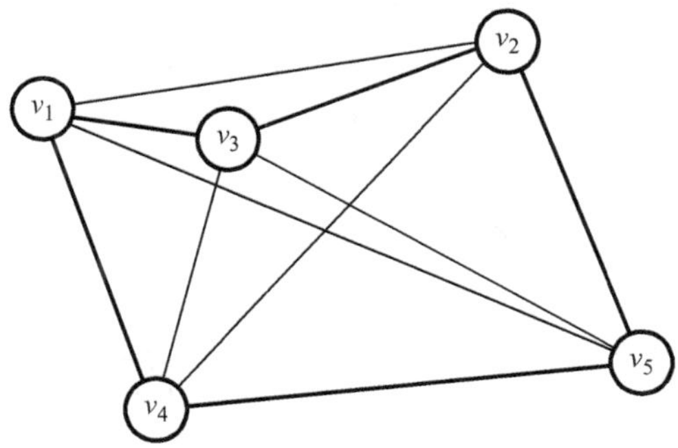

Figure 15.6 An example of TSP with 5 nodes and 10 edges with thickness of the lines showing amount of pheromone trails laid on the particular edges

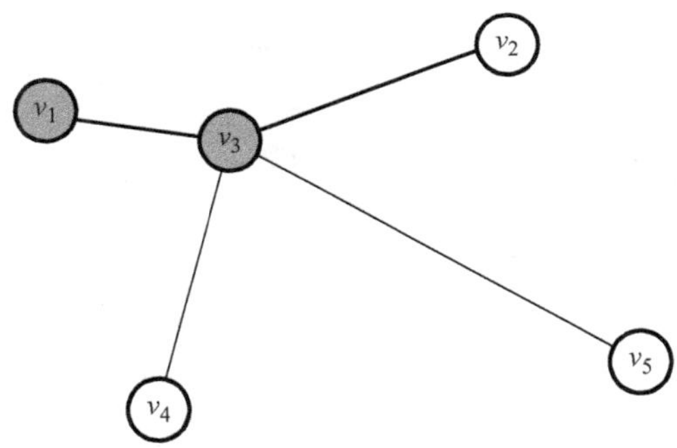

Figure 15.7 A choice that the ant has to make to expand partial solution $s^P = (v_1, v_2)$

values. Now, let us assume that v_1 is selected for starting node by the step described in row 3 in ANTCONSTRUCTTSPSOLUTION() function and partial solution is $s^P = (v_1)$. After that, let us assume that $e_{1,3}$ is selected by random-proportional rule (15.1) and the partial solution is expended to $s^P = (v_1, v_3)$. This situation is visualized in Figure 15.7. According to MMAS-TSP algorithm presented in Section 15.4.3, when the ant wants to expend partial solution $s^P = (v_1, v_3)$, it has to consider only edges $e_{3,2}$, $e_{3,4}$, and $e_{3,5}$, i.e., the set of solution components $C^P = \{e_{3,2}, e_{3,4}, e_{3,5}\}$. Alternatively interpreted, the next node to visit can be v_2, v_4, or v_5. These choices are displayed in Figure 15.7, where grey nodes represent visited cities and the white nodes represent cities that could be visited next. When values of pheromone trails from expression (15.37) are considered by the random proportional rule (15.1), the probabilities of selecting components are as follows:

$$p(c_{3,2} \mid (v_1, v_3)) = \frac{\tau_{3,2}^{\alpha} \cdot \eta_{3,2}^{\beta}}{\tau_{3,2}^{\alpha} \cdot \eta_{3,2}^{\beta} + \tau_{3,4}^{\alpha} \cdot \eta_{3,4}^{\beta} + \tau_{3,5}^{\alpha} \cdot \eta_{3,5}^{\beta}}, \tag{15.38}$$

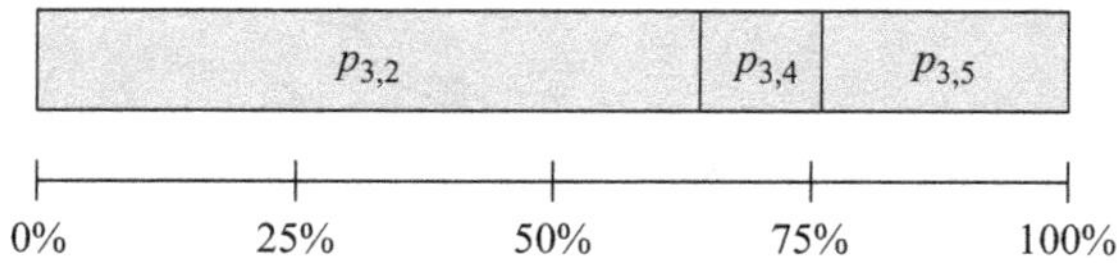

Figure 15.8 Geometrical representation of probabilities

$$p(c_{3,4} \mid (v_1, v_3)) = \frac{\tau_{3,4}^{\alpha} \cdot \eta_{3,4}^{\beta}}{\tau_{3,2}^{\alpha} \cdot \eta_{3,2}^{\beta} + \tau_{3,4}^{\alpha} \cdot \eta_{3,4}^{\beta} + \tau_{3,5}^{\alpha} \cdot \eta_{3,5}^{\beta}}, \tag{15.39}$$

$$p(c_{3,5} \mid (v_1, v_3)) = \frac{\tau_{3,5}^{\alpha} \cdot \eta_{3,5}^{\beta}}{\tau_{3,2}^{\alpha} \cdot \eta_{3,2}^{\beta} + \tau_{3,4}^{\alpha} \cdot \eta_{3,4}^{\beta} + \tau_{3,5}^{\alpha} \cdot \eta_{3,5}^{\beta}}. \tag{15.40}$$

After entering numerical values in previous expression, the probabilities for choosing particular components are calculated to be

$$p(c_{3,2} \mid (v_1, v_3)) = \frac{2}{2 + 0.2 + 0.4} \approx 76.92\%, \tag{15.41}$$

$$p(c_{3,4} \mid (v_1, v_3)) = \frac{0.2}{2 + 0.2 + 0.4} \approx 7.69\%, \tag{15.42}$$

$$p(c_{3,5} \mid (v_1, v_3)) = \frac{0.4}{2 + 0.2 + 0.4} \approx 15.38\%. \tag{15.43}$$

These probabilities $p_{3,2}, p_{3,4}, p_{3,5}$ can be visualized geometrically as shown in Figure 15.8, where together probabilities add up to 100%. To choose one component it is necessary to draw a random number between $[0, 1]$, from the generator of (pseudo)random numbers. In this case, if the random number is between $[0, 0.7692]$ then component $c_{3,2}$ is chosen and partial solution grows to $s^P = (v_1, v_3, v_2)$. Else if the random number is between $[0.7692, 0.8461]$, the component $c_{3,4}$ is chosen and partial solution grows to $s^P = (v_1, v_3, v_4)$. Else if random number is larger than 0.8461, the partial solution is expended to $s^P = (v_1, v_3, v_5)$.

15.5 Suggestions for successful applications

Implementation of successful ACO algorithm requires lots of skills and experience. The most important part is to choose appropriate model of pheromone trails. In order to achieve the best performance, it is often very useful to use some form of local optimization. Also is very important to set parameters of the algorithm to appropriate values since this can have a great influence on algorithmic performance.

There are also some opportunities to achieve greater speed of the algorithm. First, it is important to notice that it is not necessary to calculate denominator in random-proportional rule. This denominator, denoted as $D_{\texttt{r-p}}$, is equal for all the components that can be chosen in the same step of solution construction procedure. So instead of calculating probabilities $p(c|s^P)$, it is enough to calculate only numerators i.e., $D_{\texttt{r-p}} \cdot p(c|s^P)$. In that case, the geometrical representation of probability explained in

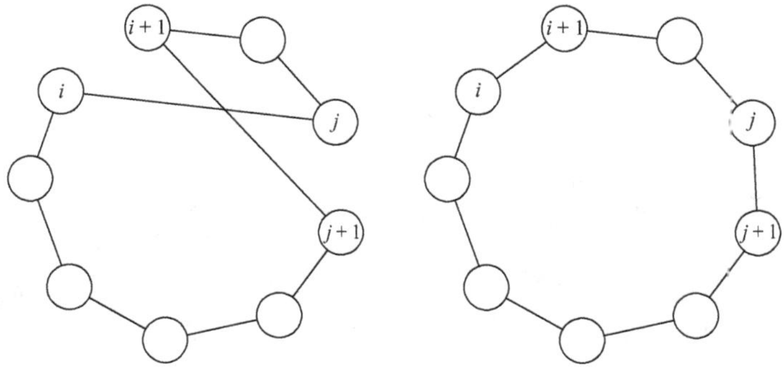

Figure 15.9 Schematic explanation of 2-exchange move for TSP

Section 15.4.5 gets scaled. The speed of the algorithm can usually be greatly improved if values $\tau^{\alpha}_{c(i)} \cdot \eta^{\beta}_{c(i)}$ are precomputed and stored into lookup tables. To achieve even greater speedups, the solution construction procedure may only consider subset of all possible components from the set C^P, this subset would usually contain only more promising components. For example, in the case of TSP, only closest cities maintained in so-called favourite lists are considered in large problem instances.

15.5.1 Local optimization

For many types of optimization problems, local optimization could be achieved by removing few solution components and then to explore different ways to finish solution with other components. In the case of removing k components, this move is called k-exchange. For example, from the solution of symmetrical TSP shown schematically on the left side of Figure 15.9, it is possible to remove edges $e_{i,j}$ and $e_{i+1,j+1}$ and instead to reconnect the solutions with edges $e_{i,i+1}$ and $e_{j,j+1}$. This 2-exchange move would yield better quality solution displayed on the right side of the same figure. Of course, not all 2-exchange moves lead to better solutions and only the moves that improve solution are actually performed. One solution might be improved more than once with different 2-exchange move. If the solution cannot be improved further with any 2-exchange move, this solution becomes 2-optimal (in general case k-optimal).

For asymmetrical TSP, 2-exchange moves are not suitable as this would reverse direction in a part of solution, but some 3-exchange moves that do preserve directions can be used instead. For some problems, like maximum clique, it is possible to use 1-exchange moves to improve the solution. Besides k-exchange moves, some other local optimization techniques like short runs of tabu search can be employed in daemon actions of ACO.

15.5.2 Parameter settings

It is very important to set parameters of ACO algorithms to appropriate values since different settings can have a significant influence on the performance of the algorithm. A larger number of parameters often allow greater control over the behaviour

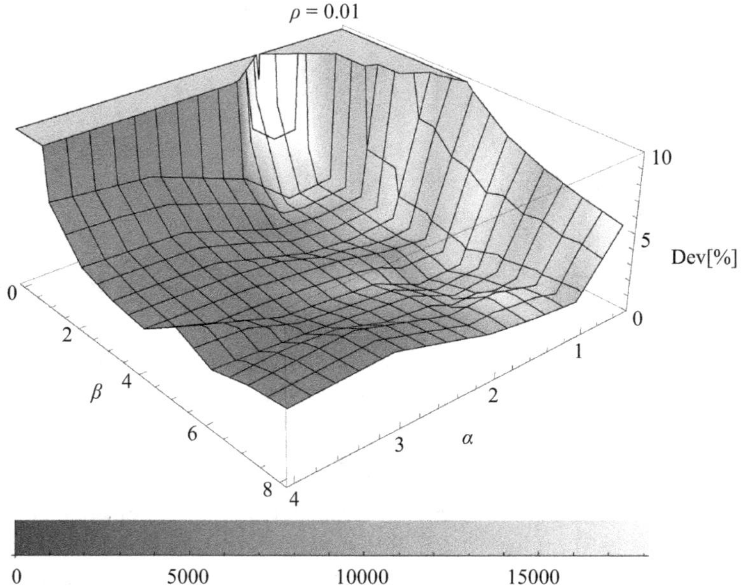

Figure 15.10 Parameter space of α and β for MMAS on TSP instance eil101

of the algorithm but also introduce the additional problem of setting parameters to suitable values. Luckily, some parameters can be calculated from other parameters, e.g., τ_0 in MMAS and TBAS. Experimental research of some problem instances can provide information about algorithmic behaviour and appropriate set of parameters. This set of parameters might provide good performance for a wide range of instances, but to achieve the best performance it is beneficial to make fine parameters tuning. The parameters of the ACO algorithms (and other metaheuristics) cannot be set independently from each other which is also observable on the following few examples of experimental research of parameter space.

Figure 15.10 shows how solution quality depends on parameters α and β, used by random-proportional rule (15.1). The other parameter are kept fixed (e.g., $\rho = 0.01$). The maximum allowed number of iterations was 20,000, but for some parameter setting the algorithm stopped producing better solution after a much lower number of iterations. The shade of grey defines the average number of iteration after which the algorithm did not find better solutions. In this case, the best solutions are obtained with parameters $\alpha = 1$ and $2 \leq \beta \leq 8$. For very small values of α and β, the algorithm exhibits the worst performance. For larger values of parameter α, the algorithm stagnate rather early, mostly after less than 1,000 iterations. For higher values of α, it is noticeable that parameter β has somewhat larger influence on performance.

Figure 15.11 shows results from the same experiment, but with dependency of solution quality on parameters α and ρ. In this case, β parameter is kept constant ($\beta = 3$). The same colour scale as in Figure 15.10 defines the average number of

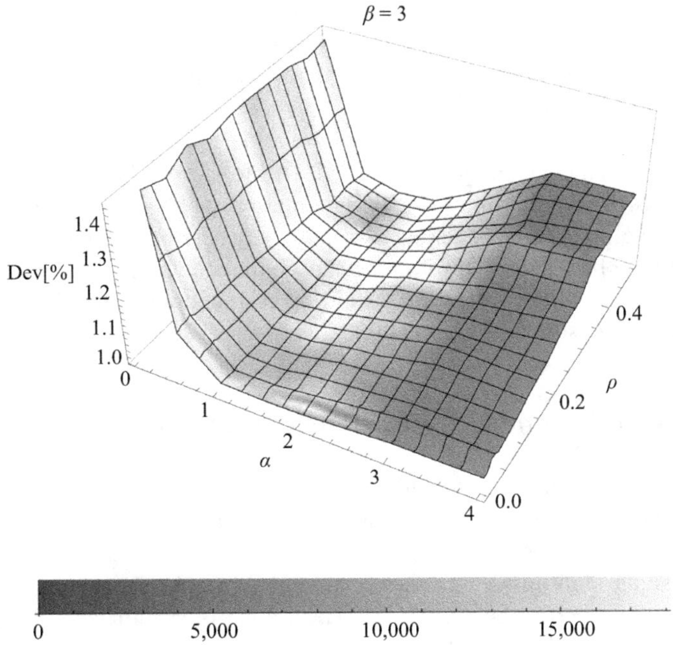

Figure 15.11 Parameter space of α and ρ for MMAS on TSP instance eil101

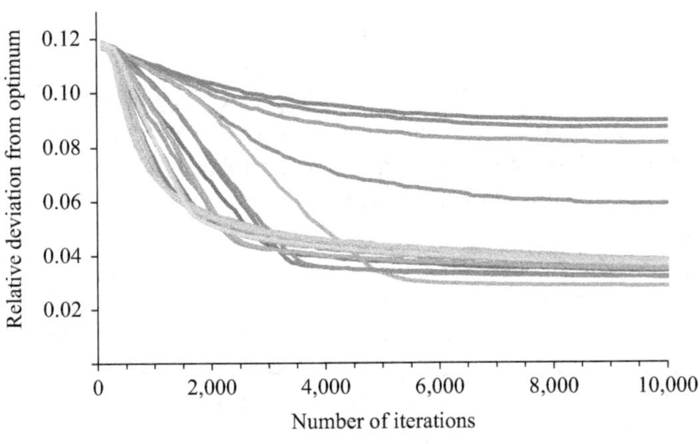

Figure 15.12 Solution quality for different reinforcement strategies in relation to number of iterations achieved by MMAS on QAP

iteration after which the algorithm did not find better solutions. In this case, the best solutions are obtained by parameter $\alpha = 1$ with a wide range of values for parameter ρ. The influence of parameter ρ on algorithmic performance is rather noticeable for higher values of parameter α.

Figure 15.12 shows solution quality in relation to number of iterations for 24 different reinforcement strategies used by MMAS to solve QAP. It is noticeable that some strategies exhibit slow improvements in solution quality while other goes quickly towards good solutions. This is also an example where it is obvious that some strategies (parameters) are better for a lower number of iterations while other are better for a larger number, e.g., the strategy that achieves best solutions after 6,000 iterations achieves rather bad solutions before 4,000 iterations.

15.6 Research trends in ant colony optimization

ACO is scientifically active field with practical applications on many hard optimization problems. The discovery of the first ACO algorithm was followed by numerous researches on application to new optimization problems, and partly to the improvement of existing algorithms, i.e., the development of new variants of ACO (many of them are described in Section 15.3). The choice of parameter settings significantly affects the efficiently of ACO algorithms. In addition to the limited parameter tuning before applying the algorithm to a specific instance of the problem, there are numerous investigations of parameter landscape and the influences of parameter settings on the behaviour of the algorithm. The most common are experimental researches of parameter space [52–56], and in lesser extent the combinations of empirical and analytical approach [57,58]. Some researches are aimed towards discovering the influence of choosing particular pheromone model on efficiency of ACO algorithm [59–62] and some try to investigate dynamics of ACO by using mathematical models [63]. It is becoming increasingly widespread recognized that experimental research in the field of ACO and other swarm intelligence and evolutionary computation algorithms should not rely on parametrical statistical methods. Nonparametric statistical methods and use of median and other quantiles instead of arithmetic mean are recommended [64,65].

In many other metaheuristical approaches, still there is some gap between theory and practice, although theoretical research is rather active. Theoretical investigations are mostly based of researching simplified variants of ACO or some special cases to answer questions about convergence [66–68], complexities for some special problems or classes of problems [69–73], influence of local optimization on algorithmic behaviour [74], etc.

References

[1] Iredi S, Merkle D, Middendorf M. Bi-Criterion Optimization with Multi Colony Ant Algorithms. In: Zitzler E, Deb K, Thiele L, *et al.*, editors. Evolutionary Multi-Criterion Optimization, First International Conference, EMO

2001, Zurich, Switzerland, March 7–9, 2001, Proceedings. vol. 1993 of Lecture Notes in Computer Science. Springer; 2001. p. 359–372.

[2] Pasia JM, Doerner KF, Hartl RF, *et al.* Solving a Bi-objective Vehicle Routing Problem by Pareto-Ant Colony Optimization. In: Stützle T, Birattari M, Hoos HH, editors. Engineering Stochastic Local Search Algorithms. Designing, Implementing and Analyzing Effective Heuristics, International Workshop, SLS 2007, Brussels, Belgium, September 6–8, 2007, Proceedings. vol. 4638 of Lecture Notes in Computer Science. Springer; 2007. p. 187–191.

[3] López-Ibáñez M, Paquete L, Stützle T. On the Design of ACO for the Biobjective Quadratic Assignment Problem. In: Dorigo M, Birattari M, Blum C, *et al.*, editors. Ant Colony Optimization and Swarm Intelligence, 4th International Workshop, ANTS 2004, Brussels, Belgium, September 5–8, 2004, Proceedings. vol. 3172 of Lecture Notes in Computer Science. Springer; 2004. p. 214–225.

[4] López-Ibáñez M, Stützle T. An experimental analysis of design choices of multi-objective ant colony optimization algorithms. Swarm Intelligence. 2012;6(3):207–232.

[5] Guntsch M, Middendorf M. Applying Population Based ACO to Dynamic Optimization Problems. In: Dorigo M, Caro GD, Sampels M, editors. Ant Algorithms, Third International Workshop, ANTS 2002, Brussels, Belgium, September 12–14, 2002, Proceedings. vol. 2463 of Lecture Notes in Computer Science. Springer; 2002. p. 111–122.

[6] Angus D, Hendtlass T. Ant Colony Optimisation Applied to a Dynamically Changing Problem. In: Hendtlass T, Ali M, editors. Developments in Applied Artificial Intelligence, 15th International Conference on Industrial and Engineering, Applications of Artificial Intelligence and Expert Systems, IEA/AIE 2002, Cairns, Australia, June 17–20, 2002, Proceedings. vol. 2358 of Lecture Notes in Computer Science. Springer; 2002. p. 618–627.

[7] Mavrovouniotis M, Yang S. A memetic ant colony optimization algorithm for the dynamic travelling salesman problem. Soft Computing. 2011;15(7): 1405–1425.

[8] Mavrovouniotis M, Yang S. Ant colony optimization with immigrants schemes for the dynamic travelling salesman problem with traffic factors. Applied Soft Computing. 2013;13(10):4023–4037.

[9] Eaton J, Yang S, Mavrovouniotis M. Ant colony optimization with immigrants schemes for the dynamic railway junction rescheduling problem with multiple delays. Soft Computing. 2016;20(8):2951–2966.

[10] Bianchi L, Gambardella LM, Dorigo M. Solving the Homogeneous Probabilistic Traveling Salesman Problem by the ACO Metaheuristic. In: Dorigo M, Caro GD, Sampels M, editors. Ant Algorithms, Third International Workshop, ANTS 2002, Brussels, Belgium, September 12–14, 2002, Proceedings. vol. 2463 of Lecture Notes in Computer Science. Springer; 2002. p. 176–187.

[11] Weyland D, Bianchi L, Gambardella LM. New Approximation-Based Local Search Algorithms for the Probabilistic Traveling Salesman Problem. In: Moreno-Díaz R, Pichler F, Quesada-Arencibia A, editors. Computer Aided

Systems Theory – EUROCAST 2009, 12th International Conference, Las Palmas de Gran Canaria, Spain, February 15–20, 2009, Revised Selected Papers. vol. 5717 of Lecture Notes in Computer Science. Springer; 2009. p. 681–688.

[12] Bianchi L, Dorigo M, Gambardella LM, *et al.* A survey on metaheuristics for stochastic combinatorial optimization. Natural Computing. 2009;8(2): 239–287.

[13] Dorigo M, Blum C. Ant colony optimization theory: a survey. Theoretical Computer Science. 2005;344(2):243–278.

[14] Dorigo M. Optimization, Learning and Natural Algorithms [in Italian]. Dipartimento di Elettronica, Politecnico di Milano, Milano; 1992.

[15] Dorigo M, Stützle T. Ant Colony Optimization. Cambridge, MA: MIT Press; 2004.

[16] Ivkovic N, Malekovic M, Golub M. Extended Trail Reinforcement Strategies for Ant Colony Optimization. In: Panigrahi BK, Suganthan PN, Das S, *et al.*, editors. Swarm, Evolutionary, and Memetic Computing – Second International Conference, SEMCCO 2011, Visakhapatnam, Andhra Pradesh, India, December 19–21, 2011, Proceedings. Part I. vol. 7076 of Lecture Notes in Computer Science. Springer; 2011. p. 662–669.

[17] Maniezzo V. Exact and approximate nondeterministic tree-search procedures for the quadratic assignment problem. INFORMS Journal on Computing. 1999;11(4):358–369.

[18] Dorigo M, Maniezzo V, Colorni A. Positive Feedback as a Search Strategy. Technical Report No. 91-016, Politecnico di Milano, Italy; 1991.

[19] Dorigo M, Maniezzo V, Colorni A. Ant system: optimization by a colony of cooperating agents. IEEE Transactions on Systems, Man, and Cybernetics, Part B. 1996;26(1):29–41.

[20] Dorigo M, Gambardella LM. Ant colony system: a cooperative learning approach to the traveling salesman problem. IEEE Trans Evolutionary Computation. 1997;1(1):53–66.

[21] Gambardella LM, Dorigo M. Ant-Q: A Reinforcement Learning Approach to the Traveling Salesman Problem. In: Prieditis A, Russell SJ, editors. Machine Learning, Proceedings of the Twelfth International Conference on Machine Learning, Tahoe City, California, USA, July 9–12, 1995. Morgan Kaufmann; 1995. p. 252–260.

[22] Bullnheimer B, Hartl RF, Strauss C. A new rank based version of the ant system: a computational study. Central European Journal for Operations Research and Economics. 1999;7(1):25–38.

[23] Stützle T, Hoos HH. MAX–MIN ant system. Future Generation Computer Systems. 2000;16:889–914.

[24] Ivkovic N, Golub M, Malekovic M. A Pheromone Trails Model for MAX–MIN Ant System. In: Hao JK, Legrand P, Collet P, *et al.*, editors. Artificial Evolution 2011 (Evolution Artificielle 2011), 10th Biennial International Conference on Artificial Evolution. Angers, France; 2011. p. 35–46.

[25] Cordón O, de Viana IF, Herrera F, *et al.* A New ACO Model Integrating Evolutionary Computation Concepts: The Best–Worst Ant System. In: Proc. of ANTS'2000. From Ant Colonies to Artificial Ants: Second International Workshop on Ant Algorithms. Brussels, Belgium; 2000. p. 22–29.

[26] Cordón O, de Viana IF, Herrera F. Analysis of the Best–Worst Ant System and Its Variants on the QAP. In: Dorigo M, Caro GD, Sampels M, editors. Ant Algorithms, Third International Workshop, ANTS 2002, Brussels, Belgium, September 12–14, 2002, Proceedings. vol. 2463 of Lecture Notes in Computer Science. Springer; 2002. p. 228–234.

[27] Cordón O, de Viana IF, Herrera F. Analysis of the best–worst ant system and its variants on the TSP. Mathware and Soft Computing. 2002;9(2–3):177–192.

[28] Guntsch M, Middendorf M. A Population Based Approach for ACO. In: Cagnoni S, Gottlieb J, Hart E, *et al.*, editors. Applications of Evolutionary Computing, EvoWorkshops 2002: EvoCOP, EvoIASP, EvoSTIM/EvoPLAN, Kinsale, Ireland, April 3–4, 2002, Proceedings. vol. 2279 of Lecture Notes in Computer Science. Springer; 2002. p. 72–81.

[29] Ivković N. Modeling, Analysis and Improvement of Ant Colony Optimization Algorithms [in Croatian]. Faculty of Electrical Engineering and Computing, University of Zagreb, Zagreb; 2014.

[30] Ivković N, Golub M. A New Ant Colony Optimization Algorithm: Three Bound Ant System. In: Dorigo M, Birattari M, Garnier S, *et al.*, editors. Swarm Intelligence – 9th International Conference, ANTS 2014, Brussels, Belgium, September 10–12, 2014. Proceedings. vol. 8667 of Lecture Notes in Computer Science. Springer; 2014. p. 280–281.

[31] Alba E, Chicano F. ACOhg: Dealing with Huge Graphs. In: Proceedings of the 9th Annual Conference on Genetic and Evolutionary Computation. GECCO '07. New York, NY, USA: ACM; 2007. p. 10–17.

[32] Blum C. Beam-ACO – hybridizing ant colony optimization with beam search: an application to open shop scheduling. Computers & Operations Research. 2005;32:1565–1591.

[33] Leguizamón G, Coello CAC. An Alternative ACO_R Algorithm for Continuous Optimization Problems. In: Dorigo M, Birattari M, Caro GAD, *et al.*, editors. Swarm Intelligence – 7th International Conference, ANTS 2010, Brussels, Belgium, September 8–10, 2010. Proceedings. vol. 6234 of Lecture Notes in Computer Science. Springer; 2010. p. 48–59.

[34] Bilchev G, Parmee IC. The Ant Colony Metaphor for Searching Continuous Design Spaces. In: Fogarty TC, editor. Evolutionary Computing, AISB Workshop, Sheffield, UK, April 3–4, 1995, Selected Papers. vol. 993 of Lecture Notes in Computer Science. Springer; 1995. p. 25–39.

[35] Monmarché N, Venturini G, Slimane M. On how pachycondyla apicalis ants suggest a new search algorithm. Future Generation Computer Systems. 2000;16(9):937–946.

[36] Dréo J, Siarry P. Continuous interacting ant colony algorithm based on dense heterarchy. Future Generation Computer Systems. 2004;20(5):841–856.

[37] Gambardella LM, Taillard ED, Dorigo M. Ant colonies for the quadratic assignment problem. Journal of the Operational Research Society. 1999; 50(2):167–176.

[38] Schoonderwoerd R, Bruten JL, Holland OE, *et al.* Ant-based load balancing in telecommunications networks. Adaptive Behavior. 1996;5(2):169–207.

[39] Caro GAD, Dorigo M. AntNet: distributed stigmergetic control for communications networks. Journal of Artificial Intelligence Research (JAIR). 1998;9:317–365.

[40] Caro GD, Ducatelle F, Gambardella LM. AntHocNet: An Ant-Based Hybrid Routing Algorithm for Mobile Ad Hoc Networks. In: Yao X, Burke EK, Lozano JA, *et al.*, editors. Parallel Problem Solving from Nature – PPSN VIII, 8th International Conference, Birmingham, UK, September 18–22, 2004, Proceedings. vol. 3242 of Lecture Notes in Computer Science. Springer; 2004. p. 461–470.

[41] Papadimitriou CH. Computational Complexity. Reading, MA: Addison-Wesley; 1994.

[42] Orponen P, Mannila H. On Approximation Preserving Reductions: Complete Problems and Robust Measures; 1987.

[43] Papadimitriou CH, Yannakakis M. The traveling salesman problem with distances one and two. Mathematics of Operations Research. 1993;18:1–11.

[44] Christofides N. Worst-Case Analysis of a New Heuristic for the Travelling Salesman Problem. Pittsburgh, PA: Carnegie-Mellon University, Management Sciences Research Group; 1976.

[45] Arora S. Polynomial time approximation schemes for Euclidean traveling salesman and other geometric problems. Journal of the ACM. 1998;45(5): 753–782.

[46] Gutin G, Punnen AP, editors. The Traveling Salesman Problem and Its Variations. Boston, MA: Kluwer Academic Publishers; 2004.

[47] Lawler EL, Lenstra JK, Kan AHGR, *et al.* The Traveling Salesman Problem: A Guided Tour of Combinatorial Optimization. New York: Wiley; 1985.

[48] Applegate DL, Bixby RE, Chvátal V, *et al.* The Traveling Salesman Problem: A Computational Study. Princeton, NJ: Princeton University Press; 2007.

[49] Davendra D, editor. Traveling Salesman Problem: Theory and Applications. London: InTech; 2010.

[50] Pardalos PM, Rendl F, Wolkowicz H. The quadratic assignment problem: a survey of recent developments. In: Proceedings of the DIMACS Workshop on Quadratic Assignment Problems, volume 16, American Mathematical Society; 1994. p. 1–42.

[51] Burkard RE, Çela E. Quadratic and Three-Dimensional Assignments: An Annotated Bibliography, Annotated Bibliographies in Combinatorial Optimization. New York: Wiley; 1997.

[52] Socha K. The Influence of Run-Time Limits on Choosing Ant System Parameters. In: Cantú-Paz E, Foster JA, Deb K, *et al.*, editors. Genetic and Evolutionary Computation – GECCO 2003, Genetic and Evolutionary Computation Conference, Chicago, IL, USA, July 12–16, 2003, Proceedings.

Part I. vol. 2723 of Lecture Notes in Computer Science. Springer; 2003. p. 49–60.

[53] Luo X, Yu F, Zhang J. Study of Parametric Relation in Ant Colony Optimization Approach to Traveling Salesman Problem. In: Huang D, Li K, Irwin GW, editors. Computational Intelligence and Bioinformatics, International Conference on Intelligent Computing, ICIC 2006, Kunming, China, August 16–19, 2006, Proceedings. Part III. vol. 4115 of Lecture Notes in Computer Science. Springer; 2006. p. 22–32.

[54] Ivkovic N. Exploring MAX–MIN Ant System Parameter Landscape for Symmetrical Traveling Salesman Problem. In: Bogunović N, Ribarić S, editors. 32nd International Convention MIPRO 2009 on Computers in Technical Systems and Intelligent Systems. CTS&CIS; 2009. p. 302–307.

[55] Maur M, López-Ibáñez M, Stützle T. Pre-scheduled and adaptive parameter variation in MAX–MIN Ant System. In: Proceedings of the IEEE Congress on Evolutionary Computation, CEC 2010, Barcelona, Spain, 18–23 July 2010. IEEE; 2010. p. 1–8.

[56] Samrout M, Kouta R, Yalaoui F, *et al.* Parameters setting of the ant colony algorithm applied in preventive maintenance optimization. Journal of Intelligent Manufacturing. 2012;18(6):663–677.

[57] Pellegrini P, Favaretto D, Moretti E. On MAX–MIN ant system's parameters. In: Dorigo M, Gambardella LM, Birattari M, *et al.*, editors. Ant Colony Optimization and Swarm Intelligence, 5th International Workshop, ANTS 2006, LNCS. vol. 4150. Brussels, Belgium. Springer; 2006. p. 203–214.

[58] Nallaperuma S, Wagner M, Neumann F. Analyzing the Effects of Instance Features and Algorithm Parameters for MAX–MIN Ant System and the Traveling Salesperson Problem. Front Robotics and AI. 2015;2:1–16.

[59] Montgomery J, Randall M, Hendtlass T. Automated selection of appropriate pheromone representations in ant colony optimization. Artificial Life. 2005;11(3):269–292.

[60] Solnon C, Fenet S. A study of ACO capabilities for solving the maximum clique problem. Journal of Heuristics. 2006;12(3):155–180.

[61] Montgomery J, Randall M, Hendtlass T. Solution bias in ant colony optimisation: Lessons for selecting pheromone models. Computers & Operations Research. 2008;35(9):2728–2749.

[62] Solnon C. Combining two pheromone structures for solving the car sequencing problem with ant colony optimization. European Journal of Operational Research. 2008;191(3):1043–1055.

[63] Merkle D, Middendorf M. Modelling the dynamics of ant colony optimization. Evolutionary Computation. 2002;10(3):235–265.

[64] Derrac J, García S, Molina D, *et al.* A practical tutorial on the use of nonparametric statistical tests as a methodology for comparing evolutionary and swarm intelligence algorithms. Swarm and Evolutionary Computation. 2011;1:3–18.

[65] Ivković N, Jakobović D, Golub M. Measuring performance of optimization algorithms in evolutionary computation. International Journal of Machine Learning and Computing. 2016;6(3):167–171.

[66] Gutjahr WJ. A Graph-based ant system and its convergence. Future Generation Computer Systems. 2000;16(8):873–888.

[67] Stützle T, Dorigo M. A short convergence proof for a class of ACO algorithms. IEEE Transactions on Evolutionary Computation. 2002;6(4):358–365.

[68] Gutjahr WJ. ACO algorithms with guaranteed convergence to the optimal solution. Information Processing Letters. 2002;82(3):145–153.

[69] Gutjahr WJ. First steps to the runtime complexity analysis of ant colony optimization. Computers & Operations Research. 2008;35(9):2711–2727.

[70] Neumann F, Sudholt D, Witt C. Analysis of different MMAS ACO algorithms on unimodal functions and plateaus. Swarm Intelligence. 2009;3(1):35–68.

[71] Neumann F, Witt C. Runtime analysis of a simple ant colony optimization algorithm. Algorithmica. 2009;54(2):243–255.

[72] Doerr B, Neumann F, Sudholt D, *et al.* Runtime analysis of the 1-ANT ant colony optimizer. Theoretical Computer Science. 2011;412(17):1629–1644.

[73] Sudholt D, Thyssen C. Running time analysis of ant colony optimization for shortest path problems. Journal of Discrete Algorithms. 2012;10:165–180.

[74] Neumann F, Sudholt D, Witt C. Rigorous Analyses for the Combination of Ant Colony Optimization and Local Search. In: Dorigo M, Birattari M, Blum C, *et al.*, editors. Ant Colony Optimization and Swarm Intelligence, 6th International Conference, ANTS 2008, LNCS. vol. 5217. Brussels, Belgium; 2008. p. 132–143.

Chapter 16
A review of particle swarm optimization for multimodal problems

Jian-Ping Li[1], Shanxin Yuan[1], Qing Sheng Li[1], Bo Li[1], and Yim Fun Hu[1]

Abstract

Particle swarm optimizations (PSOs) are population-based methods inspired from the flight of a flock of birds seeking food. After the development of over 20 years, PSOs have become a major branch of evolutionary algorithms (EAs) and have been successfully applied to solve many science and engineering optimization problems. Most of PSOs are designed to search one solution of a problem. However, many science and engineering optimization problems are complex and multimodal in nature. More and more researches are aiming to identify multiple global and local solutions of complex multimodal problems, and several competitions in recent international conferences had been set up to encourage researchers to develop more effective and efficient algorithms for exploring multiple solutions. There are several techniques in the literature which can be used by combining with an existing EA to solve multimodal optimization problems. Those techniques are called niching. The most commonly used niching techniques are crowding, fitness share, clustering and species conserving. PSO-related methods for multimodal problems are reviewed in this chapter, including hybrid PSO with other EAs. Additionally, the multimodal functions, including some challenge composition multimodal functions, are listed as references for researchers to test their new algorithms. The species conserving PSO is described in detail and used to solve some multimodal engineering optimization problems to demonstrate the power of niching in exploring multiple solutions.

16.1 Introduction

Particle swarm optimization (PSO) is a population-based metaheuristics inspired from the flight of a flock of birds seeking food [1]. Most of PSOs are developed to explore a single optimal solution of an optimization problem, and PSOs have

[1]School of Engineering, Design and Technology, University of Bradford, UK

been demonstrated to be one of efficient evolutionary algorithms (EAs), especially in solving some real-world optimization problems. Usually a real-world problem should have many constraints on resource limitations, material limitations and other design constraints, and a solution is normally located at one corner of the feasible region bounded with several constraints. Practically, all the corners are potentially becoming global and/or local solutions. When dealing with multimodal optimization problems, the techniques, called niching, are needed to preserve the diversity of populations in a PSO [2], or the algorithm may be trapped at one local solution. Different kinds of niching have been developed to group individuals in a population and to maintain the diversity of populations to let implicit/explicit niches be alive in the future generations. Those techniques can increase the possibility to explore the whole feasible region and search global solutions.

There are three types of niching techniques in the literature for dealing with multimodal optimization problems: the first one is to try to explore one of global solutions, even when there are several global solutions in the problems. This kind of algorithms can output only one global solution. If there are more than one global solution in the search space, they may output different solutions in different runs. Most of existing EAs belong to this type. The second type of technique is to explore all possible global solutions and to ignore local solutions. Finally, the third type of algorithm aims to identify all the global and local solutions. For example, Li and Wood [3] proposed an adaptive species conserving genetic algorithm (SCGA) and applied it to solve a two-dimensional Shubert function and successfully obtained 761 solutions (many text books claimed there is 760 solutions for this problem), 18 of them are global solutions. One of the tiny local solution is located at the left bottom corner of the feasible region. From the engineering design point of view, local solutions are not very interesting. However, global solutions may not be used in the real designs, because engineering design is aiming to find optimal and robust solutions. If a global solution is too sensitive and little movement in decision variables or environments will lead to a large change of the objective values, this solution will not be used. In this case, designers may select a local and more robust solution. Therefore, finding all the global and local solutions are still very useful and can help designers to understand the design space and find a better and robust solution.

After the development for decades, there are several efficient niching techniques in the literature but it is still very challenging to solve complex multimodal optimization problems with high dimensions. This is the reason why multimodal optimizations are still very active research areas in the communities of optimization and evolutionary computation. In this chapter, most important niching PSOs are reviewed, common-used multimodal benchmark functions are presented for researchers to develop some more efficient methods. The species-conserving PSO (SCPSO) is described in detail and used to solve three multimodal engineering optimization problems.

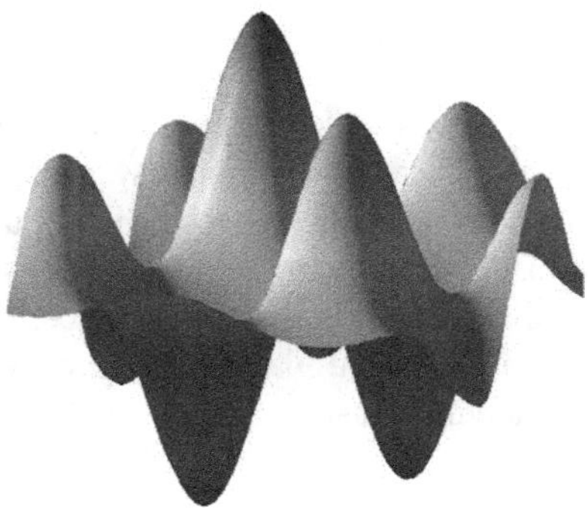

Figure 16.1 A sample multimodal function

Mathematically, a multimodal optimization problem has the same definition and description as a general optimization problem, such as

$$\begin{array}{l} \max f(\mathbf{x}) \\ \quad \mathbf{x}\ s.t. \\ g(\mathbf{x}) \leq 0 \\ h(\mathbf{x}) = 0 \end{array} \tag{16.1}$$

A multimodal problem is an optimization problem where there are more than one local and/or global solution (see Figure 16.1). Many real-world optimization problems are multimodal, for examples: data mining [4], and truss structure [5] and system identification problems [6] etc. PSOs are one part of EAs that are potentially capable of obtaining global solutions. However, classical PSOs can only explore one of multiple solutions.

When solving multimodal optimization problems, many traditional optimization algorithms may be trapped to one local solution, especially gradient-based methods can only be found one local solution that is near an initial point. Even the intrinsic parallelism of EAs suggests that EAs (including PSO, GAs) should be capable of exploring global solutions of multimodal functions; Nijssen and Back [7] claimed that those traditional EA is not suitable to obtain the global optimum of some trap functions within an acceptable fitness-evaluations or times. In most of cases, they are trapped in a local optimum.

To balance exploration and exploitation is the common and most important issue to develop an efficient EA for searching global solutions. When particles are close each other, the diversity of the population will be sharply reduced, and then the efficiency of the algorithm will drop discernibly.

The techniques that are used to deal with multimodal functions are called niching. Usually, niching will be combined with an EA to explore multiple solutions of multimodal functions. In this chapter, only related niching PSOs are discussed.

16.2 Particle swarm optimization

A PSO is a population-based algorithm [1]. A population, $S = (\mathbf{x}_1, \mathbf{x}_2, \ldots, \mathbf{x}_N)$, consists of particles, $\mathbf{x}_i = (x_{i1}, x_{i2}, \ldots, x_{in})$, also called swarms, which present potential solutions of an optimization problem, where N is the population size and n is the dimension size of an optimization problem. In the beginning of a PSO process, each particle is randomly located a position in the search space. In each generation, each particle will be relocated to a new position with an associated velocity, and the new position is mathematically calculated as

$$\mathbf{x}_i(t+1) = \mathbf{x}_i(t) + \mathbf{v}_i(t+1) \tag{16.2}$$

and its moving velocity $\mathbf{v}_i(t)$ is defined as

$$\mathbf{v}_i(t+1) = \mathbf{v}_i(t) + \phi_p * R1 * (\mathbf{x}^p - \mathbf{x}_i) + \phi_g * R2 * (\mathbf{x}^g - \mathbf{x}_i) \tag{16.3}$$

where $R1$ and $R2$ are random numbers between [0, 1]; ϕ_p and ϕ_g are, respectively, called the cognitive and social acceleration coefficient. The velocity of particle i will be affected by

1. The previous velocity $\mathbf{v}_i$ of particle i.
2. The current position of particle i.
3. The personal best position (*pBest*, $\mathbf{x}^p$) which is found by the current particle.
4. The global best (*gBest*, $\mathbf{x}^g$) which is found by all the particles. This information is shared by all the particles. So this property is also called the social communication.

Figure 16.2 illustrates the relationship of the velocity function of particle i at time $(t+1)$ with the previous velocity, the current position, $\mathbf{x}_i^l$ ($\mathbf{x}_i^p$) and $\mathbf{x}_i^g$. Equations (16.2) and (16.3) will lead all particles to move the current global best solution. If ϕ_p and ϕ_g are equal to or less than one, the algorithm may be trapped in a local solution; therefore, they are used to balance the exploration and exploitation.

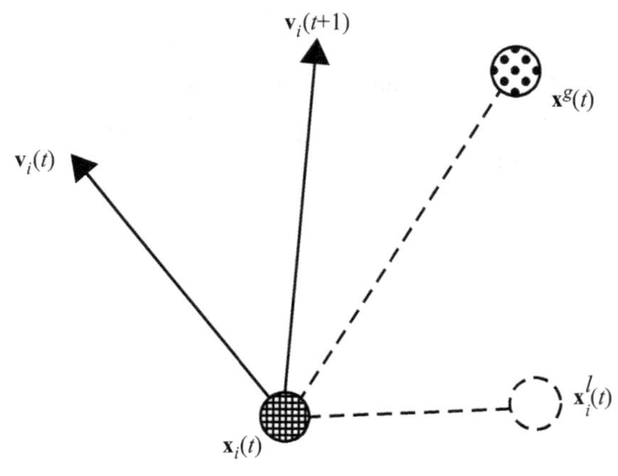

Figure 16.2 The velocity of a particle

```
Initialize a population of particles
Evaluate the fitness of all the particles.
Set each particle pBest as itself and the initial velocity as zero
Set the best particle as gBest
while (a stop criterion is not met)
  for each particle
      Change the velocity and position;
      Evaluate its fitness
      if (its fitness is better than its pBest) then
          Update pBest with the current particle
      end if
      if (its fitness is better than its gBest) then
          Update gBest with the current particle
      end if
  end for
end while
```

Figure 16.3 The procedure in a standard PSO

There are several variations of PSOs in the literature, and the typical procedure of a standard PSO can be illustrated in Figure 16.3. In the initial stage, all the particles are randomly located in the design space bounded by the lower and upper limits of design variables. The personal best (*pBest*) of each particle is set as itself. The best particle is selected from the initial population as the global best (*gBest*). The search process will be continued until the stop criterion, such as the maximum of generations or the maximum evaluations of fitness functions. In each loop, the velocity of each particle will be updated by using (16.3) and its position will be recalculated by using (16.2). If it is better than its personal best (*pBest*), then its *pBest* will be updated by using its current position. If it is better than its global best (*gBest*), its *gBest* will be updated by using its current position. Finally, *gBest* will be the solution of the solved problem.

16.3 Niching PSOs

Niching was originally developed for genetic algorithms to keep the diversity of populations and to search one solution for complex and multimodal optimization problems. A variety of niching techniques can be found in the literature, such as crowding [8,9], fitness share [10,11], clustering [12], species conserving [3,13]. More and more techniques are designed to explore multiple solutions of multimodal functions [3,13].

16.3.1 Crowding PSO

Crowding is one of the oldest niching techniques developed in EAs to give pressure to select individuals for preserving the diversity in populations and slowing down the convergence speed to find a global solution. In genetic algorithms, the fitter of individuals will have higher probability to survive in the new population. Chen

```
if f[x(t + 1)] > f[x^p] then
        x^p = x(t + 1)
end if
```

Figure 16.4 The update procedure in a standard PSO

```
if ||x(t + 1) − x^p|| ≤ threshold then
    if f[x(t + 1)] > f[x^p] then
        x^p = x(t + 1)
    end if
else
    if f[x(t + 1)] > f[x^p] then
        x^p = x(t + 1)
    end if
end if
```

Figure 16.5 The new update procedure in Chen and Montgomery (2011)

and Montgomery [14] proposed a very simple crowding strategy to maintain the diversity of particles in populations. In a standard PSO, a new position of a particle in the $(t+1)$th generation, $\mathbf{x}^{t+1}$, was obtained through the velocity function, and then its personal position will be updated. Usually, the update procedure is shown in Figure 16.4, where $\mathbf{x}^{t}_{pBest}$ is the personal best in tth iteration.

Chen and Montgomery [14] thought that each particle in a population was attracted to its *pBest*, and the operation in Figure 16.4 would cause some *pBests* to move closer and then become as a crowd. In their work, a crowding factor was introduced to specify the size of crowding, then a threshold was used to develop a new update strategy shown in Figure 16.5 to control the diversity of populations.

16.3.2 Fitness share/NichePSO

Holland [10] proposed a sharing concept in 1975, which was improved by Goldberg and Richardson [11] in 1992. Sharing has become the most known and popular method to keep the diversity of populations in a genetic algorithm to explore global solutions of complex multimodal functions.

16.3.2.1 NichePSO

Brits *et al.* [15] proposed a NichePSO algorithm to avoid premature convergence through multiple independent particles with a cognition-only-model for local search. The niches were identified by keeping records of the fitness changes of particles. When the fitness of a particle was not changed for a long time, a new niche would

```
Initialize main swarm;
Train main swarm using the cognition only model;
Update fitness of each particles in the main swarm;
for each sub-swarm do
  Train particles in the sub-swarm by using the GCPSO algorithm;
  Evaluate particles;
  Amend swarm radius;
end if
If feasible, join sub-swarms;
Move possible particles from the main swarm to a sub-swarm;
Search possible particle for partition;
If found, then build a new sub-swarm with its closest neighbor
```

Figure 16.6 The procedure of the NichePSO

be formulated with the particle and its closest particles. The Euclidean distance was applied to compute the distances between particles for the closeness check. A niche radius of the ith niche was automatically updated in each population and calculated as

$$R_i = \max_{j \in S_i}\{\|\mathbf{x}_i^g - \mathbf{x}_{ij}\|\} \tag{16.4}$$

where S_i is the collection of particles in the ith niche, $\mathbf{x}_i^g$ is the best position in the ith niche and $\mathbf{x}_{ij}$ is one particle in the ith niche.

The niche radius was used to update niches, also called as sub-swarms. For an example, moving particles to an existing niche or combining two niches into one niche if they are close to each other. A particle $\mathbf{x}_j$ will be added into a niche if

$$\|\mathbf{x}_j - \mathbf{x}_i^g\| \leq R_j \tag{16.5}$$

If the distance between two niches is less than the sum of their radius, then the two niches are joined together to formulate as one niche.

The procedure of the NichePSO is shown in Figure 16.6. Brits *et al.* [15] also used a swarm to present a niche.

The NichePSO does not need any information on the number of optimum to determine the radius value. But, a threshold parameter was required to identify when the fitness of a particle is not changed and a new niche should be formulated. A minimum distance is also needed to identify when two swarms/niches should be combined together once they are close to each other. Brits *et al.* [15] called the population as the main swarm and maintained many niches/sub-swarms. Both the main swarm and sub-swarms evolve simultaneously.

16.3.2.2 Sub-niches/swarms

Some researchers used a fixed number of explicit sub-swarms that evolve independently. Some necessary techniques are needed to prevent multiple sub-swarms to explore the same area. For an example, Chang *et al.* [16] established a parallel multi-swarm optimization (PMSO), in which many swarms evolve in parallel and independently, and each swarm explores the evenly and dynamically allocated search space with weight factors. Zhang *et al.* [17] applied the Hill-Valley function [17] to identify if a particle is a member of one existing niched. If so, the particle will be penalized. The authors claimed that this method can avoid multiple swarms from searching the same area. The penalty rule is mathematically described as

$$eval(x_i) = \begin{cases} f(x_i) & \text{if } hill\_valley(x_i, x_{best}) = 1 \\ f(x_i) - penalty(x_i) & \text{otherwise} \end{cases} \tag{16.6}$$

where $penalty(x)$ is denoted as a penalty function [18]. The Hill-Valley algorithm aims to identify whether two points/particles belong to the same valley. A series of points in a line around the tested position were generated and compared by using some heuristic methods. The best found fitness was set as the maximum while the lowest fitness was set as the minimum. If the fitness of all the points is between the minimum and the maximum, then the tested points are located in the same niche. Figure 16.7 illustrates the algorithm of the Hill-Valley function used in (16.6).

The adaptive sequential NichePSO (ASNPSO) was established by Zhang *et al.* [19] to explore optimal solutions sequentially by using multiple sub-swarm with the above valley function. Each sub-swarm scrutinizes the search space for some generations. The procedure of sequential niching is illustrated in Figure 16.8.

In the ASNPSO [19], the number of sub-swarms is given. If no prior information on a problem can be obtained, it is necessary to do some numerical experiments to

```
minfit = minimal fitness between (a,b);
maxfit = maximal fitness between (a,b);
for j=0 to samples length do
  Calculate the fitness f_j at point j between the points a and b;
  if ((f_j < minfit) or (f_j > maxfit)) then
      return 0;
  end
end
return 1;
```

Figure 16.7 The procedure of Hill-Valley function

```
N=0;
while (stop condition is not met) do
  Evolve a new sub-swarm;
  N = N + 1;
  for k = 1 to N - 1 do
      if (hill_valley(x_i, x_k) == 0) then
          Re-calculate the fitness of x_i;
      end if
  end for
  Train the sub-swarm;
end (while);
```

Figure 16.8 The procedure of the ASNPSO

guess the number of niches or use any number to try. If the number of the final solutions exceeds the number of sub-swarms, the ASNPSO will be run again with a big swarm number. The ASNPSO also needs another parameter, the number of interior points, to identify a valley used in the Hill-Valley algorithm, but unlike NichePSO, this parameter is independent to threshold parameters.

Most of existing niching techniques often require some specified parameters, which will affect the efficiency and effectiveness of the proposed algorithms. Bird and Li [20] established a niching method which could adaptively choose the number of niches in a problem and identify whether a particle is within the region of a niche. This method was called the adaptive niching PSO (ANPSO). The minimum distances (or niche distance) between particles was calculated as

$$r = \frac{\sum_{i=1}^{n} d_i}{n} \tag{16.7}$$

where $d_i = \min\{\|p_i - p_j\|, i \neq j\}$. Figure 16.9 illustrates the calculation of the minimum distance (or niche distance) in a population.

First, an undirected graph is built, which consisting of a node for each particle in the population. The above niche distance is used to search pairs of nearest particles. If their distance is less than the niche distance, they are in the same niche. The algorithm for formulating niches is illustrated in Figure 16.10.

Once the niche graph has been built, Figure 16.11 shows the procedure to identify niches by comparing each particle with its initial radius. A particle will be marked as processed if it belongs to any found niche.

Chang [21] proposed a modified PSO by splitting single population into M sub-population according to the order of particles' fitness, as shown in Figure 16.12. The strategy is very simple. For an example, assuming that the population size is

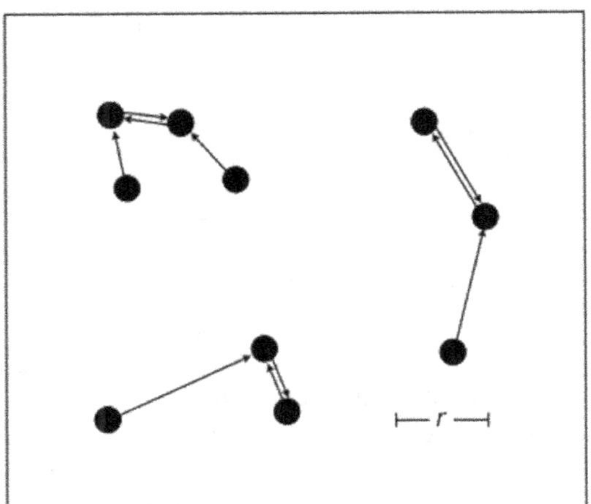

Figure 16.9 Calculating the minimum distance

```
Calculate the minimum distance (r);
Build an undirected graph G containing all particles;
for i = 1 to n−1
  for j = i+1 to n
     if ||x_i − x_j|| < r then
        s_i = s_i + 1;
        if s_i < 4 then
           s_i = 4;
        end (if)
        if s_i ≥ 2 then
           Build an edge from x_i to x_j.
        end (if)
     else
        s_i = s_i − 1;
        if s_i < 0 then
           s_i = 0;
        end (if)
     end (if)
  end (for)
end (for)
```

Figure 16.10 The algorithm of building the niche graph

```
visited = Φ;
reachable = Φ;
for i=1 to n do
  if d_i < r and p_i ∉ visited then
     reachable = reachable ∪ {p_i, p_j | p_j ∀j reachable from p_i ∈ G}
     Build a new niche s = {p_j}
     for p ∈ reachable and p_i ∉ visited do
        vivisted = vivisted ∪ {p}
        s ← {p} ∪ s
     end (for)
  end (if)
end (for)
```

Figure 16.11 Creating niches

100, so there are 100 randomly generated particles. If we want to split into five sub-populations, all the particles are ordered according to their fitness, the first 20 particles are moved to the first sub-population, the second 20 particles are placed in the second sub-population, and so on. In each generation, each group has its own *pBest* and *gBest*, then all the particles in one sub-population are updated by using the group's *pBest* and *gBest*. It looks like that there was no communication among sub-populations in Chang's algorithm.

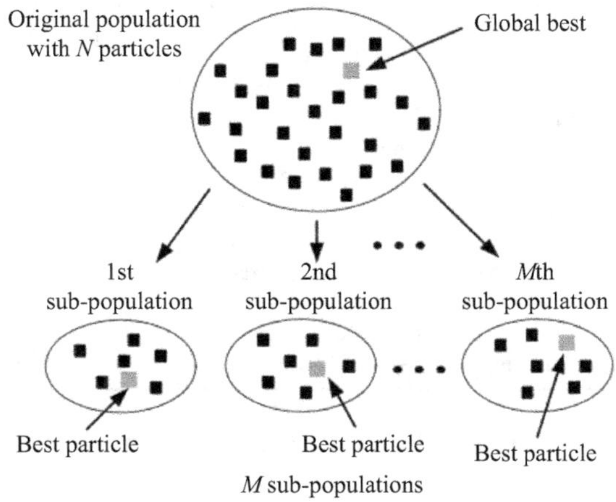

Figure 16.12 The diagram of transforming single population to M sub-population

```
Initialize k centroids;
while (stop criteria is not met) do
  Calculate the distance of each point to the k clusters;
  Allocate each point to the neighboring cluster;
  Update cluster centroid;
end (while)
```

Figure 16.13 The procedure of the k-means algorithm

16.3.3 Clustering PSO

Kennedy [12] was the first author to introduce clustering techniques into a PSO algorithm to solve multimodal optimization problems. Kennedy's method was called "stereotyping" in which the k-means algorithm in Figure 16.13 was applied to split the main swarm into k sub-swarms. Each has its own global best which is defined as the centroid of the sub-swarm.

Passaro and Starita [22] proposed another clustering approach, called k-means-based PSO (kPSO), by improving the clustering approach [12]. The k-means clustering algorithm was applied in this kPSO to find niches in the population, while a Bayesian information criterion was adaptively applied to calculate the niche number in a problem after each c iterations. The procedure of the kPSO is illustrated in Figure 16.14.

After each c generations, a niching algorithm is run to identify niches. The mean number of all particles was used as the maximum number of particles in sub-swarms. When a cluster has more particles, the worst particle in the niche will be removed. The codes for exploring a niche is shown in Figure 16.15.

```
Randomly initialize particles;
Evaluate fitness of particles;
Replace particles' pBest to their current positions;
for t = 0 to T − 1
  if t mod c = 0 then
       Identify Niches;
  end (if)
  re-calculate velocities of particles;
  Update new positions of particles;
  Re-evaluate particles' fitness;
  Amend particles' best position and neighborhood;
end (for)
```

Figure 16.14 The kPSO algorithm

```
Search particle's pBest;
Compute N_avg (the average number of particles);
Let N_u;
for each Cluster C_i do
  if N_i > N_avg then
       delete the N_i − N_avg particles from C_i;
       add N_i − N_avg to N_u;
  end (if)
end (for)
Update the neighborhood structure in C_i;
```

Figure 16.15 Algorithm for identification of niches

16.3.4 Species PSO

The species concept is an updated version of niche. A species was described as a set of particles in a population which has similar properties [13], and the particle which has the highest fitness value in the group was defined as its species seed. Mathematically, let the species seed be $\mathbf{x}^*$ of a species S_i, a particle $\mathbf{y}$ was said to belong to species S_i if

$$\|\mathbf{x}^*, \mathbf{y}\| < R_s \tag{16.8}$$

and

$$f(\mathbf{y}) \leq f(\mathbf{x}^*) \tag{16.9}$$

where R_s is defined as the species radius. In the SCGA, the species radius is a predefined and fixed number. $\|\mathbf{x}^*, \mathbf{y}\|$ is defined as the distance between two particles/individuals $\mathbf{x}^*$ and $\mathbf{y}$. An example of a distribution of species in a 2D space is illustrated in Figure 16.16. In an EA, a species consists of some particles and fills an

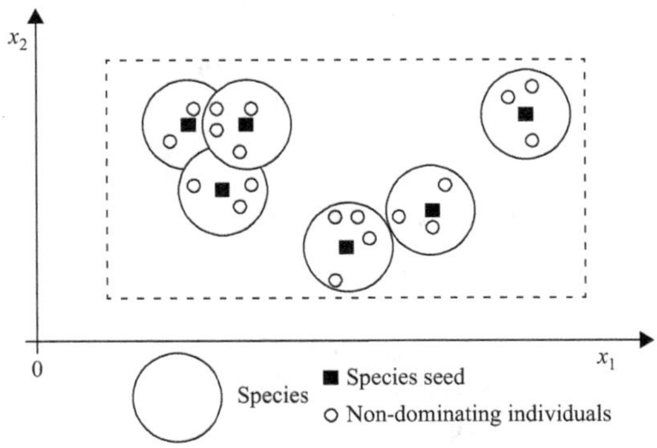

Figure 16.16 An example distribution of species in a 2D space

area of the search space. There may be some intersection among the closest species. Some particle may belong to several species. This cannot affect algorithms to search multiple solutions, because the fitness of those particles are smaller than the fitness of any species seed of their species.

The species concept was introduced to a genetic algorithm to solve multimodal functions [13]. This approach was called SCGA. Three operators were added into a traditional GA:

- Species identifier: Species are identified from the current generation. At the beginning, the species seed set X_s is set as empty and all the particles/individuals are marked as unprocessed. The best unmark individual is picked up as a species seed and casted into the species set, and the individuals within the species seed will be marked as processed. This process will be continued until no unmarked individual can be found. The procedure of identifying species is shown in Figure 16.17. This process will be added before implementing the genetic operators.
- Species conserver: After processing genetic operators, such as selection, mutation and crossover, some species may be abandoned in the population. In order to keep the diversity of populations, the found species will be copied back to the population. The algorithm of conserving species is illustrated in Figure 16.18 and will be run after implementing genetic operators.
- Solution checker: After the evolution of generations, global solutions can be explored from the saved species seed set. A solution threshold r_f $(0 < r_f < 1)$ was introduced to diagnose global solutions. A species seed $((\mathbf{x}))$ is defined as a global solution if it satisfies the condition

$$f(\mathbf{x}) \geq (f_{max} - f_{min}) * r_f \tag{16.10}$$

where f_{min} and f_{max} are the minimal and maximal values of finesses found in the current generation.

```
X_S = Φ
while (no unprocessed particle) do
    Explore the best unmarked particle x
    Mark x* ;
    found = 0
    for all x ∈ X_S do
        if dis(x, x*) ≤ R_S then
            found = 1
            Break
        end (if)
    end (for)
    if ( found = 0 ) then
        X_s = X_s ∪ {x*}
    end (if)
end (while)
```

Figure 16.17 The codes for determining species in a population

```
Mark all local best solutions as unprocessed;
for all x ∈ X_S do
    Choice the worst unmarked y from local best;
    if ( y exists) then
        y = x ;
    else
        break;
    end (if)
    Stamp y as processed;
end (for)
end
```

Figure 16.18 The codes for conserving species seeds

The species radius (R_s) has large effects on the SCGA performance. The smaller the species radius is, the more species will be found in the populations and then the efficiency of the SCGA will drop, while the larger the species radius is, the less solutions can be found. This will result that the SCGA cannot obtain all the global solutions. To beat this problem, adaptive techniques become the possible options. Average distance among members was introduced in Bird and Li's species-based PSO (SPSO) [20] as its species radius; therefore, no species parameter is required. The SPSO is illustrated in Figure 16.19.

Based on Bird and Li's SPSO [20], Yuan *et al.* [23] developed a multi-scale PSO (MPSO) to solve multimodal functions. All particles were dynamically assigned to some subgroups. Each sub-group has its own population size and simultaneously explore different spaces with different step sizes. The authors claimed that the

```
Initialize the population;
while (Termination condition is not met) do
  Calculate the fitness of all particles;
  Sort particles in descending order;
  Identify the species seeds for the population;
  Allocate species seed as gBest;
  Re-calculate the velocity and position of particles;
end while;
```

Figure 16.19 The procedure of the SPSO

possibility of exploring more solutions has greatly increased in this approach. Parmee [24] developed a cluster oriented genetic algorithms which allowed some areas which includes the highest fitness were explored for obtaining the best solution. Yao *et al.* [25] proposed a recursive middling algorithm which can identify the existence of a valley between two given points. Hua *et al.* [26] developed a detecting peak's number technique. Heuristic methods were applied to search new possible species by checking each orthogonal direction. Iwamatsu [27] developed a multi-species particle swarm optimizer for exploring multiple solutions of multimodal functions through assigning some similar particles into a cluster, called a species. Each species searches a different area and explore the global solution or local solution in the assigned area. Li and Wood [28] developed an adaptive species concept by using some heuristics. The species radius will automatically be updated according to the property of new individuals. They [28] applied the species conserving technique into a PSO and proposed a SCPSO, in which the *pBest* of a particle was assigned with its nearest species, and the *gBest* was replaced with the nearest global solution. More details will be described in the next session.

16.3.5 Hybrid PSO

"No free lunch theorems" [29] have demonstrated that any algorithm cannot be universally good. Each algorithm has its own advantage and disadvantage. For an example, if one algorithm is very good at the convergence, but it may not maintain a good diversity of populations. PSO has been hybridized with other approaches to improve its efficiency in solving multimodal problems. Pant *et al.* [30] developed a hybrid of a PSO and a differential evolution (DE). In each iteration, DE was used to generate the trial vector of a solution, and then if the trial can find a better solution, a PSO was used to generate a new solution with the PSO velocity function. The PSO can create a perturbation and help in maintain the diversity of populations.

Shelokar *et al.* [31] established a particle swarm ant colony optimization (ACO) which is a hybrid of a PSO with an ACO. First, the PSO rules are applied in a population, and the new position of a particle is updated through a velocity function. Second, the ACO evolves like a local search, in which a pheromone-guided mechanism is applied for ants to refine a new position. If the fitness value of the ant is better than the one of the corresponding particle, then the particle is replaced with the ant.

16.4 Species-conserving PSO

Li and Wood [28] proposed a SCPSO by applying the species conserving technique [13] to improve a standard PSO to search multiple solutions of multimodal functions. In this section, the details of the SCPSO is presented and used to solve some real multimodal optimization problems.

16.4.1 Identifying species

The identification algorithm [13] was applied in the SCPSO. The procedure is illustrated in Figure 16.17 for searching species seeds and implemented at each generation.

Bird and Li [20] developed a species-based PSO, in which species were searched only from the particles in the population. In a PSO, a particle's position is modified according to its local best, global best and its current positive. Therefore, a particle likely does not belong to the species of its personal best position. So in the SCPSO, species were explored among particles and their personal best positions. At the beginning of each iteration, the species set X_s is emptied Φ. In each iteration, particles and all the local best are marked as unprocessed. A best particle is selected from unmarked particles and local best solutions. If it existed, it will be marked as processed. If it existed and does not fall into any species, this particle is added to X_s. Therefore, the SCPSO has more opportunity to find species than the SPSO [20] and the SCGA [13].

16.4.2 Conserving species

After the update of a particle position by using the velocity function, the position of a particle will be changed, and its personal best may be changed as well; therefore, they may belong to the same species or some species may not survive. A special procedure will be created to copy all species back into the next population.

In the SCPSO, each species seed was assigned as the personal best position of the nearest particle. The reason is that the new position of a particle was updated through the velocity function. When the species seeds are searched from the particles and their local best solutions, all the found species can survive in the next population. The algorithm of conserving species seeds are shown in Figure 16.18.

16.4.3 Updating particle position through a velocity function

The velocity function of updating the position of particles can lead all the particles to move towards the current global solution area. This is the reason that a PSO converges faster than a GA. This also introduces some disadvantages, because it cannot maintain the diversity of populations. For multimodal functions, there are many global and local optima, this method likely guides a PSO to converge to one of solutions. The SCPSO has modified the velocity function to let particles search different regions and fly towards different solutions: the personal/local best is replaced with the nearest species seed, while the global best is set as the nearest global solution. The velocity function was changed to as

$$v_{ij}(t+1) = v_{ij}(t) + r\varphi_p[x_{ij}^{ls}(t) - x_{ij}(t)] + r\varphi_g[x_{ij}^{lg}(t) - x_{ij}(t)] \tag{16.11}$$

where $\mathbf{x}_i^{lg} = [x_{i1}^{lg}, x_{i2}^{lg}, \ldots, x_{im}^{lg}]$ is the nearest global solution (*gBest*) to the *i*th particle and $\mathbf{x}_i^{ls} = [x_{i1}^{ls}, x_{i2}^{ls}, \ldots, x_{im}^{ls}]$ is the personal best which is the nearest species seed to the *i*th particle. The algorithm in Figure 16.18 was used to update the local species seeds of particles.

16.4.4 Identifying local and global solutions

Equation (16.11) will lead particles to fly to different solutions. The species seed set X_s will consist of all the species seeds and local global and global solutions should be searched from the set X_s [13]. The local global solution, $\mathbf{x}_i^{lg}$, of a particle can be identified from the found global solutions. However, in the beginning of evolution, only a few global solutions can be found by using (16.10), and this may result the particle not to find more solutions. In the aim of keeping diversity of population, a local solution threshold r_f^l ($r_f^l < r_f$) was introduced to force the algorithm to search more space. A species seed ($\mathbf{x}$) is called as a local global solution if

$$f(x) \geq (f_{max} - f_{min}) * r_f^l + f_{min} \tag{16.12}$$

16.4.5 Procedures of the SCPSO

Figure 16.20 illustrates the procedure of the SCPSO.

Compared with a PSO, there are following changes in the SCPSO:

- A procedure is added before updating the positions of particle to explore all the species seeds from the all the particles and their personal best in the current population.
- A procedure is added after the position adjustment to copy the found species to the personal best and to update the local global solutions.
- Finally, after the evolutions of particles, a procedure is used to identify the global solutions from the species set.

```
Set t=0 and Initialize particles G(t);
Evaluate G(t)
while (termination criteria is not met) do
  Determine species seeds Xs;
  Update particle positions
  Conserve species from Xs in G(t+1);
  t=t+1;
end (while)
Determine global solutions
```

Figure 16.20 The structure of SCPSO

16.5 Multimodal testing functions

There are many mathematical multimodal functions proposed in the literature for researchers to test the efficiency and success rate of new developed algorithms in identifying all the multiple solutions. Due to the challenge and difficulty of exploring multiple solutions, there have been several "competitions on niching methods for multimodal optimization" within in IEEE Congress on Evolutionary Computation and Genetic and Evolutionary Computation Conference. The organization committee has collected and produced some popular and very challenging multimodal functions. In this session, all the multimodal functions were taken from the benchmark functions [32].

16.5.1 Simple multimodal functions

Tables 16.1 and 16.2 list eight simple multimodal functions taken from difference references in the literature. Each function has more than one global and local solution. The corresponding one-dimensional or two-dimensional graph is shown to show the nature of functions.

16.5.2 Composition multimodal function

Some very challenge multimodal functions were generated with a combination of several multimodal function by applying several transforming techniques, such as

Table 16.1 Simple multimodal functions

No.	Function	Solutions	1D/2D graph
F1	Five-uneven-peak trap $F_1(x)=\begin{cases}80(2.5-x) & \text{for } 0\le x<2.5,\\ 64(x-2.5) & \text{for } 2.5\le x<5.0,\\ 64(7.5-x) & \text{for } 5.0\le x<7.5,\\ 28(x-7.5) & \text{for } 7.5\le x<12.5,\\ 28(17.5-x) & \text{for } 12.5\le x<17.5,\\ 32(x-17.5) & \text{for } 17.5\le x<22.5\\ 32(27.5-x) & \text{for } 22.5\le x<27.5\\ 80(x-27.5) & \text{for } 27.5\le x<30\end{cases}$	2 global 4 local	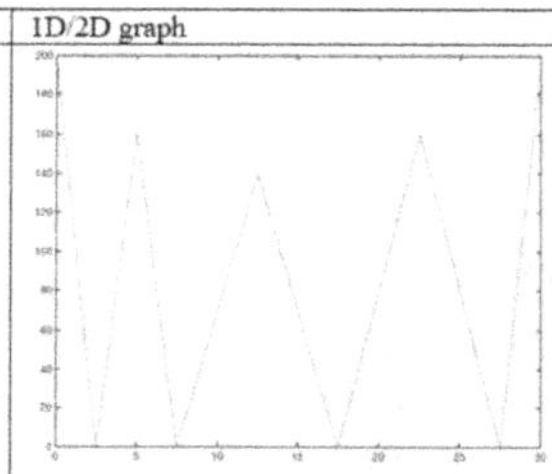
F2	Equal maximum $F_2(x)=\sin^6(5\pi x)$ where $0\le x\le 1$	5 global	
F3	Uneven decreasing maxima $F_3(x)=\exp\left[-2\log(2)\left(\frac{x-0.08}{0.854}\right)^2\right]\cdot\sin^6\left[5\pi\left(x^{3/4}-0.05\right)\right]$ where $0\le x\le 1$	1 global 4 local	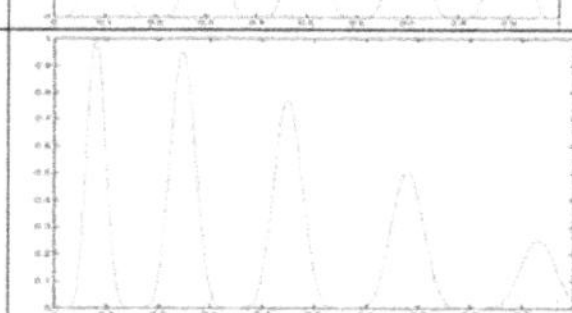
F4	Himmelblau function $F_4(x,y)=200-(x^2+y-11)^2-(x+y^2-7)^2$ where $-6\le x,y\le 6$	4 global	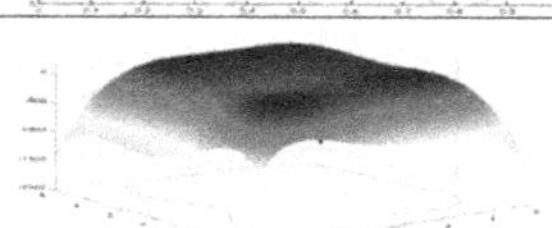

Table 16.2 Simple multimodal functions

No.	Function	Solutions	1D/2D graph
F5	Six-hump camel back $F_5(x,y) = -4\left[\left(4-2.1x^2+x^4/3\right)x^2 - 4\left[xy+\left(4y^2-4\right)\right]y^2\right]$ where $-1.9 \le x \le 1.9$ $-1.1 \le y \le 1.1$	2 global 2 local	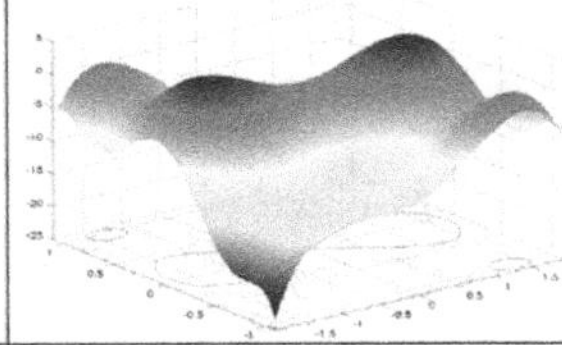
F6	Shubert function $F_6(\mathbf{x}) = -\prod_{i=1}^{D}\sum_{j=1}^{5} j\cdot\cos[(j+1)x_i + j]$ where $-10 \le \mathbf{x} \le 10$	$D\cdot 3^D$ global Many local (for 2D, 760 local)	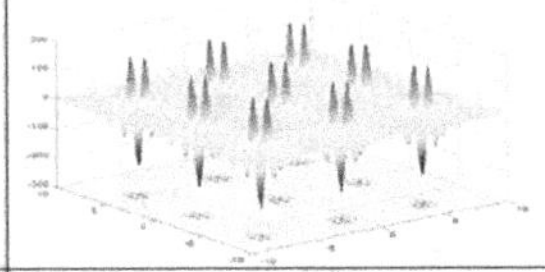
F7	Vincent function $F_7(\mathbf{x}) = \frac{1}{D}\sum_{i=1}^{D}\sin[10\log(x_i)]$ where $0.25 \le \mathbf{x} \le 10$	6^D global	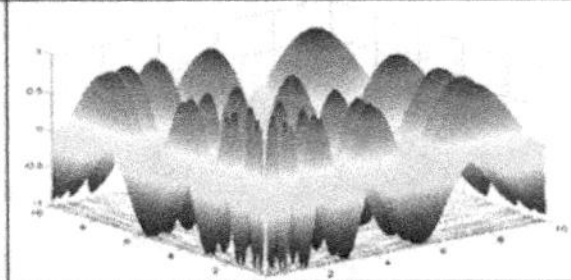
F8	Modified Rastrigin $F_8(\mathbf{x}) = -\sum_{i=1}^{D}[10 + 9\cos(2\pi k_i x_i)]$ where $0. \le \mathbf{x} \le 1$	$\prod_{i=1}^{D} k_i$ global	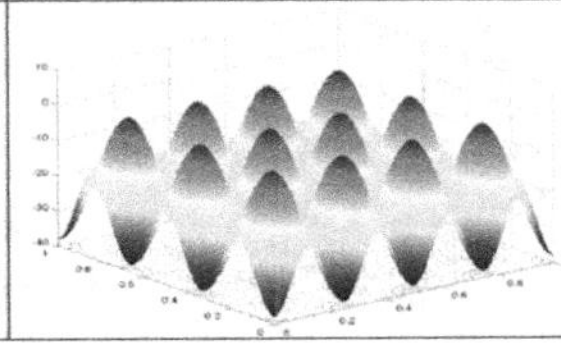

shifting, rotating, stretch/compressing and linear combination. Weight factors are used to combine several multimodal functions into a single function. Mathematically, a composition function CF is defined as

$$CF_j = \sum_{i=1}^{n} \omega_i \left\{ \hat{f}_i\left(\frac{\mathbf{x}-\sigma}{\lambda_i * M_i}\right)\right\} \tag{16.13}$$

where n is the number of basic multimodal functions, $\hat{f}_i(\mathbf{x})$ is a normalization of the ith basic function, σ_i is the new shifted optimum of function $\hat{f}_i(\mathbf{x})$, ω_i is the weight factor for the ith function, λ_i is a parameter used to stretch ($\lambda_i > 1$) or compress ($\lambda_i < 1$) the corresponding function and M_i is the linear transformation/rotation matrix of Function $\hat{f}_i(\mathbf{x})$. The weight factor ω_i was computed as

$$\omega_i = \exp\left(-\frac{1}{2D\sigma_i^2}\sum_{k=1}^{D}(x_k - o_{ik})^2\right) \tag{16.14}$$

$$\omega_i = \begin{cases} \omega_i & \omega_i = \max(\omega_i) \\ \omega_i(1-\omega_i^{10}) & \text{otherwise} \end{cases} \tag{16.15}$$

Table 16.3 Basic function

Sphere function	Grienwank's function
$f_S(\mathbf{x}) = \sum_{i=1}^{D} x_i^2$	$f_G(\mathbf{x}) = \sum_{i=1}^{D} \frac{x_i^2}{4{,}000} - \prod_{i=1}^{D} \cos\left(\frac{x_i}{\sqrt{i}}\right) + 1$
Rastrigin's function	Weierstrass function
$f_R(\mathbf{x}) = \sum_{i=1}^{D} \left[x_i^2 - 10\cos(2\pi x_i) + 10\right]$	$f_W(\mathbf{x}) = \sum_{i=1}^{D} \sum_{k=0}^{k_{max}} \alpha^k \cos[2\pi\beta^k(x_i + 0.5)] - D\sum_{k=0}^{k_{max}} \alpha^k \cos[2\pi\beta^k(0.5)]$ where $\alpha = 0.5, \beta = 3, k_{max} = 20$
Expanded Griewank's plus Rosenbrock's function (EF8F2) $EF8F2(\mathbf{x}) = F8F2(\mathbf{x}) = \mathrm{F8}(\mathrm{F2}(x_1, x_2)) + \mathrm{F8}(\mathrm{F2}(x_2, x_3)) + \mathrm{F8}(\mathrm{F2}(x_{D-1}, x_D)) + \mathrm{F8}(\mathrm{F2}(x_D, x_1))$	

The weight factor will be normalized as $\omega_i = \omega_i / \sum \omega_i$. The parameter σ_i is used to control the coverage range of each function.

Five basic functions used to create composition multimodal functions are shown in Table 16.3. The composition functions with their properties and graph of a 2D corresponding function are listed and shown in Table 16.4.

16.6 Engineering applications

16.6.1 Example 1: design of torsion rods

The design of a thin-walled cantilever torsional rod shown in Figure 16.21 aims to minimize its weight under the constraint of sinusoidal excitation [33]. Li and Wood [3] demonstrated that this is a multimodal optimization problem.

The decision variables were defined as the wall thicknesses of two elements. Without loss the generality, both elements were made from the same material; therefore, the weight of the structure will depend on the total volume, which can easily calculated as

$$w(t_1, t_2) = 720\pi(t_1^2 + t_2^2) \tag{16.16}$$

There are four constraints, two of them are related to stress requirements, in this problem:

$$g_i(t_1, t_2) = \left[\left(\frac{S_i}{S_{max}}\right)^2 - 1\right] \leq 0, \quad i = 1, 2 \tag{16.17}$$

$$t_i \geq 0.04 \text{ in.}, \quad i = 1, 2, \tag{16.18}$$

Table 16.4 Multimodal composition functions

No.	Basic functions	Parameter	2D graph
F9	*f1–f2* : Grienwank's function, *f3–f4* : Weierstrass function *f5–f6* : Sphere function	$\sigma = [1,1,1,1,1,1]$ $\lambda = [1,2,8,8,0.2,0.2]$ M_i are identity matrices	
F10	*f1–f2* : Rastrigin's function, *f3–f4* : Weierstrass function *f5–f6* : Grienwank's function, *f7–f8* : Sphere function,	$\sigma = [1,1,1,1,1,1,1,1]$ $\lambda = [1,1,10,10,0.1,0.1,1/7,1/7]$ M_i are identity matrices	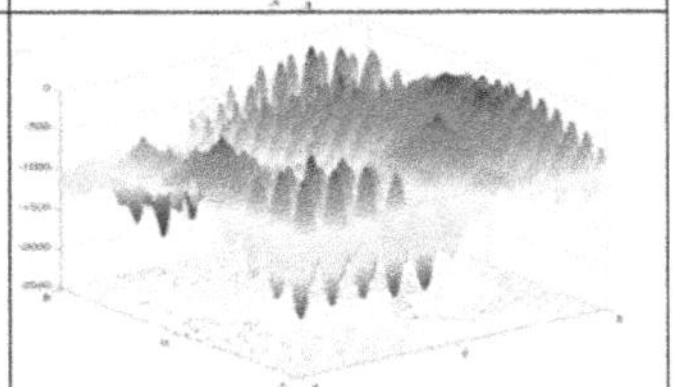
F11	*f1–f2* : EF8F2 function, *f3–f4* : Weierstrass function *f5–f6* : Griewank's function	$\sigma = [1,1,2,2,2,2]$ $\lambda = [0.25,0.1,2,1,2,5]$ M_i are the different linear transformation matrices with condition number one	
F12	*f1–f2* : Rastrigin's function, *f3–f4* : EF8F2 function *f5–f6* : EF8F2 function *f7–f8* : Griewank's function	$\sigma = [1,1,1,1,1,2,2,2]$ $\lambda = [4,1,4,1,0.1,0.2,0.1,0.025]$ M_i are the different linear transformation matrices with condition number one	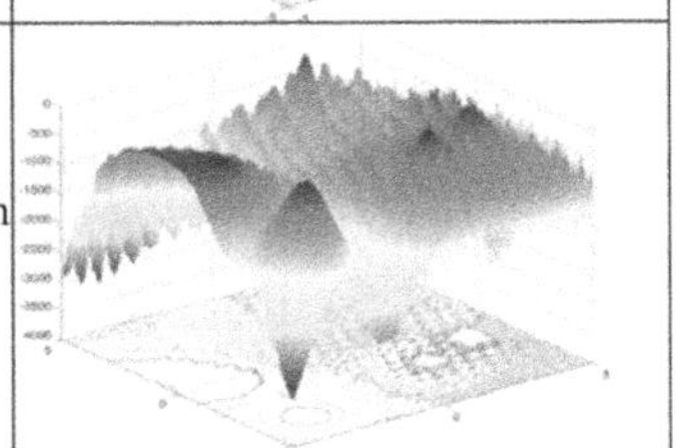

in which

$$\frac{S_1}{S_{max}} = 3T_n t_2 \left(1 - \frac{\lambda_e}{\mu}\right) \tag{16.19}$$

$$\frac{S_2}{S_{max}} = \frac{T_n[3\lambda_e t_2 + t_1(1 - 2\lambda_e)]}{\mu} \tag{16.20}$$

$$\mu = t_2[t_1(1 - 2\lambda_e)^2 - t_2(6\lambda_e - 3\lambda_e^2)] \tag{16.21}$$

where λ_e is the non-dimensional driving frequency with a value of 1/6, t_i is the thickness of the ith rod and T_n is the amplitude of the sinusoidal torsional load, with a value of 0.0085.

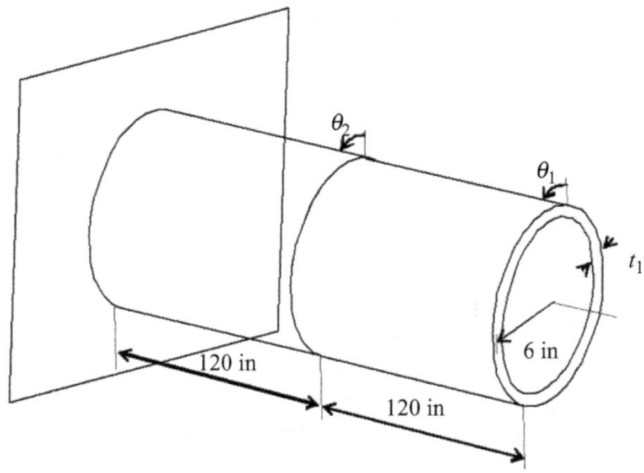

Figure 16.21 Torsion rod under sinusoidal excitation

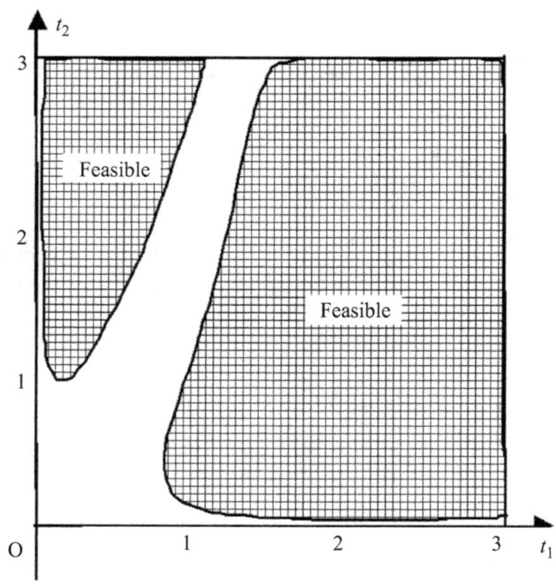

Figure 16.22 Disjoint design space for torsion-rod problem

Figure 16.22 illustrates the feasible region of this problem. It is clearly shown that the feasible region was split into two separate regions.

Using the parameter values $N = 50$ and $r_s = 0.15$, after 100 generation the SCPSO found four species that are listed in Table 16.5.

Table 16.5 Species for the torsional rod design problem

No.	Variable	Objective
1	(0.0548, 0.896)	1,830.031
2	(0.808, 0.235)	1,790.049
3	(0.819, 1.757)	13,334.561
4	(1.848, 0.474)	34,271.720

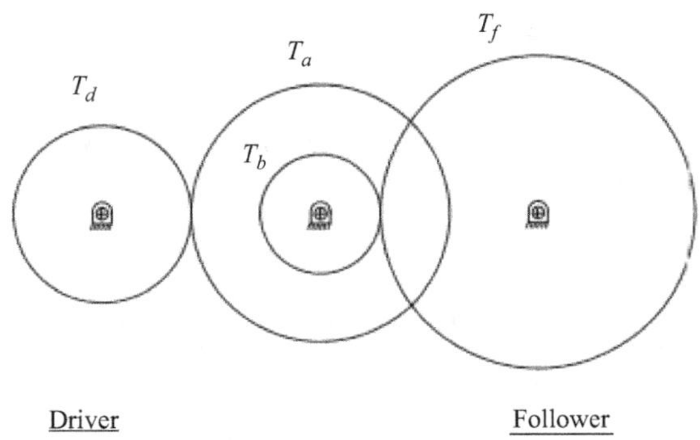

Figure 16.23 A compound gear train

16.6.2 Example 2: gear train design

A compound gear train [34] is shown in Figure 16.23. There are four gears in the system. The left one is called the driver, and the right one is the follower. The optimization design of this gear train is to find the teeth numbers of the four gears so that the error between the gear ratio and a required ration (1/6.931) [35] is minimized.

The decision variables are the teeth of gears and denoted as $\mathbf{x} = (x_1, x_2, x_3, x_4) = (T_d, T_b, T_a, T_f)$, and they must be integer numbers. The optimization problem of the compound gear train design can be described mathematically as

$$\min_{\mathbf{x}} f(\mathbf{x}) = \left[\frac{1}{6.931} - \frac{x_1 x_2}{x_3 x_4}\right]^2 \tag{16.22}$$

$$s.t.\ 12 \le x_i \le 60, \quad \text{all } x_i \text{ are integers.}$$

Using the parameter values $N = 50$ and $R_s = 1$, after 100 generation the SCPSO found 49 species and 8 of them are listed in Table 16.6. It has demonstrated that there are many solutions for this problem.

Table 16.7 lists the results obtained by different authors. It is clear that the SCPSO can find much better solution than the previous methods in the literature. The reason is that in the SCPSO, a species-conserving technique has been used to maintain the

Table 16.6 Species for the compound gear train design

No.	Variable	Objective
1	(16, 19, 43, 49)	2.7×10^{-12}
2	(12, 18, 30, 50)	7.8022×10^{-8}
3	(12, 20, 39, 43)	1.3610×10^{-6}
4	(20, 18, 41, 60)	4.2524×10^{-6}
5	(17, 19, 43, 53)	6.5050×10^{-6}
6	(15, 26, 52, 53)	7.6723×10^{-6}
7	(15, 17, 42, 43)	9.5068×10^{-6}
8	(14, 31, 58, 53)	9.5802×10^{-6}

Table 16.7 Comparison for the compound gear train design against different methods: Deb and Goyal [34], Kannan and Kramer [35], Sandgren [36]

Variables	Deb and Goyal	Kannan and Kramer	Sandgren	SCPSO (this paper)
x_1	17	13	18	16
x_2	14	15	22	19
x_3	33	33	45	43
x_4	50	41	60	49
$f(x)$	1.362×10^{-9}	2.146×10^{-8}	5.712×10^{-6}	2.7×10^{-12}
Gear ratio	0.144242	0.144124	0.146667	0.14428
Error	0.026%	0.11%	1.65%	0.0011%

diversity of the population. It can not only explore possible global and local solutions but also find better solutions than other techniques.

16.6.3 Example 3: topology of 15-members, 6-node truss

Truss optimization is commonly used in the literature as benchmark questions in engineering optimization. Truss optimization is a very challenging optimization problem, because (1) the calculation of objective functions is implemented through a finite element analysis (FEA), which is time-consumption, especially for large scales of elements; (2) there are many constraints, such as (1) stresses in each element must be within the compression and tensile strengths, or the structure will fail. The number of stress constraints is two times of the number of elements. (2) Some truss structure may have displacements and buckling requirements [37].

Usually, truss optimal designs can be grouped into three categories: (1) sizing optimization, in which the cross areas of bars in the structure are selected as decision variables. The node positions are fixed. (2) Configuration optimization, in which the coordinators of nodes are decision variables. (3) Topology optimization, in which the connections of element are decision variables. Many researches used the combination of the above three types of truss optimizations. In this session, a truss topology, in which the cross areas of existing bars and connections of nodes are decision variables, is selected as an example, because Li [5] demonstrated that it is a multimodal optimization problem.

16.6.3.1 Problem description

Figure 16.24 illustrates a 15-member and six-node truss structure with the loadings. This is a ground structure, which shows all 15 possible connections of nodes. In order to show clearly the connections of nodes, the overlapping members are illustrated with a gap, for an example, element 15 is overlap to element 13 and element 14. The design parameters are listed in Table 16.8. All the cross areas of 15 elements are design variables. There are 60 constraints in this example except of the boundaries of variables:

- 15 stress should be less than the corresponding tensile strengths
- 15 stress should be bigger than the corresponding compress strengths
- 15 x and y components of should be within allowable limits

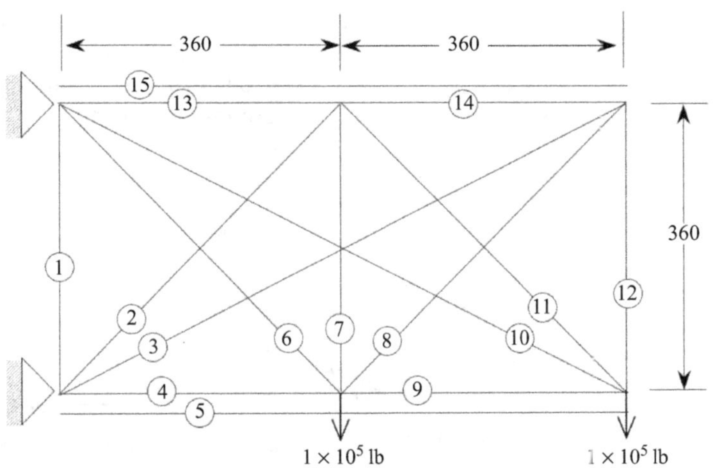

Figure 16.24 The 15-bar and six-node ground structure

Table 16.8 Parameters for Example 3

Young's modulus	104 ksi
Density	0.1 lb/in^3
Allowable compressive strength	25 ksi
Allowable tensile strength	25 ksi
Allowable displacement	2 in
Limits of cross-area	[−35, 35] in

There are several ways to model a truss topology optimization problem. In this session, Deb and Gulati's method [38] is applied to describe the topology of truss structures. A solution/truss was described as a real vector of cross-areas. The presence and absence of a bar is identified by using a critical area, ε, which is a small and positive number. If the area of a bar is bigger than the pre-defined critical area, this element is presence, or it is absence.

Element i is absence in a truss, if its cross-sectional area is less than the critical area:

$$A_i < \varepsilon \tag{16.23}$$

or element i exists if its cross-sectional area is equal or bigger than the critical area:

$$A_i \geq \varepsilon \tag{16.24}$$

In this example, the lower and upper limits of bars are designed as $A_i^{min} = -35$ m and $A_i^{max} = 35$ in., and the critical area is set as $\varepsilon = 0.01$ in. A cross area of an element will be randomly generated between -35 and 35 in. When the value of a bar's cross area is less than 0.01, the corresponding element is absent. A node will be removed from the model if there is no any element connecting to it. This method is easily implemented; however, it will introduce some disadvantage, such as some important nodes may disappear. For examples, some nodes with boundary conditions and loads should be not excluded, and a truss structure may become a machine in which some elements can move or rotate. In order to deal with such complex cases, a FEA model, illustrated in Figure 16.25, had been proposed to identify the existence

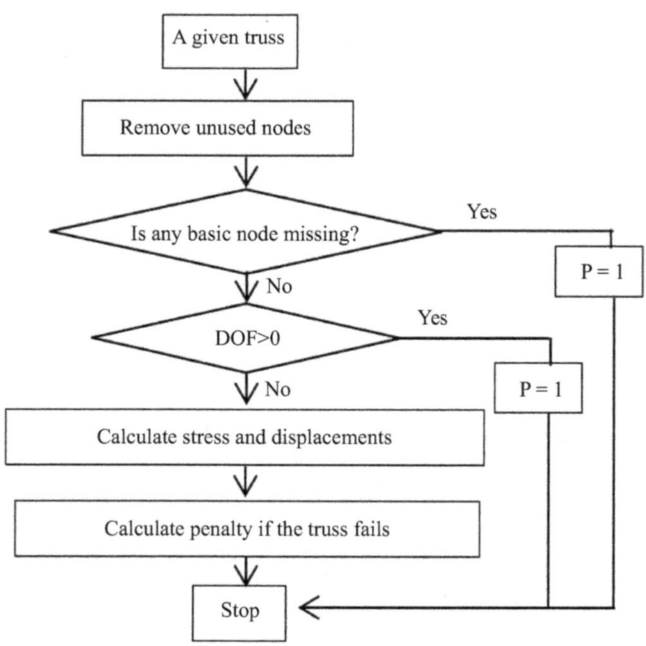

Figure 16.25 The FEA process

Table 16.9 Results of Example 3

Element	Deb and Gulati	Luh and Lin	SCGA				SCPSO	
No.			Solution 1	Solution 2	Solution 3	Solution 4	Solution 1	Solution 2
2	20.310	20.549	20.27712	21.92580	25.46163	30.24313	20.42564	26.98659
4	5.219	5.428	5.29368	16.33877	21.97115		5.03118	7.48750
5	14.593	14.308	14.33871	6.90382		20.84183	14.82824	16.58347
6	7.772	7.617	7.71425	6.07553	6.08882	0.57217	7.93105	9.79421
7						3.78584		
9				7.43428	10.37983	0.43245		0.08704
10						3.60725	21.87267	25.00552
11	20.650	20.265	21.02456	21.39010	21.93803	21.75424	26.16163	26.29662
13	28.187	28.876	28.15770	28.36948	31.20734	34.99946		
15								
Weight (lb)	4734.34039	4730.82397	4732.12081	4888.80729	5011.28834	5879.23281	4747.84	5562.89
Configuration	Figure 16.26(a)	Figure 16.26(a)	Figure 16.26(a)	Figure 16.26(b)	Figure 16.26(c)	Figure 16.26(d)	Figure 16.26(a)	Figure 16.26(e)

of basic nodes, to calculate the "degree-of-freedom" of a structure and to compute the element stresses and node displacements satisfying the given loadings and support constraints.

16.6.3.2 Optimal solutions

Table 16.9 illustrated the optimal truss found by the SCGA and SCPSO. The SCGA [13] have found four solutions on a single run, while the SCPSO found two solutions. The results are listed in Table 16.9, and the configurations of solutions are shown in Figure 16.24. The best truss (Solution 1) has the same configuration as the truss achieved by Deb and Gulati [38] and Luh and Lin [39]. Only six members in Figure 16.26 are included in the solution. Table 16.9 also demonstrated that the SCGA found more solutions than the SCPSO. The possible reason is that the GA has a mutation operator and therefore is better than the PSO in maintaining the population diversity, it got more chances to find solutions. This is also the reason that the best solution obtained by using the SCGA is better than the one from the SCPSO.

16.7 Conclusion

This chapter has briefly described the niching techniques that are used in the combination with an EA to improve the diversity of populations to achieve the aims of identifying multiple global/local solutions of multimodal problems. There are several niching techniques in the literature: crowding, fitness sharing/niche, clustering and species. The related multimodal PSOs had been viewed and discussed, especially the SCPSO was stated in detail, because it was used to solve three engineering

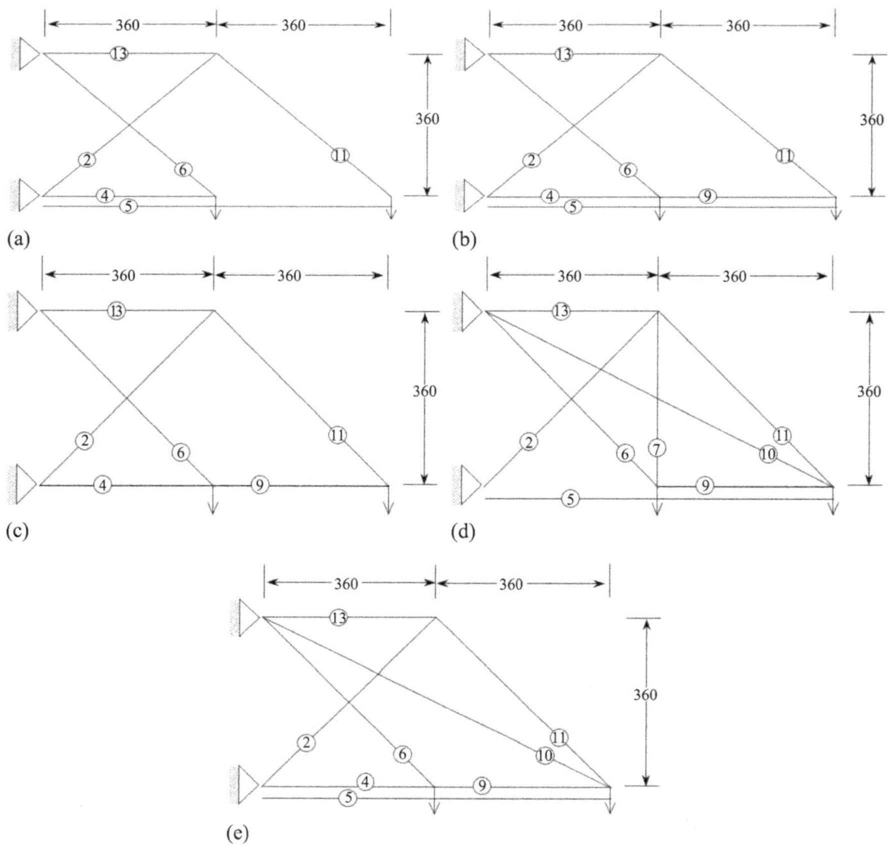

Figure 16.26 Results of truss 15-members, six-node truss

applications: design of torsion rods, design of gear train and topology of 15 members six-node truss. The numerical results demonstrated that the SCPSO can obtain multiple global/local solutions of the tested engineering design problems in a single run. Especially, the SCPSO can solve the topologies of truss structures and find different configuration with different cross area.

It should be very interesting to compare the efficiency and effectiveness of niche techniques on searching multiple solutions of multimodal problems. In the future, we are planning to design a universal platform by implementing the important PSOs for multiple solutions to compare them through solving some mathematical benchmark problems and some real engineering design problems.

References

[1] James Kennedy and Russ Eberhart. Particle swarm optimization. In *International Conference on Neural Networks*, volume 4, pages 1942–1948, 1995.

[2] Julio Barrera and Carlos A Coello Coello. A review of particle swarm optimization methods used for multimodal optimization. In *Innovations in Swarm Intelligence*, pages 9–37. Springer, 2009.
[3] Jian-Ping Li and Alastair S Wood. An adaptive species conservation genetic algorithm for multimodal optimization. *International Journal for Numerical Methods in Engineering*, 79(13):1633–1661, 2009.
[4] Min Chen, Shu-Ching Chen, Mei-Ling Shyu, and Kasun Wickramaratna. Semantic event detection via multimodal data mining. *IEEE Signal Processing Magazine*, 23(2):38–46, 2006.
[5] Jian-Ping Li. Truss topology optimization using an improved species-conserving genetic algorithm. *Engineering Optimization*, 47(1):107–128, 2015.
[6] A Naitali and F Giri. Hammerstein and wiener nonlinear models identification using a multimodal particle swarm optimizer. In *American Control Conference, 2006*. IEEE, 2006.
[7] Siegfried Nijssen and Thomas Back. An analysis of the behavior of simplified evolutionary algorithms on trap functions. *IEEE Transactions on Evolutionary Computation*, 7(1):11–22, 2003.
[8] KA De Jong. An analysis of the behavior of a class of genetic adaptive systems. PhD thesis, University of Michigan, 1975.
[9] Samir W Mahfoud. Crowding and preselection revisited. *Urbana*, 51:61801, 1992.
[10] JH Holland. Adaptation in Natural and Artificial Systems. Cambridge, MA: MIT Press, 1975.
[11] David E Goldberg, Jon Richardson. Genetic algorithms with sharing for multimodal function optimization. In *Proceedings of the Second International Conference on Genetic Algorithms on Genetic Algorithms and their Application*, pages 41–49. Hillsdale, NJ: Lawrence Erlbaum, 1987.
[12] James Kennedy. Stereotyping: Improving particle swarm performance with cluster analysis. In *Evolutionary Computation, 2000. Proceedings of the 2000 Congress on*, volume 2, pages 1507–1512. IEEE, 2000.
[13] Jian-Ping Li, Marton E Balazs, Geoffrey T Parks, and P John Clarkson. A species conserving genetic algorithm for multimodal function optimization. *Evolutionary Computation*, 10(3):207–234, 2002.
[14] Stephen Chen and James Montgomery. A simple strategy to maintain diversity and reduce crowding in particle swarm optimization. In *Australasian Joint Conference on Artificial Intelligence*, pages 281–290. Springer, 2011.
[15] Riaan Brits, Andries P Engelbrecht, and F Van den Bergh. A niching particle swarm optimizer. In *Proceedings of the 4th Asia-Pacific Conference on Simulated Evolution and Learning*, volume 2, pages 692–696. Singapore: Orchid Country Club, 2002.
[16] Shu-Kai S Fan and Ju-Ming Chang. Dynamic multi-swarm particle swarm optimizer using parallel PC cluster systems for global optimization of large-scale multimodal functions. *Engineering Optimization*, 42(5):431–451, 2010.

[17] Jun Zhang, De-Shuang Huang, and Kun-Hong Liu. Multi-sub-swarm particle swarm optimization algorithm for multimodal function optimization. In *Evolutionary Computation, 2007. CEC 2007. IEEE Congress on*, pages 3215–3220. IEEE, 2007.

[18] Özgür Yeniay. Penalty function methods for constrained optimization with genetic algorithms. *Mathematical and Computational Applications*, 10(1):45–56, 2005.

[19] Jun Zhang, De-Shuang Huang, Tat-Ming Lok, and Michael R Lyu. A novel adaptive sequential niche technique for multimodal function optimization. *Neurocomputing*, 69(16):2396–2401, 2006.

[20] Stefan Bird and Xiaodong Li. Adaptively choosing niching parameters in a pso. In *Proceedings of the 8th Annual Conference on Genetic and Evolutionary Computation*, pages 3–10. ACM, 2006.

[21] Wei-Der Chang. A modified particle swarm optimization with multiple subpopulations for multimodal function optimization problems. *Applied Soft Computing*, 33:170–182, 2015.

[22] Alessandro Passaro and Antonina Starita. Particle swarm optimization for multimodal functions: a clustering approach. *Journal of Artificial Evolution and Applications*, 2008:1–15, 2008.

[23] Xiaohui Yuan, Jing Peng, and Yasumasa Nishiura. Particle swarm optimization with multiscale searching method. In *International Conference on Computational and Information Science*, pages 669–674. Springer, 2005.

[24] Ian C Parmee. A review of the development and application of cluster oriented genetic algorithms. In *IUTAM Symposium on Evolutionary Methods in Mechanics*, pages 331–340. Springer, 2004.

[25] Jie Yao, Nawwaf Kharma, and Yu Qing Zhu. On clustering in evolutionary computation. In *Evolutionary Computation, 2006. CEC 2006. IEEE Congress on*, pages 1752–1759. IEEE, 2006.

[26] Qiang Hua, Machine Learning Center, Bin Wu, and Hao Tian. A detecting peak's number technique for multimodal function optimization. *Relation*, 11(21):12–22, 2008.

[27] Masao Iwamatsu. Multi-species particle swarm optimizer for multimodal function optimization. *IEICE Transactions on Information and Systems*, 89(3):1181–1187, 2006.

[28] Jian-Ping Li and Alastair S Wood. Species-conserving particle swarm optimisation for multimodal functions. *International Journal of Modelling, Identification and Control*, 8(4):290–300, 2009.

[29] David H Wolpert and William G Macready. No free lunch theorems for optimization. *IEEE Transactions on Evolutionary Computation*, 1(1):67–82, 1997.

[30] Millie Pant, Radha Thangaraj, Crina Grosan, and Ajith Abraham. Hybrid differential evolution-particle swarm optimization algorithm for solving global optimization problems. In *Digital Information Management, 2008. ICDIM 2008. Third International Conference on*, pages 18–24. IEEE, 2008.

[31] PS Shelokar, Patrick Siarry, Valadi K Jayaraman, and Bhaskar D Kulkarni. Particle swarm and ant colony algorithms hybridized for improved continuous optimization. *Applied Mathematics and Computation*, 188(1):129–142, 2007.

[32] Xiaodong Li, Andries Engelbrecht, and Michael G Epitropakis. Benchmark functions for CEC2013 special session and competition on niching methods for multimodal function optimization. *RMIT University, Evolutionary Computation and Machine Learning Group, Australia, Tech. Rep*, 2013.

[33] CY Lin and WT Chen. Stochastic multistage algorithms for multimodal structural optimization. *Computers & Structures*, 74(2):233–241, 2000.

[34] Kalyanmoy Deb and Mayank Goyal. A combined genetic adaptive search (geneas) for engineering design. *Computer Science and informatics*, 26:30–45, 1996.

[35] Steven N. Kramer. An augmented lagrange multiplier based method for mixed integer discrete continuous optimization and its applications to mechanical design. *Journal of Mechanical Design*, 116:405, 1994.

[36] Eric Sandgren. Nonlinear integer and discrete programming in mechanical design optimization. *Journal of Mechanical Design*, 112(2):223–229, 1990.

[37] George IN Rozvany. Difficulties in truss topology optimization with stress, local buckling and system stability constraints. *Structural Optimization*, 11(3–4):213–217, 1996.

[38] Kalyanmoy Deb and Surendra Gulati. Design of truss-structures for minimum weight using genetic algorithms. *Finite Elements in Analysis and Design*, 37(5):447–465, 2001.

[39] Guan-Chun Luh and Chun-Yi Lin. Optimal design of truss structures using ant algorithm. *Structural and Multidisciplinary Optimization*, 36(4):365–379, 2008.

Chapter 17
Decentralized control in robotic swarms

Hamed Rezaee and Farzaneh Abdollahi[1]

Abstract

Robotic swarms are groups of autonomous and self-organized mobile robots interacting under a communication network in order to accomplish cooperative tasks. This chapter will provide an overview on main outcomes on decentralized control in robotic swarms based on applications and the degree of autonomy in the view of stability and control theory. Based on application, we will categorize various problems in the area of robotic swarms such as swarming, formation, and consensus/rendezvous. First path planning and collision avoidance approaches for mobile robots will be introduced. Then, swarming and swarm formation of mobile robots as most important problems in control of multirobot systems will be investigated. Finally, the consensus and rendezvous as important problems in multirobot systems will be studied. In each category, existing structures will be classified and explained.

17.1 Introduction to decentralized control in robotic swarms

Control of autonomous groups of mobile robots such as unmanned air vehicles [1,2], unmanned ground vehicles [3,4], unmanned underwater vehicles [5,6], and unmanned surface vessels [7,8] so called *swarms* has been an important topic of research in the robotics and control communities over the past decade. The main idea of using these robots is to operate autonomously without human interaction in a lot of civilian applications such as load transportation, search and rescue missions, maintenance, fire and forest monitoring, atmospheric studies, and so on. Motivated by recent developments in communication networks, electronics, sensors technology, computing science, and control theory, control of multirobot systems recently has gained a lot of interests among researchers and engineers. These systems provide us opportunities to use simple, small, and cheap systems instead of employing a sophisticated and large single system, and they are more reliable, robust, flexible, and precise for doing tasks [2,9,10].

[1]Department of Electrical Engineering, Amirkabir University of Technology, Iran

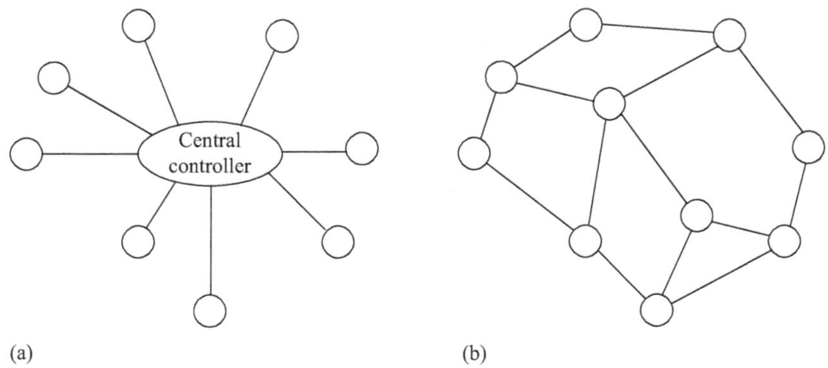

Figure 17.1 Schematics of (a) the centralized and (b) the decentralized control

To accomplish a cooperative task in a robotic swarm, interaction control laws are required. This control law can be designed and employed through a *central* computer/controller. The central controller receives the states of all the robots; then, considering all the robots as an integrated system, control signals are provided [Figure 17.1(a)]. The ability to override the system in emergency conditions and having access to global information in order to guarantee the network stability are the merits of a centralized control approach. However, the possible malfunction of the central controller affects the performance of the whole swarm, and the central controller is the single point of failure in the system. Moreover, this class of networked control systems requires high communication bandwidths, and in long maneuvers, the robots may have problems to communicate with the central controller, cohesively. To cope with the mentioned problems existing with a central controller, *decentralized* control strategies have been introduced. Under these strategies, each robot provides its own control signal locally based on the state information of other robots achieved directly through a communication network or via its own sensors [Figure 17.1(b)]. Hence, the problems associated with a central controller can be avoided [11–13]. Indeed, since the centralized control of a swarm is similar to a classic control of a multi-input multi-output system, the main objective in study of robotic swarms is decentralized control of robots when only local information is available.

In the next section, the related problems in decentralized control of robotic swarms will be introduced and the content of the chapter will be explained. In Section 17.3, preliminaries are presented. Then, in Section 17.4, we introduce motion control and path planning approaches in mobile robots. Section 17.5 is devoted to swarming problems in mobile robots. Swarming of robots considering geometric formations is studied in Section 17.6. In Section 17.7, the consensus and rendezvous problems in networked robots are investigated, and Section 17.8 concludes the chapter.

17.2 Related problems

Decentralized control in robotic swarms can be studied in various perspectives. Self-organized networks of autonomous mobile robots while keeping cohesion and collision avoidance are one of typical behaviors in decentralized control of multirobot systems called *swarming* or *flocking*. This problem is inspired by aggregation of living organisms in groups such as swarming of ants, flocking of birds, and schooling of fishes to increase the chance of finding foods or avoiding predators which recently has found a lot of engineering applications in mobile sensor networks, rescue robots, etc. Keeping cohesion with a common heading and velocity while avoiding collision (with robots and environmental obstacles) are important factors in the swarming of robots [11,14–17]. Thus, in a swarm of robots, each robot has a multi-objective behavior for weighting some difference control objectives such as target tracking, keeping cohesion, collision avoidance, and obstacle avoidance.

Since mathematical analysis of swarming (in natural living organisms) is complicated, by keeping robots in desired relative positions in specific *formations*, while satisfying the main objectives of a swarm, mathematical analysis of the robots behaviors can be simplified significantly. In other words, by decreasing the degree of freedom of the robots in a specific formation, mathematical analysis of the swarm will be more simple than the case when the robots move in a nonpredefined formation. Indeed, in a team of mobile robots, by keeping a geometric pattern as a formation, cohesion and collision avoidance, as important factors in swarming, can be guaranteed more easily and precisely [18–22]. In addition to the mentioned advantages, by keeping robots in desired formations, optimal and precise accomplishing of cooperative tasks such as coverage and exploration should be realized. Thus, maintaining relative positions among robots in desired formations is a challenging problem in swarms of autonomous robots.

Before or during each cooperative mission, it may be necessary for robots to achieve a common position/configuration or converge to a mission zone so-called *rendezvous*. Thus, one of important cooperative control problems in robotic swarms is rendezvous defined as meeting among some mobile robots at a specific place. The main objective of the rendezvous problem is convergence of robots toward a common configuration by sharing relative positions information. For instance, before a cooperative task among a team of aerial vehicles, first they may be required to converge to a common altitude [23–25]. Moreover, in addition to the rendezvous position, heading angles, velocities, etc. are other variables of interest that may be required to be identical in a swarm of robots. In these cases, the robots should achieve *consensus* on those quantities of interests via an interaction protocol [26,27].

Considering the mentioned main important problems in control of robotic swarms, this chapter will provide an overview on main outcomes in this area. The main objective of the chapter is to provide a classification and analysis of robotic swarms based on applications and the degree of autonomy in the view of stability and control theory. We introduce main important problems in this area, such as swarming, formation, and rendezvous/consensus, and explain stability challenges existing in each problem. First motion control of a single robot will be investigated. Obstacle avoidance of mobile robots based on potential fields, moving obstacles avoidance, obstacle avoidance in swarming robots, etc. are problems will be covered in this section. Then, popular techniques for swarming control of mobile robots will be investigated, and main architectures for swarm formation control of mobile robots including leader–follower and leaderless will be introduced. Finally, achieving consensus and rendezvous among a team of mobile robots will be studied. Achieving rendezvous based on consensus strategies, rendezvous considering connectivity maintenance, and rendezvous considering robots nonholonomic constraints are important problems will be studied.

Note 17.2.1 *Throughout the chapter, we assume that the robots move on a plane, and extension of the results to the three-dimensional case is straightforward.*

17.3 Preliminaries

In this section, notations and some basic concepts on graph theory are summarized.

17.3.1 Notations

Let **1** express a vector of ones with compatible dimension. $\mathbb{R}^{n\times m}$ expresses the set of $n \times m$ real matrices, and $\mathbb{R}_+$ presents the set of positive real numbers. $\mathbb{N}$ denotes the set of natural numbers. x_0 denotes the initial value of a variable x. ∇ is the gradient operator. $|\cdot|$ denotes the absolute value. $\|\cdot\|$ denotes the standard Euclidian norm. lim stands for limit. $\bigcap$ represents intersection, and $\prod$ indicates the product.

17.3.2 Graph theory

Communication among robots is expressed by a communication graph or network $\mathscr{G} = (\mathscr{V}, \mathscr{E}, \mathscr{A})$ where $\mathscr{V} = \{1, 2, \ldots, N\}$ is the set of N nodes expressing the robots, and $\mathscr{E} = \{(i,j)|i \neq j, i,j \in \mathscr{V}\}$ is the set of edges expressing communication links among the robots which an edge (i,j) (an arrow from Node i to Node j) means that the jth robot called *child* receives information from the ith robot as its *parent*. In this condition, $\mathscr{A} = [a_{ij}] \in \mathbb{R}^{N\times N}$ is defined as the adjacency matrix of the graph such that $a_{ij} \in \mathbb{R}_+$ if $(j, i) \in \mathscr{E}, i \neq j$, and $a_{ij} = 0$ otherwise. Accordingly, $\mathscr{N}_i$ is the neighboring set of the ith robot defined as a set of robots which the ith robot receives information from. Moreover, the graph is undirected if $a_{ij} = a_{ji}$, otherwise it is directed. A directed graph and an undirected graph are depicted in Figure 17.2.

A *directed tree* is a subgraph of a directed graph in which each node has exactly one parent except one node as the root of the tree which has no parents and has a directed path to each other node of the graph. In this condition, a directed *spanning tree* of a directed graph is a directed tree containing all the nodes. A directed (undirected) graph is called *strongly connected (connected)* if any two nodes of the graph are connected via a sequence of edges.

Furthermore, a set of directed graphs with the node set $\mathscr{V}$ has a *joint spanning tree* if their union has a directed spanning tree while none of them may have a spanning tree. Accordingly, a set of undirected graphs with the node set $\mathscr{V}$ is jointly connected if their union is connected [28].

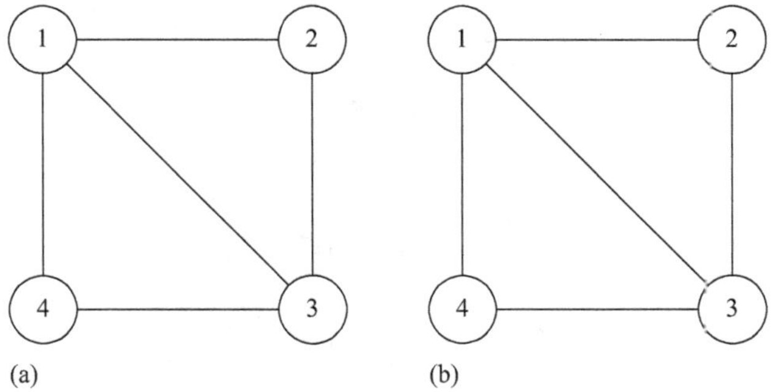

Figure 17.2 (a) A directed graph and (b) an undirected graph

Now, we define a Laplacian matrix $\mathscr{L} = [\ell_{ij}] \in \mathbb{R}^{N\times N}$ associated with $\mathscr{G}$ as follows:

$$\ell_{ij} = \begin{cases} \sum_{i=1,i\neq j}^{N} a_{ij} & i=j, \\ -a_{ij} & i \neq j \end{cases}$$

whose rows entries summations are zero. If $\mathscr{G}$ has a directed spanning tree (or if it is connected in undirected cases), $\mathscr{L}$ has a simple zero eigenvalue while other eigenvalues of $\mathscr{L}$ lie in the open right half plane. Moreover, the right and left eigenvectors associated with the single zero eigenvalue of $\mathscr{L}$ are $\mathbf{1}$ and $\xi \in \mathbb{R}^N$, respectively, where $\xi^\top \mathbf{1} = 1$ [28]. Moreover, a directed graph is called *balanced* if all of its nodes are balanced. In other words,

$$\forall i \in \mathscr{V}, \sum_{j=1,j\neq i}^{N} a_{ij} = \sum_{j=1,j\neq i}^{N} a_{ji}.$$

In this condition, $\mathbf{1}^\top \mathscr{L} = \mathbf{0}$ as well as $\mathscr{L}\mathbf{1} = \mathbf{0}$, and hence one can get $\xi = \mathbf{1}/N$[29]. Note that an undirected graph is a special form of a balanced directed graph.

17.4 Robots path planning

Before studying control of a swarm of robots, motion control of a single robot should be studied. This issue includes control of a robot toward a target while avoiding collision with obstacles and other robots. Several techniques have been introduced in the literature for path planning of mobile robots including bug algorithm [30], grid-based [31], sampling-based [32], and so on. The main problem of the mentioned approaches is that they are no efficient for online detection and collision avoidance which is an important issue for collision avoidance in robotic swarms. Using vector fields is the most popular approach for mobile robots path planning. This issue is due to ease of elegant mathematical analysis and simplicity in understand and implementation. In other words, this approach permits to online detect and avoid unknown

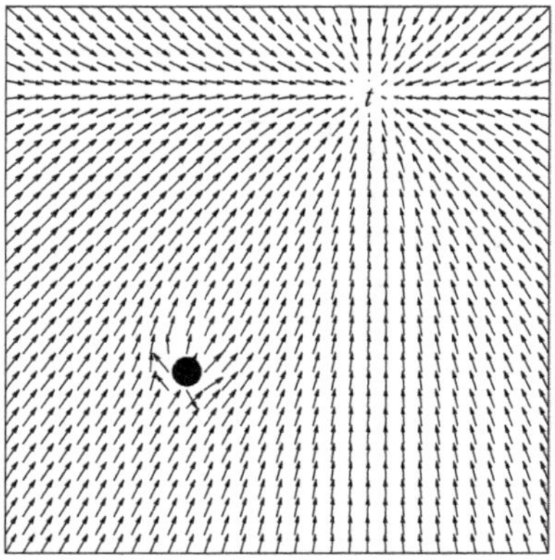

Figure 17.3 A vector field leading a robot toward a desired target (denoted by t) in a collision-free path

obstacles while seeking a desired target or trajectory. The main idea in this approach is to fill a robot working frame with artificial vectors where the robot is attracted toward its desired target/trajectory and it is repulsed from obstacles [33–37]. This technique will be investigated with more details in this section.

17.4.1 Vector field

We consider a mobile robot motion toward a target by an attractive vector field whose minimum point is a desired position during a mission. This mission can be tracking a trajectory or a desired position relative to a group of robots in a cooperative mission. Then, to lead a robot through a collision-free trajectory, we define a vector field where the obstacle is a repulsive point against the robot [37,38]. For instance, Figure 17.3 demonstrates a vector field leading a robot toward a desired target in a collision-free path. Under this vector field, the robot is attracted by a target denoted by t and an obstacle repulses it for avoiding collision. Let us define an obstacle position by $p_o = [x_o\ y_o]$. In this condition, a repulsive vector field $u_o = [u_{ox}\ u_{oy}]$ caused by the obstacle can be stated as follows:

$$\begin{aligned} u_{ox} &= k\frac{\cos\theta}{r^\rho}, \\ u_{oy} &= k\frac{\sin\theta}{r^\rho} \end{aligned} \tag{17.1}$$

in which $k \in \mathbb{R}_+$ denotes the vector field gain, $\rho \in \mathbb{N}$, r is the distance between the obstacle and the approaching robot, and θ is the angle of the vector $p - p_o$ where $p = [x\ y]$ is the robot position. Hence, according to (17.1), the repulsive vector field defined for the obstacle will be increased when the robot approaches the obstacle, and this issue guarantees obstacle avoidance. Therefore, it can be said that the repulsive vector field has a Gaussian distribution in all directions approaching the obstacle whose the standard deviation depends on ρ and the repulsion gain k.

Considering the repulsive vector field defined in (17.1), repulsion would always be employed to the robot even the times when the robot does not approach the obstacle. To cope with problem, a region of effectiveness can be defined for the repulsive vector field around the obstacle. In the case of an unknown obstacle, this region will be determined based on the sensory range of the robot (it can be any area around the obstacle when the obstacle is globally known). Hence, if we define the region of effectiveness by a circle with a radius of $r_a \in \mathbb{R}_+$, the repulsive vector field in the obstacle region of effectiveness $r \leq r_a$ can be stated as follows:

$$\begin{aligned} u_{ox} &= k\tfrac{\cos\theta}{r^\rho}\left(\tfrac{1}{r}-\tfrac{1}{r_a}\right), \\ u_{oy} &= k\tfrac{\sin\theta}{r^\rho}\left(\tfrac{1}{r}-\tfrac{1}{r_a}\right) \end{aligned}$$

where the term $((1/r)-(1/r_a))$ is useful to keep cohesion in employing the vector field. In other words, the repulsive vector field magnitude is zero on the region border. On the other hand, the vector field magnitude will be increased when the robot is approaching the obstacle.

Considering all the above-mentioned issues, the robot control input $u=[u_x\ u_y]$ can be stated as follows:

if $r \leq r_a$:

$$\begin{aligned} u_x &= u_{ox}+u_{tx}, \\ u_y &= u_{oy}+u_{ty}, \end{aligned}$$

else:

$$\begin{aligned} u_x &= u_{tx}, \\ u_y &= u_{ty} \end{aligned}$$

where $u_t=[u_{tx}\ u_{ty}]$ denotes the desired control input without obstacle avoidance to be considered for target seeking, trajectory tracking, and so on during a mission.

Instead of a point, each obstacle can be considered as an ellipse (a circle in special case) whose environment is defined by (p_o, A, B) as follows:

$$A^2(x-x_o)^2+B^2(y-y_o)^2=1$$

where $A, B \in \mathbb{R}_+$. Hence, the repulsive vectors can be provided from an ellipse enveloping the obstacle instead of the center of the gravity of the obstacle. In this condition, r will be modified as follows:

$$r=\sqrt{A^2(x-x_o)^2+B^2(y-y_o)^2-1}. \tag{17.2}$$

Therefore, it will be feasible to consider the obstacle dimensions in collision avoidance.

It should be noted that the obstacle avoidance formulation was given for a single obstacle. However, it simply can be extended to a group of obstacles labeled by $i \in \{1,2,\ldots,\ell\}$, and a group of robots labeled by $j \in \{1,2,\ldots,N\}$, by considering $u_o=\sum_{i=1}^{\ell} u_o^i+\sum_{j=1}^{N} u_r^j$ which u_o^i denotes the vector field associated with the ith obstacle and u_r^j denotes the vector field associated with the jth robot.

17.4.2 Local minima problems

A drawback of the introduced approach for obstacle avoidance of mobile robots was that if a robot moves in the exact opposite direction of the vector field, it may trap in a local minimum, and it may not be able to pass the obstacle; because the resultant vector of the robot motion and the repulsive vector may be zero in this condition. For instance, one can consider a scenario when a robot moves toward the center of the gravity of an obstacle (in the opposite direction of its vector field). In this condition, while approaching the obstacle, the vector field magnitude is increasing and there exists a distance from the obstacle where the vector field cancels the robot motion toward its target. Hence, the robot traps in a local minimum and cannot move toward its desired trajectory or target (global minimum). Another scenario is when the robot moves among some obstacles. In this condition, the resultant vector field is in the opposite direction of the robot motion. Hence, similar to the previous scenario, there exists a point on which the resultant vector field magnitude cancels the robot motion toward its target, and the robot cannot reach its desired target/trajectory.

A technique to cope with this problem is using a random input in a local minimum. If the magnitude of the robot control input is lower than a threshold $\varepsilon \in \mathbb{R}_+$, the robot control input will be a random vector with magnitude ε such that the robot changes its position and avoids local minima. In this condition, if the robot is not in the global minimum, the robot control input will be modified as follows:

$$
\begin{aligned}
u &= u_o + u_t, \quad \|u\| > \varepsilon, \\
u &= \varepsilon \left[\text{rand} \;\; \sqrt{1 - \text{rand}^2} \right]^\top, \quad \|u\| \leq \varepsilon
\end{aligned}
$$

where $0 \leq \text{rand} \leq 1$ indicates a random number.

The main problem of employing this technique is that under this technique, the vehicle speed may converge to a small value. Moreover, this technique increases the computation costs even in the global minimum. To cope with this issue, rotational vector fields are introduced in the literature [11,39,40]. The main advantage of this approach is that the repulsive forces from an obstacle are adaptive with the motion of the robot. In other words, the vector field is rotational around the obstacle whose direction depends on the motion direction of the robot. Therefore, the robot passes the obstacle while its speed is not decreased to zero. As depicted in Figure 17.4, in a case that a robot is approaching an obstacle on its left hand side, the rotational vector field around the obstacle is counter clockwise, and its direction can be clockwise when the robot approaches in the opposite direction. This issue can be distinguished by the robot by comparison of the heading angle ψ and the angle of the vector linking it to the obstacle denoted by χ. Therefore, the vector field will not be in the opposite direction of the approaching robot, and the effective control input to the mobile robot will not be zero. Therefore, the mobile robot will not trap in a local minimum. For instance, if we consider an obstacle identified as an ellipse (p_o, A, B), the vector field is rotational on ellipses $(p_o, \hat{A}, \hat{B})$ parallel to the obstacle environment such that $\hat{A}/\hat{B} = A/B$. In this condition, in a case that the mobile robot approaches the obstacle with $\psi \geq \chi$, it is proper to move in the clockwise direction. In other case that $\psi < \chi$, the vectors

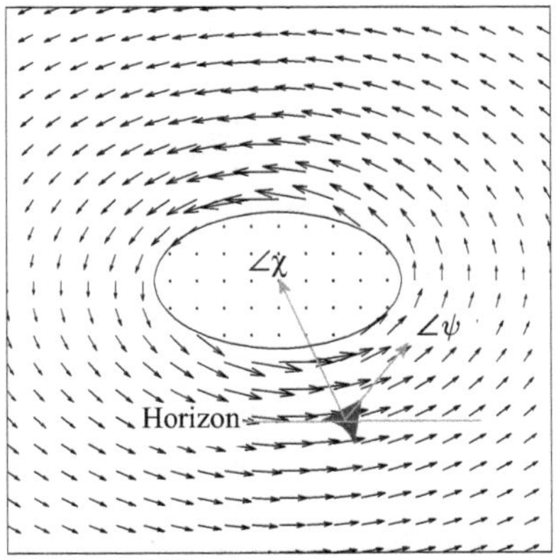

Figure 17.4 A counter-clockwise vector field around an obstacle

direction should be counter-clockwise. Considering the above-mentioned issues, the rotational vectors around the obstacle can be stated as follows:

$$u_r = [u_{rx}\ u_{ry}] \tag{17.3}$$

where

$$\begin{aligned} u_{rx} &= f_{rxc}, u_{ry} = f_{ryc} \qquad \psi \geq \chi, \\ u_{rx} &= f_{rxcc}, u_{ry} = f_{rycc} \qquad \psi < \chi, \end{aligned}$$

and $[f_{rxc}\ f_{ryc}]$ and $[f_{rxcc}\ f_{rycc}]$ show the clockwise and counter-clockwise vectors, respectively, as follows:

$$\begin{aligned} f_{rxc} &= \tfrac{\hat{B}}{\hat{A}}(y - y_o) = \tfrac{B}{A}(y - y_o), \\ f_{ryc} &= -\tfrac{\hat{A}}{\hat{B}}(x - x_o) = -\tfrac{A}{B}(x - x_o), \end{aligned} \tag{17.4}$$

and

$$\begin{aligned} f_{rxcc} &= -\tfrac{\hat{B}}{\hat{A}}(y - y_o) = -\tfrac{B}{A}(y - y_o), \\ f_{rycc} &= \tfrac{\hat{A}}{\hat{B}}(x - x_o) = \tfrac{A}{B}(x - x_o). \end{aligned} \tag{17.5}$$

The magnitudes of the vector fields expressed in (17.4) and (17.5) decrease when they get closer to the obstacle. However, it is desired to increase them when the mobile robot is approaching the obstacle. To modify (17.3) properly, let us first normalize it as follows:

$$u_{rn} = [u_{rxn}\ u_{ryn}] = \left[\frac{u_{rx}}{\|u_r\|}\ \frac{u_{ry}}{\|u_r\|}\right]$$

where u_{rn} is the normalized form of (17.3), and then multiply it by $1/r^{\rho}$ where r is corresponding to an ellipse that is tangent to the mobile robot position which is defined as (17.2). Therefore, $1/r^{\rho}$ increases when the mobile robot approaches the obstacle, and it enlarges the vector field magnitude. Considering a vector field gain k,

the following equations show the control input for obstacle avoidance considering all the above-mentioned issues:
if $r \leq r_a$:

$$\begin{aligned} u_x &= u_{tx} + \frac{ku_{rxn}}{r^\rho}\left(\frac{1}{r} - \frac{1}{r_a}\right), \\ u_y &= u_{ty} + \frac{ku_{ryn}}{r^\rho}\left(\frac{1}{r} - \frac{1}{r_a}\right), \end{aligned}$$

else:

$$\begin{aligned} u_x &= u_{tx}, \\ u_y &= u_{ty}. \end{aligned}$$

It should be noted that in any case, a vector field caused by an obstacle is in a compatible direction, and a resultant vector field never be in the opposite direction of a robot. For instance, in the case of multiple obstacles in the environment, the resultant vector field is in a direction compatible with the motion direction of the robot.

17.4.3 Avoiding moving obstacles

Another important problem in motion control of mobile robots is obstacle avoidance in dynamic environments where obstacles are moving. We consider a robot with velocity v approaching an obstacle with velocity v_o. Therefore, the velocity of the robot relative to the obstacle $v - v_o$ has a component v_{ao} on the line linking it to the obstacle and another component is perpendicular to v_{ao} indicated by $v_{ao}^{\perp}$ which are depicted in Figure 17.5. It is clear that v_{ao} is the projection of $v - v_o$ on the vector $p_o - p$ as follows [41]:

$$v_{ao} = \frac{(v - v_o)^\top (p_o - p)}{(p_o - p)^\top (p_o - p)}(p_o - p)$$

where $p = [x\, y]$ and $p_o = [x_o\, y_o]$. In this condition, to improve obstacle avoidance, an additive lateral repulsion can be employed by considering $v_{ao}^{\perp} = [v_{aox}^{\perp}\; v_{aoy}^{\perp}]$. Hence, a lateral repulsive vector field $u_o^{\perp} = [u_{ox}^{\perp}\; u_{oy}^{\perp}]$ is defined as follows:

$$\begin{aligned} u_{ox}^{\perp} &= k^{\perp} \frac{\cos\acute{\theta}}{r^\rho}\left(\frac{1}{r} - \frac{1}{r_a}\right), \\ u_{oy}^{\perp} &= k^{\perp} \frac{\sin\acute{\theta}}{r^\rho}\left(\frac{1}{r} - \frac{1}{r_a}\right) \end{aligned}$$

where

$$\begin{aligned} \cos\acute{\theta} &= \frac{v_{aox}^{\perp}}{\|v_{ao}^{\perp}\|}, \\ \sin\acute{\theta} &= \frac{v_{aoy}^{\perp}}{\|v_{ao}^{\perp}\|} \end{aligned}$$

in which $k^{\perp} \in \mathbb{R}_+$ denotes the vector field gain. Therefore, considering the repulsive vector fields u_o and $u_o^{\perp}$, the robot will be repulsed in the opposite direction of the obstacle motion.

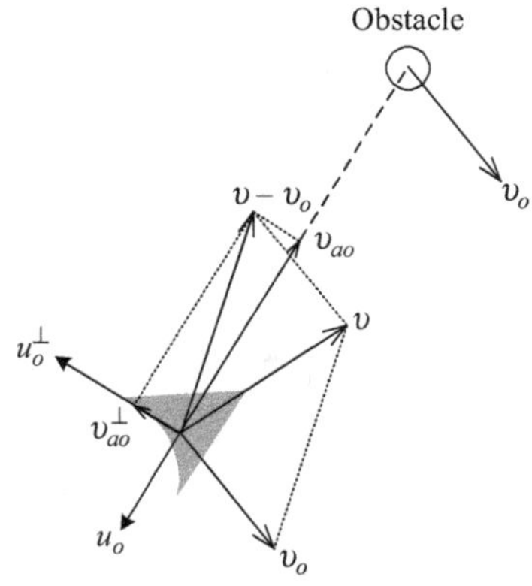

Figure 17.5 The repulsive control input configuration

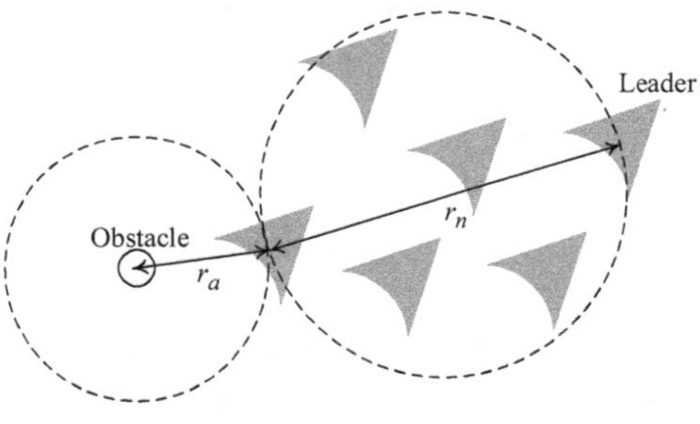

Figure 17.6 Obstacle avoidance in a group of robots

17.4.4 Obstacle avoidance in swarming robots

In this section, we extend the vector field approach for obstacle avoidance for a group of mobile robots labeled by $i \in \{1, 2, \ldots, N\}$. In some missions, there is a leader robot which determines the group trajectory. In this condition, if the leader knows the geometry of the swarm or the positions of other robots, the group obstacle avoidance is the same as a single robot which will be employed by the leader robot. In other words, the leader knows the dimensions of the group, and the effective region of the vector field will be determined by the leader such that there is no collision between the obstacle and the nearest robot to the obstacle. A simple option is to modify the effective region of the vector field as follows:

$$r_a^m = r_a + r_n$$

where r_a^m is the modified radius considered for the effective region of the vector field, and r_n is the maximum of the distance of the leader from the group members. In this condition, the sensing range of the robot should be greater than r_a^m. As shown in Figure 17.6, the leader guarantees obstacle avoidance for all the circle enveloped the group.

The introduced approach is useful when there exists a leader in the swarm and the leader has the dimension information of the group, and the obstacles are far from each other. Otherwise, obstacle avoidance should be considered by individual robots. Hence, each robot should have three behaviors before approaching an obstacle, when passing it, and after passing it. When the robots are passing the obstacle, the cooperative mission should be switched to obstacle avoidance strategy, and after passing the obstacle, the cooperative mission should be restarted. In this condition, the ith robot control law can be described as [11]

if $r_i \leq r_a$:

$$u_i = u_{ci}|_{r_i=r_a} + u_{oi},$$

else (after passing the obstacle):

$$u_i = u_{ci}|_{r_i=r_a} e^{-\tau(r_i - r_a)} + (1 - e^{-\tau(r_i - r_a)}) u_{ci}$$

where $u_{ci}|_{r_i=r_a}$ is the quantity of the cooperative control law when the formation reaches the obstacle which is considered to keep cohesion in the switching and $\tau \in \mathbb{R}_+$. The term $u_{ci}|_{r_i=r_a}$ vanishes gradually after passing the obstacle and u_{ci} will be the ith robot control law.

17.5 Swarming

Inspired by biological behaviors such as flocking of geese, schooling of fishes, bacterial colony, etc. swarming control of multirobot systems has become an interesting problem in the robotics and control communities expressed as cohesive collective motion of mobile robots in groups considering collision avoidance. In other words, the objective of a swarm of robots is cohesive collective motion with a common velocity and heading angle while considering collision avoidance among the robots. A swarm of robots can emerge as a team of ground vehicles, aerial vehicles, underwater vehicles, and so on with a wide range of applications in mobile sensory networks, surveillance and coverage robots, etc.

In 1954, in [42], a simple model of attraction and repulsion terms among members in a school of fishes was introduced. The main objective of that model was proposing an instrument to study and understand of social organisms. In 1987, in [43], a computer model of flocking birds for the animation industry was proposed under which flocking of birds with identical velocities and heading angles while considering cohesion and collision avoidance was simulated. Under that model, three main issues in flocking birds were considered:

- cohesion among the members by staying close together,
- avoid collision among the members and with obstacles,
- swarming with a common velocity.

Based on that model of swarms, a primary discrete-time model of mobile robots with limited field of sensing was introduced in [44]. Based on that model, the heading angle of each robot was updated based on the heading angles of its neighboring robots, and all the robots reached a common heading, and similar problems mainly focused on mathematical biology of swarms were studied in [45–47].

Inspired by the mentioned primary studies on swarming robots, "control and stability" of swarms has been a more challenging problem will be studied in this section.

17.5.1 Attraction/repulsion functions for swarm aggregation

Keeping cohesion among the robots while the robots avoid collision are main factors in control of robotic swarms. Using attraction/repulsion functions to attract robots toward each other while avoiding collision is a main technique for robots swarm aggregation. Thus, the following interaction protocol for mobile robots swarm aggregation can be considered [14,48]:

$$\dot{p}_i = \sum_{j=1}^{N} g(p_i - p_j), \quad i \in \{1, 2, \ldots, N\},$$

where $p_i \in \mathbb{R}^2$ is the position of the ith robot and $g(\cdot)$ denotes the function of attraction and repulsion among the robots defined as follows:

$$g(y) = -y(g_a(\|y\|) - g_r(\|y\|))$$

where $g_a(\|\cdot\|)$ is the attraction term and $g_r(\|\cdot\|)$ is the repulsion term such that there exist corresponding functions J_a and J_r where

$$\begin{aligned} \nabla_y J_a(\|y\|) &= y g_a(\|y\|), \\ \nabla_y J_r(\|y\|) &= y g_r(\|y\|). \end{aligned}$$

It should be noted that in the case of large distances among two robots, the attraction term is larger than the repulsion term, and in close distances, the repulsion one dominates the other one. In this condition, there exists a distance $\|y\| = \delta$ on which $g_a(\delta) = g_r(\delta)$. Therefore,

$$\begin{aligned} g_a(\delta) &> g_r(\delta), \|y\| > \delta, \\ g_a(\delta) &< g_r(\delta), \|y\| < \delta. \end{aligned}$$

An example of an attraction/repulsion function described above can be as follows:

$$g(y) = -y\left(a - be^{-\frac{\|y\|^2}{c}}\right)$$

where $a, b, c \in \mathbb{R}_+$. For instance, by considering $a = 1$, $b = 10$, and $c = 0.1$, $g(y)$ is depicted in Figure 17.7. Moreover, $g_a(\|y\|)$ and $g_r(\|y\|)$ are depicted in Figure 17.8 as well.

The introduced repulsion/attraction approach has the following important properties:

- The swarm center is fixed during swarming. Because the interaction among the robots is undirected, one can easily check that the derivative of $\bar{p} = (1/N)\sum_{i=1}^{N} p_i$ is zero.

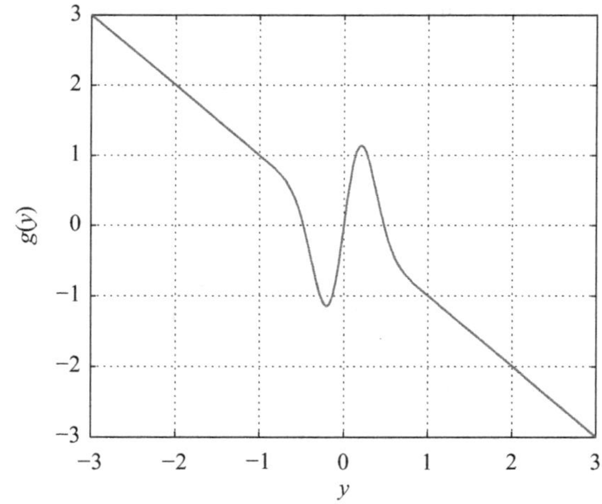

Figure 17.7 Attraction/repulsion function g(y) *when* $a = 1$, $b = 10$, *and* $c = 0.1$

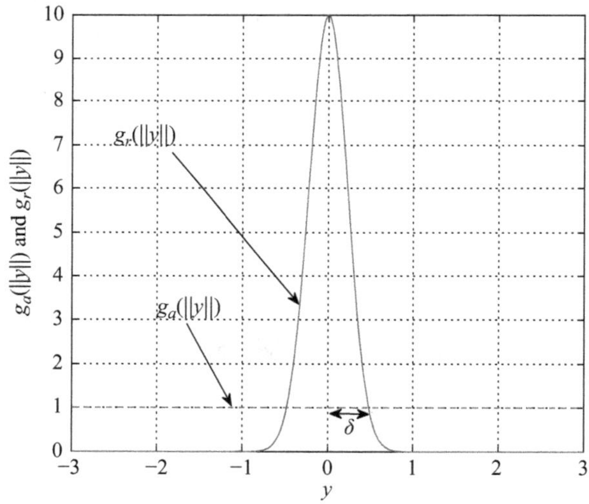

Figure 17.8 Attraction and repulsion functions $g_a(\|y\|)$ *and* $g_r(\|y\|)$ *when* $a = 1$, $b = 10$, *and* $c = 0.1$

- The robots states converge to a stationary condition such that the robots velocities converge to zero. By defining a Lyapunov candidate

  $$V = \sum_{i=1}^{N-1} \sum_{j=i+1}^{N} \left(J_a(\|p_i - p_j\|) - J_r(\|p_i - p_j\|) \right),$$

  one can check that the swarm converges to an invariant set denoted by

  $$\lim_{t\to\infty} \dot{p}_i = \mathbf{0}_{2\times 1}, \quad i \in \{1, 2, \ldots, N\}.$$

- The swarm errors, described by $e_i = p_i - \bar{p}, i \in \{1, 2, \ldots, N\}$, have a controllable bound (given later). The evolution of the Lyapunov candidate

  $$V_i = \frac{1}{2} e_i^\top e_i$$

  along the swarm trajectory implies that

  $$\|e_i\| \leq \frac{1}{aN} \sum_{j=1}^{N} g_r(\|p_i - p_j\|)\|p_i - p_j\|.$$

  Since the repulsion is bounded by b as

  $$g_r(\|p_i - p_j\|)\|p_i - p_j\| \leq b,$$

  we have

  $$\|e_i\| \leq \frac{b(N-1)}{aN} < \frac{b}{a}$$

  implying that the swarm size can be controllable by tuning a and b.

However, the mentioned approach is not applicable in the presence of a moving leader or in the presence of environmental obstacles. This problem is studied in the following section.

17.5.2 Lattice-type swarm of mobile robots with limited sensing

In this section, a swarming algorithm considering the robots limited communication ranges will be presented [15]. Keeping cohesion and collision avoidance while following a desired trajectory are considered. By letting r as the interaction range between two robots, the neighboring set of the ith robot is considered as follows:

$$\mathcal{N}_i(p) = \{j \in \mathcal{V} : \|p_j - p_i\| < r\}$$

where p describes the set of the robots positions. In this condition, the state-dependent edges associated with the network communication topology can be stated as follows:

$$\mathcal{E}(p) = \{i, j \in \mathcal{V} \times \mathcal{V}, \|p_j - p_i\| < r, \quad i \neq j\}.$$

For swarm modeling, a lattice-type structure as the desired geometry of the swarm is considered as depicted in Figure 17.9. Therefore, the objective will be arranging the robots on a plane such that the distance between each two neighbors i and j is d as

$$\|p_j - p_i\| = d.$$

In order to achieve the above-mentioned objective, to generate a smooth potential function, the following mapping is considered:

$$\|z\|_\sigma = \left(\frac{1}{\varepsilon}\sqrt{1 + \varepsilon\|z\|^2} - 1\right)$$

where $\varepsilon \in \mathbb{R}_+$ and its gradient is

$$\sigma_\varepsilon(z) = \frac{z}{\sqrt{1 + \varepsilon\|z\|^2}}.$$

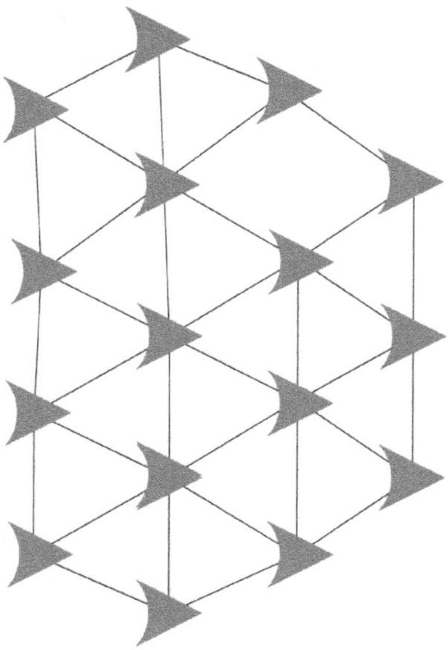

Figure 17.9 A lattice-type swarm of mobile robots

Therefore, the desired distance between each two neighboring robots can be stated in the σ-norm as

$$\|p_j - p_i\|_\sigma = \|d\|_\sigma = d_\alpha.$$

Accordingly, since the communication links are state-dependent, an adjacency matrix $A(p) = [a_{ij}(p)]$ can be described as

$$a_{ij}(p) = \rho_h(\|p_j - p_i\|_\sigma / r_\alpha) \in [0, 1], \quad j \neq i$$

where $\rho_h(\cdot)$ is a "smooth" finite cut-off function as follows:

$$\rho_h(z) = \begin{cases} 1 & z \in [0, h), \\ \frac{1}{2}\left[1 + \cos\left(\pi \frac{z-h}{1-h}\right)\right] & z \in [h, 1], \\ 0 & \text{otherwise} \end{cases}$$

in which $h \in (0, 1)$. Indeed, $\rho_h(\cdot)$ is a function which smoothly varies between 0 and 1. For instance, $\rho_h(\cdot)$ when $h = 0.7$ is depicted in Figure 17.10 which shows that the function $\rho_h(z)$ smoothly converges to zero for $z \geq 0.7$ and is zero for $z \geq 1$. In a similar way, a finite cut-off action function $\phi_\alpha(z)$ which vanishes for all $z \geq \|r\|_\sigma = r_\alpha$ is considered as follows:

$$\phi_\alpha(z) = \rho_h\left(\frac{z}{r_\alpha}\right)\phi(z - d_\alpha)$$

where $\phi(z)$ is an uneven sigmoidal function as follows:

$$\phi(z) = \frac{1}{2}[(a + b)\sigma_1(z + c) + (a - b)]$$

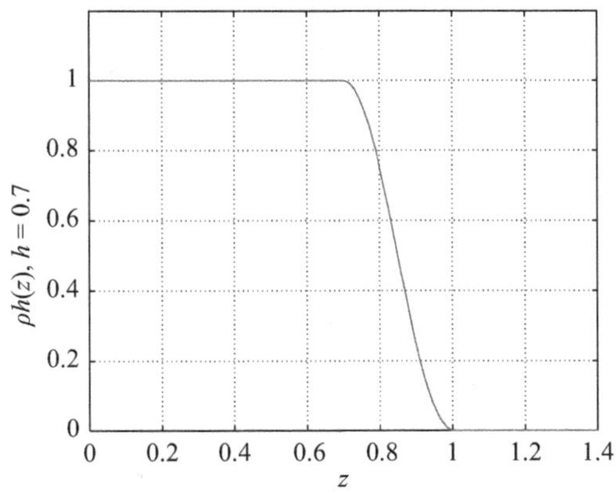

Figure 17.10 The smooth finite cut-off function $\rho_h(z)$ when $h = 0.7$

where $0 < a \le b$ and $c = |a - b|/\sqrt{4ab}$ to ensure that $\phi(0) = 0$. Therefore, an attractive/repulsive potential function can be

$$\psi_\alpha(z) = \int_{d_\alpha}^{z} \phi_\alpha(s)ds.$$

Hence, the ith robot interaction law can be designed as the following second-order differential equation:

$$\ddot{p}_i = \sum_{j \in \mathcal{N}_i(p)} \phi_\alpha(\|p_j - p_i\|_\sigma)\mathbf{n}_{ij} + \sum_{j \in \mathcal{N}_i(p)} a_{ij}(p)(p_j - p_i) + u_i^\gamma$$

where $\mathbf{n}_{ij} = \sigma_\varepsilon(p_j - p_i)$ is a vector along the line connecting p_i to p_j, and u_i^γ is tracking control law in the form

$$u_i^\gamma = -c_1(p_i - p_r) - c_2(\dot{p}_i - \dot{p}_r)$$

where $c_1, c_2 \in \mathbb{R}_+$ and $(p_r, \dot{p}_r)$ describes the desired trajectory of the swarm. Therefore, under the proposed consensus strategy, the robots move cohesively while following a desired trajectory.

17.6 Swarm formation

Another important problem in control of robotic swarms is formation control defined as coordinated motion of mobile robots while keeping desired geometric patterns. Indeed, one of main objectives of swarm formation control is that by imposing constraints on robots motion, mathematical analysis of the swarming problem is simplified. Moreover, by keeping a formation, optimal accomplishment of cooperative tasks such as coverage and exploration could be realized.

The proper strategy for formation control of multirobot systems varies based on missions, required degree of autonomy, available software/hardware, etc. In general,

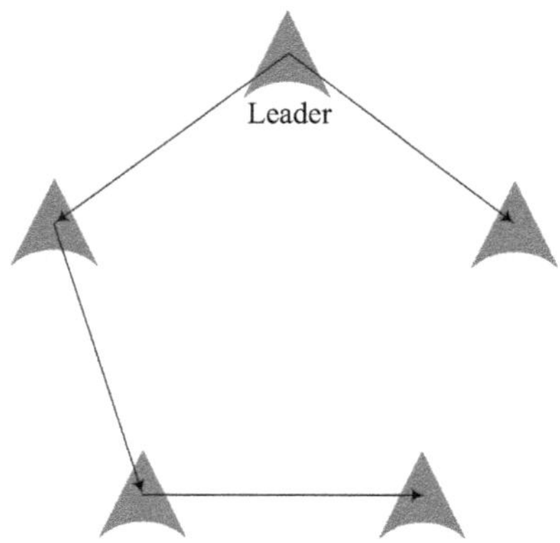

Figure 17.11 A leader–follower architecture

two classes of approaches can be considered in formation control of multirobot systems, namely, *leader-follower* and *leaderless*. This issue will be investigated in the following sections.

17.6.1 Leader–follower architecture

Most studies on formation control of multirobot systems are based on the leader–follower architecture. Indeed, ease of implementation and stability analysis are the main advantages of this architecture. As depicted in Figure 17.11, under this architecture, there exists a leader tracking a trajectory and other robots follow this leader hierarchically to keep a desired formation. In this condition, the decisions are made by the leader and the leader is the stable point of the swarm which simplifies the stability analysis of the whole swarm. However, the leader is the single point of failure in this architecture, and no feedbacks are sent from the followers to this leader, and these issues decrease the degree of autonomy of these approaches [49–52]. In this condition, if the leader is labeled as Robot 1, the following interaction protocol for the followers can be considered to achieve a formation:

$$\dot{p}_i = \sum_{j \in \mathcal{N}_i} k(p_j - p_i + d_{ji}), \quad i \in \{2, 3, \ldots, N\}, \tag{17.6}$$

where $p_i \in \mathbb{R}^2$ denotes the position of the ith robot, $k \in \mathbb{R}_+$, and $d_{ji} \in \mathbb{R}^2$ is the fixed desired position of the ith robot with respect to the jth robot. The stability analysis of the swarm is straightforward by considering formation keeping errors $e_i = \sum_{j \in \mathcal{N}_i}(-p_j + p_i - d_{ji})$. Note that since the network is decentralized, the ith robot "has no access to the derivatives of $p_j, j \in \mathcal{N}_i$, as control inputs"; because each robot control input will be calculated by the robot using the information achieved from neighboring robots, and all the robots control inputs will be determined simultaneously independent of the control inputs of other robots. In this condition, by state representation of the errors, one can get

$$\dot{e}_i = -\sum_{j\in\mathcal{N}_i} (ke_i + \dot{p}_j), \quad i \in \{2, \dots, N\}. \tag{17.7}$$

Equation (17.7) implies that if $\dot{p}_j, j \in \mathcal{N}_i$, are bounded, the formation keeping errors remains bounded. Now, since the associated communication graph is directed and no loops exist in the associated communication graph, if $\dot{p}_1$ is bounded, the formation keeping errors remain bounded as well.

Therefore, it may not be feasible for each follower to fully track the leader trajectory since $e_i, i \in \mathcal{N}_i$, may not be zero, and the formation cannot be kept rigid during swarming.

Based on the above-given arguments, the leader–follower architecture cannot guarantee fully regulation of the formation keeping errors; however, in some applications, precise formation keeping is required. To cope with the problem of the leader–follower architecture and achieve a precise formation, as depicted in Figure 17.12, the robots can track a *virtual leader* as a known reference signal. In this condition, the leader trajectory is fully known for each follower; therefore, the formation can be kept rigid and precise during maneuvers [2,53–55]. Considering the above-mentioned issue and by letting p_v as the trajectory of the virtual leader, the modified leader–follower architecture can be stated as follows:

$$\dot{p}_i = k_i(p_v - p_i + d_{vi}) + \dot{p}_v, \quad i \in \{1, 2, \dots, N\},$$

where d_{vi} is the fixed desired position of the ith robot with respect to the virtual leader. In this condition, by considering formation keeping errors $e_i = -p_v + p_i - d_{vi}, i \in \{1, 2, \dots, N\}$, we have

$$\dot{e}_i = -k_i e_i, i \in \{1, 2, \dots, N\},$$

implying the convergence of the formation keeping errors to zero. Thus, the virtual architecture is more proper when rigid formation keeping is required. However, the main problem of this strategy is that despite a real leader, a virtual leader "cannot make decision autonomously."

Due to the mentioned problems existing with the leader–follower architecture, in some applications, the robots should reach a desired formation in a leaderless architecture which will be studied in the next section.

17.6.2 Leaderless architecture

Despite leader–follower and virtual leader based approaches, under a leaderless architecture, each robot is the leader of a group of robots while it follows other robots as its leaders. In this condition, no specific leader exists in the swarm, and this issue increases the degree of autonomy in the network. Therefore, in the communication graph associated with this architecture, there is a directed path between each pair of robots. This issue has some advantages such that the robots can autonomously reach agreement on some quantities of interest such as formation centroid, velocity,

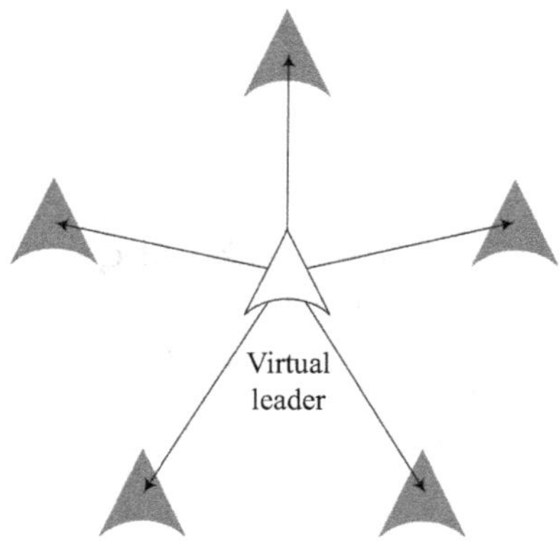

Figure 17.12 A virtual architecture

formation radius, etc. under a leaderless architecture. However, since there is no a stable leader in the swarm, the complexity of the network mathematical analysis will be increased [56–58]. In other words, by considering the interaction protocol (17.6) for a leaderless network, the stability of the formation keeping errors cannot be analyzed similar to (17.7). We consider the following interaction protocol for a leaderless network of robots:

$$\dot{p}_i = \sum_{j \in \mathcal{N}_i} a_{ij}(p_j - p_i + d_{ji}), \quad i \in \{1, 2, 3, \ldots, N\},$$

in which $d_{ji}, i, j \in \{1, 2, \ldots, N\}$, are chosen such that there exist $\tilde{p}_i, i \in \{1, 2, \ldots, N\}$, where

$$\tilde{p}_i - \tilde{p}_j = d_{ji}, \quad i, j \in \{1, 2, \ldots, N\}.$$

Indeed, based on the mentioned interaction law, the *i*th robot receives information from other robots such that the performance of each robot depends on the behaviors of its neighbors. In this condition, it is necessary to investigate the behavior of the whole network as an integrated system. By considering

$$p = \begin{bmatrix} p_1^\top - \tilde{p}_1^\top & p_2^\top - \tilde{p}_2^\top & \cdots & p_N^\top - \tilde{p}_N^\top \end{bmatrix}^\top,$$

it can be said that

$$\dot{p} = (-\mathcal{L} \otimes I_2)p$$

where $\mathcal{L}$ is Laplacian matrix associated with the network. If the network communication graph has a spanning tree, $\mathcal{L}$ has a simple zero eigenvalue and $N - 1$ stable eigenvalues implying that the states behaviors converge toward the null space of $\mathcal{L}$ (equals **1**) such that all the entries of p converge to a common value p_c which is *a priori* unknown. Therefore,

$$\lim_{t \to \infty} p_i - \tilde{p}_i = p_j - \tilde{p}_j = p_c, \quad i, j \in \{1, 2, \ldots, N\},$$

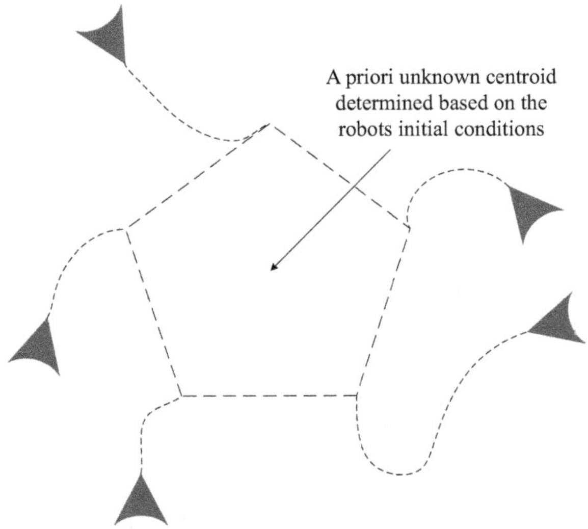

Figure 17.13 Achieving a formation in a leaderless group of mobile robots

implying convergence of formation errors $-p_j + p_i - d_{ji}, j \in \mathcal{N}_i$, to zero. The mentioned formation control strategy guarantees the convergence of the robots toward a "static" polygon formation around an *a priori* unknown centroid p_c (see Figure 17.13). The formation can be dynamic with a desired velocity v_d if the interaction protocol is modified as follows:

$$\dot{p}_i = \sum_{j \in \mathcal{N}_i} a_{ij}(p_j - p_i + d_{ji}) + v_d, \quad i \in \{1, 2, \ldots, N\}.$$

By letting p_d be an *a priori* unknown trajectory such that

$$\frac{\mathrm{d}p_d}{\mathrm{d}t} = v_d,$$

and modifying p as

$$p = \begin{bmatrix} p_1^\top - \tilde{p}_1^\top - p_d^\top & p_2^\top - \tilde{p}_2^\top - p_d^\top & \cdots & p_N^\top - \tilde{p}_N^\top - p_d^\top \end{bmatrix}^\top,$$

one can get $\dot{p} = (-\mathcal{L} \otimes I_2)p$, and thus $p_i, i \in \{1, 2, \ldots, N\}$, converge into a common value $\acute{p}$ which is *a priori* unknown. Therefore, on the one hand,

$$\lim_{t \to \infty} p_i - \tilde{p}_i - p_d = p_j - \tilde{p}_j - p_d, \quad i, j \in \{1, 2, \ldots, N\},$$

implying that the formation error $-p_j + p_i - d_{ji}, j \in \mathcal{N}_i$, converge to zero; on the other hand,

$$\lim_{t \to \infty} p_i - \tilde{p}_i - p_d = \acute{p}, \quad i \in \{1, 2, \ldots, N\}.$$

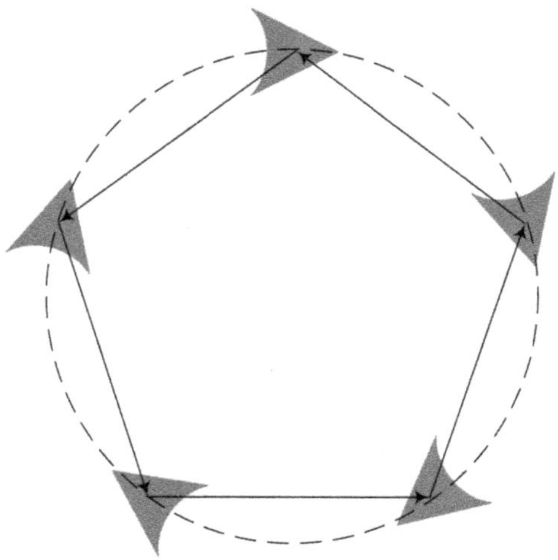

Figure 17.14 A leaderless pursuit architecture

Therefore, one can get

$$\lim_{t\to\infty} p_i = \acute{p} + \tilde{p}_i + p_d, \quad i \in \{1, 2, \ldots, N\},$$

implying convergence of the robots velocities into v_d ($\acute{p} + \tilde{p}_j$ is constant).

Inspired by the behaviors of biological organisms such as ants and beetles for searching foods or avoiding predators, in many applications, the robots may need to have a rotational swarming after achieving a formation whose centroid can be a specific position in an environment to be covered or a target to be guarded. Indeed, the rotational motion of the robots increases their searching capability and coverage range, and in this condition, perfect coverage would be feasible by employing a limited number of robots [59–61]. This architecture in which each robot pursues another leader robot under a cyclic topology is depicted in Figure 17.14. A proper control strategy on the one hand should guarantee the convergence of some formation errors to zero; on the other hand, it should guarantee the robots rotational motion as the formation errors converge to zero.

If we assume that the ith robot position is denoted by $p_i = \begin{bmatrix} x_i & y_i \end{bmatrix}^\top$, each robot pursues its leading robot with relative angle $\vartheta = 2\pi/N$ (as depicted in Figure 17.15) and angular velocity $\varpi \in \mathbb{R}$ in a regular polygon formation around a centroid $\begin{bmatrix} x_c & y_c \end{bmatrix}^\top$, if the ith robot velocity vector is defined as follows [58]:

$$\begin{bmatrix} \dot{x}_i \\ \dot{y}_i \end{bmatrix} = \begin{bmatrix} -\lambda & -\varpi \\ \varpi & -\lambda \end{bmatrix} \begin{bmatrix} x_i - x_c \\ y_i - y_c \end{bmatrix} + \begin{bmatrix} \lambda\cos\vartheta & \lambda\sin\vartheta \\ -\lambda\sin\vartheta & \lambda\cos\vartheta \end{bmatrix} \begin{bmatrix} x_{i+1} - x_c \\ y_{i+1} - y_c \end{bmatrix}$$

where $\lambda \in \mathbb{R}_+$. Indeed, by letting

$$p = \begin{bmatrix} p_1^\top & p_2^\top & \cdots & p_N^\top \end{bmatrix}^\top$$

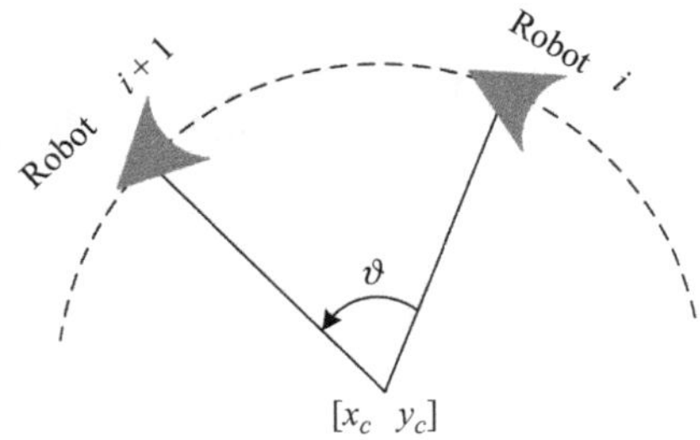

Figure 17.15 A robot pursues its leading robot with relative angle $\vartheta = 2\pi/N$

and rewriting the swarm state space equations as $\dot{p} = AP$, it can be shown that the steady-state behavior of the swarm can be described by two eigenvalues $\pm j\varpi$ and therefore pursuit of the robot with angular velocity ϖ can be concluded. Simultaneously, convergence of the following formation errors to zero can be realized:

$$e_i = \begin{bmatrix} x_i \\ y_i \end{bmatrix} - \begin{bmatrix} x_c \\ y_c \end{bmatrix} - \begin{bmatrix} \cos\vartheta & \sin\vartheta \\ -\sin\vartheta & \cos\vartheta \end{bmatrix} \begin{bmatrix} x_{i+1} - x_c \\ y_{i+1} - y_c \end{bmatrix}.$$

Remark 17.6.1. *It should be noted that in addition to the above-mentioned architectures, the robots may have* multi-objective behaviors. *Formation keeping, seeking a target, obstacle avoidance, and so on can be some weighted behaviors, and in this condition, the robots control inputs can be obtained based on the weighted control strategies associated with individual behaviors [11,62].*

17.7 Rendezvous

In a cooperative system, there are some quantities of interest such as heading angle, velocity, and so on, which should be identical for all robots. Hence, one fundamental problem in networked systems is achieving *consensus* on a quantity of interest via an interaction protocol. The main idea of consensus is to achieve agreement on a quantity of interest based on local interaction which has a wide range of applications from formation and flocking control of multivehicle systems [58,63] to data fusion in sensory networks [64,65].

The consensus problem has a long history in computer science [66,67]. However, the consensus problem for autonomous swarming mobile robots first was studied in [26] in which an algebraic graph theoretic approach for attitude alignment of a team of mobile robots was proposed. In that paper, each robot updated its state under jointly connected networks based of the states of its neighbors as follows:

$$x_i(t+1) = \frac{1}{1+n_i(t)} \left(x_i(t) + \sum_{j \in \mathcal{N}_i(t)} x_j(t) \right), \quad i \in \{1, 2, \ldots, N\},$$

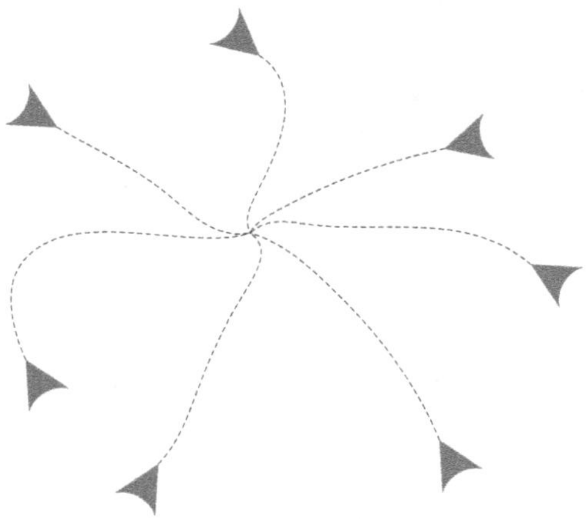

Figure 17.16 Rendezvous in a group of mobile robots

where $x_i(t)$ denotes the state and $n_i(t)$ is the number of the neighbors of the ith agent. Under that approach, the agents reached consensus upon their headings angles. However, the first systematic framework for the consensus problem in the view of control theory was proposed in [29,68] in which consensus in networks with switching strongly connected topology and identical communication delays were studied. In that paper, a consensus protocol was proposed based on relative positions measurement as follows:

$$\dot{x}_i = \sum_{j \in \mathcal{N}_i} a_{ij}(t)(x_j - x_i).$$

By employing the introduced consensus protocol, the agents reached agreement on a weighted average of their initial states as $\xi^{\top} X_0$ where $X = \begin{bmatrix} x_1 & x_2 & \cdots & x_N \end{bmatrix}^{\top}$. Under that condition, if the network topology was balanced, since $\xi = \mathbf{1}/N$, the agents reached agreement on the average of their initial states.

Note that under this condition, collision can be avoided by employing an additive potential field for collision avoidance. As mentioned in Section 17.4.4, this issue is the main assumption of similar strategies.

One of important common quantities of interest in a swarm of robots is the *rendezvous* position. Before starting swarming of a team of mobile robots, it is necessary for robots to converge to a common region, position, or configuration. Hence, one fundamental problem in robotic swarms is achieving rendezvous or *gathering* in a commonplace under an interaction protocol. Indeed, as depicted in Figure 17.16, under a rendezvous algorithm, robots should share their states information such that they eventually converge to a common position.

In the case that there exists a leader robot in the swarm, the rendezvous position will be determined by this leader. However, the rendezvous problem has been basically developed for networks of robots without any leader. This problem was introduced in [69] in which an algorithm for convergence of mobile robots toward a common point was proposed. Under the algorithm proposed in [69], each robot first observed the relative positions with respect to some visible robots as neighboring robots. Then, the next position of the robot based on the positions of the neighboring robots was determined. However, under that algorithm, the robots motions were considered instantaneously and no control errors and stability analysis were considered. Inspired by existing approaches devoted to consensus control of networked systems, approaches for rendezvous control of mobile robots have been introduced in the literature studied in this section.

17.7.1 Consensus-based rendezvous

Let us consider a group of mobile robots labeled by $i \in \{1, 2, \ldots, N\}$, whose the ith robot position is described as $p_i = \begin{bmatrix} x_i & y_i \end{bmatrix}^\top$. To achieve rendezvous in a common position, an interaction protocol is proposed in [70] as follows:

$$\begin{aligned} \dot{p}_i &= p_{i+1} - p_i, \quad i \in \{1, 2, \ldots, N-1\}, \\ \dot{p}_N &= p_1 - p_N. \end{aligned}$$

Indeed, it is supposed that the ith robot receives information from the $(i+1)$th robot, and thus similar to a leaderless formation control scheme, the performance of each robot depends on the behavior of its neighbor implying that to analyze the behavior and stability of each robot, the behavior of the whole network should be studied. Therefore, by assembling all the robots interaction protocols, one can get

$$\dot{p} = (-\mathcal{L} \otimes I_2)p$$

where

$$\mathcal{L} = \begin{bmatrix} 1 & -1 & 0 & \cdots & 0 \\ 0 & 1 & -1 & \cdots & 0 \\ \vdots & \vdots & \vdots & \ddots & \vdots \\ -1 & 0 & 0 & \cdots & 1 \end{bmatrix}$$

and $p = \begin{bmatrix} p_1^\top & p_2^\top & \cdots & p_N^\top \end{bmatrix}^\top$. The matrix $-\mathcal{L}$ has a simple zero eigenvalue and the other eigenvalues of $-\mathcal{L}$ are stable. Hence, the robot behaviors will be described by the null space of $\mathcal{L}$ which is $\mathbf{1}$ which implies convergence of the robots toward a common position. A more generalized framework is proposed in [23] such that $\mathcal{L}$ is modified as follows:

$$\mathcal{L} = \begin{bmatrix} k_1 & -k_1 & 0 & \cdots & 0 \\ 0 & k_2 & -k_2 & \cdots & 0 \\ \vdots & \vdots & \vdots & \ddots & \vdots \\ -k_N & 0 & 0 & \cdots & k_N \end{bmatrix}$$

where $k_i, i \in \{1, 2, \ldots, N\}$, should be positive real except one of them defined as k_p such that $\bar{k} < k_p$ if the lower bound satisfies

$$-\frac{\prod_{j=1,j\neq p}^{N} k_j}{\sum_{i=1,i\neq p}^{N} \prod_{j=1,j\neq i,j\neq p}^{N} k_j} = \bar{k}. \tag{17.8}$$

Indeed, the condition (17.8) implies marginal stability of the state matrix $\mathscr{L}$ and the convergence of the robots states toward the null space of $\mathscr{L}$. In the case of positive real coupling gains, in [28], the following generalized interaction protocol guarantees convergence of the robots toward a rendezvous position if the network topology dynamically changes but a robot with no neighbors does not exist during a bounded period of time, i.e., the network has a joint spanning tree in bounded time periods:

$$\dot{p}_i = \sum_{j=1}^{N} a_{ij}(t)(p_j - p_i), \quad i \in \{1, 2, \ldots, N\}.$$

In this case, it is shown that the product of transition matrices $\Phi_{t_i}, i \in \{1, 2, 3, \ldots\}$, associated with switching times $t_i, i \in \{1, 2, 3, \ldots\}$, leads the robots state vectors toward a common value.

Connectivity conditions mentioned above may not be feasible in every case, and thus achieving rendezvous with connectivity maintenance is investigated in the following section.

17.7.2 Rendezvous with connectivity maintenance

Achieving a rendezvous among a network of mobile robots with limited mobility and communication capabilities will be investigated. It is supposed that each robot can communicate with its neighbors (considered in a disk with center of the robot and radius of r), and therefore the network communication topology depends on the robots relative distance which changes over time. In [24], it is shown that under the following interaction protocol, rendezvous is guaranteed if the communication graph $\mathscr{G}$ is connected:

$$p_i(m+1) = p_i(m) + \lambda_i^*(\mathscr{CC}_i(\mathscr{M}_i) - p_i)$$

where $\mathscr{CC}_i(\mathscr{M}_i)$ is the circumcenter of the point set $\mathscr{M}_i$ comprising the ith robot and its neighbors defined as the center of the smallest radius circle enveloping the set. In this condition, by designing $\lambda_i^* < 1$ properly, connectivity maintenance among the robots can be guaranteed. To achieve this goal, λ_i^* could be obtained from the following convex problem:

$$\begin{array}{ll} \text{maximize} & \lambda_i \\ \text{subject to} & \lambda_i \leq 1, (1-\lambda_i)p_i + \lambda_i \mathscr{CC}_i(\mathscr{M}_i) \in Q_i \end{array}$$

where Q_i is a set of admissible points such that constraints such as connectivity maintenance is satisfied. The convex problem implies moving of the ith robot along the segment from p_i to $\mathscr{CC}_i(\mathscr{M}_i)$ to achieve the maximum possible distance while the position remains in the set Q_i which this solution is unique.

Note that the admissible point set should be designed considering this issue that if robots p_i and p_j are neighbors in $\mathscr{G}$; then, their subsequent positions should belong to $\overline{B}((p_i + p_j)/2, r/2)$ where $\overline{B}((p_i + p_j)/2, r/2)$ is a closed disk of radius $r/2$ centered at $(p_i + p_j)/2$. Thus, if robot p_i has its neighbors at locations $\{q_1, \dots, q_l\}$; then, its constraint set is

$$C_{p_i,r}(\{q_1,\dots,q_l\}) = \bigcap_{q\in\{q_1,\dots,q_l\}} \overline{B}\left(\frac{p_i+q}{2}, \frac{r}{2}\right).$$

The introduced strategies guarantee convergence of the robots toward a rendezvous position if the robots have no nonholonomic constraints on motion. Achieving rendezvous considering robots nonholonomic constraints is investigated in the following section.

17.7.3 Rendezvous in the presence of nonholonomic constraints

In this section, we consider a team of nonholonomic mobile robots, in which the ith one is described as

$$\begin{aligned}\dot{x}_i &= v_i \cos\psi_i,\\ \dot{y}_i &= v_i \sin\psi_i,\\ \dot{\psi}_i &= \omega_i\end{aligned} \tag{17.9}$$

where $\begin{bmatrix}x_i & y_i\end{bmatrix}$ denotes the position of the robot center, v_i is the speed, ψ_i is the heading angle, and ω_i is the heading rate. The objective is to design a control law for (v_i, ω_i) such that all the robots meet each other in a common position. In order to employ, the existing results proposed for holonomic mobile robots, in [71], the feedback linearization technique is proposed such that the ith robot described in (17.9) can be described by the following simple single-integrator model:

$$\begin{aligned}\dot{x}_{hi} &= \nu_{xi},\\ \dot{y}_{hi} &= \nu_{yi}\end{aligned} \tag{17.10}$$

where (x_{hi}, y_{hi}) is an off-axis point on the robot which can be described by

$$\begin{aligned}x_{hi} &= x_i + L_i \cos\psi_i,\\ y_{hi} &= y_i + L_i \sin\psi_i\end{aligned} \tag{17.11}$$

where $L_i \in \mathbb{R}_+$ is a constant. Indeed, by differentiation of (17.11) along (17.9) and considering (ν_{xi}, ν_{yi}) as auxiliary variables, (17.10) can be obtained as follows:

$$\begin{aligned}\dot{x}_{hi} &= v_i \cos\psi_i - L_i\omega_i \sin\psi_i = \nu_{xi},\\ \dot{y}_{hi} &= v_i \sin\psi_i + L_i\omega_i \cos\psi_i = \nu_{yi}.\end{aligned}$$

However, this approach just leads to convergence of (x_{hi}, y_{hi}), $i \in \{1, 2, \dots, N\}$, to a common position and cannot guarantee precise control of the mobile robots positions.

To consider this issue, each robot should approach its neighbors in the direction of the vector linking it to them. Thus, the robots control law will be as follows [72]:

$$\begin{aligned} \omega_i &= -\sum_{j\in\mathcal{N}_i}(\psi_i - \phi_{ij}), \\ v_i &= -\sum_{j\in\mathcal{N}_i}\big((x_i - x_j)\cos(\psi_i) + (y_i - y_j)\sin(\psi_i)\big) \end{aligned} \tag{17.12}$$

where

$$\phi_{ij} = \arctan\left(\frac{y_j - y_i}{x_j - x_i}\right).$$

Indeed, ϕ_{ij} implies the direction of the vector linking the ith robot to the jth robot. In this condition, (17.12) guarantees alignment of the ith robot heading with this direction while the robot speed is proportional to the relative position. Now, by considering a Lyapunov candidate of the consensus errors in the form $V = \frac{1}{2}\sum_{i=1}^{N}\sum_{j\in\mathcal{N}_i}(p_i - p_j)^\top(p_i - p_j)$, convergence of the consensus errors into zero along (17.12) can be concluded.

Remark 17.7.1. *It is worth mentioning that in a way similar to Section 17.6.2, by considering $p_j - p_i + d_{ji}$ instead of $p_j - p_i$, the rendezvous strategies in this section are extendable to leaderless formation control of robots.*

17.8 Discussion

Main important problems in decentralized control of robotic swarms were investigated in this chapter. First motion control of a single mobile robot was investigated. Thus, various vector field approaches for obstacle avoidance of mobile robots in static and dynamic environments as well as for cooperative robots were introduced. Then, control strategies for swarm aggregation of mobile robots were presented. Keeping cohesion among robots with velocity matching while avoiding collisions were important factors in robotic swarms which were investigated in the chapter. The results were extended to the cases when the robots kept a desired geometric formation during swarming, and various formation control architectures including leader–follower and leaderless architectures were analyzed. Accordingly, basic formulations for each architecture were presented. Finally, the rendezvous problem as an important issue before each cooperative task was investigated. Convergence of robots toward a region, connectivity maintenance, time-varying networks, etc. were important issues considered in this area.

The mentioned approaches were basic strategies in the area of robotic swarms and because of practical issues, more challenging problems have been investigated in various categories in recent years. Swarm control considering uncertain nonlinearities in robots models, swarm control of robots in the presence of time-delays, swarm formation control in the air and space, swarm formation of mobile robots based on relative distances measurement, consensus in the presence of malicious behaviors of robots, networks security in the presence of attacks, etc. have been more practical problems studied in recent years.

References

[1] M. Radmanesh and M. Kumar, "Flight formation of UAVs in presence of moving obstacles using fast-dynamic mixed integer linear programming," *Aerospace Science and Technology*, vol. 50, pp. 149–160, 2016.

[2] H. Rezaee and F. Abdollahi, "Synchronized cross coupled sliding mode controllers for cooperative UAVs with communication delays," in *Proceedings of the 51st IEEE Conference on Decision and Control*, USA, December 2012, pp. 3116–3121.

[3] Y. Hao and S. K. Agrawal, "Formation planning and control of UGVs with trailers," *Autonomous Robots*, vol. 19, no. 3, pp. 257–270, 2005.

[4] M. Saska, V. Spurny, and V. Vonasek, "Predictive control and stabilization of nonholonomic formations with integrated spline-path planning," *Robotics and Autonomous Systems*, vol. 75, pp. 379–397, 2016.

[5] Y. Wang, W. Yan, and W. Yan, "A leader-follower formation control strategy for AUVs based on line-of-sight guidance," in *Proceedings of the IEEE International Conference on Mechatronics and Automation*, China, August 2009, pp. 4863–4867.

[6] M. Mirzaei, F. Abdollahi, and N. Meskin, "Consensus of autonomous underactuated surface vessels," in *Proceedings of the American Control Conference*, USA, July 2016, pp. 6417–6422.

[7] D. Schoerling, C. V. Kleeck, F. Fahimi, C. R. Koch, A. Ams, and P. Lober, "Experimental test of a robust formation controller for marine unmanned surface vessels," *Autonomous Robots*, vol. 28, no. 2, pp. 213–230, 2010.

[8] Y. Liu and R. Bucknall, "Path planning algorithm for unmanned surface vehicle formations in a practical maritime environment," *Ocean Engineering*, vol. 97, pp. 126–144, 2015.

[9] J. L. Ramirez, M. Pavone, E. Frazzoli, and D. W. Miller, "Distributed control of spacecraft formation via cyclic pursuit: Theory and experiments," in *Proceedings of the American Control Conference*, USA, June 2009, pp. 4811–4817.

[10] F. Lin, M. Fardad, and M. R. Jovanovic, "Optimal control of vehicular formations with nearest neighbor interactions," *IEEE Transactions on Automatic Control*, vol. 57, no. 9, pp. 2203–2218, 2012.

[11] H. Rezaee and F. Abdollahi, "A decentralized cooperative control scheme with obstacle avoidance for a team of mobile robots," *IEEE Transactions on Industrial Electronics*, vol. 61, no. 1, pp. 347–354, 2014.

[12] B. Ranjbar-Sahraei, F. Shabaninia, A. Nemati, and S. D. Stan, "A novel robust decentralized adaptive fuzzy control for swarm formation of multiagent systems," *IEEE Transactions on Industrial Electronics*, vol. 59, no. 8, pp. 3124–3134, 2012.

[13] H. Rezaee and F. Abdollahi, "Motion synchronization in unmanned aircrafts formation control with communication delays," *Communications in Nonlinear Science and Numerical Simulation*, vol. 18, no. 3, pp. 744–756, 2013.

[14] V. Gazi and K. M. Passino, "Stability analysis of swarms," *IEEE Transactions on Automatic Control*, vol. 48, no. 4, pp. 692–697, 2003.
[15] R. Olfati-Saber, "Flocking for multi-agent dynamic systems: Algorithms and theory," *IEEE Transactions on Automatic Control*, vol. 51, no. 3, pp. 401–420, 2006.
[16] L. Chaimowicz and V. Kumar, "Aerial shepherds: Coordination among UAVs and swarms of robots," in *Proceedings of the International Symposium on Distributed Autonomous Robotic Systems*, France, June 2004, pp. 23–25.
[17] N. Ayanian and V. Kumar, "Decentralized feedback controllers for multiagent teams in environments with obstacles," *IEEE Transactions on Robotics*, vol. 26, no. 5, pp. 878–887, 2010.
[18] M. A. Dehghani and M. B. Menhaj, "Communication free leader-follower formation control of unmanned aircraft systems," *Robotics and Autonomous Systems*, vol. 80, pp. 69–75, 2016.
[19] F. Shamsi, F. Abdollahi, and S. K. Nikravesh, "Time varying formation control using feedback information differential game approach," in *Proceedings of the 19th Iranian Conference on Electrical Engineering*, Iran, May 2011, pp. 1–6.
[20] M. A. Dehghani and M. B. Menhaj, "Integral sliding mode formation control of fixed-wing unmanned aircraft using seeker as a relative measurement system," *Aerospace Science and Technology*, vol. 58, pp. 318–327, 2016.
[21] D. Richert and J. Cortes, "Optimal leader allocation in UAV formation pairs ensuring cooperation," *Automatica*, vol. 49, no. 11, pp. 3189–3198, 2013.
[22] F. Shamsi, F. Abdollahi, and S. K. Nikravesh, "Time varying formation control using differential game approach," in *Proceedings of the 18th IFAC World Congress*, Milan, August 2011, pp. 1114–1119.
[23] A. Sinha and D. Ghose, "Generalization of linear cyclic pursuit with application to rendezvous of multiple autonomous agents," *IEEE Transactions on Automatic Control*, vol. 51, no. 11, pp. 1819–1824, 2006.
[24] J. Cortes, S. Martinez, and F. Bullo, "Robust rendezvous for mobile autonomous agents via proximity graphs in arbitrary dimensions," *IEEE Transactions on Automatic Control*, vol. 51, no. 8, pp. 1289–1298, 2006.
[25] L. Sun and Z. Zheng, "Adaptive relative pose control for autonomous spacecraft rendezvous and proximity operations with thrust misalignment and model uncertainties," *Advances in Space Research*, vol. 59, pp. 1861–1871, 2017.
[26] A. Jadbabaie, J. Lin, and A. S. Morse, "Coordination of groups of mobile autonomous agents using nearest neighbor rules," *IEEE Transactions on Automatic Control*, vol. 48, no. 6, pp. 988–1001, 2003.
[27] H. Wang, "Position/velocity consensus in formation flying of multiple spacecraft," in *Proceedings of the 13th IFAC Symposium on Large Scale Complex Systems: Theory and Applications*, China, July 2013, pp. 141–146.
[28] W. Ren and R. Beard, *Distributed Consensus in Multi-vehicle Cooperative Control*. UK: Springer, 2008.
[29] R. Olfati-Saber, J. A. Fax, and R. M. Murray, "Consensus and cooperation in networked multi-agent systems," *Proceedings of the IEEE*, vol. 95, no. 1, pp. 215–233, 2007.

[30] J. Antich, A. Ortiz, and J. Minguez, "A bug-inspired algorithm for efficient anytime path planning," in *Proceedings of the IEEE/RSJ International Conference on Intelligent Robots and Systems*, USA, October 2009, pp. 5407–5413.

[31] G. Tanzmeister, M. Friedl, D. Wollherr, and M. Buss, "Efficient evaluation of collisions and costs on grid maps for autonomous vehicle motion planning," *IEEE Transactions on Intelligent Transportation Systems*, vol. 15, no. 5, pp. 2249–2260, 2014.

[32] S. Karaman and E. Frazzoli, "Sampling-based algorithms for optimal motion planning," *The International Journal of Robotics Research*, vol. 30, no. 7, pp. 846–894, 2011.

[33] D. An and H. Wang, "VPH: A new laser radar based obstacle avoidance method for intelligent mobile robots," in *Proceedings of the 5th World Congress on Intelligent Control and Automation*, China, June 2004, pp. 4681–4685.

[34] S. Aiushita, T. Hisanobu, and S. Kawahmura, "Fast path planning available for moving obstacle avoidance by use of Laplace potential," in *Proceedings of the IEEE/RSJ International Conference on Intelligent Robots and Systems*, Japan, July 1993, pp. 673–678.

[35] Y. S. Nam, B. H. Lee, and M. S. Kim, "View-time based moving obstacle avoidance using stochastic prediction of obstacle motion," in *Proceedings of the IEEE International Conference on Robotics and Automation*, USA, April 1996, pp. 1081–1086.

[36] T. M. Lam, H. W. Boschloo, M. Mulder, and M. M. Paassen, "Artificial force field for haptic feedback in UAV teleoperation," *IEEE Transactions on Systems, Man and Cybernetics, Part A: Systems and Humans*, vol. 39, no. 6, pp. 1316–1330, 2009.

[37] J. Borenstein and Y. Koren, "Real-time obstacle avoidance for fast mobile robots," *IEEE Transactions on Systems, Man, and Cybernetics*, vol. 19, no. 5, pp. 1179–1187, 1989.

[38] N. Y. Ko and B. H. Lee, "Avoidability measure in moving obstacle avoidance problem and its use for robot motion planning," in *Proceedings of the IEEE/RSJ International Conference on Intelligent Robots and Systems*, Japan, November 1996, pp. 1296–1303.

[39] H. Rezaee and F. Abdollahi, "Mobile robots cooperative control and obstacle avoidance using potential field," in *Proceedings of the IEEE/ASME International Conference on Advanced Intelligent Mechatronics*, Hungary, July 2011, pp. 61 –66.

[40] H. Rezaee and F. Abdollahi, "Adaptive artificial potential field approach for obstacle avoidance of unmanned aircrafts," in *Proceedings of the IEEE/ASME International Conference on Advanced Intelligent Mechatronics*, Taiwan, July 2012, pp. 1–6.

[41] S. S. Ge and Y. J. Cui, "Dynamic motion planning for mobile robots using potential field method," *Autonomous Robots*, vol. 13, no. 3, pp. 207–222, 2002.

[42] C. M. Breder, "Equations descriptive of fish schools and other animal aggregations," *Ecology*, vol. 35, no. 3, pp. 361–370, 1954.

[43] C. Reynolds, "Flocks, birds, and schools: A distributed behavioral model," *Computer Graphics*, vol. 21, no. 4, pp. 25–34, 1987.

[44] T. Vicsek, A. Czirook, E. Ben-Jacob, O. Cohen, and I. Shochet, "Novel type of phase transition in a system of self-driven particles," *Physical Review Letters*, vol. 75, no. 6, pp. 1226–1229, 1995.

[45] K. Warburton and J. Lazarus, "Tendency-distance models of social cohesion in animal groups," *Journal of Theoretical Biology*, vol. 150, no. 4, pp. 473–488, 1991.

[46] A. Okubo, "Dynamical aspects of animal grouping: Swarms, schools, flocks, and herds," *Advances in Biophysics*, vol. 22, pp. 1–94, 1986.

[47] D. Grunbaum and A. Okubo, *Modelling Social Animal Aggregations*. Berlin, Heidelberg: Springer, 1994, pp. 296–325.

[48] V. Gazi and K. M. Passino, "A class of attractions/repulsion functions for stable swarm aggregations," *International Journal of Control*, vol. 77, no. 18, pp. 1567–1579, 2004.

[49] Z. Lin, W. Ding, G. Yan, C. Yu, and A. Giua, "Leader-follower formation via complex Laplacian," *Automatica*, vol. 49, no. 6, pp. 1900–1906, 2013.

[50] M. Mesbahi and F. Hadaegh, "Formation flying control of multiple spacecraft via graphs, matrix inequalities, and switching," *Journal of Guidance, Control, and Dynamics*, vol. 24, no. 2, pp. 369–377, 2001.

[51] B. Fidan, V. Gazi, S. Zhai, N. Cen, and E. Karatas, "Single-view distance-estimation-based formation control of robotic swarms," *IEEE Transactions on Industrial Electronics*, vol. 60, no. 12, pp. 5781–5791, December 2013.

[52] H. Rezaee, F. Abdollahi, and M. B. Menhaj, "Model-free fuzzy leader-follower formation control of fixed wing UAVs," in *Proceedings of the 13th Iranian Conference on Fuzzy Systems*, Iran, August 2013, pp. 1–5.

[53] K. Tan and M. A. Lewis, "Virtual structures for high-precision cooperative mobile robotic control," in *Proceedings of the IEEE/RSJ International Conference on Intelligent Robots and Systems*, Japan, November 1996, pp. 132–139.

[54] K. D. Do, "Bounded assignment formation control of second-order dynamic agents," *IEEE/ASME Transactions on Mechatronics*, vol. 19, no. 2, pp. 477–489, 2014.

[55] H. Rezaee, F. Abdollahi, and H. A. Talebi, "$\mathscr{H}_\infty$ based motion synchronization in formation flight with delayed communications," *IEEE Transactions on Industrial Electronics*, vol. 61, no. 11, pp. 6175–6182, 2014.

[56] J. A. Marshall, M. E. Broucke, and B. A. Francis, "Formations of vehicles in cyclic pursuit," *IEEE Transactions on Automatic Control*, vol. 49, no. 11, pp. 1963–1974, 2004.

[57] Z. Chen and H. Zhang, "A remark on collective circular motion of heterogeneous multi-agents," *Automatica*, vol. 49, no. 5, pp. 1236–1241, May 2013.

[58] H. Rezaee and F. Abdollahi, "Pursuit formation of double-integrator dynamics using consensus control approach," *IEEE Transactions on Industrial Electronics*, vol. 62, no. 7, pp. 4249–4256, 2015.

[59] M. S. Klamkin and D. J. Newman, "Cyclic pursuit or the three bugs problem," *The American Mathematical Monthly*, vol. 78, no. 6, pp. 631–639, 1971.

[60] F. Behroozi and R. Gagnon, "A computer-assisted study of pursuit in a plane," *The American Mathematical Monthly*, vol. 82, no. 8, pp. 804–812, 1975.

[61] F. Behroozi and R. Gagnon, "Cylcic pursuit in a plane," *Journal of Mathematical Physics*, vol. 20, no. 11, pp. 2212–2216, 1979.

[62] L. E. Barnes, M. A. Fields, and K. P. Valavanis, "Swarm formation control utilizing elliptical surfaces and limiting functions," *IEEE Transactions on Systems, Man, and Cybernetics—Part B: Cybernetics*, vol. 39, no. 6, pp. 1434–1445, 2009.

[63] W. Ren, "Consensus strategies for cooperative control of vehicle formations," *IET Control Theory and Applications*, vol. 1, no. 2, pp. 505–512, 2007.

[64] R. Olfati-Saber, "Distributed Kalman filtering for sensor networks," in *Proceedings of the 46th IEEE Conference on Decision and Control*, USA, December 2007, pp. 5492–5498.

[65] B. Shen, Z. Wang, and Y. S. Hung, "Distributed $\mathscr{H}_\infty$ consensus filtering in sensor networks with multiple missing measurements: The finite-horizon case," *Automatica*, vol. 46, no. 10, pp. 1682–1688, 2010.

[66] V. Borkar and P. Varaiya, "Asymptotic agreement in distributed estimation," *IEEE Transactions on Automatic Control*, vol. 27, no. 3, pp. 650–655, 1982.

[67] J. N. Tsitsiklis, D. P. Bertsekas, and M. Athans, "Distributed asynchronous deterministic and stochastic gradient optimization algorithms," *IEEE Transactions on Automatic Control*, vol. 31, no. 3, pp. 803–812, 1986.

[68] R. Olfati-Saber and R. M. Murray, "Consensus problems in networks of agents with switching topology and time-delays," *IEEE Transactions on Automatic Control*, vol. 49, no. 9, pp. 1520–1523, 2004.

[69] H. Ando, Y. Oasa, I. Suzuki, and M. Yamashita, "Distributed memoryless point convergence algorithm for mobile robots with limited visibility," *IEEE Transactions on Robotics and Automation*, vol. 15, no. 5, pp. 818–828, 1999.

[70] Z. Lin, M. E. Broucke, and B. A. Francis, "Local control strategies for groups of mobile autonomous agents," *IEEE Transactions on Automatic Control*, vol. 49, no. 4, pp. 622–629, 2004.

[71] J. B. Pomet, B. Thuilot, G. Bastin, and G. Campion, "A hybrid strategy for the feedback stabilization of nonholonomic mobile robots," in *Proceedings of the IEEE International Conference on Robotics and Automation*, France, May 1992, pp. 129–134.

[72] K. D. Listmann, M. V. Masalawala, and J. Adamy, "Consensus for formation control of nonholonomic mobile robots," in *Proceedings of the IEEE International Conference on Robotics and Automation*, Japan, May 2009, pp. 3886–3891.

Chapter 18
PSO in ANN, SVM and data clustering

Terje Kristensen[1] and Fabien Guillaume[2]

Abstract

In this chapter, one gives an introduction to different kinds of particle swarm optimization (PSO) algorithms. One also introduces artificial neural networks (ANNs), support vector machines (SVMs) and evolutionary computing to show how PSO may be used to determine optimal parameters using an ANN or SVM regime, for classification of DNA strings. In addition, PSO is used in the design of an SVM-based clustering algorithm.

Ant colony optimization (ACO) algorithms are also introduced in the chapter. Using ACO algorithms has been of great success to solve many discrete optimization and non-deterministic polynomial (NP)-hard problems, for instance the travelling salesman problem. The behaviour of ants is also used, for instance, to design an algorithm for data clustering. We want to develop later a similar application based on this clustering algorithm and compare it with the SVM one using PSO.

18.1 Introduction

Imagine that you are a bird in the middle of a flock during migration. You have to fly thousands of kilometres to reach a warm destination. You have two options; you may either fly on your own or fly within the flock. So, what do you prefer? The answer should be obvious. We have all noticed the specific formation of different bird flocks in the sky, and we know this is not because of randomness. Such a formation reduces the drag. It makes it easier to fly long distances when you are in the middle of the flock. Once the birds at the edge of the flock are tired, they switched with the birds inside the flock. In this way, birds may migrate big distances.

This simple example solves a complex problem, by reducing the drag to save energy to travel long distances. The flock may be referred to as a swarm. A swarm is composed of individuals (often mobile) who communicate directly or indirectly with each other to solve complex problems. Such an interaction results in a distributive collective problem-solving strategy and is defined as swarm intelligence (SI) [1].

[1]Department of Computing, Mathematics and Physics, Western Norway University of Applied Sciences, Norway
[2]PriceWaterHouseCoopers, Luxembourg

Computational SI refers to algorithmic models of such behaviour. The idea of creating such algorithms came from the study of bird flocks and ant swarms, to reproduce their behaviours in computer models. Such simulations show great ability to explore a multi-dimensional space and quickly turned into a quite new domain of algorithmic theory exploration. A swarm can be viewed as a group of agents cooperating to achieve some goal [2]. This collective intelligence seems to emerge from groups organized in different models.

Particle swarm optimization (PSO) is a well-known algorithm for optimization [1,3,4]. It is based on the intrinsic property of swarms to execute complex tasks by a self-organization process [5]. This may be a powerful way to explore the high-dimensional search space to find near-optimal solutions. In PSO, birds are similar to particles and are flown through the hyper-dimensional space. The social tendency of individuals is used to update the velocity of the particles. In other words, each particle is influenced by the experience of other particles, its neighbours and its own knowledge. There is normally no centralized structure how the agents should behave. The local interactions between the agents often lead to emergency of global behaviour. The particles in the multi-dimensional space, the search space, represent all the feasible solutions of a given problem. Using the *fitness* function, their position is updated to find an optimal or near-optimal solution.

18.2 Global particle optimization

Let $x_i(t)$ be the position of the particle i at time step t, the position at time step $t+1$ is then given by

$$x_i(t+1) = x_i(t) + v_i(t+1) \tag{18.1}$$

where $v_i(t+1)$ is the velocity component of particle i at time step $t+1$. The velocity component is often called the *social* component (SC) of the swarm. The velocity component plays a fundamental role in the optimization process since it reflects the social information exchanged in the swarm and its own experience. The first component is often referred to as the SC while the latter is called the *cognitive* component (CC). We may say that the velocity is equal to the SC plus the CC. The two main PSO algorithms will then only be different by their SC.

$$\text{Velocity} = \text{social component} + \text{cognitive component} \tag{18.2}$$

In a global particle swarm optimization (gPSO), the neighbourhood of the particle consists of the whole swarm. This means that the SC of the velocity is obtained by gathering information from all particles. The swarm is then a fully connected graph of individuals. This social information is the global best position of the swarm. The velocity component of the best fitness of the swarm is given by

$$v_i(t+1) = v_i(t) + c_c r_c[y_i(t) - x_i(t)] + c_s r_s[y'_j(t) - x_i(t)] \tag{18.3}$$

where c_c is the cognitive componen and r_c is the social component. The values of r_c and r_s are taken from a random distribution to introduce the stochastic element of the velocity. $x_i(t)$ is the position of the particle i at time t and $y_j'(t)$ is the best known position of the swarm at time t.

If f is the fitness function and d the dimension of the search space, we may define fitness by the mapping

$$f : R^d \rightarrow R \tag{18.4}$$

The local best position for a minimization problem is then

$$y_i(t) = \begin{cases} y_i(t-1), \text{ if } f(x_i(t)) > f(y_i(t-1)) \\ x_i(t), \text{ if } f(x_i(t)) \leq f(y_i(t-1)) \end{cases} \tag{18.5}$$

The global best position of a swarm of N particles is then given by

$$y_i(t) = \min\left(f(x_1(t)), f(x_2(t)) \ldots f(x_i(t)), \ldots f(x_n(t))\right) \tag{18.6}$$

The initial population of the swarm is often given by a random distribution of N particles located in the search space. The algorithm of the global best is given by

---

Algorithm 1: Global best PSO algorithm

Gbest
Initialize a swarm of N particles
While stopping condition true **Do**
  For each particles **Do**
    If $(x_i) > (y_i)$ **Do** //Set the personal best
      $y_i = x_i$
    End if
    If f(yi) $> f(y')$ **Do** //Set the global best
      $y' = y_i$
    End if
  End for
  For each particles **Do**
      $v_i = v_i + c_c r_c[y_i - x_i] + c_s r_s[y' - x_i]$ //Update the velocity
      $x_i = x_i + v_i$ //Update the position
  End for
End while

18.3 Local particle optimization

The local particle swarm optimization (LPSO) is different from the gPSO in the way the velocity is computed. Instead of using the global best position to obtain the SC, the local best particle is used instead. A neighbourhood is defined for each particle. This may be done in two ways. The first method is based on Euclidean distance where a distance is given so that a particle i belongs to the neighbourhood of another particle j. However, such a method is computational demanding. The second method uses particle indices.

For a swarm of size N, we may define a neighbourhood of size $n_s \in \{1, \ldots, N\}$. If $n_{s=N}$, we have gPSO situation. The neighbourhood of the individual particle i is given by

$$N_i = \{y_{i-ns}(t), y_{i-ns+1}(t), \ldots, y_i(t), y_{i+ns-1}(t), y_{i+ns+1}(t)\} \tag{18.7}$$

The local best position for a maximization problem is then defined by

$$v_i(t+1) = v_i(t) + c_c r_c[y_i(t) - x_i(t)] + c_s r_s[y_j''(t) - x_i(t)] \tag{18.8}$$

where y_j'' is the best position of particle i. One advantage of using a neighbourhood based on a particle index is that the information about good solutions is spread around the swarm and is independent of their spatial proximity. The algorithm is given by

---

Algorithm 2: Local best PSO algorithm

Lbest
Initialize a swarm of N particles
While stopping condition true **Do**
  For each particle *do*
    If $(x_i) > (y_i)$ **Do** //Set the personal best
      $y_i = x_i$
    End if
    If f(y_i) $> f(y')$ **Do** //Set the global best
      $y' = y_i$
    End if
  End for
  For each particle *do*
      $v_i = v_i + c_c r_c[y_i - x_i] + c_s r_s[y'' - x_i]$ //Update the velocity
      $x_i = x_i + v_i$ //Update the position
  End for
End while

The local best version of PSO, *Lbest*, reflects the structure of the circle neighbourhood. While Lbest is slower with respect to convergence, it often results in better solutions and search of a larger part of the space. Both algorithms continue running until convergence has been reached. Usually, a PSO algorithm is executed a fixed number of iterations or evaluations of the fitness function. The process may also terminate when the velocity changes are close to zero.

18.4 PSO variations

A fundamental aspect of PSO is the trade-off between *exploration* and *exploitation*, the so-called *EE trade-off* of the search space. By exploration, we mean the ability the swarm has to leave its initial area and spread itself out through the search space. By exploitation, one means how to focus on a given area to find potential solutions of a problem. If a swarm favours first on exploration, it may miss some good areas of the search space. On the other side, if the swarm favours exploitation, it may be trapped in local minima (or maxima depending on the nature of the problem). Within the PSO algorithm, this is related to the updating of the velocity component.

There may be three different variations of the updating of the velocity; clamping, inertia weight and constriction coefficient.

18.4.1 Clamping

To limit the growth of the velocity, one introduces a *clamping* equation to limit the speed of velocity updates:

$$v_i(t+1) = \begin{cases} v_i'(t), \text{ if } v_i'(t+1) \leq V_{\text{MAX}} \\ V_{\text{MAX}}, \text{ else} \end{cases} \tag{18.9}$$

where $v_i'(t+1)$ is the result of using (18.3) and (18.8). V_{MAX} is the velocity limit. This value is important since it controls the EE trade-off. A small value increases the exploitation and a large value will encourage exploration. The problem is to find the correct value of V_{MAX} [6,7] suggested that a fraction of the domain should, for each dimension j of the search space, be given by

$$V_{\text{MAX},j} = \mu(x_{\max,j} - x_{\min,j}) \tag{18.10}$$

The coefficient μ is determined during the experiments. Use of $V_{\text{MAX},j}$ limits the step size of each particle and also the direction of the movement. Such a change of direction may also encourage the exploration but may not be able to do an optimum search. $V_{\text{MAX},j}$ does not have to be constant. One regime is to start with a big value and reduce it over time. A large value encourages exploration and then reduces it to exploit the good region found in the first stage.

18.4.2 Inertia weight

The inertia weight is the trade-off of the EE-trade-off [8]. The velocity equation has three components. The social and CC are defined by a coefficient for each of them. However, there is no momentum contribution. The role of an inertia weight is to adjust the contribution of the momentum. The updating of the velocity is then given by the inertia weight was

$$v_i(t+1) = wv_i(t) + c_c r_c[y_i(t) - x_i(t)] + c_s r_s[y_j'(t) - x_i(t)] \tag{18.11}$$

For $w > 1$, the swarm diverges and explores the search space at an accelerating velocity, but the particle cannot move back to a promising area. For $w < 1$, the particles retard until they become stationary. The optimal value of w is dependent on the nature of the problem (as for velocity clamping). However, it has been shown that w should not be less than a certain value to guarantee convergence of the swarm [9].

18.4.3 Constriction coefficient

The last control mechanism for the EE trade-off is the constriction coefficient. The idea is the same as for the inertia weight. However, the coefficient applies for the three components of the velocity update equation:

$$v_i(t+1) = \chi[v_i(t) + c_c r_c[y_i(t) - x_i(t)] + c_s r_s[y'_j(t) - x_i(t)]] \tag{18.12}$$

$$\chi = \frac{2k}{\left|2 - \phi - \sqrt{\phi(\phi - 4)}\right|} \tag{18.13}$$

where $\phi = c_c r_c + c_s r_s$ and $\phi \geq 4$ and $k \in [0,1]$. These equations have been developed using an Eigenvalue analysis [10]. We observe that $\chi \in [0,1]$, meaning that the velocity is reduced at each time step. There is no need then for velocity clamping. The parameter k controls the EE trade-off. A small value implies fast convergence with local exploitation while a large value means high exploration and slow convergence of the swarm.

18.4.4 Algorithmic aspects

18.4.4.1 Velocity models

- *The cognitive-only model:* The SC of the velocity update equation is removed. The resulting model tends to perform local search around the initial position and displays poor performance in general. However, it is well suited for niching algorithms.
- *The social-only model:* The CC is now removed from the velocity update equation. The swarm converges faster than for the cognitive-only model and demonstrates better result for dynamic environments [11].
- *The selfless model:* This model is similar to the social-only model with one exception. The current particle is not allowed to become the best solution of its neighbourhood.

18.4.4.2 Initialization

The first step of any PSO algorithm is to initialize the control parameters and the swarm itself. A general approach is to uniformly cover the search space, using a generator of a random particle within the search space. To ensure a good cover of the search space at initialization, the Sobol and Faure sequences have been used [12,13].

18.4.4.3 Termination

Another aspect of PSO is the stopping condition. It has to satisfy two conditions. First, it should not stop the PSO to converge too fast, since it may find a suboptimal solution. Second, it should not require frequent computing of the fitness function to minimize the number of computations needed. Five major stopping conditions have been used.

- *Iterations:* The PSO stops when a certain number of iterations are reached. If this limit is too small, the swarm will find a suboptimal solution. If it is too large, the PSO may be doing useless computations.
- *Acceptable solution:* The PSO stops when an 'acceptable' solution has been found. A threshold is defined. When a particle achieves a fitness value within this threshold, the algorithm stops.
- *No more improvements:* The PSO stops when it fails to improve the fitness value over a certain number of iterations. If the best solution of the swarm stays stuck in a small window over a predefined number of iterations, we may consider that PSO has converged.
- *Normalized swarm radius close to zero:* The PSO stops when the radius of the swarm becomes small. The normalized swarm radius is obtained using the following equations:

$$R_{norm} = \frac{R_{\max}}{diameter(S)} \tag{18.14}$$

  where

$$R_{\max} = \|x_m - y'\| \tag{18.15}$$

  x_m being the particle with largest distance from y'. Diameter (S) is the diameter of the swarm at time step zero. This approach defines the global best at the centre of the swarm. If this method is used in a particle clustering algorithm, it will stop when the slope of the objective function is close to zero and creates a single cluster of $\in$ percent of the swarm.
  By letting $\in = 0.9$, the PSO stops when 90% of the particles is centred at the best solution. It gives us more control over the spatial distribution of the swarm to stop the search.
- *Small objective function slope:* The PSO will stop when the slope of the objective function is close to zero. We obtain the slope at time step t by

$$f'(t) = \frac{f(y'(t)) - f(y'(t-1))}{f(y'(t))} \tag{18.16}$$

Algorithm 3: Particle clustering algorithm

Particle Clustering Algorithm
Initialize Cluster $C = \{y'\}$
For about k times **Do**
    Calculate the centroid of C

$$x' = \frac{\sum_{i=1, x_i \in C}^{|C|} x_i}{|C|}$$

    For each particles **Do**
        If $\|x_i - x'\| < \in$ **Do**
            $C \leftarrow C \cup \{x_i\}$
        End if
    End for
End for

We recognize the derivative of the objective function in the formula above. If $f'(t)$ stays below a certain threshold $\in$ for a number of iterations, we may consider that the swarm is not making any progress and let PSO stop. However, this solution might cause the PSO to be trapped in a local minimum.

18.5 PSO parameters

The parameter values of PSO may vary by

The size of the swarm: A large initial population implies a large diversity and a good space coverage. However, the larger the swarm, the larger number of the computation per iteration we need. A big swarm may obtain an optimal solution in fewer iterations. Usually, a swarm should have between 10 and 30 particles [14] even if good results may be obtained by using smaller swarms.

The number of iterations: This depends on the external parameter values. The neighbourhood size influences the interaction between particles. A small neighbourhood implies less interactions and vice versa.

The acceleration coefficient: The acceleration coefficient, the social and the cognitive coefficient do not need to be constant. An adaptive acceleration coefficient has been proposed by Clerc [15].

18.6 Evolutionary computing and PSO

A PSO algorithm may use a genetic algorithm (GA) approach [16]. This type of algorithm is inspired by the genetic evolution of chromosomes. They maintain a population of genes (equivalent to a particle in PSO). The main idea is to take this initial population, mutate it and reproduce certain of the individuals. Mutation and

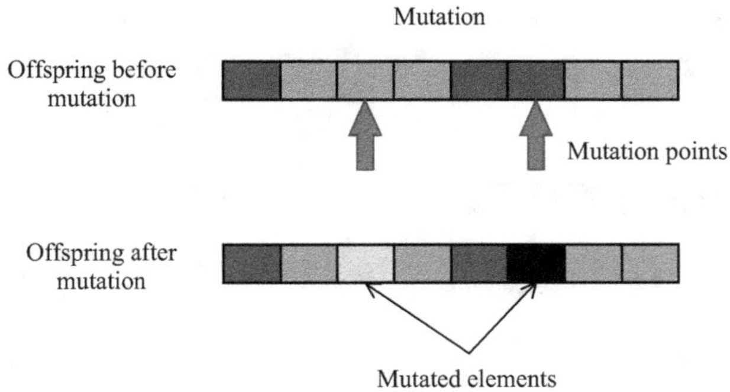

Figure 18.1 The mutation operator

reproduction are also used in a biological context. We may then replace a part of the initial population by the genes created. The major steps of any GA are as follows:

- Parent selection
- Crossover (or reproduction) strategies
- Mutation strategies
- Replacement strategies

There are different selection operators as follows:

- Random selection: each individual has a probability of $1/N$ to be selected where N is the total number of individuals.
- Proportional selection: the selection is biased towards the most fitted individuals. The higher the value of the fitness, the higher is the probability to be selected.
- Tournament selection: a random group of the population is selected, and the best individuals of this group are then returned. If two parents are needed, this strategy is applied twice.

Many variations exist. For more information, see [17].

18.6.1 Mutation

Mutation operators are used to introduce new genetic material in the population [9]. In other words, it introduces diversity. Along with the reproduction operators, it ensures a strong diversity. Mutation is defined by a certain probability p_m for each offspring that influences the selection of the offsprings (Figure 18.1).

A random value between 0 and 1 is added to the mutation points. If m_i is a mutation point of an individual, the mutated value is defined as

$$m_i = m_i + U(0, 1) \tag{18.17}$$

where (0,1) is the Gaussian (normalized) distribution between 0 and 1.

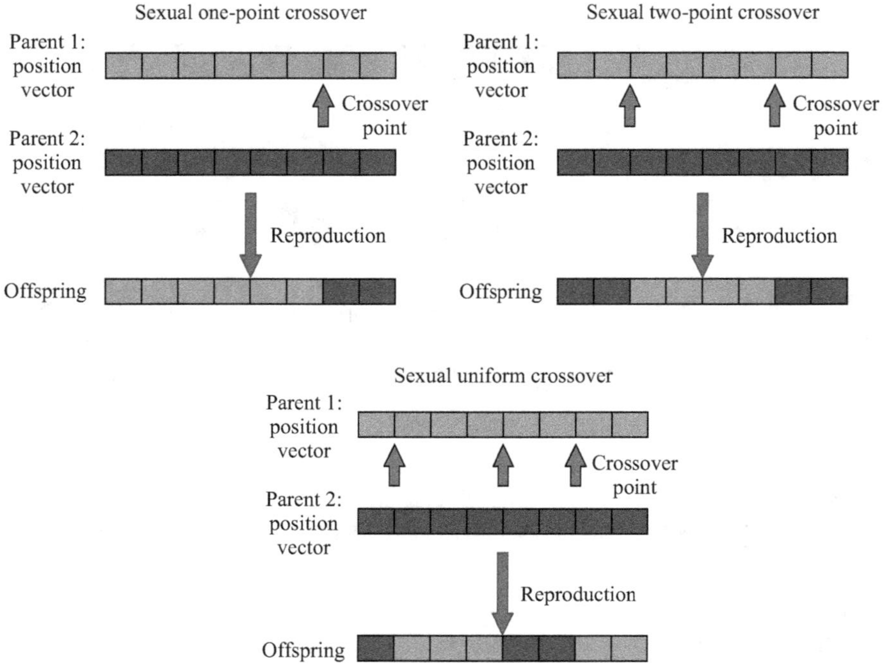

Figure 18.2 Different crossover points

18.6.2 Crossover

The crossover operators are divided into three categories (Figure 18.2):

- Asexual: only one parent is used for the reproduction.
- Sexual: two parents are needed for reproduction. In this case, the same parent can be selected twice depending on what kind of selection operator is used.
- Multi-recombination: more than two parents can be used for the reproduction.

Each of these categories may be divided further, such as

- One-point crossover: each parent is split in two; the offspring consists of a part of each parent.
- Two-point crossover: each parent is split in three ways, each part is then swapped to create a new offspring.
- Uniform crossover: the same principle, but with more than two crossover points.

These strategies have been used in the design of the custom PSO algorithm developed by the authors and described in Algorithm 6.

18.6.3 Summary

To briefly summarize, the GA is repeated by the following steps until certain termination conditions are satisfied:

- Selecting parents
- Applying the crossover operator to generate new individuals
- Applying the mutation operator to ensure diversity without destroying good individuals
- Integrating the new individuals in the current population

18.6.4 A selection-based PSO algorithm

In the selection-based PSO algorithm, the algorithm is combined with a GA. Algorithm 4 is executed before the velocity update of either local or global PSO. Their personal best position is saved. A selection-based PSO improves the local search capabilities. However, the diversity is reduced because half of the swarm is removed. This may be overcome by applying a mutation regime. The worst half of the swarm is then replaced by a mutated version that maintains the diversity of the swarm.

The reproduction process involves selection of particles that will generate offsprings. In Clerc [15], one allows a particle to generate a new one, kill itself or modify the inertia and acceleration coefficient. If no improvement in the neighbourhood is observed, a new particle is generated in the neighbourhood. By using this approach, a large swarm reduces the probability to generate a particle, and a small swarm increases this probability. In [8], another approach was proposed to spawn a particle in the neighbourhood of the global best particle to reduce the complexity. The global best particle will then generate an offspring if it gets trapped in a local minimum. The Gaussian mutator is introduced by Higashi and Iba [9] and may be defined by the following:

Let $x_i'(t+1)$ be the position of a particle after the velocity updating, and P the probability to mutate. For each component $j \in \{1 \dots d\}$, if $U(0,1) < P$, the component $x_{ij}'(t+1)$ is mutated by

$$x_{ij}(t+1) = x_{ij}'(t+1) + N(0,\sigma)x_i'(t+1) \tag{18.18}$$

where $(0, \sigma)$ is the Gaussian distribution and σ is given by

$$\sigma = \alpha(x_{max} - x_{min}) \text{ and } \alpha \in [0, 1] \tag{18.19}$$

Different Gaussian or Cauchy mutators have been proposed as in Secrest and Lamont [18], Wei *et al.* [19] and Yao *et al.* [11]. The Cauchy mutators are used on the Cauchy distribution instead on the Gaussian. Algorithm 4 presents the pseudo-code for the selection-based PSO.

18.6.5 Replacement

Once the offspring is created and mutated, it is integrated into the population. The main replacement strategies are as follows:

- Worst replacement, the worst individuals are replaced by the new offsprings.

- Random replacement, individuals are randomly selected to be replaced by the offspring.
- Parent–offspring, offsprings replace their parents.
- Elitist strategy, the best individuals are protected from the replacement.

Algorithm 4: A selection-based PSO algorithm

Selection-based PSO
Initialize Cluster $C = \{y'\}$
For each particle **Do**
  Select n_s particles
  Evaluate the performance of the current particle relative to the n_s selected particles
 End for
 For each particle **Do**
 Order the particles according to their scores
 Replace the worst half of the swarm by the best half of swarm without changing the personal best positions
End for

18.7 A multi-phase PSO

The sub-swarm concept has been developed by Løvberg *et al.* [12] and Al-Kazemi and Mohan [13]. In a multi-phase PSO, the swarm is split in subgroups where each subswarm exhibits different behaviour and performs different tasks. The individuals can be allowed to migrate between groups. A common approach is to split the initial swarm in two groups and randomly assigned individuals to the groups. The research strategy alternates between two phases. First, a group in the *attraction* phase means that particles within the group will move towards the global best position. The second phase is the *repulsion* phase where individuals move away from the global best position. The velocity component is defined by

$$v_i(t+1) = wv_i(t) + c_c x_i(t) + c_s y'(t) \tag{18.20}$$

In order to change between phases, the triplet representing the inertia weight, the cognitive and the social coefficient (w, c_c, c_s) is changed. The *attraction* phase pushes the particle towards the global best particle; i.e. they tend to ignore their CC and follow the social one. The triplet becomes then (1,−1,1). For the *repulsion* phase, individuals do the opposite; they ignore the SC and follow their personal best position. The triplet is then (1,1, −1). A group can change the phase after a user-defined number of iterations or when there is no improvement of the fitness value. In order to detect an improvement or a stagnation of the fitness, different termination conditions may be selected. Furthermore, the velocity vector of every particle is periodically randomly reset.

A multi-phase PSO can exhibit more than two phases. The life-cycle PSO uses three phases. A regular PSO behaviour updating the velocity using either the LPSO or the GPSO. The second phase consists of a GA [20] where particles reproduce with each other, mutate and finally the best offsprings are selected. Finally, a hill climber phase is established, where a particle's position is updated only if the new position is better. They become solitary stochastic hill climbers. In the life-cycle PSO, the three types of individuals may exist together in the same swarm. The conditions for change of the phase are the same for each sub-swarm PSO. In Algorithm 5, the pseudo-code of the life-cycle algorithm is presented.

Algorithm 5: The life-cycle PSO

```
Life-Cycle PSO
Repeat
For each particle Do
   evaluate fitness
   if no improvement
     change the phase
   End if
  End for
 For each PSO particle Do
  Update the velocity vector
  Update the position
 End for
For each GA individual do
  Reproduce
  Mutate
  Select the new position
End for
For each Hill Climber Do
  Find the new neighbouring positions
  Evaluate their fitness
  Move toward the best one using a specified probability
 End for
Until stopping condition is true
```

18.8 Custom PSO

The custom PSO tries to combine several types of PSO. The first step is to split the swarm in two sub-swarms, each one in a different phase. Both attraction and repulsion are used in multi-phase PSO. Each group is following a global PSO approach and allows big diversity in the swarm. At every iteration, each group updates its velocity and position. The algorithm then evaluates the probability of the worst sub-swarm to mutate. If the random number generated is greater than the predefined mutation probability, the whole sub-swarm is mutated, and the phase is changed. This is the approach used in the selection-based PSO. However, the selection-based PSO tends to reduce the diversity to avoid introduction of Gaussian mutation and changing the phase of the group. Finally, at later stages of the search process, the algorithm switches the behaviour of the best swarm to a hill climbing behaviour (defined earlier). The process stops after a predefined number of iterations have been reached. The pseudo-code of this algorithm is presented in Algorithm 6. A simple iteration condition is selected to stop the search process since the goal is to explore the search space. The custom PSO is applied in this chapter together with some PSO variations.

Algorithm 6: The custom PSO algorithm

```
Custom PSO
Initialize the swarm using a uniform distribution
Split the swarm in two equal groups, one in attraction phase and one in
repulsion phase
For each group do
  Update the fitness of the group // also set local and personal best
  For each particle Do
    Update the velocity using (18.18)
    Update the position
  End for
If random probability > mutation_probability and current group is the
worst Do
  order the group based on fitness
  Replace the worst half by the top half (applying Gaussian mutation
  and keeping the personal best)
  Switch the Phase (Attraction or Repulsion)
 End for
End if
if reached 80% of max number of iterations Do
  Switch the best swarm to Hill climbing behaviour
 End if
End For (the max number of iterations reached)
```

18.9 Artificial neural network

18.9.1 The biological neuron

The biological neuron consists of dendrites, synapse and axon. The axon transfers nerve signals to the other neurons. It is about 0.1 mm–1 m big. The dendrites receive incoming signals from the other neurons. The synapse consists of the synaptic cleft which contains neurotransmitter molecules. The building block of the brain is the

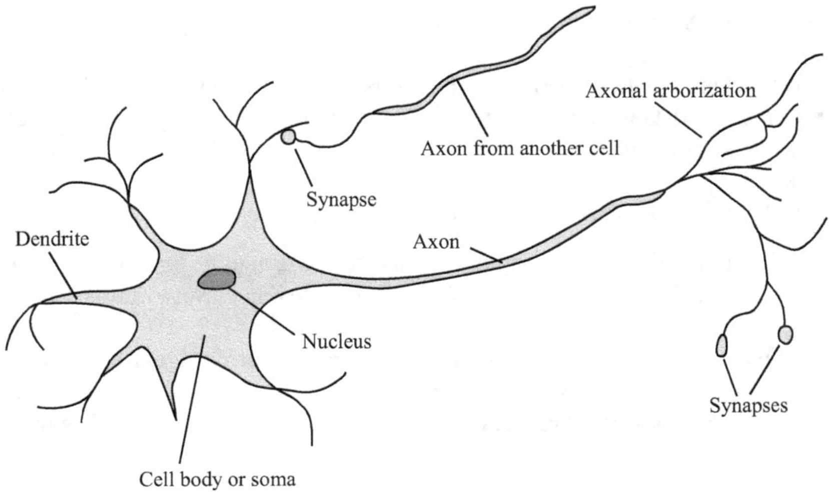

Figure 18.3 The biological neuron

neuron. The transport of signal in the brain consists of two types of signals as follows (Figure 18.3):

- Chemical signals across the synapse
- Electrical signals inside the neuron

The rest potential of a neuron is about −470 mV compared to its surroundings. Outside the neuron, the potential is 0 mV. Each signal from the other neurons gives a transient change in the resting potential. When the resting potential is about −50 mV, the resting potential changes rapidly which results in a positive signal towards its surroundings of about +50 mV. This recharging impulse (spike) propagates along the axon away from the nucleus towards the synapses.

18.9.2 The multi-layer perceptron

The multi-layer perceptron (MLP) artificial neural network (ANN) has been known for a long time and is inspired by the design of the biological neural network in the brain. The brain is composed of biological neurons connected by synapses and organized in different layers [21]. The MLP has three basic features:

- Each unit (neuron) includes a differentiable nonlinear activation function
- The network contains one or more hidden layers
- The network is highly connected: each unit of a layer is usually fully connected to every unit of the next layer.

These features make the theoretical analysis of the network difficult for two reasons. First, the high connectivity and the nonlinearity of the network make the analysis hard to do. Second, the presence of hidden layers makes the learning process difficult to visualize.

The backpropagation (BP) algorithm is used to train the network. The training requires a set of data associated with their corresponding target vector. This means, we know to which cluster the data belong, to allow us to use them in the training process.

The BP algorithm consists of two phases.

- *Feed-forward phase*: the synaptic weights are fixed, and the input signal is propagated through the network layer by layer, from the input layer towards the output layer.
- *Backpropagation phase*: The network computes an error signal by comparing the output of the network to the desired output (target vector). The errors are propagated backward, from the output layer towards the input layer. The process stops when the global error is reaching a certain threshold.

MLP uses two types of signals:

- *Function signals*, defined as input signals, propagating forward and emerging at the output layer
- *Error signals*, defined by output neurons using an error-dependent function. The errors propagate backward through the network

18.9.2.1 The artificial neuron

An artificial neuron is an entity that receives one or several signals as input. Each input signal arrives through a synapse. It simply assigns a weight to the original signal, representing the strength of the signal. Each unit computing the weighted sum, called the induced local field:

$$y = \sum_{i=0}^{n} x_i w_i \tag{18.21}$$

A sigmoid function (or logistic function) is used to compute the output signal of the current unit (Figure 18.4):

$$\sigma(y) = \frac{1}{1 + e^{-y}} \tag{18.22}$$

The input neuron is the only unit that does not compute anything. It just transmits the input signal through its synapses. x_0 is a bias that may be applied to each layer. Using a bias adds one more dimension to the input space and gives the network greater flexibility.

18.9.2.2 The layers

An MLP is composed of three types of layers:

- An input layer that transmits the input data through its synapses to the others layers.

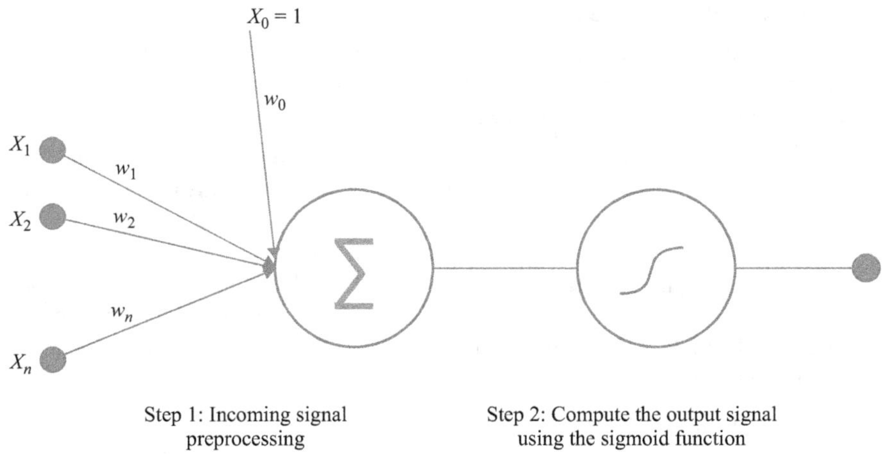

Figure 18.4 Sigmoid unit

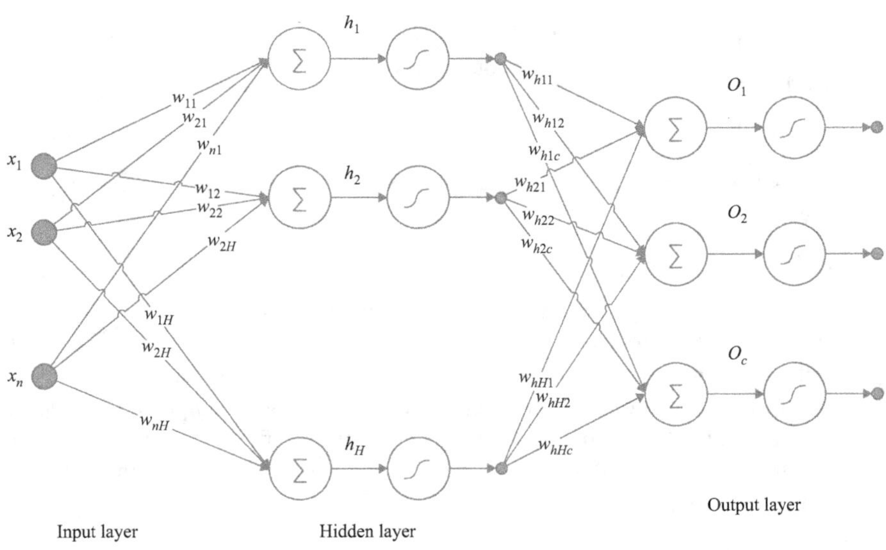

Figure 18.5 The MLP architecture

- One or several hidden layers that receive the signals coming from the input layer and computes the signals using the sigmoid function and sends the signals to another hidden layer or the output layer.
- An output layer, where the classification is done.
- Figure 18.5 shows a typical architecture of a MLP with one hidden layer.

18.9.2.3 The backpropagation algorithm

For a classification problem with c clusters (categories), we may use l–2 hidden layers. We may define a network with L layers as the depth of the network.

The BP algorithm consists of five steps:

1. Initialization: the weight matrix is initialized using a uniform distribution
2. Each training example is presented and the network performs one forward and one backward propagation step
   The structure of MLP is given in Figure 18.5.
3. Forward propagation: a training example $\{x(t), d(t)\}$, $x(t)$ is presented as input, and $d(t)$ is the target vector. The signal of each unit is computed using the induced local field equation. The induced local field $v_j^l(t)$ for iteration step t, neuron j and layer l is given by

$$v_j^l(t) = \sum_i w_{ji}^{(l)}(t) y_j^{l-1}(t) \tag{18.23}$$

   where $y_j^{l-1}(t)$ is the output signal of unit j of layer $L-1$ and $w_{ji}^{(l)}(t)$ is the weight of synapse between unit j in layer l and unit i in layer $l-1$ at iteration t. Assuming a sigmoid function, we have

$$y_j^l(t) = \begin{cases} x_j(t) & \text{if } j = 0, \text{ i.e. input layer} \\ o_j(t) & \text{if } j = L, \text{ i.e. output layer} \\ & \dfrac{1}{1 + e^{-v_j^l(t)}} \end{cases} \tag{18.24}$$

   Once every induced local field and function signal are computed, the error signal is evaluated as

$$e_j(t) = d_j(t) - o_j(t) \tag{18.25}$$

   where d_j is the jth element of the desired output vector $d(t)$, and o_j is the jth element of the real output vector $o(t)$ obtained from the output layer. When the error vector has been computed, forward phase propagation finishes.
4. Backward propagation: it uses the error vector previously computed to compute the local gradient. Indeed, the backpropagation tries to find the steepest vector in the error function in order to minimize the error for the next iteration. The local gradient δ is defined by:

$$\delta_j^{(l)}(t) = \begin{cases} e_j^{(L)}(t)\varphi_j'\left(v_j^{(L)}(t)\right) & \text{for neuron } j \text{ in output layer } L \\ \varphi_j'\left(v_j^{(L)}(t)\right) \sum_k \delta_k^{(l+1)}(t) w_{kj}^{(l+1)}(t) & \text{for neuron } j \text{ in hidden layer } l \end{cases} \tag{18.26}$$

   where φ' is the derivative of the sigmoid function with respect to the argument and k is varying through the connected neurons of the jth neuron. Finally, the synaptic weights in layer l are adjusted using the equation:

$$w_{ji}^{(l)}(t+1) = w_{ji}^{(l)}(t) + \alpha[w_{ji}^{(l)}(t)] + \eta\delta_j^{(l)}(t) y_i^{(l-1)}(t) \tag{18.27}$$

   where α is the learning rate and η is the momentum.

5. Iteration: forward and backward propagation are repeated until the stopping criterion is reached

Once the BP algorithm terminates, the weights are fixed and the network can be used to classify unseen data. A major difficulty with MLP is to find the right value for the different parameters of the network, that is the number of hidden units, number of hidden layers, learning rate and momentum. Another aspect is the long time it may take to train the network. The network can be over-trained if the network learning is performing too well. This means that the network may perform well on the training data, but not so well on unseen data. For a complete proof of the BP algorithm, see [22].

18.10 Support vector machine

A support vector machine (SVM) [23], is a binary learning machine (theoretical) to be able to classify a pattern in two different classes. The core idea is

> Given some training samples, the SVM constructs a hyper plane (decision surface) to discriminate between positive and negative samples based on the principle of maximizing the separation of margins, Figure 18.6 illustrates the concept.

Support vectors (SVs) are a subset of data points selected by the learning algorithm and are the most difficult to classify. The SVM is based on the *inner product kernel* between a SV and a vector drawn from the training data. Because of this property, the SVM algorithm is referred to as a *kernel method.* A kernel method designs an

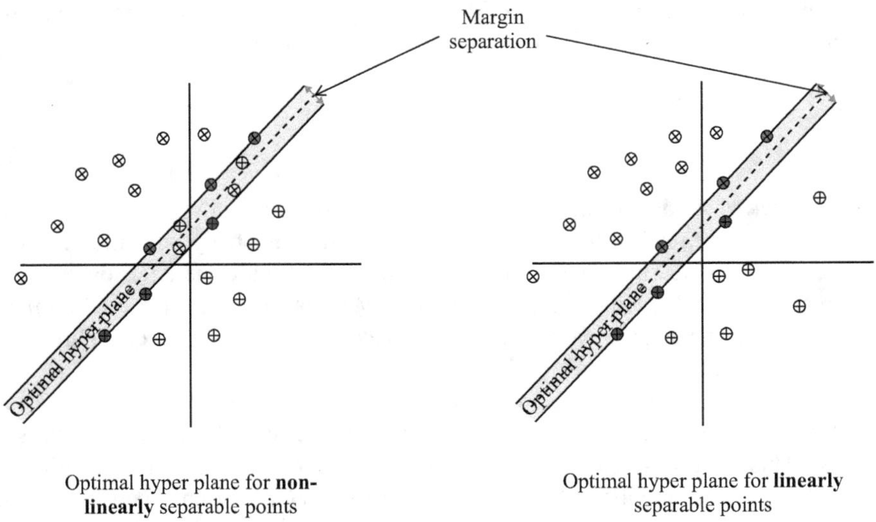

Figure 18.6 The margin separation of SVM

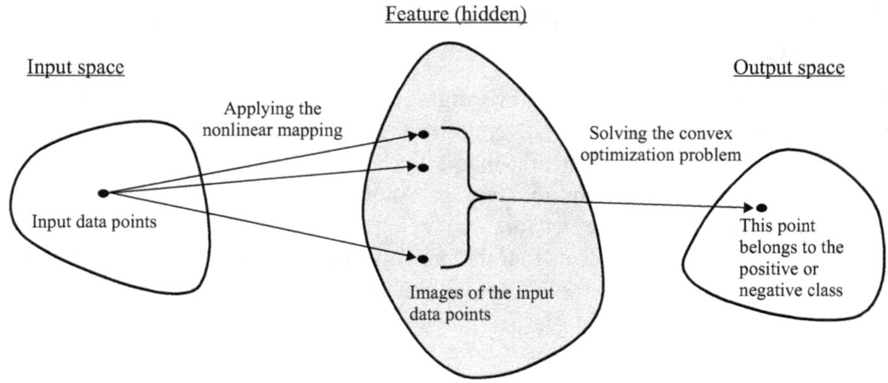

Figure 18.7 The mappings of the SVM

optimal decision surface. This is a convex optimization problem [24,25]. However, such a property comes at the price of high computational complexity, but the *kernel trick* solves this problem.

The SVM algorithm projects the training points into a feature space with a much higher dimension than the input space. Such projection is nonlinear, but according to the Cover Theorem [21], the data may become linearly separable in the feature space. This means that the data points can be separated by the hyper plane in the feature space. The nonlinear mapping is unknown and is computationally heavy and sometimes become impossible to compute. The kernel trick is then used to compute the images of the training points in the feature space. By using this trick, one may avoid the concrete computation of the nonlinear mapping. Let us be more formal.

A nonlinear mapping $\Phi : R^D \rightarrow F$ where F represents the feature space and k be a Mercer's Kernel [26] is given. The inner product of Φ is defined by

$$\Phi : R^D \rightarrow F \tag{18.28}$$

$$k(x, x') = \Phi^T(x')\Phi(x) \tag{18.29}$$

Mercer's kernel is known mathematical functions (polynomial, sigmoid etc.), and therefore we can calculate the inner product of Φ without actualy knowing it. In the feature space, the learning algorithm selects the SVs to build the decision surface. This mapping is achieved by first solving a convex optimization problem and then applying a linear mapping from the feature space to the output space. Figure 18.7 illustrates the concept of SVM.

SVM has been initially designed to classify binary data. However, a multi-class SVM has been designed by allowing classification of a finite number of classes [27,28]. This kind of learning assumes a priori knowledge of the data (the different classes). However, an algorithm for SV clustering (SVC) [29] has been developed using the core idea of SVM. We will explore the SVM kernel clustering algorithm more in details.

18.10.1 The Kernel

The introduction of the SVM has increased the popularity of the Kernel algorithm since the 1990s, due to its high performance in both supervised classification and regression analysis [30]. It has also been successfully applied in unsupervised clustering. The key behind it is the Cover Theorem:

> *A complex pattern-classification problem, cast in a high-dimensional space nonlinearly, is more likely to be linearly separable than in a low-dimensional space, provided that the space is not densely populated.*

Given a set $X = \{x_1, \ldots, x_j, \ldots, x_N\}$ with $x_j \in R^d$ of input pattern and a nonlinear map $\Phi : R^D \to F$ where F represents a feature space of arbitrarily high dimension. X is mapped into F through Φ so that a linear algorithm can be performed. The cover theorem is illustrated in Figure 18.8.

A problem soon appears. As the dimension increases, so does the complexity of the problem. This is called the *curse of dimension* [31]. This difficulty is overcome using the kernel trick from Mercer's theorem [26]. See also Duda *et al.* [32].

> *Let $k(x, x')$ be a continuous symmetric kernel that is defined in the closed interval $a \leq x \leq b$, and likewise for x'. The kernel $k(x, x')$ can be expanded in the series:*

$$k(x, x') = \sum_{i=1}^{\infty} \lambda_i \phi_i(x) \phi_i(x') \tag{18.30}$$

> *With positive coefficients $\lambda_i > 0$ for all i. For this expansion to be valid and to converge, it is necessary and sufficient that the condition*

$$\int_b^a \int_b^a k(x, x') \psi(x) \psi(x') dx dx' \geq 0 \tag{18.31}$$

> *holds for all $\psi(\cdot)$ for which we have*

$$\int_b^a \psi^2(x) dx < \infty \tag{18.32}$$

> *where a and b are the constants of integration.*

Calculating Φ is time-consuming and often infeasible. However, Mercer Theorem [26] allows us to avoid this computation, and there is no need to explicitly describe the non-linear mapping Φ neither the image points in the feature space F. This technique is known as the Kernel trick.

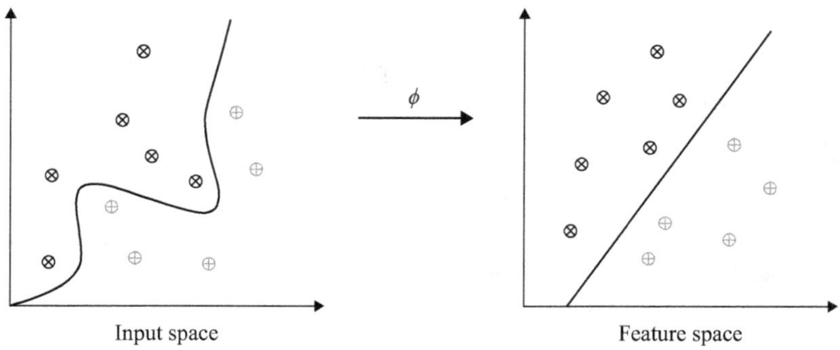

Figure 18.8 The cover theorem

18.10.2 Definition of Kernels

The function $k(x,x')$ is called an inner product kernel, or kernel. It is a function that computes the inner product of images produced in the feature space under Φ of two data points in the input space [23]. Two properties are derived from this definition:

Property 1: $k(x,x')$ is symmetric about the centre x':
The maximum is reached for $x = x'$
Property 2: The volume under $k(x,x')$

However, this is the general definition of a kernel. To use the kernel trick, the function $k(x,x')$ has to satisfy Mercer's theorem. In the scope of this observation, only four types of kernel remain as listed below.

The polynomial kernel:

$$k(x,x') = (x * x' + 1)^p \tag{18.33}$$

The Gaussian kernel:

$$k(x,x') = \exp\left(-\frac{1}{2\sigma^2}\|x - x'\|^2\right) \tag{18.34}$$

where σ is the width parameter or margin separation. In the case of clustering, it controls the granularity of the clustering.

The sigmoid kernel:

$$k(x,x') = \tan\text{h}(\theta_0(x + x') + \theta_1) \tag{18.35}$$

where θ_0 and θ_1 are user-specified parameters. Furthermore, the sigmoid kernel satisfies Mercer's theorem only for specific value of θ_0 and θ_1. In SVC, the Gaussian kernel is the most used because the polynomial kernel does not allow tight contours of the clusters to be created [33].

18.10.3 Quadratic programming

Quadratic programming (QP) [34] is a class of problems that are close to linear programming problems, with the difference that the objective function contains products of pairs of variables (called quadratic terms). In SVC, such problems are more abstract. Fortunately, this constitutes a well-known class of problems, and the solution does exist. They belong to the field of convex programming [25,35]. Although solving such problems are beyond the scope of this book, it is necessary to give an overview of QP here. QP problems are usually minimization problems of the following form:

$$\begin{aligned}&\text{minimize } c^T x + \frac{1}{2}x^t Qx\\&\text{subject to } Ax \geq b \text{ and } x \geq 0\end{aligned} \tag{18.36}$$

This is solved by calculating the dual problem by introducing the Lagrangian. The dual problem is obtained using the Karush–Kuhn–Tucker (KKT) complementarity conditions [36]:

$$\begin{aligned}&\text{maximize } b^T y\\&-\frac{1}{2}x^t Qx \text{ subject to } A^T y + z - Qx = c \quad \text{and} \quad y, z \geq 0\end{aligned} \tag{18.37}$$

18.10.4 KKT conditions

The KKT complementarity conditions are defined beneath by letting $\lambda \in R^m_+$. The following Lagrangian function is defined as

$$L(X, \lambda) = f(X) + \sum_{i=1}^{m} \lambda_i g_i(X) \tag{18.38}$$

where λ_i is the Lagrangian multiplier.

The KKT conditions combine two conditions: the stationary condition and the complementary slackness condition:

To maximize the Lagrangian L with respect to X, the first order derivative needs to be zero:

$$\frac{\partial f}{\partial X}(X) + \sum_{i=1}^{m} \lambda_i \frac{\partial g_i}{\partial X}(X) = 0 \quad \text{where} \quad \frac{\partial}{\partial X} \text{ is the gradient} \tag{18.39}$$

The complementary slackness conditions or positivity conditions are stated as

$$min[\lambda_i, g_i(X)] = 0, \forall i \in \{1 \ldots m\} \tag{18.40}$$

For more information, see [37]. The literature about this topic is very rich.

18.11 DNA classification

18.11.1 Presentation of the problem

One of the main research goals of molecular biology has been to determine a complete genetic description of any organism. In the Human Genome Project [38], the goal was to decipher the exact sequence of about 3 billion nucleotides in the 46 human chromosomes. An important part of the genome project is the computational processing of data. Since the birth of the Human Genome Project, sequence analysis as a computational method has been used to infer biological information from the sequenced data.

The classical approach for analysing sequences is by the use of sequence matching, using either single or multiple alignment techniques [39]. With these techniques, one seeks to determine whether sequences are significantly similar or not. Another approach is to use theories from neural computing or statistical learning theory to detect genetic information on the DNA sequences.

Neural networks have been applied to various tasks such as DNA sequence recognition. A neural network trained by a BP algorithm may learn to categorize between different types of bacteria cells related to the structure of their DNA-sequences. Such a method is based on pattern recognition analysis and is built on the assumption that some underlying characteristics of the DNA-sequence can be used to identify its bacteria type. Other neural network paradigms than an MLP network may also be used to analyse DNA sequences.

In this chapter, however, we will focus on how to use both a MLP and an SVM to distinguish between eukaryotic and prokaryotic DNA-sequences on basis of their nucleotide frequency structure. Cells can be divided into two major groups, prokaryotic and eukaryotic cells. All prokaryotic cells are uni-cellular organisms and consist mostly of bacteria. The genome of a prokaryotic cell consists of one double helix DNA strand, floating freely in the cell. This double helix strand is often circular. The genome of the bacterium *E. coli*, for instance, consists of a circular strand of five and a half million bases.

A nucleotide sequence can be viewed as a language, based on an alphabet of four letters: A, G, C and T where the number of As is the same as the number of Ts, and the number of Cs is the same as the number of Gs. However, the relation of A(T) and G(C) can vary tremendously and depends on the actual species that are studied. Most eukaryotes are multi-cellular, but some are uni-cellular. The main difference between prokaryotic cells and eukaryotic cells is that eukaryotic cells contain a nucleus that is surrounded by a membrane. Prokaryotic cells do not have such a nucleus. In such cells, the frequency distribution of pairs of nucleotides is different from those in prokaryotic cells. We want to train an MLP and an SVM combined with PSO and compare the results with our previous research [40]. But first, we need to introduce our DNA recognition theory.

18.11.2 DNA recognition theory

Statistical analysis of several DNA sequences has shown that the distribution of nucleotides is far from random. Some dinucleotide combinations in prokaryotic DNA sequences are more dominating than in eukaryotic cells. We will anticipate that this simple difference in data occurrence might be sufficient to allow species identification. We may then train, for instance, an MLP network, to use the differences in the nucleotide distribution to discriminate between eukaryotic and prokaryotic cells.

A C A T G A T G C T A ...

A C A T G A T G C T A ...

A C A T G A T G C T A ...

A C A T G A T G C T A ...

Figure 18.9 Illustration of a sliding window of size 2

We assume therefore that the identification of the DNA sequence is based solely on the frequency of nucleotide sub-sequences.

A sliding window is used to count the number of nucleotide sub-sequences of the DNA sequence. In general, the size of the window may vary, from one base wide to a user defined number w. By choosing a window of length one, we simply count the number of the different bases of the DNA sequence. The result will be four different frequencies, one for each base. A window of length two will give 16 different ordered sub-sequences. The frequency of each sub-sequence is computed by counting the occurrence of each nucleotide pair in the DNA sequence.

The number of triplets or *codon* units of the DNA sequence may be estimated by using a window of three bases wide. This results in 4^3 or 64 ordered triplets. This is maybe the most relevant sub-sequence to study since the codon itself has important meanings in the DNA sequence. In general, a window of w bases results in 4^w sub-sequences of length w to be counted.

18.11.3 Feature extraction

For a sliding window of length two, the frequency of sub-sequence AA is denoted as f_{AA}, for AC as f_{AC} and so on. These numbers are collected in a vector $\mathbf{F_n}$, where n denotes the number of DNA sub-sequences. For a sliding window of size two, n is equal to 16. The counting of the different nucleotide pairs is illustrated in Figure 18.9. In Figure 18.9 the counting of the nucleotide pair AC is shown. After counting the pair AC, the window is moved one letter to the right to cover the next nucleotide pair. This is done to the end of the DNA sequence. The relative frequencies of each nucleotide pair are put into a vector and normalized. This method for feature extraction has been developed by Kristensen [40].

18.11.3.1 Normalization

The DNA sequences obtained from on online databases have different sequence lengths (Figure 18.10). The frequency of the different nucleotide pairs has to be normalized before presenting them to the MLP or SVM network. The normalization condition of the frequency vector F_n is given by

$$S_n = \frac{F_n}{|F_n|} \tag{18.41}$$

Here $|F_n|$ means the Euclidean norm of the frequency vector. This nonlinear transformation conserves the direction of the vector and enhances the differences among

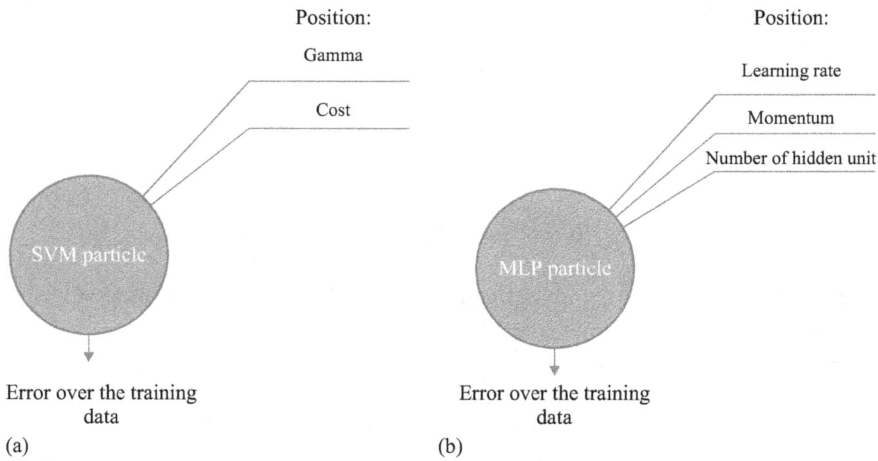

Figure 18.10 An SVM/MLP particle

the input vectors. The geometric interpretation of the transformation is that the vector F_n is moved onto the hyper unit sphere. In our case, we will use a sliding window of size three and incorporate the redundancy of the genetic code. It means that the representation of the genetic code may be built into the neural network by use of the frequency vector.

18.11.4 Use of PSO

The BP algorithm tries to minimize an error function. This could be done by using PSO [41]. This is a common approach; instead of using the BP algorithm, swarm optimization can be used to find the weight matrix that minimizes the error function. However, we will use the PSO to find the other parameters of the network, as well. Indeed, a particle will represent a possible configuration of the network. In other words, the position of a particle will be given by the following vector: learning-rate, momentum, number of hidden neuron [42]. The weight matrix is determined by using the back-propagation algorithm.

PSO is used to determine the optimal parameters of the MLP and the SVM networks. The SVM uses a Gaussian Kernel to determine two parameters: gamma and the cost. The MLP needs three parameters to be determined: the learning rate, the momentum and the number of hidden units (the input and output layer are already determined).

The implementation of the MLP was done using the Neuroph library, while the SVM was implemented by using LIBSVM [43].

The program works in several steps:

1. The data is stored in text files containing the DNA sequences. The ‘IOManager’ class calls the ‘TextReader’ class to extract the DNA sequences from our text files and converts them into the ‘Sequence’ class.

```
SQ   Sequence 691 BP; 135 A; 243 C; 192 G; 121 T; 0 other;
     CCTGAACCCG GTGTCCCCGG GTGGGGGGTG GGGACGCCAC GGCCGAAGCA GCTAGCTCCG        60
     TTCGTGATCC GGGAGCCTGG TGCCAGCGAG ACCTGGAATT TCCGGTCTGG TTGGTCTGGG       120
     GCCCCGCGGA GCCAGGTTGA TACCCTCACC TCCCAACCCC AGGCCCTCGG ATGCCCAGAA       180
     CCTGTAGGCC GCACCGTGGA CTTGTTCTTA ATCGAGGGGC ACTTCTACCC TAGCCGGGCC       240
     CAGCCCCCGA GCAGTGCAGC CTCCCGAGTG CAGAGTGCAG CCCCTGCCCG CCCTGGCCCA       300
     GCTGCCCATG TCTACCCTGC TGGATCCCAA GTAATGATGA TCCCTTCCCA GATCTCCTAC       360
     CCAGCCTCCC AGGGGGCCTA CTACATCCCT GGACAGGGGC GTTCCACATA CGTTGTCCCG       420
     ACACAGCAGT ACCCTGTGCA GCCAGGAGCC CCAGGCTTCT ATCCAGGTGC AAGCCCTACA       480
     GAATTTGGGA CCTACGCTGG CGCCTACTAT CCAGCCCAAG GGGTGCAGCA GTTTCCCACT       540
     GGCGTGGCCC CCGCCCCAGT TTTGATGAAC CAGCCACCCC AGATTGCTCC CAAGAGGGAG       600
     CGTAAGACGA TCCGAATTCG AGATCCAAAC CAAGGAGGAA AGGATATCAC AGAGGAGATC       660
     ATGTCTGGGG CCCGCACTGC CTCCACACCC A          A          A                691
```

Figure 18.11 A database entry of eukaryotic DNA sequence from the EMBL database found on the Internet. The length of the sequence is 691 bases. The DNA sequence belongs to a protein found in human beings

2. When the sequences are loaded, the 'IOManager' computes the frequency vector of each sequence using the 'Alphabet' class, which contains the representation of the codons of the genetic code to identify triplets (amino acids).
3. The fitness of either the MLP or the SVM network is initialized, followed by the generation of the swarm.
4. The PSO is executed and the data retrieved.

18.11.5 Experiments

The dataset we used was composed of 54 sequences: 28 eukaryotic sequences and 26 prokaryotic sequences. Figure 18.13 shows the type of file we have to deal with. The process required to automate the extraction of the DNA sequence within the file. Every DNA sequences were tagged with 'SQ', and the extraction was a straight forward string analysis of the file. The common approach is to split the data set in two, one part to train the SVM or the MLP, another to test their performance. However, in our case, the data set is too small to be split. The MLP does not perform well with too few training data. We therefore decided to evaluate each classifier only on the training data. In order to limit the training time of the MLP, the number of iterations was set to 10,000 (Figure 18.11 and Table 18.1).

The swarm was composed of 20 particles separated in two sub-swarms. A fixed number of iterations (10) have been used. In total, the MLP has been trained about 2,000 times more than SVM.

18.11.5.1 Results

The PSO performed well to find the configuration of the neural network. The parameters are given in the table beneath. The parameters of SVM are given in Table 18.2:

The results in the SVM table look better than MLP. However, the experiment did not work so well. The custom-PSO algorithm we have used was not able to optimize properly the SVM classifier. It converged after only two iterations.

Table 18.1 Final MLP results

Learning rate	Momentum	Hidden neurons	Error	Running time (min)
0.83	0.3	59	0.0096	11.5

Table 18.2 SVM results

Gamma	Cost	Error	Running time (s)
0.58	975	0.074	3.6

Table 18.3 MLP results

Learning rate	Momentum	Hidden neurons	Error	Running time (min)
0.78	0.82	75	0.0094	17

Table 18.4 SVM results

Gamma	Cost	Error	Running time (s)
0.8	624	0.074	5

To confirm the results, the experiments were reproduced using the global PSO. The swarm had the following parameters:

- Initial size: 25 particles
- Inertia coefficient: 0.95
- Cognitive coefficient: 0.9
- Social coefficient: 0.9
- Number of iterations: 10

The results are similar. For MLP, we observe a high learning rate in both cases. A large hidden layer, in the second experiments, was associated with high momentum. Both experiments gave a low error considering such a small number of DNA sequences. However, the downside is a long training time for the MLP network. In the second experiments of SVM, gamma was bigger with a lower cost, but the error was identical, and the running time still small. The experiments show the ability of PSO to optimize an MLP and an SVM network. However, it also may show some limitations of the SVM for the DNA classification problem. The MLP performed better than SVM but has a longer training time (Tables 18.3 and 18.4).

We want to come back to this DNA classification problem more thoroughly in a later version of the book and use this feature extraction method to classify between virus, bacteria and fungi, based on their complete genome. We are working on this problem now and the preliminary results are very good. By using PSO to estimate optimal values for the parameters of SVM recognition, we were able to get nearly 100% correct classification of the species.

18.12 Support vector clustering

SI has been used to find the parameters of an SVM [24]. We are now going to see how we may use this technique in clustering. In this case, PSO can be used to determine a near-optimal Eigenvector of the SVC method. We know that SVM learning results in a quadratic optimization problem to solve. The classical approach involves numerical methods or more advanced algorithms such as the Sequential Minimal Optimization (SMO) algorithm [44]. The SMO version of SVC exists and involves heavy numerical methods to be run, but it is fast to execute.

18.12.1 Theory

Over the last years, various Kernel clustering algorithms have been developed. They may be grouped in three families. The first type is based on *Kernelising the Metric*, where the metric is computed by means of Mercer's kernel in the feature space. The second type implements the K-means method [20] in the feature space, using *Kernel Methods*. The last one is based on SVM including *Kernel Methods*.

In SVC, the data points are mapped from the data space into a high-dimensional feature space by means of a Mercer's kernel (here the Gaussian kernel). We then look for the smallest hyper-sphere capturing all the image data points. When mapped back into the input space, the contour of the sphere describes the boundaries of the cluster. Finally, data points enclosed by the same contours are assigned to a cluster by computing an adjacency matrix.[1] Let us be more formal:

Given a set $X = \{x_1, \ldots, x_j, \ldots, x_N\}$ where$x_j \in R^d$ is an input pattern and a nonlinear mapping is given by $\Phi : R^D \rightarrow F$. The goal is to find the smallest hyper-sphere in F capturing all the data points of X after the nonlinear mapping is done. This means, we need to find the hyper-sphere H with the minimal radius R such as

$$\|\Phi(x_j) - \alpha\|^2 \leq R^2, \forall j \tag{18.42}$$

where $\|.\|$ is the Euclidian distance and α is the centre of H. The introduction of the slack variables ξ_j incorporates soft constraints. The goal is to find a hyper-sphere with a minimal radius R such as

$$\|\Phi(x_j) - \alpha\|^2 \leq R^2 + \xi_j, \xi_j \geq 0 \tag{18.43}$$

[1]An adjacency matrix is a matrix representing the connection of nodes in a graph. It is just one of many representations of a graph.

where $\xi_j \geq 0$, is a *slack* variable and α is the centre of the hyper-sphere. The primal problem [35] is solved in its dual form by introducing the Lagrangian.

$$L = R^2 - \sum_j (R^2 + \xi_j - \|\Phi(x_j) - \alpha\|^2)\beta_j$$
$$- \sum_j \xi_j \mu_j + C \sum_j \xi_j \tag{18.44}$$

$\beta_j \geq 0$ and $\mu_j \geq 0$ are the Lagrange multipliers, $C \sum_j \xi_j$ is a penalty term and C a regularization constant. The stationary condition of L is given by setting the derivative of L to zero with respect to R, α and ξ_j. This leads to

$$\sum_j \beta_j = 1 \tag{18.45}$$

$$a = \sum_j \beta_j \Phi(x_j) \tag{18.46}$$

$$\beta_j = C - \mu_j \tag{18.47}$$

This is a quadratic optimization problem [25] which involves the KKT complementarity conditions [36]. We then have

$$\xi_j \mu_j = 0 \tag{18.48}$$

$$(R^2 + \xi_j - \|\Phi(x_j) - \alpha\|^2)\beta_j = 0 \tag{18.49}$$

We observe from (18.49) that the image of a point x_j with $\beta_j > 0$ and $\xi_j > 0$ is outside the hyper-sphere H. According to (18.48), $\mu_j = 0$. It follows now from (18.47) that $\beta_j = C$. This point is called a *Bounded Support Vector*. A point x_j where $0 < \beta_j < C$ we see from $\mu_j > 0$ that $\xi_j = 0$. Such a point lies on the feature space sphere and is called Support Vectors (SV). All other points are inside the sphere.

SVs lie on cluster boundaries, while BSVs are outside, the rest are then inside the clusters. Using these relations, we can remove R, α and μ_j and rewrite into its dual expression:

$$W = \sum_j \Phi(x_j)^2 \beta_j - \sum_{ij} \beta_i \beta_j \Phi(x_i).\Phi(x_j) \tag{18.50}$$

The variables μ_j are replaced by the following constraints:

$$0 < \beta_j < C, \quad j = 1, \ldots, N \tag{18.51}$$

At this stage, we still neither know the nonlinear map nor the Lagrange Multipliers. However, according to Mercer's theorem, the dot product $\Phi(x_i).\Phi(x_j)$ may be replaced by a Mercer kernel $K(x_i, x_j)$ such as

$$K(x_i, x_j) = \Phi(x_i).\Phi(x_j) \tag{18.52}$$

This replacement is known as *the Kernel trick*. The dual Lagrangian W is now written as

$$\max_{\beta_j} W = \sum_j K(x_j, x_j)\beta_j - \sum_{ij} \beta_i \beta_j K(x_i, x_j), \quad \text{where } \beta \text{ is the eigenvector.} \tag{18.53}$$

And subject to the constraints

$$0 < \beta_j < C \tag{18.54}$$

$$\sum_j \beta_j = 1 \quad \text{for } j = 1, \ldots, N \tag{18.55}$$

As previously mentioned, the appropriate kernel for SVC is the Gaussian kernel with the width parameter q:

$$K(x_i, x_j) = e^{-q\|x_i - x_j\|^2} \tag{18.56}$$

We define now the distance of each image point from the centre of the sphere:

$$R^2(x) = \|\Phi(x) - \alpha\|^2 \tag{18.57}$$

Using the definition of Mercer kernel, we have

$$R^2(x) = K(x,x) - 2\sum_j \beta_j K(x_j, x) - \sum_{ij} \beta_i \beta_j K(x_i, x_j) \tag{18.58}$$

SV lies on the surface of the sphere, i.e.

$$R = \{R(x_i) | x_i \in SV\} \tag{18.59}$$

And vice versa, the set of contours C that defines the cluster is given by

$$C = \{x | R(x) = R\} \tag{18.60}$$

The set C establish the cluster boundaries. One last step is necessary to complete the algorithm. We need to assign a cluster to each data points and label them. We observe that the path connecting two given data points belonging to different clusters must exceed the radius of the hyper-sphere of the feature space. Such a path contains a point y where $R(y) > R$. This defines the adjacency matrix A_{ij} between two points whose images $\Phi(x_i)$ and $\Phi(x_j)$ lie within or on the sphere (Figure 18.12):

$$A_{ij} = \begin{cases} 1 \text{ if for all } y \text{ on the line segment connecting } x_i \text{ and } x_j, R(y) \leq R \\ 0 \text{ otherwise} \end{cases} \tag{18.61}$$

A graphical illustration may show this better.

By using this matrix, clusters are defined as the connected components of the graph induced by the matrix A. BSVs are not being classified by using this procedure. A possible solution is to assign them to the closest cluster. The PSO will try to maximize the Lagrangian function by computing the Eigenvectors. Once the PSO procedure has terminated, the SVC resumes the clustering process. The kernel matrix

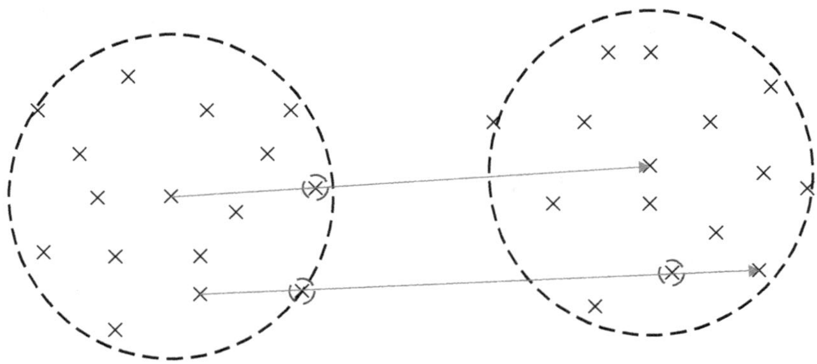

Figure 18.12 Line segments connecting data points

is computed once since it does not depend on the Eigenvectors. We therefore do not need demanding computations every time the position of a particle is evaluated. The PSO sends its output solution to SVC, and the result of the clustering process is returned.

18.12.2 SVC–PSO relation

The connection between the two algorithms is made by using a controller module. Using a controller, the SVC calls directly on PSO to find a near-optimal Eigenvalue. Then the adjacency matrix is computed and exported into the software *Gephi* using the Gephi API [45].

18.12.3 Data set

A data set called 'Flame' [46] of two dimensions has been used for benchmarking. It is composed of 240 vectors and contains two clusters. The scatter diagram beneath represents the data set. No clusters are labelled as no analysis has been done yet on the data points. The scatter diagram is just the actual data points of a two-dimensional plot, based on raw data. Two clusters are clearly visible (Figure 18.13).

18.12.4 The experiment

We have tried different values for the cost and gamma parameters. However, the output was always the same. After several tests, a swarm was defined by using 20 particles and 100 iterations. The same evolution of the swarm occurred in every experiment. Currently, we do not know why.

Figure 18.14 shows what is happening. The swarm always was decreasing its level in the fitness landscape and tends to stay at this low fitness region for about half the time. Then, the whole swarm increases its performances, and afterwards slowly decreases.

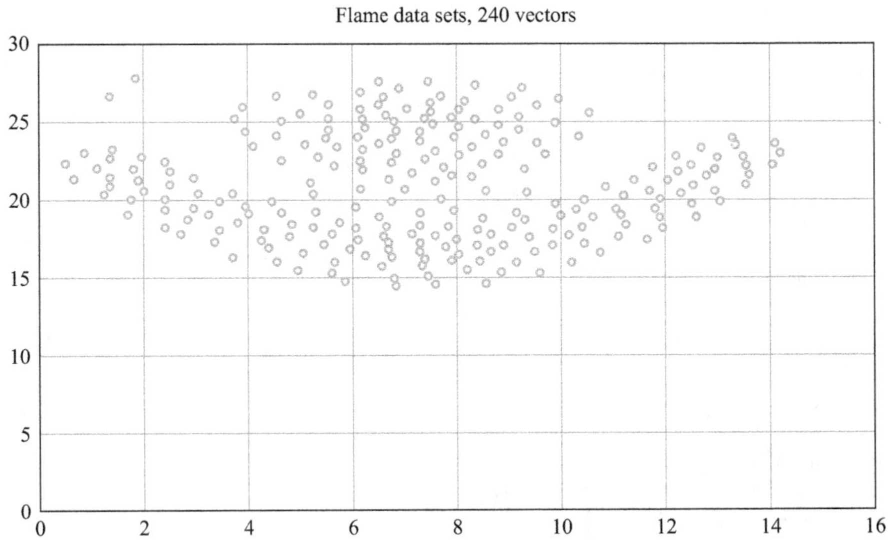

Figure 18.13 Clustering of the flame set

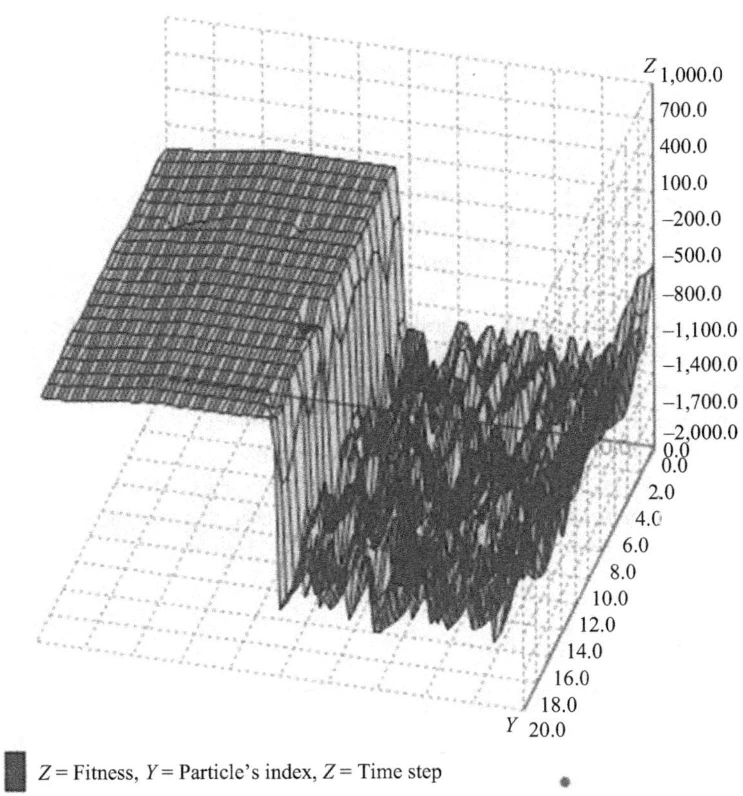

Figure 18.14 A PSO plot of SVC

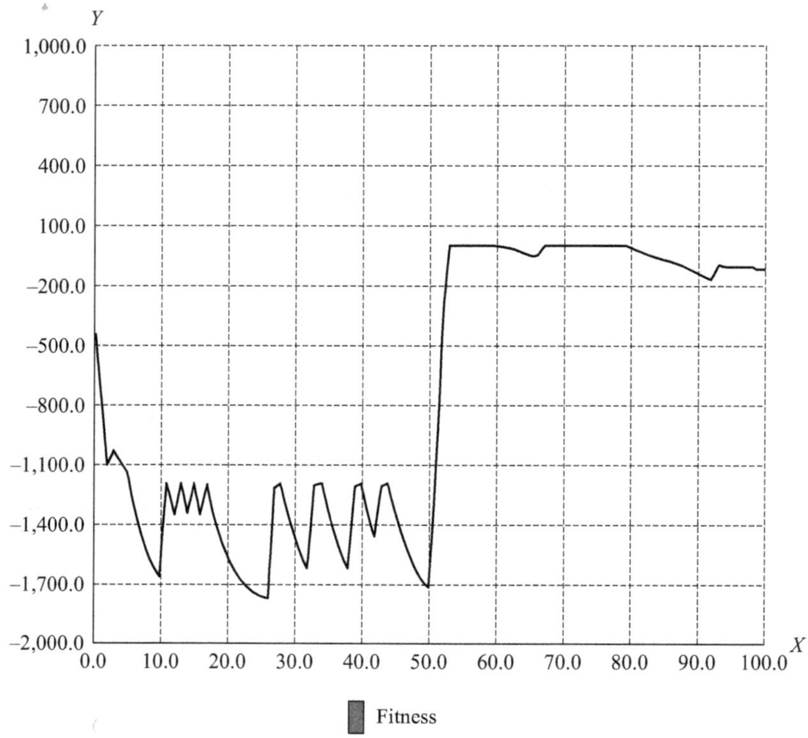

Figure 18.15 The evolution of the fitness of a SVC particle

Figure 18.15 shows the behaviour of an arbitrary single particle through the optimization process. Several experiments have been done, and the same pattern always occurred. Increasing the length of the optimization process did not change the output. The swarm still has the same behaviour. We used different behaviours of a particle such as an *only-cognitive* model or an *only-social* model, but the fitness was always the same.

The adjacency matrix computed from the PSO's solution was nearly a fully connected graph. Such a graph has only one strongly connected component leading to only one cluster by the algorithm. Unfortunately, this means that PSO is unable to find the Eigenvector of the SVC algorithm. The reason we believe is that the SVC procedure requires an exact solution of the QP equation, while the PSO algorithm provides only an approximation solution.

18.13 Ant colony systems

Ant colony optimization (ACO) [47,48] is another direction with great success in SI that focuses on discrete optimization problems. ACO has been applied to a great number of NP-hard [35] discrete optimization problems. An artificial Ant Colony

System may be defined as an agent-based system. It may be used to simulate the natural behaviour of ants by use of cooperation and learning. The new heuristics ACO have shown to be robust and can be used to solve a wide range of combinatorial optimization problems. The main idea of ACO is to define the problem as a search for a minimum cost of the path in the graph [49]. The ants are walking on this graph in costly or cheaper manner. Cheaper paths are an emergent property of global cooperation between ants in the ant colony. The behaviour of artificial ants is defined by which edges they are selecting with respect to the probabilities of the pheromone trials. Artificial ants are living in a discrete world (a graph) and are moving from node to node. The distributed behaviour of an ant colony is called the *stigmergy*. The concept of artificial stigmergy is defined by Dorigo and Di Caro [48] as

> *indirect communication mediated by numerical modifications of environment states which are only locally accessible by the communication agents'*

When modelling ant colonies we want to find mathematical models that describe the stigmergetic characteristics of the corresponding ant individuals.

18.13.1 The pheromone

One important task for an ant colony is the food collection, the dropping of pheromone and how the movement of ants may be modelled. The ants have always the ability to find the shortest path between the nest and the food source. Different experiments have been carried out to study the behaviour of the ants. The experiments may be made by making paths of different length between the nest and the food source. It has been observed that more and more ants are following the shorter paths. Such a behaviour may be explained by the dropping of pheromone of each ant. Each ant drops pheromone deposits on the path and from it search for food and returning back to the nest with the collected food. The ants are following the path given by the biggest pheromone concentration. The shortest paths will have stronger pheromone deposits than longer paths. In addition, the pheromone concentration evaporates with time. The strength will decrease more quickly on longer paths than on shorter ones.

Although the ant colony may exhibit very complex adaptive behaviour, a single ant follows very simple behaviours. We may interpret an ant as a stimulus-response agent. An ant observes the concentration of the pheromone and produces an action based on this. We may therefore consider an ant as a simple computational entity or agent. Such a system makes a simple production system of rules.

18.13.2 Optimization: travelling salesman problem

ACO may be used to find an optimal tour of the travelling salesman problem (TSP). Given a set of n cities and a set of distances between them, the problem is to find a minimum tour of the cities and return at the end to where it starts. The travel should not include a city more than once. At each city, the task of the ant is to select the next city on the route, based on a probability that depends on the pheromone deposits.

The pheromone amount is randomly initialized to a small value. According to a classical ant system [50], the probability of an ant k in city i visits the city j is given by

$$p_{i,j,k} = \begin{cases} \dfrac{\tau_{ij}^{\alpha}\eta_{ij}^{\beta}}{\sum_{h \in \mathrm{N}_{i,k}} \tau_{ih}^{\alpha}\eta_{ih}^{\beta}} & \text{if } j \in \mathrm{N}_{i,k} \\ 0 & \text{otherwise} \end{cases} \tag{18.62}$$

where τ_{ij} is the pheromone intensity on the edge (i,j). α and β are parameters that represent the importance of pheromone trail and heuristic information, $\mathrm{N}_{i,k}$ represents the set of cities that can be visited by the ant k, i.e. the set of feasible cities. In (18.62), shorter edges with greater amount of pheromone are favoured by multiplying the pheromone on the edge (i,j) with a heuristic value $\eta(i,j)$. The parameters α and β determine the relative importance of pheromone versus cost. $\eta(i,j)$ is defined by

$$\eta(i,j) = 1/d_{ij} \tag{18.63}$$

where d_{ij} is the Euclidean distance between city i and city j. Each route T_k is constructed by an ant k. The pheromone intensity τ_{ij} on the edge of that route is updated as

$$\tau_{ij}(t+1) = (1-\rho)\tau_{ij}(t) + \Delta\tau_{ij}(t) \tag{18.64}$$

where

$$\Delta\tau_{ij}(t) = \sum_{k=1}^{m} \Delta\tau ij, k(t) \tag{18.65}$$

Algorithm 7: ACO algorithm of TSP

ACO Algorithm for TSP
Initialize the pheromone deposits on the edge, i.e. τ(i,j) to small positive random values
$\tau_{ij}(0) \sim U(0,max)$
For each ant k = 1, 2, 3, . . . ,m on the start city **Do**
  T+ is the shortest trip, L+ is the length of the trip
   For t =1 to t_{max} **Do**
    For each ant build the trip $T_k(t)$ by choosing the next city
        n − 1 times with probability $p_{ij}(t)$ in (59)
        //n is the number of cities
     Compute the length of the route $L_k(t)$ for each ant
     If improved route is found update T^+ and L^+
     update deposits on each edge using (61)
       End if
    End for
    Output the shortest route $\mathbf{T}^+$
   End for
End for

---

is the total deposits of the pheromone $\Delta\tau ij, k(t)$ for each ant k defined beneath. m is the total number of ants:

$$\Delta\tau ij, k(t) = \begin{cases} Q/L_k(t) & \text{if } (i,j) \in T_k(t) \\ 0 & \text{if } (i,j) \notin T_k(t) \end{cases} \tag{18.66}$$

Q is a constant and $L_k(t)$ is the length of the tour travelled by the ant k. Possible terminating conditions are

- a maximum number of iterations
- an acceptable solution has been found
- all the ants are following the same path

18.14 Ant colonies clustering

Ants may be useful for solving discrete optimization problems. In this section, we will explain how the clustering and sorting behaviour of ants may be used to design an algorithm for clustering of data. Experimental work has shown that a collection of ants that are initially randomly placed will after some hours be distributed into clusters.

The general idea is that a number of ants (or agents) walking around in a search space will pick up items and drop the items based on some probability measure. Suppose that the environment contains only one object that is distributed uniformly across the search space. The probability for an agent to pick up an object randomly may be given by

$$p_p = \left(\frac{f}{k_1 + f}\right)^2 \tag{18.67}$$

where f is a fraction of objects the agent sees in his neighbourhood where k_1 is a constant. When there are not many objects in the neighbourhood, we may suppose that $f \ll k_1$. In this case, we see that p_p is approaching 1 and therefore of high probability for being picked up. The other case may be that $f \gg k_1$ and p_p will approach 0. The agent will then pick up an object with little probability.

The agent wearing an object has the probability of dropping the carried object given by

$$p_d = \left(\frac{f}{k_2 + f}\right)^2 \tag{18.68}$$

where k_2 is a constant. When $f \gg k_2$, the p_d will approach 1 and the probability of dropping an item is high. If $f \ll k_2$, the p_d will then approach 0 and has a small probability for dropping an item.

Such an approach has been utilized within robotics. We now want to use the same algorithmic approach for data clustering. The ant colony clustering method is similar to self-organizing maps (SOM) [5] where the input vectors are mapped on a two-dimensional grid space conserving the topological structure of the input space. Agents may move around randomly on the grid consisting of an area of s^2 sites. The surrounding area is a square neighbourhood $N_{s\times s}(r)$ of the $s \times s$ around the position r of the agent. Assuming at time t, the agent in position r finds a data vector z_i. We may then define 'local' density $f(z_i)$ within the neighbourhood of an agent as

$$f(z_i) = \begin{cases} \dfrac{1}{s^2} \sum\limits_{z_j \in Neigh(s\times s)(r)} \left[1 - \dfrac{d(z_i, z_j)}{\alpha}\right] & \text{when} f > 0 \\ 0 & \text{otherwise} \end{cases} \tag{18.69}$$

$f(z_i)$ is here a measure of the average similarity of other data in the neighbourhood. The constant α determines the degree of dissimilarity and decides when two data vectors may be grouped together or not.

The probability of picking up or dropping a data vector may be expressed as

$$p_p(z_i) = \left(\frac{k_1}{k_1 + f(z_i)}\right)^2 \tag{18.70}$$

$$p_d(z_i) = \begin{cases} 2f(z_i) \text{ when} f(z_i) < k_2 \\ 1 \text{ when} f(z_i) \geq k_2 \end{cases} \tag{18.71}$$

Algorithm 8: Ant colonies clustering algorithm

Ant Colonies Clustering
Place initially randomly each data vector z_i on the grid
Place initially agents randomly on selected sites
Define the values k_1, k_2, α, s, t_{max}
For t =1 to t_{max} **Do**
  For each agent **Do**
    If current agent unladen and the site is occupied by item z_i
      compute f($\mathbf{z_i}$), and p_p (**zi**) according to the formulas above
    else
      If U(0, 1) ≤ p_p ($\mathbf{z_i}$)
        drop data vector $\mathbf{z_i}$
     else
              pick randomly a neighbouring site not occupied by an ant
      End if
    End if
  End for
End for

A summary of the ant colony algorithm for data clustering is given in [51]. In the next version of the book, we may implement this clustering algorithm in the same way that was done for PSO and visualize the solution using for instance the Gephi API [52]. We want to compare this kind of clustering with PSO clustering on some benchmark data sets to see how the algorithms differ.

References

[1] J. Kennedy, R. Eberhart. Particle Swarm Optimization, Proceedings of IEEE International Conference on Neural Networks, 1995, pp. 1942–1948.

[2] J. Kennedy, R.C. Eberhart. A Discrete Binary Version of the Particle Swarm Algorithm, Proceedings of the 1997 Conference on Systems, Man, and Cybernetics, IEEE Service Center, Piscataway, NJ, 1997, pp. 4104–4109.

[3] A. Abraham, C. Grosan, V. Ramos. Swarm Intelligence and Data Mining, Studies in Computational Intelligence, Springer Verlag, Germany, 2006, pp. 270.

[4] J. Kennedy, R. Eberhart, Y. Shi. Swarm Intelligence, Morgan Kaufmann Academic Press, San Francisco, CA, 2001.

[5] T. Kohonen. Self-Organizing Maps, Springer Series in Information Sciences, vol 30, Springer-Verlag, Berlin, Heidelberg, 1995.

[6] Y. Shi, R. Eberhart. Empirical Study of Particle Swarm Optimization, Proceedings of the IEEE Congress on Evolutionary Computation, 1999.

[7] Y. Shi, R. Eberhart. A Modified Particle Swarm Optimizer, Proceeding of the IEEE Congress on Evolutionary Computation, 1998.

[8] C. Koay, D. Srinivasan. Particle Swarm Optimization-Based Approach for Generator Maintenance Scheduling, Proceedings of the IEEE Swarm Intelligence Symposium, 2003, pp. 167–173.

[9] H. Higashi, H. Iba. Particle Swarm Optimization with Gaussian Mutation, Proceedings of the IEEE Swarm Intelligence Symposium, 2003, pp. 72–79.

[10] M. Clerc, J. Kennedy. The Particle Swarm – Explosion, Stability and Convergence in a Multidimensional Complex Space. IEEE Transactions on Evolutionary Computation, 6 (1), 2002, pp. 58–73.

[11] X. Yao, Y. Liu, G. Liu. Evolutionary Programming Made Faster. IEEE Transactions on Evolutionary Computation, 2 (3), 1999, pp. 82–102.

[12] M. Løvberg, T. Rasmussen, T. Krink. Hybrid Particle Swarm Optimizer with Breeding and Subpopulation, Proceedings of Genetic and Evolutionary Computation, 2001, pp. 469–476.

[13] B. Al-Kazemi, C. Mohan. Multi-phase Generalization of the Particle Swarm Optimization Algorithm, Proceedings of the IEEE Congress on Evolutionary Computation, 2002, pp. 489–494.

[14] R. Brits, A. Engelbrecht, F. van den Bergh. A Niching Particle Swarm Optimizer, Proceedings of the Fourth Asia-Pacific Conference on Simulated Evolution and Learning, 2002, pp. 692–696.

[15] M. Clerc. Think Locally, Act Locally: The Way of Life of Cheap-PSO, An Adaptive PSO, 2001.

[16] M. Dorigo, L.M. Gambardella. Ant Colony System: A Cooperative Learning Approach to he Traveling Salesman Problem. IEEE Transactions Evolutionary Computing, 1, 1997, pp. 53–66.

[17] A.P. Engelbrecht. Computational Intelligence. An Introduction. Chichester: Wiley, 2007.

[18] B. Secrest, G. Lamont. Visualizing Particle Swarm Optimization – Gaussian Particle Swarm Optimization, Proceedings of the IEEE Swarm Intelligence Symposium, 2003, pp. 198–204.

[19] C. Wei, Z. He, Y. Zheng, W. Pi. Swarm Direction Embedded in Fast Evolutionary Programming, Proceedings of the IEEE Congress on Evolutionary Computation, volume 2, 2002, pp. 1278–1283.

[20] T. Kristensen. Computational Intelligence, Evolutionary Computing and Evolutionary Clustering Algorithms. Sharjah: Bentham Science Publishers, 2016.

[21] S. Haykin, Neural Network and Learning Machine. Upper Saddle River, NJ: Pearson International Edition, Hamilton, 2009.

[22] T.M. Mitchell. Machine Learning. Pittsburgh, PA: Carnegie Mellon University, McGraw-Hill International Editions, 1997.

[23] J. Shaw-Taylor, N. Christianini. Kernel Methods for Pattern Analysis. Cambridge: Cambridge University Press, 2004.

[24] B. Schölkopf, A.J. Smola. Learning with Kernels: Support Vector Machines, Regularization, Optimization and Beyond. Cambridge, MA: MIT Press, 2002.

[25] S. Boyd, L. van den Bergh. Convex Optimization. New York: Cambridge University Press, 2004.

[26] J. Mercer. Functions of Positive and Negative Type, and Their Connection with the Theory of Integral Equations. Transactions of London Philosophical Society, 209 (441–458), 1909, pp. 415–446.

[27] R. Chattamvelli. Data Mining Algorithms, Alpha Science International Ltd. Oxford, UK, 2011.

[28] T.M. Huang, V. Kechman, I. Kopriva. Kernel Based Algorithms for Mining Huge Data Sets. Supervised, Semi-supervised and Unsupervised Learning. Berlin, Heidelberg: Springer-Verlag, 2006.

[29] R. Xu, I.C. Wunsch. Clustering. Hoboken, NJ: Wiley & Sons, 2009.

[30] N.V. Vapnik. The Nature of Statistical Learning Theory, 2nd ed. New York: Springer-Verlag, 2000.

[31] R. Bellman. Dynamic Programming. Princeton, NJ: Princeton University Press, 1957.

[32] R. Duda, P. Hart, D. Stork. Pattern Classification. New York: John Wiley & Sons, 2001.

[33] D.M.J. Tax, R.P.W. Duin. Support Vector Domain Description. Pattern Recognition letters, 20, 2004, pp. 1191–1199.

[34] R.J. Vanderbei. Linear Programming: Foundations and Extensions, 3rd ed. New York: Springer, 2008.

[35] D.P. Williamson, D.B. Shmoys. The Design of Approximation Algorithms, Cambridge, MA: Cambridge University Press, 2011.

[36] R. Fletcher, Practical Methods of Optimization. Chichester: Wiley-Interscience, 1987.

[37] F. van den Bergh. An Analysis of Particle Swarm Optimizers, PhD thesis, Pretoria, South Africa, 2002.

[38] F. Collins, D. Gallas. A New Five-Year Plan for U.S. Human Genome Project. Science, 262, 1993, pp. 43–46.
[39] D.F. Feng, F. Dolittle. Progressive Alignment of Acid Sequences and Construction of Phylogenetic Trees from Them. Methods in Enzymology, 266, 1996, pp. 368–382.
[40] T. Kristensen, F. Guillaume. Classification of DNA Sequences by a MLP and SVM Network, Biocomp'13 – International Conference of Bioinformatics and Computational Biology, Las Vegas, USA, 2013.
[41] S. Dehuri, S. Ghosh, S.-B. Cho. Integration of Swarm Intelligence and Artificial Neural Network, vol. 78, World Scientific Publishing, 2011.
[42] J. Salerno. Using Particle Swarm Optimization Technique to Train a Recurrent Neural Model, Proceeding of the IEEE International Conference on Tools with Artificial Intelligence, 1997.
[43] LiBSVM – a library for support vector machines. Available: http://www.csie.ntu.edu.tw/~cjlin/libsvm/.
[44] J. Platt. Sequential Minimal Optimization. A Fast Algorithm for Training Support Vector Machines. Microsoft, 1998.
[45] Gephi. Gephi: makes graphs handy. Available: https://gephi.org/, 2009.
[46] L. Fu, E. Medico. FLAME, a Novel Fuzzy Clustering Method for Analysis of DNA Microarray Data. BMC Bioinformatics, 1 (8), 2007, pp. 3.
[47] M. Dorigo, L. Gambardella. Ant Colonies for the Travelling Salesmen Problem. Biosystems, 43 (2), 1997, pp. 73–81.
[48] M. Dorigo, G. Di Caro. The Ant Colony Optimization Meta-Heuristics, New Ideas in Optimization. Maidenhead: McGraw-Hill, 1999.
[49] M. Dorigo, T. Styzle. Ant Colony Optimization. Cambridge, MA: MIT Press, 2004.
[50] B. Everitt, Cluster Analysis, 2nd ed. Social Science Research Council, London, 1993.
[51] E. Lumer, B. Faieta. Diversity and Adaption in Populations of Clustering Ants, Proceedings of Third International Conference on Simulation of Adaptive Behavior: From Animals to Animates, MIT Press, volume 3, 1994, pp. 499–508.
[52] M. Bastian, S. Heyman, M. Jacomy. Gephi: An Open Source Software for Exploring and Manipulating Networks. International AAAI Conference on Weblogs and Social Media, Paris, France, 2009.

Chapter 19
Modelling of interaction in swarm intelligence focused on particle swarm optimization and social networks optimization

A. Niccolai[1], F. Grimaccia[1], M. Mussetta[1] and R.E. Zich[1]

Abstract

Swarm intelligence is an important and popular branch of computational intelligence. This technique is based on the analysis of the collective behaviour of swarms in nature and of social phenomena, and it is aimed to solve highly non-linear problems. The optimization performances of these techniques are highly affected by the specific choice of the working parameters. In this chapter, the modelling of interaction in swarm intelligence is done by means of different models, and the space of the working parameter is finally divided accordingly to the specific dynamic behaviour of the swarm for each point of this space. In this analysis, the attention has been focused in two optimization algorithms: the well-known particle swarm optimization and the recently developed social network optimization.

19.1 Introduction

In this chapter, the modelling of the interaction in swarm intelligence (SI) will be analysed using differential and discrete models.

Among the several computational intelligence techniques recently developed, SI is one of the most popular ones [1,2]. Based mainly on the development of particle swarm optimization (PSO) [3] and ant colony optimization [4], SI focuses on the collective behaviour of swarms in nature and social phenomena, to solve highly non-linear optimization problems.

The aim of this chapter is to provide an overview of the general interaction models in SI algorithms. In particular, the presented interaction analysis will focus on the

[1]Dipartimento di Energia, Politecnico di Milano, Italy

well-known PSO technique and it will be applied considering a recently developed population-based algorithm, namely the social network optimization (SNO).

The chapter is structured as follows: Section 19.2 will introduce the definition of interaction among agents in swarms; Section 19.3 will introduce the swarm topologies and the algorithm used for numerical results; a parameter sensitivity analysis will be presented in Section 19.4; Section 19.5 will discuss the trajectory of particles in SI algorithms and their convergence will be analysed in Section 19.6, before drawing the conclusions in Section 19.7.

19.2 Swarm interaction definition

The first application of swarm interaction in PSO has been done in 1995 by Kennedy and Eberhart [3]. Their algorithm, PSO, has been used in many fields, giving an empirical evidence of its effectiveness [5–7].

The PSO has undergone many modifications [8] and inceptions [9,10] to improve and to increase the optimization capability of the algorithm. In more recent years, other optimization algorithms based on swarm interaction have been implemented [11,12], such as the SNO [13], bee algorithms [14], fireworks algorithm [15] and imperialist competitive algorithm [16].

All these algorithms have been classified [17] under the class of swarm algorithm because all of them try to reproduce the interaction between different elements. Even if, considering performances, computational complexity and specific operators, these algorithms are different, they are all based on the same principle that here will be defined as *swarm interaction*.

The swarm interaction is the aggregating principle of the algorithm: it gives the powerful exploitation capability that is typical of all these algorithms. Extracting from all the swarm algorithms the common feature, it is possible to express by word the swarm interaction with the two following statements:

- *I go on my way*: in almost all the literature, this principle is identified with the term of 'personal inertia', it is the capability of the specific algorithm feature to keep part of its previous behaviour even if it is subjected to external forces;
- *I can change my mind*: this is the capability of a specific algorithmic feature to be influenced by the behaviour of other features.

The first element, the *inertia*, is common to all the algorithms with similar shape, while on the other one there are many differences from an algorithm to another. In fact, it can be declined in different ways, for example as Personal and Global Best in standard PSO or as influencing posts in SNO. All these differences are in the selection of the type of force that acts on the particle (mechanical force, swarm behaviour, social forces, etc.) but the contribution to the overall swarm interaction is almost the same. In the next subsection, several notations will be analysed.

19.2.1 Traditional notation

In the previous section, the two main terms of the swarm interaction have been investigated from a general point of view. Now their mathematical implementation in the algorithm will be shown.

First of all, it is important to define the space in which the equations are applied. In all these algorithms, it is possible to define a specific feature that will be considered as a candidate solution. This is a point in the space of the solutions. The change of the position of this point in the space of the solutions can be considered as its speed.

The swarm interaction can be defined with two equations: the first characterizes the change of the speed and the second the change in the position [18]:

$$\vec{v}_{t+1} = \vec{a} \cdot \vec{v}_t + \sum_{i=1}^{M} \vec{b}_i \cdot \vec{r}_i \cdot (\vec{p}_i - \vec{x}_t) \tag{19.1}$$

$$\vec{x}_{t+1} = \vec{c} \cdot \vec{x}_t + \vec{d} \cdot \vec{v}_{t+1} \tag{19.2}$$

where

- $\vec{x}$ is the position of a candidate solution in the variable space. In general, an N-dimensional space is considered; thus, $\vec{x}$ is a vector. The subscript t refers to the actual value, while $t+1$ refers to the value that will be updated for the following iteration;
- $\vec{v}$ is an indicator of how a candidate solution will move in the variable space. For this variable, the subscripts have the same meaning of the variable $\vec{x}$;
- $\{\vec{p}_i\}$ is a set of the M attracting points in the variable space. Their choice and meaning makes the difference among the several algorithms in this class;
- $\vec{a}$ is the inertia parameter. In the most general case, it is a vector;
- $\vec{b}_i$ is the weight of each attracting position: they are sometime called *accelerating coefficients* [19];
- $\vec{c}$ and $\vec{d}$ are the weights of the attracting forces applied to the candidate solution;
- $\vec{r}$ is a random vector used to increase the exploration capability of these pseudo-stochastic algorithms.

Observing carefully (19.1) and (19.2), it is possible to notice that the evolution of each variable is independent of the others. This brings to two considerations. The first one is that the working principles of SI algorithms are not affected by the size of the problem under investigations. This means that having a problem with some non-influencing variables do not affect the optimization behaviour. The second consideration regards the fact that, if 1D swarm interaction is analysed, all the considerations found can be applied to the N-D. This decrease a lot the effort of analysing the behaviour of swarm interaction, first because working with scalars is much easier than working with vectors, and finally because it is possible to easily visualize the trajectories of the particles.

Having done these considerations, it is possible to rewrite (19.1) and (19.2) in a scalar form:

$$v_{t+1} = a \cdot v_t + \sum_{i=1}^{M} b_i \cdot r_i \cdot (p_i - x_t) \tag{19.3}$$

$$x_{t+1} = c \cdot x_t + d \cdot v_{t+1} \tag{19.4}$$

In the next two subsections, these two equations will be simplified in order to get to a final model that can be easily analysed.

19.2.2 Deterministic model

The first step to simplify the analysis is to transform the swarm interaction model in a deterministic model [18]. By deterministic model, we mean a model in which there are no stochastic variables. These variables are important in the optimization process as they give the algorithm exploration capability and so the possibility to work as a global optimizer. Nevertheless, the analysis of the stochastic model requires statistic tools that can increase the effort.

All these problems can be avoided substituting to all the stochastic variables in the model their expected value. In this way, the deterministic model resulting will be equivalent to the average behaviour in a stochastic model.

In our model, there are M stochastic variables, where M is the number of attracting positions. The stochastic variables present are uniformly distributed between 0 and 1, so their expected value is

$$E[U(0, 1)] = \frac{1}{2} \tag{19.5}$$

Due to this consideration, it is possible to rewrite (19.3) using the expected value of the stochastic variables:

$$v_{t+1} = a \cdot v_t + \frac{1}{2} \sum_{i=1}^{M} b_i \cdot (p_i - x_t) \tag{19.6}$$

The constant value of $1/2$ can be incorporated in the constants b_i getting an equivalent deterministic model:

$$v_{t+1} = a \cdot v_t + \sum_{i=1}^{M} b_i \cdot (p_i - x_t) \tag{19.7}$$

In the next subsection, the model will be rewritten in order to simplify it and to reduce the number of parameters.

19.2.3 Equivalent simplified notations

In this subsection, the model will be analysed with the aim to get a simplified version of it and to reduce the number of free parameters [18]. In fact, it is possible to demonstrate, and it will be done, that not all the parameters a, b_i, c, and d are needed to have a completely equivalent model.

By inserting (19.7) into (19.4),

$$x_{t+1} = c \cdot x_t + d \cdot \left(a \cdot v_t + \sum_{i=1}^{M} b_i \cdot (p_i - x_t) \right) \tag{19.8}$$

it is possible to split the summation into two parts:

$$x_{t+1} = c \cdot x_t + d \cdot \left(a \cdot v_t + \sum_{i=1}^{M} b_i \cdot p_i - \sum_{i=1}^{M} b_i \cdot x_t \right) \tag{19.9}$$

Doing the products,

$$x_{t+1} = c \cdot x_t + d \cdot a \cdot v_t + d \cdot \sum_{i=1}^{M} b_i \cdot p_i - d \cdot \sum_{i=1}^{M} b_i \cdot x_t \tag{19.10}$$

Putting the constant valued in the summations,

$$x_{t+1} = c \cdot x_t + d \cdot a \cdot v_t + \sum_{i=1}^{M} d \cdot b_i \cdot p_i - \sum_{i=1}^{M} d \cdot b_i \cdot x_t \tag{19.11}$$

Writing (19.4) for the time t,

$$x_t = c \cdot x_{t-1} + d \cdot v_t \tag{19.12}$$

it is possible to get the value of v_t:

$$v_t = \frac{1}{d} x_t - \frac{c}{d} \cdot x_{t-1} \tag{19.13}$$

Using this last result in (19.11):

$$x_{t+1} = c \cdot x_t + a x_t - a \cdot c \cdot x_{t-1} + \sum_{i=1}^{M} d \cdot b_i \cdot p_i - \sum_{i=1}^{M} d \cdot b_i \cdot x_t \tag{19.14}$$

For the sake of clarity, it is possible to rewrite this equation as

$$x_{t+1} + \left(\sum_{i=1}^{M} d \cdot b_i - c - a \right) \cdot x_t + a \cdot c \cdot x_{t-1} = \sum_{i=1}^{M} d \cdot b_i \cdot p_i \tag{19.15}$$

From this last equation, it is clear that the time behaviour of a single particle depends on the product of $d \cdot b_i$ and never on the single value of d or b_i, so it is possible to assume, without losing the generality of the model, that

$$d \equiv 1 \tag{19.16}$$

In other words, any sequence of particle positions generated by the original swarm interaction model can be generated also with the simplified one, with a proper choice of the value of the constants b_i.

The second simplification of the model can be done expecting the convergence of the particles in the optimal position [18]. So, if we consider p as the optimal solution of the problem and the only attracting point ($M = 1$), we want that

$$\lim_{t \to \infty} x_t = p \tag{19.17}$$

This condition can be applied to (19.15)

$$p + (d \cdot b - c - a)p + a \cdot c \cdot p = d \cdot b \cdot p \tag{19.18}$$

thus,

$$p + (-c - a)p + a \cdot c \cdot p = 0 \tag{19.19}$$

Finally we can get

$$(a - 1) \cdot (c - 1) = 0 \tag{19.20}$$

This equation has two solutions:

$$(a = 1) \vee (c = 1) \tag{19.21}$$

Due to the fact that (19.15) is symmetric with respect to a and c, it is possible to choose freely the solution. To better simplify the model, we assume

$$c \equiv 1 \tag{19.22}$$

The assumptions made on c and d values are justified because, in this way, the set of v_t according to (19.4) are effectively true 'velocities' with respect to the position x_t.

19.3 Different swarm topologies and social network optimization

In this section, some swarm interaction topologies will be described and finally a brief description of the optimization algorithm used for the sensitivity test will be provided.

Swarm topologies are different ways to define the attracting point or the set of points between which the attracting points can be chosen. Often the set of points of possible attractor are called *neighbours*. The spatial arrangement of the neighbours is the *topology* of the algorithm.

The neighbourhood size is an important factor in the convergence rate of the algorithm [3]. Small size neighbourhoods provide a better global behaviour, while larger neighbourhoods provide a faster convergence, giving more importance to the exploitation [20].

It is possible to identify different methods for the definition of the neighbourhood [21]. The first possibility [17] in the definition of the neighbours is to define then as the n closest particles at the beginning of the algorithm and then to maintain them fixed during the optimization process.

When the optimization stagnates, a possibility is to define the neighbours randomly [22]. Also the entire swarm can be defined as the neighbourhood of all the particles.

One of the most common choice in PSO is the definition of the global best and personal best attractors [23]. Other possibilities for the choice of the neighbours are related to distance metrics [24].

19.3.1 Social network optimization

In order to perform numerical analysis, in this chapter we will consider the recently developed SNO [25].

SNO is a population-based evolutionary algorithm. The idea of this algorithm is to emulate the main characteristics of social networks [26]. The population of SNO (candidate solutions of the problem) represents a set of people. Each person lives a real life, in which he lives situations and then these situations are shared on the social network. The real life of each person is also characterized by his character, considered as an inclination to make choices.

The social network is the virtual space in which people interact informing the other individual on his status and commenting ('evaluating') the performance of the others. The evaluation of a situation is divided into two parts: the first one is the evaluation by the person that lived the situation, the second one is by some of the other members of the social.

Some people of the social network are called explorers, because they are not attracted by other people and their attraction capability is much higher than the one of the average person.

Figure 19.1 shows the pseudo-code of SNO. In the outer loop, the first process is the creation of a new memory, recalling that the most of lived situations are stored in the memory of the social network. The choice of which situation store and which not is first based on the evaluation given to that situation. The second criterion to decide is the 'age' of the post. Posts shared in past are more suitably substituted by new posts. The inner loop is the heart of the algorithm: the process of group division, influencer selection and interaction.

```
Initialize a random population
Create an empty memory
while not termination criterion do
    Create the new memory
    for all the population do
        if Is an explorer then
            Random change of the character
        else
            Creation of group of friend
            Creation of peer group
            Selection of the influencers
            Interaction between people
        end if
        Change of the situation
        Evaluation of the cost
    end for
end while
```

Figure 19.1 Pseudo-code of SNO

People, in SNO, interact in groups: friends groups and peers groups. Groups of friends collect people that have similar situations. The actual situation of a person is compared with the situations memorized in the memory. If these are similar, the situation memorized can influence the person, and so this person and the person who has lived the situation memorized are considered friends. Each person has a group of friends that is different from groups of friends of other people. The similarity between two situations is evaluated as a distance in the space of the candidate solutions.

Groups of peers collect people with the same most predominant feature of the character. Similarly for friends' groups, for peers' groups are compared the actual character of a person with the character of a person who has lived a situation contained in the memory. All groups change during time: actual situations and characters of people change so also solutions contained in the memory change.

The probability that a person can be chosen as influencer depends on the position in the ranking of the group. In fact, in all groups a ranking is created, where people are ranked by the evaluation of their situation. Each person has a probability greater than zero to be chosen as influencer. This probability is based with a non-linear relation with the rank of each person in the groups.

After having selected some influencers, people interact with them: the interaction takes into account the tendency of a person to maintain his character, the influence of the influencer and the type of groups from which the influencer is chosen.

The interaction of a person is driven by the swam interaction model where the individual is attracted by one of his friends and one of his peers.

19.4 Parameter sensitivity analysis

This section is aimed to demonstrate the importance of the analysis on the swarm interaction formula. In fact, all swarm algorithms are characterized by a great dependence from the choice of the parameters in the interaction.

Many papers are aimed to find the best parameter set for a specific problem or for a set of specific problems [18,27]. In reality the problem of the right choice of the parameters cannot be solved completely, but it can be drastically reduced by means of the analysis that will be explained in the next section.

In this section, we want to give an overview of the performance of one swarm interaction algorithm, the SNO, on three different fitness function changing the parameters of the interaction.

The test is a standard sensitivity analysis on the algorithm changing the parameters. Due to the strong dependence of the parameters of the interaction, it is mainly meaningless to perform an analysis one at time.

To highlight the parameters involved in the parametric analysis, it is necessary to briefly analyse the implementation of the swarm interaction in SNO. In SNO, there are two attractors extracted from some particular groups or sets of candidate solutions.

Considering the mono-dimensional definition given in (19.3), the SNO interaction formula is

$$v_{t+1} = a \cdot v_t + b_1 \cdot (p_1 - x_t) + b_2 \cdot (p_2 - x_t) \tag{19.23}$$

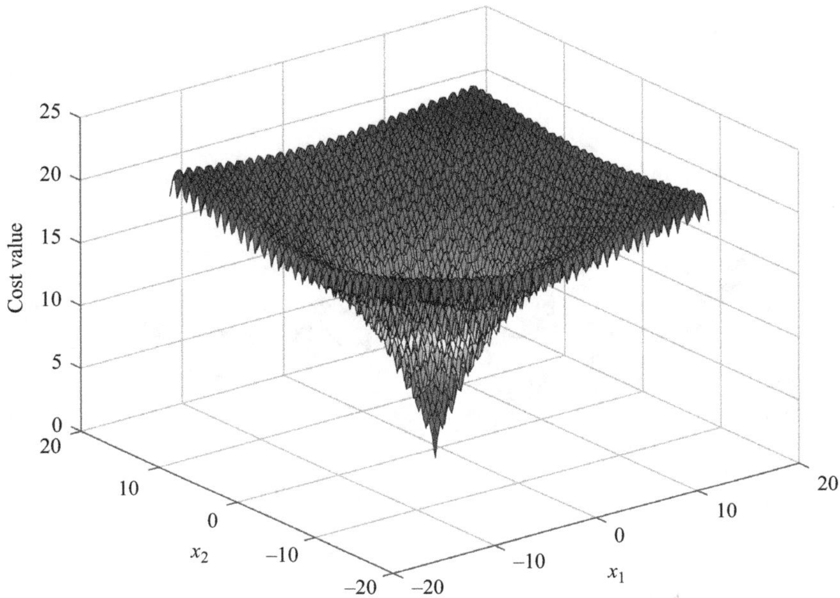

Figure 19.2 Plot of 2D Ackley function

and, after some manipulations, it becomes

$$v_{t+1} = a \cdot v_t - (b_1 + b_2) \cdot x_t + b_1 \cdot p_1 + b_2 \cdot p_2 \tag{19.24}$$

In order to reduce the number of parameters in the analysis, and consequently the computational time required, it is possible to make the assumption that $b_1 = b = b_2$. With this assumption, the formula becomes

$$v_{t+1} = a \cdot v_t - 2b \cdot x_t + b \cdot (p_1 + p_2) \tag{19.25}$$

In this way, we have just only two parameters a and b. This is good because the number of trials that should be done in the parametric analysis increases with A^{N_P} where N_P is the number of parameters, and A is the number of sampling point for each parameter.

Three different fitness functions have been used to perform this sensitivity analysis. These functions have been chosen in order to have three different typologies of landscape: the Ackley function (see Figure 19.2) is a multimodal, noisy function; the Penalty-1 function (see Figure 19.3) is a multimodal function with penalty definition of the constraints of the domain; finally, the Schwefel-226 function (see Figure 19.4) is a multimodal function with low frequency variations.

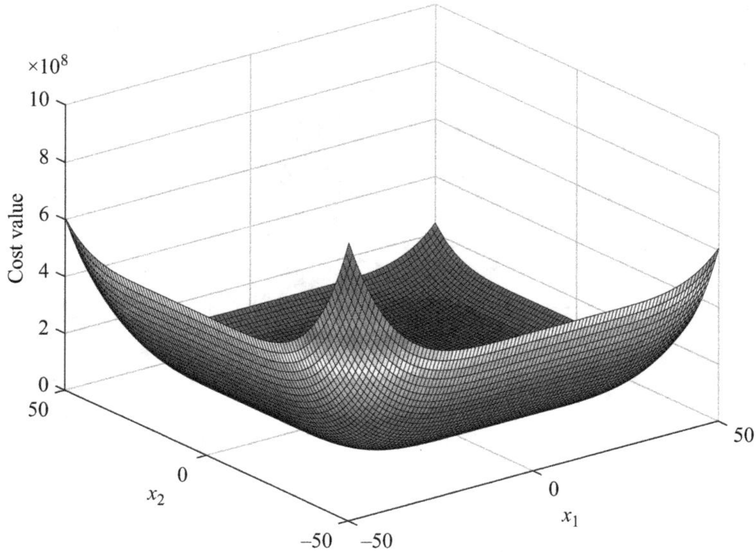

Figure 19.3 Plot of 2D Penalty-1 function

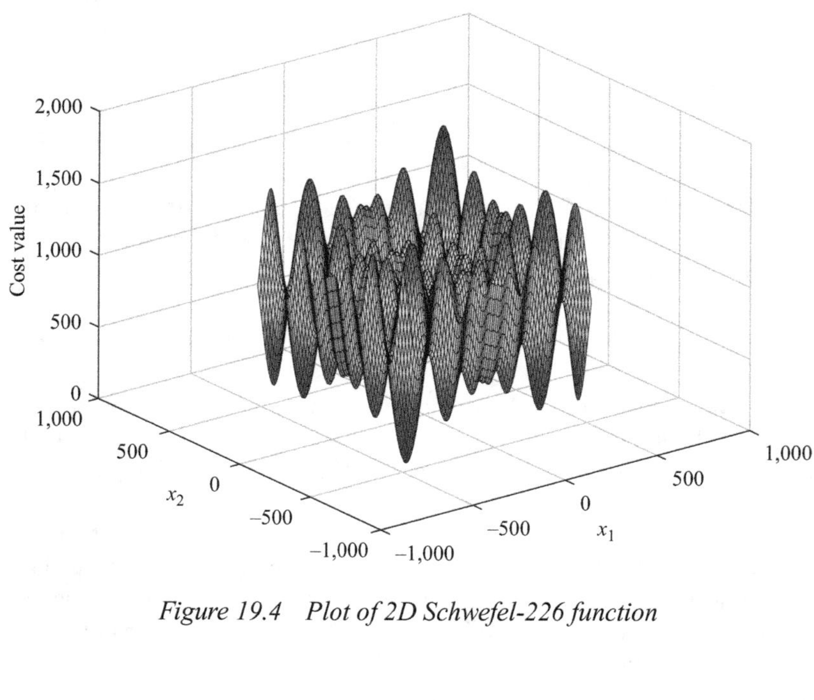

Figure 19.4 Plot of 2D Schwefel-226 function

The fitness function mathematical formulation is the following, derived from [17].

- Ackley function:

$$f(\overline{x}) = 20 + e^1 - 20 \cdot e^{-0.2\sqrt{\sum (x_i - x_0)^2/N}} - e^{\sum \cos(2\pi(x_i - x_0))/N} \tag{19.26}$$

  where N is the length of $\overline{x}$ and $x_0 = -2$ and the domain of definition is $-15 \leq x_i \leq 15$.
- Penalty-1 function:

$$f(\overline{x}) = g(\overline{x}) + h(\overline{x}) \tag{19.27}$$

$$g(\overline{x}) = \sum \begin{cases} 100(x_i - x_0 - 10)^4 & (x_i - x_0) > 10 \\ 100(-x_i - x_0 - 10)^4 & (x_i - x_0) < -10 \\ 0 & (x_i - x_0) = 10 \end{cases} \tag{19.28}$$

$$h(\overline{x}) = 10\sin^2\left[\pi\left(1 + \frac{x_1 - x_0 + 1}{4}\right)\right] + \left(1 + \frac{x_N - x_0 + 1}{4} - 1\right)^2 \cdot \frac{\pi}{30} + \sum_{i=1}^{N-1}\left[1 + \frac{x_i - x_0 + 1}{4} - 1\right]^2 \cdot \left(1 + 10\sin^2\left[\pi\left(1 + \frac{x_{i+1} - x_0 + 1}{4}\right)\right]\right)\frac{\pi}{30} \tag{19.29}$$

  where $x_0 = 7$ and its domain is $-50 \leq x_i \leq 50$
- Schwefel-226 function:

$$f(\overline{x}) = 418.9829 \cdot N + \sum x_i \cdot \sin\sqrt{|x_i|} \tag{19.30}$$

  where N is the length of $\overline{x}$ and its domain is $-512 \leq x_i \leq 512$.

The dimensionality of all the fitness function has been chosen to be 25-D in order to have a problem enough complex for SNO to be able to appreciate the differences in the choice of the parameters. To have a good statistical analysis, 50 independent trials have been done for each couple of parameters and then the optimal results found have been averaged.

It is possible to draw some considerations from the results of the parametric tests shown in Figure 19.5. The first consideration regards the fact that for all the functions there is a common area in which choice of the parameters gives a good optimization capability. Considering the different shape of the adopted benchmark functions, this means that a set of parameters chosen in this area can give good performances on a wide range of problems. Consequently, it is possible to perform the optimization without performing another sensitivity analysis on the specific problem. The second consideration regards the fact that the area of convergence is close to the region that in Section 19.6 will be demonstrated to be unstable. This means that to have good performances the algorithms should be close to instability. This is due to the fact that in this way it is possible to have a good compromise between exploration and exploitation.

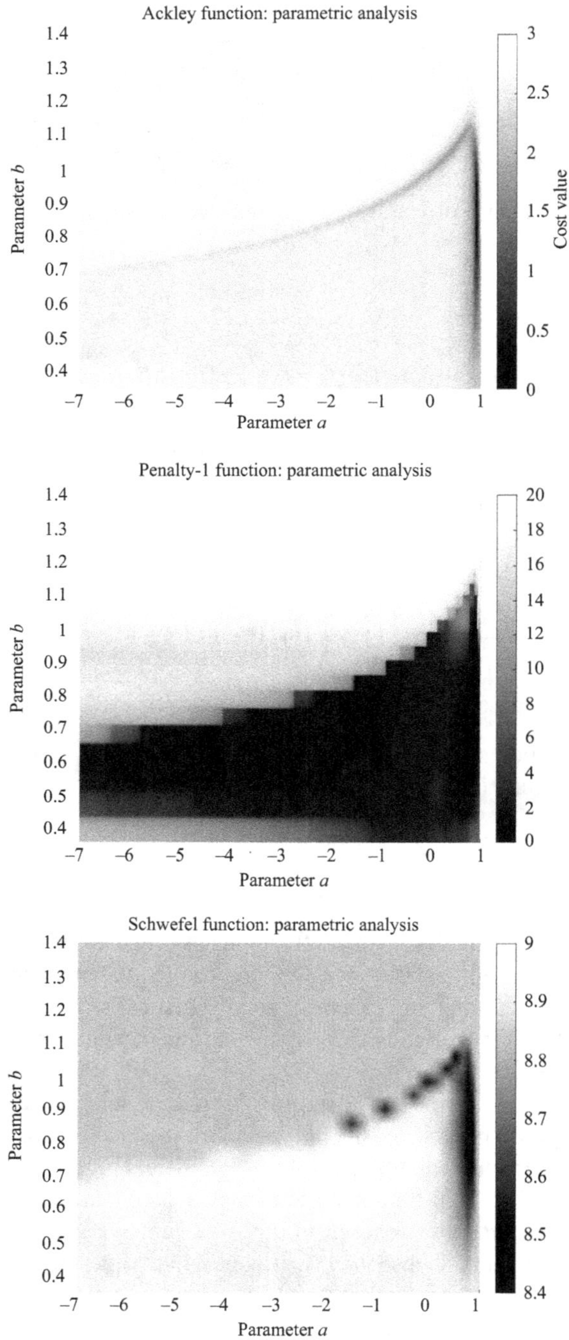

Figure 19.5 Parametric analysis for a and b. The fitness is proportional to the colour, the darker the better

19.5 Particle trajectory analysis

In this section, the trajectory of the particles will be analysed with theoretical and experimental approaches.

The first part is the definition of a simplified model in which it is easy to define the trajectories behaviour. Then the real model will be approached and the two approaches will be consequently compared. The last part of this section will be devoted to the experimental approach in which the behaviour of the entire swarm of particles will be analysed.

19.5.1 Differential equation interpretation

In this section, an interpretation of the swarm interaction model will be given. This interpretation is not completely equivalent to the real model, but due to its intuitiveness and similarities with other systems, it is worthy to analyse.

As described in Section 19.2.3, it is possible to express the interaction formula of swarm interaction in the 1D case as follows:

$$v_{t+1} = a \cdot v_t + \sum_{i=1}^{M} b_i \cdot (p_i - x_t) \tag{19.31}$$

$$x_{t+1} = x_t + v_{t+1} \tag{19.32}$$

From (19.32), it is possible to get to

$$x_{t+1} - x_t = v_{t+1} \tag{19.33}$$

So, it is possible to interpret it as a difference quotient:

$$v = \dot{x} \tag{19.34}$$

Subtracting v_t from both sides of (19.31):

$$v_{t+1} - v_t = a \cdot v_t + \sum_{i=1}^{M} b_i \cdot (p_i - x_t) - v_t \tag{19.35}$$

Rearranging the equation, the result is

$$v_{t+1} - v_t = (a-1) \cdot v_t + \sum_{i=1}^{M} b_i \cdot (p_i - x_t) \tag{19.36}$$

Interpreting again $v_{t+1} - v_t$ as a difference quotient, it is possible to write

$$v_{t+1} - v_t = \ddot{x} \tag{19.37}$$

So the equation of the velocity update (19.31) becomes

$$\ddot{x} = (a-1) \cdot \dot{x} + \sum_{i=1}^{M} b_i \cdot (p_i - x) \tag{19.38}$$

This last interpretation is not equivalent to the original model because this is a discrete interpretation of (19.38): they have different stability regions, and for some parameters they describe different trajectories.

Nonetheless, this model is really easy to be used since it is a second-order differential equation.

Equation (19.38) can be written as

$$\ddot{x} - (a-1)\cdot\dot{x} + \sum_{i=1}^{M} b_i \cdot x = \sum_{i=1}^{M} b_i \cdot p_i \tag{19.39}$$

From this equation, it is possible to make two preliminary considerations. The first one is that the behaviour of the swarm and the stability of the system do not depend on the choice of the attracting points. The second one is that the attracting points influence only the steady-state solution of the system.

The behaviour of the swarm can be analysed focusing the attention only on the homogeneous part of the differential (19.39):

$$\ddot{x} - (a-1)\cdot\dot{x} + \sum_{i=1}^{M} b_i \cdot x = 0 \tag{19.40}$$

It is possible to redefine the parameters of this equation to get to some parameters that are typical of the analysis of the second-order differential equations: the adimensional damping ratio h and the natural frequency of the system ω_0.

The equation written with these parameters is

$$\ddot{x} + 2h\omega_0 + \omega_0^2 x = 0 \tag{19.41}$$

Comparing (19.40) and (19.41), it is possible to correlate the parameters of one equation with the ones of the other:

$$\omega_0^2 = \sum_{i=1}^{M} b_i \tag{19.42}$$

$$h = \frac{1-a}{2\omega_0} \tag{19.43}$$

This model can be easily analysed because the frequency of oscillation of the system almost depends only on ω_0 and the reduction of the oscillation amplitude is due to the damping ratio. Thus, it is possible to evaluate the analytical solution of (19.41). The two eigenvalues are as follows:

$$\lambda_{1,2} = -h\omega_0 \pm \omega_0\sqrt{h-1} \tag{19.44}$$

With this model, it is trivial to identify oscillating solution to statically converging solutions. In fact, the solutions oscillate when the eigenvalues have an imaginary part not equal to zero. This condition can be described as follows: if $h < 1$ then the solution is oscillating.

The complete solution of the homogeneous differential equation, considering the eigenvalues calculated in (19.44), is

$$x(t) = A_1 \cdot e^{-h\omega_0 t} \cdot e^{-\omega_0\sqrt{h-1}t} + A_2 \cdot e^{-h\omega_0 t} \cdot e^{\omega_0\sqrt{h-1}t} \tag{19.45}$$

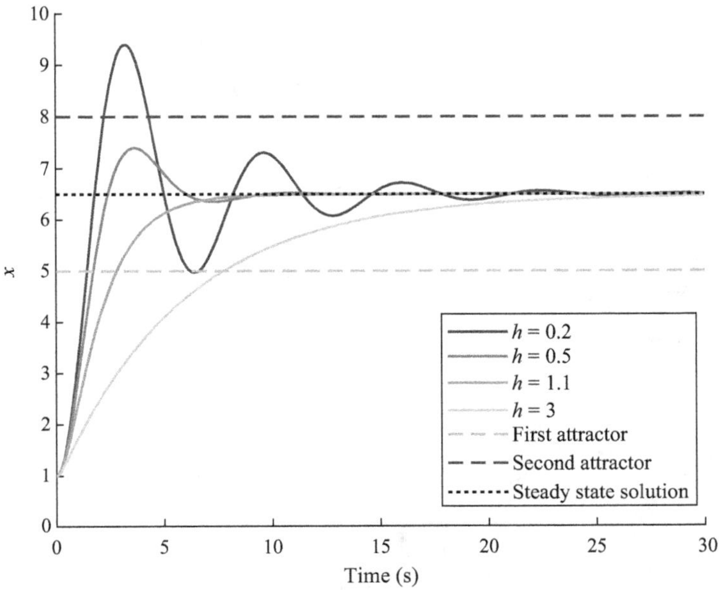

Figure 19.6 Effect of the parameter h on particle trajectories

The solution of the particular integral is

$$x_\infty = \sum_{i=1}^{M} b_i \cdot p_i \tag{19.46}$$

and this is the steady-state solution of the overall differential equation.

The value of the two constants A_1 and A_2 can be found from the initial conditions.

It has been proved [28] that the initial speed in swarm interaction has no influence on the optimization capability of the algorithms. Thus, the initial conditions are

$$\dot{x}(t=0) = 0 \tag{19.47}$$

$$x(t=0) = x_0 \tag{19.48}$$

The two coefficients A_1 and A_2 can be calculated:

$$A_1 = \frac{(x_0 - x_\infty)}{1 - \frac{-h\omega_0 + \omega_0\sqrt{h-1}}{-h\omega_0 - \omega_0\sqrt{h-1}}} = (x_0 - x_\infty)\frac{(h + \sqrt{h-1})}{2\sqrt{h-1}} \tag{19.49}$$

$$A_1 = \frac{(x_0 - x_\infty)}{1 - \frac{-h\omega_0 - \omega_0\sqrt{h-1}}{-h\omega_0 + \omega_0\sqrt{h-1}}} = (x_0 - x_\infty)\frac{(-h + \sqrt{h-1})}{2\sqrt{h-1}} \tag{19.50}$$

With these two values, the analytical solution of the differential equation is complete and can be plotted. Figures 19.6 and 19.7 show some solution changing the parameters.

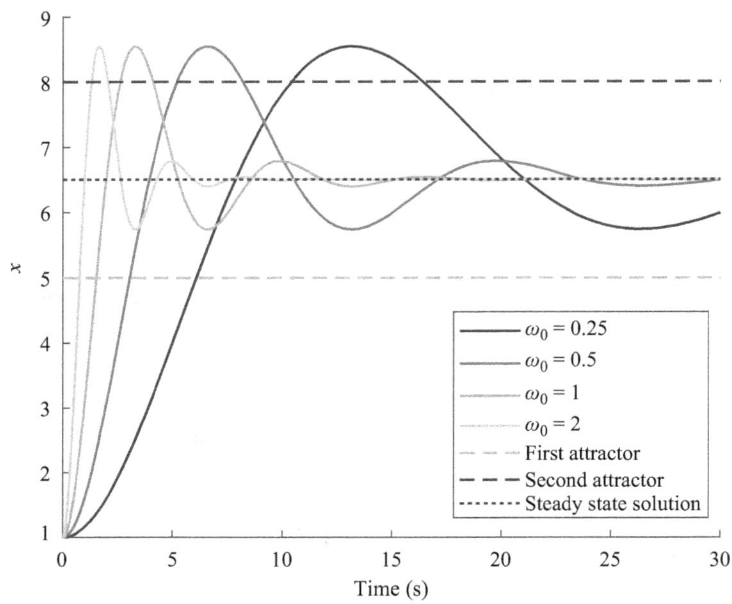

Figure 19.7 Effect of the parameter ω_0 on particle trajectories

This method can help in the choice of the right parameters for the swarm interaction model. In fact, solutions with $h > 1$ should be avoided: the convergence of the solutions is slower than for $h = 1$ and in both cases the swarm individuals do not explore the space of the solutions. Generally, for swarm interaction algorithms, the choice is a value of h, quite small, because these types of algorithms tend to be lack of exploration, so letting big oscillations is a way to increase the exploration capability of the algorithm.

Trajectory types can be analysed considering the results obtained with this model:

- Oscillating trajectories, when the eigenvalues are complex numbers, so when $h < 1$;
- Statically converging trajectories, when the eigenvalues are real, so when $h \geq 1$.

19.5.2 Discrete model interpretation

In this subsection, the discrete model will be analysed [18] and the types of trajectories that can be obtained are shown.

The discrete model is based directly the two original equations: (19.32) can be combined with (19.31) to get a system of two equations where the values at $t + 1$ are function of the values at t:

$$x_{t+1} = x_t + a \cdot v_t + \sum_{i=1}^{M} b_i \cdot (p_i - x_t) \tag{19.51}$$

Collecting the term x_t:

$$x_{t+1} = \left(1 - \sum_{i=1}^{M} b_i\right) x_t + a \cdot v_t + \sum_{i=1}^{M} b_i \cdot p_i \tag{19.52}$$

Calling

$$b = \sum_{i=1}^{M} b_i \tag{19.53}$$

and

$$p = \sum_{i=1}^{M} b_i \cdot p_i \tag{19.54}$$

it is possible to simplify the notation:

$$x_{t+1} = (1 - b)\, x_t + a \cdot v_t + p \tag{19.55}$$

Equation (19.31) can be rewritten using the same simplifications in the notations used before:

$$v_{t+1} = a \cdot v_t - b x_t + p \tag{19.56}$$

Considering the last two equations, the matrix form is

$$\begin{bmatrix} x_{t+1} \\ v_{t+1} \end{bmatrix} = \begin{bmatrix} 1 - b & a \\ -b & a \end{bmatrix} \cdot \begin{bmatrix} x_t \\ v_t \end{bmatrix} + \begin{bmatrix} p \\ p \end{bmatrix} \tag{19.57}$$

The behaviour of the individuals in this model is driven by the eigenvalues of the matrix, they are as follows:

$$\lambda_{1,2} = \frac{-(b - a - 1) \pm \sqrt{(b - a - 1)^2 - 4a}}{2} \tag{19.58}$$

Adopting this model, it is possible to have a larger number of trajectory types if compared with the approximated differential model.

The trajectories can be oscillating, non-oscillating, zigzagging and zigzagging with oscillations.

These trajectories are driven by two elements [18]:

- trajectories are oscillating if the eigenvalues are complex values with non-null imaginary part;
- trajectories are oscillating when the real part of at least one of the eigenvalues is positive.

It is possible to represent the conditions of the different trajectory types.

The condition for oscillation can be expressed by the inequality:

$$(b - a - 1)^2 - 4a < 0 \tag{19.59}$$

so

$$b^2 + a^2 - 2ab - 2b - 2a + 1 < 0 \tag{19.60}$$

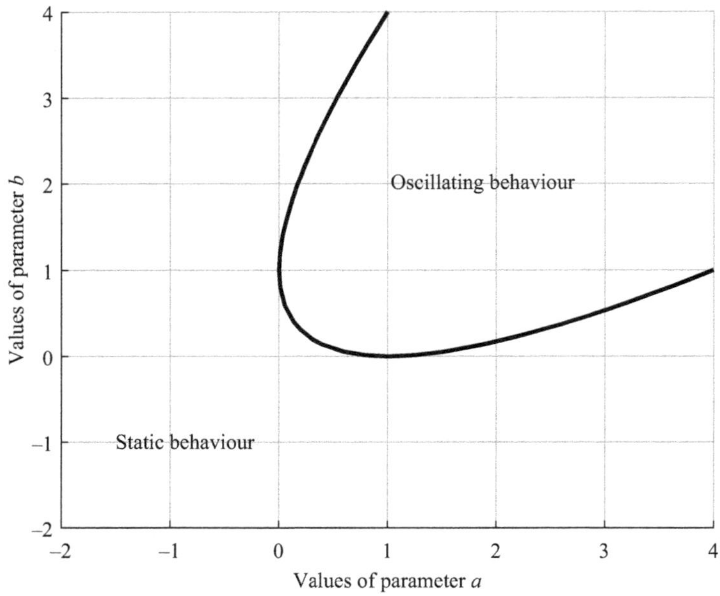

Figure 19.8 Oscillating behaviour in the b–a plane

Figure 19.8 shows the condition of oscillating behaviour in the *b–a* plane. This curve is not related with the convergence or divergence of the particle trajectories to the attractor, but just on how particle movement.

The second possible particle behaviour is the zigzagging. This behaviour appears when at least one of the eigenvalues has a real negative part. This condition can be analysed in the two cases: the first case is when the eigenvalues are complex, the other one is when the eigenvalues are both real. This mathematical consideration reflect the fact that oscillatory behaviour and zigzagging can be combined.

When the eigenvalues are complex, the condition on zigzagging is expressed as

$$-(b - a - 1) < 0 \tag{19.61}$$

so

$$a - b + 1 < 0 \tag{19.62}$$

On the other hand, when eigenvalues are real, it is possible to analyse the smaller:

$$-(b - a - 1) - \sqrt{(b - a - 1)^2 - 4a} < 0 \tag{19.63}$$

so

$$\sqrt{(b - a - 1)^2 - 4a} > -(b - a - 1) \tag{19.64}$$

Taking the square and performing some calculations:

$$a > 0 \tag{19.65}$$

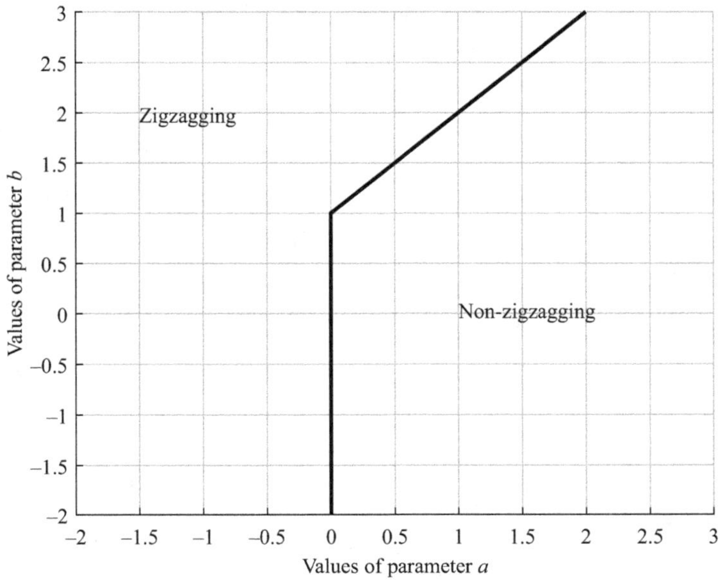

Figure 19.9 Zigzagging behaviour in the b–a plane

Figure 19.9 shows the condition of zigzagging behaviour in the *b–a* plane. The two lines correspond to the inequalities (19.62) and (19.65).

19.5.3 Complete algorithmic behaviour

In this section, the complete algorithmic behaviour will be analysed. In order to study the algorithmic behaviour, it is needed to choose a specific algorithm; as said before, the SNO algorithm has been chosen.

The analysis is composed of three steps with increasing complexity. In the first one, a simplified 1D version of SNO with fixed attractors will be compared with the analytic solution of the differential equation. Then the complete SNO will be considered analysing one single particle. Finally, the compete SNO swarm will be analysed.

The simplified version of SNO used in the first comparison, is a 1D SNO with only one person in the population and always the same two influencers. Figures 19.10 and 19.11 show the results of this comparison for different values of h and ω_0. The two dashed lines represent the position of the two influencers and the dotted one represents the steady-state solution.

The continuous line is the analytic solution of the differential equation while the crosses are the positions of the person of the simplified SNO.

Then, it is possible to observe the trajectory of one person of a complete SNO on the Ackley function. Figure 19.12 shows with the circles the behaviour of one person of SNO and with the crosses the position of the two influencers at each iteration.

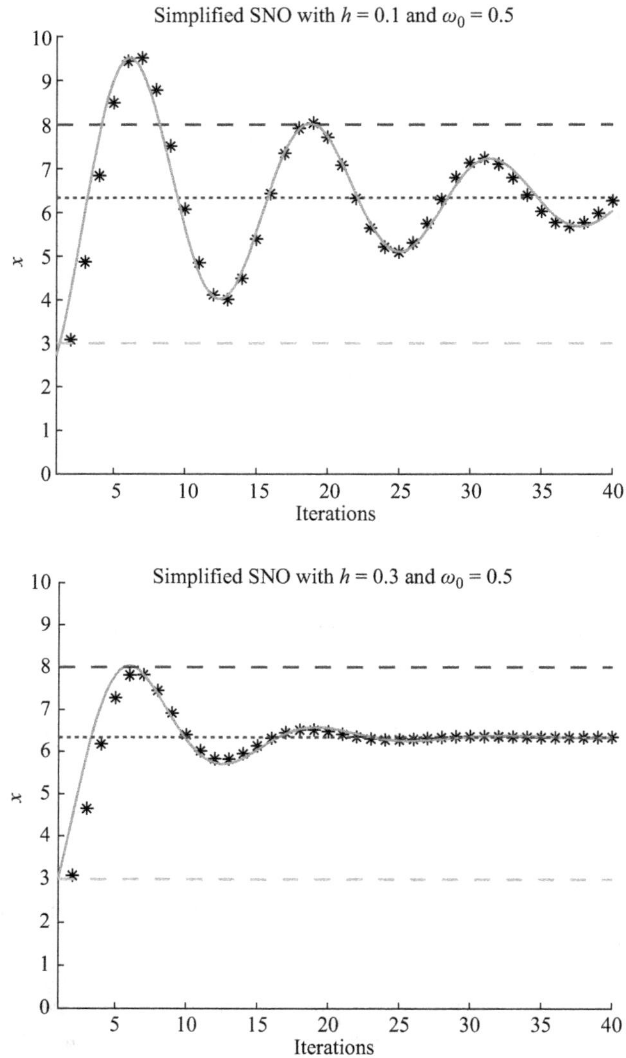

Figure 19.10 Comparison between differential equations with simple 1D SNO – Part I. The dashed lines correspond to the two attractors, the dotted line is the steady state solution, the continuous line is the analytical solution of the differential equation, while the crosses are the positions of the person of the simplified SNO

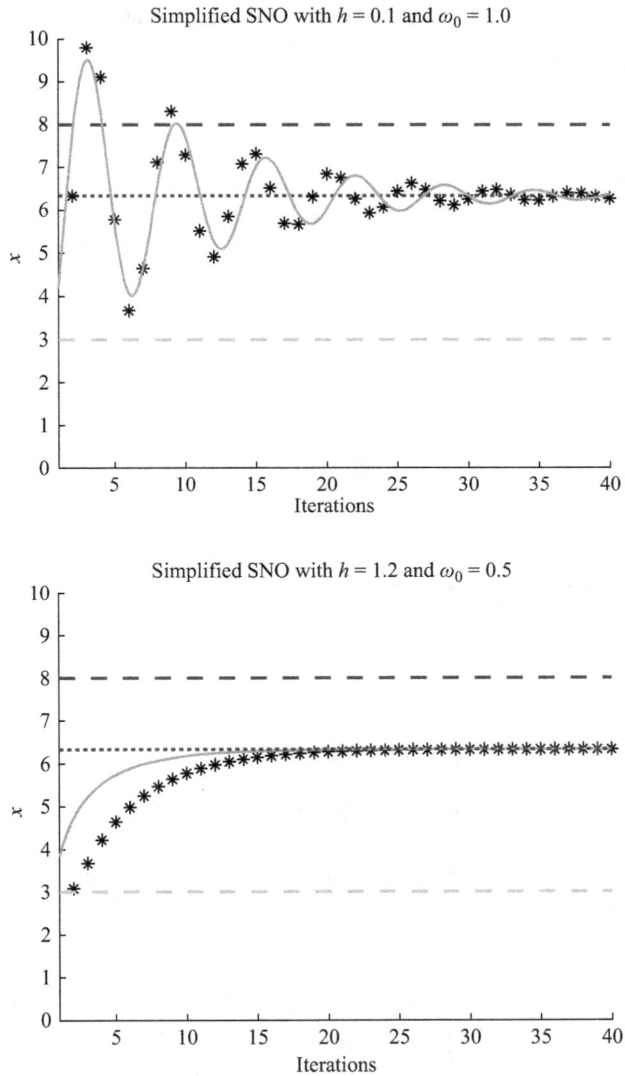

Figure 19.11 Comparison between differential equations with simple 1D SNO – Part II. The dashed lines correspond to the two attractors, the dotted line is the steady state solution, the continuous line is the analytical solution of the differential equation, while the crosses are the positions of the person of the simplified SNO

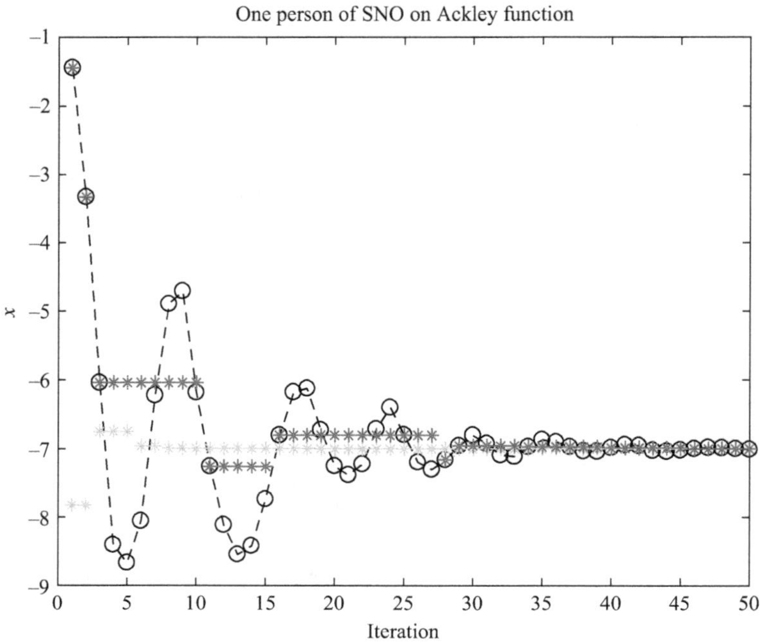

Figure 19.12 Behaviour of one person of real SNO

In this case, the trajectory is a bit less regular because at each iteration the influencers change.

Finally Figure 19.13 shows the convergence of all the population and of the memory of SNO on one of the 20 dimensions of an Ackley function. The colours of the points depend on the fitness of the person. Lighter the better. In this figure, the behaviour of the single person cannot be seen clearly. The main trend that is visible is the attraction of all the people to the minimum. Moreover, it is possible to see that there are some people that do not converge: they are the explorers that continuously try to find a new good solution, different from the one that is attracting all the population.

19.6 Stability analysis

In this section, the stability analysis of the different models of the swarm interaction will be provided.

Analysing the stability is an important study for the understanding of the optimization capability of the algorithm itself [29,30]. It is not strictly required that at each iteration the behaviour is stable, but over long run the stability is required for the most of the step to guarantee that, if the attracting point is the global optimal solution, the algorithm is able to reach that solution.

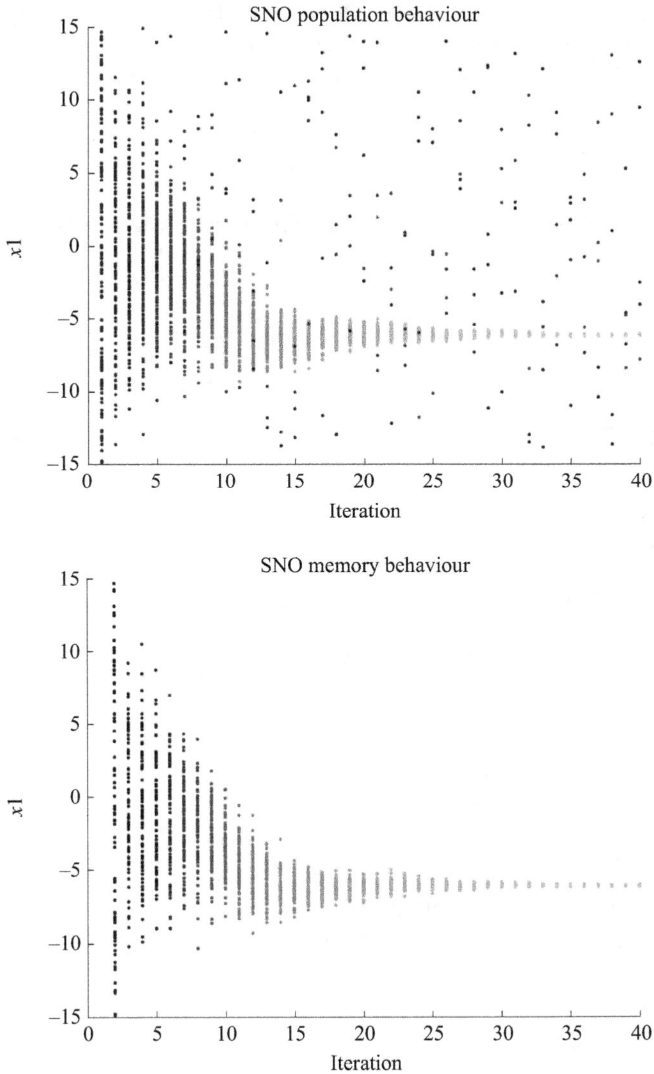

Figure 19.13 Behaviour of the population and of the memory of SNO: colours depend on the fitness of the person. Light colours mean good fitness, darker colours a worst fitness

19.6.1 Stability of differential equation

In this section, the convergence of the differential equation is analysed.

The analysis of the differential equation is a preliminary step for the complete stability analysis. In fact, it is easy to demonstrate that if the differential model is stable, the discrete one is also stable. On the other hand, there are some regions in which the differential is stable, but the discrete not.

We have seen that it is possible to write the differential model as

$$\ddot{x} - (a - 1) \cdot \dot{x} + \sum_{i=1}^{M} b_i \cdot x = \sum_{i=1}^{M} b_i \cdot p_i \tag{19.66}$$

The convergence can be analysed only considering the homogeneous part of the differential equations:

$$\ddot{x} - (a - 1) \cdot \dot{x} + \sum_{i=1}^{M} b_i \cdot x = 0 \tag{19.67}$$

Calling

$$b = \sum_{i=1}^{M} b_i \tag{19.68}$$

the equation becomes

$$\ddot{x} - (a - 1) \cdot \dot{x} + b \cdot x = 0 \tag{19.69}$$

The eigenvalues of the system are as follows:

$$\lambda_{1,2} = \frac{a - 1 \pm \sqrt{(a-1)^2 - 4b}}{2} \tag{19.70}$$

To have convergence, it is necessary that

$$Re(\lambda_i) < 0 \tag{19.71}$$

for all the values of i.

The first case is the oscillating behaviour, when $(a-1)^2 - 4b < 0$. The two eigenvalues are complex conjugate. Both their real parts are equal to

$$Re(\lambda_i) = a - 1 \tag{19.72}$$

The stability condition is

$$a < 1 \tag{19.73}$$

On the other hand, when $(a-1)^2 - 4b > 0$, the critical eigenvalue is positive:

$$\lambda_1 = \frac{a - 1 + \sqrt{(a-1)^2 - 4b}}{2} \tag{19.74}$$

The stability condition becomes

$$a - 1 + \sqrt{(a-1)^2 - 4b} < 0 \tag{19.75}$$

so

$$\sqrt{(a-1)^2 - 4b} < 1 - a \tag{19.76}$$

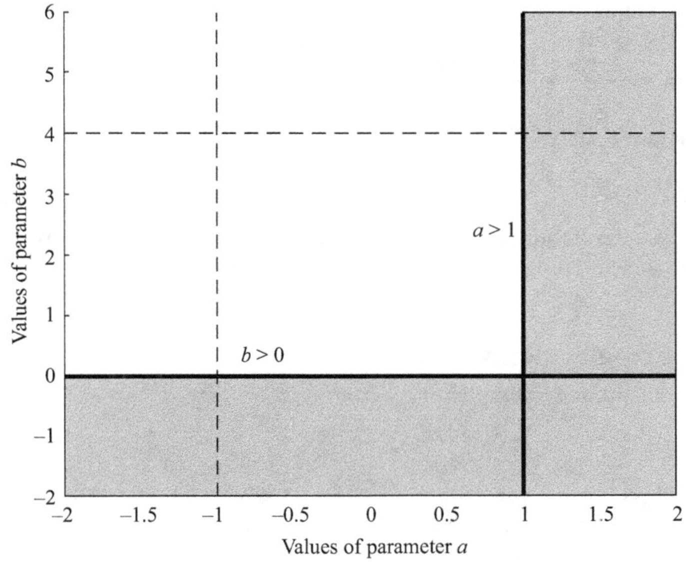

Figure 19.14 Stability of differential equation in the b–a plane

Finally,

$$(a-1)^2 - 4b < (1-a)^2 \tag{19.77}$$

so

$$b > 0 \tag{19.78}$$

The two conditions of stability can now be plotted in the plane b–a, as shown in Figure 19.14.

19.6.2 Stability with standard model

As seen many times before, the standard swarm interaction model is described by the two equations:

$$v_{t+1} = a \cdot v_t + \sum_{i=1}^{M} b_i \cdot (p_i - x_t) \tag{19.79}$$

$$x_{t+1} = x_t + v_{t+1} \tag{19.80}$$

Doing the same math seen in Section 19.5.2, it is possible to get to the matrix form:

$$\begin{bmatrix} x_{t+1} \\ v_{t+1} \end{bmatrix} = \begin{bmatrix} 1-b & a \\ -b & a \end{bmatrix} \cdot \begin{bmatrix} x_t \\ v_t \end{bmatrix} + \begin{bmatrix} p \\ p \end{bmatrix} \tag{19.81}$$

The two eigenvalues are as follows:

$$\lambda_{1,2} = \frac{-(b-a-1) \pm \sqrt{(b-a-1)^2 - 4a}}{2} \tag{19.82}$$

The system is stable if

$$\max(|\lambda_1|, |\lambda_2|) < 1 \tag{19.83}$$

To analyse the stability, it is convenient to split the analysis into two conditions: the first one is when the eigenvalues are imaginary and the second one is when they are both real.

19.6.2.1 Complex eigenvalues

The two eigenvalues are complex if

$$(b-a-1)^2 - 4a < 0 \tag{19.84}$$

In this case, the two eigenvalues λ_1 and λ_2 are two complex conjugate numbers. So their modules are equal and they can be calculated as

$$|\lambda_1| = |\lambda_2| = \sqrt{Re(\lambda_1)^2 + Im(\lambda_1)^2} \tag{19.85}$$

The stability condition is

$$\sqrt{Re(\lambda_1)^2 + Im(\lambda_1)^2} < 1 \tag{19.86}$$

Taking the square of this expression,

$$Re(\lambda_1)^2 + Im(\lambda_1)^2 < 1 \tag{19.87}$$

Before doing the evaluation of the stability, it is better to underline the imaginary part of the eigenvalues:

$$\lambda_{1/2} = \frac{-(b-a-1) \pm i\sqrt{4a - (b-a-1)^2}}{2} \tag{19.88}$$

Thus,

$$\frac{[-(b-a-1)]^2 + [4a - (b-a-1)^2]}{4} < 1 \tag{19.89}$$

Eventually, it results in

$$a < 1 \tag{19.90}$$

This is the first condition for the stability. This condition should be taken into account dealing with time-varying coefficients [31].

19.6.2.2 Real eigenvalues

The second case is when the eigenvalues are both real. This condition is satisfied when

$$(b-a-1)^2 - 4a > 0 \tag{19.91}$$

The stability condition is, as always,

$$\max(|\lambda_1|, |\lambda_2|) < 1 \tag{19.92}$$

In this case, it is required to find the largest eigenvalue.

When the square root is taken with the positive sign, the stability condition becomes

$$-(b-a-1) + \sqrt{(b-a-1)^2 - 4a} < 2 \tag{19.93}$$

On the other hand, when the square root is taken with the negative sign, the stability condition becomes

$$-(b-a-1) - \sqrt{(b-a-1)^2 - 4a} > -2 \tag{19.94}$$

Starting the analysis from (19.93),

$$\sqrt{(b-a-1)^2 - 4a} < b - a + 1 \tag{19.95}$$

taking the square

$$(b-a-1)^2 - 4a < (b-a+1)^2 \tag{19.96}$$

Doing some math, it is possible to reach the condition:

$$b > 0 \tag{19.97}$$

Analysing (19.94),

$$a - b + 3 > \sqrt{(b-a-1)^2 - 4a} \tag{19.98}$$

taking the square,

$$(a-b+3)^2 > (b-a-1)^2 - 4a \tag{19.99}$$

It is possible to reach the condition:

$$2a - b - 2 > 0 \tag{19.100}$$

19.6.2.3 Stability analysis

The three stability conditions of stability seen before are as follows:

$$a < 1 \tag{19.101}$$

$$b > 0 \tag{19.102}$$

$$2a - b - 2 > 0 \tag{19.103}$$

The representation of these conditions in the space of the parameters *b–a* is depicted in Figure 19.15.

It is possible to see the effect of the parameters plotting the eigenvalues in the real-imaginary plane. It can be first done fixing the value of *a*. The result is shown in Figure 19.16. The figure shows the behaviour of the eigenvalues changing the value of *b*, with a fixed value of $a = 0.5$.

The stability condition is equivalent to the unitary module circle depicted in the figure. While both the two eigenvalues are in the circle, the system results stable. The

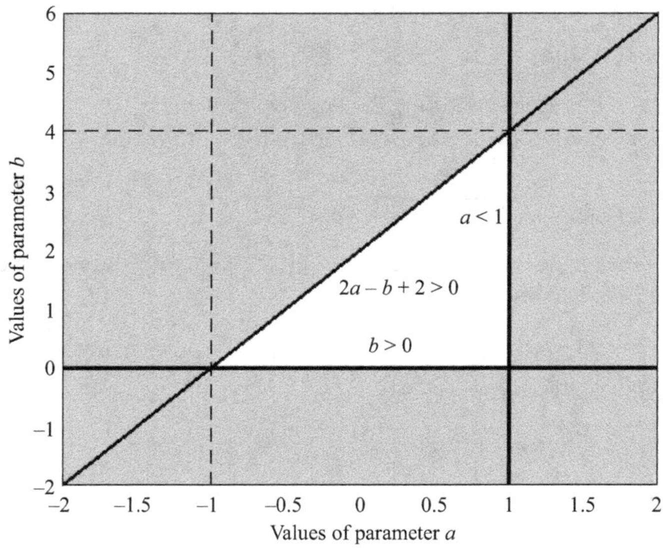

Figure 19.15 Stability analysis in the b–a plane

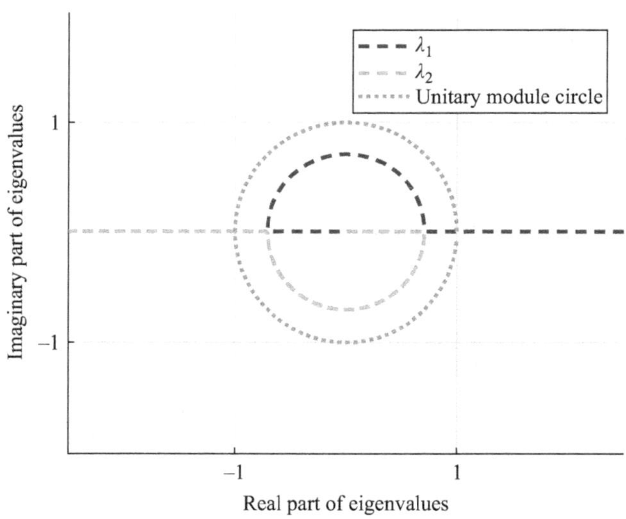

Figure 19.16 Effect of b on the eigenvalues

oscillatory behaviour is found when the eigenvalues have an imaginary part, different from zero.

It is possible to analyse the effect of the parameter a on one eigenvalues (just for sake of simplicity). Analysing Figure 19.17, we can observe that the value of a dilates the trajectories of the eigenvalues.

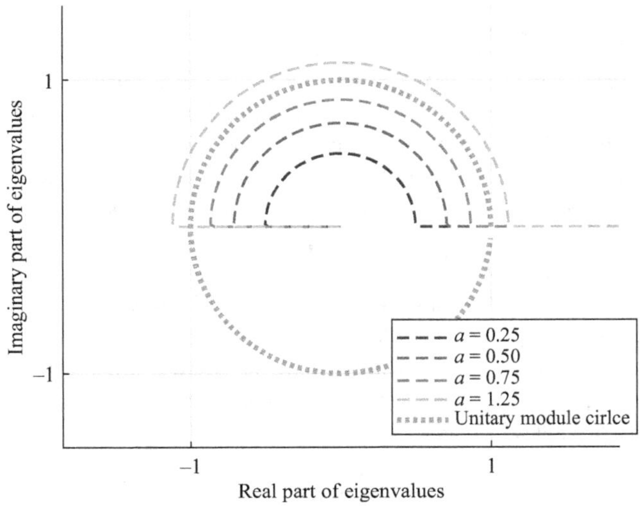

Figure 19.17 Effect of a on one eigenvalue

19.7 Conclusions

In this chapter, an overview of the general interaction in SI algorithms has been provided, analysing their behaviour using differential and discrete models. In particular, the presented interaction analysis has been focused on the well-known PSO technique and applied also to a recently developed algorithm, namely the SNO.

From the presented parametric analysis, it has been demonstrated that, for all the considered benchmark functions, there is a common set of parameters for which the algorithm convergence is optimal. Considering the different shape of the functions, this means that the choice of the parameters associated with inertia and attraction does not depend on the specific optimization problem.

This numerical and experimental finding has also been compared with theoretical analysis, showing that the above-mentioned region of optimal parameter values corresponds to oscillating trajectories close to instability. The main explanation for this outcome may be due to the fact that, in this way, it is possible to have a good balance between exploration and exploitation, resulting in overall performance maximization.

Finally, the presented stability analysis has shown that the constrains found for inertia and attraction parameters in SI are critical aspects to be considered in research and applications, in particular when dealing with time-varying and adaptive coefficients.

References

[1] C. Blum and X. Li, "Swarm intelligence in optimization," *Swarm Intelligence*, pp. 43–85, 2008.

[2] A. P. Engelbrecht. Fundamentals of Computational Swarm Intelligence. Hoboken, NJ: John Wiley & Sons, 2006.
[3] R. Eberhart and J. Kennedy, "A new optimizer using particle swarm theory," in *Micro Machine and Human Science, 1995. MHS'95, Proceedings of the Sixth International Symposium on*. IEEE, 1995, pp. 39–43.
[4] M. Dorigo, V. Maniezzo, and A. Colorni, "Ant system: optimization by a colony of cooperating agents," *IEEE Transactions on Systems, Man, and Cybernetics, Part B (Cybernetics)*, vol. 26, no. 1, pp. 29–41, 1996.
[5] R. C. Eberhart and Y. Shi, "Comparison between genetic algorithms and particle swarm optimization," in *International Conference on Evolutionary Programming*. Springer, 1998, pp. 611–616.
[6] J. Robinson and Y. Rahmat-Samii, "Particle swarm optimization in electromagnetics," *IEEE Transactions on Antennas and Propagation*, vol. 52, no. 2, pp. 397–407, 2004.
[7] A. Janecek and Y. Tan, "Using population based algorithms for initializing nonnegative matrix factorization," in *International Conference in Swarm Intelligence*. Springer Berlin Heidelberg, 2011, pp. 307–316.
[8] M. Ruello, A. Niccolai, F. Grimaccia, M. Mussetta, and R. E. Zich, "Blackhole PSO and SNO for electromagnetic optimization," in *2014 IEEE Congress on Evolutionary Computation (CEC)*, July 2014, pp. 1912–1916.
[9] S. Selleri, M. Mussetta, P. Pirinoli, R. E. Zich, and L. Matekovits, "Some insight over new variations of the particle swarm optimization method," *IEEE Antennas and Wireless Propagation Letters*, vol. 5, no. 1, pp. 235–238, 2006.
[10] A. Gandelli, F. Grimaccia, M. Mussetta, P. Pirinoli, and R. E. Zich, "Development and validation of different hybridization strategies between GA and PSO," in *Evolutionary Computation, 2007. CEC 2007. IEEE Congress on*. IEEE, 2007, pp. 2782–2787.
[11] B. K. Panigrahi, Y. Shi, and M.-H. Lim, *Handbook of swarm intelligence: concepts, principles and applications*. Springer Science & Business Media, 2011, vol. 8.
[12] E. Bonabeau, M. Dorigo, and G. Theraulaz, *Swarm intelligence: from natural to artificial systems*. New York: Oxford University Press, 1999, no. 1.
[13] R. Albi, F. Grimaccia, M. Mussetta, A. Niccolai, and R. Zich, "A new algorithm for antenna optimization in aerospace applications: the sounding rocket test case," in *Electromagnetics in Advanced Applications (ICEAA), 2013 International Conference on*. IEEE, 2013, pp. 1537–1540.
[14] D. Karaboga and B. Akay, "A survey: algorithms simulating bee swarm intelligence," *Artificial Intelligence Review*, vol. 31, no. 1–4, pp. 61–85, 2009.
[15] Y. Tan and Y. Zhu, "Fireworks algorithm for optimization," in *International Conference in Swarm Intelligence*. Springer, 2010, pp. 355–364.
[16] E. Atashpaz-Gargari and C. Lucas, "Imperialist competitive algorithm: an algorithm for optimization inspired by imperialistic competition," in *Evolutionary Computation, 2007. CEC 2007. IEEE Congress on*. IEEE, 2007, pp. 4661–4667.

[17] D. Simon, *Evolutionary optimization algorithms*. Hoboken, NJ: John Wiley & Sons, 2013.

[18] I. C. Trelea, "The particle swarm optimization algorithm: convergence analysis and parameter selection," *Information Processing Letters*, vol. 85, no. 6, pp. 317–325, 2003.

[19] F. Van Den Bergh and A. P. Engelbrecht, "A study of particle swarm optimization particle trajectories," *Information Sciences*, vol. 176, no. 8, pp. 937–971, 2006.

[20] J. Zhang, K. Liu, Y. Tan, and X. He, "Allocation of local and global search capabilities of particle in canonical PSO," in *Proceedings of the 10th Annual Conference on Genetic and Evolutionary Computation*. ACM, 2008, pp. 165–166.

[21] S. B. Akat and V. Gazi, "Particle swarm optimization with dynamic neighborhood topology: three neighborhood strategies and preliminary results," in *Swarm Intelligence Symposium, 2008. SIS 2008. IEEE*. IEEE, 2008, pp. 1–8.

[22] M. Clerc, "Stagnation analysis in particle swarm optimisation or what happens when nothing happens," Technical Report CSM-460, University of Essex, 2006.

[23] R. Poli, J. Kennedy, and T. Blackwell, "Particle swarm optimization," *Swarm Intelligence*, vol. 1, no. 1, pp. 33–57, 2007.

[24] O. Olorunda and A. P. Engelbrecht, "Measuring exploration/exploitation in particle swarms using swarm diversity," in *Evolutionary Computation, 2008. CEC 2008.(IEEE World Congress on Computational Intelligence). IEEE Congress on*. IEEE, 2008, pp. 1128–1134.

[25] A. Niccolai, F. Grimaccia, M. Mussetta, P. Pirinoli, V. H. Bui, and R. E. Zich, "Social network optimization for microwave circuits design," *Progress In Electromagnetics Research C*, vol. 58, pp. 51–60, 2015.

[26] F. Grimaccia, M. Mussetta, A. Niccolai, P. Pirinoli, and R. E. Zich, "Recently developed social-based algorithms for antennas optimization," in *2014 International Conference on Numerical Electromagnetic Modeling and Optimization for RF, Microwave, and Terahertz Applications (NEMO)*, May 2014, pp. 1–4.

[27] R. C. Eberhart and Y. Shi, "Comparing inertia weights and constriction factors in particle swarm optimization," in *Evolutionary Computation, 2000. Proceedings of the 2000 Congress on*, vol. 1. IEEE, 2000, pp. 84–88.

[28] S. Helwig and R. Wanka, "Theoretical analysis of initial particle swarm behavior," in *International Conference on Parallel Problem Solving from Nature*. Springer, 2008, pp. 889–898.

[29] M. Clerc and J. Kennedy, "The particle swarm-explosion, stability, and convergence in a multidimensional complex space," *IEEE Transactions on Evolutionary Computation*, vol. 6, no. 1, pp. 58–73, 2002.

[30] C. W. Cleghorn and A. P. Engelbrecht, "A generalized theoretical deterministic particle swarm model," *Swarm Intelligence*, vol. 8, no. 1, pp. 35–59, 2014.

[31] X. Cai, Y. Cui, and Y. Tan, "Predicted modified PSO with time-varying accelerator coefficients," *International Journal of Bio-Inspired Computation*, vol. 1, no. 1–2, pp. 50–60, 2009.

Chapter 20

Coordinating swarms of microscopic agents to assemble complex structures

Bruce J. MacLennan[1]

Abstract

This chapter addresses the problem of coordinating very large swarms of microscopic agents to assemble complex, hierarchically structured physical systems. The agents might be microscopic robots or genetically engineered microorganisms. The approach we use is a form of artificial morphogenesis, which studies the self-organized morphogenetic processes in the developing embryo, by which billions of cells cooperate to create physical form and abstracts these processes to control microscopic agents to assemble desired physical structures. We use an approach based on the description of morphogenetic processes by partial differential equations, which ensures that our morphogenetic algorithms will scale up to arbitrarily large swarms (millions or more). We present several simulation experiments demonstrating the coordination of massive swarms to construct complex objects.

20.1 The challenge

20.1.1 Assembling complex hierarchical structures

We have limited means for structuring systems at the microscopic level. For example, we use photolithography for creating very large scale integration (VLSI) circuitry, but this is limited to two dimensions. (Three-dimensional VLSI is limited to stacking a few layers and does not achieve the same fineness of structure in the third dimension.) To some extent, we can design materials that self-assemble from the nanoscale up to the macroscale, but such techniques are limited to regular patterns analogous to crystals. Such techniques are inadequate for future technological opportunities, which will require macroscopic systems with complex structures at many hierarchical levels down to a microscopic scale.

For example, consider the challenge of designing a robot with physical competence and intelligence comparable to a mammal. It will need an artificial nervous

[1]Department of Electrical Engineering and Computer Science, University of Tennessee, USA

system, perhaps implemented with many dense neural networks intricately interconnected as they are in animals' nervous systems. How can we assemble artificial neural networks comprising millions or perhaps billions of artificial neurons with specific patterns of interconnection and many thousands of connections each, as we find in mammalian brains? Our robot may require complex sense organs, for example, artificial eyes with millions of receptors. Competent physical action may require force, torque, and pressure sensors distributed throughout the robot's body and its artificial muscles. More broadly, we may need to organize many millions or even billions of microscopic components in very specific configurations in order to assemble a macroscopic object. Our solution is to employ massive swarms of microscopic agents to accomplish this task.

20.1.2 Artificial morphogenesis approach

One might reasonably wonder if this goal is achievable, that is, if it is possible to coordinate massive swarms of agents to assemble complex hierarchical structures. Fortunately, we have an existence proof, for in the course of embryological development, a very large number of cells—with limited intelligence—self-organize to create an incredibly complex living body. In particular, through self-organized *morphogenesis*, cells create physical forms and shapes by moving, dividing, pushing, and pulling on one another and on the substrates they have created. Therefore, embryological morphogenesis provides a model for *artificial morphogenesis* as an approach for coordinating microscopic agents to assemble complex physical structures. It is a step toward *programmable matter*, that is, toward complex systems whose physical properties are systematically controllable (programmable) [1].

Other research projects have been inspired by embryological morphogenesis as a basis for self-assembly. For example, there are active research programs in *morphogenetic engineering*, which are inspired more or less closely by natural morphogenesis [2]. Other work inspired by embryological morphogenesis includes [3–8]. Fortunately, a small number of morphogenetic processes (about 20) seem to be adequate to account for embryological development [9–13]. In our approach to artificial morphogenesis, we follow embryological morphogenesis quite closely, because we know it works for microscopic agents and scales up to very large numbers (up to 10^{12}). The consequences of this strategy are discussed in the following sections.

20.1.3 Distinctive properties of microscopic agents

Our goal is to control swarms of microscopic agents to assemble complex hierarchical structures, and so we must be cognizant of the nature of these agents. The two most likely kinds of agents are microrobots (by which we mean robots less than a millimeter in size) and genetically engineered microorganisms (10–500 μm). Such small agents will be subject to many unavoidable perturbations, including Brownian motion, contamination, and irregularities in the medium through which or on which they move. We cannot treat such disturbances as afterthoughts but must consider them from the beginning. Therefore, we are operating in a stochastic context.

Because our agents are very small, they will have limited capabilities. We cannot assume they have much power, either for movement or for communication. We cannot have complex moving parts (although it is reasonable to assume effectors such as cilia, flagella, and simple actuators for controlling adhesion and connection). Microscopic agents might not have precise control over their position or velocity. We cannot assume that the agents can be guided by complex programs, and therefore analog control is more appropriate since it is simpler to implement [14]. We have discussed reasonable assumptions about microscopic assemblers in more detail in [1,15].

As mentioned, our intended agents may be manufactured or living but there are differences. Advantages of using living agents include the facts that they are capable of self-reproduction, have efficient metabolic systems for powering their behavior, and are capable of chemical signaling. On the other hand, we have more control over the design of manufactured agents (but progress in synthetic biology is shifting this trade-off). Hybrid agents, which incorporate living cells in a manufactured vehicle, are also possible.

Embryological morphogenesis is coordinated by living cells, but also involves nonliving substances as substrates, structures, and communication media. Likewise, artificial morphogenesis makes use of microscopic *passive components* as well as assembler agents (which are considered *active components*). Passive components are especially needed when the agents are manufactured and therefore incapable of reproduction, for then growth takes place by assembling passive components (provided from an external source) rather than by the reproduction of the agents themselves. Passive components also include things such as signaling molecules (called *morphogens*), which are used for coordinating assembly rather than being part of the final assembled structure.

20.1.4 Scalability

A fundamental goal of our approach to artificial morphogenesis is scalability: we are seeking algorithms that will scale to very large numbers of microscopic assemblers (millions or even billions). Therefore, we limit our attention to processes that can be demonstrated to work for sufficiently large swarms of agents. Whether the process works for modest numbers of agents (say, hundreds to tens of thousands) is not especially important for our purposes.

A further goal of our approach in an independence of scales, by which we mean that the size of structures assembled by the agents should be independent of the size of the components (so long as the components are small enough compared to those structures). For example, our morphogenetic processes should be capable of generating structures with features of desired dimensions independent of the size of the components. This permits different components (e.g., newer smaller assemblers) to implement the same process, much as a conventional computer algorithm can be executed on various hardware platforms.

20.2 Massive swarms

In this section, we explain the approach we use for describing and controlling massive swarms of microscopic assemblers. In brief, we treat them from the perspective of continuum mechanics, that, as an infinite number of infinitesimal components.

20.2.1 Continuum mechanics approach

There are several reasons continuum mechanics is an appropriate theoretical framework for artificial morphogenesis. First, it is worth noting that developmental biologists often use partial differential equations (PDEs) for describing morphogenetic processes; PDEs abstract away from discrete cells, treating tissues and fluids as continuous materials [16,17]. Second, biologists have identified the regime of *soft matter* or *viscoelastic materials* as especially relevant to embryological development [10,18,19]. Tissues stretch, bend, and flow viscously in development, and continuum mechanics is the appropriate tool for describing the processes. Third, continuum mechanics provide important mathematical concepts and tools for treating the behavior of substances comprising vast numbers of microscopic particles. It describes substances such as solids, liquids, and gasses—including soft matter—that are phenomenologically continuous but composed of discrete molecules or atoms.

20.2.2 Mathematical framework

20.2.2.1 Intensive quantities

We begin by defining some fundamental quantities for any artificial morphogenetic system. Suppose we have k_e species of elementary or "atomic" components, which might be active or passive. These elementary components can join together into various small complexes or "molecules," and we let k be the total number of possible species, both elementary and complex. Let n_i be the number of instances of species i in the system, and let $n_s = \sum_{i=1}^{k} n_i$ be the total number of components (elementary and complex) in the system. We define $X_i = n_i/n_s$ to be the *number fraction* or *relative concentration* of species i in the system.

If μ_i is the mass of elementary component i, then we can compute the mass of each complex μ_j for each complex species j. For example, suppose species 3 is a complex of elementary species 1 and 2; perhaps species 1 is a transporter agent and species 2 is a passive component being transported. Then $\mu_3 = \mu_1 + \mu_2$. The total mass of species i in the system is $m_i = n_i\mu_i$, and the *mass fraction* of species i is $Y_i = m_i/m$, where $m = \sum_{i=1}^{k} m_i$ is the total mass of components in the system.

In order to describe morphogenetic algorithms independently of the size of the components, it is preferable to express them in terms of *intensive quantities* that are independent of volume, such as mass density and number density. Therefore, let n_i be the number of instances of species i in a volume V. We define the concentration (partial number density) of this species as $\xi_i = n_i/V$. The partial (mass) density is $\rho_i = m_i/V$.

20.2.2.2 Balance equations

Movement and interaction of the components obey a number of balance equations. Consider an arbitrary but fixed control volume V. Let $\boldsymbol{Q}(t)$ be the total of some scalar or vector quantity in V at time t, and let $\boldsymbol{q}(\boldsymbol{x}, t)$ be the density of that quantity at location $\boldsymbol{x} \in V$. Therefore, $\boldsymbol{Q}(t) = \int_V \boldsymbol{q}(\boldsymbol{x}, t)\mathrm{d}V$. For example, $\boldsymbol{q}(\boldsymbol{x}, t)$ might be the number density or flux of some component species at $\boldsymbol{x}$.

The quantity $\boldsymbol{Q}$ may be changing as a consequence of various factors. First, there may be a flux of $\boldsymbol{Q}$ across the surface A of volume V; let $\boldsymbol{\phi}_q(\boldsymbol{x}, t)$ be the flux at surface location $\boldsymbol{x} \in A$. If $\boldsymbol{n}$ is an outward-pointing normal to the surface at $\boldsymbol{x}$, then the quantity leaving the volume at that point is $\boldsymbol{\phi}_q \cdot \boldsymbol{n}$, and the total outflow due to flux across the boundary is $\int_A \boldsymbol{\phi}_q \cdot \boldsymbol{n}\mathrm{d}A$.

There may also be local sources (and sinks) for $\boldsymbol{Q}$ within the volume; for example, the number of free elementary components may decrease or increase by being incorporated into complex components or released from them, respectively. Likewise, the number of complex components can increase when they assemble from elementary components, or decrease when they disassemble. Living components may die or reproduce, thereby changing their number. Therefore, let $\boldsymbol{\sigma}_q(\boldsymbol{x}, t)$ be the amount of $\boldsymbol{Q}$ being produced (or consumed, if negative) at location $\boldsymbol{x}$. These internal sources and sinks contribute $\int_V \boldsymbol{\sigma}_q \mathrm{d}V$ to $\boldsymbol{Q}(t)$. Combining the changes in the interior and across the boundary,

$$\frac{\partial \boldsymbol{Q}}{\partial t} = -\int_A \boldsymbol{\phi}_q \cdot \boldsymbol{n}\mathrm{d}A + \int_V \boldsymbol{\sigma}_q \mathrm{d}V$$

Since $\boldsymbol{Q} = \int_V \boldsymbol{q}\mathrm{d}V$, the rate of change of $\boldsymbol{Q}$ is

$$\frac{\partial \boldsymbol{Q}}{\partial t} = \int_V \frac{\partial \boldsymbol{q}}{\partial t}\mathrm{d}V = -\int_A \boldsymbol{\phi}_q \cdot \boldsymbol{n}\mathrm{d}A + \int_V \boldsymbol{\sigma}_q \mathrm{d}V$$

By Gauss's law, $\int_A \boldsymbol{\phi}_q \cdot \boldsymbol{n}\mathrm{d}A = \int_V \nabla \cdot \boldsymbol{\phi}_q \mathrm{d}V$; hence, we have a balance equation:

$$\int_V \frac{\partial \boldsymbol{q}}{\partial t}\mathrm{d}V = -\int_V \nabla \cdot \boldsymbol{\phi}_q \mathrm{d}V + \int_V \boldsymbol{\sigma}_q \mathrm{d}V$$

Since the control volume was arbitrary, the balance is independent of it:

$$\frac{\partial \boldsymbol{q}}{\partial t} = -\nabla \cdot \boldsymbol{\phi}_q + \boldsymbol{\sigma}_q$$

This accounts for the local change in a quantity due to both production/consumption and motion.

It will be helpful to consider further the sources and sinks of components. In the case of living components, sources include locations where the components are reproducing, and there are sinks wherever they are dying. In the case of nonliving components, sources may include locations where components of various species are supplied from outside the system. Similarly, sinks may be locations from which unused or waste components are removed from the system so that they can be recycled or otherwise disposed.

Less obvious sources and sinks are locations where components change their species. For example, a component may change from one species to another by altering its characteristics or behavior in significant ways; this is analogous to cell differentiation in biological morphogenesis. In this case, the source for a species i is the sink for a species j: $\sigma_i(\boldsymbol{x}, t) = -\sigma_j(\boldsymbol{x}, t)$. The conversion is often controlled by some components, either those converting or some third species. Therefore, the rate of conversion $\nu_{ij}(\boldsymbol{x}, t) = \sigma_i(\boldsymbol{x}, t)$ often depends on other quantities, for example, the number density of the species that is changing: $\nu_{ij} \propto \xi_j$.

Another common case is when species are combining, such as when an agent attaches a passive component, or when they are separating, as when an agent releases its cargo. We can think of these as reactions, with reaction rates that depend on the concentration of the reacting components, but perhaps on other factors as well (since we are dealing with active agents). For example, suppose that species 1 (unladen agent) combines with species 2 (cargo) to produce species 3 (laden agent). This reaction is characterized by $\sigma_3 = -\sigma_1 = -\sigma_2$. For this process to occur, there have to be both unladen agents and cargo available, and the rate of production of species 3 depends on the number densities of species 1 and 2: $\sigma_3 \propto \xi_1\xi_2$. The constant of proportionality might depend on the species 1 agents; for example, they might do quorum sensing and begin forming species 3 complexes only if their number density is above a threshold.

20.2.2.3 Lagrangian reference frame

As in continuum mechanics, it is often convenient to express artificial morphogenesis in a Lagrangian reference frame. That is, rather than adopting a fixed (Eulerian) spatial reference frame, through which material particles flow, we take the perspective of the material particles flowing through space. A Lagrangian reference frame, then, is a more agent-oriented perspective, and more directly corresponds to the control programs of the agents.

The Lagrangian approach describes continuous *bodies* (tissues, masses, fluids, etc.) that change their configuration (shape, position, volume, etc.) through time. Bodies are composed of *material points* or *particles*, which are infinitesimal elements or parcels of these continua. In artificial morphogenesis, particles do not correspond to individual components, but to small volumes (parcels) containing multiple components. This may seem counter-intuitive, but it is exactly parallel with the interpretation in continuum mechanics. For example, the material points (particles) in a solid or fluid do not represent individual atoms or molecules but idealized infinitesimal volumes or parcels containing many atoms or molecules of the material.

> The continuum theory regards matter as indefinitely divisible. Thus, within the theory, one accepts the idea of an infinitesimal volume of materials referred to as a particle in the continuum, and in every neighborhood of a particle there are always materials present. [20, p. 1]

Likewise, in artificial morphogenesis, the particles represent tiny parcels with specific densities of components.

We think of each material point or particle as having a label or identifier P by which we can track it (mathematically) as it moves through space (conventionally, the initial position of the point in space is used as its label). A body $\mathscr{B}$ is a continuum of particles, $P \in \mathscr{B}$. At any given time t, each particle has a position $\boldsymbol{p} = C_t(P)$, and therefore $C_t : \mathscr{B} \to \mathbb{R}^3$ describes the configuration of the body at that time. Particles can have other properties, such as density, concentration, velocity, and acceleration, which are bound to the particles as they move in space.

Mathematically a variable or property Q can be considered either a function of spatial location $q(\boldsymbol{p})$ (thus, in a spatial or Eulerian reference frame), or a function of a particle $Q(P)$ as it moves through space (i.e., in a material or Lagrangian reference frame). The time derivative of such a quantity may be measured while holding location constant (a spatial derivative) $\partial q/\partial t$, or while holding the particle constant (a *substantial* or *material derivative*), $\mathrm{D}Q/\mathrm{D}t$. The substantial derivative is related to the spatial derivative as follows. Since $\boldsymbol{p} = C_t(P)$, in Cartesian coordinates,

$$
\begin{aligned}
\frac{\mathrm{D}Q(P,t)}{\mathrm{D}t} &= \left.\frac{\partial q(\boldsymbol{p},t)}{\partial t}\right|_{P \text{ fixed}} \\
&= \frac{\partial q}{\partial x}\frac{\partial x}{\partial t} + \frac{\partial q}{\partial y}\frac{\partial y}{\partial t} + \frac{\partial q}{\partial z}\frac{\partial z}{\partial t} + \left.\frac{\partial q}{\partial t}\right|_{p \text{ fixed}} \\
&= \frac{\partial q}{\partial t} + \boldsymbol{v} \cdot \nabla q
\end{aligned}
$$

Here, $\boldsymbol{v}$ is the velocity of the material point or particle P. In words, the Lagrangian (substantial) rate of change is the Eulerian (spatial) rate of change plus the convective rate of change. (Although we have shown this for a Cartesian coordinate system, it is valid for any coordinate system.)

The substantial acceleration of a particle, which is its acceleration as a result of its interactions with other particles and its own motive forces, is the substantial derivative of its velocity, $\boldsymbol{A} = \mathrm{D}\boldsymbol{v}/\mathrm{D}t$. It can be determined by writing the velocity in Cartesian coordinates, $\boldsymbol{v} = v_1\boldsymbol{e}_1 + v_2\boldsymbol{e}_2 + v_3\boldsymbol{e}_3$. The substantial derivative of each of these components is $\mathrm{D}v_j/\mathrm{D}t = \partial v_j/\partial t + \boldsymbol{v} \cdot \nabla v_j$. Since this applies for each component of the velocity vectors, we can write $\boldsymbol{A} = \partial \boldsymbol{v}/\partial t + \boldsymbol{v} \cdot \boldsymbol{\nabla v}$. Technically, $\boldsymbol{\nabla v}$ is a second-order tensor; in Cartesian coordinates, $(\boldsymbol{\nabla v})_{jk} = \partial v_j/\partial x_k$, so $\boldsymbol{A} = \partial \boldsymbol{v}/\partial t + \boldsymbol{v}^{\mathrm{T}}\boldsymbol{\nabla v}$. Hence, $\boldsymbol{A} = \boldsymbol{a} + \boldsymbol{v}^{\mathrm{T}}\boldsymbol{\nabla v}$; that is, the Lagrangian acceleration is the Eulerian acceleration plus the convective acceleration.

In the context of continuum mechanics, a particle represents a small amount of material, an infinitesimal parcel of the substance (active and passive components, in our case). As a continuous body deforms through time and space, the volume of these parcels can change, but a particle's mass—the amount of material—cannot. If ρ is the density of a particle and $\mathrm{d}V$ is its volume, then its constant mass is expressed by $\mathrm{D}(\rho \mathrm{d}V)/\mathrm{D}t = 0$. Therefore,

$$
\rho\frac{\mathrm{D}(\mathrm{d}V)}{\mathrm{D}t} + \mathrm{d}V\frac{\mathrm{D}\rho}{\mathrm{D}t} = 0
$$

and hence

$$\frac{\rho}{\mathrm{d}V}\frac{\mathrm{D}(\mathrm{d}V)}{\mathrm{D}t}+\frac{\mathrm{D}\rho}{\mathrm{D}t}=0$$

Observe that $(1/\mathrm{d}V)\mathrm{D}(\mathrm{d}V)/\mathrm{D}t$ is the rate of change of the volume relative to the volume; it can be shown that this equals the divergence of the velocity:

$$\frac{1}{\mathrm{d}V}\frac{\mathrm{D}(\mathrm{d}V)}{\mathrm{D}t}=\nabla\cdot\boldsymbol{v}$$

Therefore, we have the equation of continuity (conservation of mass):

$$\rho\nabla\cdot\boldsymbol{v}+\frac{\mathrm{D}\rho}{\mathrm{D}t}=0$$

In the case of an incompressible material, $\mathrm{D}\rho/\mathrm{D}t=0$, and therefore also $\nabla\cdot\boldsymbol{v}=0$.

20.2.3 Global-to-local compilation

A central problem in swarm intelligence is *global-to-local compilation*, that is, given desired swarm behavior, how do we derive from it control programs for the individual agents that will produce the desired behavior? Our approach to this problem proceeds in several stages. Often we can define PDEs in an Eulerian reference frame that will result in the desired three-dimensional structures. In particular, they may be inspired by biological morphogenesis. To achieve a more agent-oriented perspective, the equations for agent swarms can be transformed to a Lagrangian frame, which takes the perspective of particles moving through space and interacting with their neighbors. The terms in these equations may be divided into those that depend on the physical characteristics of the agents (their hardware) and those that depend on their control processes (their software), a distinction discussed in Section 20.3.3. As PDEs, these control processes are expressed in terms of infinitesimal particles that nevertheless represent multiple agents (this is the mathematical fiction of continuum mechanics, as discussed in Section 20.2.2.3). These control relationships need to be adjusted to apply to single agents of finite size.

Some aspects of this issue are addressed in the context of the examples in Section 20.4, but a systematic solution is the subject of ongoing research. For example, *smoothed particle hydrodynamics* (SPH) is a mesh-free particle simulation method for continuous materials [21,22]; it has been applied to coordinating modest-sized robot swarms [23–26]. SPH and similar mesh-free particle methods are a promising approach to deriving single agent behavior from continuum-mechanics descriptions of swarm behavior.

20.3 Morphogenetic programming notation

Fundamentally, our approach to controlling agent swarms to accomplish artificial morphogenesis is based on using PDEs to describe the behavior of the agents. Therefore, our programming language is based on ordinary mathematical notation, but we have made a few extensions and have formalized the notation in some ways to make it more convenient for artificial morphogenesis.

20.3.1 Change equations

PDEs are the primary constructs for describing the behavior of substances and agent swarms in our approach to artificial morphogenesis. However, especially for simulation purposes, we want to allow discrete-time as well as continuous-time descriptions of these processes. Therefore, we use *change equations*, which can be interpreted ambiguously as either differential equations or as finite difference equations. The change equation $ĐX = F(X, Y, Z, \ldots)$, which we read, "the change in X is $F(X, Y, Z, \ldots)$," can be interpreted either as the differential equation $\mathrm{D}X/\mathrm{D}t = F(X, Y, Z, \ldots)$ or as the finite difference equation $\Delta X/\Delta t = F(X, Y, Z, \ldots)$, in which the time increment Δt is an implicit symbolic quantity. The rules of derivation for change equations respect both the continuous-time and discrete-time interpretations. Within change equations, we can use any of the ordinary mathematical operators, including vector operations such as "∇" and "$\nabla\cdot$."

Change equations with several terms can be broken up into an *extended equation*. For example, an equation

$$ĐX = F(X, Y, Z) + G(X, Y, Z) - H(X, Y, Z)$$

can be broken into a main equation and several extensions, such as this:

$$\begin{aligned} ĐX &= F(X, Y, Z) \\ ĐX &\mathrel{+}= G(X, Y, Z) \\ ĐX &\mathrel{-}= H(X, Y, Z) \end{aligned}$$

This allows a very long equation to be broken across several lines, improving readability. The extensions, however, do not need to follow the main equation but can be located elsewhere. This permits the various contributions to a variable to be defined in more appropriate places (e.g., with related equations for other variables). For example, the invariable physical properties of components can be separated from their controllable behavior (see Section 20.3.3).

In the control of agents, it is often useful to have them behave differently depending on whether some quantity is above or below some threshold. Therefore we have a conditional notation to handle these situations; it is defined in terms of the Heaviside or unit step function $u(x)$, which is 1 for $x > 0$ and 0 for $x < 0$. Since physical realizations are continuous or stochastic, we adopt the convention $u(0) = 1/2$. Conditional factors (which have the value 1 if true and 0 if false) are written as bracketed Boolean expressions defined as follows:

$$\begin{aligned} [x > \theta] &= u(x - \theta) \\ [x < \phi] &= u(\phi - x) \\ [\theta < x < \phi] &= u(x - \theta)u(\phi - x) \\ [x > \theta \wedge y > \phi] &= u(x - \theta)u(y - \phi) \end{aligned}$$

and so forth. They are analogous to epigenetic switches that enable or disable particular regulatory programs in cells.

As explained previously, in the microscopic realm random effects must be expected. These are treated as random processes driven by the Wiener derivative $\text{Đ}\mathbf{W}^n$, which is an n-dimensional normally distributed random vector with zero mean and unit variance. The resulting stochastic differential equations are understood in accord with the Itō calculus, which is required for consistent continuous-time and discrete-time interpretations [11,27,28].

20.3.2 Substance definitions

Our approach to artificial morphogenesis uses continuum mechanics to describe massive collections of animate and inanimate microscopic objects as continuous substances. These substances are classified according to their common properties and behaviors as defined by PDEs. As a matter of programming syntax, classes of similarly behaving components are described by *substance definitions*, which are analogous to class definitions in object-oriented programming languages. The basic syntax of a substance definition is as follows:

substance *name*:
    variable definitions

    behavior:

        equations

The variables are scalar and vector properties of the substance, and in particular are properties of any physical body composed of the substance. The behavioral equations are typically change equations defining relations among the properties of the substance and properties of other substances with which it may interact.

Variable declarations can take several forms. The simplest is a scalar field defined across any three-dimensional volume of the substance; it is understood as a substantial or material variable bound to each particle of the substance (i.e., in a Lagrangian frame). A common example is the mass density:

scalar field ρ || mass density

(Comments follow the "||" symbol.) Vector and tensor fields are defined similarly:

vector field $\boldsymbol{j}$ || flux
order-2 **field** $\boldsymbol{D}$ || diffusion tensor

Vector fields are assumed to be three-dimensional. Multiple variables of the same type (e.g., scalar field, vector field) can be declared together, for example,

scalar fields:
    ρ || mass density
    ξ || concentration

Behavior is defined by equations (often change equations) defined over the variables of the substance and other substances. For example,

behavior:

$$\begin{aligned} \boldsymbol{j} &= \mu\xi - \nabla \cdot (\boldsymbol{D}\xi) && \| \text{ flux from drift and diffusion} \\ \text{Đ}\xi &= -\nabla \cdot \boldsymbol{j} && \| \text{ change in concentration} \end{aligned}$$

Every substance has two automatically defined variables that are vector fields (i.e., defined for every particle of a body composed of the substance). The variable $\boldsymbol{p}$ is automatically bound to the position (in a global Eulerian reference frame) of the material point at any time. Specifically, if $P \in \mathscr{B}$ is any material point in the body, then its position variable is defined $\boldsymbol{p} = C_t(P)$, where C_t is the configuration of the body at that time. In addition, each material point has a velocity, which is the value of a predefined vector, $\boldsymbol{v} = \text{Đ}\boldsymbol{p}$. The velocity of a point is determined by defining either $\boldsymbol{v}$ or $\text{Đ}\boldsymbol{p}$ in the behavior equations. Specifying $\boldsymbol{v} = 0$ in a substance definition makes any bodies composed of that substance rigid and immobile, and it thus functions as a spatially fixed (Eulerian) frame.

Similar to classes in object-oriented languages, substances can be subclasses of other substances. Subclass substances inherit all the properties and behaviors of their parent substances but may define additional variables and behaviors. For example, the parent might define a general diffusible substance, and its children could further specify the particular properties of particular diffusible substances. The syntax for defining a substance as a subclass of another substance is

substance morphogen **is** diffusible_substance **with**:
additional variable definitions

behavior:

additional equations

Some properties might not have behavioral definitions in the parent substance because they are intended to be provided in the child substances (they are *virtual* properties in the terminology of object-oriented programming).

20.3.3 Physical vs. controllable substances

Artificial morphogenesis deals with both active and passive components. Passive components have properties that depend on their material, size, mass, etc. and that are constant for their substance. Active components (agents) have these fixed properties, but they also have additional, programmable behaviors, which we control in order to have swarms behave as intended. In the case of microrobots, these behaviors are implemented by internal control systems; in the case of genetically engineered cells, they are implemented through genetic regulatory networks [14]. Therefore, in programming an artificial morphogenesis system, we distinguish between *physical substances* and *controllable substances*.

Physical and controllable substances are not distinguished in the substance definition syntax, because the distinction is more a matter of intent than formal definition.

A physical substance definition can be thought of as a description of properties and behaviors that are consequences of the physical nature of components and are thus relatively fixed; they are analogous to hardware. In contrast, a controllable substance has properties that are programmable in some way appropriate to its constituent active components (agents). It is by programming these components that the goals of an artificial morphogenetic program are achieved, and thus controllable substances are more akin to software.

By way of example, we might have a physical substance representing agents with certain physical characteristics (size, mass, etc.), including the ability to exert motive force on the surrounding fluid medium. The physical substance reflects the agents' hardware, which might include its sensors and actuators, including motive devices. Given the motive forces exerted by the agents, we would know the flux of the physical substance, but we might want these forces to be programmable. That is, the control processes that make use of the sensors and actuators would be software that can be changed independently of the hardware. Moreover, we might have several different classes of agents that use the same hardware but different control programs. In such a case, we could define various controllable substances to be subclasses of this physical substance, so that the different controllable substances could define different control programs for this motive force (e.g., chemotaxis).

20.3.4 Bodies

Substance definitions specify the properties and behaviors of abstract classes of similar active and passive components, but they do not define any specific swarms or masses of these components. Therefore, a specific artificial morphogenesis system comprises a number of *bodies*, each composed of a massive swarm or ensemble of components of the same type (substance). (This is analogous to instantiating an object as a member of a class in object-oriented programming.) In our approach, a body is treated as a continuum of infinitesimal components (material points). Depending on the properties and behavior of the substance, a body might have the characteristics of a fluid, a rigid solid, a viscoelastic mass, or a field (such as an electrical field), and we may refer to them as tissues, bodies, fluids, fields, etc. A morphogenetic process is initialized by preparing one or more bodies in simple configurations, which then self-organize into the desired complex structures. In addition to initial masses or swarms, bodies may include fixed sources and sinks for components.

A body is defined by a declaration that determines the initial values of its properties over a specific region of space. For example, the following defines a cubical body with a spherical concentration of morphogen in the center:

body Attractant **of** morphogen:
  for $|\boldsymbol{p}\cdot\boldsymbol{x}| \leq 1 \wedge |\boldsymbol{p}\cdot\boldsymbol{y}| \leq 1 \wedge |\boldsymbol{p}\cdot\boldsymbol{z}| \leq 1$:
    $\boldsymbol{j} = 0$ ‖ initial zero flux
    $\boldsymbol{\mu} = 0$ ‖ drift vector
    $\boldsymbol{D} = 0.05\boldsymbol{I}$ ‖ isotropic diffusion
  for $\|\boldsymbol{p}\| \leq 0.001$: $\phi = 1$ ‖ initial density inside sphere
  for $\|\boldsymbol{p}\| > 0.001$: $\phi = 0$ ‖ zero density outside sphere

Such definitions determine initial conditions; as the morphogenetic process progresses, a body may change its shape or move. Since bodies may be liquids, gasses, or fields as well as solids, several bodies may occupy the same region of space.

If substances have properties with the same name, then we can qualify the property with the name of the body, as is done in object-oriented languages. For example, "Attractant.*j*" refers to the flux of the attractant body.

20.4 Examples

In this section, we discuss two examples of artificial morphogenesis, presenting the morphogenetic programs and simulation results.

20.4.1 Fiber bundle routing

In this section, we show how an algorithm for controlling a swarm of discrete agents can be scaled to control a spatially continuous swarm.

20.4.1.1 The U-machine

The *U-machine*, which is named for the mathematician Pavel Urysohn (1898–1924), is a machine model capable of massively parallel generalized computation, including both analog and digital computations [29]. It intended as an architecture exploiting large numbers of nanoscale computational elements. Configuring a U-machine for specific computations involves routing bundles comprising many fibers from one region in a hemispherical shell to others, somewhat analogous to axonal (nerve fiber) bundles within the brain. Due to the small scale of the structures and the large number of fibers in a bundle, the process has to be self-organized except for the selection of the origin and destination regions.

20.4.1.2 Modified flocking algorithm

To implement fiber bundle routing, we have developed the algorithm in three stages. In the first stage, each fiber bundle is treated as a single path from a chosen origin to a chosen destination. An attractant diffuses from the destination and a repellent diffuses from already created pathways. The assembler, which is analogous to a "neural growth cone," emerges from the origin and follows up the gradient of the difference of the attractant and repellent, creating the neural pathway as it goes. In this way, it creates a pathway to the destination without colliding with other pathways (see Figure 20.1). The exact algorithm is described in [12,15].

For the second stage, we use a modified bird flocking algorithm [30,31]. Each growth cone is subject to a number of weighted "urges," including

1. a *destination urge* ($w_d \boldsymbol{u}_d$), which is directed up the attractant gradient
2. an *avoidance urge* ($w_a \boldsymbol{u}_a$) to avoid already generated bundles
3. a *velocity urge* ($w_v \boldsymbol{u}_v$) to match the average velocity of nearby agents
4. a *spacing urge* ($w_s \boldsymbol{u}_s$) to avoid collisions with nearby agents

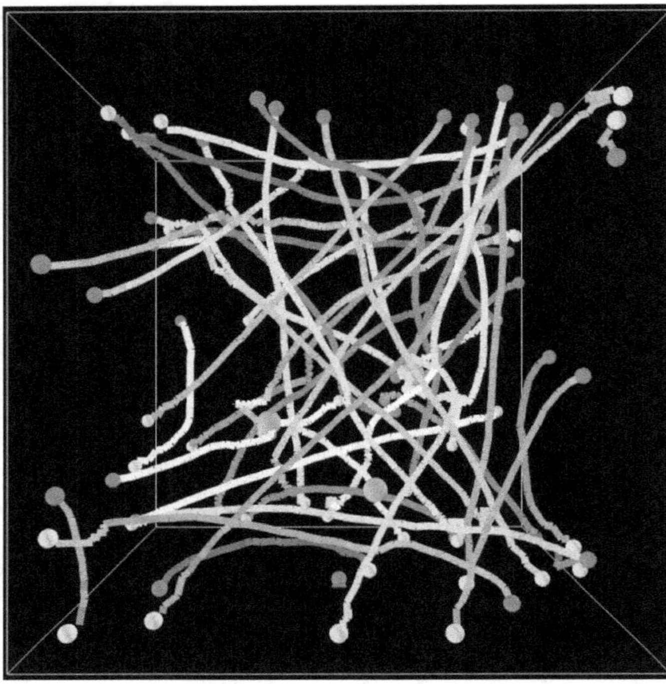

Figure 20.1 Simulation of nerve fiber growth by single agents. Forty paths between randomly chosen origins (darker shade) and destinations (lighter) on the back five surfaces of a cube. In this simulation, there were five collisions (marked by larger spheres)

5. a *center urge* ($w_c\boldsymbol{u}_c$) to steer toward the centroid of nearby agents and keep the group cohesive
6. a random *wander urge* ($w_w\boldsymbol{u}_w$) to break deadlocks and keep agents from getting stuck.

An agent's acceleration, then, is the sum of these six weighted urges:

$$\boldsymbol{a} = w_d\boldsymbol{u}_d + w_a\boldsymbol{u}_a + w_v\boldsymbol{u}_v + w_s\boldsymbol{u}_s + w_c\boldsymbol{u}_c + w_w\boldsymbol{u}_w$$

Figure 20.2 shows a typical simulation of this algorithm with a swarm of 5,000 agents creating six bundles, each comprising 5,000 fibers, between randomly chosen origins and destinations. (The space itself becomes too crowded for much larger numbers of bundles.) Significantly, the same parameters are effective in controlling swarms from 5 to 5,000 assemblers, that is, the algorithm scales over three orders of magnitude. To date, we have performed simulations with up to 20,000 assemblers in a swarm and the same parameters, and these swarms perform correctly. The model is implemented in NetLogo 6.0 and was adapted from a flocking model by Wilensky [32,33].

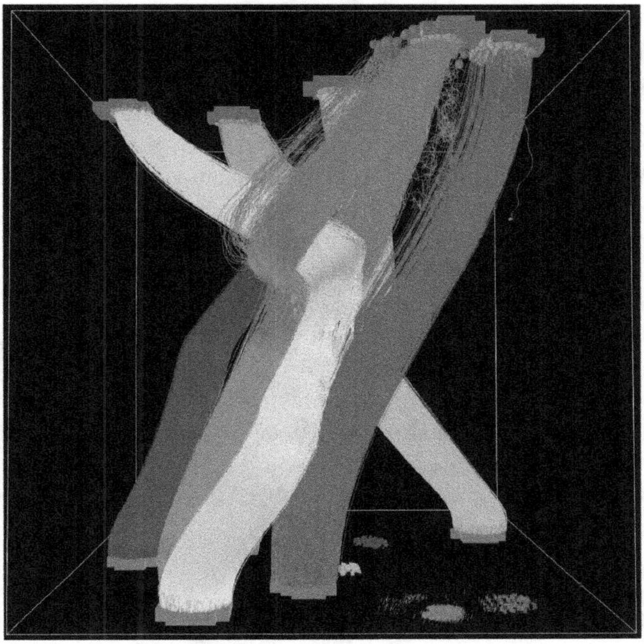

Figure 20.2 *Simulation of fiber bundle routing. Six paths (shaded arbitrarily for clarity) are created sequentially between randomly chosen origins on the lower surface and randomly chosen destinations on the upper surface. Each path is created by a cohort of 5,000 agents that together create a bundle of 5,000 fibers. Notice how bundle had to split to find its way past the lightest bundle. As a result, several agents lost their way (one can be viewed near the righthand surface). Small spheres on the upper surface indicate incorrect connections; in this example, there were 162 incorrect out of 30,000 (99.5 percent correct)*

20.4.1.3 Continuous version

The third stage in the algorithm development, which ensures that it scales to an arbitrarily large number of agents, is to express it in PDEs. In this case, we take the behavior of the agents in the modified flocking algorithm and transfer it to infinitesimal particles representing small parcels of agents; that is, we approximate a massive agent swarm by a continuous mass of agents. This is, in effect, a continuous flocking algorithm, such as explored in [34–36].

In the first stage of algorithm development, we avoided already constructed paths by means of a repellent morphogen. Growth cones followed the gradient of the difference between the attractant, which diffused very rapidly and decayed slowly, and the repellent, which diffused very slowly and decayed rapidly (thus allowing the agents to skirt existing paths while approaching the goal). In the second stage of algorithm development, we computed the avoidance urge by summing individual avoidance vectors from points within a certain radius containing repellent above a threshold. In this third stage, we use a simpler and more effective method, which accomplishes the same end by having existing paths quickly degrade or absorb the attractant, so they are effectively attractant sinks. Since the attractant decreases in absolute concentration with distance from the destination, a repellent diffusing from obstacles near the destination would have less of an effect on the effective gradient (the gradient of attractant minus repellent) than would obstacles further away from the destination. Therefore, it is impossible to find an appropriate balance between the attractant and repellent on the gradient. On the other hand, by having the obstacles absorb attractant, they have the same relative effect on the gradient independent of their distance from the destination (see Figure 20.3).

Let G be a three-dimensional field representing the concentration of goal (destination) material, which is a source of attractant A, which diffuses with diffusion constant D_A and decays at a rate given by the time constant τ_A. The goal region produces attractant at a rate κ_G and saturates at $A = 1$. In addition, let P represent the

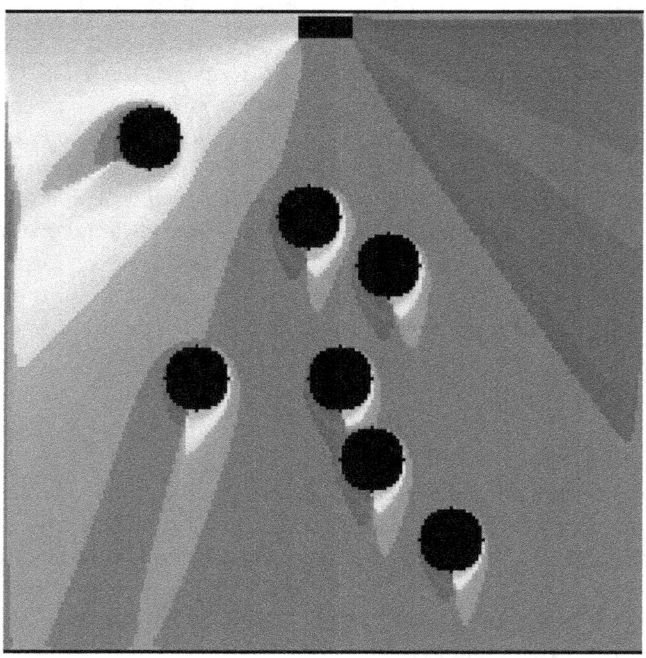

Figure 20.3 Normalized attractant gradient. The attractant is diffusing from the destination at the center top and being absorbed by the seven obstacles shown as black circles. Shading indicates the (quantized) direction of the normalized gradient

three-dimensional concentration of path material, which absorbs the attractant with time constant τ_P. Then, the concentration of the attractant is given by the PDE:

$$ÐA = D_A \nabla^2 A + \kappa_G G(1 - A) - A/\tau_A - PA/\tau_P$$

We want the particles to steer up the attractant gradient ∇A, which will lead it to the goal and also avoid existing paths. This gradient will be stronger near the goal and weaken with increasing distance, and so the algorithm will be more robust if the *direction* of movement is given by the gradient but *rate* of movement is constant (κ_C). Thus, one component of swarm velocity is the versor $\nabla A/\|\nabla A\|$, but we must beware of a zero gradient; more realistically, we want to ignore very weak gradients, which might be noise. Therefore, let $S = \|\nabla A\|$ be the strength of the gradient, and θ be the threshold below which the gradient is ignored. Then, the first term of the swarm velocity is $\kappa_C[S > \theta](\nabla A)/S$.

The second term serves a function similar to the velocity, spacing, and center urges in the flocking algorithm: to keep the swarm cohesive and avoid collisions. In the continuous case, we define a potential function U on the swarm concentration (number density) C that is minimal at the optimal swarm concentration. Mathematically, this may defined by a function such as $U(C) = (C - C_0)^2$, where C_0 is the desired swarm density. One way to implement such a potential function is for the agents to emit a slowly diffusing, quickly decaying morphogen c, which then becomes a surrogate for the density C. In such a case, in addition to steering up the attractant gradient, the swarm should steer down the $U(C)$ gradient in order to maintain an optimal density, which is indicated by $c_{\text{upb}} > c > c_{\text{lwb}}$. Alternatively, we might define

$$U(c) = (u(c_{\text{lwb}} - c) + [c - c_{\text{upb}}]^+)(u(c - c_{\text{upb}}) + [c_{\text{lwb}} - c]^+)$$

which is zero in the allowed range of morphogen concentrations and increases linearly outside of it. We can assemble these considerations into behavioral equations for the swarm velocity, using a parameter λ to control the relative influence of the attractant and density:

$$\begin{aligned} W &= A - \lambda U(C) \\ S &= \|\nabla W\| \\ \mathbf{v} &= [S > \theta](\nabla W)/S \end{aligned}$$

Figures 20.4 and 20.5 show a two-dimensional simulation of the continuous flocking applied to coordinating an agent swarm to create a massive fiber bundle. The quadratic potential function is used to control swarm density and cohesion.

20.4.2 Clock and wavefront process

Our goal is to develop morphogenetic programs—often inspired by embryological morphogenesis—that can be used to coordinate the activity of very large swarms of agents to assemble complex structures. Therefore, we have been exploring how controllable morphogenetic processes in nature can be exploited in artificial morphogenesis to serve similar functions to what they serve in nature but also for different

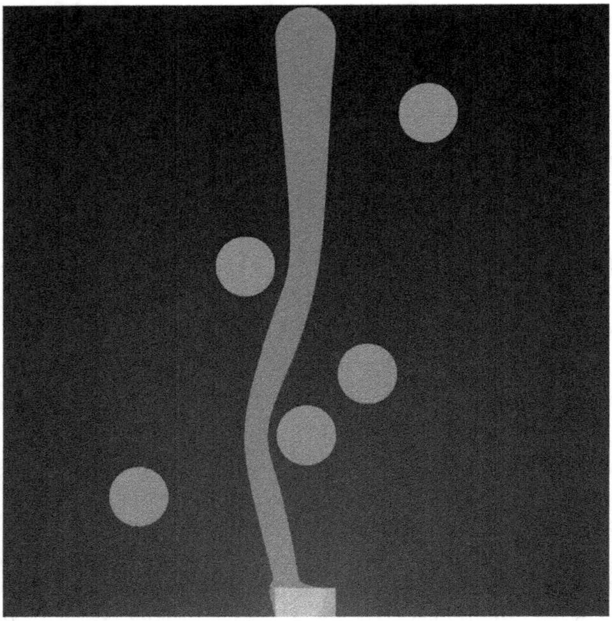

Figure 20.4 *Simulation of fiber bundle routing by continuous flocking. Origin is at top, destination at bottom, obstacles shown as disks. The swarm finds its way around obstacles representing previously assembled bundles*

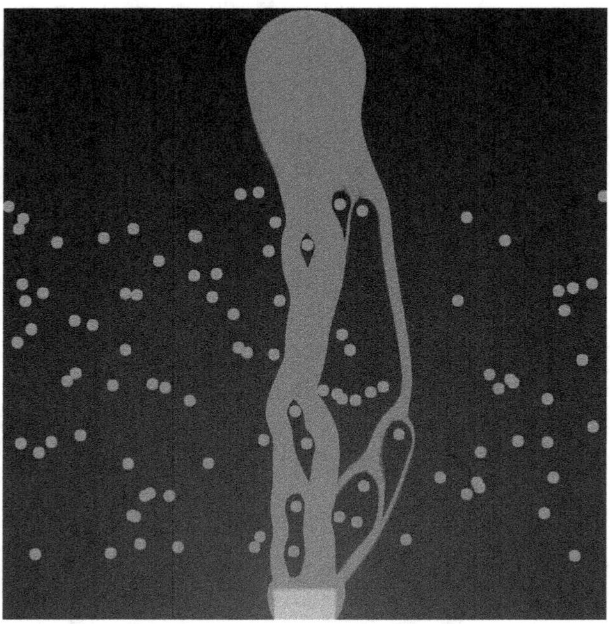

Figure 20.5 *Simulation of fiber bundle routing by continuous flocking. The large number of obstacles and small passages between them cause the swarm to divide into multiple streams*

purposes. In particular, the *clock-and-wavefront process* controls *somitogenesis* in embryos: the segmentation of the spinal precursor into somites, the number and size of which is characteristic of the species (e.g., the 33 vertebrae of human beings) [37,38]. The sequence of segments defines a kind of endogenous reference frame for later developmental processes (e.g., determining the location of the limbs and internal organs).

We have applied this process to a sample problem in artificial morphogenesis: the assembly of a frame that might be the basis for an insect-like robot. The frame has a sort of spine for which we expect to be able to specify the size and number of segments; this application of the clock-and-wavefront process is similar to its function in vertebrate development. We also want our robot frame to have a pair of segmented legs on each spinal segment, and we want to be able to control the position of the legs on each segment and to specify the number and size of the leg segments. This application is a redeployment of the clock-and-wavefront process, which is not involved in the segmentation of insect legs. We will summarize recent experiments simulating the application of the clock-and-wavefront process to the assembly of this robot frame; prior experiments and more detailed information can be found in [11–13,15].

20.4.2.1 Biological model

In vertebrate development, the clock-and-wavefront process operates through the interaction of three morphogens (chemicals controlling development). The embryo grows in length through cell proliferation at the tail end, and the somites differentiate one at a time in order from the head end. A *caudal morphogen* diffuses from the tail end, and a *rostral morphogen* diffuses from already differentiated somites toward the head end. The next somite will differentiate in a region sufficiently far from both the tail bud and the differentiated somites, as determined by relatively low concentrations of these two morphogens. However, differentiation does not take place until a segmentation signal propagates forward from the tail bud. Pacemaker cells in the tail bud produce periodic pulses of a *segmentation morphogen*, which stimulate nearby cells to produce a pulse of this morphogen and then enter a refractory period. The consequence is that a wave of segmentation morphogen propagates through the cells toward the head of the embryo. When this wave passes through the cells that are ready for differentiation (as indicated by rostral and caudal morphogens), they differentiate, forming the next somite. The number of segments is determined by the product of the pacemaker frequency and the growth duration; the length of the segments is determined by the ratio of the tail growth rate and the pacemaker frequency.

20.4.2.2 Spine assembly

To apply artificial morphogenesis to the assembly of the robot frame, we define the required substances and the initial states of the bodies. The primary substance is "medium," which represents the mass of agents and constitutes the body of the robot; at any point in space, M is the concentration of "medium" agents. In addition to its undifferentiated state, "medium" may be in two major states of differentiation. The first is as "terminal" tissue, which corresponds to the tail bud; the variable T is the relative concentration (number density) of terminal agents, so $T = 1$ (maximum density) in the tail region and $T = 0$ elsewhere. The terminal agents are responsible for the growth of the structure and the periodic emission of the segmentation morphogen. The remainder of the agents constituting the structure also belong to the "medium" substance and can be in either of two states, undifferentiated tissue ($S \approx 0$) or differentiated segment tissue ($S \approx 1$).

The structure grows through the creation of M tissue just anterior to the T tail tissue. In an embryo, this is the consequence of cell proliferation, and if the agents in artificial morphogenesis are living cells and capable of reproduction, undifferentiated tissue could grow in the same way. If the agents are microrobots and incapable of reproduction, undifferentiated agents must be provided externally and migrate to the growth site anterior to the tail tissue (attracted, e.g., by the caudal morphogen). To cover both possibilities, our current simulation has the new undifferentiated tissue simply appear behind the tail tissue as it moves in a caudal direction; it doesn't model how the new tissue gets there.

The initial preparation for the morphogenetic processes is a cylinder of already differentiated somite tissue (the "head") contiguous with a cylinder of terminal tissue. Assuming the spinal axis is oriented along the X-axis, the following body declarations define the preparation:

body Spine **of** medium:
  for $0 \le \boldsymbol{p}\cdot\boldsymbol{x} \le 2 \wedge (\boldsymbol{p}\cdot\boldsymbol{y})^2 + (\boldsymbol{p}\cdot\boldsymbol{z})^2 \le 1$: $M = 1$ || initial medium tissue

body Head **of** medium:
  for $0 \le \boldsymbol{p}\cdot\boldsymbol{x} \le 1 \wedge (\boldsymbol{p}\cdot\boldsymbol{y})^2 + (\boldsymbol{p}\cdot\boldsymbol{z})^2 \le 1$:
    $S = 1$ || initial head somite
    $T = 0$ || no tail agents

body Tail **of** terminal:
  for $1 \le \boldsymbol{p}\cdot\boldsymbol{x} \le 2 \wedge (\boldsymbol{p}\cdot\boldsymbol{y})^2 + (\boldsymbol{p}\cdot\boldsymbol{z})^2 \le 1$:
    $T = 1$ || initial tail bud
    $S = 0$ || no segmented tissue

The caudal morphogen C is produced in the tail bud, from which it diffuses and decays at rates defined in its physical substance definition, which is omitted here, but has this behavioral equation:

$$ÐC = D_C \nabla^2 C - C/\tau_C \tag{20.1}$$

Notice that there is no source term, since it is provided in the definition of the controllable substance "terminal", which includes this incremental extension of the previous equation (Equation (20.1)):

$$ÐC \mathrel{+}= \kappa_C T(1 - C)$$

This is the missing source term, which describes how terminal tissue ($T \approx 1$) produces caudal morphogen up to saturation ($C = 1$).

Similarly, the rostral morphogen R is produced from differentiated somites (initially, just the head), from which it diffuses and decays, as defined by the physical behavioral equation:

$$\text{Đ}R = D_R \nabla^2 R - R/\tau_R \tag{20.2}$$

The definition of the controllable substance "medium" provides an incremental extension to (20.2), which describes the production of rostral morphogen from differentiated somites ($S \approx 1$):

$$\text{Đ}R \mathrel{+}= \kappa_R S(1 - R)$$

Hence, differentiated segments produce rostral morphogen up to saturation ($R = 1$). Figure 20.6 shows a two-dimensional simulation of the concentrations of the R and C morphogens after eight segments have differentiated.

A pulse of segmentation morphogen α is produced under either of two conditions (details can be found in [13,15]): (1) Growth has not been completed, and the pacemaker is in the correct phase; this condition is indicated by the control variable $\psi = 1$. (2) The agent senses the local concentration of α to be above a threshold, and the agent is not in its refractory period; this condition is indicated by the control variable $\phi = 1$. The generation and propagation of the segmentation morphogen is defined:

$$\text{Đ}\alpha = \kappa_\psi \psi(1 - \alpha) + \kappa_\phi \phi(1 - \alpha) + D_\alpha \nabla^2 \alpha - \alpha/\tau_\alpha$$

Since the control variables in the first two terms are only briefly nonzero, they provide a pulse of morphogen, which then diffuses and decays. For clarity, we have written $\text{Đ}\alpha$ as a single PDE, but it is better broken into three parts. The last two terms are part of the definition of the physical substance "segmentation morphogen" and describe its diffusion and decay:

$$\text{Đ}\alpha = D_\alpha \nabla^2 \alpha - \alpha/\tau_\alpha$$

The ψ and ϕ terms are better provided as incremental extensions in the definitions of the "terminal" and "medium" substances, respectively. In other words, these controllable substances have the capability of emitting α.

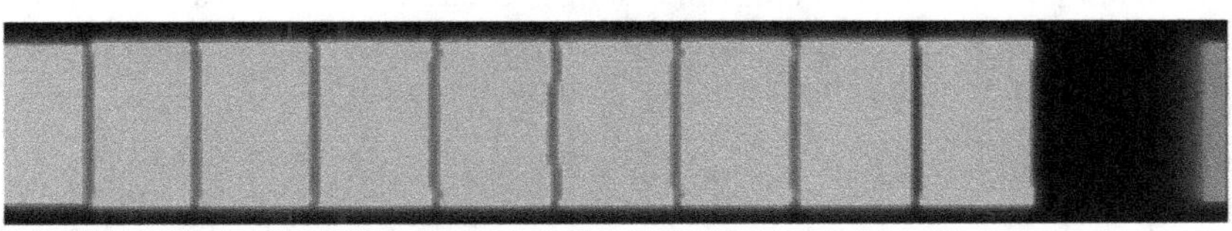

Figure 20.6 Concentrations of rostral and caudal morphogens during spinal segmentation. Lighter rectangles represents rostral morphogen emitted from nine already differentiated segments. Band at extreme right represents caudal morphogen emitted from tail tissue

Therefore, the behavior of the "terminal" substance includes these extensions, which control its production of the segmentation morphogen:

$$\psi = [G > \vartheta_G \wedge K > \vartheta_K]T$$
$$\text{Đ}\alpha \mathrel{+}= \kappa_\psi \psi(1 - \alpha)$$

The pulse ψ is emitted by terminal tissue ($T > 0$) when the pacemaker cells are in a certain phase (indicated by $K > \vartheta_K$) so long as growth continues (indicated by $G > \vartheta_G$). This is implemented by an oscillator such as the following (in which ω is the pacemaker frequency):

$$\text{Đ}K = -\omega^2 L$$
$$\text{Đ}L = [G > \vartheta_G]K$$

The growth duration is controlled typically by the exponential decay of the growth timer G:

$$\text{Đ}G = -\frac{G}{t_G} Gt_G$$

where for convenience we initialize $G = 1$ and set $\vartheta_G = 1/e$. The number of segments n is determined by the product of the frequency (in cycles per unit time) and growth duration t_G, specifically, $n = t_G\omega/2\pi$. The length of the segments λ_S is determined by the ratio of the growth rate r to the clock frequency, specifically, $\lambda_S = 2\pi r/\omega$.

The other source of segmentation morphogen is the "medium" substance, which includes this behavior:

$$\phi = [\alpha > \vartheta_\alpha \wedge \rho < \vartheta_\rho]M$$
$$\text{Đ}\alpha \mathrel{+}= \kappa_\phi \phi(1 - \alpha)$$

The variable ϕ represents the density of "medium" tissue with local morphogen concentration above threshold ($\alpha > \vartheta_\alpha$) but not in a refractory state ($\rho < \vartheta_\rho$). The refractory state ensures that the segmentation signal propagates in a single direction; the refractory period is initialized and timed by the following equation:

$$\text{Đ}\rho = \phi - \rho/\tau_\rho$$

which is also part of the definition of "medium."

Differentiation of the medium into somite tissue ($S \approx 1$) takes place when a sufficiently strong segmentation morphogen wave passes through a region where the rostral and caudal morphogens are below certain thresholds:

$$\text{Đ}S = \kappa_S S(1 - S) + [\alpha > \alpha_{\text{lwb}} \wedge R < R_{\text{upb}} \wedge C < C_{\text{upb}}]c_\varsigma(1 - S)$$

If α is above its threshold, and R and C are below their thresholds, then a differentiation trigger c_ς (a constant) precipitates self-reinforcing differentiation, which goes to saturation [$S(1 - S)$]. A two-dimensional simulation of this process is shown in Figure 20.7.

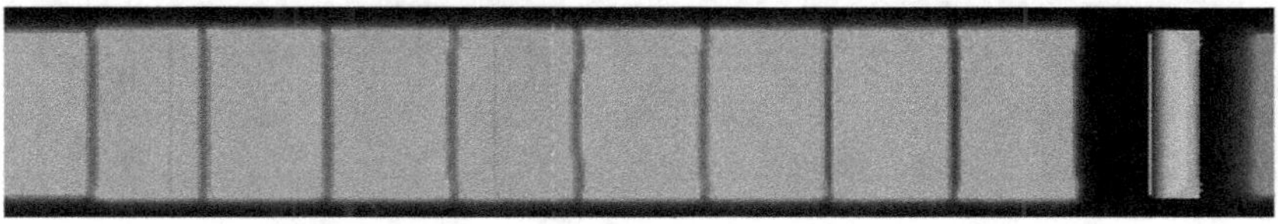

Figure 20.7 *Simulation of spinal segmentation. This shows a stage in the process in which eight new segments have differentiated (indicated by light rectangles), and a ninth segment is in the process of differentiation. The first (left-most) or head segment is initialized; the other nine are differentiating through the clock-and-wavefront process. The narrow lines represent the waves of the* α *segmentation morphogen, which propagate from the tail to the head. The rightmost is differentiating the tenth segment as it passes through undifferentiated tissue; three previous waves are visible passing through already differentiated tissue*

20.4.2.3 Leg assembly

The second phase of the assembly process is to grow a pair of segmented legs on each spinal segment. To this end, we want to control the position of the legs along the spinal segment, which requires us to differentiate the anterior and posterior ends of each segment. Anterior segment tissue will be indicated by a differentiation state $A = 1$, and posterior segment tissues will be indicated by $P = 1$ (see Figure 20.8). Anterior tissue develops when a sufficiently strong segmentation wave passes through differentiated segment tissue, and the rostral morphogen is in a particular range:

$$\text{Đ}A = \kappa_A SA(1 - A) + [0.5R_{\text{upb}} > R > 0.25R_{\text{upb}} \wedge \alpha > \alpha_{\text{lwb}}]c_A - A/\tau_A$$

Similarly, caudal tissue develops when a sufficiently strong segmentation wave passes through differentiated segment tissue, and the caudal morphogen is in a particular range.

$$\text{Đ}P = \kappa_P SP(1 - P) + [0.95C_{\text{upb}} > C > 0.8C_{\text{upb}} \wedge \alpha > \alpha_{\text{lwb}}]c_P - P/\tau_P$$

The anterior and posterior tissues each emit morphogens (a and p, respectively), which diffuse from these tissues and provide a reference frame for determining distance along the segment. Their defining equations are

$$\text{Đ}a = [A > \vartheta_A]\kappa_a S(1 - a) + D_a \nabla^2 a - a/\tau_a \qquad (20.3)$$

$$\text{Đ}p = [P > \vartheta_P]\kappa_p S(1 - p) + D_p \nabla^2 p - p/\tau_p \qquad (20.4)$$

The bracketed conditionals ensure that the morphogens are emitted only when the fraction of differentiated anterior and posterior tissue is above specified thresholds $(\vartheta_A, \vartheta_P)$.

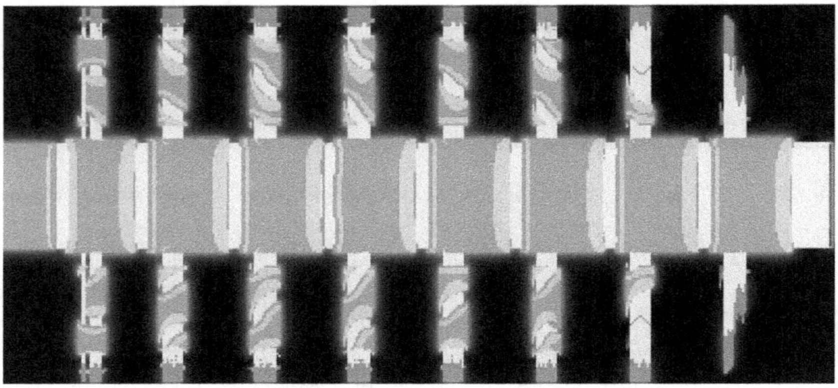

Figure 20.8 Anterior and posterior tissues during spine and leg growth. Lighter bands at left represents anterior segment tissue ($A \approx 1$) and lighter bands at right represents posterior segment tissue ($P \approx 1$). (Within the legs, anterior and posterior mean proximal and distal, respectively.) Darker shading at tips represents terminal tissue ($T \approx 1$), that is, the "tail" and the "feet"; lightest shade represents undifferentiated, intersegmental tissue

In insect embryological development, legs form out of groups of cells called *imaginal disks*. Similarly, we will select a group of agents to form an imaginal disk on which a new leg will be assembled. Agents differentiate into the imaginal state ($I = 1$) when the anterior and posterior morphogens are in the appropriate range to place the disk where desired, and the agents are on the surface (indicated by a control variable $E = 1$):

$$ÐI = [a_{\text{upb}} > a > a_{\text{lwb}} \wedge p_{\text{upb}} > p > p_{\text{lwb}}]ES(1 - I) \tag{20.5}$$

(Specific means for accomplishing this are described in [12,13,15].) Figure 20.9 illustrates the use of the a and p morphogens in the placement of the imaginal disks.

The rapid differentiation of the imaginal disk agents into the I state (signaled by $ÐI > \vartheta_{ÐI}$) causes them to orient outward (i.e., down the gradient ∇S) and to enter the terminal ($T = 1$) state. Now each imaginal disk is in the initial state for the clock-and-wavefront process (proximal $S = 1$, distal $T = 1$); the S and T agents begin emitting the R and C morphogens, respectively, and the T agents begin assembling new undifferentiated "medium" tissue and producing periodic pulses of α. Therefore, the segmented legs grow just as the spine did (see Figures 20.10 and 20.11).

The "terminal" agents in the imaginal disks can have a different growth duration and pacemaker frequency than those in the tail bud, so the number and length of the leg segments is independent of the number and length of spinal segments. (The number of leg segments is $t_{\text{A}}\omega_{\text{A}}/2\pi$, where t_{A} is the duration of leg growth and ω_{A} is the leg pacemaker frequency; the length of the leg segments is $2\pi r/\omega_{\text{A}}$, where r is the growth rate.)

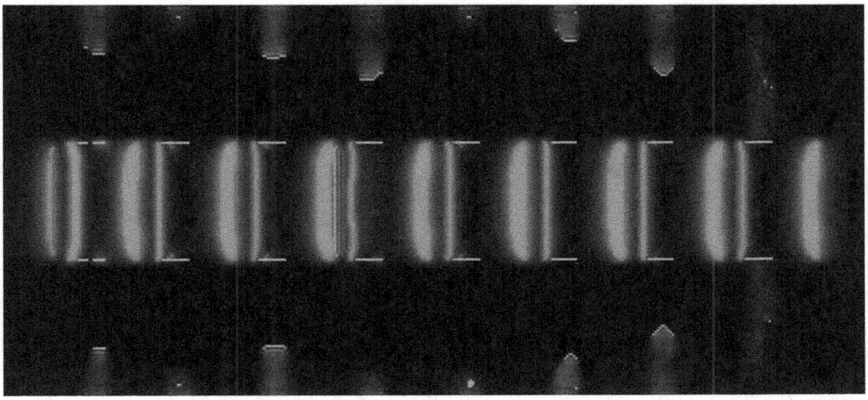

Figure 20.9 *Morphogens involved in leg placement and formation. Wider, ovoid bands represents the "a" morphogen, which marks the anterior (rostral) ends of spinal segments, and narrower bands represents the "p" morphogen, which marks the posterior (caudal) ends. Small lines represent imaginal disks. Streaks on both sides represents the α segmentation morphogen, which is propagating both in the spine and the legs*

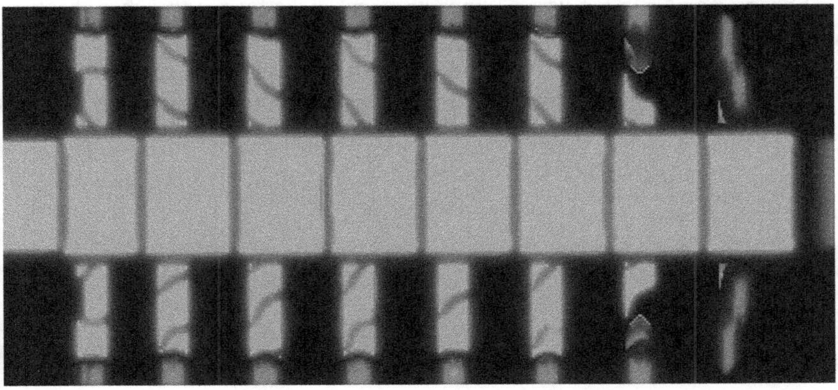

Figure 20.10 *Assembly of legs. Lighter shading represents differentiated segments, and darker shading is C morphogen. Several pulses of α segmentation morphogen are visible in the legs. There are irregularities in the leg formation, in part because the simulation space is too narrow and the growing legs have reached the edges of the simulation space*

There are some differences between the processes in the spine and legs, since otherwise the leg segments would form imaginal disks of their own, and little "leglets" would grow out of the legs. There are various ways to prevent formation of imaginal disks, but perhaps the simplest is to suppress emission of the anterior and posterior

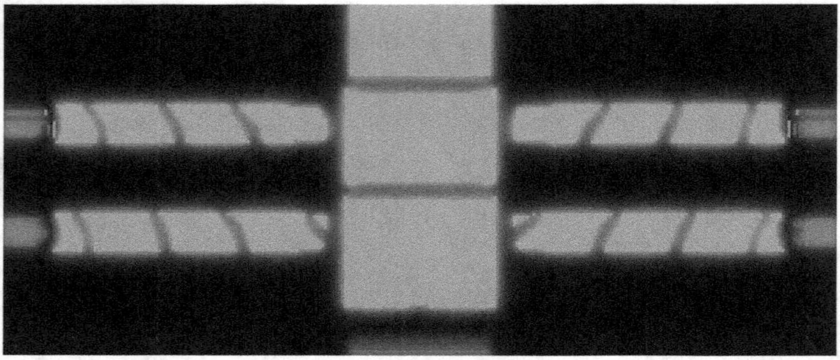

Figure 20.11 Assembly of legs. The first three segments of the first two pairs of legs have formed, and the fourth segments are developing. Darker shades represents caudal (C) and rostral (R) morphogens, but in the case of legs, they represent the distal and proximal orientations. The angled leg joints are a result of a greater concentration of R in the rostral direction

morphogens a and p in leg tissue. Since imaginal disks form only when these morphogens are in appropriate ranges [Equation (20.5)], the disks will not form in the absence of the morphogens. The a and p morphogens are produced at rates κ_a and κ_p, respectively [Equations (20.3) and (20.4)], and so these rates are set to zero in the imaginal disks:

$$\text{Đ}\kappa_a = -\frac{\kappa_a I}{\tau_I}$$

$$\text{Đ}\kappa_p = -\frac{\kappa_p I}{\tau_I}$$

In addition, terminal tissue has to copy these rates into newly assembled "medium" tissue, so that the tissue that grows out of the imaginal disks inherits the zero rates.

20.5 Conclusions

Creating complex hierarchical structures that are organized from the microscale up to the macroscale will be facilitated by coordinating massive swarms of microscopic agents to assemble themselves or passive components into definite shapes. One promising approach is artificial morphogenesis, which adapts morphogenetic processes in embryological development that coordinate the movement, assembly, and differentiation of billions of cells into a living organism. This requires the development of swarm intelligence algorithms that will scale up to unlimited numbers of microscopic components (active assemblers and passive elements). We accomplish this objective by going to the continuum limit, that is, by treating swarms as though composed of an infinite number of infinitesimal components. The mathematical framework for this is continuum mechanics, which has been used to describe the behavior of fluids, solids, and viscoelastic materials in embryological development. To this end, we have designed an artificial morphogenesis oriented programming notation, which uses PDEs to describe the coordination of massive swarms of active and passive components. Swarm types with common properties and behaviors are organized into substances and continuous bodies, which are analogous to classes and objects, respectively, in object-oriented programming languages.

We illustrated artificial morphogenesis with simulations of two different processes. In the first, we showed how to coordinate large swarms of assemblers to route bundles of many fibers between chosen origins and destinations; such a procedure might be used to establish massive connection pathways in a neuromorphic computer architecture. This algorithm was developed in stages from a single assembler to swarms of thousands of assemblers and culminating in continuous swarms of assemblers. The second example showed how a natural morphogenetic process, the clock-and-wavefront model of spinal morphogenesis, could be applied to an artificial morphogenesis problem: the swarm construction of an insect-like frame with a segmented spine and a pair of segmented legs on each spine segment.

The foregoing are first steps in the development of a morphogenetic programming approach to the coordination of massive swarms of microscopic agents (microrobots or genetically engineered microorganisms). In addition to the development of conceptual, programming, and simulation tools, ongoing research is investigating the adoption of natural developmental processes to the creation of complex hierarchical structures.

Acknowledgment

The author grateful to Allen McBride for programming the continuous flocking simulation and producing the images illustrating it.

References

[1] MacLennan BJ. The Morphogenetic Path to Programmable Matter. Proceedings of the IEEE. 2015;103(7):1226–1232.

[2] Doursat R, Sayama H, Michel O, editors. Morphogenetic Engineering: Toward Programmable Complex Systems. Springer, Berlin, Heidelberg; 2012.

[3] Kitano H. Morphogenesis for Evolvable Systems. In: Sanchez E, Tomassini M, editors. Towards Evolvable Hardware: The Evolutionary Engineering Approach. Berlin: Springer; 1996. p. 99–117.

[4] Nagpal R, Kondacs A, Chang C. Programming Methodology for Biologically-Inspired Self-Assembling Systems. In: AAAI Spring Symposium on Computational Synthesis: From Basic Building Blocks to High Level Functionality; 2003. Available from: http://www.eecs.harvard.edu/ssr/papers/aaaiSS03-nagpal.pdf.

[5] Spicher A, Michel O, Giavitto J. Algorithmic Self-assembly by Accretion and by Carving in MGS. In: Proc. of the 7th International Conference on Artificial Evolution (EA '05). No. 3871 in Lecture Notes in Computer Science. Berlin: Springer-Verlag; 2005. p. 189–200.

[6] Murata S, Kurokawa H. Self-Reconfigurable Robots: Shape-Changing Cellular Robots Can Exceed Conventional Robot Flexibility. IEEE Robotics & Automation Magazine. 2007 March; p. 71–78.

[7] Doursat R. Organically Grown Architectures: Creating Decentralized, Autonomous Systems by Embryomorphic Engineering. In: Würtz RP, editor. Organic Computing. Springer, Berlin, Heidelberg; 2008. p. 167–200.

[8] Bourgine P, Lesne A, editors. Morphogenesis: Origins of Patterns and Shapes. Berlin: Springer; 2011.

[9] Salazar-Ciudad I, Jernvall J, Newman S. Mechanisms of Pattern Formation in Development and Evolution. Development. 2003;130:2027–2037.

[10] Forgacs G, Newman SA. Biological Physics of the Developing Embryo. Cambridge, UK: Cambridge University Press; 2005.

[11] MacLennan BJ. Morphogenesis as a Model for Nano Communication. Nano Communication Networks. 2010;1(3):199–208.

[12] MacLennan BJ. Molecular Coordination of Hierarchical Self-Assembly. Nano Communication Networks. 2012 June;3(2):116–128.

[13] MacLennan BJ. Embodied Computation: Applying the Physics of Computation to Artificial Morphogenesis. Parallel Processing Letters. 2012;22(3).

[14] Teo JJY, Woo SS, Sarpeshkar R. Synthetic Biology: A Unifying View and Review Using Analog Circuits. IEEE Transactions on Biomedical Circuits and Systems. 2015 August;9(4):453–474.

[15] MacLennan BJ. Coordinating Massive Robot Swarms. International Journal of Robotics Applications and Technologies. 2014;2(2):1–19.

[16] Beysens DA, Forgacs G, Glazier JA. Cell Sorting is Analogous to Phase Ordering in Fluids. Proceedings of the National Academy of Sciences of the United States of America. 2000;97:9467–9471.

[17] Meinhardt H. Models of Biological Pattern Formation. London: Academic Press; 1982.

[18] de Gennes PG. Soft Matter. Science. 1992;256:495–497.

[19] Taber LA. Nonlinear Theory of Elasticity: Applications in Biomechanics. Singapore: World Scientific; 2004.

[20] Lai WM, Rubin D, Krempl E. Introduction to Continuum Mechanics. Revised edition in SI/Metric units. Oxford: Pergamon Press; 1978.

[21] Gingold RA, Monaghan JJ. Smoothed Particle Hydrodynamics: Theory and Application to Non-spherical Stars. Monthly Notices of the Royal Astronomical Society. 1977 November;181:375–389.

[22] Liu GR, Liu MB. Smoothed Particle Hydrodynamics: A Meshfree Particle Method. Singapore: World Scientific; 2003.

[23] Perkinson JR, Shafai B. A Decentralized Control Algorithm for Scalable Robotic Swarms Based on Mesh-free Particle Hydrodynamics. In: Proc. of the IASTED Int. Conf. on Robot. and Applications; 2005. p. 102–107.

[24] Pimenta LCA, Mendes ML, Mesquita RC, *et al.* Fluids in Electrostatic Fields: An Analogy for Multirobot Control. IEEE Transactions on Magnetics. 2007 April;43(4):1765–1768. Available from: http://ieeexplore.ieee.org/document/4137791/.

[25] Pimenta LCA, Pereira GAS, Michael N, *et al.* Swarm Coordination Based on Smoothed Particle Hydrodynamics Technique. IEEE Transactions on Robotics. 2013 April;29(2):383–399. Available from: http://ieeexplore.ieee.org/document/6407937/.

[26] Bandala AA, Dadios EP. Dynamic Aggregation Method for Target Enclosure Using Smoothed Particle Hydrodynamics Technique: An Implementation in Quadrotor Unmanned Aerial Vehicles (QUAV) Swarm. Journal of Advanced Computational Intelligence. 2016;20(1):84–91.

[27] MacLennan BJ. Preliminary Development of a Formalism for Embodied Computation and Morphogenesis. Knoxville, TN: Department of Electrical Engineering and Computer Science, University of Tennessee; 2009. UT-CS-09-644.

[28] MacLennan BJ. Artificial Morphogenesis as an Example of Embodied Computation. International Journal of Unconventional Computing. 2011;7(1–2): 3–23.

[29] MacLennan BJ. The U-Machine: A Model of Generalized Computation. International Journal of Unconventional Computing. 2010;6(3–4):265–283.

[30] Reynolds CW. Flocks, Herds and Schools: A Distributed Behavioral Model. ACM SIGGRAPH Computer Graphics. 1987;21(4):25–34.

[31] Spector L, Klein J, Perry C, *et al.* Emergence of Collective Behavior in Evolving Populations of Flying Agents. Genetic Programming and Evolvable Machines. 2005;6(1):111–125.

[32] Wilensky U. NetLogo. Evanston, IL: Center for Connected Learning and Computer-Based Modeling, Northwestern University; 1999. Available from: http://ccl.northwestern.edu/netlogo/.

[33] Wilensky U. NetLogo Flocking 3D Alternate model. Evanston, IL: Center for Connected Learning and Computer-Based Modeling, Northwestern University; 2005. Available from: http://ccl.northwestern.edu/netlogo/models/Flocking3DAlternate.

[34] Topaz CM, Bertozzi AL, Lewis MA. A Nonlocal Continuum Model for Biological Aggregation. Bulletin of Mathematical Biology. 2006;68:1601–1623.

[35] Chuang YL, D'Orsogna MR, Marthaler D, *et al.* State Transitions and the Continuum Limit for a 2D Interacting, Self-propelled Particle System. Physica D. 2007;232:33–47.

[36] Carrillo JA, Fornasier M, Toscani G, *et al.* Particle, Kinetic, and Hydrodynamic Models of Swarming. In: Naldi G, Pareschi L, Toscani G, editors.

Mathematical Modeling of Collective Behavior in Socio-Economic and Life Sciences. Boston: Birkhāuser; 2010. p. 297–336.
[37] Cooke J, Zeeman EC. A Clock and Wavefront Model for Control of the Number of Repeated Structures During Animal Morphogenesis. Journal of Theoretical Biology. 1976;58:455–476.
[38] Dequéant ML, Pourquié O. Segmental patterning of the vertebrate embryonic axis. Nature Reviews Genetics. 2008;9:370–382.

Author index

Subject index

www.ingramcontent.com/pod-product-compliance
Lightning Source LLC
LaVergne TN
LVHW011854200926
848484LV00002B/451

* 9 7 8 1 7 8 5 6 1 6 2 7 3 *